T0335841

Physics of the Atmosphere and Climate

Murry Salby's new book provides an integrated treatment of the processes controlling the Earth-atmosphere system developed from first principles through a balance of theory and applications. This book builds on Salby's previous book *Fundamentals of Atmospheric Physics*. The scope has been expanded to include climate, while streamlining the presentation for undergraduates in science, mathematics, and engineering. Advanced material, suitable for graduate students and researchers, has been retained but distinguished from the basic development. The book offers a conceptual yet quantitative understanding of the controlling influences integrated through theory and major applications. It leads readers through a methodical development of the diverse physical processes that shape weather, global energetics, and climate. End-of-chapter problems of varying difficulty develop student knowledge and its quantitative application, supported by answers and detailed solutions online for instructors.

MURRY SALBY is Chair of Climate Science at Macquarie University, Sydney, Australia. He was previously a Professor at the University of Colorado, where he served as Director of the Center for Atmospheric Theory and Analysis. Previously, he was a researcher at the U.S. National Center for Atmospheric Research and at Princeton University. Professor Salby has authored more than 100 scientific articles in major international journals, as well as the textbook *Fundamentals of Atmospheric Physics* (1996). His research focuses on changes of the atmospheric circulation in relation to global structure, energetics, and climate. Involving large-scale computer simulation and satellite data, Salby's research has provided insight into a wide range of phenomena in the Earth-atmosphere system.

Praise for *Physics of the Atmosphere and Climate*

"Salby's book is a graduate textbook on Earth's atmosphere and climate that is well balanced between the physics of the constituent materials and fluid dynamics. I recommend it as a foundation for anyone who wants to do research on the important open questions about aerosols, radiation, biogeochemisty, and ocean-atmosphere coupling."

–Professor Jim McWilliams, University of California, Los Angeles

"Salby's book provides an exhaustive survey of the atmospheric and climate sciences. The topics are well motivated with thorough discussion and are supported with excellent figures. The book is an essential reference for researchers and graduate and advanced undergraduate students who wish to have a rigorous source for a wide range of fundamental atmospheric science topics. Each chapter ends with an excellent selection of additional references and a challenging set of problems. Atmospheric and climate scientists will find this book to be an essential one for their libraries."

–Associate Professor Hampton N. Shirer, Pennsylvania State University

"Murry Salby presents an informative and insightful tour through the contemporary issues in the atmospheric sciences as they relate to climate. *Physics of the Atmosphere and Climate* is a valuable resource for educators and researchers alike, serving both as a textbook for the graduate or advanced undergraduate student with a physics or mathematics background and as an excellent reference and refresher for practitioners. It is a welcome addition to the field."

–Professor Darin W. Toohey, University of Colorado at Boulder

Salby's earlier book is a classic. As a textbook it is unequaled in breadth, depth, and lucidity. It is the single volume that I recommend to all of my students in atmospheric science. This new version improves over the previous version, if that is possible, in three aspects: beautiful illustrations of global processes (e.g. hydrological cycle) from newly available satellite data, new topics of current interest (e.g. interannual changes in the stratosphere and the oceans), and a new chapter on the influence of the ocean on the atmosphere. These changes make the book more useful as a starting point for studying climate change."

–Professor Yuk Yung, California Institute of Technology

PHYSICS OF THE ATMOSPHERE AND CLIMATE

MURRY L. SALBY
Macquarie University

CAMBRIDGE
UNIVERSITY PRESS

CAMBRIDGE
UNIVERSITY PRESS

University Printing House, Cambridge CB2 8BS, United Kingdom

One Liberty Plaza, 20th Floor, New York, NY 10006, USA

477 Williamstown Road, Port Melbourne, VIC 3207, Australia

314-321, 3rd Floor, Plot 3, Splendor Forum, Jasola District Centre, New Delhi - 110025, India

79 Anson Road, #06-04/06, Singapore 079906

Cambridge University Press is part of the University of Cambridge.

It furthers the University's mission by disseminating knowledge in the pursuit of
education, learning and research at the highest international levels of excellence.

www.cambridge.org
Information on this title: www.cambridge.org/9780521767187

© Murry L. Salby 2012

This publication is in copyright. Subject to statutory exception
and to the provisions of relevant collective licensing agreements,
no reproduction of any part may take place without the written
permission of Cambridge University Press.

First published 2012
Reprinted 2016

A catalogue record for this publication is available from the British Library

Library of Congress Cataloging in Publication data
Salby, Murry L.
Physics of the atmosphere and climate / Murry L. Salby. – 2nd ed.
p. cm.
Revised ed. of: Fundamentals of atmospheric physics. 1996.
Includes bibliographical references and index.
ISBN 978-0-521-76718-7 (hardback)
1. Atmospheric physics. I. Salby, Murry L. Fundamentals of atmospheric physics. II. Title.
QC861.3.S257 2011
551.5–dc23   2011033950

ISBN 978-0-521-76718-7 Hardback

Additional resources for this publication at www.cambridge.org/salby

Cambridge University Press has no responsibility for the persistence or
accuracy of URLs for external or third-party internet websites referred to in
this publication, and does not guarantee that any content on such websites is,
or will remain, accurate or appropriate.

To Alon

Strive for the Best

Contents

See colour plate section between pages 252 and 253

Preface

Global measurements from space, coupled with large-scale computer models, have widened the perspective of atmospheric science along with its subdiscipline, the study of climate. Supporting those tools are proxy records of previous climate upon which rest interpretations of the current state of the Earth-atmosphere system. While opening new avenues of investigation, these modern tools have introduced increasingly complex questions. Many concern the tools themselves. Uncertainties surround the interpretation of observations, especially proxy records of previous climate, how key physical processes are represented in Global Climate Models (GCMs), and discrepancies between those models. These uncertainties make an understanding of the controlling physical processes and limitations that surround their description essential for drawing reliable insight into the Earth-atmosphere system.

Emerging simultaneously with technological advances has been growing concern over the role of humans in global climate. Buttressed by wide-ranging claims of environmental consequences, such concern has been pushed into the limelight of major national and international policy. The popular ascent of climate research has not been without criticism. Notable are concerns over rigor and critical analysis, whereby (i) proxies of previous climate, relied upon in interpretations of current climate, are often remote and ambiguous and (ii) insight into underlying physical mechanisms has been supplanted by models which, although increasingly complex, remain, in many respects, primitive and poorly understood. Despite technological advances in observing the Earth-atmosphere system and in computing power, strides in predicting its evolution reliably – on climatic time scales and with regional detail – have been limited. The pace of progress reflects the interdisciplinary demands of the subject. Reliable simulation, adequate to reproduce the observed record of climate variation, requires a grasp of mechanisms from different disciplines and of how those mechanisms are interwoven in the Earth-atmosphere system.

Historically, students of the atmosphere and climate have had proficiency in one of the physical disciplines that underpin the subject, but not in the others. Under the fashionable umbrella of climate science, many today do not have proficiency in even one. What is today labeled climate science includes everything from archeology of the Earth to superficial statistics and a spate of social issues. Yet, many who embrace the label have little more than a veneer of insight into the physical processes that actually control the Earth-atmosphere system, let alone what is necessary to simulate its evolution reliably. Without such insight and its application to resolve major uncertainties, genuine progress is unlikely.

The atmosphere is the heart of the climate system, driven through interaction with the sun, continents, and ocean. It is the one component that is comprehensively observed. For this reason, the atmosphere is the central feature against which climate simulations must ultimately be validated.

This book builds on a forerunner, *Fundamentals of Atmospheric Physics*. It has been expanded to include climate, while streamlining the presentation for undergraduates in science, mathematics, and engineering. Advanced material, suitable as a resource for graduate students and researchers, has been retained (distinguished by shading). The treatment focuses upon physical concepts, which are developed from first principles. It integrates five major themes:

1. Atmospheric Thermodynamics;
2. Hydrostatic Equilibrium and Stability;
3. Radiation, Cloud, and Aerosol;
4. Atmospheric Dynamics and the General Circulation;
5. Interaction with the Ocean and Stratosphere.

Cornerstones of modern research, these themes are developed in a balance of theory and applications. Each is illustrated with manifestations on an individual day, the same day used to illustrate other themes. In this fashion, the Earth-atmosphere system is dissected in contemporaneous properties, revealing interactions among them. Supporting the development are detailed solutions to selected problems.

Chapter 1 presents an overview of the Earth-atmosphere system, describing its composition, structure, and energetics. It culminates in a discussion of global mean temperature, its relationship to atmospheric composition, and issues surrounding uncertainties in instrumental and proxy records of climate. Chapters 2–5 are devoted to atmospheric thermodynamics. Developed from a Lagrangian perspective, the discussion concentrates on heterogeneous systems that figure in considerations of cloud and its interaction with radiation, as well as the role of water vapor in the global energy budget. Hydrostatic equilibrium and stability are treated in Chapters 6 and 7, which develop their roles in convection and its influence on thermal and humidity structure. Chapters 8 and 9 focus on atmospheric radiation, cloud, and aerosol. After developing the laws governing radiative transfer, the presentation moves to the energetics of radiative and radiative-convective equilibrium. It then considers climate feedback mechanisms, which are discussed in relation to major contributors to the greenhouse effect, and their simulation in GCMs. Chapters 10–16 are devoted to atmospheric dynamics and the general circulation. The perspective is then transformed, via Reynolds' transport theorem, to the Eulerian description of behavior. Large-scale motion is first treated in terms of geostrophic and hydrostatic equilibrium and then extended to vorticity dynamics and quasi-geostrophic motion. The general circulation is motivated by a zonally symmetric model of heat transfer, setting the stage for baroclinic instability. Supporting it is a treatment of thermal properties of the Earth's surface, persistent

features of the circulation, and interannual fluctuations that comprise climate vari-
ablity. The presentation then turns in Chapter 17 to the ocean, its structure, dynamics,
and how it influences the atmosphere. The book closes with a treatment of the stratosphere, issues surrounding ozone, and interactions with the troposphere.

This book has benefited from interaction with numerous colleagues and students.
In addition to those received earlier, contributions and feedback were generously provided by W. Bourke, J. Frederiksen, R. Madden, E. Titova, D. Toohey, and J. Wu. Figures
were skillfully prepared by J. Davis and D. Oliver. Lastly, I am grateful for the understanding and encouragement of my son, without which this book would not have been
completed.

Murry L. Salby

Prelude

The most fruitful areas for growth of the sciences are those between established fields. Science has been increasingly the task of specialists, in fields which show a tendency to grow progressively narrower. Important work is delayed by the unavailability in one field of results that may have already become classical in the next field. It is these boundary regions of science that offer the richest opportunities to the qualified investigator.

Norbert Wiener

CHAPTER ONE

The Earth-atmosphere system

1.1 INTRODUCTION

The Earth's atmosphere is the gaseous envelope surrounding the planet. Like other planetary atmospheres, it figures centrally in transfers of energy between the sun, the Earth, and deep space. It also figures in transfers of energy from one region of the globe to another. By maintaining thermal equilibrium, such transfers determine the Earth's climate. However, among neighboring planets, the Earth's atmosphere is unique because it is related closely to ocean and surface processes that, together with the atmosphere, form the basis for life.

Because it is a fluid system, the atmosphere is capable of supporting a wide spectrum of motions. These range from turbulent eddies of a few meters to circulations with dimensions of the Earth itself. By rearranging mass, air motion influences other atmospheric components such as water vapor, ozone, and cloud, which figure prominently in radiative and chemical processes. Such influence makes the atmospheric circulation a key ingredient of the global energy budget.

1.1.1 Descriptions of atmospheric behavior

The mobility of a fluid system makes its description complex. Atmospheric motion redistributes mass and constituents into a variety of complex configurations. Like any fluid system, the atmosphere is governed by the laws of continuum mechanics. They can be derived from the laws of mechanics and thermodynamics that govern a discrete fluid body by generalizing those laws to a continuum of such systems. In the atmosphere, the discrete system to which these laws apply is an infinitesimal fluid element, or *air parcel*, which is defined by a fixed collection of matter.

1

Two frameworks are used to describe atmospheric behavior. The *Eulerian description* represents atmospheric behavior in terms of field properties, such as the instantaneous distributions of temperature, motion, and constituents. Governed by partial differential equations, the field description of atmospheric behavior is convenient for numerical applications. The *Lagrangian description* represents atmospheric behavior in terms of the properties of individual air parcels (e.g., in terms of their instantaneous positions, temperatures, and constituent concentrations). Because it focuses on transformations of properties within an air parcel and on interactions between that system and its environment, the Lagrangian description offers conceptual as well as certain diagnostic advantages. For this reason, the basic principles governing atmospheric behavior are developed in this text from a Lagrangian perspective.

In the Lagrangian framework, the system considered is an individual air parcel moving through the circulation. Although it may change in form through deformation and in composition through thermodynamic and chemical transformations, this system is uniquely identified by the matter comprising it initially. Mass can be transferred across the boundary of an air parcel through molecular diffusion and turbulent mixing. However, such transfers are slow enough to be ignored for many applications. An individual parcel can then change only through interaction with its environment and through internal transformations that alter its composition and state.

1.1.2 Mechanisms influencing atmospheric behavior

Of the factors influencing atmospheric behavior, gravity is the single most important. Even though it has no upper boundary, the atmosphere is contained by the gravitational field of the Earth, which prevents atmospheric mass from escaping to space. Because it is such a strong body force, gravity determines many atmospheric properties. Most immediate is the geometry of the atmosphere. Atmospheric mass is concentrated in the lowest 10 km – less than 1% of the Earth's radius. Gravitational attraction has compressed the atmosphere into a shallow layer above the Earth's surface in which mass and constituents are stratified vertically: They are layered.

Through stratification of mass, gravity imposes a strong kinematic constraint on atmospheric motion. Circulations with dimensions greater than a few tens of kilometers are quasi-horizontal. Vertical displacements of air are then much smaller than horizontal displacements. Under these circumstances, constituents such as water vapor and ozone fan out in layers or "strata." Vertical displacements are comparable to horizontal displacements only in small-scale circulations such as convective cells and fronts, which have horizontal dimensions comparable to the vertical scale of the mass distribution.

The compressibility of air complicates the description of atmospheric behavior by enabling the volume of a parcel to change as it experiences changes in surrounding pressure. Therefore, concentrations of mass and constituents for the parcel can change, even though the number of molecules remains fixed. The concentration of a chemical constituent can also change through internal transformations, which alter the number of a particular type of molecule. For example, condensation decreases the abundance of water vapor in an air parcel that passes through a cloud system. Photodissociation of O_2 will increase the abundance of ozone in a parcel that passes through a region of sunlight.

Exchanges of energy with its environment and transformations between one form of energy and another likewise alter the properties of an air parcel. By expanding, an air parcel exchanges energy mechanically with its environment through work that it performs on the surroundings. Heat transfer, as occurs through absorption of radiant energy and conduction with the Earth's surface, represents a thermal exchange of energy with a parcel's environment. Absorption of water vapor by an air parcel (e.g., through contact with a warm ocean surface) has a similar effect. When the vapor condenses, latent heat of vaporization carried by the vapor is released to the surrounding molecules of dry air that comprise the parcel. If the condensed water then precipitates back to the Earth's surface, this process leads to a net transfer of heat from the Earth's surface to the parcel.

The Earth's rotation, like gravity, exerts an important influence on atmospheric motion and, hence, on distributions of atmospheric properties. Because the Earth is a noninertial reference frame, the conventional laws of mechanics do not hold; they must be modified to account for its acceleration. Apparent forces introduced by the Earth's rotation are responsible for properties of the large-scale circulation, in particular, the flow of air around centers of low and high pressure. Those forces also inhibit meridional, e.g., NS (North-South) motion. Consequently, they inhibit transfers of heat and constituents between the equator and poles. For this reason, rotation tends to stratify properties meridionally, just as gravity tends to stratify them vertically.

The physical processes described in the preceding paragraphs do not operate independently. They are interwoven in a complex fabric of radiation, chemistry, and dynamics that govern the Earth-atmosphere system. Interactions among these can be just as important as the individual processes themselves. For instance, radiative transfer controls the thermal structure of the atmosphere, which determines the circulation. Transport by the circulation, in turn, influences the distributions of radiatively active components such as water vapor, ozone, and cloud. In view of their interdependence, understanding how one of these processes influences behavior requires an understanding of how that process interacts with others. This feature makes the study of the Earth-atmosphere system an eclectic one, involving the integration of many different physical principles. This book develops the most fundamental of these.

1.2 COMPOSITION AND STRUCTURE

The Earth's atmosphere consists of a mixture of gases, mostly molecular nitrogen (78% by volume) and molecular oxygen (21% by volume); see Table 1.1. Water vapor, carbon dioxide, and ozone, along with other minor constituents, comprise the remaining 1% of the atmosphere. Although present in very small abundances, trace species such as water vapor and ozone play a key role in the energy balance of the Earth through their involvement in radiative processes. Because they are created and destroyed in particular regions and are linked to the circulation through transport, these and other minor species are highly variable. For this reason, trace species are treated separately from the primary atmospheric constituents, which are referred to simply as "dry air."

1.2.1 Description of air

The starting point for describing atmospheric behavior is the ideal gas law

$$
\begin{aligned}
pV &= nR^*T \\
&= \frac{m}{M}R^*T \\
&= mRT,
\end{aligned}
\tag{1.1}
$$

Table 1.1. Atmospheric Composition. Constituents are listed with volume mixing ratios representative of the Troposphere or Stratosphere, how the latter are distributed vertically, and controlling processes

Constituent	Tropospheric Mixing Ratio	Vertical Distribution (Mixing Ratio)	Controlling Processes
N_2	.7808	Homogeneous	Vertical Mixing
O_2	.2095	Homogeneous	Vertical Mixing
*H_2O	≤0.030	Decreases sharply in Troposphere Increases in Stratosphere Highly Variable	Evaporation, Condensation, Transport Production by CH_4 Oxidation
Ar	.0093	Homogeneous	Vertical Mixing
*CO_2	380 ppmv	Homogeneous	Vertical Mixing Production by Surface and Anthropogenic Processes
*O_3	10 ppmv$^\$$	Increases sharply in Stratosphere Highly Variable	Photochemical Production in Stratosphere; secondarily through pollution in troposphere Destruction at Surface Transport
*CH_4	1.8 ppmv	Homogeneous in Troposphere Decreases in Middle Atmosphere	Production by Surface Processes Oxidation Produces H_2O
*N_2O	320 ppbv	Homogeneous in Troposphere Decreases in Middle Atmosphere	Production by Surface and Anthropogenic Processes Dissociation in Middle Atmosphere Produces NO Transport
*CO	70 ppbv	Decreases in Troposphere Increases in Stratosphere	Production Anthropogenically and by Oxidation of CH_4 Transport
NO	0.1 ppbv$^\$$	Increases Vertically	Production by Dissociation of N_2O Catalytic Destruction of O_3
*CFC–11	0.2 ppbv	Homogeneous in Troposphere	Industrial Production
*CFC–12	0.5 ppbv	Decreases in Stratosphere	Mixing in Troposphere
*HFC–134A	30 ppt		Photo-dissociation in Stratosphere

* Radiatively active
$ Stratospheric value

which constitutes the equation of state for a pure (single-component) gas. In (1.1), p, T, and M denote the pressure, temperature, and molar weight of the gas, and V, m, and $n = \frac{m}{M}$ refer to the volume, mass, and molar abundance of a fixed collection of matter (e.g., an air parcel). The *specific gas constant R* is related to the *universal gas constant R** through

$$R = \frac{R^*}{M}. \tag{1.2}$$

Equivalent forms of the ideal gas law that do not depend on the dimension of the system are

$$p = \rho RT$$
$$pv = RT, \tag{1.3}$$

where ρ and $v = \frac{1}{\rho}$ (also denoted α) are the density and *specific volume* of the gas.

Because it is a mixture of gases, air obeys similar relationships. So do its individual components. The *partial pressure* p_i of the *i*th component is that pressure the *i*th component would exert in isolation at the same volume and temperature as the mixture. It satisfies the equation of state

$$p_i V = m_i R_i T, \tag{1.4.1}$$

where R_i is the specific gas constant of the *i*th component. Similarly, the *partial volume* V_i is that volume the *i*th component would occupy in isolation at the same pressure and temperature as the mixture. It satisfies the equation of state

$$pV_i = m_i R_i T. \tag{1.4.2}$$

Dalton's law asserts that the pressure of a mixture of gases equals the sum of their partial pressures

$$p = \sum_i p_i. \tag{1.5}$$

Likewise, the volume of the mixture equals the sum of the partial volumes[1]

$$V = \sum_i V_i. \tag{1.6}$$

The equation of state for the mixture can be obtained by summing (1.4) over all of the components

$$pV = T \sum_i m_i R_i.$$

Then, defining the mean specific gas constant

$$\overline{R} = \frac{\sum_i m_i R_i}{m} \tag{1.7}$$

yields the equation of state for the mixture

$$pV = m\overline{R}T. \tag{1.8}$$

The mean molar weight of the mixture is defined by

$$\overline{M} = \frac{m}{n}. \tag{1.9}$$

Because the molar abundance of the mixture is equal to the sum of the molar abundances of the individual components,

$$n = \sum_i \frac{m_i}{M_i},$$

(1.9) may be expressed

$$\overline{M} = \frac{R^* m}{\sum_i m_i \left(\frac{R^*}{M_i}\right)}.$$

---

[1] These are among several consequences of the Gibbs-Dalton law, which relates the properties of a mixture to properties of the individual components (e.g., Keenan, 1970).

Then applying (1.2) for the ith component together with (1.7) leads to

$$\overline{R} = \frac{R^*}{\overline{M}}.$$ (1.10)

Equation (1.10) is analogous to (1.2) for a single-component gas.

Because of their involvement in radiative and chemical processes, variable components of air must be quantified. The "absolute concentration" of the ith species is measured by its density ρ_i or, alternatively, by its *number density*

$$[i] = \left(\frac{N_A}{M_i}\right)\rho_i$$ (1.11)

(also denoted n_i), where N_A is Avogadro's number and M_i is the molar weight of the species. Partial pressure p_i and partial volume V_i are other measures of absolute concentration.

The compressibility of air makes absolute concentration an ambiguous measure of a constituent's abundance. Even if a constituent is passive, namely if the number of molecules inside an individual parcel is fixed, its absolute concentration can change through changes of volume. For this reason, a constituent's abundance is more faithfully described by the "relative concentration," which is referenced to the overall abundance of air or simply dry air. The relative concentration of the ith species is measured by the *molar fraction*

$$N_i = \frac{n_i}{n}.$$ (1.12)

Dividing (1.4) by (1.8) and applying (1.2) for the ith component leads to

$$N_i = \frac{p_i}{p} = \frac{V_i}{V}.$$ (1.13)

Molar fraction uses as a reference the molar abundance of the mixture, which can vary through changes of individual species. A more convenient measure of relative concentration is mixing ratio. The *mass mixing ratio* of the ith species

$$r_i = \frac{m_i}{m_d},$$ (1.14)

where the subscript d refers to dry air, is dimensionless. It is expressed in g kg^{-1} for tropospheric water vapor and in parts per million by mass (ppmm, or simply ppm) for stratospheric ozone. Unlike molar abundance, the reference mass m_d is constant for an individual air parcel. If the ith species is passive, namely if it does not undergo a transformation of phase or a chemical reaction, its mass m_i is also constant. The mixing ratio r_i is then fixed for an individual air parcel.

For a trace species, such as water vapor or ozone, the mixing ratio is closely related to the molar fraction

$$N_i \cong \frac{r_i}{\epsilon_i},$$ (1.15.1)

where

$$\epsilon_i = \frac{M_i}{M_d},$$ (1.15.2)

because the mass of air in the presence of such species is nearly identical to that of dry air. The *volume mixing ratio* provides similar information. It is distinguished from mass mixing ratio by dimensions such as parts per million by volume (ppmv) for stratospheric ozone (Probs. 1.2, 1.3). From (1.13) and (1.12), it follows that the volume mixing ratio is approximately equal to the molar fraction. Each measures the relative abundance of molecules of the ith species.

As noted, the mixing ratio of a passive species is fixed for an individual air parcel. A property that is invariant for individual parcels is said to be *conserved*. Although constant for individual air parcels, a conserved property is generally not constant in space and time. Unless that property happens to be homogeneous, its distribution must vary spatially and temporally, as parcels with different values exchange positions. A conserved property is a *material tracer* because particular values track the motion of individual air parcels. Thus, tracking particular values of r_i provides a description of how air is rearranged by the circulation and, therefore, of how all conserved species are redistributed.

1.2.2 Stratification of mass

By confining mass to a shallow layer above the Earth's surface, gravity exerts a profound influence on atmospheric behavior. If vertical accelerations are ignored, then Newton's second law of motion applied to the column of air between some level at pressure p and a level incrementally higher at pressure $p + dp$ (Fig. 1.1) reduces to a balance between the weight of that column and the net pressure force acting on it

$$pdA - (p + dp)dA = \rho g dV,$$

where g denotes the acceleration of gravity, or

$$\frac{dp}{dz} = -\rho g.$$ (1.16)

This simple form of mechanical equilibrium is known as *hydrostatic balance*. It is a good approximation even if the atmosphere is in motion because, for large-scale circulations, vertical displacements of air are small. This feature renders vertical acceleration two to three orders of magnitude smaller than the forces in (1.16). Applying the same analysis between the pressure p and the top of the atmosphere (where p vanishes) illustrates the origin of atmospheric pressure: The pressure at any level must equal the weight of the atmospheric column of unit cross-sectional area above that level.

Owing to the compressibility of air, the density in (1.16) is not constant. It depends on the air's pressure through the gas law. Eliminating ρ with (1.3) and integrating from the surface to an altitude z yields

$$\frac{p}{p_s} = e^{-\int_{z_s}^{z} \frac{dz'}{H(z')}},$$ (1.17.1)

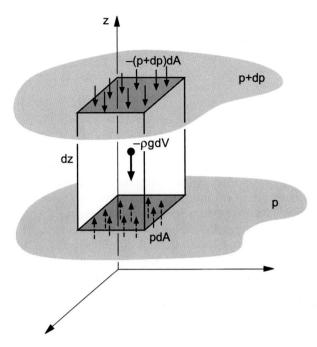

Figure 1.1 Hydrostatic balance for an incremental atmospheric column of cross-sectional area dA and height dz, bounded vertically by isobaric surfaces at pressures p and $p + dp$.

where

$$H(z) = \frac{RT(z)}{g} \tag{1.17.2}$$

is the pressure *scale height* and p_s is the surface pressure. The scale height represents the characteristic vertical dimension of the mass distribution. A function of altitude, it varies from about 8 km near the Earth's surface to 6 km in very cold regions of the atmosphere.

As illustrated by Fig. 1.2, global-mean pressure and density decrease with altitude approximately exponentially. Pressure decreases from about 1000 hPa or 10^5 Pascals (Pa) at the surface to only 10% of that value at an altitude of 15 km (two scale heights).[2] According to hydrostatic balance, 90% of the atmosphere's mass then lies beneath this level. Pressure decreases by another factor of 10 for each additional 15 km of altitude. Density decreases with altitude at about the same rate, from a surface value of about 1.2 kg m^{-3}. The sharp upward decrease of pressure implies that isobaric surfaces, along which $p = $ const, are quasi-horizontal. Deflections of those surfaces introduce comparatively small horizontal variations of pressure that drive atmospheric motion.

---

[2] The historical unit of pressure, millibar (mb), has been replaced by its equivalent in the Standard International (SI) system of units, the hectoPascal (hPa), where 1 hPa = 100 Pa = 1 mb. See Appendix A for conversions between the SI system of units and others.

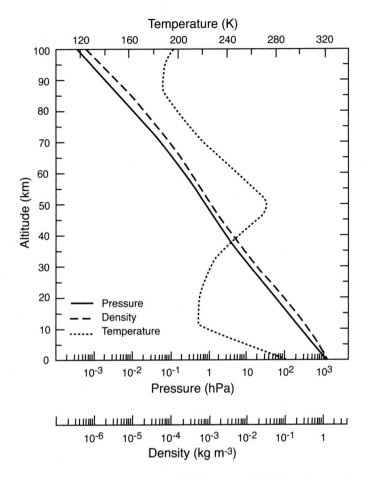

Figure 1.2 Global-mean pressure (solid), density (dashed), and temperature (dotted), as functions of altitude. *Source: U.S. Standard Atmosphere* (1976).

Above 100 km, pressure and density also decrease exponentially (Fig. 1.3), but at a rate which differs from that below and which varies gradually with altitude. The distinct change of behavior near 100 km marks a transition in the processes that control the stratification of mass and the composition of air. The mean free path of molecules is determined by the frequency of collisions. It varies inversely with air density. Consequently, the mean free path increases exponentially with altitude, from about 10^{-7} m at the surface to of order 1 m at 100 km. Because it controls molecular diffusion, the mean free path determines properties of air such as viscosity and thermal conductivity. Diffusion of momentum and heat supported by those properties dissipate atmospheric motion by destroying gradients of velocity and temperature.

Below 100 km, the mean free path is short enough for turbulent eddies in the circulation to be only weakly damped by molecular diffusion. At those altitudes, bulk transport by turbulent air motion dominates diffusive transport of atmospheric constituents. Turbulence stirs different gases with equal efficiency. Mixing ratios of

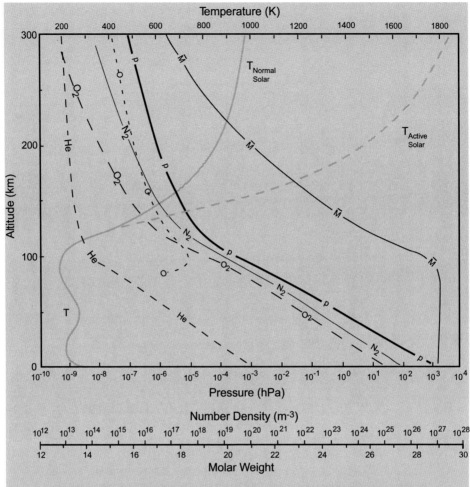

Figure 1.3 Global-mean pressure (bold), temperature (stippled), mean molar weight (solid), and number densities of atmospheric constituents as functions of altitude. *Source: U.S. Standard Atmosphere* (1976).

passive constituents are therefore homogeneous in this region. Those constituents are said to be "well mixed." The densities of passive constituents then all decrease with altitude at the same exponential rate. This gives air a homogeneous composition, with constant mixing ratios $r_{N_2} \cong 0.78$, $r_{O_2} \cong 0.21$, and the constant gas properties[3]

$$M_d = 28.96 \ g \ mol^{-1} \tag{1.18.1}$$

$$R_d = 287.05 \ J kg^{-1} K^{-1}. \tag{1.18.2}$$

The well-mixed region below 100 km is known as the *homosphere*.

---

[3] Properties of dry air are tabulated in Appendix B, along with other thermodynamic constants.

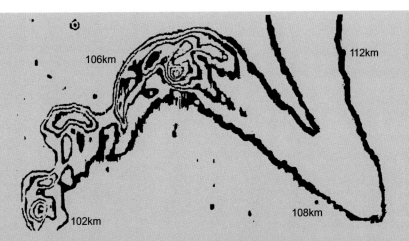

Figure 1.4 Constant-density contours of a chemical vapor trail released by a rocket traversing the turbopause. Beneath 107 km, the vapor trail is distorted by an array of turbulent eddies that form in the wake of the rocket. Above 107 km, the vapor trail remains laminar, reflecting the absence of turbulence, and expands under the action of molecular diffusion. Adapted from Roper (1977).

Above 100 km, the mean free path quickly becomes larger than turbulent displacements of air. As a result, turbulent air motion there is strongly damped by diffusion of momentum and heat. Diffusive transport then becomes the dominant mechanism for transferring properties vertically. The transition from turbulent transport to diffusive transport occurs at the *homopause* (also known as the *turbopause*), which has an average altitude of about 100 km. Figure 1.4 shows a rocket vapor trail traversing this region. Below 100 km, the trail is marked by turbulent eddies. Produced in the wake of the rocket, they homogenize different constituents with equal efficiency. Those eddies are conspicuously absent above 107 km. At this altitude, molecular diffusion suppresses turbulent air motion, becoming the prevailing form of vertical transport.

The region above the homopause and below 500 km is known as the *heterosphere*. In the heterosphere, air flow is nearly laminar, dominated by molecular diffusion. Because it operates on gases according to their molar weights, molecular diffusion stratifies constituents so that the heaviest species O_2 decreases with altitude more rapidly than the second heaviest species N_2 and so forth (Fig. 1.3). For this reason, the composition of air changes with altitude in the heterosphere, as is evidenced by the mean molar weight in Fig. 1.3. Constant below the homopause, $\overline{M}$ changes abruptly near 100 km, above which it decreases upward monotonically.

Diffusive separation is primarily responsible for the stratification of constituents in the heterosphere. However, photo-dissociation also plays a role. Energetic ultraviolet (UV) radiation incident at the top of the atmosphere dissociates molecular oxygen. This process provides an important source of atomic oxygen at these altitudes. In fact, not far above the homopause, O becomes the dominant form of oxygen.

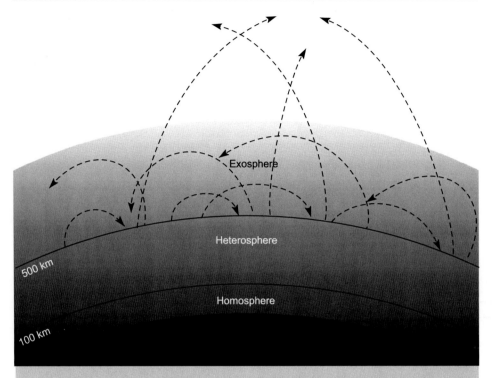

Figure 1.5 Schematic cross section of the atmosphere illustrating the homosphere, heterosphere, and the exosphere, in which molecular trajectories are shown.

Photo-dissociation of H_2O at lower altitudes by less energetic UV radiation liberates atomic hydrogen. H produced in this manner is gradually mixed and diffused to higher altitude. By dissociating such molecules, energetic radiation is filtered from the solar spectrum that penetrates to lower levels.

In the homosphere and heterosphere, atmospheric molecules interact strongly through frequent collisions. Their interaction becomes infrequent above an altitude of about 500 km, known as the *critical level*. At this altitude, molecular collisions are so rare that a significant fraction of the molecules pass out of the atmosphere without experiencing a single collision. Known as the *exosphere*, the region above the critical level contains molecules that leave the denser atmosphere and move out into space. As is illustrated in Fig. 1.5, these molecules follow ballistic trajectories that are determined by the molecular velocity at the critical level and the gravitational attraction of the Earth. Most of the molecules in the exosphere are captured by the Earth's gravitational potential. They return to the denser atmosphere along parabolic trajectories. However, some molecules have velocities great enough to escape the Earth's gravitational potential entirely. Those are lost to deep space.

The escape velocity v_e is determined by the kinetic energy adequate to liberate a molecule from the potential well of the Earth's gravitational field. That energy equals the work performed to displace a molecule from the critical level to infinity

$$\frac{1}{2}mv_e^2 = \int_a^\infty mg_0 \left(\frac{a}{r}\right)^2 dr, \tag{1.19}$$

where a is the Earth's radius, g_0 is the gravitational attraction averaged over the surface of the Earth, $m \cdot g_0 \left(\frac{a}{r}\right)^2$ is the molecule's local weight, dr is its local displacement, and the difference between a and the radial distance to the critical level is negligible. The escape velocity follows from (1.19) as

$$v_e = \sqrt{2g_0 a}. \tag{1.20}$$

For Earth, v_e has a value of about 11 km s^{-1}. Independent of molecular weight, v_e is the same for all molecules. However, different molecules do not all have the same distribution of velocities. Energy is equipartitioned in a molecular ensemble (see, e.g., Lee, Sears, and Turcotte, 1973). Lighter molecules thus have faster velocities than heavier ones. For this reason, lighter atmospheric constituents escape to space more readily than do heavier constituents.

The critical level of the Earth's atmosphere lies near 500 km. The temperature at this altitude is about 1000 K under conditions of normal solar activity. However, it can reach 2000 K under disturbed solar conditions (Fig. 1.3). Figure 1.6 shows for O and H the Boltzmann distributions of a molecular ensemble (see, e.g., Lee, Sears, and Turcotte, 1973), as functions of molecular velocity:

$$\frac{dn}{n} = \frac{4}{\sqrt{\pi}} \frac{v^2}{v_0^3} e^{-\left(\frac{v}{v_0}\right)^2} dv, \tag{1.21}$$

where $\frac{dn}{n}$ represents the fractional number of molecules having velocities in the range $(v, v + dv)$,

$$v_0 = \sqrt{\frac{2kT}{m}} \tag{1.22}$$

is the most probable velocity, m is the mass of the molecules, and k is the Boltzmann constant. The fraction of molecules with velocities exceeding the escape velocity is

$$\frac{\Delta n_e}{n} = \int_{v_e}^\infty \frac{4}{\sqrt{\pi}} \frac{v^2}{v_0^3} e^{-\left(\frac{v}{v_0}\right)^2} dv$$
$$\cong \frac{2}{\sqrt{\pi}} \left(\frac{v_e}{v_0}\right) e^{-\left(\frac{v_e}{v_0}\right)^2} \tag{1.23}$$

for $v_0 \ll v_e$.

For atomic oxygen, the most probable velocity is $v_0 = 1.02$ km s^{-1}. The fraction of O molecules with velocities faster than v_e is only about 10^{-45}. A lower bound on the time to deplete all O molecules initially at the critical level (e.g., neglecting production of O molecules locally by photo-dissociation and diffusive transport from below) is given by the mean time between collisions divided by the fraction of molecules moving upward with $v > v_e$. Near 500 km, the mean time between collisions is of order 10 s. Hence the time for all O molecules initially at the critical level to escape the Earth's gravitational field is greater than 10^{46} s. This is far greater than the 4 billion years that the Earth has existed. Heavier species are captured by the Earth's gravitational field even more effectively.

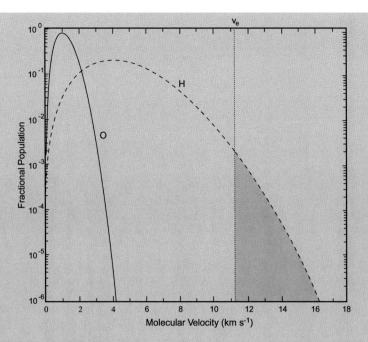

Figure 1.6 Boltzmann distribution of velocities for a molecular ensemble of oxygen atoms and hydrogen atoms. Escape velocity v_e for Earth also indicated.

The situation for hydrogen differs significantly. The population of H is distributed over much faster velocities, so many more molecules exceed the escape velocity v_e. The most probably velocity is 4.08 km s^{-1}, whereas the fraction of molecules with velocities faster than v_e is about 10^{-4}. By previous reasoning, a significant fraction of H molecules initially at the critical level is lost to deep space after only 10^5 s -1 day. Under conditions of disturbed solar activity, temperatures at the critical level are substantially greater. Hydrogen molecules are then boiled off of the atmosphere even faster. The rapid escape of atomic hydrogen from the Earth's gravitational field explains why H is found in the atmosphere only in very small abundances, despite its continual production by photo-dissociation of H_2O.

1.2.3 Thermal and dynamical structure

The atmosphere is categorized according to its thermal structure, which determines dynamical properties of individual regions. The simplest picture of atmospheric thermal structure is the vertical profile of global-mean temperature in Fig. 1.2. From the surface up to about 10 km, temperature decreases upward at a nearly constant *lapse rate*, which is defined as the rate at which temperature "decreases" with altitude. This layer immediately above the Earth's surface is known as the *troposphere*, which means "turning sphere." Its title is symbolic of the convective overturning that characterizes this region. Having a global-mean lapse rate of about 6.5 K km^{-1}, the troposphere contains most of the behavior identified with weather. It is driven ultimately by surface heating, which is then transferred to the atmosphere. The upper boundary of the troposphere, or "tropopause," lies at an altitude of about 10 km (100 hPa), marked by a minimum of temperature and a sharp change of lapse rate.

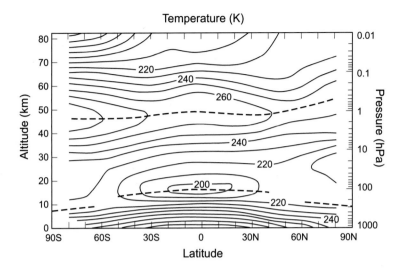

Figure 1.7 Zonal-mean temperature during northern winter as a function of latitude and altitude. Adapted from Fleming et al. (1988).

The region from the tropopause to an altitude of about 85 km is known as the *middle atmosphere*. Temperature first remains nearly constant and then increases upward in the *stratosphere*, which means "layered sphere." Its title is symbolic of properties at these altitudes, which experience little vertical mixing. In the stratosphere, temperature increases upward (negative lapse rate). This thermal structure reflects ozone heating; the latter results from the absorption of solar UV by O_3, which increases upward above the tropopause. Contrary to the troposphere, the stratosphere involves only weak vertical motions. It is dominated by radiative processes. The upper boundary of the stratosphere, or "stratopause," lies at an altitude of about 50 km (1 hPa), where temperature reaches a maximum.

Above the stratopause, temperature again decreases with altitude in the *mesosphere*, where ozone heating diminishes. Convective motion and radiative processes are both important in the mesosphere. Meteor trails form in this region of the atmosphere; so do lower layers of the ionosphere during daylight hours. The "mesopause" lies at an altitude of about 85 km (.01 hPa), where a second minimum of temperature is reached.

Above the mesopause, temperature increases steadily in the *thermosphere* (cf. Fig. 1.3). Unlike lower regions, the thermosphere cannot be treated as an electrically neutral continuum. Ionization of molecules by energetic solar radiation produces a plasma of free electrons and ions. Oppositely charged, those particles interact differently with the Earth's electric and magnetic fields. As is apparent in Fig. 1.3, this region of the atmosphere is influenced strongly by variations of solar activity. Below the mesopause, however, the influence of solar variations is limited.

A more complete picture of the thermal structure of the atmosphere is provided by the zonal-mean temperature $\overline{T}$, where $\overline{(\ )}$ denotes the longitudinal average. $\overline{T}$ is plotted in Fig. 1.7, as a function of latitude and altitude, during northern winter. In the troposphere, temperature decreases with altitude and latitude. The tropopause, which is characterized by an abrupt change of lapse rate, is highest in the tropics (~16 km), where temperature has decreased to 200 K. It is lowest in polar regions

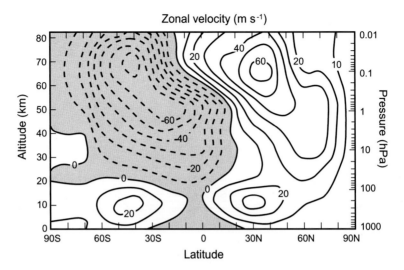

Figure 1.8 As in Fig. 1.7, but for the zonal component of velocity.

(~8 km), where temperature is 10–20 K warmer, even in perpetual darkness that pre-vails over the winter pole (*polar night*). A sharp change of zonal-mean lapse rate is not observed at mid-latitudes, symbolized by a break in the tropopause. Above the tropopause, temperature increases with altitude. It is warmest over the summer pole and decreases steadily to coldest temperature over the winter pole. In the mesosphere, where temperature again decreases with altitude, the horizontal temperature gradient is reversed. Temperature is actually coldest over the summer pole, which lies in perpet-ual sunlight (*polar day*). It increases steadily to warmest temperature over the winter pole, which lies in perpetual darkness. This peculiarity of the temperature distribution is incongruous with radiative considerations. It illustrates the importance of dynamics to establishing observed thermal structure.

The thermal structure in Fig. 1.7 is related closely to the zonal-mean circulation $\bar{u}$, which is shown in Fig. 1.8 at the same time of year. In the troposphere, the zonal (EW) velocity u is characterized by subtropical *jet streams*. They strengthen upward, maximizing near the tropopause. These jets describe circumpolar motion that is west-erly in each hemisphere.[4] Above the subtropical jets, zonal-mean wind first weakens with altitude. It then intensifies, but with opposite sign in the two hemispheres. In the winter hemisphere, westerlies intensify above the tropopause, forming the *polar-night jet*. There, wind speed reaches 60 m s^{-1} in the lower mesosphere. In the summer hemisphere, westerlies weaken above the tropopause, replaced by easterlies that inten-sify up to the mesosphere. This easterly circulation attains speeds somewhat stronger than the westerly circulation in the winter hemisphere. At lower altitudes, it joins weak easterlies that prevail in the tropical troposphere.

On individual days, the circulation is more complex and variable than is represented in the zonally- and time-averaged distributions in Figs. 1.7 and 1.8. Figure 1.9a shows, for an individual day, the instantaneous circulation on the 500-hPa isobaric surface: that locus of points where the pressure equals 500 hPa. Contoured over the circulation is the height of the 500 hPa isobaric surface. It reflects the horizontal distribution of pressure.

---

[4] In meteorological vernacular, *westerly* refers to motion from the west, whereas *easterly* refers to motion from the east.

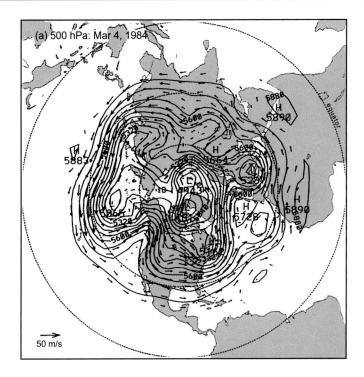

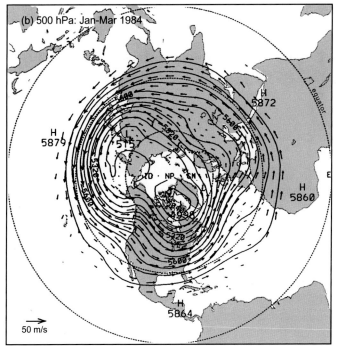

Figure 1.9 (a) Height (contours) of and horizontal velocity (vectors) on the 500-hPa isobaric surface for March 4, 1984. (b) Time-mean height of and velocity on the 500-hPa isobaric surface for January–March, 1984. Isobaric heights shown in meters. From National Meteorological Center (NMC) analyses.

Hydrostatic equilibrium (1.17) requires pressure and altitude to maintain a single-valued relationship (Fig. 1.2). Therefore, if one is known, so is the other. The altitude of the 500-hPa isobaric surface may then be interpreted analogously to the pressure on a surface of constant altitude $z \cong 5.5$ km, the mean height of the 500-hPa surface. Because pressure decreases upward monotonically, low altitude of the 500-hPa surface corresponds to low pressure on that surface of constant altitude (cf. Fig. 6.2). Similarly, high altitude of the 500-hPa surface corresponds to high pressure on that surface of constant altitude. Thus, centers of low altitude that punctuate the height of the 500-hPa surface in Fig. 1.9a imply centers of low pressure on the surface of constant altitude, $z \cong 5.5$ km. Centers of high altitude imply the reverse.

The 500-hPa surface slopes downward toward the pole, in the direction of decreasing tropospheric temperature (Figs. 1.9; 1.7). The circulation at 500 hPa is characterized by westerly circumpolar flow. It corresponds to the zonal-mean subtropical jet in Fig. 1.8. However, on the individual day shown, the jet stream is far from zonally symmetric. It is disturbed by half a dozen depressions of the 500-hPa surface. They are associated with *synoptic weather systems*. Those unsteady disturbances deflect the air stream meridionally (i.e., NS), distorting the circulation into a wavy pattern that meanders around the globe. Although highly disturbed, the circumpolar flow remains nearly tangential to contours of isobaric height.

In addition to synoptic weather systems, disturbances of global dimension also appear in the 500-hPa circulation. Known as *planetary waves*, these disturbances are manifested by a displacement of the circumpolar flow out of zonal symmetry, as is apparent at high latitude. They introduce a gradual undulation of the jet stream about latitude circles. Planetary waves are more evident in the time-mean circulation, shown in Fig. 1.9b. In it, unsteady synoptic weather systems have been filtered out – because they fluctuate on short time scales. The time-mean circulation exhibits more zonal symmetry. It therefore possesses greater correspondence to zonal-mean structure in Fig. 1.8 (e.g., the subtropical jet) than does the instantaneous circulation in Fig. 1.9a. Nonetheless, the time-mean flow is still disturbed, only on larger scales. Steady planetary waves are not removed by time averaging. They too displace the circumpolar flow out of zonal symmetry, deflecting the air stream across latitude circles. In addition, the time-mean circulation contains locally intensified jets. They mark the *North Atlantic storm track* and the *North Pacific storm track*. Accompanied by a steep meridional gradient of height, they are preferred sites of cyclone development. By determining these and other longitudinally dependent features of the time-mean circulation, planetary waves shape the climate of individual regions.

In the stratosphere, the circulation is also unsteady. However, the synoptic disturbances that prevail in the troposphere are not evident in the instantaneous circulation at 10 hPa, shown in Fig. 1.10a for the same day as in Fig. 1.9a. Instead, only planetary-scale disturbances appear at this level. The time-mean circulation at 10 hPa (Fig. 1.10b) is characterized by strong westerly flow. It corresponds to the polar-night jet in Fig. 1.8 and to a circumpolar vortex, the *polar-night vortex*. Like motion at 500 hPa, the time-mean circulation is still disturbed from zonal symmetry by steady planetary waves, which are not eliminated by averaging over time. By comparison, the instantaneous circulation in Fig. 1.10a is much more disturbed. Counterclockwise motion about the low, comprising the polar-night vortex, has been displaced off the pole, replaced by clockwise motion about a high that has amplified and temporarily invaded the polar cap. This disturbance deflects the air stream meridionally, which therefore undergoes

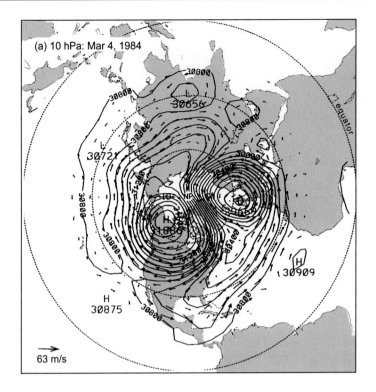

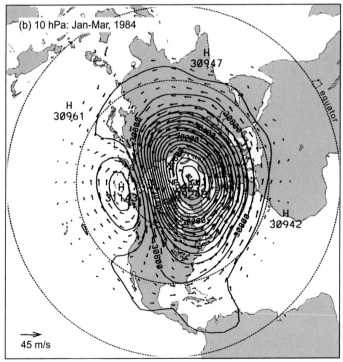

Figure 1.10 As in Fig. 1.9, but for the 10-hPa isobaric surface.

Figure 1.11 Satellite image of wavy cloud patterns found downwind of mountainous terrain. From Scorer (1986). Reproduced with permission of Ellis Horwood Ltd.

large excursions in latitude. Air is then transported from one radiative environment to another (e.g., from the sunlit tropics into polar darkness).

In addition to large-scale features, like those described in the preceding paragraphs, the circulation is also disturbed on smaller dimensions that are not resolved in the global analyses shown in Figs. 1.9 and 1.10. Known as *gravity waves*, these small-scale disturbances owe their existence to buoyancy and the stratification of mass. Gravity waves are manifested in wavy patterns that appear in layered clouds. They are often observed from the ground and in satellite imagery like that shown in Fig. 1.11. Like planetary waves, gravity waves contain both transient and steady components. Consequently, they are present even in time-mean fields.

1.2.4 Trace constituents

Beyond its primary constituents, air contains a variety of trace species. Although present in minor abundances, several play key roles in radiative and chemical processes. Perhaps the simplest is CO_2 because it is chemically inert away from the Earth's surface and therefore well-mixed throughout the homosphere. Like N_2 and O_2, carbon dioxide has a nearly uniform mixing ratio, $r_{CO_2} \cong 380$ ppmv in 2010. However, unlike the primary constituents of air, CO_2 is coupled to human activities.

Carbon Dioxide

Involved in chemical and biological processes, CO_2 is produced and destroyed naturally near the Earth's surface. It is produced through oxidation processes, like cell respiration and soil respiration. The latter is associated with the bacterial decomposition of organic matter under aerobic conditions (aerobic fermentation). These processes absorb oxygen and convert it into carbon dioxide. CO_2 is also produced under anaerobic conditions, at deeper soil levels. Involving similar pathways, anaerobic fermentation likewise involves the bacterial decomposition of organic matter, whereby carbohydrates are oxidized to produce CO_2 and hydrocarbons like methane. CO_2 is destroyed through reverse pathways, reduction processes – notably photosynthesis. Those processes absorb carbon dioxide and convert it into carbohydrate and back into oxygen. Together, these natural processes lead to emission of ~150 gigatons of carbon per year (150 GtC/yr), offset by about as much absorption (see Fig. 17.11). Almost half of the O_2 emitted to and CO_2 absorbed from the atmosphere are estimated to come from *phytoplankton*, marine biota that form the underpinnings of the food chain and cycle CO_2 into O_2.

The relationship of CO_2 to temperature is documented in prehistorical records of the Earth's climate. Glacial ice cores, drilled from great depths, provide a record of atmospheric composition dating far into the past. In concert with isotopic information on atmospheric temperature, that record suggests a link between atmospheric CO_2 and global temperature.

During previous climates of the Earth, these quantities varied in a systematically related fashion. Figure 1.12a compares records of CO_2 and temperature, inferred from an Antarctic ice core that extends back 400,000 years. Minima in CO_2 (dashed) were achieved when its mixing ratio had decreased below 200 ppmv. Those features coincide with anomalously-cold conditions, when, simultaneously, temperature (solid) had decreased by some 8 K. Both coincide with *glacial periods*, when ice volume (Fig. 1.12b) maximized some 20,000 140,000, 250,000, and 350,000 years ago. The periods of ice maxima were followed by periods of warming, *interglacial periods*, when CO_2 and temperature increased sharply. During interglacials, there was a sharp recession of ice volume. Compared with the gradual approach to glacial maxima, changes of CO_2, temperature, and ice during the swings to glacial minima are abrupt. Close inspection of the records in Fig. 1.12a also indicates a small but repeated separation between contemporaneous features, with CO_2 lagging temperature by 500–1000 years. The implied relationship of CO_2 to temperature is manifest even on much shorter time scales (Sec. 1.6.2).

The veracity of proxy records like those in Fig. 1.12 is clouded by uncertainties (ibid). Notable is the long-term stability of gases that are trapped inside ice, along with their diffusion between layers. Those uncertainties limit temporal resolution, which hampers the discrimination to individual periods. They also act to homogenize properties, limiting excursions in older portions of the record.[5] Equally important is the issue of causality. While illustrating the interdependence of CO_2 and temperature, these proxies provide little insight into which property produced changes in the other. Perhaps relevant is that they also leave undocumented contemporaneous changes in the most important radiatively-active species, water vapor and cloud, which are far more influential (Chap. 8).

More recent records evidence a human contribution to the budget of CO_2. Since the dawn of the industrial era (late eighteenth-century), the combustion of fossil fuel has steadily increased the rate at which carbon dioxide is introduced into the atmosphere. Augmenting that source is biomass destruction, notably, in connection

---

[5] The signature of diffusion is apparent in Fig. 1.12, wherein older sections of the record become increasingly blurred.

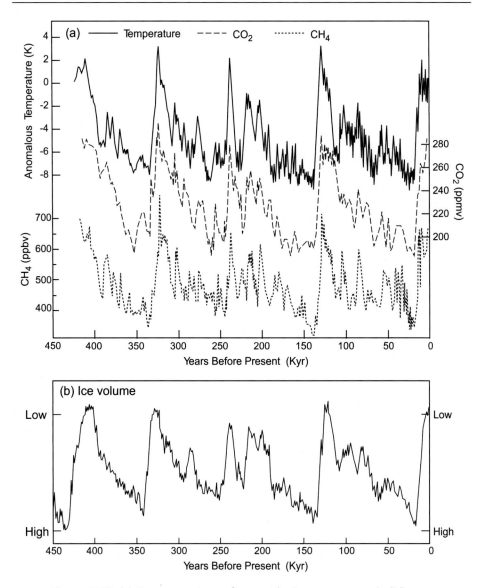

Figure 1.12 (a) Reconstructions of atmospheric temperature (solid), carbon dioxide (dashed), and methane (dotted) as functions of time before present, inferred from ice core drilled at Vostok Antarctica. (b) Reconstruction of global ice volume, inferred from isotopic analysis of ^{18}O in sedimentary foraminifera. *Sources*: Petit et al. (1999); Lisiecki and Raymo (2005); http://www.globalchange.gov.

with the clearing of dense tropical rainforest for timber and agriculture. (This process produces CO_2 either directly, through burning of vegetation, or indirectly, through its subsequent decomposition.) Interactions with the ocean and the biosphere make the budget of CO_2 complex. Nevertheless, the involvement of human activities is strongly suggested by observed changes.

Plotted in Fig. 1.13 is the record of CO_2 mixing ratio over the last 1000 years, inferred from ice cores over Antarctica (symbols). According to this proxy, r_{CO_2} varied

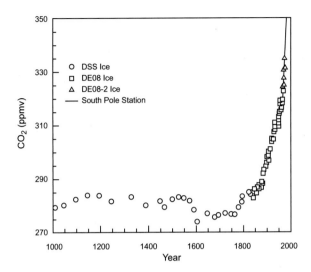

Figure 1.13 Mixing ratio of proxy CO_2 (ppmv) during the preceding millenium, inferred from Antarctic ice cores (symbols). Superimposed is the more
recent instrumental record of atmospheric CO_2 (solid). After Ethridge et al.
(1996).

during most of the last millenium by less than 10 ppmv. It hovered about a nearly
constant value of 280 ppmv. After ~1800, however, r_{CO_2} began a steep and monotonic
increase, one that has prevailed to date. The proxy evidence is consistent with nearby
instrumental measurements of CO_2, which became available in the twentieth century
(solid). Jointly, these records describe a modern increase that has brought r_{CO_2} to
values in excess of 380 ppmv, about 35% higher than pre-industrial values in the proxy
record.

The ice record of CO_2 in Fig. 1.12a is expanded in Fig. 1.14, which plots, at higher
resolution, r_{CO_2} over the last 150 years (squares). Data have been smoothed to exclude
variability with periods shorter than two decades. The monotonic increase dates back
to around 1850, with interruptions during the 1880s and 1890s and, more conspicuously, during the 1940s and 1950s. If anything, r_{CO_2} during those intervals decreased.
The reconstructed evolution is, like the longer proxy record (Fig. 1.13), consistent with
more recent instrumental measurements of CO_2 (triangles).

The upward trend of CO_2 is commonly ascribed to emission by human activities. Support for this interpretation comes from isotopes of carbon. Carbon 13,
like carbon 12, is stable. It represents about 1% of the isotopic composition of CO_2.
However, its concentration varies between reservoirs of carbon. Vegetation and ances-
tral carbon, fossil fuel, are slightly leaner in ^{13}C than is the atmosphere.[6] Also plotted
in Fig. 1.14 is the relative concentration of atmospheric ^{13}C,

$$\delta^{13}C = \frac{(^{13}C/^{12}C)}{(^{13}C/^{12}C)_{ref}} - 1$$

in parts per thousand ($^0/_{00}$), referenced against a standard value (solid circles). Proxy
evidence of $\delta^{13}C$ is more variable than that of r_{CO_2}. It also has little overlap with the

---

[6] Reflecting increased efficiency of photosynthesis with the lighter form of carbon.

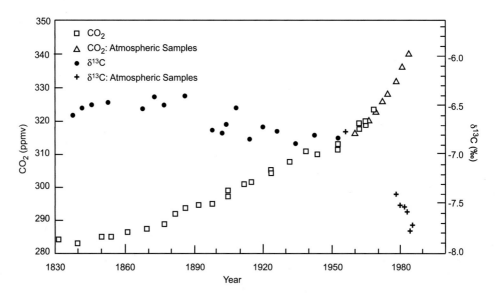

Figure 1.14 Mixing ratio of proxy CO_2 (ppmv) during the last two centuries, inferred from Antarctic ice cores (squares). Superimposed is the contemporaneous instrumental record of r_{CO_2} over the South Pole (triangles). Also plotted is the relative concentration of isotopic carbon (per thousand) $\delta^{13}C = \frac{(^{13}C/^{12}C)}{(^{13}C/^{12}C)_{ref}} - 1$ (solid circles), along with recent instrumental observations (crosses). *Sources*: Etheridge et al. (1996); Friedli et al. (1986).

more recent instrumental record (crosses). Nevertheless, reconstructed $\delta^{13}C$ decreased over the last two centuries, mirroring the contemporaneous increase of r_{CO_2}.

The decrease of $\delta^{13}C$, together with the increase of r_{CO_2}, reflects the addition of CO_2 that is ^{13}C lean. This feature is consistent with the combustion of fossil fuel, as well as biomass destruction. It is equally consistent, however, with the decomposition of organic matter derived from vegetation. Thus, associating the decrease of $\delta^{13}C$ to the combustion of fossil fuel requires the exclusion of other sources that are ^{13}C lean. In particular, it relies on CO_2 emission from the ocean, which overshadows other sources of CO_2 (Sec. 17.3), having the same isotopic composition as the atmosphere (which would then be left unchanged). Only then can the decrease of $\delta^{13}C$ be isolated to continental sources, which are weaker and, in particular, to the combustion of fossil fuel, which is an order of magnitude weaker. Yet, the isotopic composition of marine organic matter is influenced by a variety of biological and environmental factors (Francois et al., 1993; Goericke et al., 1994). Through those factors, $\delta^{13}C$ in the upper ocean varies significantly. Along with transport from the deep ocean, which is likewise uncertain, they leave the magnitude and composition of ocean emission poorly understood (Sec. 1.6.2).

The instrumental record is expanded in Fig. 1.15, which plots r_{CO_2} observed at Mauna Loa during the latter half of the twentieth century. Situated in the central Pacific and several kilometers above sea level, Mauna Loa measurements are removed from continental and urban sources. Accordingly, they reflect CO_2 that has become well mixed across the troposphere. An annual variation of about ± 3 ppmv is synchronized

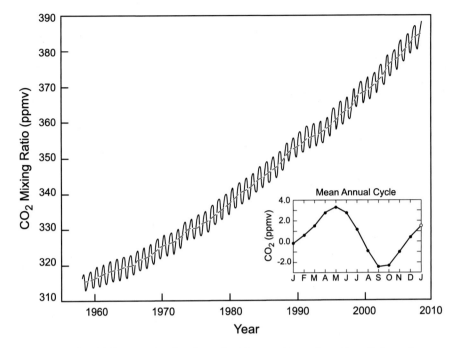

Figure 1.15 Mixing ratio of carbon dioxide over Mauna Loa, Hawaii. Superimposed is the mean annual variation. *Source*: www.esrl.noaa.gov/gmd.

with the growing season of the Northern Hemisphere: CO_2 decreases during May–September, reflecting absorption by vegetation. The annual variation, however, is not cyclic. More CO_2 is emitted during part of the year, when sources prevail over sinks and r_{CO_2} increases, than is absorbed during the other part of the year, when sinks prevail over sources and r_{CO_2} decreases. This yields a hysteresis in the annual variation: CO_2 in one January is ~1.5 ppmv greater than CO_2 in the preceding January. During successive years, the hysteresis accumulates to form a steady upward trend, the same trend apparent over Antarctica. Through that trend, CO_2 has increased by more than 20% in just half a century.

The rapid increase of CO_2 during the industrial era has raised concern over global warming because of the role carbon dioxide plays in trapping radiant energy near the Earth's surface. Developed in Chap. 8, this role makes carbon dioxide a *greenhouse gas*. The implication to temperature of increasing CO_2 was first noted by the Swedish chemist, Arrhenius (1894). For Arrhenius, global warming was, ironically, but a curiosity. His real interest was to explain the ice ages, behavior he pursued through a decrease of CO_2. In a crude but challenging calculation of radiative energy transfer, Arrhenius estimated that halving CO_2 would lead to cooling of 4–5 K, sufficient to form an ice age. Through the same calculation, he estimated that a doubling of CO_2 would lead to warming of 5–6 K. For the rate at which CO_2 is currently increasing, ~20% during the last 5 decades, this translates into a warming rate of about 0.25 K/decade.

Arrhenius was unconcerned by the latter possibility because, at the rate of emission then, he estimated that such levels of CO_2 would not be achieved for millennia. The possibility became more tangible in his subsequent book (Arrhenius, 1908), wherein

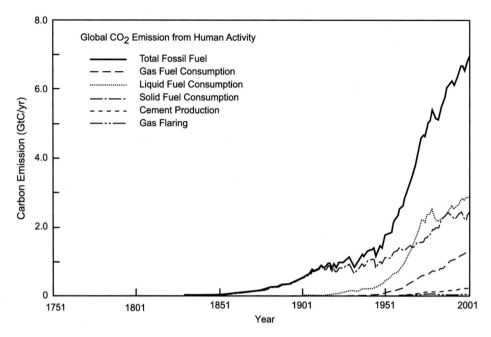

Figure 1.16 Anthropogenic emission of CO_2, in gigatons of carbon per year (GtC/yr), as total and from major sources. *Source*: Marland et al. (2008).

CO_2 was foreseen to double in only centuries. Other researchers doubted that it would ever occur. This position followed from radiative considerations: (1) Absorption by CO_2 was thought to involve the same wavelengths of radiation as water vapor, which is far more abundant and fully absorbs radiation at those wavelengths. (2) The atmosphere already contained sufficient CO_2 to absorb all of the radiation possible; increased CO_2 would therefore be inconsequential. Even exclusive of radiative considerations, most of the CO_2 emission at the turn of the nineteenth century was thought to be absorbed by the ocean. What had not been evaluated was the capacity of the ocean to keep pace with an increasing rate of CO_2 emission.[7]

Plotted in Fig. 1.16 is anthropogenic emission of CO_2 in gigatons of carbon per year (GtC/yr). Around 1850, CO_2 emission ramped up exponentially, approaching 1 GtC/yr at the time of Arrhenius. Following World War I and during the Great Depression, it increased slower. Then, after World War II, CO_2 emission increased sharply, chiefly through combustion of liquid fuel that supports transportation. Almost as great is emission from the combustion of solid fuel in power generation and industry. It is noteworthy that, during the 1940s and 1950s, when emission from fossil fuel accelerated, proxy CO_2 inferred from the ice-core record (Fig. 1.14) did not. If anything, r_{CO_2} during that period decreased. Current rates of anthropogenic emission exceed 7 GtC/yr. Although increasing, this is still only about 4% of total CO_2 emission, which is dominated by natural sources; cf. Fig. 17.11. Consequently, even a minor imbalance

---

[7] The growth of anthropogenic emission parallels the expansion of humans, who share basic needs of food, heat, and transportation. Those needs all require energy. At the time of Arrhenius, the human population was 1.5 billion. Today, it exceeds 6 billion. By the second half of the twenty-first century, it is projected to exceed 10 billion.

between natural sources and sinks can overshadow the anthropogenic component of CO_2 emission.

The concern over increasing CO_2 is supported, in part, by large-scale numerical simulations. Global Climate Models (GCMs) are used to study climate by including a wide array of physical processes. Many can be represented only crudely, with more than a few represented through ad hoc treatment. Although central to many of the considerations, those processes must be treated through crude parameterization because either (1) they cannot be resolved in the computation, (2) they are not adequately observed, or (3) the governing equations that describe them are not even known. Owing to such uncertainties, GCMs differ widely in detail. Nevertheless, on basic features, they are in general agreement. Current models predict that a doubling of CO_2 would lead to warming of global-mean temperature by 2–6 K (IPCC, 2007). Those projections are broadly consistent with Arrhenius' simple calculation, which considered only fundamental issues surrounding the transfer of radiative energy.

Although differing between models, projected changes of temperature are sufficiently great to imply important changes to the Earth's climate (e.g., by melting of continental ice and increasing sea level). In the stratosphere, where CO_2 dominates infrared (IR) cooling to space, temperature decreases as large as 10 K have been suggested. They, in turn, could alter other radiatively active constituents such as ozone. Distinguishing the anthropogenic component of such changes from the natural component will be essential to correctly interpret observed changes and to understand how they are likely to evolve. This is especially true for changes of regional climate, for which model projections are far less reliable than for global-mean temperature (Sec. 8.7.3).

Water vapor

The uniform composition of dry air in the homosphere is not to be confused with the distributions of trace species such as water vapor and ozone. These constituents are highly variable because they are not merely redistributed by atmospheric motion, which would eventually homogenize them. Instead, water vapor and ozone are produced in some regions and destroyed in others. By transporting them from their source regions to their sink regions, the circulation exerts an important influence on these species and makes their distributions dynamic.

Owing to its involvement in radiative processes, cloud formation, and in exchanges of energy with the ocean, water vapor is the single most important trace species in the atmosphere. Its strong absorptivity in the IR, developed in Chap. 8, makes it also the single most important greenhouse gas. The zonal-mean distribution of water vapor is shown in Fig. 1.17 as a function of latitude and altitude. Water vapor is confined almost exclusively to the troposphere. Its zonal-mean mixing ratio $\bar{r}_{H_2O}$ decreases steadily with altitude, from a maximum of about 20 g kg^{-1} (0.020) at the surface in the tropics to a minimum of less than 1 ppm at the tropopause. The absolute concentration of water vapor, or *absolute humidity* $\bar{\rho}_{H_2O}$ (shaded), decreases with altitude even more rapidly. From (1.14), the density of the ith constituent is just its mixing ratio times the density of dry air

$$\rho_i = r_i \rho_d. \tag{1.24}$$

Because ρ_d decreases exponentially with altitude, water vapor tends to be concentrated in the lowest 2 km of the atmosphere. The zonal-mean mixing ratio also decreases with

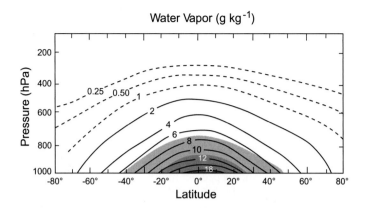

Figure 1.17 Zonal-mean mixing ratio of water vapor (contoured) and density of water vapor or absolute humidity (shaded), as functions of latitude and pressure. Shading levels correspond to 20%, 40%, and 60% of the maximum value. *Source*: Oort and Peixoto (1983).

latitude, falling to under 5 g kg^{-1} poleward of 60°. These characteristics of water vapor reflect its production over warm tropical ocean at the Earth's surface, redistribution by the atmospheric circulation, and destruction at altitude and at middle and high latitudes through condensation and precipitation.

Owing to those production and destruction mechanisms and the rapid transport of air between source and sink regions, tropospheric water vapor is short-lived. A characteristic lifetime, the time for r_{H_2O} inside an individual air parcel to change significantly, is of order days. Every few days, an air parcel encounters a warm ocean surface, where it absorbs moisture through evaporation, or a region of cloud, where it loses water vapor through condensation and precipitation.

Most of the water vapor in Fig. 1.17 originates near the equator from warm ocean surface. Consequently, transport by the circulation plays a key role in determining the mean distribution $\bar{r}_{H_2O}$. Vertical and horizontal air motion are referred to respectively as *convection* and *advection*. Each contributes to the redistribution of r_{H_2O}. Introduced at the surface of the tropical atmosphere, water vapor is carried upward by deep convective cells. It is carried horizontally by large-scale eddies, which disperse r_{H_2O} across the globe. Some bodies of air escape production and destruction long enough for r_{H_2O} to be conserved. Mixing ratio is then rearranged as a tracer.

Figure 1.18 presents, for the day shown in Fig. 1.9, an image from the 6.3-μm water vapor channel of the geostationary satellite Meteosat-2, which observed the Earth from above the Greenwich meridian (see Fig. 1.29 for geographical landmarks). The gray scale in Fig. 1.18 represents cold emission temperature (high altitude) as bright and warm emission temperature (low altitude) as dark. The water vapor column is optically thick at this wavelength. Outgoing radiation therefore derives from H_2O emission only at the highest levels of the water vapor column. For this reason, behavior in Fig. 1.18 characterizes the top of the moisture layer. Bright regions indicate moisture at high altitude, corresponding to deep convective displacements of surface air. Conversely, dark regions indicate moisture that remains close to the Earth's surface. Together, those structures reflect the horizontal distribution of r_{H_2O}.

Figure 1.18 Water vapor image on March 4, 1984, from the 6.3 μm channel of Meteosat-2, which is in geostationary orbit over the Greenwich meridian. Gray scale displays equivalent blackbody temperature from warmest (black) to coldest (white). (See Fig. 1.29 for geographical landmarks). Supplied by the European Space Agency.

Unlike its mean distribution in Fig. 1.17, which is fairly smooth, the global distribution of water vapor on an individual day is quite variable. Especially dynamic is the distribution in the tropics, where the moisture pattern is granular. There, water vapor has been displaced upward by deep convective cells that have horizontal dimensions as short as tens of kilometers and time scales of only hours. At middle and high latitudes, the pattern is smoother, but still complex. Swirls of light and dark mark bodies of air that are rich and lean in water vapor, respectively. They reflect air which originated in tropical and extratropical regions but which has been rearranged by the circulation. The local abundance reflects the history of the air parcel residing at that location, namely where that parcel has been and what processes influencing water vapor have acted on it. A plume of high humidity stretches northeastward from deep convection over the Amazon basin (cf. Fig. 1.29). It extends across the Atlantic and into Africa, where it joins a plume of drier air that is being drawn southward behind a cyclone in the eastern Atlantic (cf. Fig. 1.9a). In the Southern Hemisphere, a band of high humidity is sharply delineated from neighboring dry air along a front that spans the south Atlantic.

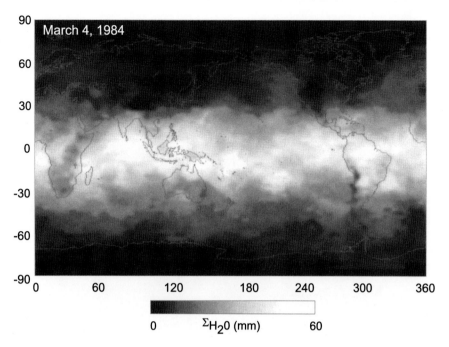

Figure 1.19 Global distribution of the column abundance of water vapor, or *total precipitable water vapor*, on March 4, 1984, derived from the TIROS Operational Vertical Sounder (TOVS). Data courtesy of I. Wittmeyer and T. Vonderharr.

More relevant to radiative processes than the local concentration is the total abundance of a species over a particular horizontal position. The *column abundance*

$$\Sigma_i = \int_0^\infty \rho_i dz$$
$$= \frac{1}{g} \int_0^{p_s} r_i dp,$$

(1.25)

describes the mass of the ith species contained by an atmospheric column of unit cross-sectional area. Figure 1.19 displays the distribution of Σ_{H_2O}, referred to as *total precipitable water vapor*. Expressed in millimeters of liquid water, it is presented for the same day as in Fig. 1.18. Σ_{H_2O} is sharply confined to the tropics, resembling the distribution of temperature. It is distributed more uniformly than r_{H_2O} (Fig. 1.18). In part, this feature of precipitable water vapor follows from the concentration of ρ_{H_2O} in the lowest 2 km. But it also indicates that, at higher altitude, even deep convective towers that punctuate Fig. 1.18 contain comparatively little water in vapor phase.

Together, the zonal-mean and horizontal distributions illustrate the prevailing mechanisms that control atmospheric water vapor. Large values of $\bar{r}_{H_2O}$ at the Earth's surface in the tropics reflect production of water vapor through evaporation of warm tropical ocean. Only inside convective towers and in analogous features of greater dimension are large mixing ratios found far above the ground. Even there, vertical transport of H_2O is limited by thermodynamic constraints. Developed in Chap. 5, they prevent water vapor from reaching great altitude, where it would suffer photodissociation by UV and eventually be lost to space through escape of atomic hydrogen (Sec. 1.2.2).

Ozone

O_3 is another radiatively active trace gas. It plays an essential role in supporting life at the Earth's surface. By intercepting harmful UV radiation, ozone allows life as we know it to exist.[8] The evolution of the Earth's atmosphere and the formation of the ozone layer are, in fact, thought to be closely related to the development of life on Earth.

Geological evidence suggests that primitive forms of plant life developed *in aqua*, deep in the ocean, at a time when the Earth's atmosphere contained little or no oxygen. Damaging UV radiation then passed freely to the Earth's surface. Through photosynthesis, these early forms of life are thought to have liberated oxygen, which then passed to the surface. There, it was photo-dissociated by UV radiation in the reaction

$$O_2 + h\nu \to 2O. \tag{1.26}$$

Atomic oxygen produced by (1.26) could then recombine with O_2 to form ozone in the thermolecular reaction

$$O_2 + O + M \to O_3 + M, \tag{1.27}$$

where M represents a third body needed to carry off excess energy liberated by the combination of O and O_2. Ozone created in (1.27) is dissociated by UV radiation according to the reaction

$$O_3 + h\nu \to O_2 + O. \tag{1.28}$$

If third bodies are abundant, atomic oxygen produced by (1.28) recombines almost immediately with O_2 in (1.27) to re-form ozone. Thus reactions (1.27) and (1.28) constitute a "closed cycle," one that involves no net loss of components. Because the only result is the absorption of solar energy, this cycle can process UV radiation efficiently. By removing harmful UV from the solar spectrum, ozone is thought to have allowed life to advance upward to the ocean surface. There, it had greater access to visible radiation, could produce more oxygen through photosynthesis, and, consequently, was able to evolve into more sophisticated forms.

Ozone also absorbs IR radiation, a property that makes it a greenhouse gas. Its contribution to increased warming of the Earth's surface is thought to be about 20% of the contribution from CO_2 (Fig. 8.30). The increased warming from O_3 follows from tropospheric ozone, which, like CO_2, is increasing. It represents as much as a third of the total ozone column (Fig. 18.2).

The zonal-mean distribution of ozone mixing ratio is plotted in Fig. 1.20 as a function of latitude and altitude. Whereas atmospheric water vapor is confined chiefly to the troposphere, ozone is concentrated in the stratosphere. Ozone mixing ratio increases sharply above the tropopause, reaching a maximum of about 10 ppmv near

---

[8] UV exposure is strongly correlated to common forms of skin cancer, basal cell and squamous cell carcinoma, which have low rates of mortality. Its correlation to melanoma, which has a high rate of mortality, is weaker and remains under debate (Osterlind et al, 2006; Cascinelu and Marchesini, 2008; Bauer, 2010).

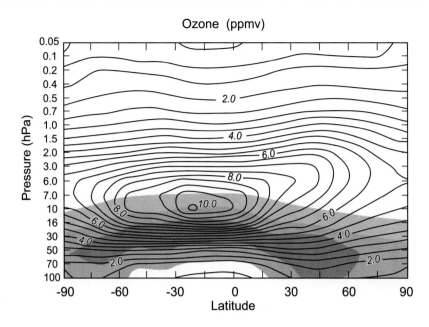

Figure 1.20 Zonal-mean mixing ratio of ozone (contoured) and density of ozone (shaded) averaged over January–February 1979, as functions of latitude and pressure, obtained from the Limb Infrared Monitor of the Stratosphere (LIMS) on board Nimbus-7. Shading levels correspond to 20%, 40%, and 60% of the maximum value.

30 km (10 hPa). The zonal-mean ozone mixing ratio $\bar{r}_{O_3}$ is largest in the tropics, where the flux of UV and photodissociation of O_2 are large.

The photochemical lifetime of ozone varies sharply with altitude. In the lower stratosphere, ozone has a photochemical lifetime of several weeks. Because this is long compared with the characteristic time scale of air motion ($\sim$1 day), r_{O_3} behaves as a tracer at these altitudes. For the same reason, its distribution there is controlled by dynamical influences. The lifetime of ozone decreases upward, shortening to of order 1 day by 30 km and only an hour by the stratopause. For this reason, the distribution of ozone in the upper stratosphere and mesosphere is controlled by photochemical influences.

The troposphere serves as a sink of stratospheric ozone. Should ozone find its way into the troposphere, it is quickly destroyed. Its water-solubility enables O_3 to be absorbed partially by convective systems. Through vertical transport, they transfer ozone to the Earth's surface, where it is consumed in a variety of oxidation processes.

Even though its mixing ratio $\bar{r}_{O_3}$ maximizes near 30 km, atmospheric ozone is concentrated in the lower stratosphere (Fig. 1.20). Because air density decreases exponentially with altitude, the density of ozone $\bar{\rho}_{O_3}$ (shaded) is concentrated at altitudes of 10–20 km (1.24), not far above the tropopause. The largest values are found in a shallow layer near 30 hPa in the tropics, which descends and deepens in extratropical regions. The column abundance, or *total ozone* Σ_{O_3}, is expressed in Dobson units (DU). One DU measures, in thousandths of a centimeter, the depth the ozone column

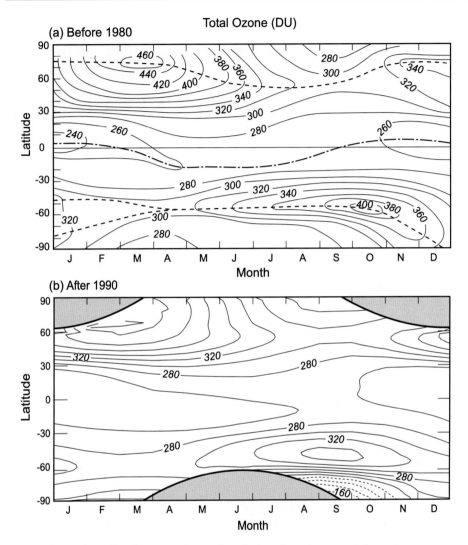

Figure 1.21 Zonal-mean column abundance of ozone, or *total ozone*, as a function of latitude and month. (a) Based on the historical record prior to 1980. From London (1980). (b) Based on satellite observations during the 1990s, after the formation of the Antarctic ozone hole, from the Total Ozone Mapping Spectrometer (TOMS). Values less than 220 DU (dotted) characterize the ozone hole. Observations not available in regions of darkness (shaded).

would assume if brought to standard temperature and pressure. Figure 1.21a plots the zonally averaged column abundance of ozone, as a function of latitude and season, based on observations before 1980. Values of $\overline{\Sigma}_{O_3}$ range from about 250 DU near the equator to in excess of 400 DU at high latitude. At standard temperature and pressure, the entire ozone column measures less than one half of one centimeter! Even though most stratospheric ozone is produced in the tropics, the greatest column abundance is found at middle and high latitudes. This peculiarity of the ozone distribution illustrates the importance of dynamics in shaping observed composition and structure.

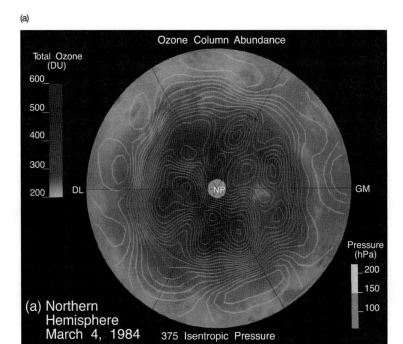

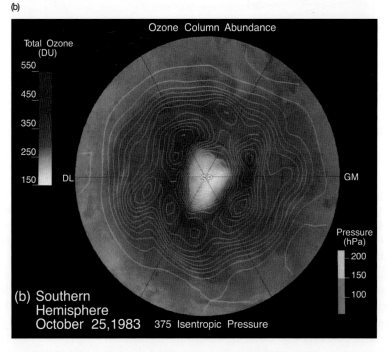

Figure 1.22 Distribution of total ozone (color) from the Total Ozone Mapping Spectrometer (TOMS) on board Nimbus-7 and pressure (contours) on the 375°K isentropic surface (see Sec. 2.4.1) over (a) the Northern Hemisphere on March 4, 1984 and (b) the Southern Hemisphere on October 25, 1983. See color plate section: Plate 1.

Compared in Fig. 1.21b is the same information, but from satellite observations during the 1990s (available only in sunlit regions). Distinguishing the behavior is anomalously low $\overline{\Sigma}_{O_3}$ over the Antarctic during Austral spring. Values less than 220 DU (dotted) then characterize the Antarctic ozone hole, which appeared around 1980. Reflecting strong depletion, they take a bite out of the annual cycle of Antarctic ozone, which, before 1980, achieved values nearly twice as great.

As is true for water vapor, the circulation plays a key role in determining the mean distribution of O_3. It makes the global distribution on individual days dynamic. Figure 1.22 presents the distribution of Σ_{O_3} over the Northern and Southern Hemispheres on individual days, as observed by the Total Ozone Mapping Spectrometer (TOMS). The distribution over the Northern Hemisphere (Fig. 1.22a) has been arranged by the circulation into several anomalies in which Σ_{O_3} varies by as much as 100%. Figure 1.22a, which is contemporaneous with the 500-hPa circulation in Fig. 1.9a, reveals a strong correspondence between variations of Σ_{O_3} and synoptic weather systems in the troposphere. Both evolve on a time scale of a day. The distribution over the Southern Hemisphere (Fig. 1.22b) is distinguished by column abundance less than 200 DU (white), which delineates the "Antarctic ozone hole." Anomalously low Σ_{O_3} develops over the South Pole each year during Austral spring. Forming inside the polar-night vortex, it then disappears some 2 months later. The formation of the ozone hole, which emerged in the 1980s, follows from increased levels of atmospheric halogens, like chlorine and bromine. The ozone hole's disappearance each year occurs through dynamics.

Several other trace gases also figure importantly in radiative and chemical processes. These include species that are produced naturally, such as methane (CH_4), and species that are produced exclusively by human activities, such as chlorofluorocarbons (CFCs).

Methane

CH_4 forms through bacterial and surface processes that occur naturally. It is produced by digestion in grazing animals, by insects, and through anaerobic fermentation, the decomposition of organic matter under anoxic conditions (chiefly in wetlands). Methane is also produced by living plants, a process that accelerates with temperature (Keppler et al, 2006; Reid and Qaderi, 2009). Like natural processes that emit CO_2, these natural sources of CH_4 are poorly documented. Methane is also produced by a variety of human activities such as mining, agriculture, landfills, industrial processes, and biomass destruction. Collectively, those anthropogenic sources may account for as much as half of CH_4 emission.

Methane is long-lived and therefore well mixed in the troposphere. There, it has a uniform mixing ratio of $r_{CH_4} \cong 1.7$ ppmv (Fig. 1.23). In the stratosphere, r_{CH_4} decreases with altitude as a result of oxidation. This process ultimately leads to the formation of stratospheric water vapor. It is thought to be responsible for the upward increase of r_{H_2O} in the stratosphere.

Methane also varied in prehistoric times. The record in ice cores is superimposed in Fig. 1.12 (dotted). Like carbon dioxide (dashed), the variation of methane tracks the variation of temperature (solid), at least on time scales longer than millennial. Consequently, CO_2 and CH_4 do not vary independently. Changes of one are attended by changes of the other. In the ice core record of prehistoric composition, the percentage change of CH_4 is 2.0-2.5 times the percentage change of CO_2.

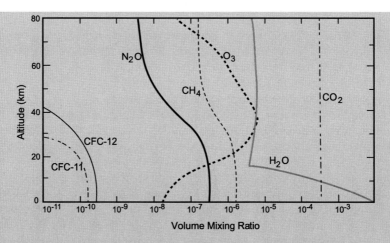

Figure 1.23 Mixing ratios of radiatively active trace species as functions of altitude. *Source*: Goody and Yung (1989).

Similar behavior characterizes changes during the modern era. As for carbon dioxide, methane concentrations are steadily increasing. The record inferred from ice cores over Antarctica and Greenland is presented in Fig. 1.24. It implies that, over much of the last 1000 yrs, r_{CH_4} hovered about a value of $\sim$700 ppbv, when r_{CO_2} hovered about a value of $\sim$280 ppmv (Fig. 1.13). However, like CO_2, methane increased sharply after about 1850, now having values more than twice as great. In relation to anthropogenic emission, why CH_4 should increase contemporaneously with CO_2 is paradoxical. Methane has only a tenuous connection to the combustion of fossil fuel. Its production by agriculture has certainly increased. However, that increase has been in place for at least 7000 years, since the domestication of cattle. It is noteworthy that the decomposition of organic matter, which is intrinsic to the cycling

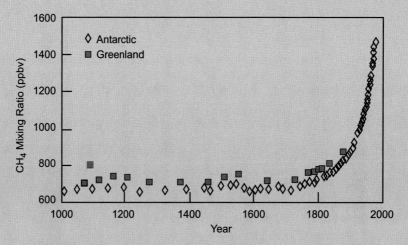

Figure 1.24 Mixing ratio of proxy CH_4 (ppbv) during the last millennium, inferred from ice cores from Antarctic and Greenland (symbols). *Source*: http://cdiac.ornl.gov, Etheridge et al. (CSIRO).

between oxygen and carbon dioxide, produces both methane and CO_2. For many pathways, such as anaerobic fermentation, the concentration of emitted CH_4 is two to three times that of CO_2.

Modern values of r_{CH_4} represent an increase over pre-industrial values of about 100%. This is three times the percentage increase of CO_2 during the same period. In the ice core record of prehistoric composition (Fig. 1.12), the relative change of CH_4 to CO_2 is 2.0–2.5. Although the ratio of modern changes exceeds that in the prehistoric record, the two are broadly consistent.

Like CO_2, CH_4 is a greenhouse gas. It is less abundant than carbon dioxide, but more effective at trapping IR radiant energy. Per unit mass, methane is some 25 times more effective. Despite its much smaller abundance, this feature enables methane to contribute about a fourth as much increased warming of the Earth's surface as follows from elevated carbon dioxide (Fig. 8.30). Methane is also involved in the complex photochemistry of ozone (Sec. 18.2).

After the turn of the century, the rate at which methane is increasing slowed significantly. Estimates of anthropogenic emission did not (http://www-naweb.iaea. org/nafa/aph/stories/2008-atmospheric-methane.html; 14.11.10). In addition to ongoing production, large reserves of methane are trapped in frozen tundra beneath Arctic ice, notably over Siberia. It has been suggested that the release of those reserves, if Arctic ice recedes, would sharply magnify CH_4 emission. The attendant warming would, in turn, reinforce the melting of Arctic ice, representing a positive feedback. How this suggestion fits with the seasonality of methane emission over Arctic tundra is unclear. During the months of autumn, when tundra freezes, CH_4 emission is as great as during the entire unfrozen period of summer (Mastepanov et al., 2008).

Halocarbons

Industrial halocarbons such as the chlorofluorocarbons CFC-10 (CCl_4), CFC-11 ($CFCl_3$), and CFC-12 (CF_2Cl_2) have been used widely as aerosol propellants, in refrigeration, and in a variety of manufacturing processes. As shown in Fig. 1.25, the production of these gases increased sharply after World War II. Their rate of increase abated after 1970, in the wake of heightened environmental concern and reduced demand. Following the Montreal Protocol, which led to a multilateral ban on CFCs, their production ground to a halt in the mid 1990s.

These anthropogenic species are stable in the troposphere, where their water-insolubility makes them immune to normal scavenging processes associated with precipitation. Because they are long-lived, CFCs are well-mixed in the troposphere, with CFC-11 and CFC-12 having nearly uniform mixing ratios there of $r_{CFC-11} \cong 0.2$ ppbv and $r_{CFC-12} \cong 0.3$ ppbv, respectively.

The greatest interest in CFCs surrounds their impact on the ozone layer. Since they are long-lived in the troposphere, chlorofluorocarbons are eventually transported into the stratosphere. There, they are photo-dissociated by UV radiation. Although uniform in the troposphere, mixing ratios of CFC-11 and −12 therefore decrease in the stratosphere (Fig. 1.23). Free chlorine that is released via dissociation of CFCs can destroy ozone. So can free bromine, which is released in similar fashion. Reservoir species for these free radicals are collected in the Equivalent

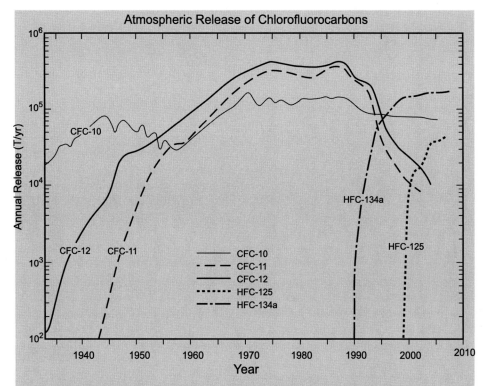

Figure 1.25 Production and eventual emission of chlorofluorocarbons (CFCs) and hydrofluorocarbons (HFCs). *Sources*: http://www.afeas.org, WMO (2006).

Effective Stratospheric Chlorine (EESC). EESC increased during the 1970s and 1980s – by more than 100% (see Fig. 18.36). After peaking in the mid 1990s, EESC began a gradual decline in which values have decreased by about 15%.

Replacing CFCs are hydrofluorocarbons, such as HFC-134a (CF_3CFH_2) and HFC-125 (CF_3CF_2H). HFCs have lifetimes in the troposphere that are shorter than those of CFCs. For this reason, HFCs suffer greater removal before they can reach the stratosphere. Nonetheless, like their predecessors, HFCs are steadily increasing (Fig. 1.25).

Photodissociation of CFCs in the stratosphere produces free radicals of chlorine and bromine, which are then said to be "activated." Through a series of reactions, Cl and Br support catalytic destruction of O_3, which leads to the formation of the Antarctic ozone hole in Fig. 1.22b. Those free radicals are but one of several ingredients involved in the formation of the ozone hole. Very high cloud in the stratosphere, which rarely forms except over Antarctica, also plays a key role. So does the stratospheric circulation, which controls temperature and the spring breakdown of the polar-night vortex, when the ozone hole disappears.

Other concerns regarding CFCs and HFCs surround their involvement in the Earth's energy budget. Like CO_2 and O_3, industrial halocarbons interact with radiation. Consequently, they too are greenhouse gases. Collectively, their contribution to increased warming of the Earth's surface is thought to be comparable to that of methane (Sec. 8.7).

Nitrogen compounds

Oxides of nitrogen, such as nitrous oxide (N_2O) and nitric oxide (NO), are also relevant to the photochemistry of ozone. Nitrous oxide is produced primarily by natural means relating to bacterial processes in soils. Anthropogenic sources of N_2O include nitrogen fertilizers, combustion of fossil fuel, and biomass destruction. They may account for as much as 25% of its total production. These sources have altered the natural cycle of nitrogen by introducing it in the form of N_2O instead of N_2. Like methane, N_2O is long-lived and therefore well-mixed in the troposphere. There, it has a nearly uniform mixing ratio of $r_{N_2O} \cong 300$ ppbv (Fig. 1.23).

In the stratosphere, r_{N_2O} decreases with altitude because of dissociation of nitrous oxide. That, in turn, represents the primary source of stratospheric NO. Like free chlorine, NO can destroy ozone catalytically. Nitric oxide is also produced as a by-product of inefficient combustion (e.g., in aircraft exhaust). Nitrous oxide has increased steadily in recent years, similar to other gases with anthropogenic sources. Beyond its relevance to ozone, N_2O is also a greenhouse gas. Its contribution to increased warming of the Earth's surface is thought to be about half as large as that of methane.

Atmospheric aerosol

Suspensions of liquid and solid particles are relevant to radiative as well as chemical processes. Aerosol particles range in size from thousandths of a micron to several microns. They promote cloud formation, making them vital to atmospheric behavior. Aerosol particles serve as condensation nuclei for water droplets and ice crystals, which do not form readily in their absence. These small particulates are produced naturally (e.g., as dust, sea salt, and volcanic debris). They are also produced anthropogenically through combustion and industrial processes. An important source of aerosol is *gas-to-particle conversion*. It occurs through chemical reactions involving, among other precursors, the gaseous emission sulfur dioxide (SO_2). Sulfur dioxide is produced by industry and naturally by volcanos.

Aerosol concentrations are high in urban areas and near industrial complexes (Fig. 1.26). Reflecting anthropogenic sources, their concentrations there are as much as an order of magnitude greater than over maritime regions. Even higher concentrations accompany windblown silicates during desert dust storms (cf. Figs. 9.43; 9.44) and smoke from forest fires and biomass burning (Cover). In addition to their role in cloud formation, aerosol particles interact with radiation: They scatter solar radiation at visible wavelengths. They absorb IR radiation that is emitted by the Earth's surface and atmosphere. These radiative properties involve aerosol in the global energy balance, where it has a role analogous to greenhouse gases. In that capacity, its influence is thought to be as great as that of CO_2 (Fig. 8.30).

Aerosol also plays a key role in chemical processes. One of the more notable is heterogeneous chemistry, which requires the presence of multiple phases. Reactions

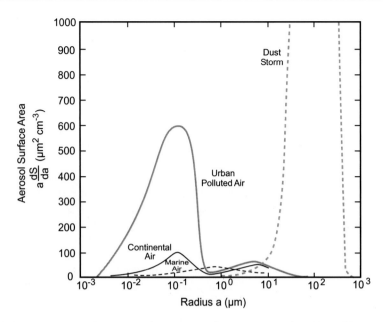

Figure 1.26 Size spectrum of aerosol surface area, as a function of particle radius. The representation is area-preserving: equal areas under the curves represent equal number densities of aerosol particles. *Source*: Slinn (1975).

involving chlorine take place on the surfaces of cloud particles. They lie at the heart of the chain of events that culminates in the formation of the Antarctic ozone hole (Fig. 1.22b).

Beyond this role, aerosol is thought to figure in fluctuations of climate, possibly in the evolution of the atmosphere. Sporadic increases of aerosol produced by major volcanic eruptions have been linked to changes in thermal, optical, and chemical properties of the atmosphere. Figure 1.27a plots the change of global-mean temperature in the lower troposphere following the eruption of Mt. Pinatubo in 1991. Pinatubo introduced volcanic gas and debris well into the stratosphere. Its enhancement of aerosol there remained long enough to modify transfers of radiative energy, in particular, to diminish solar radiation that reaches the Earth's surface. Global-mean temperature decreased following the eruption, attaining a maximum depression of ~0.5 K one and a half years later. The depression remained for another 2 years, while global-mean temperature gradually returned to its unperturbed value. Accompanying the depression of temperature was a depression of the column abundance of water vapor (Fig. 1.27b). Precipitable water vapor decreased in much the same fashion, attaining a maximum depression at about the same time of ~0.5 mm. Much of the reduction in the column abundance of water vapor derived from humidity changes in the middle and upper troposphere. Their resemblance to the evolution of temperature in the lower troposphere points to a modification of the mechanism by which the troposphere is humidified. Involving convection, such behavior is developed in Chaps. 5 and 7.

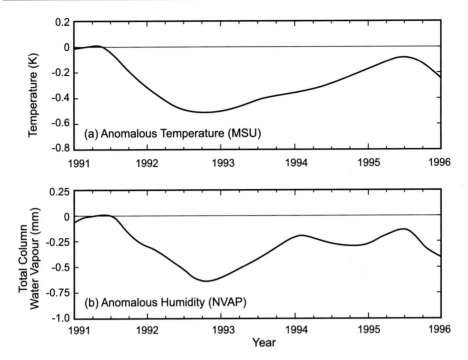

Figure 1.27 (a) Anomalous temperature observed by the Microwave Sounding Unit (MSU) and (b) anomalous column abundance of water vapor from the NASA Water Vapor Project (NVAP) following the eruption of Pinatubo. *Source*: Soden et al. (2002).

1.2.5 Cloud

One of the most striking features of the Earth, when viewed from space, is its extensive coverage by cloud. At any instant, about half of the planet is cloud-covered. Cloud appears with a wide range of shapes, sizes, and microphysical properties. Like trace species, the cloud field is highly dynamic because of its dependence on the circulation. Beyond these more obvious characteristics, cloud and related phenomena figure prominently in a variety of atmospheric processes.

The reason cloud is so striking in satellite imagery is that it reflects back to space a large fraction of incident solar radiation, which is concentrated at visible wavelengths. Owing to its high reflectance at visible wavelengths, cloud shields the planet from solar radiation. This is one important role that cloud plays in the Earth's energy budget. Cloud also plays a major role in the budget of terrestrial radiation, which is emitted in the IR by the Earth's surface and atmosphere to offset solar heating. Because particles of water and ice absorb strongly in the IR, cloud increases the atmosphere's opacity to terrestrial radiation and hence its trapping of radiant energy near the Earth's surface.

Its optical properties enable cloud to be observed from space in measurements of visible and IR radiation. Figure 1.28 shows the IR image from the 11-μm channel of Meteosat-2 that is contemporaneous with the water vapor image in Fig. 1.18. The gray scale in Fig. 1.28 emphasizes the highest objects emitting in the IR. Bright

Figure 1.28 Infrared image at 1200 UT on March 4, 1984, from the 11-μm channel of Meteosat-2. Gray scale displays equivalent blackbody temperature from warmest (black) to coldest (white). (See Fig. 1.29 for geographical landmarks). Supplied by the European Space Agency.

areas correspond to cold high cloud, whereas dark areas indicate warm surface under cloud-free conditions. Many features in the IR image are well correlated with features in the water vapor image. This is especially true of high convective cloud in the tropics, which extends upward to the tropopause. Having horizontal dimensions of tens to a few hundred kilometers, that cloud marks deep convective tower that displaces surface air upward on time scales of hours. On the other hand, the IR image clearly reveals the subtropical Sahara and Kalahari deserts in dark areas that, in the water vapor image, are masked.

High cloud at mid-latitudes also corresponds well with features in the water vapor image. Most striking is the band of high cirrus that extends northeastward from South America, across the Atlantic, and into Africa (cf. Fig. 1.29 for geographical landmarks). Coinciding with the plume of high humidity in Fig. 1.18, this high cloud cover is part of a frontal system that accompanies the cyclone in the eastern Atlantic (cf. Fig. 1.9a). In Figs. 1.18 and 1.28, the cyclone appears as a spiral of water vapor and low cloud. As it advances, the cyclone draws humid tropical air poleward ahead of the front, entraining it with drier mid-latitude air that has been drawn equatorward behind the front.

Cloud also appears in visible imagery, like that shown in Fig. 1.29, which is contemporaneous with the IR and water vapor images. Unlike the latter two images, which

Figure 1.29 Visible image at 1200 UT on March 4, 1984, from Meteosat-2. Supplied by the European Space Agency.

follow from emission of IR radiation, the visible image represents reflected solar radiation. For this reason, it does not depend on the temperature of individual features. Low and comparatively warm stratiform cloud therefore appears just as prominently as high and cold convective cloud. The prevalence of stratiform features in Fig. 1.29 indicates that even shallow cloud is an efficient reflector of solar radiation. Its extensive coverage of the globe makes such cloud an important consideration in the Earth's radiation budget. Swirls of light and dark in the water vapor image, which are signatures of horizontal motion, are only weakly evident in the IR image. They are completely absent in the visible image.

The roles cloud plays in the budgets of solar and terrestrial radiation make it a key ingredient of climate. In fact, the influence cloud cover exerts on the Earth's energy balance is an order of magnitude greater than that of CO_2. Its dependence on the circulation, thermal structure, and distribution of water vapor make cloud an especially interactive component of the climate system.

With the exception of shallow stratus, most cloud in Figs. 1.28 and 1.29 develops through vertical motion. Two forms of convection are distinguished in the atmosphere. *Cumulus convection* is often implied by the term convection alone. It involves thermally-driven circulations that operate on horizontal dimensions of order 100 km and smaller. Deep tropical cloud in Fig. 1.28 is a signature of cumulus convection. It displaces surface air upward on small horizontal dimensions. *Sloping convection* is associated with forced lifting, when one body of air overrides another. It occurs coherently over large horizontal dimensions. The band of high cloud preceding the cyclone in the eastern Atlantic is a signature of sloping convection.

Beyond its involvement in radiative processes, convection plays a key role in the dynamics of the atmosphere and in its interaction with the ocean. Deep convection in the tropics releases large quantities of latent heat when water vapor condenses and precipitates back to the Earth's surface. Derived from heat exchange with the ocean, latent heating in the tropics represents a major source of energy for the atmosphere. For this reason, deep cumulus cloud is often considered as a proxy for atmospheric heating.

Figure 1.30a shows a nearly instantaneous image of the global cloud field, as constructed from 11-μm radiances measured aboard six satellites. The highest (brightest) cloud is found in the tropics, inside a narrow band of cumulus convection. Marking the *Inter-Tropical Convergence Zone* (ITCZ), this band of organized convection is oriented parallel to the equator. Exceptional are the tropical landmasses: South America, Africa, and the *maritime continent* of Indonesia and the surrounding archipelago, where the zone of convection widens. Inside the ITCZ, deep convection is supported by the release of latent heat when moisture condenses. Vertical transfers of moisture and energy make the ITCZ important to the tropical circulation and to interactions between the atmosphere and ocean.

The ITCZ emerges conspicuously in the time-mean cloud field, shown in Fig. 1.30b. Over maritime regions, the time-mean ITCZ appears as a narrow strip parallel to the equator. It reflects the convergence of surface air from the two hemispheres inside the Hadley circulation (see Fig. 1.35). Over tropical landmasses, time-mean cloud cover expands due to the additional influence of surface heating, which triggers convection diurnally. There, as throughout the tropics, the unsteady component of the cloud field is as large as the time-mean component. In addition to varying on short time scales, tropical convection migrates north and south annually with the sun. Its annual swing culminates during the solstices in the monsoons over southeast Asia and northern Australia.

Time-mean cloud also reveals the North Atlantic and North Pacific storm tracks. Coinciding with intensified zonal wind (Fig. 1.9a), they are preferred sectors of longitude, where convection is organized by synoptic weather systems. Individual systems are evident in the instantaneous cloud field in Fig. 1.30a. By contrast, in the Southern Hemisphere, such systems are distributed almost uniformly in longitude, inside an unbroken storm track that circumscribes the Antarctic. The distinction between hemispheres is apparent in Fig. 1.30b: Time-mean cloud at mid-latitudes of the Northern Hemisphere is concentrated in the North Atlantic and North Pacific Oceans. However, at mid-latitudes of the Southern Hemisphere, it is almost circumpolar. This hemispheric asymmetry reflects the relative absence of major orographic features in the Southern Hemisphere, limited to the Andes. In the Northern Hemisphere, which includes the Himalayas, the Alps, and the Rockies, such features are substantial. They excite amplified planetary waves that disturb the zonal circulation, introducing strong longitudinal dependence.

Cloud is also important in chemical processes. Condensation and precipitation constitute the primary removal mechanism for many chemical species. Gaseous pollutants that are water soluble are absorbed in cloud droplets, eventually eliminated when those droplets precipitate to the Earth's surface. Referred to as *rain out*, this mechanism also

Brightness Temperature

(a) 1200 March 4. 1984

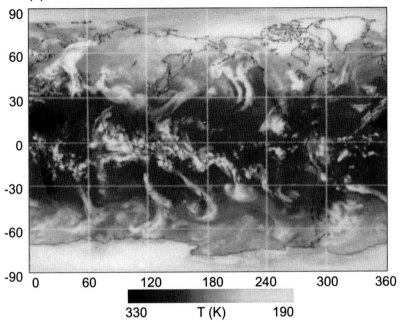

(b) Time–Mean

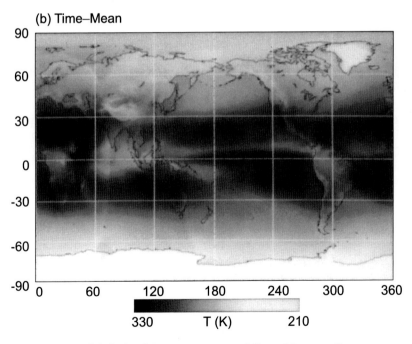

Figure 1.30 Global cloud image constructed from 11-μm radiances measured aboard six satellites simultaneously viewing the Earth. (a) Instantaneous cloud field at 1200 UT on March 4, 1984. (b) Time-mean cloud field for January–March, 1984. Gray scale displays equivalent blackbody temperature as indicated.

scavenges aerosol pollutants, which serve as condensation nuclei for cloud droplets and ice crystals. Although improving air quality, these scavenging mechanisms operate at the expense of surface hydrology. They transfer pollutants to the Earth's surface, often manifest as "acid rain."

Another chemical process in which cloud figures importantly pertains to the ozone hole in Fig. 1.22b. Because moisture is sharply confined to the troposphere, cloud forms in the stratosphere only under exceptionally cold conditions. The Antarctic stratosphere is one of the coldest sites in the atmosphere. As a result, it is populated by a rare form of cloud. Tenuous and very high, Polar Stratospheric Cloud (PSC) is common over the Antarctic during winter. On the surface of PSC, particles operate heterogeneous chemical reactions that involve chlorine and bromine (Sec. 1.2.4). They figure centrally in the formation of the ozone hole each year during Austral spring.

1.3 RADIATIVE EQUILIBRIUM OF THE EARTH

The driving force for the atmosphere is the absorption of solar energy at the Earth's surface. Over time scales long compared with those involved in the redistribution of energy, the Earth-atmosphere system is in thermal equilibrium. The net energy gained must then vanish. Solar radiation is concentrated at visible wavelengths, termed *shortwave* (SW) *radiation*. Terrestrial radiation, that emitted by the Earth's surface and atmosphere, is concentrated at IR wavelengths, termed *longwave* (LW) *radiation*. For thermal equilibrium, the absorption of SW radiation must be balanced by emission to space of LW radiation. This basic principle leads to a simple estimate of the mean temperature of the Earth.

The Earth intercepts a beam of SW radiation of cross-sectional area πa^2 and flux F_s (energy/area·time), as illustrated in Fig. 1.31. A fraction of the intercepted radiation, the *albedo* $\mathcal{A}$, is reflected back to space by the Earth's surface and components of the atmosphere. The remainder of the incident SW flux: $(1 - \mathcal{A})F_s$, is then absorbed by the Earth-atmosphere system. It is distributed across the globe as it spins in the line of the SW beam.

To maintain thermal equilibrium, the Earth and atmosphere must re-emit to space LW radiation at exactly the same rate. Also referred to as *Outgoing Longwave Radiation* (OLR), the emission to space of terrestrial radiation is described by the Stefan-Boltzmann law

$$\pi B = \sigma T^4, \tag{1.29}$$

where πB represents the energy flux integrated over wavelength that is emitted by a blackbody at temperature T and σ is the Stefan-Boltzmann constant. Integrating the emitted LW flux over the Earth and equating the result to the SW energy absorbed obtains the simple energy balance

$$(1 - \mathcal{A})F_s \pi a^2 = 4\pi a^2 \sigma T_e^4, \tag{1.30.1}$$

where T_e is the equivalent blackbody temperature of the Earth. Then

$$T_e = \left[\frac{(1 - \mathcal{A})F_s}{4\sigma} \right]^{\frac{1}{4}} \tag{1.30.2}$$

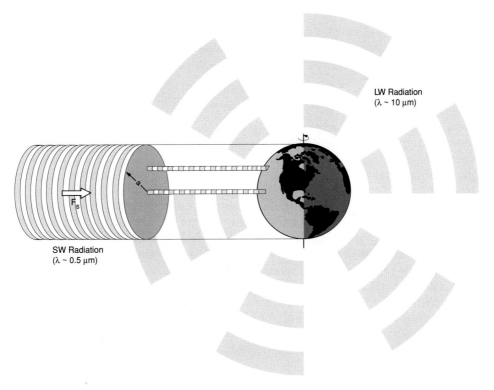

LW Radiation
($\lambda \sim 10\ \mu m$)

SW Radiation
($\lambda \sim 0.5\ \mu m$)

Figure 1.31 Schematic of SW radiation intercepted by the Earth and LW radiation emitted by it.

provides a simple estimate of the Earth's temperature. An incident SW flux of $F_s = 1372$ W m^{-2} and an albedo of $A = 0.30$ lead to an equivalent blackbody temperature for Earth of $T_e = 255$ K. This value is some 30 K colder than the global-mean surface temperature, $T_s = 288$ K (Fig 1.2).

The discrepancy between T_s and T_e follows from the different ways the atmosphere processes SW and LW radiation. Although nearly transparent to SW radiation (wavelengths $\lambda \sim 0.5\ \mu m$), the atmosphere is almost opaque to LW radiation ($\lambda \sim 10\ \mu m$) that is re-emitted by the Earth's surface. For this reason, SW radiation passes relatively freely to the Earth's surface, where it can be absorbed. However, LW radiation emitted by the Earth's surface is captured by the overlying air, chiefly by the major LW absorbers: water vapor and cloud. Energy absorbed in an atmospheric layer is re-emitted, half upward and half back downward. The upwelling re-emitted radiation is absorbed again in overlying layers, which subsequently re-emit that energy in similar fashion. This process is repeated until LW energy is eventually radiated beyond all absorbing components of the atmosphere and rejected to space. By inhibiting the transfer of energy from the Earth's surface, repeated absorption and emission by intermediate layers of the atmosphere traps LW energy, elevating surface temperature over what it would be in the absence of an atmosphere.

The elevation of surface temperature that results from the atmosphere's different transmission characteristics to SW and LW radiation is known as the *greenhouse effect*. The greenhouse effect is controlled by the IR opacity of atmospheric constituents, which radiatively insulate the Earth's surface. In the atmosphere, the primary absorbers are water vapor and cloud. Carbon dioxide, ozone, methane, and nitrous oxide are also radiatively active at wavelengths of LW radiation, as are aerosol and halo-carbons.

1.4 THE GLOBAL ENERGY BUDGET

Because it follows from a simple energy balance, the equivalent blackbody temperature of the Earth provides some insight into where LW radiation is ultimately emitted to space. The value $T_e = 255$ K corresponds to the middle troposphere, above most of the water vapor and cloud. Most of the energy received by the atmosphere is supplied from the Earth's surface, where SW radiation is absorbed. Transfers of energy from the surface constitute a heat source for the atmosphere. Conversely, LW emission to space by the middle troposphere constitutes a heat sink for the atmosphere. Representing heating and cooling, these energy transfers drive the atmosphere into motion. It then functions as a global heat engine in the energy budget of the Earth, as described in Chap. 3.

Radiative energy absorbed at the Earth's surface must be transmitted to the middle troposphere, where it is rejected to space as LW radiation. From the time it is absorbed at the surface as SW radiation until it is eventually rejected to space, energy assumes a variety of forms. Most of the energy transfer between the Earth's surface and the atmosphere occurs through LW radiation. In addition, energy is transferred through thermal conduction, which is referred to as the transfer of *sensible heat*. It is also transferred through *latent heat*, when water vapor that is absorbed by the atmosphere condenses, releasing its latent heat of vaporization, and then the resulting condensate precipitates back to the Earth's surface.

In addition, energy is stored internally in the atmosphere, in thermal and mechanical forms. The thermal or internal energy of air is measured by temperature. It represents the random motion of molecules. Mechanical energy is represented in the distribution of atmospheric mass within the Earth's gravitational field (potential energy), as well as in the motion of air (kinetic energy). These mechanical forms of energy are involved in the redistribution of energy within the atmosphere. However, they do not contribute to globally integrated transfers between the Earth's surface, the atmosphere, and deep space.

1.4.1 Global-mean energy balance

The globally averaged energy budget is illustrated in Fig. 1.32. Incoming solar energy is distributed across the Earth. Therefore, the global-mean SW flux incident on the top of the atmosphere is given by $\bar{F}_s = \frac{F_s}{4} = 343$ W m^{-2}, where the factor 4 represents the ratio of the surface area of the Earth to the cross-sectional area of the intercepted beam of SW radiation (Fig. 1.31). Of the incident 343 W m^{-2}, a total of 106 W m^{-2} (approximately 30%) is reflected back to space: 21 W m^{-2} by air, 69 W m^{-2} by cloud, and 16 W m^{-2} by the surface. The remaining 237 W m^{-2} is absorbed in the Earth-atmosphere system. Of this, 68 W m^{-2} (about 20% of the incident SW flux) is absorbed by the atmosphere: 48 W m^{-2} by atmospheric water vapor, ozone, and aerosol and 20 W m^{-2} by cloud. This leaves 169 W m^{-2} to be absorbed by the surface – nearly 50% of that incident on the top of the atmosphere.

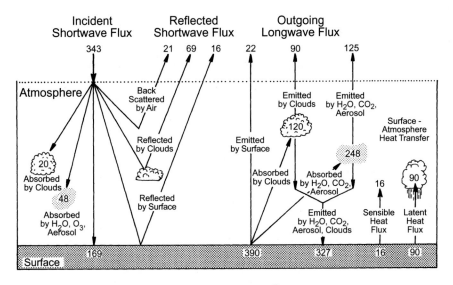

Figure 1.32 Global-Mean Energy Budget (W m^{-2}). Updated from *Understanding Climate Change* (1975) with recent satellite measurements of the Earth's radiation budget. Additional sources: Ramanathan (1987), Ramanathan et al. (1989).

The 169 W m^{-2} absorbed at the ground must be re-emitted to maintain thermal equilibrium of the Earth's surface. At a global-mean surface temperature of $T_s = 288$ K, the surface emits 390 W m^{-2} of LW radiation according to (1.29). This is far more energy than it absorbs as SW radiation. Excess LW emission must be balanced by transfers of energy to the surface from other sources. Owing to the greenhouse effect, the surface also receives LW radiation that is emitted downward by the atmosphere, in the amount 327 W m^{-2}. Collectively, these contributions result in a net transfer of radiative energy to the Earth's surface:

SW Absorption	+	LW Absorption from Atmosphere	−	LW Emission	=	Net Radiative Forcing of Surface
169 W m^{-2}	+	327 W m^{-2}	−	390 W m^{-2}	=	+106 W m^{-2} .

The surplus in absorption of 106 W m^{-2} represents net "radiative heating of the surface." For equilibrium, it must be balanced by transfers of sensible and latent heat to the atmosphere. Sensible heat transfer accounts for 16 W m^{-2} through conduction with the atmosphere. Latent heat transfer accounts for the remaining 90 W m^{-2}. It follows from evaporative cooling of the ocean: Heat absorbed from the ocean to evaporate water is transferred to the atmosphere when that water vapor condenses, releases the latent heat of vaporization, and precipitates back to the Earth's surface. Were it not for these auxiliary forms of energy transfer, the Earth's surface would have to be some 50 K warmer for its emission of LW energy to balance its absorption of SW energy.

The atmosphere's energy budget must likewise balance to zero to maintain thermal equilibrium. The atmosphere receives 68 W m^{-2} directly through absorption of SW radiation. More important, however, is its absorption of LW radiation emitted by the

Earth's surface. Of the 390 W m^{-2} of LW radiation that is emitted by the Earth's surface, nearly all of it is absorbed by overlying atmosphere: Only 22 W m^{-2} passes freely through the atmosphere and is rejected to space. The remaining 368 W m^{-2} is absorbed by the atmosphere: 120 W m^{-2} by cloud and 248 W m^{-2} by water vapor, CO_2, and aerosol. The atmosphere loses 327 W m^{-2} through LW emission to the Earth's surface. Another 215 W m^{-2} of LW radiation is rejected to space: 90 W m^{-2} emitted by cloud and 125 W m^{-2} emitted by water vapor, CO_2, and other minor constituents. Collecting these contributions gives the net transfer of radiative energy to the atmosphere:

$$
\begin{array}{ccccccc}
\text{SW Absorption} & + & \text{LW Absorption} & - & \text{LW Emission} & - & \text{LW Emission} & = & \text{Net Radiative} \\
 & & \text{from Surface} & & \text{to Surface} & & \text{to Space} & & \text{Forcing of} \\
 & & & & & & & & \text{Atmosphere}
\end{array}
$$

$$
68 \text{ W m}^{-2} + 368 \text{ W m}^{-2} - 327 \text{ W m}^{-2} - 215 \text{ W m}^{-2} = -106 \text{ W m}^{-2}.
$$

The deficit in absorption of 106 W m^{-2} represents net "radiative cooling of the atmosphere." That cooling is balanced by mechanical heating: transfers of sensible and latent heat from the Earth's surface.

1.4.2 Horizontal distribution of radiative transfer

The preceding discussion focuses on vertical transfers involved in the global-mean energy balance. Were the absorption of SW energy and the emission of LW energy uniform across the Earth, those vertical transfers could accomplish most of the energy exchange needed to preserve thermal equilibrium. However, geometrical considerations, variations in surface properties, and cloud cover make the horizontal distribution of radiative energy transfer nonuniform.

Even if optical properties of the Earth-atmosphere system were uniform, the absorption of solar energy would not be. At low latitude, SW radiation arrives nearly perpendicular to the Earth (cf. Fig. 1.31). A pencil of radiation incident on the top of the atmosphere is distributed there across a perpendicular cross section. Consequently, the flux of energy crossing the top of the atmosphere at low latitude equals that passing through the pencil. On the other hand, SW radiation at high latitude arrives at an oblique angle. A pencil of radiation incident on the top of the atmosphere is distributed there across an oblique cross section. The latter has area greater than that of a perpendicular cross section. Consequently, the flux of energy crossing the top of the atmosphere at high latitude is less than that passing through the pencil and, therefore, less than the flux arriving at low latitude.

The daily-averaged SW flux incident on the top of the atmosphere defines the *insolation*. Insolation depends on the *solar zenith angle*, which is measured from local vertical. It also depends on the length of day. Both vary with location and season. Figure 1.33 presents the insolation as a function of latitude and month. During equinox, insolation is a maximum on the equator, decreasing poleward like the cosine of latitude. At solstice, the maximum insolation occurs over the summer pole, with values uniformly high across much of the summer hemisphere. In the winter hemisphere, insolation decreases sharply with latitude. It vanishes at the *polar-night terminator*, beyond which the Earth is not illuminated.[9]

---

[9] The slight asymmetry between the hemispheres evident in Fig. 1.33 follows from the eccentricity of the Earth's orbit, which brings the Earth closest to the sun during January and farthest from the sun during July.

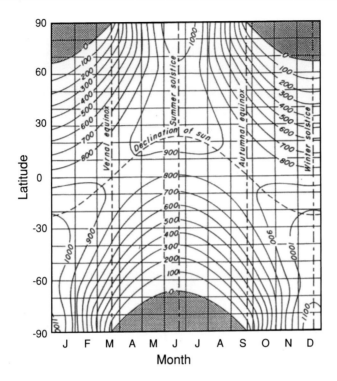

Figure 1.33 Average daily SW flux incident on the top of the atmosphere (cal cm^{-2}day^{-1}), as a function of latitude and time of year. *Source:* List (1958).

Optical properties of the atmosphere also lead to nonuniform heating of the Earth. At low latitude, SW radiation passes almost vertically through the atmosphere, so the distance traversed through absorbing constituents is minimized. However, at middle and high latitudes, SW radiation traverses the atmosphere along a slant path, which involves a much longer distance and therefore results in greater absorption. A similar effect is introduced through scattering of SW radiation by atmospheric aerosol, which increases sharply with solar zenith angle. Each of these atmospheric optical effects leads to more SW energy reaching the Earth's surface at low latitude than at high latitude.

Optical properties of the Earth's surface introduce similar effects. They emerge in OLR and reflected SW radiation that were measured by the Earth Radiation Budget Experiment (ERBE). The findings of ERBE are collected in Fig. 1.34 for northern winter. Figure 1.34a plots the distribution of albedo. Low latitudes, which account for much of the Earth's surface area, have extensive coverage by ocean. Those regions are characterized by very low albedo and thus absorb most of the SW energy incident on them. High latitudes, particularly in the winter hemisphere, have extensive coverage by snow and ice. Having high albedo, those surfaces reflect back to space much of the SW energy incident on them. This is especially true for large solar zenith angle (intrinsic to high latitudes), which increases scattering. Cloud cover adds to the high albedo at extratropical latitudes, for example, in the storm tracks, where cloud is organized by synoptic weather systems. High albedo is also evident in the tropics, where deep

(a) Albedo

(b) Longwave (Wm⁻²)

(c) Net Radiation (Wm⁻²)

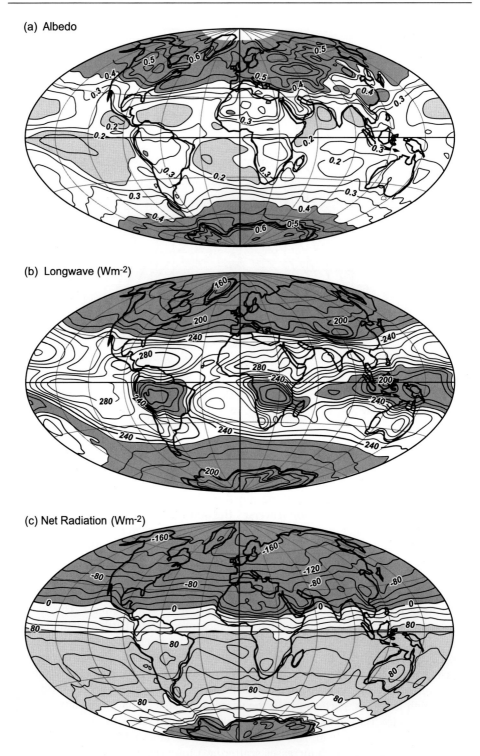

Figure 1.34 Top-of-the-atmosphere radiative properties for December–February derived from the Earth Radiation Budget Experiment (ERBE), which was on board the ERBS and NOAA-9 satellites: (a) albedo, (b) outgoing long-wave radiation, (c) net radiation. Adapted from Hartmann (1993).

convective cloud leads to maxima over the Amazon Basin, tropical Africa, and Indonesia. Flanking those regions are areas of low albedo. They correspond to cloud-free oceanic regions in the subtropics and between tropical landmasses.

Outgoing longwave radiation (Fig. 1.34b) is dominated by the poleward decrease of temperature, which causes OLR to decrease steadily with latitude (1.29). At middle and high latitudes, anomalously low OLR is found over the Tibetan Plateau and Rocky Mountains, where cold surface temperature reduces LW emission (cf. Fig. 1.30b). Anomalously-low OLR is also found near the equator, over the convective centers of South America, Africa, and Indonesia. High cloud in those regions is very cold. Its LW emission is therefore substantially weaker than emission by warm surface in neighboring cloud-free regions (1.29). In fact, subtropical latitudes are marked by distinct maxima of OLR in regions that flank organized convection. Sinking motion there compensates rising motion inside the ITCZ. It inhibits cloud formation and precipitation, maintaining arid zones that typify the subtropics.

The difference between the SW radiation absorbed by the Earth-atmosphere system and the LW radiation it emits to space defines the *net radiation*. In the global mean, net radiation must vanish for thermal equilibrium (1.30). Locally, however, this need not be the case. The distribution of net radiation observed by ERBE is plotted in Fig. 1.34c. Low latitudes are characterized by a surplus of net radiation: They experience radiative heating. High latitudes, on the other hand, are characterized by a deficit of net radiation: They experience radiative cooling. During northern winter, the entire Northern Hemisphere poleward of 15° absorbs less energy in the form of SW radiation than it emits to space as LW radiation. At the same time, most of the Southern Hemisphere absorbs more energy in the form of SW radiation than it emits to space as LW radiation. Only over Antarctica, where permanent snow cover maintains high albedo, is net radiation negative. Net radiation is positive and large almost uniformly across the tropics and in the subtropics of the summer hemisphere, where strong insolation and cloud-free conditions prevail. Notice that the strong zonal asymmetries that punctuate the SW and LW components individually (Figs. 1.34a,b) are absent. Although individually, albedo and OLR exhibit large geographical variations across the tropics, a cancellation between these components in regions of convection nearly eliminates those anomalies in the distribution of net radiation. Net radiative forcing of the Earth-atmosphere system is therefore almost zonally symmetric.

The optical properties just described lead to nonuniform heating of the Earth-atmosphere system. Low latitudes experience radiative heating, whereas middle and high latitudes experience radiative cooling. To preserve thermal equilibrium, the surplus of radiative energy received by low latitudes must be transferred to middle and high latitudes, where it offsets the deficit of radiative energy. This meridional transfer of energy is accomplished by the general circulation, about 60% of it by the atmosphere.

1.5 THE GENERAL CIRCULATION

The term *general circulation* refers to the aggregate of motions controlling transfers of heat, momentum, and constituents. Broadly speaking, the general circulation also includes interactions with the Earth's surface (e.g., transfers of latent heat and moisture from the ocean) because atmospheric behavior, especially that operating on time scales longer than a month, is influenced importantly by such interactions. For this reason, properties of the Earth's surface, like sea surface temperature (SST), figure importantly in the general circulation of the atmosphere.

The general circulation is maintained against frictional dissipation by a conversion of potential energy, associated with the distribution of atmospheric mass, to kinetic energy, associated with the motion of air. Radiative heating acts to expand the atmospheric column at low latitude, according to hydrostatic equilibrium (1.17). It thus acts to raise the column's center of mass. By contrast, radiative cooling acts to compress the atmospheric column at middle and high latitudes and lower its center of mass. The uneven horizontal distribution of mass that results introduces an imbalance of pressure forces. It drives a meridional overturning, with air rising at low latitude and sinking at middle and high latitudes.

The simple meridional circulation implied in the preceding paragraph is modified importantly by the Earth's rotation. As is evident from the instantaneous circulation at 500 hPa (Fig. 1.9a), the large-scale circulation remains nearly tangential to contours of isobaric height (e.g., tangential to isobars on a constant-height surface). Net radiative heating in Fig. 1.34c tends to establish time-mean thermal structure in which isotherms and contours of isobaric height are oriented parallel to latitude circles. The time-mean circulation at 500 hPa (Fig. 1.10b) is therefore almost westerly and circumpolar at middle and high latitudes. Characterized by a nearly zonal jet stream, the time-mean circulation possesses only a small meridional component to transfer heat between the equator and poles. A similar conclusion applies in the stratosphere, where time-mean motion is strongly zonal (Fig. 1.10b).

Asymmetries in the instantaneous circulation involve meridional deflections of air. Consequently, they play a key role in transferring heat between the equator and poles. In the troposphere, much of the heat transfer is accomplished by unsteady synoptic weather systems. They transport heat poleward through sloping convection, which exchanges cold polar air with warm tropical air. Ubiquitous in the troposphere, those disturbances contain much of the kinetic energy at mid-latitudes. They develop preferentially in the North Pacific and North Atlantic storm tracks and in the continuous storm track of the Southern Hemisphere. By rearranging air, synoptic disturbances also shape the distributions of water vapor and other constituents that are introduced at the Earth's surface.

In the stratosphere and mesosphere, synoptic disturbances are absent (Fig 1.10). Planetary waves, which propagate upward from the troposphere, play a role at these altitudes, similar to the role played by synoptic disturbances in the troposphere. Generated near the Earth's surface, these global-scale disturbances force the middle atmosphere mechanically. By deflecting air across latitude circles, planetary waves transport heat and constituents between low latitudes and high latitudes. Such transport is responsible for the greatest abundances of ozone being found at middle and high latitudes (Fig. 1.21), despite its production at low latitude.

At low latitude, the Earth's rotation exerts a weaker influence on air motion. Kinetic energy there is associated primarily with *thermally direct circulations*, in which air rises in regions of heating and sinks in regions of cooling. Thermally direct circulations in the tropics are forced by the geographical distribution of atmospheric heating (e.g., as implied by the distributions of net radiation in Fig. 1.34c and time-mean cloud in Fig. 1.30b).

Latent heat release inside the ITCZ drives a meridional overturning that prevails within 30° of the equator. Depicted in Fig. 1.35, the *Hadley circulation* involves upwelling near the equator and downwelling at subtropical latitudes. (A meridional section of observed motion is presented in Fig. 15.5.) Further poleward are westerlies, which intensify to form the subtropical jets (Fig. 1.8). At low latitude, equatorward

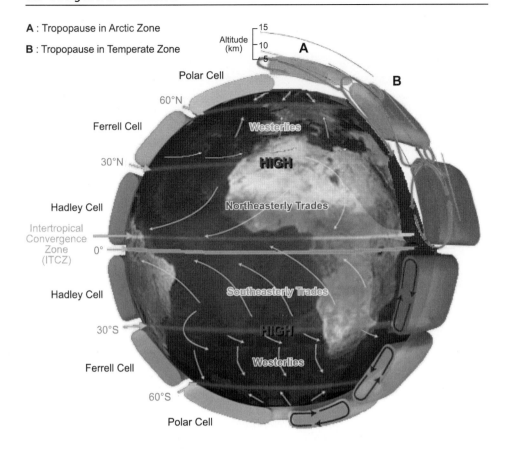

A : Tropopause in Arctic Zone

B : Tropopause in Temperate Zone

Figure 1.35 Schematic of the mean circulation, which is comprised of Hadley cells equatorward of 30°, Ferrell cells at mid-latitudes, and polar cells at higher latitude. The lower branch of the Hadley cell involves northeasterly and southeasterly Trade Winds, which meet in the InterTropical Convergence Zone (ITCZ), where air rises to form deep cumulus convection and heavy rainfall. *Source*: sealevel.jpl.nasa.gov. See color plate section: Plate 2.

flow near the Earth's surface is manifested by easterly *trade winds*, northeasterly in the Northern Hemisphere and southeasterly in the Southern Hemisphere (cf. Fig. 15.9). They converge into the ITCZ. Upwelling inside the ITCZ is compensated by downwelling at subtropical latitudes. Comprising the descending branch of the Hadley circulation, the latter forms a *subtropical high* in each hemisphere. By suppressing cloud and precipitation, downwelling of the Hadley circulation maintains deserts (cf. Fig. 1.34b). It is for this reason that all major deserts on Earth are found in the subtropics.

Nonuniform heating across the tropics also drives zonal overturning. Known as a *Walker circulation*, the latter involves upwelling at longitudes of heating and downwelling at longitudes of cooling. The nonuniform distribution of land and sea, along with resulting asymmetries in atmospheric heating, lead to Walker circulations along the equator. The concentration of latent heating over the maritime continent (Fig. 1.34b) forces the Pacific Walker circulation, which is illustrated in Fig. 1.36. This circulation reinforces easterly trade winds across the equatorial Pacific. Its descending

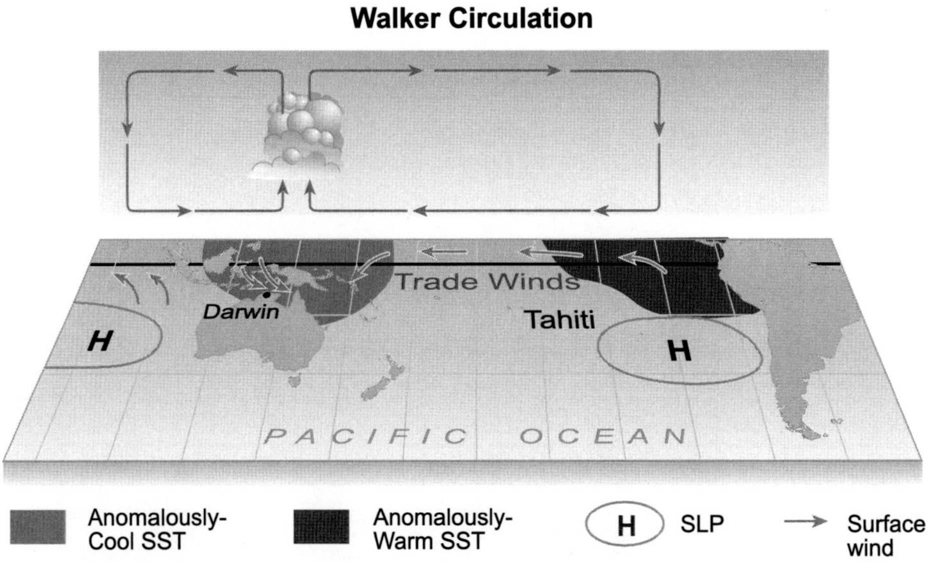

Walker Circulation

Figure 1.36 Schematic of the Pacific Walker circulation. *Source*: http://www.bom.gov.au/lam/climate/levelthree/analclim. See color plate section: Plate 3.

branch, found to the east, maintains the arid climate that typifies the eastern Pacific (cf. Figs. 1.34a,b).

In addition to driving quasi-steady circulations, latent heat release excites unsteady wave motions. Those disturbances are another form of asymmetry in the circulation. Wave motions are also excited mechanically at the Earth's surface. By displacing atmospheric mass vertically, orographic features such as the Alps, the Himalayas, and the Rocky Mountains excite planetary waves and gravity waves that radiate away from those source regions. Like wave activity generated by unsteady heating, orographically-excited disturbances propagate the influence of their forcing to other regions.

Kinetic energy associated with the general circulation is damped by frictional dissipation. Frictional damping occurs principally through turbulence. Involving eddies with a wide range of scales, it homogenizes large-scale gradients. About half of the kinetic energy of the large-scale circulation is dissipated in the lowest kilometer of the atmosphere. Inside the *planetary boundary layer*, small-scale turbulence extracts energy from large-scale motion and cascades it to small dimensions, where it is dissipated by molecular diffusion. In the *free atmosphere*, namely, away from the surface, the large-scale circulation suffers frictional dissipation through convective motion and dynamical instability. Both generate turbulence from large-scale organized motion. The large-scale circulation is also damped by thermal dissipation. LW cooling to space damps atmospheric motion indirectly, by destroying the accompanying temperature anomaly.

1.6 HISTORICAL PERSPECTIVE: GLOBAL-MEAN TEMPERATURE

This chapter closes with an overview of the salient property of climate, global-mean surface temperature, along with evidence of how it has evolved. Surface temperature dictates many features of climate. Among them is the amount of water that is sequestered in glacial ice. The latter, in turn, influences sea level.

Like any property of climate, Global Mean Temperature (GMT) is comprised of two components: (1) *Climate variability* involves fluctuations that enter through influences that are internal to the climate system, like transient exchanges of heat between the atmosphere and ocean, and influences that are external to the climate system, like variations of solar luminosity. Involving a wide range of time scales (years, decades, and longer), they represent natural fluctuations of the climate system. (2) The *secular* variation of climate evolves systematically in one direction. Unlike climate variability, it is not offset by subsequent swings in the opposite direction. It is in the secular variation that anthropogenic changes should be concentrated.

1.6.1 The instrumental record

Routine measurements of air temperature began in the nineteenth century. They eventually formed a network of ground-based weather observations, which contribute today to the operations of major weather centers to produce short-term forecasts. Beyond surface measurements, the operational network makes vertical soundings of temperature, humidity, and wind via *rawinsonde*. Launched synchronously at 0000 and 1200 Universal Time (UT = local time at Greenwich England), these balloon-borne instruments relay back local conditions during their ascent (nominally reaching 50 hPa). The rawinsonde network, parts of which date back to the 1930s, is comprised of about 1000 stations over the globe, displayed in Fig. 1.37 (open circles). It provides coverage that is dense over populated continents, sparse over remote landmasses like Africa, South America, Greenland, and Antarctica, and almost nonexistent over the oceans. Augmenting the rawinsonde network are numerous other reporting stations, especially for surface measurements, which are superimposed in Fig. 1.37 (dots). Nonetheless, even collectively, the ground-based network suffers from the same nonuniform sampling that limits the rawinsonde network.

Because it is discriminated to continental regions, such sampling can introduce a systematic error or bias into the record of global-mean temperature. Changes over continental regions are inadvertently magnified relative to changes over maritime regions, which are left unaccounted for. The respective error in global-mean temperature, however, appears to be small, a couple of percent – at least as inferred from a climate simulation (Madden and Meehl, 1993).

Exacerbating the sampling error are two related features of continental temperature: Owing to thermal properties of the Earth's surface, diurnal and seasonal temperature extremes are large over land, but small over ocean (Sec. 15.4). The diurnal cycle of temperature, because it is controlled by water vapor and cloud, is not a robust feature of climate simulations (e.g., as have been relied on to evaluate sampling). In addition, many stations in the ground network are, for historical reasons, situated near population centers (Fig. 1.37). Over the course of the instrumental record, those centers expanded through urban development. With their expansion was an amplification of the *urban heat island effect*, wherein urban centers are distinctly warmer than their surroundings.[10] The heat island effect is greatest at night, when rural areas cool faster

---

[10] The heat island effect develops through several mechanisms, notably through the greater heat capacity of concrete structures and their interruption of LW cooling to space.

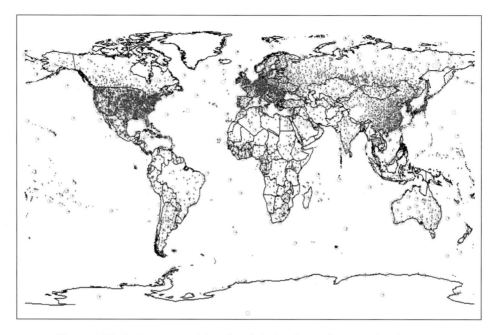

Figure 1.37 Stations comprising the global radiosonde network, where vertical profiles of atmospheric structure are measured twice daily at 0000 and 1200 UT (open circles), and the ground network of surface temperature measurements (dots). *Sources:* Gruber and Haimberger (2008), Peterson and Vose (1997). See color plate section: Plate 4.

and, therefore, achieve colder temperatures than urban centers. It emerges conspicuously in temperature trends.

Figure 1.38a compares records of annual-mean temperature between stations that are in the same region (California) but neighboring disparate populations at the end of the century.[11] The average of stations neighboring population centers that exceed 1,000,000 (upper) has a mean over the twentieth century of $\sim 62°F$ ($17°C$) and a warming trend of some $0.3°F$/decade. Population centers of less than 100,000 experienced far less growth and urban development. The average of stations neighboring those centers (lower) has a mean over the twentieth century that is some $6°F$ cooler. It exhibits a warming trend, but one that is an order of magnitude weaker. Hence, more-urbanized regions are warmer and they warmed faster than other regions.

Superimposed in Fig. 1.38a is the record of Global Mean Temperature, derived from the network of surface stations (dotted). The trend exhibited by GMT is closer to that reported by large population centers. The variability of GMT, however, is an order of magnitude weaker than that in the records of regional temperature. Close inspection reveals little correspondence with either. Changes of regional temperature are strongly correlated between neighboring centers of different population. However, they are virtually uncorrelated to changes of GMT. Hence the comparatively large changes that dominate regional climate have little relationship to those of GMT.

The lack of dependence on GMT follows from the spatial coherence of regional changes. Plotted in Fig. 1.38b is the same information for SE Australia, a region at

[11] Notice that, even in annual means, regional temperature exhibits climate variability of 1–2 K.

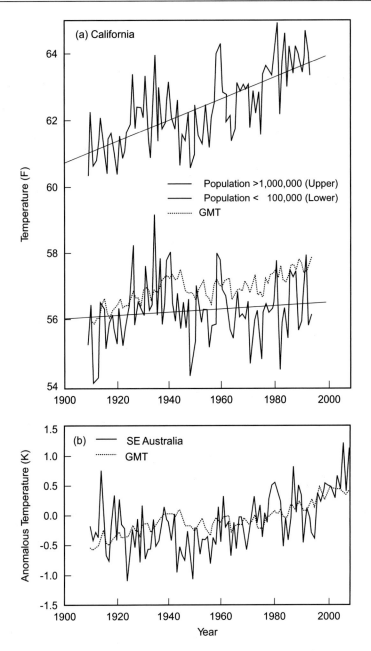

Figure 1.38 (a) Record of annual-mean temperature in California, averaged over population centers exceeding 1,000,000 (upper) and of less than 100,000 (lower). Superimposed is the record of Global Mean Temperature (GMT) from the network of surface stations (dotted). (b) Record of annual-mean temperature over SE Australia (undiscriminated by population), a region at conjugate latitude, of comparable area, and proportionate population growth (solid), and GMT (dotted). (c) Histogram of observed temperature trend over California, as a function of population. *Sources:* Goodridge (1996); Robinson et al. (1998).

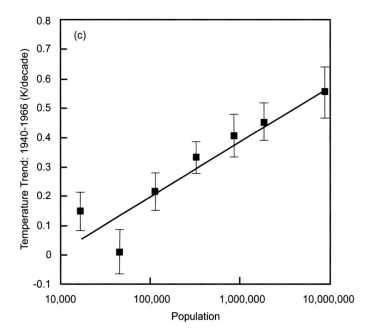

Figure 1.38 (*continued*)

conjugate latitude, of comparable area, and proportionate population growth. The picture (undiscriminated by population) is much the same: Even in annual means, changes of regional temperature exhibit little correspondence to changes of GMT, which are an order of magnitude weaker. The records have a correlation of 0.29. Thus GMT accounts for only about 8% of the variance of regional temperature, rendering one largely independent of the other. Instead, regional temperature is dominated by climate variability, fluctuations that operate on a wide range of time scales. Included are decadal drifts of regional climate. For example, during the first half of the twentieth century, temperature over SE Australia declined, whereas GMT then increased. During the last two decades of the twentieth century, both increased. Afterward, the records diverged again, GMT no longer increasing.

The correlation between changes over SE Australia and California is even smaller, 0.24. Only about 6% of the variance operates coherently between those regions. It follows that changes of regional temperature, even in annual means, have limited spatial coherence. When averaged over the Earth, those spatially incoherent changes cancel. Left behind are much weaker changes that are globally coherent and therefore contribute to changes of GMT. Changes of regional temperature that operate on shorter time scales, for example, represented in seasonal means, exhibit even greater variance, with even less dependence on GMT. So do episodes of extreme weather, which develop through anomalous regional conditions. As is apparent in Figs. 1.38a and 1.38b, such conditions are dominated by climate variability, which is largely independent of GMT.

The long-term evolution of station reports in Fig. 1.38a suggests a correspondence to population growth. That relationship is borne out more generally. Plotted in Fig. 1.38c is a histogram of the temperature trend reported by stations, as a function of neighboring population at the end of the twentieth century. The reported trend is strongly correlated with population in the surroundings. Stations in large urban centers that underwent major development report magnified trends. They are unrepresentative of trends from smaller population centers in the wider surroundings. The distorted contribution to global-mean temperature is intrinsic to populated landmasses, which are sampled by the ground network (Fig. 1.37). It is not representative of remote landmasses and maritime regions, which are left unsampled.

Circumventing the sampling limitations of the ground network are satellite observations. The satellite era began around 1979, when atmospheric structure and composition began to be monitored from space by several operational and research platforms. Most are in a polar orbit, which, on time scales longer than a day, provides uniform and nearly complete coverage of the Earth. However, the benefit of satellite observations in horizontal sampling is offset by their limitation in vertical sampling. Unlike surface measurements and rawinsondes, which provide temperature in the lowest 100 m, satellite retrievals describe temperature that is averaged over a comparatively deep layer of the atmosphere, characterizing the lowest 5 km. Accordingly, sharp horizontal changes of surface temperature (e.g., associated with abrupt differences between land and ocean) are lost, blended with more gradual horizontal variations that prevail at upper levels. Nonetheless, the coverage and continuity afforded by the satellite record makes it unrivalled in considerations of global-mean temperature.

Plotted in Fig. 1.39, by month, is the satellite record of GMT (dashed) recovered from measurements of channel 2 of the Microwave Sounding Unit (MSU) and channel 5 of the Advanced Microwave Sounding Unit (AMSU). Those measurements represent temperature in the lower troposphere. Superimposed is the corresponding record derived from surface measurements (solid). The two records exhibit much the same behavior, deviating only in detail. Both evidence an upward trend. GMT warmed since 1982, by some 0.19 K/decade in the surface record and by 0.16 K/decade in the satellite record. About the trend are interannual fluctuations of order ±0.20 K. They represent the climate variability of GMT. Conspicuous among interannual fluctuations is the sharp warming during 1997–1998. It coincides with the El Nino during that period (Chap. 17). El Nino is an oceanic phenomenon. Its signature is only half as strong in the surface record of GMT, which is discriminated to continent, as it is in the satellite record, which has uniform coverage of the Earth. A signature of the 1982–1983 El Nino is less evident. Nevertheless, it exhibits a similar discrepancy between the surface and satellite records of GMT. Also evident is influence from the eruption of Pinatubo in 1991 (cf. Fig. 1.27). After declining for about 2 years, global temperature rebounded, increasing for about a decade. Global-mean temperature then leveled off around 2002, followed by several years of general decline.

In the surface record, much of the warming trend derives from continental regions (Fig. 1.40a); that is, after all, where those observations are concentrated (cf. Fig. 1.38). It is especially strong at polar sites of the Northern Hemisphere, which are glaciated during winter. The trend over continental regions follows principally from nighttime temperature, which exhibits faster warming than daytime temperature (Easterling et al., 1997). The trend derived from the satellite record (Fig. 1.40b) is more uniform longitudinally, with weaker geographical dependence. Nonetheless, it shares major features with the trend in surface record, notably, the strongest warming being found at polar latitudes of the Northern Hemisphere. Offsetting the latter is a cooling trend over the Southern Ocean and Antarctica. Notice that, due to to scant coverage, it is virtually unrepresented in the surface record (Fig. 1.40a). Smoother horizontal structure

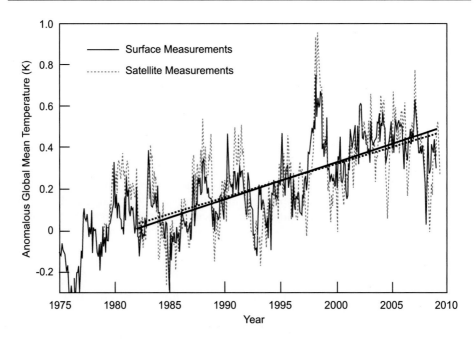

Figure 1.39 Record of Global Mean Temperature (GMT) from station measurements at the Earth's surface (solid) and from satellite measurements by Channel 2 of MSU and Channel 5 of AMSU (dashed). Both evidence a warming trend
since 1982, of 0.19 K/decade in the surface record and 0.16 K/decade in the
satellite record. (Over 1979–2009, the trend in MSU is ∼0.125 K/decade.)
The satellite record derives from the retrieval by University of Alabama at
Huntsville (Christy, Spencer, and Braswell, 2000), in association with National
Aeronautics and Space Administration (NASA). A different retrieval of the same
measurements has been performed by Remote Sensing Systems (Mears et al.,
2003), in association with National Oceanic and Atmospheric Administration
(NOAA) – (not shown). It is very similar, albeit with a decadal trend closer to
that of the surface record. However, that retrieval omits data from Antarctica,
which, during the same interval, exhibits a cooling trend.

intrinsic to the satellite trend reflects the comparatively deep layer that is represented
by those measurements. In the vertical average over 5 km, sharp horizontal differences of surface temperature, for example, between land and ocean, are dissolved
into structure that is more zonally uniform – smooth structure at upper levels that
is maintained by horizontal air motion. Although it blurs geographical features over
large horizontal dimension, this process should have little impact on global-mean temperature (Prob. 1.18). It is for this reason that the surface and satellite records of GMT
track as closely as they do (Fig. 1.39).

In Fig. 1.41, the ground-based record is extended back to 1850, along with the
respective uncertainty. During the latter half of the nineteenth century, GMT mean-
dered, with deviations of a few tenths of a Kelvin. Between 1880 and 1910, it declined by
about 0.5 K. Global-mean temperature then increased steadily for three decades, during
1910–1940, also by about 0.5 K. During the 1940s and early 1950s, GMT declined again,
followed by two decades when it simply fluctuated. It is noteworthy that the decline

Temperature Trend

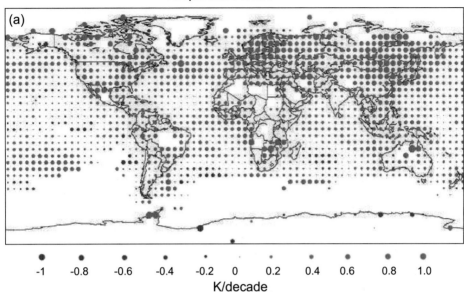

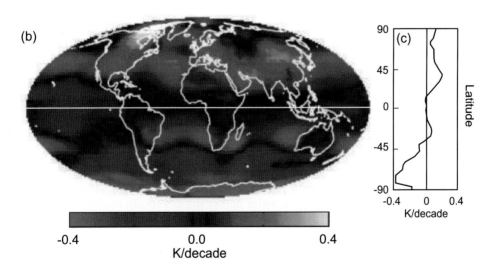

Figure 1.40 Temperature trend during the satellite era, as function of geographical position, derived from (a) the surface record and (b) the satellite record. (c) Zonal-mean trend in the satellite record, which has global coverage. *Sources:* IPCC (2007), Mears et al. (2003). See color plate section: Plate 5.

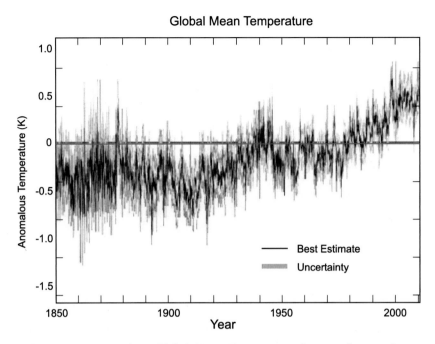

Figure 1.41 Anomalous Global Mean Temperature from surface stations (solid), relative to the mean over 1960–1990, along with estimates of uncertainty (shaded). *Source*: http://hadobs.metoffice.com/hadcrut3.

of temperature during the 1940s and 1950s coincides with a plateau in atmospheric CO_2 (Fig. 1.14). r_{CO_2} then actually decreased, even though anthropogenic emission accelerated during the period (Fig. 1.16). Then, in the late 1970s, GMT again increased steadily. That warming continued until the close of the twentieth century, after which GMT plateaued, declining slightly. This evolution is mirrored in the satellite record (Fig. 1.39).

Overall, the twentieth century was marked by three major episodes: 3 decades of steady warming, followed by 4 decades of steady cooling or no systematic change at all, followed by 2 decades of renewed steady warming. Together, those episodes led to net warming during the twentieth century of nearly 1.0 K. That deviation is twice as great as the one recorded during the latter half of the nineteenth century (albeit when measurements were sparse).

The systematic variation of GMT is modulated substantially by its climate variability. During the 1940s and 1950s, GMT decreased. Systematic warming over the twentieth century was then clearly overshadowed by climate variability. During the early twentieth century and late twentieth century, when GMT increased, it was, with equal likelihood, reinforced by climate variability. For this reason, the trend during an individual episode, even one lasting a couple of decades, represents not the secular variation or climate variability, but both. Only over a long enough interval for opposing swings of climate variability to cancel does a trend reflect the true secular variation of GMT. Between 1910 and 2010, GMT increased at a rate of ∼0.09 K/decade. The mean trend over the twentieth century is ∼0.075 K/decade (IPCC, 2007). These values are only about half of the warming trend that appeared during the last of the

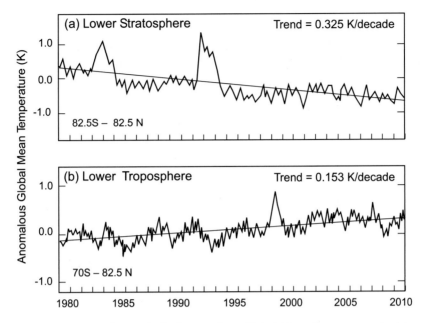

Figure 1.42 Satellite record of anomalous Global Mean Temperature in (a) lower stratosphere and (b) lower troposphere. Note that, in (b), the contribution from Antarctica is omitted. *Source*: www.ssmi.com/msu.

major episodes in Fig. 1.39 (e.g., during the 1980s and 1990s). It is about a third of the rate implied by Arrhenius' calculation, which did not account for adjustments of the climate system to increasing CO_2.

Compared in Fig. 1.42 are the satellite records of GMT in the troposphere and stratosphere. In the lower troposphere (Fig. 1.42b), GMT is punctuated by warming during the 1997–1998 El Nino. A counterpart signature of the 1983 El Nino is less conspicuous. The tropospheric record exhibits a steady warming trend, of order 0.15 K/decade. (Noteworthy, however, is the exclusion of contributions from Antarctica, where the trend is negative; Fig. 1.40b,c.) In the lower stratosphere (Fig. 1.42a), even the 1997–1998 El Nino is absent. Instead, GMT in the stratosphere is punctuated by episodes of warming following the eruptions of El Chichon (1982) and Pinatubo (1991). Those episodes of warming are scarcely visible in the tropospheric record, which exhibits cooling following the eruptions. (cf. Fig. 1.27). In contrast to the warming trend in the troposphere, GMT in the stratosphere exhibits a cooling trend. Of order −0.30 K/decade, it is opposite to and about twice as strong as the warming trend in the lower troposphere.

1.6.2 Proxy records

Instrumental records before the nineteenth century are not available. In place of temperature measurements is a host of proxies, geological and other evidence that is used to infer temperature through a variety of techniques. These records are relied upon by students of *paleoclimate*, for periods before written records were maintained. Such records are limited by coverage and uncertainties, which cloud their interpretation.

Figuring prominently among proxy records are ice cores (Fig. 1.12). Many come from Antarctica and Greenland, where deep cores extend far into the past. Through isotopic analysis, those cores provide estimates of prior temperature. Bubbles of air trapped inside the ice also provide estimates of atmospheric composition, including the concentrations of trace gases. Records like that in Fig. 1.12 indicate that, on long time scales, temperature and CO_2 vary dependently. They imply further that CO_2 lags temperature by 500–1000 years (Fischer et al., 1999; Indermuhle et al., 2000; Monnin et al., 2001). The lag suggests that, at least on millennial time scales, changes of temperature lead to changes of CO_2 (e.g., through changes in production at the Earth's surface and in exchanges with the ocean).

A close relationship between temperature and CO_2 is also evident on much shorter time scales, short enough to be resolved in the instrumental record. That relationship is hinted at by the plateau of r_{CO_2} during the 1940s and 1950s (Fig. 1.14), which coincided with a decline of GMT (Fig. 1.41). The budget of carbon is complex (Sec. 17.3). Poorly documented, individual sources and sinks may never be resolved in detail. Yet, their collective impact on atmospheric carbon is consolidated in a single property, r_{CO_2}.

The rate of increase $\frac{d}{dt} r_{CO_2} = \dot{r}_{CO_2}$ must equal the net emission rate of CO_2, collected from all sources and sinks. Plotted in Fig. 1.43a is the net emission rate of CO_2 (solid), evaluated from the record at Mauna Loa (Fig. 1.15). The behavior has been lowpass filtered to changes that occur on time scales longer than 2 years. By doing so, it discriminates to the hysteresis in the annual cycle of emission and absorption (Fig. 1.15) – the part that accumulates over successive years. Net emission is uniformly positive, consistent with CO_2 increasing monotonically. However, it varies substantially from one year to the next. About a mean of ~1.5 ppmv/yr, net emission changes by almost 100%: from twice as great in some years to nearly zero in others. These large changes reflect differences between years in how much CO_2 is emitted during one phase of its annual cycle versus how much is absorbed during the opposite phase. Of order 50–100%, the large change of net emission represents an imbalance between sources and sinks. By comparison, the contribution from fossil fuel emission varies only gradually (Rotty, 1987); cf. Fig. 1.16. Superimposed in Fig. 1.43 is the satellite record of global-mean temperature (Fig. 1.39), likewise lowpass filtered (dashed). Net emission of CO_2 closely tracks the evolution of GMT. Achieving a correlation of 0.80, the variation of GMT accounts for most of the variance in CO_2 emission.

Plotted in Fig. 1.43b is the rate of change in isotopic composition, $\frac{d}{dt} \delta^{13}C = \dot{\delta}^{13}C$ (solid).[12] Its mean is negative, consistent with the long-term decline of $\delta^{13}C$ in ice cores (Fig. 1.14). However, like emission of CO_2, differential emission of $^{13}CO_2$ varies substantially from one year to the next. It too tracks the evolution of GMT – just out of phase. When GMT increases, emission of $^{13}CO_2$ decreases and vice versa. The records achieve a correlation of −0.86. Hence the variation of GMT, which accounts for most of the variance in emission of CO_2, also accounts for most of the variance in differential emission of $^{13}CO_2$.

---

[12] $\frac{d}{dt} \delta^{13}C$ is approximately equal to the percentage rate of change of $^{13}CO_2$ minus that of $^{12}CO_2$. It thus measures the rate of differential emission of $^{13}CO_2$.

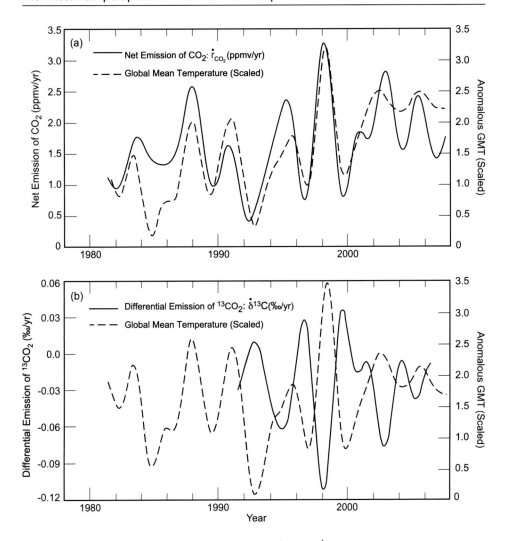

Figure 1.43 (a) Net emission rate of CO_2, $\dot{r}_{CO_2} = \frac{d}{dt}r_{CO_2}$ (ppmv/yr), derived from the Mauna Loa record (Fig. 1.15), lowpass filtered to changes that occur on time scales longer than 2 years (solid). Superimposed is the satellite record of anomalous Global Mean Temperature (Fig. 1.39), lowpass filtered likewise and scaled by 0.225 (dashed). Trend in GMT over 1979–2009 (not included) is ∼0.125 K/decade. (b) As in (a), but for rate of change in isotopic carbon fraction, $\dot{\delta}^{13}C = \frac{d}{dt}\delta^{13}C$, likewise from Mauna Loa (cf. Fig. 1.14). $\frac{d}{dt}\delta^{13}C$ is approximately equal to the percentage rate of change of $^{13}CO_2$ minus that of $^{12}CO_2$. It thus measures the rate of differential emission of $^{13}CO_2$.

The out-of-phase relationship between r_{CO_2} and $\delta^{13}C$ in the instrumental record (Fig. 1.43) is the same one evidenced on longer time scales by ice cores (Fig. 1.14). The out-of-phase relationship in ice cores is regarded as a signature of anthropogenic emission, subject to uncertainties (Sec. 1.2.4). The out-of-phase relationship in the instrumental record, however, is clearly not anthropogenic. Swings of GMT following the eruption of Pinatubo and during the 1997–1998 El Nino were introduced through natural mechanisms (cf. Figs 1.27; 17.19, 17.20). Changes in Fig. 1.43 reveal that net emission of CO_2, although ^{13}C lean, is accelerated by increased surface temperature. Outgassing from ocean, which increases with temperature (Sec. 17.3), is consistent with the observed relationship – if the source region has anomalously low $\delta^{13}C$. So is the decomposition of organic matter derived from vegetation. Having $\delta^{13}C$ comparable to that of fossil fuel, its decomposition is likewise accelerated by increased surface temperature.

Conspicuous in the records is the warming during the 1997–1998 El Nino, when GMT increased sharply. Intrinsic to the ocean, that perturbation to the climate system led to a surge in emission of CO_2 and a simultaneous reduction in differential emission of $^{13}CO_2$ – each by more than 100%. Such changes of global temperature cannot be explained by changes of carbon dioxide. Coherent with large swings in the net emission of CO_2 and $^{13}CO_2$, they imply that net transfer of carbon to the atmosphere involves processes that are sharply modulated by surface temperature. The respective sensitivity of those processes to surface temperature is established by coherent changes of net emission (Sec. 8.7.1). The dependence of CO_2 emission on temperature is poorly understood. It is not accounted for in GCMs, wherein atmospheric CO_2 is simply specified.

Older records of temperature and composition rest on proxy evidence, which is limited by coverage and uncertainty. Atmospheric properties inferred from ice cores are subject to a number of unknowns. At shallow depth, their layering resolves individual years. However, at greater depth, such features are blurred by diffusion, which limits temporal resolution. The isolation and stability of air trapped inside ice is a matter of ongoing debate.

So is the dating of features, which is pursued through half a dozen different techniques (e.g., Ruddiman and Raymo, 2003; Petit et al., 1999; Jouzel et al., 1996; Lorius et al., 1985; Sowers et al., 1993). Discrepancies in timing between those techniques range from centuries at shallow depth to millennia at greater depth. Those uncertainties are not allayed by cores from neighboring sites, which reveal similar discrepancies (Parrenin et al., 2007).

Dating of tree rings, *dendrochronology*, also figures prominently in reconstructions of previous climate. It relies on the width and number of annual growth cycles, in concert with radiocarbon dating of fossil specimens. The width of individual rings is used to infer temperature. Like ice cores, however, inferences drawn from those records are clouded by uncertainties; see Moberg et al. (2008) for an overview. In particular, temperature is but one of several environmental factors that influence annual tree growth (Bednarz and Ptak, 1990). Annual growth is also influenced by moisture, disease, and infestation. Another influence may be related more directly to its interpretation. Tree growth during the twentieth century differs distinctly from earlier growth (Briffa, 2000). It then diverges from the instrumental record – about the time that direct measurements of temperature became widespread. Why is not understood. Regardless of its origin, this feature seriously complicates the inference of temperature from annual tree growth. Supporting tree-ring evidence are sedimentary records from lakes and rivers. Because they form differently under frozen and unfrozen conditions, they too are clouded by uncertainties.

Changes in temperature inferred for previous climates are accompanied by changes in glaciation. Figure 1.12 compares, during the last 400,000 years, the record of temperature inferred from the Vostok core over Antarctica against global ice volume (reversed) inferred from sedimentary cores that were collected from around the globe. On time scales of millennia and longer, the two are strongly correlated. Gradual cooling

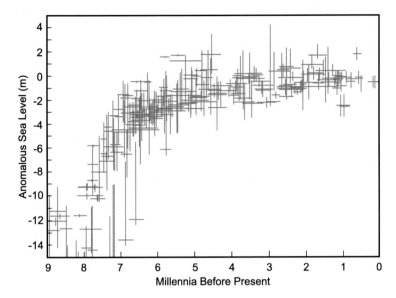

Figure 1.44 Reconstruction of sea level during the Holocene epoch, inferred from geological proxies at several sites around the Earth. The estimates rely on corrections for the vertical displacement of continents, which introduces one of several major uncertainties in dating (horizontal bar) and value (vertical bar). *Sources*: Fleming et al. (1998), Fleming (2000), and Milne et al. (2009).

toward temperature minima corresponds well with glacial periods. The comparatively abrupt warming to temperature maxima corresponds equally well with the approach to interglacial periods. Included is the current one, which defines the *Holocene* epoch.

The swings of ice volume in Fig. 1.12 imply reverse swings in sea level. To leading order, maritime ice does not contribute to such changes (Prob. 1.17). Sea level can then be influenced only by melting of continental ice. Plotted in Fig. 1.44 is a reconstruction of sea level during the last 10,000 years. It has been inferred from a variety of geological proxies, including sediments and coral skeletons that flourish at particular depths; Milne et al. (2009) provides a tabulation. Following the last glacial maximum 20,000 years ago, sea level rose sharply, at some 10 m/millennium. It then leveled off about 7000 years ago, although it continued to increase gradually. The attainment of those levels was contemporaneous with the development by humans of agriculture; the rise of major civilizations in the fertile crescent, Egypt, and Indus valley; and the onset of historical records. Since then, sea level appears to have risen by a couple of meters. However, the scatter of proxy evidence is noteworthy: about a meter.

Overall, the proxy record indicates a gradual increase of sea level during the last 5000 years, at ~4 cm/century. By comparison, sea level is currently rising at some 10-20 cm/century (Douglas, 1997). Most of this rise derives not from glacial melting, but from thermal expansion of the ocean (IPCC, 2007). The latter involves changes of temperature in the ocean interior, for which observations are scant (Chap. 17). It is noteworthy that other records during the same period as Fig. 1.44 reveal short-term fluctuations of sea level that are not resolved in such reconstructions (e.g., Siddall et al., 2003). As large as a couple of meters, those fluctuations operate on centennial time scales. A similar event some 14,000 years ago, the so-called *meltwater pulse*, involved a

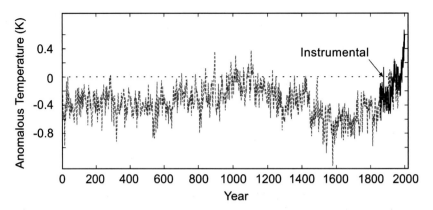

Figure 1.45 A reconstruction of Global Mean Temperature during last two millennia, inferred from a variety of proxies. After Moberg et al. (2005). Reprinted by permission of Macmillan Publishers Ltd: *Nature* (Moberg et al., 2005), Copyright (2005).

sea level rise of more than 10 m in just a couple of centuries (Webster, 2004; Stanford et al., 2006).

Presented in Fig. 1.45 is a reconstruction of temperature during the last two millennia based on a variety of proxies from the Northern Hemisphere, including tree rings. The record presented is but one of several such reconstructions, which are technique dependent (Moberg et al., 2008). Nonetheless, it illustrates major features that have long been recognized. After about the year 600, temperature warmed steadily, maximizing during the so-called *Medieval Warm Period* that prevailed during 1000–1300 (Lamb, 1965). The period is well documented in anecdotal evidence from the North Atlantic and Europe, in Norse explorations that were unhindered by ice, in the Viking colonization of Greenland, and in the northward advance of agriculture that enabled a wine industry to flourish in England (see, e.g., Lamb, 1982). The Medieval Warm Period is also evidenced by records of the tree line in the Alps, which ascended then by some 100 m.

Following the Medieval Warm Period was a temperature swing in the opposite direction. After the twelfth century, temperature decreased steadily – for several centuries. Minimum values were attained during the so-called *Little Ice Age*, which prevailed between the sixteenth and nineteenth centuries. The Little Ice Age is chronicled in anecdotal evidence from Europe and North America, surrounding the southward advance of pack ice into the North Atlantic; the expansion of glaciers, which descended some 100 m lower than their extent in the late twentieth century; and the freezing of major bodies of water like the Thames, New York harbor, and parts of the Baltic and Bosphorus. Then, in the nineteenth century; temperature began increasing. It eventually merged with the increase that appears in the instrumental record, which commenced later that century. How much temperature rebounded, like so many features in proxy evidence, depends on which reconstruction is adopted.

Whether the Medieval Warm Period and Little Ice Age were global events or even existed has been questioned. Uncertainties over their existence surround the veracity and interpretation of tree-ring records. However, reconstructions exclusive of tree-ring evidence produce much the same pattern as in Fig. 1.45, which is a robust feature of the proxy record (Loehle, 2007; Loehle and McCulloch, 2008). The Mevieval Warm Period and Little Ice Age have also been documented in records worldwide, including moraine deposits, sedimentary records, and ice cores (e.g., Wilson et al., 1979; Grove and Switsur, 1994; Broecker, 2000; Kreutz et al., 1997; Adhikari and Kumon, 2001; Cook et al., 2002; Hall and Denton, 2002; McGann, 2008; Williams et al., 2007; Araneda et al., 2007). Those records evidence a global retreat of glaciers during the Medieval Warm Period, followed by their systematic advance during the approach to the Little Ice Age.

How much temperature was elevated during the Medieval Warm Period, like many features of temperature reconstructions, is a matter of debate. The reconstruction in Fig. 1.45 implies that temperature then was about 0.5 K warmer than previously. Other reconstructions place temperature then 1.0–2.0 K warmer (Lamb, 1965; Keigwin, 1996; Huang and Pollack, 1997; Loehle, 2007). That would make the Medieval Warm Period as warm or warmer than temperature during the late twentieth century.

Similar uncertainty surrounds how fast temperature rebounded following the Little Ice Age. Because, on long time scales, CO_2 is coupled to temperature, the same uncertainty applies to its increase. The reconstruction in Fig. 1.45 implies warming over the last two centuries of 0.03–0.05 K/decade (e.g., Moberg et al., 2005). Reconstructions free of tree-ring uncertainty imply similar warming since the Little Ice Age (Loehle, 2007). Representing the average trend over two centuries, those estimates of natural warming are about half of the warming trend that appears in the instrumental record during the twentieth century.

SUGGESTED REFERENCES

The Boltzmann distribution is developed along with other components of molecular kinetics in *Statistical Thermodynamics* (1973) by Lee, Sears, and Turcotte.
Aeronomy of the Middle Atmosphere (1986) by Brasseur and Solomon includes a comprehensive treatment of anthropogenic trace gases.
The Intergovernmental Panel Report on Climate Change (2007) provides an overview of the climate problem, numerical modeling of it, and historical evidence.
A comprehensive treatment of the global energy budget and its relationship to the general circulation is presented in *Global Physical Climatology* (1994) by Hartmann.

PROBLEMS

1. Derive expression (1.15) for the molar fraction of the ith species in terms of its mass mixing ratio.
2. Derive an expression for the volume mixing ratio of the ith species in terms of its mass mixing ratio.
3. Relate the volume mixing ratio of the ith species to the fractional abundance of i molecules present.
4. Show that the sum of partial volumes equals the total volume occupied by a mixture of gases (1.6).
5. Demonstrate that 1 atm of pressure is equivalent to that exerted by (a) a 760 mm column of mercury, (b) a 32' column of water. (The density of Hg at 273 K is 1.36 10^4 kg m^{-3}.)
6. Consider moist air at a pressure of 1 atm and a temperature of 20°C. Under these conditions, a relative humidity of 100% corresponds to a vapor pressure

of 23.4 hPa. At that vapor pressure, air holds the maximum abundance of water vapor possible under the foregoing conditions and is said to be *saturated*. For this mixture, determine (a) the molar fraction N_{H_2O}, (b) the mean molar weight $\overline{M}$, (c) the mean specific gas constant $\overline{R}$, (d) the absolute concentration of vapor ρ_{H_2O}, (e) the mass mixing ratio of vapor r_{H_2O}, (f) the volume mixing ratio of vapor.

7. From the distribution of water vapor mixing ratio shown in Fig. 1.17, plot the vertical profile of water vapor number density $[H_2O](z)$ at (a) the equator, (b) 60°N.

8. In terms of the vertical profile of water vapor mixing ratio $r_{H_2O}(z)$, derive an expression for the total precipitable water vapor Σ_{H_2O} in mm.

9. If

$$r_{H_2O}(\phi, z) = r_0(\phi)e^{-\frac{z}{h}},$$

where $h = 3$ km and $r_0(\phi)$ is the zonal-mean mixing ratio at the surface and latitude ϕ in Fig. 1.17, use the results of Prob. 1.8 to compute the total precipitable water vapor at (a) the equator, (b) 60°N.

10. From the mixing ratio of ozone r_{O_3} shown in Fig. 1.20, plot the vertical profile of ozone number density $[O_3](z)$ above 100 hPa at (a) the equator, (b) 60°N.

11. As in Prob. 1.10a, but plot the contribution to total ozone $\Sigma_{O_3}(z)$ from 100 hPa to an altitude z over the equator.

12. Cloud contributes much of the Earth's albedo, which has a value of about 0.30. Calculate the change of equivalent blackbody temperature corresponding to a 5% reduction of albedo.

13. The Earth's equivalent blackbody temperature, $T_e = 255$ K, corresponds to a level in the middle troposphere, above the layer in which water vapor is concentrated. This also corresponds to the level from which most of the outgoing LW radiation is emitted to space. Suppose atmospheric water vapor increases to raise the top of the (optically thick) water vapor layer by 1 km, while the albedo remains unchanged. (a) What is the corresponding change of the Earth's equivalent blackbody temperature? (b) If a lapse rate of 6.5 K km^{-1} is maintained inside the water vapor layer, by how much will the Earth's mean surface temperature change?

14. By appealing to the hydrostatic relationship (1.17), contrast the vertical spacing of isobaric surfaces ($p = $ const) at a horizontal site where air is cold from one where air is warm.

15. The mean pressure observed in a planetary atmosphere varies with height as

$$p(z) = \frac{p_0}{1 + \left(\frac{z}{h}\right)^2}.$$

Determine the mean variation of temperature with height.

16. Venus has a mean distance from the sun of 0.7 that of Earth and an albedo of 0.75. Calculate the radiative-equilibrium temperature of Venus. Compare the result against the observed surface temperature of 750 K.

17. (a) Apply Archemides Principle to argue that maritime ice, if it melts, does not alter sea level. (b) The density of water is actually greatest at 4°C (1000 kg/m^3), some 8% greater than at 0°C (916.8 kg/m^3). If maritime ice occupies 10% of the surface area of ocean and if the ocean has a mean depth of 2 km, estimate the mean rise

of sea level that would result if all maritime ice underwent a transformation of state from $0°C$ to the temperature $4°C$ of surrounding sea water. (c) Maritime ice is almost pure water, which, at $4°C$, has a density some 2.5% less than sea water (1027 kg/m^3). Under the same circumstances, repeat the calculation in (b).

18. Lower-tropospheric temperature measured from space by the microwave instruments MSU and AMSU represents, not temperature at the earth's surface, but temperature that is averaged vertically over the lowest ~5 km of atmosphere. As horizontal structure is smoother at upper levels than at the surface, vertical averaging blurs features that are present in the horizontal distribution of surface temperature. Suppose that the blurring of horizontal structure is described by diffusion-like equation

$$\frac{\partial T}{\partial z} = k(z)\nabla^2 T,$$

where $T = T(z)$ is the temperature at altitude z, ∇ denotes horizontal gradient, and $k(z)$ is the effective diffusivity. According to this relation, T changes with altitude in proportion to the horizontal curvature of T, which is thereby damped with increasing altitude. Show that, although horizontal structure becomes increasingly blurred with altitude, its global mean does not change.

CHAPTER TWO

Thermodynamics of gases

The link between the circulation and transfers of energy from the Earth's surface is thermodynamics. Thermodynamics deals with internal transformations of the energy of a system and exchanges of energy between that system and its environment. Here, we develop the principles of thermodynamics for a discrete system, namely an air parcel moving through the circulation. In Chap. 10, these principles are generalized to a continuum of such systems, which represents the atmosphere as a whole.

2.1 THERMODYNAMIC CONCEPTS

A *thermodynamic system* refers to a specified collection of matter (Fig. 2.1). Such a system is said to be "closed" if no mass is exchanged with its surroundings. Otherwise it is "open." The air parcel that will serve as our system is, in principle, closed. In practice, however, mass can be exchanged with the surroundings through entrainment and mixing across the system's boundary, which is referred to as the *control surface*. In addition, trace constituents such as water vapor can be absorbed through diffusion across the control surface. Above the planetary boundary layer, such exchanges are slow compared with other processes that influence an air parcel. The system may therefore be treated as closed.

The thermodynamic state of a system is defined by the various properties characterizing it. In a strict sense, all of those properties must be specified to define the system's thermodynamic state. However, that requirement is simplified for many applications, as is discussed next.

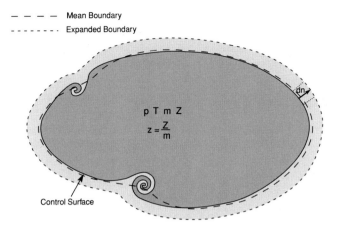

Figure 2.1 A specified collection of matter defining an infinitesimal air parcel. The system's thermodynamic state is characterized by the extensive properties m and Z and by the intensive properties p, T, and z.

2.1.1 Thermodynamic properties

Two types of properties characterize the state of a system. A property that does not depend on the mass of the system is said to be *intensive*. Otherwise it is *extensive*. Intensive and extensive properties are usually denoted with lower and upper case symbols, respectively. Pressure and temperature are examples of intensive properties. Volume is an extensive property. An intensive property z may be defined from an extensive property Z, by referencing the latter to the mass m of the system

$$z = \frac{Z}{m}. \tag{2.1}$$

The intensive property is then referred to as a *specific property*. The specific volume $v = \frac{V}{m}$ is an example. If a system's properties do not vary in space, it is said to be *homogeneous*. Otherwise it is *heterogeneous*. Because an air parcel is of infinitesimal dimension, it is by definition homogeneous (so long as it involves only gas phase). On the other hand, stratification of density and pressure make the atmosphere as a whole a heterogeneous system.

A system can exchange energy with its surroundings through two fundamental mechanisms. It can perform work on its surroundings, which represents a mechanical exchange of energy with the environment. In addition, heat can be transferred across the control surface, which represents a thermal exchange of energy with the environment.

2.1.2 Expansion work

The system relevant to the atmosphere is a compressible gas, perhaps containing an aerosol of liquid and solid particles. For this reason, the primary mechanical means of exchanging energy with the environment is expansion work (see Fig. 2.1). If a pressure imbalance exists across the control surface, the system is out of mechanical equilibrium. It will automatically expand or contract to relieve the mechanical imbalance. By

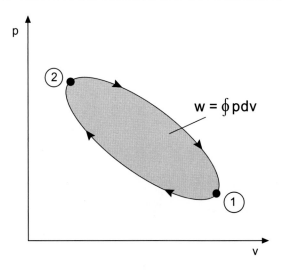

Figure 2.2 Expansion work performed by the system during a cyclic process, wherein it is restored to its initial state.

adjusting to the environmental pressure, the system performs expansion work or has such work performed on it. The incremental expansion work δW performed by displacing a section of control surface $d\mathbf{S} = dS\mathbf{n}$, with unit normal $\mathbf{n}$, perpendicular to itself by $d\mathbf{n}$ is

$$\delta W = (pd\mathbf{S}) \cdot d\mathbf{n}$$
$$= pdV, \qquad (2.2)$$

where $dV = dSdn$ is the incremental volume displaced by the section dS. Integrating (2.2) between the initial and final volumes yields the net work performed "by" the system

$$W_{12} = \int_{V_1}^{V_2} pdV. \qquad (2.3)$$

Dividing both sides by the mass of the system then gives the specific work

$$w_{12} = \int_{v_1}^{v_2} pdv. \qquad (2.4.1)$$

For a cyclic variation, namely one in which the initial and final states of the system are identical,

$$w = \oint pdv. \qquad (2.4.2)$$

The cyclic work (2.4.2) is represented in the area enclosed by the path of the system in the $p - v$ plane (Fig. 2.2). As the system is restored to its initial state, the net change of properties is zero. However, the same is not true of the work performed during the cycle. Unless the system returns to its initial state along the same path it followed out of that state, net work is performed during a cyclic variation.

On a fluid system, another form of work can be performed, one analogous to stirring. *Paddle work* corresponds to a rearrangement of fluid, as would occur by turning a paddle immersed in the system. The reaction of the fluid against the paddle must be overcome by a torque, which in turn performs work when exerted across an angular displacement. Although secondary to expansion work, paddle work has applications to dissipative processes associated with turbulence (cf. Fig. 2.1).

2.1.3 Heat transfer

Energy can also be exchanged thermally, via heat transfer Q across the system's control surface. If an air parcel moves into an environment of a higher temperature, it will absorb heat from its surroundings (e.g., through diffusion or thermal conduction). If the system is open, a similar process can occur through the absorption of water vapor from the surroundings, followed by condensation and the release of latent heat. Heat can also be exchanged through radiative transfer. If it interacts radiatively with surroundings at a higher temperature, an air parcel will absorb more radiant energy than it emits.

For many applications, heat transfer is secondary to processes that are introduced through motion, which operates on time scales of a day and shorter. This is especially true above the boundary layer, where turbulent mixing between bodies of air is relatively weak. In the free atmosphere and away from cloud, the prevailing form of heat transfer is radiative. Operating on a time scale of order 2 weeks, radiative transfer is slow by comparison with expansion work, which influences an air parcel on time scales of only a day.

If no heat is exchanged between a system and its environment, the control surface is said to be *adiabatic*. Otherwise, it is *diabatic*. Because, for many applications, heat transfer is slow compared with other processes that influence a parcel, adiabatic behavior is a good approximation. Obviously, heat transfer must be central to processes that operate on long time scales, as it is ultimately responsible for driving the atmosphere into motion. For this reason, diabatic effects prove to be important for the long-term maintenance of the general circulation, even though they can be ignored for behavior that operates on shorter time scales.

2.1.4 State variables and thermodynamic processes

In general, describing the thermodynamic state of a system requires specifying all of its properties. However, that requirement can be relaxed for gases and other substances at normal temperatures and pressures. A *pure substance* is one whose thermodynamic state is uniquely determined by any two intensive properties. For such a system, intensive properties are called *state variables*. From any two state variables z_1 and z_2, a third z_3 may be determined through an *equation of state* of the form

$$f(z_1, z_2, z_3) = 0,$$

or, equivalently,

$$z_3 = g(z_1, z_2).$$

In this sense, a pure substance has two *thermodynamic degrees of freedom*. Any two state variables uniquely specify its thermodynamic state. An ideal gas is a pure substance. It has as its equation of state the ideal gas law (1.1). Because no more than two intensive properties are independent, the state of a pure substance may be specified in the plane of any two of its state variables (z_1, z_2). The latter represents the *state space* of the system. It defines the collection of all possible thermodynamic states. For instance, the state space of an ideal gas may be represented by the $v - T$ plane. Any third state variable is then described by a surface, as a function of the two independent state variables. For example, $p = p(v, T)$, as illustrated in Fig. 2.3 for an ideal gas.

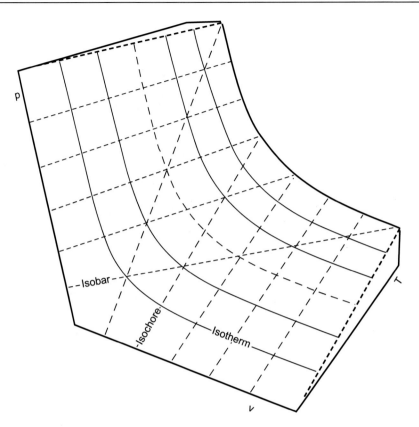

Figure 2.3 State space of an ideal gas, illustrated in terms of p as a function of the state variables v and T. Superposed are contours of constant pressure (*isobars*), constant temperature (*isotherms*), and constant volume (*isochores*).

A homogeneous system is said to be in *mechanical equilibrium* if there exists at most an infinitesimal pressure difference between the system and its surroundings (unless its control surface happens to be rigid). Likewise, a homogeneous system is said to be in *thermal equilibrium* if there exists at most an infinitesimal temperature difference between the system and its surroundings (unless its control surface happens to be adiabatic). If it is in mechanical equilibrium and in thermal equilibrium, a homogeneous system is said to be in *thermodynamic equilibrium*.

Thermodynamics addresses how a system evolves from one state to another (e.g., from one position in state space to another position). The transformation of a system between two states is referred to as a *thermodynamic process*. It describes a path in state space. As depicted in Fig. 2.4, two thermodynamic states are connected by infinitely many paths. Consequently, defining a process requires specifying the particular path in state space along which the system evolves. In contrast, the change of a state variable depends on only the initial and final states of the system. It is therefore "path independent." As illustrated in Fig. 2.2, a cyclic process may be decomposed into two legs: a forward leg, from the initial state to an intermediate state, and a reverse leg, from that intermediate state back to the initial state. Because it is path independent, the change of a state variable z along the reverse leg must equal minus the change along the forward leg. Therefore, the cyclic integral of a state variable vanishes

$$\oint dz = 0. \tag{2.5}$$

The incremental change of a quantity $z(x, y)$ that satisfies (2.5) represents an exact differential

$$dz = \frac{\partial z}{\partial x}dx + \frac{\partial z}{\partial y}dy, \tag{2.6}$$

a description which holds only under certain conditions.

Exact Differential Theorem: Consider two continuously differentiable functions $M(x, y)$ and $N(x, y)$ in a simply-connected region of the $x - y$ plane. The contour integral between two points (x_0, y_0) and (x, y)

$$\int_{(x_0,y_0)}^{(x,y)} M(x', y')dx' + N(x', y')dy' \tag{2.7}$$

(e.g., along a specified curve $g(x, y)$ that relates x and y) is path independent if and only if

$$\frac{\partial M}{\partial y} = \frac{\partial N}{\partial x}. \tag{2.8}$$

Under these circumstances, the quantity

$$M(x, y)dx + N(x, y)dy = dz$$

represents an exact differential. That is, there exists a function z such that

$$M(x, y) = \frac{\partial z}{\partial x}$$
$$N(x, y) = \frac{\partial z}{\partial y}. \tag{2.9.1}$$

The contour integral (2.7) then reduces to

$$\int_{z(x_0,y_0)}^{z(x,y)} dz = z(x, y) - z(x_0, y_0) \tag{2.9.2}$$
$$= \Delta z$$

or

$$z = z(x, y) + c. \tag{2.9.3}$$

The variable defined by (2.9) is a *point function*: It depends only on the evaluation point (x, y). Similarly, its net change along a contour depends only on the initial and final points of the contour. Because it is unique only up to an additive constant, only the change of a point function is significant. Also referred to as a *potential function*, $z(x, y)$ defines an *irrotational* vector field

$$\boldsymbol{v} = \nabla z$$
$$= M(x, y)\boldsymbol{i} + N(x, y)\boldsymbol{j}, \tag{2.10.1}$$

which satisfies

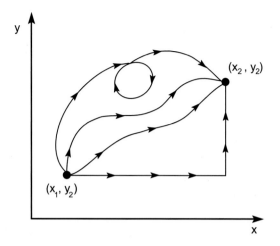

Figure 2.4 Possible thermodynamic processes between two states: (x_1, y_1) and (x_2, y_2).

$$\nabla \times \boldsymbol{v} \equiv 0.^{[1]} \tag{2.10.2}$$

An irrotational vector field may be represented as the gradient of a scalar potential z. Gravity g is an irrotational vector field: $\nabla \times g \equiv 0$. The work performed to displace a unit mass between two points in a gravitational field is independent of path. It defines the gravitational potential Φ.

Thermodynamic state variables are point functions. Properties of the system, they depend only on the system's state, not on its history. By contrast, the work performed by the system and the heat transferred into it during a thermodynamic process are not properties of the system. Work and heat transfer are, in general, *path functions*. They depend on the path in state space that is followed by the system. For this reason, the thermodynamic process must be specified to unambiguously define those quantities.

Path dependent, work and heat transfer can differ along the forward and reverse legs of a cyclic process. Consequently, the net work and heat transfer during a cyclic variation of the system need not vanish, as does the net change of a state variable (2.5). The path dependence of work and heat transfer produce a *hysteresis* in the $w - q$ plane, illustrated in Fig. 2.5. During a cyclic variation of the system, the cumulative work and heat

$$w = \int \delta w$$

$$q = \int \delta q$$

do not return to their original positions after the system has been restored to its initial state. The discrepancy $\Delta w = \oint \delta w$ after completing a cycle equals the area in the $p - v$ plane enclosed by the contour in Fig. 2.2. Due to hysteresis, successive cycles lead to a drift of the above quantities in the $w - q$ plane. The drift reflects the net work performed by the system and the net heat transferred into it.

Under special circumstances, the work performed by a system or the heat transferred into it "are" independent of path. Work and heat transfer then vanish for a

---

[1] If it is a force, the vector field $\boldsymbol{v}$ is said to be *conservative* because the net work performed along a cyclic path vanishes (2.5).

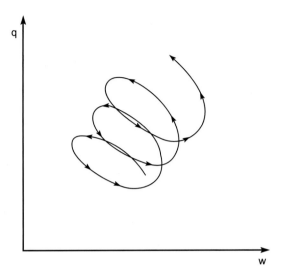

Figure 2.5 Cumulative work and heat transfer during successive thermodynamic cycles of the system. The path dependence of w and q introduces a hysteresis into those quantities, even though the system (i.e., the collection of state variables) is restored to its initial state after each cycle.

cyclic process. Each is therefore a point function. This special form of work or heat transfer may be used to define a state variable. For instance, the displacement work in a gravitational field is path independent. That work can therefore be used to define the gravitational potential Φ, which is a state variable (Sec. 6.2).

2.2 THE FIRST LAW

2.2.1 Internal energy

The First Law of Thermodynamics is inspired by an empirical finding: The work performed on an adiabatic system is independent of the process. That is, under adiabatic conditions, the work is independent of the path in state space followed by the system. For an adiabatic process, expansion work depends only on the initial and final states of the system. It therefore behaves as a state variable. The *internal energy u* is defined as that state variable whose change, under adiabatic conditions, equals the work performed on the system or minus the work performed "by" the system.

$$\Delta u = -w_{ad}. \qquad (2.11.1)$$

For an incremental process,

$$du = -\delta w_{ad}, \qquad (2.11.2)$$

where δ describes an incremental change that is, in general, "path dependent" and d to one that is "path independent" (i.e., of a state variable). Under the special circumstances defined in the preceding paragraph, work is path independent. The net work performed

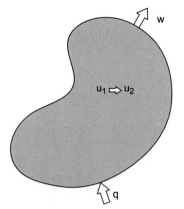

Figure 2.6 An air parcel undergoing a process between states 1 and 2, during which it performs work w and absorbs heat q.

during a cyclic process therefore vanishes. The system must then return to its initial state along the same path that it followed out of that state.

2.2.2 Diabatic changes of state

Consider now a diabatic process, as illustrated schematically in Fig. 2.6. Because heat is exchanged with the environment,

$$w \neq w_{ad} = -\Delta u.$$

The work performed by the system w will differ from that performed under adiabatic conditions, w_{ad}. The difference, $w - w_{ad} = q$, equals the energy transferred "into" the system through heat exchange (e.g., it measures the departure from adiabatic conditions). The change of internal energy is then described by

$$\Delta u = q - w. \tag{2.12.1}$$

Equation (2.12.1) constitutes the *First Law of Thermodynamics*. For an incremental process and for a system capable of expansion work only, the First Law may be expressed

$$du = \delta q - pdv. \tag{2.12.2}$$

In (2.12), pdv represents the work performed "by" the system on its surroundings and q the heat transfer "into" the system. The change of internal energy between two states is path independent. However, the same is not true of the work performed by the system and the heat transferred into it during the change of state. Except under adiabatic conditions, the work performed by the system during a change of state depends on the thermodynamic process, namely, the path in state space along which the system evolves. The same applies to the heat transferred into the system during the change of state. Because each depends on path, neither the work nor the heat transfer vanishes for a cyclic process. However, the change of internal energy does vanish for a cyclic process. The First Law therefore reduces to

$$\oint pdv = \oint \delta q. \tag{2.13}$$

During a cyclic process, the net work performed by the system (energy out) is balanced by the net heat absorbed by the system (energy in).

A closed system that performs work through a conversion of heat that is absorbed by it is a *heat engine*. Conversely, a system that rejects heat through a conversion of work that is performed on it is a *refrigerator*. In Chap. 6, we will see that individual air parcels comprising the circulation of the troposphere behave as a heat engine. By absorbing heat at the Earth's surface, through transfers of radiative, sensible, and latent heat, individual parcels perform net work as they evolve through a thermodynamic cycle (2.13). Ultimately realized as kinetic energy, the heat absorbed maintains the circulation against frictional dissipation. It makes the circulation of the troposphere thermally driven.

In contrast, the circulation of the stratosphere behaves as a radiative refrigerator. For motion to occur, individual air parcels must have work performed on them. The kinetic energy produced is eventually converted to heat and rejected to space through LW cooling. It makes the circulation of the stratosphere mechanically driven. Gravity waves and planetary waves that propagate upward from the troposphere are dissipated in the stratosphere. Their absorption exerts an influence on the stratosphere analogous to paddle work. By forcing motion that rearranges air, it drives the stratospheric circulation out of radiative equilibrium, which results in net LW cooling to space.

It is convenient to introduce the state variable *enthalpy*

$$h = u + pv. \tag{2.14}$$

In terms of enthalpy, the First Law becomes

$$dh = \delta q + v\,dp. \tag{2.15}$$

Enthalpy is useful for diagnosing processes that occur at constant pressure. Under those circumstances, the First Law reduces to a statement that the change in enthalpy equals the heat transferred into the system.

2.3 HEAT CAPACITY

Observations indicate that the heat absorbed by a homogeneous system which is maintained at constant pressure or at constant volume is proportional to the change of the system's temperature. The constants of proportionality between heat absorption and temperature change define the *specific heat capacity at constant pressure*

$$c_p = \frac{\delta q_p}{dT} \tag{2.16.1}$$

and the *specific heat capacity at constant volume*

$$c_v = \frac{\delta q_v}{dT}, \tag{2.16.2}$$

where the subscripts denote processes that are *isobaric* (p = const) and *isochoric* (v = const).

The specific heat capacities are related closely to the internal energy and enthalpy of the system. Because they are state variables, u and h may be expressed in terms of any two other state variables. For instance, $u = u(v, T)$, in which case

$$du = \left(\frac{\partial u}{\partial v}\right)_T dv + \left(\frac{\partial u}{\partial T}\right)_v dT$$

Incorporating this transforms the First Law (2.12.2) into

$$\left(\frac{\partial u}{\partial T}\right)_v dT + \left[\left(\frac{\partial u}{\partial v}\right)_T + p\right] dv = \delta q.$$

For an isochoric process, this reduces to

$$\left(\frac{\partial u}{\partial T}\right)_v = \frac{\delta q_v}{dT} \tag{2.17.1}$$

$$= c_v.$$

Thus, the specific heat capacity at constant volume measures the rate at which internal energy increases with temperature during an isochoric process. In similar fashion, the enthalpy may be expressed $h = h(p, T)$, with which the First Law (2.15) becomes

$$\left[\left(\frac{\partial h}{\partial T}\right)_T - v\right] dp + \left(\frac{\partial h}{\partial T}\right)_p dT = \delta q.$$

For an isobaric process, this reduces to

$$\left(\frac{\partial h}{\partial T}\right)_p = \frac{\delta q_p}{dT} \tag{2.17.2}$$

$$= c_p.$$

The specific heat capacity at constant pressure measures the rate at which enthalpy increases with temperature during an isobaric process.

In a strict sense, c_p and c_v are state variables. Hence they too depend on pressure and temperature. However, over ranges of pressure and temperature relevant to the atmosphere, the specific heats may be regarded as constant. Therefore, the change of internal energy during an isochoric process is proportional to the change of temperature alone (c_v) and similarly for the change of enthalpy during an isobaric process (c_p). For an ideal gas, the relationships (2.17) turn out to hold irrespective of process.

Consider the internal energy of an ideal gas in the form $u = u(p, T)$. Joule's experiment, which is pictured in Fig. 2.7, demonstrates that u is then a function of temperature alone. Two ideal gases that are initially isolated and at pressures p_1 and p_2 are brought into contact and allowed to equilibrate (e.g., by rupturing a diaphragm that separates them). Observations indicate that, during this process, no heat transfer takes place with the environment. The First Law then reduces to a statement that the change of internal energy equals minus the work performed. However, the volume of the system (that occupied collectively by both gases) does not change. Consequently, the work also vanishes, leaving $\Delta u = 0$. Yet, the final equilibrated pressure clearly differs from the initial pressures of the individual gases in isolation. Because the internal energy of the system does not change, it follows that u is not a function of pressure. Hence, for an ideal gas, $u(p, T)$ reduces to

$$u = u(T).$$

As it involves only state variables, this relationship must hold irrespective of process. Incorporating it and the ideal gas law into (2.14) demonstrates that the enthalpy is likewise a function of temperature alone

$$h = h(T).$$

Involving only state variables, it too holds irrespective of process.

Joule's Experiment

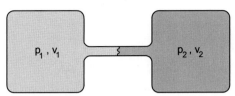

Figure 2.7 Schematic of Joule's experiment: Two ideal gases in states 1 and 2 are brought into contact by rupturing a diaphragm that separates them initially. No heat transfer with the surroundings is observed.

These conclusions imply that, for an ideal gas, equations (2.17) are valid in general. Integrating with respect to temperature yields finite values of internal energy and enthalpy, which are unique up to constants of integration. It is customary to define u and h so that they vanish at a temperature of absolute zero. With that convention,

$$u = c_v T \qquad (2.18.1)$$

$$h = c_p T. \qquad (2.18.2)$$

It follows that

$$c_p - c_v = R. \qquad (2.19)$$

According to statistical mechanics, the specific heat at constant volume is given by

$$c_v = \frac{3}{2}R \qquad (2.20.1)$$

for a monotomic gas, and by

$$c_v = \frac{5}{2}R \qquad (2.20.2)$$

for a diatomic gas (e.g., Lee, Sears, and Turcotte, 1973). These values are confirmed experimentally over a wide range of pressure and temperature relevant to the atmosphere. Taking air to be chiefly diatomic, together with the value of R_d in (1.18), yields the specific heats for dry air

$$c_{vd} = 717.5 \text{ Jkg}^{-1}\text{K}^{-1}$$
$$c_{pd} = 1004.5 \text{ Jkg}^{-1}\text{K}^{-1} \qquad (2.21.1)$$

and the dimensionless constants

$$\gamma = \frac{c_p}{c_v} = 1.4 \qquad (2.21.2)$$

$$\kappa = \frac{R}{c_p}$$
$$= \frac{\gamma - 1}{\gamma} \qquad (2.21.3)$$
$$\cong 0.286.$$

With the aforementioned definitions, the First Law may be expressed in the two equivalent forms

$$c_v dT + p dv = \delta q \qquad (2.22.1)$$

$$c_p dT - v dp = \delta q. \qquad (2.22.2)$$

For isochoric and isobaric processes, these expressions reduce to

$$\delta q_v = c_v dT \tag{2.23.1}$$

$$\delta q_p = c_p dT, \tag{2.23.2}$$

respectively. Because the right-hand sides of (2.23) involve only state variables, the same must be true of the left-hand sides. Hence, under these special circumstances, heat transfer behaves as a state variable. Although generally path dependent, heat transfer during an isochoric process or during an isobaric process is uniquely determined by the change of temperature.

2.4 ADIABATIC PROCESSES

For an adiabatic process, the First Law reduces to

$$c_v dT + p dv = 0 \tag{2.24.1}$$

$$c_p dT - v dp = 0. \tag{2.24.2}$$

Dividing through by T and introducing the gas law transforms (2.24) into

$$c_v d\ln T + R d\ln v = 0 \tag{2.25.1}$$

$$c_p d\ln T - R d\ln p = 0. \tag{2.25.2}$$

Equations (2.25) may be integrated to obtain the identities

$$T^{c_v} v^R = \text{const} \tag{2.26.1}$$

$$T^{c_p} p^{-R} = \text{const}. \tag{2.26.2}$$

A third identity, which relates p and v, may be derived from (2.25) with the aid of a differential form of the ideal gas law

$$d\ln p + d\ln v = d\ln T, \tag{2.27}$$

which follows from (1.1). Using (2.27) to eliminate T from (2.25.2) gives

$$c_v d\ln p + c_p d\ln v = 0. \tag{2.28}$$

Integrating then yields

$$p^{c_v} v^{c_p} = \text{const}. \tag{2.29}$$

The three identities (2.26.1), (2.26.2), and (2.29), may be expressed in terms of dimensionless constants as

$$Tv^{\gamma-1} = \text{const} \tag{2.30.1}$$

$$Tp^{-\kappa} = \text{const} \tag{2.30.2}$$

$$pv^\gamma = \text{const}. \tag{2.30.3}$$

Known as *Poisson's equations*, (2.30) define adiabatic paths in the state space of an ideal gas. Each describes the evolution of a state variable during an adiabatic process in terms of only one other state variable. Thus, the change of a single state variable, together with the condition that the process be adiabatic, is sufficient to determine the change of a second state variable. That, in turn, determines the change of thermo- dynamic state. For this reason, an adiabatic system possesses only one independent state variable and thus only one thermodynamic degree of freedom.

Because the state space of a pure substance is represented by the plane of any two intensive properties z_1 and z_2, a thermodynamic process describes a contour

$$g(z_1, z_2) = \text{const.}$$

Poisson's equations (2.30) are of this form. They describe a family of contours, known as *adiabats*, in the plane of any two of the state variables p, T, and v (Fig. 2.8). In similar fashion, isobaric processes describe a family of *isobars* ($p = $ const), isothermal processes describe a family of *isotherms* ($T = $ const), and isochoric processes describe a family of *isochores* ($v = $ const), which are superposed in Fig. 2.8. Each of the latter paths is characterized by invariance of a certain state variable. The same applies to adiabats.

2.4.1 Potential temperature

Poisson's relation between pressure and temperature motivates the introduction of a new state variable, one that is preserved during an adiabatic process. The *potential temperature* θ is defined as that temperature the system would assume were it com- pressed or expanded adiabatically to a reference pressure of $p_0 = 1000$ hPa. According to (2.30.2), an adiabatic process from the state (p, T) to the reference state (p_0, θ) satisfies

$$\theta p_0^{-\kappa} = T p^{-\kappa}.$$

Hence, the potential temperature is described by

$$\frac{\theta}{T} = \left(\frac{p_0}{p} \right)^{\kappa}. \tag{2.31}$$

A function of pressure and temperature, θ is a state variable. According to (2.31) and Poisson's relation (2.30.2), θ is invariant along an adiabatic path in state space.

Adiabatic behavior of individual air parcels is a good approximation for many atmospheric applications. Above the boundary layer and outside of cloud, the time scale of heat transfer is of order 2 weeks. It is thus long compared with the character- istic time scale of motion, which influences an air parcel through changes of pressure and expansion work. For instance, vertical motion of air and accompanying changes of pressure and volume occur inside cumulus convection on a time scale of minutes to hours. Even in motions of large horizontal dimension (e.g., in sloping convection associated with synoptic weather systems), air displacements occur on a characteristic time scale of order 1 day. Thus, for a wide range of motions, the time scale for an air parcel to adjust to changes of pressure and to perform expansion work is short compared with the characteristic time scale of heat transfer.

Under these circumstances, the potential temperature of individual air parcels is approximately conserved. An air parcel descending to greater pressure experi- ences an increase of temperature according to (2.30.2) due to compression work that is performed on it. However, its increase of temperature is in such proportion to its increase of pressure as to preserve the parcel's potential temperature through (2.31).

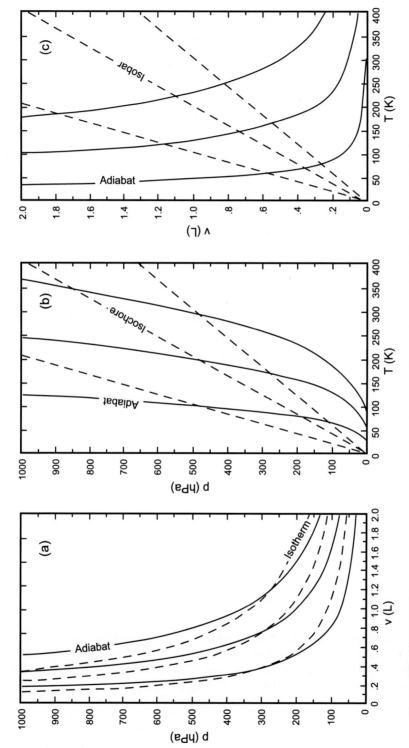

Figure 2.8 Changes of pressure, volume, or temperature in terms of one other state variable during an adiabatic process, which define *adiabats* in the state space of an ideal gas. Also shown are the corresponding changes during an isothermal process ($T =$ const), an isochoric process ($v =$ const), and an isobaric process ($p =$ const), which define *isotherms*, *isochores*, and *isobars*, respectively.

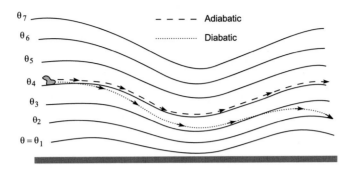

Figure 2.9 Surfaces of constant potential temperature θ. An air parcel remains coincident with a particular θ surface under adiabatic conditions (dashed), whereas it drifts across θ surfaces under diabatic conditions (dotted).

Similar considerations apply to an air parcel that is ascending. It follows that, under adiabatic conditions, θ is a conserved quantity. It therefore behaves as a tracer of air motion. On time scales for which individual parcels are adiabatic, particular values of θ track the movement of those bodies of air. Conversely, a collection of air parcels that has a particular value of θ remains coincident with the corresponding isopleth of potential temperature. The locus $\theta = \text{const}$ thus provides a constraint on how air moves.

The distribution of θ in the atmosphere is determined by the distributions of pressure and temperature. Because pressure decreases upward sharply, (2.31) implies that surfaces of constant θ tend to be quasi-horizontal like isobaric surfaces (Fig. 2.9). On time scales characteristic of air motion, a parcel initially coincident with a certain θ surface must remain coincident with that surface. Deflections of these quasi-horizontal surfaces therefore describe the vertical motion of individual bodies of air. Under steady conditions, air is constrained to move along those surfaces.

Magnified values of ozone column abundance that appear in Fig. 1.22 are attributable in part to vertical motion along θ surfaces. Contoured over Σ_{O_3} is the pressure on the 375 K potential temperature surface, which lies just above the tropopause. Above tropospheric cyclones (cf. Fig. 1.9a), that surface is deflected downward to greater pressure. Air moving along the θ surface then descends, experiencing compression. That, in turn, increases the absolute concentration ρ_{O_3}, at the expense of ozone in the surroundings, increasing Σ_{O_3} (1.25).

Another example is provided by the distribution of nitric acid on the 470 K potential temperature surface (Fig. 2.10), which also lies in the lower stratosphere. Like ozone, HNO_3 is long-lived at this level, so its mixing ratio is conserved. Therefore, r_{HNO_3} is passively rearranged by the circulation on potential temperature surfaces, along which air moves. The distribution of r_{HNO_3} on the $\theta = 470$ K surface illustrates how air is rearranged quasi-horizontally along that surface. r_{HNO_3} has been distorted into a pentagonal pattern by synoptic weather systems in the troposphere, which are distributed almost uniformly in longitude along the continuous storm track of the Southern Hemisphere (cf. Fig. 1.30a). In fact, the pattern of r_{HNO_3} suggests that air is being overturned horizontally by those synoptic disturbances (like amplifying waves in their approach to a beach).

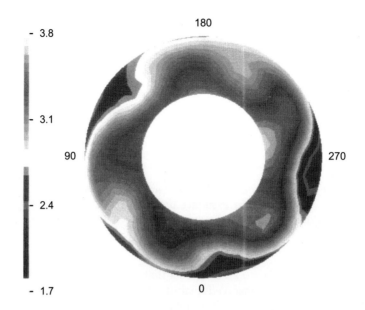

Figure 2.10 Distribution of nitric acid mixing ratio on the $\theta = 470$ K potential temperature surface over the Southern Hemisphere on January 3, 1979, as observed by Nimbus-7 LIMS. Adapted from Miles and Grose (1986).

2.4.2 Thermodynamic behavior accompanying vertical motion

According to (2.30.2), the temperature of an air parcel moving vertically changes due to expansion work. However, it does so in such proportion to its pressure as to preserve its potential temperature. An expression for the rate at which a parcel's temperature T' changes with its altitude z' under adiabatic conditions follows from (2.25.2), in combination with hydrostatic equilibrium (1.16). With the gas law, the change with altitude of environmental pressure p is described by

$$dlnp = -\frac{dz}{H},\qquad(2.32)$$

where H is given by (1.17.2). To preserve mechanical equilibrium, the parcel's pressure p' adjusts automatically to the environmental pressure: $p' = p$. Equation (2.32) therefore also describes how the parcel's pressure changes with altitude. Then, for the parcel, (2.25.2) becomes

$$c_p dT' + g dz' = 0.$$

Thus, the temperature of a displaced air parcel evolving adiabatically decreases with its altitude at a constant rate

$$-\frac{dT'}{dz'} = \frac{g}{c_p}$$
$$= \Gamma_d,\qquad(2.33)$$

which defines the *dry adiabatic lapse rate*. The linear profile of temperature followed by a parcel moving vertically under adiabatic conditions (Fig. 2.11) is associated with a uniform profile of potential temperature (i.e., along which θ' for the parcel remains constant). Both describe the evolution of a displaced air parcel that does not interact

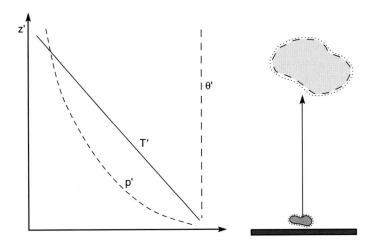

Figure 2.11 Profiles of pressure, temperature, and potential temperature for an individual air parcel ascending adiabatically.

thermally with its environment. The only interaction between the parcel and its environment is then mechanical.

Γ_d has a value of approximately 9.8 K km^{-1}. It applies to a parcel of dry air undergoing vertical motion. For reasons developed in Chap. 5, (2.33) also applies to an air parcel that is moist, but outside of cloud. Even then, heat transfer remains slow enough, compared with expansion work, to be ignored. Inside cloud, the simple behavior predicted by adiabatic considerations does not apply. It is invalidated by the release of latent heat. Accompanying the condensation of water vapor, that diabatic process occurs on the same time scale as vertical motion. Although the behavior of an air parcel is then more complex, a generalization of (2.33) applies under those circumstances (Sec. 5.4.3).

2.5 DIABATIC PROCESSES

Adiabatic conditions are violated in certain locations and over long time scales. Near the surface, thermal conduction and turbulent mixing become important because they operate on short time scales. A similar conclusion holds inside cloud, where the release of latent heat operates on the same time scale as vertical motion. For changes that occur on time scales longer than a week, radiative transfer becomes important.

Under diabatic conditions, an air parcel interacts with its environment thermally as well as mechanically. Its potential temperature is then no longer conserved. Instead, θ changes in proportion to the heat transferred into the parcel, where θ and other unprimed variables are hereafter understood to refer to an individual air parcel. Taking the logarithm of (2.31) gives a differential relation among the variables θ, p, and T

$$dln\theta - dlnT = -\kappa\, dlnp. \tag{2.34}$$

Likewise, dividing the First Law (2.22.2) by T and incorporating the gas law leads to a similar relation

$$dlnT - \kappa\, dlnp = \frac{\delta q}{c_p T}. \tag{2.35}$$

Combining (2.34) and (2.35) then yields

$$dln\theta = \frac{\delta q}{c_p T}. \tag{2.36}$$

Potential temperature increases in direct proportion to the heat transferred into the parcel (e.g., in proportion to its departure from adiabatic conditions). An air parcel will then drift across potential temperature surfaces, according to the net heat exchanged with its environment (Fig. 2.9).

Equation (2.36) may be regarded as an alternate and more compact expression of the First Law. Expansion work there has been absorbed into the state variable θ. In this light, θ for a compressible fluid, like air, is an analogue of T for an incompressible fluid, like water. Each increases in direct proportion to the heat absorbed by the system.

2.5.1 Polytropic processes

Most of the energy exchanged between the Earth's surface and the atmosphere and between one atmospheric layer and another is accomplished through radiative transfer (cf. Fig. 1.32). In fact, outside of the boundary layer and cloud, radiative transfer is the primary diabatic influence. It is sometimes convenient to model radiative transfer as a *polytropic process*

$$\delta q = cdT, \tag{2.37}$$

wherein the heat transferred into a parcel is proportional to its change of temperature. The constant of proportionality c is the *polytropic specific heat capacity*. *Newtonian cooling* damps the departure from an equilibrium temperature. It may be regarded as a polytropic process with $c < 0$. For deep temperature anomalies, as are characteristic of planetary waves, Newtonian cooling provides a convenient approximation of LW cooling to space.

For a polytropic process, the two expressions of the First Law (2.22) reduce to

$$(c_v - c)dT + pdv = 0 \tag{2.38.1}$$

$$(c_p - c)dT - vdp = 0. \tag{2.38.2}$$

Equations (2.38) resemble the First Law for an adiabatic process, but with modified specific heats. Consequently, previous formulae valid for an adiabatic process hold for a polytropic process with the transformation

$$c_p \rightarrow (c_p - c)$$
$$c_v \rightarrow (c_v - c). \tag{2.39}$$

Then (2.39) may be used in (2.31) to define a *polytropic potential temperature* $\hat{\theta}$, which is conserved during a polytropic process (Prob. 2.16). Analogous to potential temperature, $\hat{\theta}$ is the temperature assumed by an air parcel if compressed polytropically to 1000 hPa. It characterizes a family of polytropes in state space.

In terms of potential temperature, the First Law becomes

$$d\ln\theta = \left(\frac{c}{c_p}\right) d\ln T. \tag{2.40}$$

If c is negative (e.g., Newtonian cooling), the effective specific heats $(c_p - c)$ and $(c_v - c)$ are increased from their adiabatic values. A given change of a parcel's pressure or volume in (2.38) then leads to a diminished change of its temperature. For an increase of temperature (e.g., introduced through compression), warming is diminished. The reduction from adiabatic warming reflects heat loss to the environment (e.g., through turbulent mixing or LW cooling). The heat loss is accompanied by a decrease of θ (2.40). These circumstances apply to a displaced air parcel that finds itself warmer than its environment. For a decrease of temperature (e.g., introduced through expansion), cooling is diminished. The reduction from adiabatic cooling implies heat absorption from the environment. The heat absorption is accompanied by an increase of θ. These circumstances apply to a displaced air parcel that finds itself cooler than its environment.

SUGGESTED REFERENCES

A detailed discussion of thermodynamic properties is given in *Atmospheric Thermodynamics* (1981) by Irabarne and Godson.

Statistical Thermodynamics (1973) by Lee, Sears, and Turcotte provides a clear treatment of specific heats and other gas properties from the perspective of statistical mechanics.

PROBLEMS

1. A plume of heated air leaves the cooling tower of a power plant at 1000 hPa with a temperature of 30°C. If the air may be treated as dry, to what level will the plume ascend if the ambient temperature varies with altitude as (a) $T(z) = 20 - 8z$ (°C), (b) $T(z) = 20 + z$ (°C), with z in km.

2. Suppose that the air in (1) contains moisture, but that its effect is negligible beneath a cumulus cloud, which forms overhead between 900 hPa and 700 hPa. (a) What is the potential temperature inside the plume at 950 hPa? (b) How is the potential temperature inside the plume at 800 hPa related to that at 950 hPa? (c) How is the buoyancy of air at 800 hPa related to that were the air perfectly dry?

3. One mole of water is vaporized at 100°C and 1 atm of pressure (1.013×10^5 Pa), with an observed increase in volume of 3.02×10^{-2} m^3. (a) How much work is performed by the water? (b) How much work would be performed if the water behaved as an ideal gas and accomplished the same change of volume isobarically? (c) The enthalpy of vaporization for water is 4.06×10^4 J mol^{-1}. What heat input is necessary to accomplish the above change of state? Why does the heat input differ from the work performed?

4. Demonstrate (a) that $h = h(T)$, (b) relation (2.19).

5. Demonstrate that θ is conserved during an adiabatic process.

6. A large helium-filled balloon carries an instrumented payload from sea level into the stratosphere. If the balloon ascends rapidly enough for heat transfer across its surface to be negligible, what relative increase in volume must be accommodated

for the balloon to reach 10 hPa? (Note: The balloon exerts negligible surface tension – it serves merely to isolate the helium.)

7. Through sloping convection, dry air initially at 20°C ascends from sea level to 700 hPa. Calculate (a) its initial and final specific volumes, (b) its final temperature, (c) the specific work performed, (d) changes in its specific energy and enthalpy.

8. Commercial aircraft normally cruise near 200 hPa, where $T = -60$°C. (a) Calculate the temperature air would be if compressed adiabatically to the cabin pressure of 800 hPa. (b) How much specific heat must then be added/removed to maintain a cabin temperature of 25°C?

9. Surface wind blows down a mountain range during a *Chinook*, which is an Indian term for "snow eater." If the temperature at 14,000′ along the continental divide is −10°C and if the sky is cloud-free leeward of the divide, what is the surface temperature at Denver, which lies at approximately 5,000′?

10. An air parcel descends from 10 hPa to 100 hPa. (a) How much specific work is performed on the parcel if its change of state is accomplished adiabatically? (b) if its change of state is accomplished isothermally at 220 K? (c) Under the conditions of (b), what heat transfer must occur to achieve the change of state? (d) What is the parcel's change of potential temperature under the same conditions?

11. Derive (2.27) for an ideal gas.

12. A commercial aircraft is en route to New York City from Miami, where the surface pressure was 1013 hPa and its altimeter was adjusted at takeoff to indicate 0 feet above sea level. If the surface pressure at New York is 990 hPa and the temperature is −3°C, estimate the altimeter reading when the aircraft lands if no subsequent adjustments are made.

13. A system comprised of pure water is maintained at 0°C and 6.1 hPa, at which the specific volumes of ice, water, and vapor are: $v_i = 1.09 \ 10^{-3}$ m^3 kg^{-1}, $v_w = 1.00 \ 10^{-3}$ m^3 kg^{-1}, $v_v = 206$ m^3 kg^{-1}. (a) How much work is performed by the water substance if 1 kg of ice melts? (b) How much work is performed by the water substance if 1 kg of water evaporates? (c) Contrast the values in (a) and (b) in light of the heat transfers required to accomplish the preceding transformations of phase: $Q_{fusion} = 0.334 \ 10^6$ J and $Q_{vaporization} = 2.50 \ 10^6$ J.

14. A scuba diver releases a bubble at 30°C at a depth of 32′, where the absolute pressure is 2 atm. Presume that the bubble remains intact and is large enough for heat transfer across its surface to be negligible, and that condensational heating in its interior can be ignored. Beginning from hydrostatic equilibrium and the first law, (a) derive an expression for the bubble's temperature as a function of its elevation z above its initial elevation. (b) Determine the bubble's temperature upon reaching the surface under the foregoing conditions. (c) How would these results be modified under fully realistic conditions? (Note: Normal respiration processes only about 30% of the oxygen available, so the bubble may be treated as air.)

15. Describe physical circumstances under which polytropic heat transfer would be a useful approximation to actual heat transfer.

16. Use the transformation (2.29) to define a *polytropic potential temperature* $\hat{\theta}$, which is conserved during a polytropic process. How is $\hat{\theta}$ related to θ?

THREE

The Second Law and its implications

The First Law of Thermodynamics describes how the state of a system changes in response to work that it performs and heat that it absorbs. The Second Law of Thermodynamics deals with the direction of thermodynamic processes and the efficiency with which they occur. Because these characteristics control how a system evolves out of a given state, the Second Law also governs the stability of thermodynamic equilibrium.

3.1 NATURAL AND REVERSIBLE PROCESSES

A process for which the system can be restored to its initial state without leaving a net influence on the system or on its environment is said to be *reversible*. A reversible process is actually an idealization: a process that is free of friction and for which changes of state occur slowly enough for the system to remain in thermodynamic equilibrium. By contrast, a *natural process* is one that proceeds spontaneously. It can be stimulated by an infinitesimal perturbation. Because the system is then out of thermodynamic equilibrium with its surroundings, a natural process cannot be reversed entirely, namely, without leaving a net influence on either the system or its environment. A natural process is therefore inherently "irreversible."

Irreversibility arises whenever the system is out of thermodynamic equilibrium. This occurs during rapid changes of state. Performing work across a finite pressure difference, wherein the system is out of mechanical equilibrium, is irreversible. So is transferring heat across a finite temperature difference, wherein the system is out of thermal equilibrium. A process involving friction (e.g., turbulent mixing, which damps large-scale kinetic energy) is also irreversible. So are transformations of phase when different phases of water are not in equilibrium with one another. Such phase transformations occur in states away from saturation.

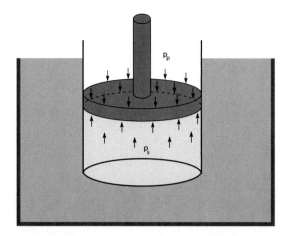

Figure 3.1 A gas at the system pressure p_s that is acted upon by a piston with pressure p_p and is maintained at constant temperature through contact with a heat reservoir.

An example of irreversible work follows from a gas that is acted upon by a piston, while being maintained at a constant temperature through contact with a heat reservoir (Fig. 3.1). Suppose the gas is first expanded isothermally at temperature T_{12} from state 1 to state 2 and subsequently restored to state 1 through isothermal compression at the same temperature (Fig. 3.2). If the cycle is executed slowly and without friction, then the system pressure p_s is uniform throughout the gas and equal to that exerted by

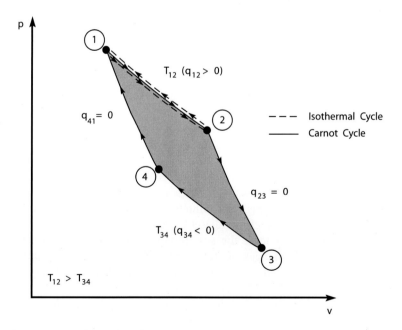

Figure 3.2 Isothermal cycle (dashed) and Carnot cycle (solid) in the $p - v$ plane.

the piston p_p at each stage of the process. The work performed by the system during isothermal expansion is thus given by

$$
\begin{aligned}
w_{12} &= \int_1^2 p_p dv \\
&= \int_1^2 \frac{RT_{12}}{v} dv \\
&= RT_{12} \ln\left(\frac{v_2}{v_1}\right).
\end{aligned}
\tag{3.1.1}
$$

Likewise, the work performed on the system during isothermal compression is

$$
\begin{aligned}
-w_{21} &= -\int_2^1 p_p dv \\
&= RT_{12} \ln\left(\frac{v_2}{v_1}\right),
\end{aligned}
\tag{3.1.2}
$$

equal to the work performed by the system during expansion. Hence, the net work performed during the cycle vanishes. Because $\oint du = 0$, (2.13) implies

$$
\oint \delta q = \oint p dv = 0.
$$

The net heat absorbed during the cycle also vanishes. Consequently, both the system and the environment are restored to their original states. The isothermal expansion from state 1 to state 2 is entirely reversible. The same holds for the isothermal compression thereafter and hence for the cycle as a whole.

If the cycle is executed rapidly, some of the gas is accelerated. Pressure is then no longer uniform across the system, so p_s does not equal p_p. During rapid compression, the piston must exert a pressure which exceeds that which occurs if the compression is executed slowly (e.g., to offset the reaction of the gas being accelerated). Therefore, the work performed on the system exceeds that performed under reversible conditions.[1] During rapid expansion, the piston must exert a pressure which is smaller than that when the expansion is executed slowly. The work performed by the system is then less than that performed under reversible conditions. Thus, executing the cycle rapidly results in

$$
\oint p dv = \oint \delta q < 0.
$$

Net work is performed on the system. It must be compensated by net rejection of heat to the environment.

This example illustrates that executing a transition between two states reversibly minimizes the work that must be performed on a system. Alternatively, it maximizes the work that is performed by the system. These results may be extended to a system that performs work through a conversion of heat transferred into it, viz. to a heat engine. With (2.13), it follows that the cyclic work performed by such a system is maximized when the cycle is executed reversibly. Irreversibility reduces the net work performed by the system over a cycle. By the First Law, the deficit in work must be

---

[1] Kinetic energy generated by the excess work is eventually dissipated and lost through heat transfer to the environment.

associated with some of the heat that is absorbed by the system being lost to the environment.

3.1.1 The Carnot cycle

Consider a cyclic process that is comprised of two isothermal legs and two adiabatic legs (Fig. 3.2). During isothermal expansion and compression, the system is maintained at constant temperature through contact with a thermal reservoir, which serves as a heat source or heat sink. During adiabatic compression and expansion, the system is thermally isolated. If executed reversibly, this process describes a *Carnot cycle*.

The leg from state 1 to state 2 is isothermal expansion at temperature T_{12}, so

$$\Delta u_{12} = 0.$$

Then the heat absorbed during that leg is given by

$$q_{12} = w_{12}$$

$$= \int_1^2 p\,dv \tag{3.2.1}$$

$$= RT_{12} \ln\left(\frac{v_2}{v_1}\right).$$

The leg from state 2 to state 3 is adiabatic, so

$$q_{23} = 0$$

and

$$-w_{23} = \Delta u_{23}$$

$$= c_v(T_{34} - T_{12}). \tag{3.2.2}$$

Similarly, the leg between states 3 and 4 is isothermal compression, so the heat transfer and work are given by

$$q_{34} = w_{34} = RT_{34} \ln\left(\frac{v_4}{v_3}\right). \tag{3.2.3}$$

The closing leg between states 4 and 1 is adiabatic, so it involves work

$$w_{41} = c_v(T_{12} - T_{34}). \tag{3.2.4}$$

If the cycle is executed in reverse, the work and heat transfer during each leg are exactly opposite to those during the corresponding leg of the forward cycle. All exchanges of energy during the two cycles then cancel. The Carnot cycle is thus reversible.

During the two adiabatic legs (3.2.2) and (3.2.4), the work performed cancels. The heat transfer vanishes identically. Hence, the net work and heat transfer over the cycle follow exclusively from the isothermal legs. The volumes in (3.2.1) and (3.2.3) are related to the changes of temperature along the adiabatic legs. They follow from Poisson's identity (2.30.1)

$$\frac{T_{12}}{T_{34}} = \left(\frac{v_3}{v_2}\right)^{\gamma-1} = \left(\frac{v_4}{v_1}\right)^{\gamma-1}.$$

Thus,

$$\frac{v_2}{v_1} = \frac{v_3}{v_4}. \tag{3.3}$$

Collecting the heat transferred during the isothermal legs yields

$$\oint \delta q = RT_{12} \ln\left(\frac{v_2}{v_1}\right) + RT_{34} \ln\left(\frac{v_4}{v_3}\right)$$
$$= R(T_{12} - T_{34}) \ln\left(\frac{v_2}{v_1}\right). \tag{3.4}$$

Heat transfer into the system is positive if $T_{12} > T_{34}$ and $v_2 > v_1$, namely, if isothermal expansion proceeds at high temperature and isothermal compression proceeds at low temperature. Under these conditions,

$$\oint p\,dv = \oint \delta q > 0. \tag{3.5}$$

The system then performs net work, which is converted from heat that it absorbs during the cycle.

The Carnot cycle is a paradigm of a heat engine. For net work to be performed by the system, heat must be absorbed at high temperature T_{12} (during isothermal expansion) and rejected at low temperature T_{34} (during isothermal compression). According to (3.2), more heat is absorbed at high temperature than is rejected at low temperature. The First Law (2.13) then implies that the net heat absorbed during the cycle is balanced by net work that is performed by the system.

If the Carnot cycle is executed in reverse, the system constitutes a refrigerator. Heat transfer into the system is negative. More heat is rejected at high temperature than is absorbed at low temperature. Then (2.13) implies that the net heat rejected during the cycle is balanced by work that must be performed on the system.

Net work and heat transfer over the Carnot cycle are proportional to the temperature difference $T_{12} - T_{34}$ between the heat source and heat sink. Not all of the heat absorbed by the system during expansion is converted into work. Some of that heat is lost during compression. The efficiency of a heat engine is therefore limited – even for a reversible cycle.

3.2 ENTROPY AND THE SECOND LAW

In the development of the First Law, we observed that, under adiabatic conditions, work is independent of path. This empirical finding permitted the introduction of a state variable, internal energy (2.11). The Second Law of Thermodynamics is inspired by the observation that, under reversible conditions, the quantity $\frac{q}{T}$ is independent of path. As for the First Law, this empirical finding permits the introduction of a state variable, *entropy*. Like internal energy, entropy is defined in terms of a quantity that, in general, is path dependent.

Consider the Carnot cycle. According to (3.4),

$$\oint \frac{\delta q}{T} = R\left[\ln\left(\frac{v_2}{v_1}\right) + \ln\left(\frac{v_4}{v_3}\right)\right] = 0.$$

This is equivalent to the identity

$$\frac{q_{12}}{T_{12}} + \frac{q_{41}}{T_{41}} = 0. \tag{3.6}$$

It turns out that this relationship holds under rather general circumstances.

Carnot's Theorem: The identity (3.6) holds for any reversible cycle between two heat reservoirs at temperatures T_{12} and T_{41}, irrespective of details of the cycle.

It can be shown that any reversible cycle can be represented as a succession of infinitesimal Carnot cycles (see, e.g., Keenan, 1970). Therefore, Carnot's theorem implies the identity

$$\oint \left(\frac{\delta q}{T}\right)_{rev} = 0, \tag{3.7}$$

irrespective of path.

By the Exact Differential Theorem (Sec. 2.1.4), it follows that the quantity $(\frac{\delta q}{T})_{rev}$ represents a point function. That is, $\int_{(x_0,y_0)}^{(x,y)} (\frac{\delta q}{T})_{rev}$ depends only on the thermodynamic state (x, y), not on the path along which the system evolved to that state. We define the state variable *entropy*

$$ds = \left(\frac{\delta q}{T}\right)_{rev}, \tag{3.8}$$

which constitutes a property of the system.

Carnot's theorem and the identity (3.7) apply to a reversible process. Under more general circumstances, the following inequality holds.

Clausius Inequality: For a cyclic process,

$$\oint \frac{\delta q}{T} \leq 0, \tag{3.9}$$

where equality applies if the cycle is executed reversibly.

One of several statements of the Second Law, the Clausius inequality has the following consequences that pertain to the direction of thermodynamic processes:

(1) Heat must be rejected to the environment somewhere during a cycle.

(2) Under reversible conditions, more heat is absorbed at high temperature than at low temperature.

(3) Irreversibility reduces the net heat absorbed during a cycle.

The first consequence precludes the possibility of a process that converts heat from a single source entirely into work: a *perpetual motion machine of the 2nd kind.* Some of the heat absorbed by a system that performs work must be rejected. Representing a thermal loss, that heat rejection limits the efficiency of any heat engine, even one operated reversibly. The second consequence of (3.9) implies that net work is performed by the system during a cycle (viz. it behaves as a heat engine) if heat is absorbed at high temperature and rejected at low temperature. Conversely, net work must be performed on the system during the cycle (viz. it behaves as a refrigerator) if heat is absorbed at low temperature and rejected at high temperature. The third consequence of (3.9) implies that, in the case of a heat engine, irreversibility reduces the net work

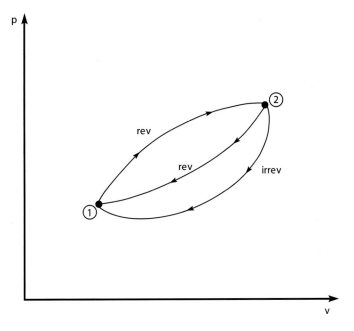

Figure 3.3 Cyclic processes between two states, consisting of (1) two reversible legs and (2) one reversible and one irreversible leg.

performed by the system. In the case of a refrigerator, irreversibility increases the net work that must be performed on the system.

A differential form of the Second Law, which applies to an incremental process, may be derived from the Clausius inequality. Consider a cycle comprised of a reversible and an irreversible leg between two states 1 and 2 (Fig. 3.3). A cycle comprised of two reversible legs may also be constructed between those states. For the first cycle, the Clausius inequality yields

$$\int_1^2 \left(\frac{\delta q}{T}\right)_{rev} + \int_2^1 \left(\frac{\delta q}{T}\right)_{irrev} \leq 0,$$

whereas for the second cycle

$$\int_1^2 \left(\frac{\delta q}{T}\right)_{rev} + \int_2^1 \left(\frac{\delta q}{T}\right)_{rev} = 0.$$

Subtracting gives

$$\int_2^1 \left(\frac{\delta q}{T}\right)_{irrev} - \int_2^1 \left(\frac{\delta q}{T}\right)_{rev} \leq 0$$

or with (3.8)

$$\int_2^1 ds \geq \int_2^1 \left(\frac{\delta q}{T}\right)_{irrev}.$$

Because it holds between two arbitrary states, the inequality must also apply to the integrands. Thus,

$$ds \geq \frac{\delta q}{T},$$ (3.10)

where equality holds if the process is reversible.

Equation (3.10) is the most common form of the Second Law. In combination with (3.8), it implies an upper bound to the heat that can be absorbed by the system during a given change of state:

$$\delta q \leq T ds,$$

namely the heat absorbed when that process is executed reversibly. If the process is executed irreversibly, some heat is lost to the environment. This reduces δq and the net heat absorbed. That, in turn, reduces the net work performed by the system during a cycle (2.13).

Also represented in (3.10) is the direction of thermodynamic processes. Through the inequality, the Second Law restricts the paths along which a system can evolve. A process for which the change of entropy satisfies (3.10) is possible. If (3.10) is satisfied through equality, that process is reversible. If it is satisfied through inequality, that process is irreversible (e.g., a natural process). By contrast, a process that satisfies the reverse inequality is impossible.

3.3 RESTRICTED FORMS OF THE SECOND LAW

The entropy of a system can either increase or decrease, depending on the heat transfer needed to achieve the same change of state under reversible conditions. For certain processes, the change of entropy implied by the Second Law is simplified.

For an adiabatic process, (3.10) reduces to

$$ds_{ad} \geq 0,$$ (3.11)

so the entropy can then only increase. It follows that irreversible work increases a system's entropy. Letting the control surface of a hypothetical system pass to infinity eliminates heat transfer to the environment. This leads to the conclusion that the entropy of the universe can only increase. For a reversible adiabatic process, (3.11) implies that $ds = 0$. Such a process is therefore *isentropic*.

For an isochoric process, expansion work vanishes. The First Law (2.23.1) then transforms (3.8) into

$$ds = c_v \left(\frac{dT}{T} \right)_{rev}.$$

Because it involves only state variables, this expression must hold irrespective of whether the process is reversible. Hence,

$$ds_v = c_v \left(\frac{dT}{T} \right)_v.$$ (3.12)

In the absence of work, entropy can either increase or decrease, depending on the sign of dT. It follows that, unlike work, heat transfer can either increase or decrease a system's entropy.

According to (3.11) and (3.12), changes of entropy follow from

(1) irreversible work
(2) heat transfer

Irreversible work only increases s. Conversely, heat transfer (irreversible or otherwise) can either increase or decrease s. In the absence of work, the change of entropy equals $\frac{\delta q}{T}$, irrespective of path (e.g., whether or not the process is reversible).

3.4 THE FUNDAMENTAL RELATIONS

Substituting the Second Law (3.10) into the two forms of the First Law (2.22) leads to the inequalities

$$du \leq T ds - p dv \qquad (3.13.1)$$

$$dh \leq T ds + v dp \qquad (3.13.2).$$

For a reversible process, these reduce to

$$du = T ds - p dv \qquad (3.14.1)$$

$$dh = T ds + v dp. \qquad (3.14.2)$$

Because they involve only state variables, the equalities (3.14) cannot depend on path. Consequently, they must hold irrespective of whether the process is reversible. Known as the *fundamental relations*, these identities describe the change in one state variable in terms of the changes in two other state variables. Although generally valid, the fundamental relations cannot be evaluated easily under irreversible conditions. The values of p and T in (3.14) refer to the pressure and temperature "of the system." The latter can be specified only under reversible conditions, wherein they equal "applied values." Under irreversible conditions, they are unknown. The relationship among these variables then reverts to the inequalities (3.13), wherein p and T denote applied values, which can be specified.

It is convenient to introduce two new state variables, which are referred to as auxiliary functions. The *Helmholtz function* is defined by

$$f = u - T s. \qquad (3.15.1)$$

The *Gibbs function* is defined by

$$\begin{aligned} g &= h - T s \\ &= u + p v - T s. \end{aligned} \qquad (3.15.2)$$

In terms of these variables, the fundamental relations become

$$df = -s dT - p dv \qquad (3.16.1)$$

$$dg = -s dT + v dp. \qquad (3.16.2)$$

The Helmholtz and Gibbs functions are each referred to as the *free energy* of the system: f for an isothermal process and g for an isothermal-isobaric process, because they reflect the energy available for conversion into work under those conditions (Prob. 3.10).

3.4.1 The Maxwell Relations

Variables appearing in (3.14) and (3.16) are not entirely independent. Involving only state variables, each fundamental relation has the form of an exact differential. According to the Exact Differential Theorem (2.8), these relations can hold only if the cross derivatives of the coefficients on the right hand sides are equal. Applying that condition to (3.14) and (3.16) yields the identities

$$\left(\frac{\partial T}{\partial v}\right)_s = -\left(\frac{\partial p}{\partial s}\right)_v$$

$$\left(\frac{\partial s}{\partial v}\right)_T = \left(\frac{\partial p}{\partial T}\right)_v$$

$$\left(\frac{\partial T}{\partial p}\right)_s = \left(\frac{\partial v}{\partial s}\right)_p$$

$$\left(\frac{\partial s}{\partial p}\right)_T = -\left(\frac{\partial v}{\partial T}\right)_p,$$

(3.17)

which are known as *the Maxwell relations*.

3.4.2 Noncompensated heat transfer

The inequalities in (3.13) account for additional heat rejection to the environment that occurs through irreversibility. Those inequalities may be eliminated in favor of equalities by introducing the *noncompensated heat transfer* $\delta q'$, defined by

$$\delta q = T\,ds - \delta q',$$

(3.18)

where $\delta q' > 0$. The noncompensated heat transfer represents the additional heat that is lost to the environment (and hence cannot be converted to work) as a result of irreversibility (e.g., that associated with frictional dissipation of kinetic energy).

Substituting (3.18) into the First Law (2.22.1) yields

$$du = T\,ds - p\,dv - \delta q',$$

(3.19)

where p and T are understood to refer to applied values. An expression for the noncompensated heat transfer can be derived from (3.14) and (3.19), which hold for applied values under reversible and irreversible conditions, respectively. Subtracting gives

$$\delta q' = (T - T_{rev})ds - (p - p_{rev})dv,$$

(3.20)

where T_{rev} and p_{rev} refer to applied values under equilibrium conditions (i.e., those assumed by the system when the process is executed reversibly). According to (3.20), noncompensated heat transfer results from (1) thermal dis-equilibrium of the system, which is represented in the difference $T - T_{rev}$, and (2) mechanical dis-equilibrium of the system, which is represented in the difference $p - p_{rev}$.

3.5 CONDITIONS FOR THERMODYNAMIC EQUILIBRIUM

By restricting the direction of thermodynamic processes, the Second Law implies whether or not a path out of a given thermodynamic state is possible. Because this determines the likelihood of the system remaining in that state, the Second Law characterizes the stability of thermodynamic equilibrium.

Consider a system in a given thermodynamic state. An arbitrary infinitesimal process out of that state is referred to as a *virtual process*. The system is said to be in *stable* or *true thermodynamic equilibrium* if no virtual process out of that state is a natural process, namely, if all virtual paths out of that state are either reversible or impossible. If all virtual paths out of the state are natural processes, the system is said to be in *unstable equilibrium*. An infinitesimal and arbitrary perturbation will then result in a finite change of state. If only some of the virtual processes out of the state are natural, the system is said to be in *metastable equilibrium*. A small perturbation then may or may not result in a finite change of state, depending on the nature of the perturbation.

These definitions may be combined with the Second Law to determine conditions that characterize thermodynamic equilibrium. The relations

$$du \leq T\,ds - p\,dv$$
$$dh \leq T\,ds + v\,dp$$
$$df \leq -s\,dT - p\,dv$$
$$dg \leq -s\,dT + v\,dp,$$

hold for a reversible process, in the case of equality, and for an irreversible (e.g., natural) process, in the case of inequality. For a state to represent thermodynamic equilibrium, the reverse must be true, namely,

$$du \geq T\,ds - p\,dv$$
$$dh \geq T\,ds + v\,dp$$
$$df \geq -s\,dT - p\,dv \tag{3.21}$$
$$dg \geq -s\,dT + v\,dp,$$

where inequality describes processes that are impossible. Inequalities (3.21) provide criteria for thermodynamic equilibrium. If they are satisfied by all virtual processes out of the current state, the system is stable. If only some virtual paths out of the state satisfy (3.21), the system is metastable.

Under special circumstances, simpler criteria for thermodynamic equilibrium exist. For an adiabatic enclosure, (3.10) reduces to

$$ds \geq 0,$$

where inequality corresponds to a natural process. A criterion for thermodynamic equilibrium is then

$$ds_{ad} \leq 0, \tag{3.22}$$

which describes processes that are reversible or impossible. According to (3.22), a stable state for an adiabatic system coincides with a local maximum of entropy

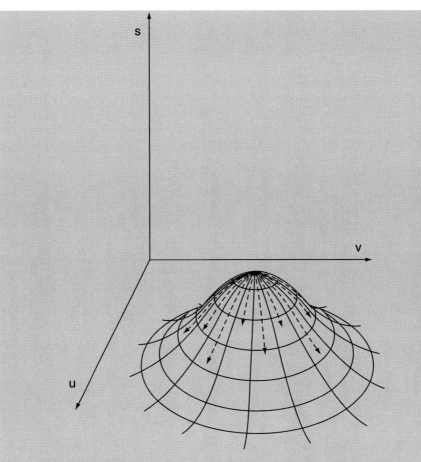

Figure 3.4 Local entropy maximum in the $u - v$ plane, symbolizing a state that corresponds to thermodynamic equilibrium.

(Fig. 3.4). Therefore, an adiabatic system's entropy must increase as it approaches thermodynamic equilibrium.

Choosing processes for which the right-hand sides of (3.21) vanish yields other criteria for thermodynamic equilibrium:

$$du_{s,v} \geq 0$$
$$dh_{s,p} \geq 0$$
$$df_{T,v} \geq 0$$
$$dg_{T,p} \geq 0,$$

(3.23)

which must be satisfied by all virtual paths out of a state for it to be stable. For the processes defined, (3.23) implies that a stable state coincides with local minima in the properties u, h, f, and g, respectively. Internal energy, enthalpy, Helmholtz function, and Gibbs function must therefore decrease as a system approaches thermodynamic equilibrium.

3.6 RELATIONSHIP OF ENTROPY TO POTENTIAL TEMPERATURE

In Chap. 2, we saw that the increase of potential temperature is proportional to the heat absorbed by an air parcel. Under reversible conditions, (3.10) implies that the same is true of entropy. Substituting (2.36) into the Second Law yields

$$d\,ln\theta \leq \frac{ds}{c_p}, \tag{3.24}$$

where equality holds for a reversible process. Because it involves only state variables, that equality must hold irrespective of whether the process is reversible. Hence

$$d\,ln\theta = \frac{ds}{c_p}. \tag{3.25}$$

The change of potential temperature is related directly to the change of entropy. Despite its validity, (3.25) applies to properties of the system (e.g., to p and T through θ). Thus like the fundamental relations, it can be evaluated only under reversible conditions.

3.6.1 Implications for vertical motion

If a process is adiabatic, $d\theta = 0$ and $ds \geq 0$. The entropy remains constant or it can increase through irreversible work (e.g., that associated with turbulent dissipation of large-scale kinetic energy). In the case of an air parcel, the conditions for adiabatic behavior are closely related to those for reversibility. Adiabatic behavior requires not only that no heat be transferred across the control surface, but also that no heat be exchanged between one part of the system and another (see, e.g., Landau and Lifshitz 1980). The latter requirement excludes turbulent mixing, which is the principal form of mechanical irreversibility in the atmosphere. It also excludes irreversible expansion work because such work introduces internal motions that eventually result in mixing.

Because they exclude the important sources of irreversibility, the conditions for adiabatic behavior are tantamount to conditions for isentropic behavior. Thus, adiabatic conditions for the atmosphere are equivalent to requiring isentropic behavior for individual air parcels. Under these circumstances, potential temperature surfaces: $\theta = $ const, coincide with isentropic surfaces: $s = $ const. An air parcel coincident initially with a certain isentropic surface remains on that surface. Because those surfaces tend to be quasi-horizontal, adiabatic behavior implies no net vertical motion (cf. Fig. 2.9). Air parcels may ascend and descend along isentropic surfaces, which undulate spatially. However, they experience no systematic vertical displacement.

Under diabatic conditions, an air parcel moves across isentropic surfaces, in proportion to the heat exchanged with its environment (2.36). Consider an air parcel moving horizontally through different thermal environments, like those represented in the distribution of net radiation in Fig. 1.34. Because radiative transfer varies sharply with latitude, this occurs whenever the parcel's motion is deflected across latitude circles. Figure 3.5 shows a wavy trajectory followed by an air parcel that is initially at latitude ϕ_0. Symbolic of the disturbed circulations in Figs. 1.9 and

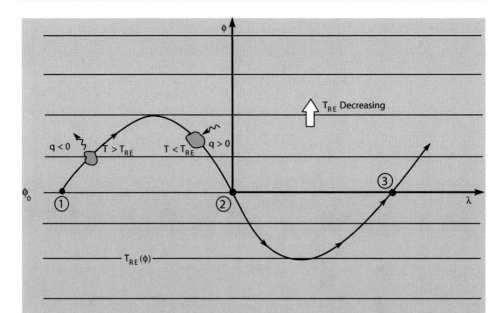

Figure 3.5 A wavy trajectory followed by an air parcel, which symbolizes the disturbed circulations in Figs. 1.9a and 1.10a. When displaced poleward, the parcel is warmer than the local radiative-equilibrium temperature, so it rejects heat. The reverse process occurs when the parcel moves equatorward.

1.10, that motion advects air through different radiative environments. Also indicated in Fig. 3.5 is the distribution of radiative-equilibrium temperature $T_{RE}(\phi)$, at which air emits radiant energy at the same rate as it absorbs it. That thermal structure is achieved if the motion is everywhere parallel to latitude circles, because air parcels then have infinite time to adjust to local thermal equilibrium.

Suppose the disturbed motion in Fig. 3.5 is sufficiently slow for the parcel to equilibrate with its surroundings at each point along the trajectory. The parcel's temperature then differs from T_{RE} only infinitesimally (Fig. 3.6), so the parcel remains in thermal equilibrium. Heat transfer along the trajectory then occurs reversibly. Between two successive crossings of the latitude ϕ_0, the parcel absorbs heat such that

$$\int_1^2 c_p d \ln\theta = \int_1^2 \frac{\delta q}{T}. \qquad (3.26)$$

If the heat exchange depends only on the parcel's temperature, e.g.,

$$\delta q = T df(T),$$

which is symbolic of radiative transfer, then (3.26) reduces to

$$c_p \ln\left(\frac{\theta_2}{\theta_1}\right) = \Delta f = 0,$$

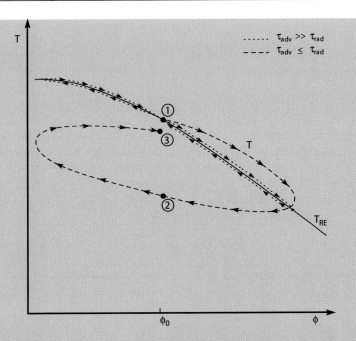

Figure 3.6 Thermal history of the idealized air parcel in Fig. 3.5. If advection occurs on a time scale much longer than radiative transfer (dotted), the parcel remains in thermal equilibrium with its surroundings (i.e., it assumes the local radiative-equilibrium temperature). The heat it absorbs and rejects is then reversible. If advection occurs on a time scale comparable to or shorter than radiative transfer (dashed), the parcel is driven out of thermal equilibrium with its surroundings. The heat it absorbs and rejects is then irreversible, which introduces a hysteresis into the parcel's thermodynamic state as it is advected through successive cycles of the disturbance (cf. Fig. 2.5).

because $T_2 = T_1 = T_{RE}(\phi_0)$. Thus $\theta_2 = \theta_1$ and the parcel is restored to its initial thermodynamic state when it returns to latitude ϕ_0.

While moving poleward, the parcel is infinitesimally warmer than the local radiative-equilibrium temperature. It therefore emits more radiant energy than it absorbs. Rejection of heat results in the parcel drifting off its initial isentropic surface toward lower θ, which, for reasons developed in Chap. 7, corresponds to lower altitude. While moving equatorward, the parcel is infinitesimally colder than the local radiative-equilibrium temperature. It therefore absorbs more radiant energy than it emits. Absorption of heat then results in the parcel ascending to higher θ, just enough to restore the parcel to its initial isentropic surface when it returns to the latitude ϕ_0. Successive crossings of the latitude ϕ_0 therefore result in no net vertical motion. The parcel's evolution is then perfectly cyclic.

Now suppose the motion is sufficiently fast to carry the parcel between radiative environments before it has equilibrated to the local radiative-equilibrium temperature. During the excursion poleward of ϕ_0, the parcel is out of thermal equilibrium. Heat transfer along the trajectory therefore occurs irreversibly. Because its temperature then lags that of its surroundings, the parcel returns to the latitude ϕ_0 with a temperature different from that initially (Fig. 3.6). By the foregoing analysis, the parcel's potential temperature θ_2 also differs from that initially

$$c_p \ln \left(\frac{\theta_2}{\theta_1} \right) = \Delta f \neq 0,$$

because $T_2 \neq T_1$. Thus the parcel is not restored to its initial isentropic surface. Rather, it remains displaced vertically after returning to the latitude ϕ_0. Similar reasoning shows that heat transfer during the excursion equatorward of ϕ_0 (e.g., between positions 2 and 3) does not exactly cancel net heat transfer during the poleward excursion. Hence a complete cycle results in net heat transfer and therefore a net vertical displacement of the parcel from its initial isentropic surface.

Whether the parcel returns above or below that isentropic surface depends on the radiative-equilibrium temperature and on details of the motion, which control the history of heating and cooling. In either event, irreversible heat transfer introduces a hysteresis (cf. Fig. 2.5). Through it, successive cycles yield a vertical drift of air across isentropic surfaces. Because advection operates on time scales much shorter than those of radiative transfer, disturbed horizontal motion invariably drives air out of thermal equilibrium. That introduces irreversible heat transfer, which in turn drives vertical motion.

In the troposphere, irreversible heat transfer enters through quasi-horizontal mixing by synoptic weather systems. Amplifying through instability (Chap. 16), they represent exaggerated forms of the wavy pattern in Fig. 3.5. Trajectories then steepen and are eventually overturned (as in a breaking wave); cf. Figs. 2.10; 16.6. Air that is rearranged in latitude is driven out of radiative equilibrium, so it experiences irreversible heat transfer. In the stratosphere, the circulation is more often wavelike (Fig. 3.5). Through hysteresis (Fig. 3.6), it too introduces irreversible heat transfer that drives air across isentropic surfaces.[2] In each case, vertical motions generated in this manner contribute to mean meridional circulations, which transfer heat, moisture, and chemical constituents.

SUGGESTED REFERENCES

An illuminating treatment of the Second Law and its consequences to mechanical and chemical systems is given in *The Principles of Chemical Equilibrium* (1971) by K. Denbigh.

Statistical Physics (1980) by Landau and Lifshitz provides a clear discussion of reversibility and its implications for fluid systems.

PROBLEMS

1. 200 g of mercury at 100°C is added to 100 g of water at 20°C. If the specific heat capacities of water and mercury are 4.18 J kg^{-1}g^{-1} and 0.14 J kg^{-1}K^{-1}, respectively, determine (a) the limiting temperature of the mixture, (b) the change of entropy for the mercury, (c) the change of entropy for the water, (d) the change of entropy for the system as a whole.

---

[2] These sources of vertical motion are coupled to a mean poleward drift of air, which is forced by the absorption of wave activity (Sec. 18.3).

2. For reasons developed in Chap. 9, many clouds are supercooled: They contain droplets at temperatures below 0°C. Consider 1 mole of supercooled water in the metastable state: $T = -10$°C and $p = 1$ atm. Following a perturbation, the water freezes spontaneously, with the ice and its surroundings eventually returning to the original temperature. If the heat capacities of water and ice remain approximately constant with the values: 75 J K^{-1} mol^{-1} and 38 J K^{-1} mol^{-1}, respectively, and if the enthalpy of fusion at 0°C is 6026 J mol^{-1}, calculate (a) the change of entropy for the water, (b) the change of entropy for its environment.

3. Consider a parcel crossing isentropic surfaces. If radiative cooling is just large enough to maintain the parcel on a fixed isobaric surface, what is the relative change of its temperature when the parcel's potential temperature has decreased to 90% of its original value?

4. Air moving inland from a cooler maritime region warms through conduction with the ground. If the temperature and potential temperature increase by 5% and 6%, respectively, determine (a) the fractional change of surface pressure between the parcel's initial and final positions, (b) the heat absorbed by the parcel.

5. During a cloud-free evening, LW heat transfer with the surface causes an air parcel to descend from 900 hPa to 910 hPa and its entropy to decrease by 15 J kg^{-1} K^{-1}. If its initial temperature is 280 K, determine the parcel's (a) final temperature, (b) final potential temperature.

6. Derive expression (3.20) for the noncompensated heat transfer.

7. Air initially at 20°C and 1 atm is allowed to expand freely into an evacuated chamber to assume twice its original volume. Calculate the change of specific entropy. (Hint: How much work is performed by the air?)

8. The thermodynamic state of an air parcel is represented conveniently on the *pseudo-adiabatic chart* (Chap. 5), which displays altitude in terms of the variable p^κ. Show that the work performed during a cyclic process equals the area circumscribed by that process in the $\theta - p^\kappa$ plane.

9. One mole of water at 0°C and 1 atm is transformed into vapor at 200°C and 3 atm. If the enthalpy of vaporization for water is 4.06 10^4 J mol^{-1} and if the specific heat of vapor is approximated by

$$c_p = 8.8 - 1.9 \ 10^{-3} T + 2.2 \ 10^{-6} T^2 \qquad \text{cal K}^{-1} \text{mol}^{-1},$$

calculate the change of (a) entropy, (b) enthalpy.

10. The Helmholtz and Gibbs functions are referred to as *free energies* or *thermodynamic potentials*. Use the definitions (3.15) together with the First and Second Laws for a closed system to show that (a) under isothermal conditions, the decrease of Helmholtz function describes the maximum total work which can be performed by the system:

$$w_{max} = -\Delta f,$$

(b) under isothermal-isobaric conditions, the decrease of Gibbs function describes the maximum work, exclusive of expansion work, which can be performed by the system:

$$w'_{max} = -\Delta g,$$

where $\delta w' = \delta w - p dv$.

Heterogeneous systems

The thermodynamic principles developed in Chaps. 2 and 3 apply to a homogeneous system, which can involve only a single phase. For it to be in thermodynamic equilibrium, a homogeneous system must be in thermal equilibrium: At most, an infinitesimal temperature difference exists between the system and its environment. It must also be in mechanical equilibrium: At most, an infinitesimal pressure difference exists between the system and its environment.

A heterogeneous system can involve more than one phase. For such a system, thermodynamic equilibrium imposes an additional constraint. The system must also be in *chemical equilibrium*: No conversion of mass occurs from one phase to another. Analogous to thermal and mechanical equilibrium, chemical equilibrium requires a certain state variable to have, at most, an infinitesimal difference between the phases present.

4.1 DESCRIPTION OF A HETEROGENEOUS SYSTEM

For a homogeneous system, two intensive properties describe the thermodynamic state. Conversely, only two state variables may be varied independently. A homogeneous system therefore has two thermodynamic degrees of freedom. For a heterogeneous system, each phase may be regarded as a homogeneous subsystem, one that is "open" due to exchanges with the other phases present. Consequently, the number of intensive properties that describe the thermodynamic state of a heterogeneous system is proportional to the number of phases present. Were they independent, those properties would constitute additional degrees of freedom for a heterogeneous system. However, thermodynamic equilibrium between phases introduces additional constraints. They reduce the degrees of freedom of a heterogeneous system – even below those of a homogeneous system.

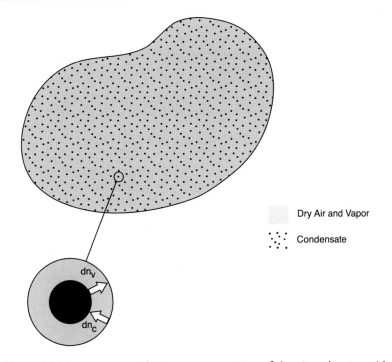

Figure 4.1 Two-component heterogeneous system of dry air and water, with the water component present in vapor and one condensed phase. The water component is transformed incrementally from condensate to vapor in amount dn_v and from vapor to condensate in amount dn_c.

The system we consider is a two-component mixture of dry air and water, with the latter existing in vapor and possibly one condensed phase (Fig. 4.1).[1] This heterogeneous system is described conveniently in terms of extensive properties, which depend on the abundance of each species present. An extensive property Z for the total system follows from the contributions from the individual phases

$$Z_{tot} = Z_g + Z_c, \tag{4.1}$$

where the subscripts g and c refer to the gas and condensate subsystems. The gas subsystem includes both dry air and vapor. The extensive property for it is specified by its pressure and temperature and the molar abundances of dry air and vapor

$$Z_g = Z_g(p, T, n_d, n_v).$$

If the condensed phase is a pure substance, the extensive property for it is determined by the pressure and temperature and the molar abundance of condensate

$$Z_c = Z_c(p, T, n_c).$$

---

[1] The term *vapor* refers to water in gas phase. Invisible, it should not be confused with condensate, such as aerosol or cloud.

Then changes of the extensive property Z for the individual subsystems are described by

$$dZ_g = \left(\frac{\partial Z_g}{\partial T}\right)_{pn} dT + \left(\frac{\partial Z_g}{\partial p}\right)_{Tn} dp + \left(\frac{\partial Z_g}{\partial n_d}\right)_{pTn} dn_d + \left(\frac{\partial Z_g}{\partial n_v}\right)_{pTn} dn_v \qquad (4.2.1)$$

$$dZ_c = \left(\frac{\partial Z_c}{\partial T}\right)_{pn} dT + \left(\frac{\partial Z_c}{\partial p}\right)_{Tn} dp + \left(\frac{\partial Z_c}{\partial n_c}\right)_{pT} dn_c, \qquad (4.2.2)$$

where the subscript n denotes holding fixed all molar values except the one being varied.

It is convenient to introduce a state variable that measures how the extensive property Z of the total system changes with an increase in one of the components (e.g., through a conversion of mass from one phase to another). It will be seen in Sec. 4.5 that a change of phase that occurs isobarically also occurs isothermally. Consequently, the foregoing state variable is expressed most conveniently for processes that occur at constant pressure and temperature. The *partial molar property* is defined as the rate an extensive property changes, under isobaric and isothermal conditions, with a change in the molar abundance of the kth species

$$\overline{Z}_k = \left(\frac{\partial Z}{\partial n_k}\right)_{pTn}, \qquad (4.3.1)$$

where the subscript k refers to a particular component and phase. Similarly, the *partial specific property* is defined as

$$\overline{z}_k = \left(\frac{\partial Z}{\partial m_k}\right)_{pTm}. \qquad (4.3.2)$$

In general, the partial molar and partial specific properties differ from the molar and specific properties for a pure substance:

$$\tilde{z}_k = \frac{Z}{n_k}$$

$$z_k = \frac{Z}{m_k}, \qquad (4.4)$$

due to interactions with the other components present.[2] However, for a system of dry air and water in which the latter is present only in trace abundance, such differences are small enough to be ignored. Hence, to a good approximation, the partial properties may be replaced by the molar and specific properties

$$\overline{Z}_d \cong \tilde{z}_d \quad \overline{z}_d \cong z_d$$

$$\overline{Z}_v \cong \tilde{z}_v \quad \overline{z}_v \cong z_v \qquad (4.5)$$

$$\overline{Z}_c \cong \tilde{z}_c \quad \overline{z}_c \cong z_c.$$

If the system is closed, the abundance of individual components is also preserved, so

$$dn_d = 0 \qquad (4.6.1)$$

$$d(n_v + n_c) = 0. \qquad (4.6.2)$$

---

[2] For example, adding air to to a mixture of air and water at constant pressure and temperature results in a change of the extensive property Z which differs from that brought about by adding the same amount of air to a system of pure air if, in the former case, some of that air passes into solution with the water.

Adding (4.2) and incorporating (4.6) and (4.5) yields the change of the extensive property Z for the total system

$$dZ_{tot} = \left(\frac{\partial Z_{tot}}{\partial T}\right)_{pn} dT + \left(\frac{\partial Z_{tot}}{\partial p}\right)_{Tn} dp + (\breve{z}_v - \breve{z}_c)\, dn_v, \qquad (4.7.1)$$

where

$$Z_{tot} = n_d \breve{z}_d + n_v \breve{z}_v + n_c \breve{z}_c. \qquad (4.7.2)$$

In terms of specific properties, the change of Z for the total system is given by

$$dZ_{tot} = \left(\frac{\partial Z_{tot}}{\partial T}\right)_{pm} dT + \left(\frac{\partial Z_{tot}}{\partial p}\right)_{Tm} dp + (z_v - z_c)\, dm_v, \qquad (4.8.1)$$

where

$$Z_{tot} = m_d z_d + m_v z_v + m_c z_c. \qquad (4.8.2)$$

These expressions represent generalizations of the exact differential relation (2.6) for a homogeneous system to account for exchanges of mass between the phases of a heterogeneous closed system.

4.2 CHEMICAL EQUILIBRIUM

We now develop a criterion for two phases of the system to be at equilibrium with one another. In addition to thermal and mechanical equilibrium, those phases must also be in chemical equilibrium. The latter is determined by the diffusion of mass from one phase to the other, which, for reasons to become apparent, is closely related to the Gibbs function. The *chemical potential* for the kth species, μ_k, is defined as the partial molar Gibbs function:

$$\mu_k = \overline{G}_k = \left(\frac{\partial G}{\partial n_k}\right)_{pTn}, \qquad (4.9)$$

which is tantamount to the molar Gibbs function $\breve{g}_k$. For the gas phase, (4.2.1) gives the change of Gibbs function

$$dG_g = \left(\frac{\partial G_g}{\partial T}\right)_{pn_v n_d} dT + \left(\frac{\partial G_g}{\partial p}\right)_{Tn_v n_d} dp + \mu_d dn_d + \mu_v dn_v.$$

In a constant n_v, n_d process (e.g., one not involving a phase transformation), this expression must reduce to the fundamental relation (3.16.2) for a homogeneous closed system. Accordingly, we identify

$$\left(\frac{\partial G_g}{\partial T}\right)_{pn_v n_d} = -S_g$$

$$\left(\frac{\partial G_g}{\partial p}\right)_{Tn_v n_d} = V_g. \qquad (4.10)$$

Because they involve only state variables, expressions (4.10) must hold irrespective of path (e.g., whether or not a phase transformation is involved). Thus,

$$dG_g = -S_g dT + V_g dp + \mu_d dn_d + \mu_v dn_v \qquad (4.11)$$

describes the change of Gibbs function for the gas phase. Similar analysis leads to the relation

$$dG_c = -S_c dT + V_c dp + \mu_c dn_c \tag{4.12}$$

for the condensed phase. Adding (4.11) and (4.12) and incorporating (4.6.2) yields the change of Gibbs function for the total system

$$dG_{tot} = -S_{tot} dT + V_{tot} dp + (\mu_v - \mu_c) dn_v, \tag{4.13.1}$$

where

$$G_{tot} = n_d \tilde{g}_d + n_v \tilde{g}_v + n_c \tilde{g}_c. \tag{4.13.2}$$

Equation (4.13.1) is a generalization to a heterogeneous system of the fundamental relation (3.16.2).

For the heterogeneous system to be in thermodynamic equilibrium, the pressures and temperatures of the different phases present must be equal. Further, there can be no conversion of mass from one phase to another. Corresponding to the fundamental relation (4.13) is the inequality

$$dG_{tot} \leq -S_{tot} dT + V_{tot} dp + (\mu_v - \mu_c) dn_v, \tag{4.14}$$

where p, T, and n_v refer to applied values and where equality and inequality correspond to reversible and irreversible processes, respectively. For the system to be in equilibrium, all virtual processes emanating from the state under consideration must be either reversible or impossible. Hence, thermodynamic equilibrium is characterized by the reverse inequality

$$dG_{tot} \geq -S_{tot} dT + V_{tot} dp + (\mu_v - \mu_c) dn_v, \tag{4.15}$$

which must hold for all virtual paths out of the state. Consider a transformation of phase that occurs at constant pressure and temperature. Then, for equilibrium,

$$dG_{tot} \geq (\mu_v - \mu_c) dn_v$$

must hold for all virtual paths. As dn_v is arbitrary, the requirement can be satisfied only if

$$\mu_v = \mu_c. \tag{4.16}$$

For chemical equilibrium, the chemical potentials of different phases of the water component must be equal. This criterion is analogous to the requirements that, under thermal and mechanical equilibrium, the temperatures and pressures of those phases must be equal.

The chemical potential determines the diffusive flux of mass from one species to another. In that respect, μ_k may be regarded as a diffusion potential from the kth species.[3] Under the circumstances described by (4.16), the flux of mass from one phase of water to another is exactly balanced by a flux in the opposite sense. The net diffusion of mass therefore vanishes. If the chemical potential differs between the phases, there

---

[3] Chemical potential is the ultimate determinant of diffusion. It supersedes concentration in certain applications; see Denbigh (1971).

will exist a net diffusion of mass from the phase of high μ to the phase of low μ. This transformation of mass will proceed until the difference of chemical potential between the phases has been eliminated, establishing chemical equilibrium.

4.3 FUNDAMENTAL RELATIONS FOR A MULTI-COMPONENT SYSTEM

In a manner similar to that used to develop the change of Gibbs function, all of the fundamental relations for a homogeneous system can be generalized to a heterogeneous system of C chemically distinct components and P phases. Expanding U, H, F, and G as above leads to the relations

$$dU = TdS - pdV + \sum_{j=1}^{P}\sum_{i=1}^{C} \mu_{ij} dn_{ij}$$

$$dH = TdS + Vdp + \sum_{j=1}^{P}\sum_{i=1}^{C} \mu_{ij} dn_{ij}$$

$$dU = -SdT - pdV + \sum_{j=1}^{P}\sum_{i=1}^{C} \mu_{ij} dn_{ij} \tag{4.17}$$

$$dG = -SdT + Vdp + \sum_{j=1}^{P}\sum_{i=1}^{C} \mu_{ij} dn_{ij},$$

where we have introduced alternate expressions for the chemical potential of the ith component and jth phase:

$$\mu_{ij} = \left(\frac{\partial U}{\partial n_{ij}}\right)_{SVn} = \left(\frac{\partial H}{\partial n_{ij}}\right)_{Spn} = \left(\frac{\partial F}{\partial n_{ij}}\right)_{TVn} = \left(\frac{\partial G}{\partial n_{ij}}\right)_{pTn} \tag{4.18}$$

in terms of the variables in (4.17). As will be seen shortly, the definitions in (4.18) are equivalent. However, only the last corresponds to an isobaric-isothermal process. Consequently, the first three do not represent partial molar properties according to the definition (4.3). Because transformations of phase at constant pressure also occur at constant temperature, the last expression in (4.18) is the most convenient form of chemical potential. For the same reason, the last of the fundamental relations in (4.17) is the most convenient description of such transformations.

For a closed system, the mass of each component is preserved, so

$$\sum_{j=1}^{P} dn_{ij} = 0, \quad i = 1, 2, 3, \dots C. \tag{4.19}$$

For the ith component, (4.19) implies

$$\sum_{j=1}^{P} \mu_{ij} dn_{ij} = \sum_{j=2}^{P} \mu_{ij} dn_{ij} + \mu_{i1} dn_{i1}$$

$$= \sum_{j=2}^{P} \left(\mu_{ij} - \mu_{i1}\right) dn_{ij}, \tag{4.20}$$

where $j = 1$ denotes an arbitrary reference phase for that component. Then (4.17) may be expressed

$$dU = T\,dS - p\,dV + \sum_{j=2}^{P}\sum_{i=1}^{C}\left(\mu_{ij} - \mu_{i1}\right)dn_{ij}$$

$$dH = T\,dS + V\,dp + \sum_{j=2}^{P}\sum_{i=1}^{C}\left(\mu_{ij} - \mu_{i1}\right)dn_{ij}$$

$$dF = -S\,dT - p\,dV + \sum_{j=2}^{P}\sum_{i=1}^{C}\left(\mu_{ij} - \mu_{i1}\right)dn_{ij} \tag{4.21}$$

$$dG = -S\,dT + V\,dp + \sum_{j=2}^{P}\sum_{i=1}^{C}\left(\mu_{ij} - \mu_{i1}\right)dn_{ij},$$

which represent generalizations of the fundamental relations (3.14) and (3.16) to a heterogeneous closed system. According to (4.21), each of the definitions of chemical potential in (4.18) implies the same statement of chemical equilibrium (4.16). However, only the expression involving Gibbs function applies to a process executed at constant pressure and temperature (e.g., to an isobaric transformation of phase).

4.4 THERMODYNAMIC DEGREES OF FREEDOM

Each phase of a heterogeneous system constitutes a homogeneous subsystem. Its state is therefore specified by two intensive properties. Thus, the number of intensive properties that describe the thermodynamic state of a heterogeneous system is proportional to the number of phases present.

Consider a single-component system involving two phases (e.g., pure water and vapor). The state of each homogeneous subsystem is specified by two intensive properties. Four intensive properties then specify the state of the total system. However, thermodynamic equilibrium requires each subsystem to be in thermal, mechanical, and chemical equilibrium with the other subsystem. Consequently, the heterogeneous system must also satisfy three constraints:

$$T_1 = T_2$$
$$p_1 = p_2 \tag{4.22}$$
$$\mu_1 = \mu_2.$$

The heterogeneous system is left with only one independent state variable.

Thus, a one-component system involving two phases at equilibrium with one another possesses only one thermodynamic degree of freedom. Such a system must therefore possess an equation of state of the form

$$p = p(T). \tag{4.23}$$

Involving a single independent state variable, the equation of state of the heterogeneous system has the same form as that governing adiabatic changes of a homogeneous system (2.30). It describes a family of curves, along which the heterogeneous system evolves reversibly (i.e., during which it remains in thermodynamic equilibrium). According to (4.23), fixing the temperature of a single-component mixture of two phases also fixes its pressure and vice versa. Consequently, an equilibrium transformation of phase that occurs at constant pressure also occurs at constant temperature. This feature makes the last expression in (4.18) the most useful definition of chemical potential and the change of Gibbs function the most convenient of the fundamental relations (4.21).

If all three phases are present, six intensive properties describe the state of a single-component heterogeneous system. For the system to be in thermodynamic equilibrium, those properties must also satisfy six independent constraints, like those in (4.22). Consequently, a single-component system involving all three phases possesses no thermodynamic degrees of freedom. Such a system can exist in only a single state, referred to as the *triple point*.

In general, the number of thermodynamic degrees of freedom possessed by a heterogeneous system is described by the following principle.

Gibbs' Phase Rule: The number of independent state variables for a heterogeneous system involving C chemically distinct but non-reactive components and P phases is given by

$$N = C + 2 - P. \qquad (4.24)$$

According to (4.24), the degrees of freedom possessed by a heterogeneous system increases with the number of chemically distinct components but decreases with the number of phases present.[4]

4.5 THERMODYNAMIC CHARACTERISTICS OF WATER

The remainder of this chapter focuses on the thermodynamic behavior of a single-component system of pure water. This system may involve one, two, or possibly all three phases at equilibrium with one another (Fig. 4.2). Because water is a pure substance, its equation of state may, in general, be expressed

$$p = p(v, T), \qquad (4.25)$$

regardless of how many phases are present (Sec 2.1.4). However, the particular form of (4.25) depends on which phases are present. If only vapor is present, (4.25) is given by the ideal gas law (1.1). If two phases are present, the equation of state must reduce to the form of (4.23).

Like any pure substance, water has an equation of state that describes a surface over the plane of two intensive properties, illustrated in Fig. 4.3 for p as a function of v and T. Portions of that surface describe states in which only a single phase is present and the system is homogeneous. In those regions of state space, the system possesses two thermodynamic degrees of freedom. Any two state variables may be varied independently. For instance, in the region of vapor, the behavior is that of an ideal gas (cf. Fig. 2.3). Both v and T are then required to specify p and the thermodynamic state.

---

[4] A generalization of the phase rule to reactive components can be found in Denbigh (1971).

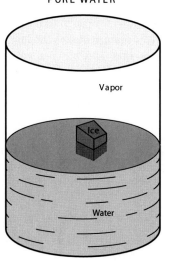

Figure 4.2 Single-component heterogeneous system of pure water.

At pressures and temperatures below the *critical point:* (p_c, T_c), the system can assume heterogeneous states. Multiple phases then coexist at equilibrium with one another.[5] If two phases are present, the system possesses only one thermodynamic degree of freedom (4.24). Consequently, in such regions, the surface in Fig. 4.3 assumes a simpler form. It is discontinuous in slope at the boundary with regions where the system is homogeneous. For instance, in the region of vapor and water, isotherms coincide with isobars. Specifying one of those properties determines the other and thus the thermodynamic state of the pure substance. Similar behavior is found in the regions of vapor and ice and water and ice.

At temperatures below the critical point, the system can evolve from a homogeneous state of one phase to a homogeneous state of another phase only by passing through intermediate states that are heterogeneous. Due to a change in the number of degrees of freedom, the path describing this process is discontinuous in slope at the boundary between homogeneous and heterogeneous states. Consider isobaric heat rejection (cooling), indicated in Fig. 4.3. From an initial temperature above the critical point, where the system is entirely gaseous, temperature and volume both decrease. Those changes reflect the two degrees of freedom possessed by a homogeneous system. Eventually, the process encounters the boundary separating homogeneous states, wherein only vapor is present, from heterogeneous states, wherein vapor and water coexist at equilibrium. At that state, the slope of the process changes discontinuously – because the temperature of the system can no longer decrease. Instead, isobaric cooling results in condensation of vapor and a sharp reduction of volume – all at constant temperature. This simplified behavior continues until the vapor has been converted

---

[5] At temperatures above T_c, the system remains homogeneous and its properties vary smoothly across the entire range of pressure. Condensed phases are then not possible at any pressure. For those temperatures, water substance is referred to as *gas*, to distinguish it from *vapor*, which can coexist with condensed phases.

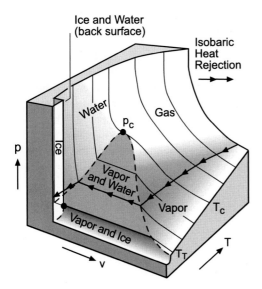

Figure 4.3 State space of a single-component system of pure water, illustrated in terms of its pressure as a function of its specific volume and temperature. Heterogeneous states occur below the *critical point* (p_c, T_c). A process corresponding to isobaric heat rejection (bold) is also indicated. In the heterogeneous region of vapor and water, isobars coincide with isotherms. Also shown is the temperature of the *triple point* (p_T, T_T, v_T), where all three phases coexist.

entirely into water. At that point, the system is again homogeneous. The slope of the process then changes discontinuously a second time. Beyond that state, heat rejection results in a decrease of both temperature and volume. Those changes reflect the two degrees of freedom that are again possessed by the system.

In a heterogeneous state, different phases coexist at equilibrium only if their temperatures, pressures, and chemical potentials are equal. The individual phases are then said to be *saturated*, because the net flux of mass from one phase to another vanishes. If one of those phases is vapor, the pressure of the heterogeneous system represents the *equilibrium vapor pressure* with respect to water or ice, denoted p_w and p_i, respectively. Should the heterogeneous system have a pressure below the equilibrium vapor pressure (e.g., below p_w), the chemical potential of the vapor will be less than that of the condensed phase. Mass will then diffuse from the condensed phase to the vapor phase, increasing the system's pressure until $p = p_w$. At that point, the net flux of mass between the two subsystems vanishes: Chemical equilibrium has been established. Conversely, a pressure above the equilibrium vapor pressure will result in a conversion of mass from vapor to condensate. Accompanied by a decrease of the system's pressure, the conversion will proceed until the difference of chemical potential between the phases has been eliminated: Chemical equilibrium is likewise established.

According to Gibbs' phase rule, there exists a single state at which all three phases coexist at equilibrium. Defined by the intersection of surfaces where water and vapor coexist, where vapor and ice coexist, and where water and ice coexist, the triple point for water is given by

$$
\begin{aligned}
p_T &= 6.1 \text{ mb} \\
T_T &= 273 \text{ K} \\
v_{Tv} &= 2.06 \ 10^5 \text{ m}^3 \text{ kg}^{-1} \\
v_{Tw} &= 1.00 \ 10^{-3} \text{ m}^3 \text{ kg}^{-1} \\
v_{Ti} &= 1.09 \ 10^{-3} \text{ m}^3 \text{ kg}^{-1},
\end{aligned}
\tag{4.26}
$$

where the subscripts v, w, and i refer to vapor, water, and ice, respectively.

4.6 EQUILIBRIUM PHASE TRANSFORMATIONS

According to Sec. 2.3, heat transferred during an isobaric process between two homogeneous states of the same phase is proportional to the change of temperature. Under those circumstances, the constant of proportionality is the specific heat at constant pressure c_p. By contrast, heat transfer during an isobaric process between two heterogeneous states of the same two phases involves no change of temperature (Fig. 4.3). Instead, heat transfer results in a conversion of mass from one phase to the other. That transformation of phase is associated with a change of internal energy and with work being performed when the heterogeneous system's volume changes. Conversely, an isobaric transformation of phase can occur only if it is accompanied by a transfer of heat to support the change of internal energy and the work performed by the heterogeneous system during its change of volume.

4.6.1 Latent heat

The preceding effects are collected in the change of enthalpy, which equals the heat transfer into the system during an isobaric process (2.15). Analogous to the specific heat capacity at constant pressure, the specific *latent heat* of transformation is defined as the heat absorbed by the system during an isobaric transformation of phase

$$l = \delta q_p$$
$$= dh,$$

(4.27.1)

which, by the First Law, equals the change of enthalpy during the phase transformation. In (4.27.1), the mass in l refers only to the substance undergoing the transformation of phase (e.g., to the water component in a mixture with dry air). The specific latent heats of vaporization, fusion (solid $\rightarrow$ liquid), and sublimation (solid $\rightarrow$ vapor) are denoted l_v, l_f, and l_s, respectively. They are related as

$$l_s = l_f + l_v.$$

(4.27.2)

Like specific heat capacity, latent heat is a property of the system. Thus, l depends on the thermodynamic state. For a heterogeneous system of two phases, it may be expressed $l = l(T)$. The dependence on temperature of l may be established by considering a transformation from one phase to another and how that transformation varies with temperature. Consider a homogeneous state wherein the system is entirely in phase a and another homogeneous state wherein it is entirely in phase b. By (2.23.2), an isobaric process between two homogeneous states that involve only phase a results in a change of enthalpy

$$dh_a = \left(\frac{\partial h_a}{\partial T}\right)_p dT$$
$$= c_{pa} dT.$$

Likewise, an isobaric process between two homogeneous states that involve only phase b results in

$$dh_b = \left(\frac{\partial h_b}{\partial T}\right)_p dT$$
$$= c_{pb} dT.$$

Subtracting gives an expression for how the difference of enthalpy between phases a and b changes with temperature

$$d(\Delta h) = (c_{pb} - c_{pa})dT.$$

Now the difference of enthalpy, at a given temperature, equals the latent heat of transformation between phase a and phase b (4.27.1). Therefore,

$$\frac{dl}{dT} = \Delta c_p, \tag{4.28}$$

where Δc_p refers to the difference of specific heat between the phases.

Known as *Kirchoff's equation*, (4.28) provides a formula for calculating the latent heat, as a function of temperature, in terms of the difference of specific heat between the two phases. Of the three latent heats, only l_f varies significantly over a range of temperature relevant to the atmosphere. At $0°C$, the latent heats of water have the values

$$l_v = 2.50 \ 10^6 \ \mathrm{J \ kg^{-1}}$$
$$l_f = 3.34 \ 10^5 \ \mathrm{J \ kg^{-1}} \tag{4.29}$$
$$l_s = 2.83 \ 10^6 \ \mathrm{J \ kg^{-1}}.$$

Because it is defined for an isobaric process, l describes the change of enthalpy during a transformation of phase at constant pressure. The corresponding change of internal energy follows from the First Law. Consider the system in a heterogeneous state and undergoing an isobaric transformation of phase. Then (2.12.2) becomes

$$du = l - pdv. \tag{4.30}$$

For fusion, dv is negligible, so $du = l$. The change of internal energy is then just the latent heat, which also represents the change of enthalpy under those circumstances. For vaporization and sublimation,

$$dv \cong v_v, \tag{4.31}$$

where v_v refers to the volume of vapor produced. Incorporating the ideal gas law for the vapor gives the change of internal energy

$$du = l - R_v T \tag{4.32}$$

during isobaric vaporization and sublimation.

4.6.2 Clausius-Clapeyron Equation

In states involving two phases, the system of pure water possesses only one thermodynamic degree of freedom. Thus, specifying its temperature determines the system's pressure and hence its thermodynamic state. Under those circumstances, the equation of state (4.25) reduces to the form of (4.23), which describes the simplified behavior of the water surface in Fig. 4.3. The equation of state describing those heterogeneous regions may be derived from the fundamental relations, subject to conditions of chemical equilibrium.

Consider two phases a and b and a transformation between them that occurs reversibly (e.g., wherein the system remains in thermodynamic equilibrium). The heat transfer during such a process equals the latent heat of transformation. Then, with (4.27), the Second Law becomes

$$ds = \frac{l}{T}. \tag{4.33}$$

For the system to be in chemical equilibrium, the chemical potential of phase a must equal that of phase b, so by (4.5)

$$g_a = g_b.$$

This condition must be satisfied throughout the process. Therefore, the change of Gibbs function for one phase must track that of the other

$$dg_a = dg_b.$$

Applying the fundamental relation (3.16.2) to each homogeneous subsystem then implies

$$-(s_b - s_a)\,dT + (v_b - v_a)\,dp = 0$$

or

$$\frac{dp}{dT} = \frac{\Delta s}{\Delta v}, \tag{4.34}$$

where Δ refers to the change between the phases. Incorporating (4.33) yields a relationship between the equilibrium pressure and temperature

$$\frac{dp}{dT} = \frac{l}{T\Delta v}, \tag{4.35}$$

where l is the latent heat appropriate to the phases present. Known as the *Clausius-Clapeyron equation*, (4.35) relates the equilibrium vapor pressure (e.g., $p = p_w$ or p_i) to the temperature of the heterogeneous system. It thus constitutes an equation of state for the heterogeneous system when two phases are present, describing the simplified surfaces in Fig. 4.3 that correspond to such states.

The Claussius-Clapeyron equation may be specialized to each of the heterogeneous regions in Fig. 4.3. For water and ice, l corresponds to fusion. Under those circumstances, (4.35) is expressed most conveniently in inverted form

$$\frac{dT}{dp} = \frac{T\Delta v}{l}.$$

It describes the influence on melting temperature exerted by a change of pressure. Because the change of volume during fusion is negligible, the equation of state in the region of water and ice reduces to

$$\left(\frac{dT}{dp}\right)_{fusion} \cong 0. \tag{4.36}$$

Changing the pressure has only a negligible effect on the temperature at which water is at equilibrium with ice. Consequently, the surface of water and ice in Fig. 4.3 is vertical.

For vapor and a condensed phase, l corresponds to the latent heat of vaporization or sublimation. Under those circumstances, the change of volume is approximately equal to that of the vapor produced (4.31). Thus

$$\Delta v \cong \frac{R_v T}{p},$$

which transforms (4.35) into

$$\left(\frac{d\ln p}{dT}\right)_{\substack{vaporization\\sublimation}} = \frac{l}{R_v T^2}. \tag{4.37}$$

For $l \cong$ const, (4.37) gives

$$\ln\left(\frac{p}{p_0}\right) = \frac{l}{R_v}\left(\frac{1}{T_0} - \frac{1}{T}\right). \tag{4.38}$$

Using the value of l_v in (4.29) yields the equilibrium vapor pressure with respect to water

$$log_{10} p_w \cong 9.4041 - \frac{2.354\ 10^3}{T}, \tag{4.39}$$

where p_w is in hPa and $T_0 = 0$ C is used as a reference state. Similarly, the value of l_s in (4.29) yields the equilibrium vapor pressure with respect to ice

$$log_{10} p_i \cong 10.55 - \frac{2.667\ 10^3}{T}. \tag{4.40}$$

It differs from (4.39) only modestly.

Equations (4.39) and (4.40) describe the simplified surfaces in Fig. 4.3 that correspond to vapor being in chemical equilibrium with a condensed phase (e.g., to the system pressure equaling the equilibrium vapor pressure p_w or p_i). If the system's pressure is below the equilibrium vapor pressure for the temperature of the system, water will evaporate or sublimate until the system's pressure reaches the equilibrium vapor pressure. A pressure above the equilibrium vapor pressure will result in the reverse transformation. Owing to the exponential dependence in (4.39) and (4.40), p_w and p_i vary sharply with temperature. In the presence of a condensed phase, much more water can exist in vapor phase at high temperature than at low temperature. It will be seen in Chap. 5 that the principles governing a single-component heterogeneous system carry over to a two-component system of dry air and water. The equilibrium vapor pressure then represents the maximum amount of vapor than can be supported by air at a given temperature.

The exponential dependence of p_w on T has an important implication for exchanges of water between the Earth's surface and the atmosphere. According to (4.39), warm tropical ocean with high SST can transfer much more water into the atmosphere than can colder extratropical ocean. For this reason, tropical oceans serve as the primary source of water vapor for the atmosphere. After being introduced in the tropics, water vapor is redistributed over the globe by the circulation (cf. Fig. 1.18). Much of the water vapor absorbed by the tropical atmosphere is precipitated back to the Earth's surface in organized convection inside the ITCZ (Fig. 1.30). However, latent heat that is released during condensation remains in the overlying atmosphere. Cyclic transfer of water between the ocean surface and the tropical troposphere thus results in a net transfer of heat to the atmosphere. Eventually converted into work, that heat generates kinetic energy. Along with radiative transfer from the Earth's surface (Fig. 1.32), it maintains the general circulation against frictional dissipation.

SUGGESTED REFERENCES

The Principles of Chemical Equilibrium (1971) by K. Denbigh includes a thorough treatment of phase equilibria in heterogeneous systems.

A detailed description of water substance is presented in *Atmospheric Thermodynamics* (1981) by Iribarne and Godson.

PROBLEMS

1. The Gibbs-Dalton law implies that the partial pressure of vapor at equilibrium with a condensed phase of water is the same in a mixture with dry air as it would be were the water component in isolation. Because it corresponds to the abundance of vapor at which no mass is transformed from one phase to another, this vapor pressure describes the state at which air is *saturated*. For a lapse rate of 6.5 K km^{-1}, representative of thermal structure in the troposphere (Fig. 1.2), calculate (a) the equilibrium vapor pressure as a function of altitude, and (b) the corresponding mixing ratio of water vapor as a function of altitude. (c) Apply the result to explain why water vapor is distinguished from other trace gases in Fig. 1.23.

2. A more accurate version of (4.39) is given by

$$log_{10} p_w = -\frac{2937.4}{T} - 4.9283 log_{10} T + 23.5471 \text{ (hPa)}.$$

    Estimate the boiling temperature for water at the altitude of (a) Denver: 5000′, (b) the North America continental divide: 14,000′, (c) the summit of Mt. Everest: 29,000′.

3. Present-day Venus contains little water, which is thought to have been absorbed by its atmosphere and eventually destroyed during the planet's evolution (Sec. 8.7). (a) For present conditions, wherein Venus has a surface pressure of 9×10^6 Pa, at what surface temperature does water vapor become the atmosphere's primary constituent? Compare this with Venus' present surface temperature of 750 K. (b) Were these conditions to prevail to altitudes where energetic UV radiation is present, what would be the implication for the abundance of water on the planet? (c) Contrast these circumstances with present-day conditions on Earth.

4. Cloud seldom forms in the stratosphere because air there is very dry. Exceptional is Polar Stratospheric Cloud (PSC), which forms over the Antarctic and, less frequently, over the Arctic. The thicker form of such cloud (Type II PSC), which is still quite tenuous compared with tropospheric cloud, is composed of ice. (a) For a mixing ratio at 80 hPa of 3 ppmv, representative of water vapor in the lower stratosphere (cf. Fig. 18.8), calculate the temperature at which ice cloud forms. (b) Referring to Fig. 1.7 and to the discussion in Sec 18.3.2, where is such cloud likely to form?

5. The average precipitation rate inside the ITCZ is of order 10 mm day^{-1} (Fig. 9.41). (a) In W m^{-2}, calculate the average column heating rate inside the ITCZ. (b) Compare this value with the LW radiative flux emitted to the atmosphere (and largely absorbed by it) if the surface behaves as a blackbody at a temperature of 300 K (1.29).

6. Derive the fundamental relations (4.17) for a mixture involving multiple phases.

7. Derive alternate expressions for chemical equilibrium under (a) isothermal-isochoric conditions, (b) reversible-adiabatic and isobaric conditions.

8. Consider a mixture of dry air and water. Describe the state space of this generally heterogeneous system and the geometry its graphical representation assumes, noting the number of thermodynamic degrees of freedom in regions where 1, 2, and 3 phases are present.

9. Use (4.39) to estimate the latent heat of vaporization for water.

FIVE

Transformations of moist air

The atmosphere is a mixture of dry air and water in varying proportion. Although its abundance varies widely, water vapor seldom represents more than a few percent of air by mass. We shall consider a two-component system comprised of these species, with water appearing in possibly one condensed phase. According to the *Gibbs-Dalton law* (which is accurate at pressures below the critical point), an individual component of a mixture of gases behaves the same as if the other components were absent. Consequently, the abundance of vapor at equilibrium with a condensed phase in a mixture of water and dry air is the same as if the water component were in isolation.[1] For this reason, concepts established in Chap. 4 for a single-component system of pure water carry over to a two-component system of dry air and water.

5.1 DESCRIPTION OF MOIST AIR

5.1.1 Properties of the gas phase

For the moment, we focus on the gas phase of this system, irrespective of whether condensate happens to be present. The vapor is in solution with dry air. It is represented by the partial pressure of vapor, e, which obeys the equation of state

$$ev_v = R_v T. \qquad (5.1.1)$$

In (5.1.1),

$$v_v = \frac{V}{m_v} \qquad (5.1.2)$$

---

[1] Implications of the Gibbs-Dalton law are developed in Keenan (1970).

is the specific volume of vapor (not to be confused with the partial volume),

$$
\begin{aligned}
R_v &= \frac{R^*}{M_v} \\
&= \frac{1}{\epsilon_v} R_d,
\end{aligned}
\tag{5.1.3}
$$

and $\epsilon_v \cong 0.622$ is the ratio of molar weights defined by (1.15.2).

The *absolute humidity* $\rho_v = \frac{1}{v_v}$ measures the absolute concentration of vapor. The relative concentration of vapor is measured by the *specific humidity*

$$
q = \frac{\rho_v}{\rho} = \frac{m_v}{m},
\tag{5.2}
$$

which equals the ratio of the masses of vapor and mixture. Closely related is the mass *mixing ratio* $r = m_v/m_d = \rho_v/\rho_d$, which is referenced to the mass of dry air (1.14). Because

$$
m = m_d + m_v,
\tag{5.3}
$$

the mixing ratio is approximately equal to the specific humidity

$$
\begin{aligned}
r &= \frac{q}{1 - q} \\
&\cong q
\end{aligned}
\tag{5.4}
$$

because vapor is present in only trace abundance (Table 1.1). Hence, to a good approximation, q and r may be used interchangeably. Both are conserved for an individual air parcel outside regions of condensation. By contrast, measures of absolute concentration like e and ρ_v change for an individual air parcel through changes of pressure, even if the mass of vapor remains constant.

Despite the advantages of relative concentration, chemical equilibrium of the water component is controlled by the absolute concentration of vapor. For this reason, it is convenient to express the mixing ratio r in terms of the vapor pressure e. The dry air component of the mixture obeys the equation of state

$$
p_d v_d = R_d T,
\tag{5.5}
$$

where p_d and v_d denote the partial pressure and specific volume of dry air. Dividing (5.1) by (5.5) obtains

$$
\left(\frac{e}{p_d} \right) \left(\frac{v_v}{v_d} \right) = \frac{1}{\epsilon}.
\tag{5.6}
$$

Now

$$
\begin{aligned}
\left(\frac{v_v}{v_d} \right) &= \left(\frac{m_d}{m_v} \right) \\
&= \frac{1}{r}.
\end{aligned}
$$

Because vapor exists only in trace abundance,

$$
\begin{aligned}
p &= p_d + e \\
&\cong p_d.
\end{aligned}
$$

Then (5.6) reduces to

$$\frac{r}{\epsilon} = \frac{e}{p}$$

$$= N_v,$$

(5.7)

where the molar fraction of vapor N_v is defined by (1.12).

Because the composition of air varies with the abundance of water vapor, so too do composition-dependent properties such as the specific gas constant R. It is convenient to absorb such variations into state variables and, instead, deal with the fixed properties of dry air. From (1.7) and (5.3), the specific gas constant of the mixture may be expressed

$$R = (1 - q)R_d + qR_v$$

$$= (1 - q)R_d + \frac{q}{\epsilon}R_d$$

$$= \left[1 + \left(\frac{1}{\epsilon} - 1\right)q\right]R_d,$$

or

$$R = (1 + .61q)R_d.$$

(5.8)

Then the equation of state for the mixture becomes

$$pv = (1 + .61q)R_d T.$$

(5.9)

The moisture dependence in (5.9) can be absorbed into the *virtual temperature*

$$T_v = (1 + .61q)T.$$

(5.10)

Then the equation of state for the gas phase of the system is simply

$$pv = R_d T_v,$$

(5.11)

which involves only the fixed specific gas constant of dry air. In practice, $q = O(.01)$, so, to good approximation, T may be used in place of T_v.

The specific heats of air also depend on moisture content:

$$c_v = (1 + .97q)c_{vd}$$

$$c_p = (1 + .87q)c_{pd},$$

(5.12)

from which the dimensionless quantities γ and κ (2.21) follow directly. Like virtual temperature, c_v, c_p, γ, and κ differ only slightly from the constant values for dry air.

5.1.2 Saturation properties

Consider now the gas phase of the system in the presence of a condensed phase of water. If the vapor is in chemical equilibrium with the condensed phase, it is said to be *saturated*. Corresponding to this condition, and for a given pressure and temperature, are particular values of the foregoing moisture variables. They are referred to as *saturation values*. According to the Gibbs-Dalton law, the *saturation vapor pressure with respect to water* e_w is identical to the equilibrium vapor pressure p_w of a single-component system of vapor and water (Sec. 4.6). Likewise, the saturation vapor pressure with respect to ice e_i is identical to p_i for a single-component system.[2]

The saturation vapor pressure e_c, where c denotes either of the condensed phases, is a function of temperature alone. Described by the Claussius-Clapeyron relations (4.39) and (4.40), e_c increases sharply with temperature. The *saturation specific humidity* q_c and the *saturation mixing ratio* r_c, which follows from e_c through (5.7), also describe the abundance of vapor at equilibrium with a condensed phase. Like the saturation vapor pressure, these quantities are state variables. But, because they describe the mixture and not just the vapor, q_c and r_c also depend on pressure (in accord with Gibbs' phase rule (4.24) for a two-component system involving two phases). However, the strong temperature dependence of $e_c(T)$ in the Claussius-Clapeyron equation is the dominant influence on $q_c(p, T)$ and $r_c(p, T)$. Therefore, a decrease of temperature introduced by adiabatic expansion sharply reduces the saturation values q_c and r_c. Just the reverse results from an increase of temperature introduced by adiabatic compression.

Unlike saturation values (e.g., r_c), which change with the thermodynamic state of the system, the abundance of vapor actually present (e.g., r) changes only through a transformation of phase. If no condensed phase is present (namely, under unsaturated conditions), the abundance of vapor is preserved. A decrease of temperature then results in a decrease of r_c, but no change of r. On the other hand, if the system is saturated, $r = r_c(p, T)$. A change of state in which the system remains saturated must then result in a change of both $r_c(p, T)$ and r. The vapor must adjust to maintain chemical equilibrium with the condensate present.

Two other quantities are used to describe the abundance of water vapor. The *relative humidity* is defined as

$$RH = \frac{N_v}{N_{vc}}$$
$$= \frac{e}{e_c} \cong \frac{r}{r_c}, \tag{5.13}$$

where N_{vc} denotes the saturation molar abundance of vapor, and equilibrium with respect to water is implicit. The *dew point temperature* T_d is defined as that temperature to which the system must be cooled "isobarically" to achieve saturation. If saturation occurs below 0°C, that temperature is the *frost point temperature* T_f. The *dew point spread* is given by the difference $(T - T_d)$. For a given temperature T, a high dew point implies a small dew point spread. Each reflects a large abundance of vapor, which requires only a small depression of temperature to achieve saturation.

Neither relative humidity nor dew point spread are direct measures of vapor concentration. Rather, they describe how far the system is from saturation. Because saturation values increase sharply with temperature, inferring moisture content from the aforementioned quantities is misleading. For instance, r_w at 1000 hPa and 30°C is

---

[2] Strictly, the water component of the two-component system does not behave exactly as it would in isolation. Discrepancies with that idealized behavior stem from (1) near saturation, departures of the vapor from the behavior of an ideal gas; (2) the condensed phase being acted upon by the total pressure and not just that of the vapor; and (3) some of the air passing into solution with the water. However, these effects introduce discrepancies that are smaller than 1%, so they can be ignored for most applications; see Iribarne and Godson (1981) for a detailed treatment.

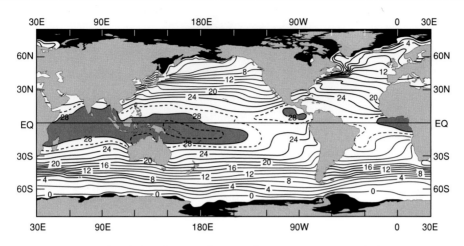

Figure 5.1 Global distribution of sea surface temperature (SST) for March. Temperatures warmer than 28°C are shaded. From Shea et al. (1990).

nearly 30 g kg^{-1}. However, at the same pressure but 0°C, it is only 4 g kg^{-1}. Thus, a relative humidity of 50% implies an abundance of vapor of 2 g kg^{-1} at 30°C but of 15 g kg^{-1} at 0°C - more than seven times greater!

5.2 IMPLICATIONS FOR THE DISTRIBUTION OF WATER VAPOR

Saturation values describe the maximum abundance of vapor that can be supported in solution with dry air at a given temperature and pressure. For that abundance, diffusion of mass from vapor to condensate is exactly balanced by diffusion of mass in the opposite sense. If the system is heterogeneous and has a vapor abundance below the saturation value (e.g., an unsaturated air parcel in contact with warm ocean surface), vapor will be absorbed until the difference of chemical potential between the phases of water has been eliminated. For this transformation to occur, the water component must absorb heat equal to the latent heat of vaporization. Conversely, if the system is heterogeneous and has a vapor abundance slightly above the saturation value (e.g., a supersaturated air parcel containing an aerosol of droplets), vapor will condense to relieve the imbalance of chemical potential. For this transformation to occur, the water component must reject heat equal to the latent heat of vaporization.

At temperatures and pressures representative of the atmosphere, the saturation vapor pressure seldom exceeds 60 hPa. The saturation mixing ratio seldom exceeds 30 g kg^{-1} or 0.030. It is for this reason that water vapor exists only in trace abundance in the atmosphere. One should note that the foregoing moisture quantities refer only to vapor - not to the total water content of the system. Condensation results in a reduction of q and r, but a commensurate increase of condensate according to (4.6.2). Unless condensate precipitates out of the system, the total water content of an air parcel is preserved.

According to the Clausius-Clapeyron equation, saturation vapor pressure increases exponentially with temperature. Thus, air can support substantially more vapor in solution at high temperature than at low temperature. For this reason, water vapor is produced efficiently in the tropics, where warm SST corresponds to a high equilibrium vapor pressure (Fig. 5.1). For the same reason, its mixing ratio decreases

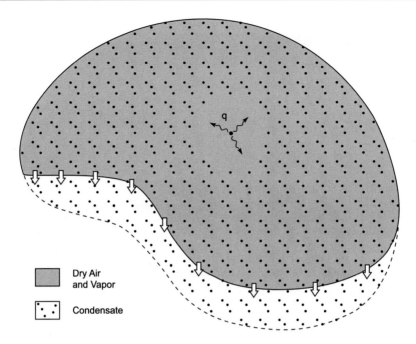

Figure 5.2 Schematic illustration of a saturated air parcel undergoing condensation, in which latent heat q is released to the gas phase and some of the condensate precipitates out of the parcel.

poleward (Fig. 1.17). Conversely, water vapor is destroyed aloft through condensation and precipitation, after displaced air parcels have cooled through adiabatic expansion work (2.33) and suffered a sharp reduction of saturation mixing ratio. Condensation also occurs at high latitude, after displaced air parcels have cooled through radiative and conductive heat transfer.

Production of vapor at an ocean surface can occur only if the water component absorbs latent heat to support the transformation of phase. Absorbed from the ocean (which then experiences evaporative cooling), that latent heat is then transferred to the air with which the vapor passes into solution. When the water re-condenses, the latent heat is released to the air surrounding the condensate (Fig. 5.2). It remains in the atmosphere after the condensate has precipitated back to the Earth's surface.[3] The preceding cycle of water results in no net exchange of mass but a net transfer of heat from the ocean to the atmosphere (cf. Fig. 1.32).

5.3 STATE VARIABLES OF THE TWO-COMPONENT SYSTEM

Under unsaturated conditions, the two-component system involves only a single phase. By Gibbs' phase rule (4.24), the system then possesses three thermodynamic degrees

---

[3] The respective heat transfer is substantial. The latent heat that is released inside a strong hurricane in one day exceeds the collective energy that is generated by the electrical grid of the United States in an entire year.

of freedom. Thus three intensive properties are required to specify the system's state if no condensate is present. Usually, the state of moist air is specified by pressure, temperature, and a humidity variable like mixing ratio that describes the abundance of vapor, e.g., $\theta = \theta(p, T, r)$. Under saturated conditions, two phases are present. One of the independent state variables is then eliminated by the constraint of chemical equilibrium. For instance, the abundance of water vapor is already determined by other state variables: $r = r_c(p, T)$. This leaves two degrees of freedom – the same number as for a homogeneous single-component system. Thus, under saturated conditions, a state variable can be specified as $\theta = \theta(p, T)$.

5.3.1 Unsaturated behavior

Under unsaturated conditions, thermodynamic processes for moist air occur much as they would for dry air because thermal properties are modified only slightly by the trace abundance of vapor. A change of temperature leads to changes of internal energy and enthalpy given by (2.18), but with slightly modified specific heats (5.12). The saturation mixing ratio r_c, which measures the capacity of air to support vapor in solution, varies with the system's pressure and temperature. By contrast, the mixing ratio r of vapor that is actually present remains constant – so long as the system is unsaturated.

The same is approximately true of the potential temperature. Under adiabatic conditions, pressure and virtual temperature change in such proportion to preserve the *virtual potential temperature*

$$\frac{\theta_v}{T_v} = \left(\frac{p_{00}}{p}\right)^{\kappa_d},$$

$$(5.14)$$

where κ_d denotes the value for dry air. Like virtual temperature, θ_v is nearly identical to its counterpart for dry air, θ. Therefore, virtual properties of moist air like T_v and θ_v will hereafter be referred to by their counterparts for dry air, but the former will be understood to apply in a strict sense.

5.3.2 Saturated behavior

Under saturated conditions (e.g., in the presence of droplets), the aforementioned relationships no longer hold. Because the vapor must then be at equilibrium with a condensed phase, $e = e_c$. A change of thermodynamic state that alters e_c must then also alter e. It must therefore result in a transformation of mass from one phase of the water component to another. Accompanying that transformation of mass is an exchange of latent heat between the condensed and gas phases of the heterogeneous system (Fig. 5.2). That heat exchange alters the potential temperature of the gas phase via (2.36).

State variables describing the two-component heterogeneous system must account for these changes. For a closed system, (4.7) gives the change of total enthalpy

$$dH = \left(\frac{\partial H}{\partial T}\right)_{pm} dT + \left(\frac{\partial H}{\partial p}\right)_{Tm} dp + (h_v - h_c)dm_v.$$

$$(5.15)$$

Because

$$H = m_d h_d + m_v h_v + m_c h_c,$$

$$\left(\frac{\partial H}{\partial T}\right)_{pm} = m_d \left(\frac{\partial h_d}{\partial T}\right)_{pm} + m_v \left(\frac{\partial h_v}{\partial T}\right)_{pm} + m_c \left(\frac{\partial h_c}{\partial T}\right)_{pm} \tag{5.16}$$

$$= m_d c_{pd} + m_v c_{pv} + m_c c_{pc}.$$

Similarly,

$$\left(\frac{\partial H}{\partial p}\right)_{Tm} = m_d \left(\frac{\partial h_d}{\partial p}\right)_{Tm} + m_v \left(\frac{\partial h_v}{\partial p}\right)_{Tm} + m_c \left(\frac{\partial h_c}{\partial p}\right)_{Tm}.$$

For components of the gas phase, (2.18) gives

$$\left(\frac{\partial h_d}{\partial p}\right)_{Tm} = \left(\frac{\partial h_v}{\partial p}\right)_{Tm} = 0.$$

For the condensed phase, the corresponding change of enthalpy can be expressed

$$\left(\frac{\partial h_c}{\partial p}\right)_{Tm} = v_c(1 - T\alpha_p),$$

where

$$\alpha_p = \frac{1}{v}\left(\frac{\partial v}{\partial T}\right)_p$$

is the *isobaric expansion coefficient* (e.g., Denbigh, 1971). Because α_p is small for condensed phases,

$$\left(\frac{\partial h_c}{\partial p}\right)_{Tm} \cong v_c.$$

The contribution to (5.15) from this pressure term can be shown to be negligible compared with the corresponding contribution from temperature (5.16). The specific heat of the heterogeneous system is

$$c_p = \frac{m_d c_{pd} + m_v c_{pv} + m_c c_{pc}}{m}. \tag{5.17}$$

The latent heat is just the difference of enthalpy between the phases of water

$$l = h_v - h_c. \tag{5.18}$$

Incorporating the above into (5.15) then obtains the change of specific enthalpy for the system

$$dh = c_p dT + l \frac{dm_v}{m} \tag{5.19}$$

$$\cong c_p dT + l dr.$$

The last expression holds exactly if the specific enthalpy refers to a unit mass of dry air, derivation of which is left as an exercise. In (5.19), the term

$$l dr = -\delta q$$

represents the heat transferred to the gas phase (mostly dry air) from the water component when the latter undergoes condensation. It reflects an internal heat source for a saturated parcel undergoing condensation. Under adiabatic conditions, wherein heat transfer between the parcel and its environment vanishes, the First Law (2.15) reduces to

$$dh = 0.$$

Equation (5.19) then gives

$$c_p dT = -l dr.$$

A decrease in vapor leads to an increase in temperature of the parcel, which is composed chiefly of dry air.[4]

If l is treated as constant (Sec. 4.6.1), (5.19) may be integrated to yield an expression for the absolute enthalpy of the two-component heterogeneous system:

$$h = c_p T + lr + h_0, \tag{5.20}$$

where h_0 denotes the enthalpy at a suitably defined reference state.

Expressions for the internal energy and entropy of the system follow in similar fashion:

$$u = c_v T + lr + u_0 \tag{5.21}$$

$$\frac{s}{c_p} = lnT - \kappa_d lnp + \frac{lr}{c_p T} + \frac{s_0}{c_p}$$

$$= ln\theta + \frac{lr}{c_p T} + \frac{s_0}{c_p}. \tag{5.22}$$

Like (5.19), the expressions for absolute enthalpy, internal energy, and entropy of the two-component system are exact if referenced to a unit mass of dry air.

5.4 THERMODYNAMIC BEHAVIOR ACCOMPANYING VERTICAL MOTION

The thermodynamic state of a moist air parcel changes through vertical motion. Vertical displacement alters the parcel's environmental pressure, which varies hydrostatically according to (1.17). To preserve mechanical equilibrium, the parcel expands or contracts, which results in work being performed. Compensating that work is a change of internal energy. It alters the parcel's temperature and, hence, the saturation vapor pressure of the two-component system.

5.4.1 Condensation and the release of latent heat

From (4.38), the change of saturation vapor pressure between a reference temperature T_0 and a temperature T may be expressed

$$ln\left(\frac{e_c}{e_{c0}}\right) = -\frac{l}{R_v}\left(\frac{1}{T} - \frac{1}{T_0}\right). \tag{5.23}$$

---

[4] The effect is analogous to the combustion of fuel in a combustion engine. Like water vapor, vaporized fuel has a mixing ratio of only a few percent. Its combustion heats the dry air in which it has been dissolved, which then expands and performs work.

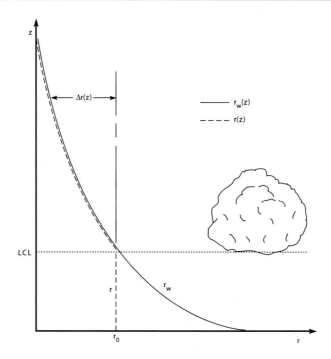

Figure 5.3 Vertical profiles of mixing ratio r and saturation mixing ratio wrt water $r_c = r_w$ for an ascending air parcel below and above the Lifting Condensation Level (LCL).

Then (5.7) implies that the saturation mixing ratio varies with pressure and temperature as

$$\frac{r_c}{r_{c0}} = \frac{e^{-\frac{l}{R_v}\left(\frac{1}{T} - \frac{1}{T_0}\right)}}{\left(\frac{p}{p_0}\right)}. \tag{5.24}$$

According to (5.24), the saturation mixing ratio increases with decreasing pressure, as accompanies upward motion. However, r_c decreases sharply with decreasing temperature, which likewise accompanies upward motion. Therefore, even though an ascending parcel's pressure decreases exponentially with its altitude, the temperature dependence in (5.24) prevails. For this reason, the parcel's saturation mixing ratio decreases monotonically with altitude.

Consider a moist air parcel ascending in thermal convection. Under unsaturated conditions, the parcel's mixing ratio and saturation mixing ratio satisfy

$$r < r_c.$$

As it rises, the parcel performs work at the expense of its internal energy. Its temperature therefore decreases at the dry adiabatic lapse rate Γ_d. From (5.24), the decrease of temperature is attended by a reduction of saturation mixing ratio r_c (Fig. 5.3). By contrast, the parcel's actual mixing ratio r remains constant, equal to its initial value $r = r_0$. The same applies to its potential temperature $\theta = \theta_0$. After sufficient upward displacement, the saturation mixing ratio will have been reduced to the actual

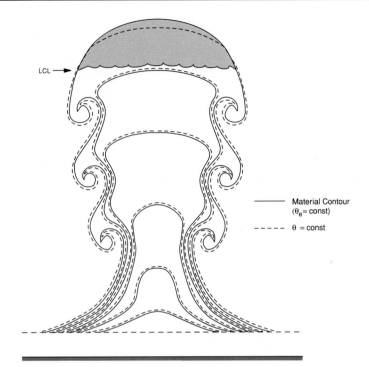

Figure 5.4 An ascending plume of moist air that develops from surface air that has become positively buoyant. Material lines (solid), which are defined by a fixed collection of air parcels, remain coincident with isentropes ($\theta =$ const) below the LCL. However, they drift to higher θ and higher altitude above the LCL, where air warms through latent heat release. Although they depart from isentropes, material lines remain coincident with pseudo-isentropes ($\theta_e =$ const) above the LCL.

mixing ratio:

$$r_c = r.$$

At that point, the parcel is saturated. The altitude where this first occurs is the Lifting Condensation Level (LCL). Because convective cloud forms through this process, the LCL defines the base of cumulus cloud, which is fueled by buoyant air that originates at the surface.

Below the LCL, the parcel's thermodynamic behavior may be regarded as adiabatic because the characteristic time scale for vertical displacement (e.g., from minutes inside cumulus convection to 1 day in sloping convection) is small compared with the characteristic time scale for heat transfer with the surroundings. Therefore, the parcel evolves in state space along a dry adiabat, which is described by Poisson's equations (2.30) and characterized by a constant value of θ.

In physical space, the parcel is actually part of a layer that is displaced vertically (Fig. 5.4). *Material contours*, which correspond to a fixed locus of air parcels, buckle to form a plume of moist air that ascends through positive buoyancy. Folds at the wall of the plume reflect turbulent entrainment with surrounding air. This exchange dilutes the plume's buoyancy with that of the environment. The associated mixing introduces diabatic effects that are ignored in the present analysis. However, those effects play a key role in damping cumulus cloud (Chap. 9).[5] Under unsaturated adiabatic conditions, θ and r are both conserved for individual parcels. Therefore, a material contour that coincides initially with a certain isentrope $\theta = \theta_0$ and mixing ratio contour $r = r_0$ remains coincident with those isopleths – so long as they reside beneath the LCL.

Above the LCL, the foregoing behavior breaks down. Continued ascent and expansional cooling reduce the saturation mixing ratio r_c below the mixing ratio r of vapor present. To restore chemical equilibrium, some of the vapor must condense – just enough to maintain chemical equilibrium:

$$r = r_c.$$

Thus, above the LCL, r and r_c both decrease with altitude (Fig. 5.3). The decrease of vapor is compensated by an increase of condensate (4.6.2).

In this fashion, ascent above the LCL wrings vapor out of solution with dry air, producing condensate (cloud droplets). The production of condensate is attended by a release of latent heat to the gas phase of the system. This heats the gas phase, which is comprised chiefly of dry air that accounts for nearly all of the parcel's mass. Latent heating thus serves as an internal heat source for the ascending parcel. Through (2.36), it increases the parcel's potential temperature. Simultaneously, it adds positive buoyancy, which reinforces ascent.

A displacement Δz above the LCL results in a decrease of water vapor mixing ratio

$$\Delta r(z) = r_c[p(z), T(z)] - r_0,$$

where z, p, and T are understood to refer to the displaced parcel with surface mixing ratio r_0. Because r_c decreases monotonically with altitude, the greater the displacement above the LCL, the less is the vapor that remains and the greater is the condensate produced and the latent heat that has been released. Cloud droplets produced in this manner grow (through mechanisms described in Chap. 9) until they can no longer be supported by the updraft. At that point, they precipitate out of the parcel.

Because θ and r are no longer conserved, the material contour coincident initially with the isopleths $\theta = \theta_0$ and $r = r_0$ deviates from those isopleths (Fig. 5.4). The release of latent heat increases θ (2.36). Consequently, the material contour advances to isentropes of greater potential temperature. For reasons developed in Chap. 7, those isentropes lie at higher altitude than the original isentrope. Similarly, condensation reduces r. The material contour therefore moves to isopleths of smaller mixing ratio, which likewise lie at higher altitude.

Saturated air that is descending undergoes just the reverse behavior. Adiabatic compression then increases the internal energy and temperature of the gas phase. That increases the saturation mixing ratio r_c over the actual mixing ratio r. Condensate can then re-evaporate to restore chemical equilibrium: $r = r_c$. Evaporation of condensate is attended by absorption of latent heat from the gas phase. This cools the parcel, which is chiefly dry air. Evaporative cooling also introduces negative buoyancy, which reinforces

---

[5] The behavior depicted in Fig. 5.4 is familiar to the sailplane pilot. The updraft fueling a growing cumulus cloud (lift) is often surrounded by a downdraft (a ring of sink), which must be penetrated to reach the core of a thermal. Above the LCL, the surrounding downdraft involves cloudy air that has been detrained from the updraft. It experiences compressional warming during its descent and evaporates.

descent. Such cooling relies on the presence of condensate that was produced during ascent. Any condensate that precipitated out of the parcel is not available to re-absorb latent heat that it released during ascent. That latent heat therefore remains in the gas phase.

The foregoing process confines water vapor to a neighborhood of the Earth's surface. Introduced over warm ocean, water vapor is extracted when moist air ascends above its LCL and experiences condensation. After cloud particles have become sufficiently large, water is removed altogether when the resulting condensate precipitates back to the Earth's surface. This process maintains upper levels of the atmosphere very dry. Even inside convective tower, mixing ratio decreases upward, because chemical equilibrium requires r to equal r_c, which decreases with altitude. By restricting vertical transport, thermodynamics prevents water vapor from reaching high altitude. There, it would be photo-dissociated by energetic radiation and eventually destroyed, when the free hydrogen produced escaped to space (Sec. 1.2.2).

Due to exchange of latent heat with the condensed phase, the gas phase of an air parcel is not adiabatic above the LCL – even though the total system may still be. Consequently, the potential temperature of the gas phase is no longer conserved. Because the parcel's mass is dominated by dry air, the change of mass during condensation exerts only a minor influence on the energetics of the gas phase. However, the accompanying release of latent heat exerts a major influence because it serves as an internal heat source for the system. Latent heat released to the gas phase during condensation offsets adiabatic cooling due to expansion work that is performed by an ascending parcel. Conversely, latent heat absorbed from the gas phase during evaporation offsets adiabatic warming due to compression work that is performed on a descending parcel. Owing to the transfer of latent heat, the parcel's temperature no longer varies with altitude at the dry adiabatic lapse rate. Instead, under saturated conditions, it varies with altitude more slowly.

5.4.2 The pseudo-adiabatic process

If expansion work occurs fast enough for heat transfer with the environment to remain negligible and if none of its moisture precipitates out, the parcel is closed. Its behavior above the LCL is then described by a reversible saturated adiabatic process. That process depends weakly on the abundance of condensate present (e.g., on how much of the system's enthalpy is represented by condensate). It therefore depends on the LCL of the parcel. However, because condensate is present only in trace abundance, its variation unnecessarily complicates the description of a saturated parcel. This complication is averted by approximating the parcel's behavior as a *pseudo-adiabatic process*: An ascending parcel is then treated as open, with condensate removed immediately after it is produced. For a descending parcel, condensate is introduced immediately before it is evaporated. Because the water component accounts for only a small fraction of the system's mass, the pseudo-adiabatic process is nearly identical to a reversible saturated adiabatic process.

A pseudo-adiabatic change of state may be constructed in two legs:

(1) Reversible saturated adiabatic expansion (compression), which results in the production (evaporation) of condensate of mass dm_c with a commensurate release (absorption) of latent heat to (from) the gas phase.

(2) Removal (introduction) of condensate of mass dm_c.

Because the phase transformation occurs adiabatically and reversibly, this process is isentropic

$$ds = 0.$$

As just enough vapor condenses to maintain chemical equilibrium: $r = r_c$, the change of water vapor mixing ratio is described by

$$dr = dr_c.$$

Then (5.22) implies

$$dln\theta = -d\left(\frac{lr_c}{c_p T}\right). \tag{5.25}$$

With the removal of condensate, (5.25) describes the change of potential temperature of the gas phase in terms of the transformation of the water component. A decrease of r_c, as accompanies ascent and expansional cooling, releases latent heat that increases θ. Conversely, an increase of r_c, as accompanies descent and compressional warming, absorbs latent heat that decreases θ.

Integrating (5.25) obtains

$$\theta e^{\left(\frac{lr_c}{c_p T}\right)} = \text{const}, \tag{5.26}$$

which describes a family of paths in the state space of saturated air. It is analogous to the family of adiabats described by Poisson's equation (2.30) for unsaturated air. The latter was used to introduce the state variable potential temperature, which is preserved during an adiabatic process. The identity (5.26) is now used to introduce another state variable, which is preserved during a pseudo-adiabatic process.

Evaluating (5.26) at a reference state of zero pressure (toward which r_c approaches zero faster than does the parcel's temperature) yields

$$\frac{\theta_e}{\theta} = e^{\left(\frac{lr_c}{c_p T}\right)}, \tag{5.27}$$

which defines the *equivalent potential temperature*. θ_e represents the maximum temperature the parcel could assume, namely, if it was raised to $z = \infty$, whereupon all of its water vapor condensed and released its latent heat, the resulting condensate was removed, and the parcel was then lowered adiabatically to the Earth's surface. According to (5.26), θ_e is constant during a pseudo-adiabatic process. Condensation is accompanied by a reduction of r_c and an increase of θ. However, the two vary in such proportion as to preserve θ_e. The same holds under evaporation.

In physical space, the material contour in Fig. 5.4 now separates from the isentrope ($\theta = $ const) with which it coincided initially. Because of the release of latent heat, it advances to isentropes of higher θ. Yet, it remains coincident with the isopleth of θ_e with which it coincided initially.

Like θ, $\theta_e = \theta_e(p, T, r)$ is a state variable. θ is conserved along an adiabat in state space below the LCL. θ_e is conserved along a *pseudo* or *saturated adiabat* in state space above the LCL. An adiabatic process involving no transformation of phase is also pseudo-adiabatic. Therefore, θ_e is conserved under unsaturated conditions as well. However, the definition (5.27) can be applied only under saturated conditions because only then does $r = r_c$, as is implicit in the derivation of θ_e. Alternatively, because it is conserved, θ_e can be calculated with r in place of r_c if T is replaced by the parcel's temperature at the LCL, where it just becomes saturated.

5.4.3 The Saturated Adiabatic Lapse Rate

The temperature of a dry air parcel decreases with its altitude at the dry adiabatic lapse rate Γ_d. To a good approximation, the same holds for a moist air parcel under unsaturated conditions – because the trace abundance of water vapor modifies thermal properties of air only slightly. Under saturated conditions, the adiabatic description of air breaks down due to the release of latent heat that accompanies the transformation of water from one phase to another. Latent heat exchanged with the gas phase then offsets adiabatic cooling and warming, which accompany ascending and descending motion.

An approximate description of how the temperature of a saturated parcel changes with altitude can be derived from the First Law with the aid of (5.19). From (2.35), the First Law for the gas phase may be expressed

$$c_p d\ln T - R d\ln p = \frac{\delta q}{T}. \tag{5.28}$$

If the parcel is unsaturated, (5.28) recovers the dry adiabatic lapse rate (2.33). If it is saturated, the heat transferred to the gas phase is given by

$$\delta q = -l \, dr_c. \tag{5.29}$$

Then (5.28) becomes

$$c_p d\ln T - R d\ln p = -\frac{l}{T} dr_c,$$

where l is treated as constant. With hydrostatic equilibrium (1.16), this reduces to

$$c_p dT + g dz = -l \, dr_c. \tag{5.30}$$

Strictly, the saturation mixing ratio depends on both pressure and temperature, so

$$dr_c = \left(\frac{\partial r_c}{\partial T}\right)_p dT + \left(\frac{\partial r_c}{\partial p}\right)_T dp.$$

However, the strong dependence on temperature conveyed from the Claussius-Clapeyron equation through (5.7) is the dominant influence on r_c. If its dependence on pressure is ignored, the change of saturation mixing ratio can be written

$$dr_c = \frac{dr_c}{dT} dT,$$

where T and z are understood to refer to the displaced parcel. Then (5.30) reduces to

$$c_p dT + g dz = -l \frac{dr_c}{dT} dT$$

or

$$\left(c_p + l \frac{dr_c}{dT}\right) dT + g dz = 0.$$

In terms of the dry adiabatic lapse rate, this may be expressed

$$\left(1 + \frac{l}{c_p}\frac{dr_c}{dT}\right)dT + \Gamma_d dz = 0. \tag{5.31}$$

Then the saturated parcel's temperature decreases with altitude according to

$$-\frac{dT}{dz} = \frac{\Gamma_d}{1 + \frac{l}{c_p}\frac{dr_c}{dT}} \tag{5.32}$$

$$= \Gamma_s,$$

which defines the *saturated adiabatic lapse rate*.

Unlike Γ_d, Γ_s varies with the parcel's altitude as a result of the nonlinear dependence on T of r_c. However, because $\frac{dr_c}{dT} > 0$, (5.32) implies

$$\Gamma_s < \Gamma_d. \tag{5.33}$$

A parcel's temperature decreases with altitude slower under saturated conditions than under unsaturated conditions. This property of saturated vertical motion follows from the release of latent heat to the gas phase, which offsets cooling associated with adiabatic expansion. Although variable, the saturated adiabatic lapse rate has a value of $\Gamma_s \cong 5.5$ K km^{-1} for conditions representative of the troposphere. This is only slightly smaller than the global-mean lapse rate of the troposphere (Fig. 1.2). No accident, the resemblance follows from dynamical processes that are developed in Chaps. 6 and 7.

5.5 THE PSEUDO-ADIABATIC CHART

Thermodynamic processes associated with vertical motion are represented conveniently on a diagram of the state space of moist air. Shown in Fig. 5.5 is the *pseudo-adiabatic chart*, which displays, as functions of T and $-p^\kappa$,

(1) Adiabats: $\theta = $ const (heavy solid),
(2) Pseudo Adiabats: $\theta_e = $ const (heavy dashed), and
(3) Isopleths of Saturation Mixing Ratio $r_w = $ const (light solid).[6]

The ordinate $-p^\kappa$ reflects altitude (1.17), which is indicated at the right for the US Standard Atmosphere. Adiabats, along which a parcel evolves under unsaturated conditions and its temperature varies with altitude at the dry adiabatic lapse rate Γ_d, appear as straight lines in this representation (2.34). Pseudo-adiabats, along which a parcel evolves under saturated conditions and its temperature varies with altitude at the saturated adiabatic lapse rate Γ_s, are curved – but only weakly. Consistent with (5.33), pseudo-adiabats ($\theta_e = $ const) have a slope with respect to altitude that is everywhere smaller than that of adiabats ($\theta = $ const). Both are labeled in degrees K, which correspond to the constant values of θ and θ_e that characterize those paths.

---

[6] If ice is present, moisture properties actually depend on r_i. However, according to (4.39) and (4.40), isopleths of r_i differ from those of r_w only modestly.

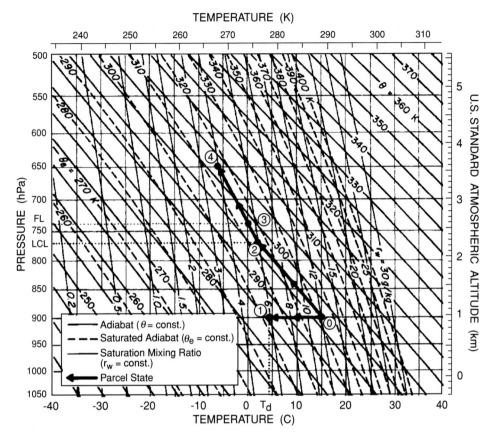

Figure 5.5 Pseudo-adiabatic chart illustrating thermodynamic processes for an ascending air parcel. Shown as functions of temperature and elevation are (1) adiabats (heavy solid), which are labeled by the constant values of θ characterizing those lines; (2) pseudo-adiabats (heavy dashed), which are labeled by the constant values of θ_e characterizing those curves; and (3) isopleths of saturation mixing ratio (light solid), which are labeled by the constant values of r_w defining those curves.

The use of the pseudo-adiabatic chart is best illustrated with an example. Consider conditions leading to the formation of a cumulus cloud. The cloud is fueled by moist air that ascends in a buoyant thermal, which is driven by surface heating (Fig. 5.4). At the surface, which is located at 900 hPa, air has a temperature of $T_0 = 15°C$ and a mixing ratio $r_0 = 6.0$ g kg^{-1}. Because air inside the thermal originates at the surface, thermodynamic properties inside the cloud, as well as those beneath the cloud, can be inferred from the evolution of an air parcel that is initially at the ground. Once the parcel's initial state (state 0) has been located on the pseudo-adiabatic chart, individual properties follow by allowing the system to evolve along certain paths in the state space of the parcel (indicated in Fig. 5.5).

Surface relative humidity

The saturation mixing ratio at the surface follows from the isopleth of r_w passing through state 0, which gives

$$r_{w0} = 12 \ g \ kg^{-1}.$$

The initial relative humidity of surface air is thus

$$RH_0 = \frac{6}{12} = 50\%.$$

Surface potential temperature

The parcel's initial potential temperature is determined by the adiabat passing through state 0, which is characterized by the constant value

$$\theta = \theta_0 = 297 \text{ K}.$$

Surface dew point

The dew point of surface air follows from isobaric cooling out of state 0. During that process, the parcel's state evolves along the isobar $p = 900$ hPa, which crosses isopleths of saturation mixing ratio toward lower values of r_w. Eventually, state 1 is reached, at which $r_w = r = 6$ g kg^{-1}. The parcel is then saturated. The temperature at which this occurs defines the dew point temperature of surface air

$$T_{d0} = 4 \text{ C}.$$

The above process is responsible for the formation of ground fog. The dew point spread of surface air, 11°C in this example, provides an indirect measure of the base of convective cloud. It reflects the amount of adiabatic cooling necessary to achieve saturation (Prob. 5.26).

Cumulus cloud base

The base of convective cloud corresponds to the LCL of surface air. The latter may be determined by displacing the parcel upward adiabatically. During that process, the parcel's state evolves along the adiabat passing through state 0 ($\theta = 297$ K). It too crosses isopleths of saturation mixing ratio toward lower values of r_w. Eventually, state 2 is reached, at which $r_w = r$ and the parcel is again saturated. The altitude where this occurs corresponds to the LCL

$$p_{LCL} = 770 \text{ hPa}.$$

The parcel's temperature at this altitude is 13 K colder than its initial temperature. However, its temperature is only 2 K colder than its initial dew point temperature. (It equals the parcel's dew point temperature at this altitude.) The small change of T_d between the surface and LCL reflects the weak pressure dependence of r_c.

Equivalent potential temperature at the surface

The equivalent potential temperature is determined once the LCL is located. Through state 2 (and along subsequent states) passes a pseudo adiabat. It defines θ_e for the parcel. Because θ_e is conserved under both saturated and unsaturated conditions, that value is also the equivalent potential temperature of the parcel below the LCL, so

$$\theta_{e0} = 315 \ K$$

at 900 hPa. Note that, even though it is conserved throughout, θ_e must be inferred at and above the LCL, for reasons discussed in Sec. 5.4.2.

Freezing level of surface air

At the LCL, the parcel's temperature is greater than 0°C. Consequently, the freezing level lies higher, inside the cloud. Lower portions of the cloud contain water droplets, whereas higher portions contain ice particles.[7] Conditions inside the cloud may be determined by displacing the parcel above the LCL. The parcel's state then evolves along the pseudo adiabat passing through state 2. From there, the $\theta_e = 315$ K pseudo adiabat crosses adiabats toward higher θ and isopleths of saturation mixing ratio toward lower r_w. Eventually, it reaches state 3, where the temperature is $0°$C. Defining the freezing level, that condition is achieved at

$$p_{FL} = 740 \text{ hPa}.$$

Liquid water content at the freezing level

Because the parcel is saturated, its mixing ratio must equal the saturation mixing ratio at the freezing level. The isopleth of saturation mixing ratio passing through state 3 gives for the mixing ratio there

$$r_{FL} = 5.5 \text{ g kg}^{-1}.$$

If no precipitation occurs, the total water content of the parcel is preserved. Therefore, the liquid water content at the freezing level is given by

$$r_l = 0.5 \text{ g kg}^{-1}.$$

Approximately 10% of the parcel's moisture has condensed by this altitude.

Temperature inside cloud at 650 hPa

At 650 hPa, the parcel has evolved along the $\theta_e = 315$ K pseudo adiabat to state 4, where its temperature is

$$T_{650} = -6 \text{ C}.$$

Were the parcel perfectly dry, it would have continued to evolve above the LCL (770 hPa) along the 315 K adiabat. Without the release of latent heat to offset adiabatic cooling, the parcel's temperature would then have decreased more rapidly, resulting in a temperature at 650 hPa of $-12°$C.

Mixing ratio inside cloud at 650 hPa

Because the parcel remains saturated, the isopleth of saturation mixing ratio that passes through state 4 gives for the mixing ratio at 650 hPa

$$r_{650} = 4.0 \text{ g kg}^{-1}.$$

By this level, one third of the parcel's water vapor has been transformed into condensate.

This example illustrates the strong constraint on water vapor imposed by thermodynamics and hydrostatic stratification. Inside convective tower, which transports moisture upward from the Earth's surface, the abundance of vapor that can be supported in solution with dry air decreases with altitude (Fig. 5.3). By 500 hPa, less than 30% of the surface mixing ratio of water inside the parcel under consideration remains as vapor. According to (1.24), the absolute humidity ρ_v decreases with altitude even faster. Less than 15% of the absolute concentration of water vapor remains after

---

[7] Ice forms only in the presence of a special type of aerosol particle, referred to as a *freezing nucleus* (Chap. 9). Because freezing nuclei are comparatively rare, many cloud droplets are actually "supercooled": They remain liquid at temperatures below $0°$C.

the parcel has been displaced to the middle troposphere. It is for this reason that deep convection increases the column abundance of water vapor in Fig. 1.19 only modestly. Much of Σ_{H_2O} resides in the lowest 1–2 km, where water vapor is distributed in the horizontal more uniformly than it is at higher altitude (cf. Fig. 1.18).

Through this process, water vapor that is transported upward inside convective tower is systematically removed. Condensation leads to a decrease with altitude of mixing ratio and, via precipitation, a similar decrease of total water content. Outside convective tower, air is even drier – because it has not recently been in contact with the source of water vapor at the Earth's surface or because it has been dehydrated inside convective tower. By restricting vertical transport of water, thermodynamics, together with hydrostatic stratification, confines moisture to a shallow neighborhood of the Earth's surface. It thereby isolates water from photo-dissociating radiation at higher altitude, a process that has preserved water on the Earth.

SUGGESTED REFERENCES

A complete discussion of the Gibbs-Dalton law, along with its implications to a multi-component system, is given in *Thermodynamics* (1970) by J. Keenan.

Atmospheric Thermodynamics (1981) by Iribarne and Godson contains a detailed treatment of moisture-dependent properties and a survey of thermodynamic charts.

The Principles of Chemical Equilibrium (1971) by K. Denbigh discusses the thermodynamics of condensed phases.

PROBLEMS

The pseudo-adiabatic chart in Appendix F is to be used only for those problems in which it is explicitly indicated.

1. Downslope winds in North America are called a *chinook*, an Indian term for "snow eater." A chinook often occurs with the mountains blanketed in cloud but with clear skies leeward. Consider the following synoptic situation: Moist air originating in the eastern Pacific is advected over the western United States, where it is forced over the continental divide. On the windward side, the surface lies at 800 hPa, where the temperature and mixing ratio are 20°C and 15 g kg^{-1}, respectively. If the summit lies at 600 hPa and if any condensate that forms precipitates out, determine (a) the surface air temperature at Denver, which lies at 830 hPa, (b) the mixing ratio at Denver, (c) the relative humidity at Denver.

2. A refrigerator having an interior volume of 2.0 m^3 is sealed and switched on. If the air initially has a temperature of 30°C and a relative humidity of 0.50, determine (a) the temperature at which condensation forms on the walls, (b) how much moisture will have condensed when the temperature reaches 2°C, (c) how much heat must be rejected to the surroundings to achieve the final state in (b).

3. State the number of thermodynamic degrees of freedom for (a) moist unsaturated air, (b) moist saturated air. (c) Explain the numbers in (a) and (b) and why they differ.

4. Compare θ and θ_v for saturated conditions at 1000 hPa and a temperature of (a) 10°C, (b) 20°C, (c) 30°C.

5. Warm moist air leaves an array of cooling towers at a power plant situated at 825 hPa. A cloud forms directly overhead. If the ambient temperature profile is isothermal, with $T = 5$°C, and if the initial temperature and mixing ratio of air leaving the towers are 30°C and 25 g kg^{-1}, respectively, determine (a) the relative

humidity immediately above the towers, (b) the dew point immediately above the towers, (c) the virtual potential temperature immediately above the towers; compare this value with the potential temperature and discuss it in relation to typical differences between θ and θ_v, (d) the pressure at cloud base, (e) the equivalent potential temperature at 800 hPa, (f) the mixing ratio at 700 hPa, (g) the pressure at cloud top. A pseudo-adiabatic chart is provided in Appendix F.

6. Room temperature on a given day is 22°C, whereas the outside temperature at 1000 hPa is 2°C. Calculate the maximum relative humidity that can be accommodated inside without room windows fogging, if the windows can be treated as having a uniform temperature.

7. (a) Under the same conditions as in (6), but for an aircraft cabin that is pressurized to 800 hPa. (b) As in (a), but noting that the interior pane is thermally isolated from the exterior pane.

8. Evaporation is an efficient means of cooling air and, through contact with other media, of rejecting heat to the environment. An evaporative cooler processes ambient air at 35°C and 900 hPa to produce room air with a relative humidity of 0.60. Calculate the temperature inside if the relative humidity outside is (a) 0.10, (b) 0.50.

9. A cooling tower situated at 1000 hPa processes ambient air to release saturated air at a temperature of 35°C and at a rate of 10 $m^3 s^{-1}$. Calculate the rate heat is rejected if ambient air has a temperature of 20°C and a relative humidity of (a) 0.20, (b) 0.80.

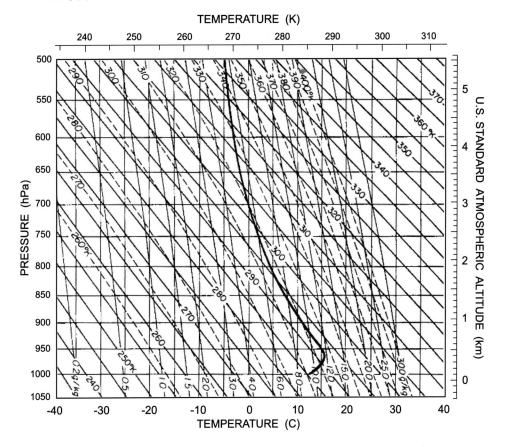

10. A morning temperature sounding is plotted above. Through absorption of SW radiation at the ground, the surface inversion ($\Gamma < 0$) that developed during the night is replaced during the day by adiabatic thermal structure in the lowest half kilometer. A cumulus cloud then forms over an asphalt parking lot at 1000 hPa, where the mixing ratio is 10 g kg^{-1} and the relative humidity is 50%. For the air column above the parking lot, determine the (a) surface air temperature, (b) pressure at the cloud base, (c) potential temperature at 900 hPa, (d) potential temperature at 700 hPa, (e) equivalent potential temperature at the surface, (f) mixing ratio at 700 hPa, (g) pressure at the cloud top, (i) mixing ratio at the cloud top. A pseudo-adiabatic chart is provided in Appendix F.

11. Moist air moves inland from a maritime region, where it has a temperature of 16°C and a relative humidity of 66%. Through contact with the ground, the air cools. Calculate the temperature at which fog forms.

12. Use the pseudo-adiabatic chart in Appendix F to determine the LCL above terrain at 1000 hPa for (a) moist surface conditions representative of the eastern US: $T = 30°C$ and $RH = 70\%$, (b) arid surface conditions representative of the southwestern US: $T = 30°C$ and $RH = 10\%$. (c) Contrast the heights of cumulus cloud bases under these conditions in relation to the likelihood of precipitation reaching the surface.

13. An altitude chamber is used to simulate a sudden decompression from normal cabin pressure: 800 hPa, to ambient pressure at 18,000′. If the initial temperature is 22°C, how small must the relative humidity be to avoid spontaneous cloud formation during the decompression? A pseudo-adiabatic chart is provided in Appendix F.

14. Outside air has a temperature of −10°C and a relative humidity of 0.50. (a) What is the relative humidity indoors if the room temperature is 22°C and if air is simply heated, without humidification? (b) What mass of water vapor must be added to a room volume of 75 m^3 to elevate its relative humidity to 40%? (c) How much energy is required to achieve this state?

15. On a given day, the lapse rate and relative humidity are constant, with $\Gamma = 8°K$ km^{-1} and $RH = 0.80$. If the surface temperature at 1000 hPa is 20°C, (a) estimate the total precipitable water vapor. (b) Below what height is 90% of the water vapor column represented? (c) Now calculate the precipitable water vapor inside a column of ascending air in a cumulonimbus tower that extends to 100 hPa. A pseudo-adiabatic chart is provided in Appendix F.

16. A morning temperature sounding over Florida reveals the profile

$$T = \begin{cases} 10 - 6(z - 1) \,(°C) & z \geq 1 \text{ km} \\ 20 - 10z \,(°C) & z < 1 \text{ km.} \end{cases}$$

That afternoon, surface conditions are characterized by a temperature of 30°C and a mixing ratio of 20 g kg^{-1}. (a) To what height will cumulus clouds develop if air motion is nearly adiabatic? (b) Plot the vertical profile of mixing ratio from the surface to the top of a cumulus cloud. (c) Over the same altitude range, plot the vertical profile of mixing ratio but neglecting the explicit pressure dependence of saturation mixing ratio. (d) Discuss how the results in (a) and (b) would be modified if air inside the ascending plume is not adiabatic but, rather, mixes with surrounding air in roughly equal proportion. A pseudo-adiabatic chart is provided in Appendix F.

17. For the temperature sounding and surface conditions in (16), determine the height of convection and the vertical profile of temperature inside an ascending plume were the air perfectly dry. A pseudo-adiabatic chart is provided in Appendix F.

18. Graph the dew point temperature as a function of relative humidity for 1000 hPa and 20°C.

19. For reasons developed in Chap. 9, many clouds are supercooled (e.g., they contain liquid droplets at temperatures below 0°C). For cloud that is saturated with respect to water, calculate the relative humidity with respect to ice at a temperature of (a) -5°C, (b) -20°C.

20. Inspection of the pseudo-adiabatic chart reveals that, as $z \to \infty$, pseudo-adiabats ($\theta_e = $ const) approach adiabats ($\theta = $ const). Why?

21. The development in Sec. 5.1 for a mixture of dry air and water considers the presence of only one condensed phase. Why and under what conditions is this permissible?

22. A cold front described by the surface

$$z = h(x - ct)$$

moves eastward with velocity c and undercuts warmer air ahead of it. Far ahead of the front, undisturbed air is characterized by the temperature and mixing ratio profiles $T_\infty(z)$ and $r_\infty(z)$, respectively. If air is lifted in unison over the frontal surface, for a fixed location x, derive an expression for (a) the variation of temperature with altitude above the frontal surface but beneath the LCL as a function of time, (b) the variation of mixing ratio with altitude above the frontal surface but beneath the LCL as a function of time, (c) an expression for the altitude of the LCL if, far from the frontal zone, pressure decreases with altitude approximately as $e^{-\frac{z}{H}}$, with $H = $ const.

23. Show that the exponent in (5.26) vanishes for an air parcel that is displaced to the top of the atmosphere.

24. Revisit Prob. 2.14, recognizing now that the bubble is surrounded by water. If the bubble is saturated throughout its ascent, yet remains adiabatic, calculate the bubble's temperature on reaching the surface.

25. An air parcel drawn into a cumulonimbus tower becomes saturated at 900 hPa, where its temperature is 20°C. Calculate the fractional internal energy represented by condensate when the parcel has been displaced to 100 hPa. A pseudo-adiabatic chart is provided in Appendix F.

26. The base of convective cloud may be estimated from the dew point spread at the surface. (a) Derive an expression for the dew point temperature T_d of an ascending parcel as a function of its height in terms of the surface mixing ratio r_0, presuming environmental conditions to be isothermal. (b) Show that T_d for the parcel varies with height approximately linearly in the lowest few kms. (c) Use the lapse rate of T_d in (b) to derive an expression for the height of the LCL in terms of the dew point spread at the surface: $(T - T_d)_0$. (d) Use the result of (c) to estimate the cloud base under the conditions in (16).

CHAPTER SIX

Hydrostatic equilibrium

Changes of thermodynamic state that accompany vertical motion follow from the distribution of atmospheric mass, which is determined ultimately by gravity. In the absence of motion, Newton's second law applied to the vertical reduces to a statement of *hydrostatic equilibrium* (1.16). Gravity is then balanced by the vertical pressure-gradient force. This simple form of mechanical equilibrium is accurate even in the presence of motion because the acceleration of gravity is, almost invariably, much greater than vertical acceleration of individual air parcels. Only inside deep convective towers and other small-scale phenomena is vertical acceleration large enough to invalidate hydrostatic equilibrium.

Because it is such a strong body force, gravity must be treated with some care. Complications arise from the fact that the gravitational acceleration experienced by an air parcel does not act purely in the vertical. It also varies with location. According to the preceding discussion, gravity is large enough to overwhelm other contributions in the balance of vertical forces. The same holds for the balance of horizontal forces. Horizontal components of gravity that are introduced by the Earth's rotation and other sources must be balanced by additional horizontal forces. Unrelated to air motion, those additional forces unnecessarily complicate the description of atmospheric motion.

6.1 EFFECTIVE GRAVITY

In the reference frame of the Earth, the gravitational acceleration experienced by an air parcel involves three basic contributions:

(1) radial gravitation by the Earth's mass,

(2) centrifugal acceleration due to rotation of the reference frame, and

(3) anisotropic contributions,

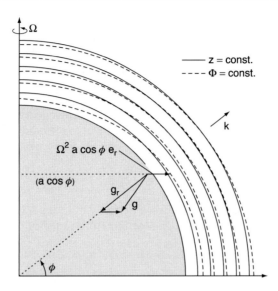

Figure 6.1 Effective gravity g illustrated in relation to its contributions from radial gravitation and centrifugal acceleration. Also shown are surfaces of constant geometric altitude z and corresponding surfaces of constant geopotential Φ, to which g is orthogonal.

as illustrated in Fig. 6.1. Radial gravitation by the Earth's mass is the dominant contribution. Under idealized circumstances, that component of g acts perpendicular to surfaces of constant geometric altitude z, which are concentric spheres. The Earth's rotation introduces another contribution. Because it is noninertial, the reference frame of the Earth includes a centrifugal acceleration. It acts perpendicular to and away from the axis of rotation. Lastly, departures of the Earth from sphericity and homogeneity (e.g., through variations of surface topography) introduce anisotropic contributions to gravity. They must be determined empirically.

Collectively, these contributions determine the *effective gravity*:

$$g(\lambda, \phi, z) = \frac{a^2}{(a+z)^2} g_0 k + \Omega^2 (a+z) cos\phi\, e_r + \epsilon(\lambda, \phi, z), \tag{6.1}$$

where a is the mean radius of the Earth, g_0 is the radial gravitation at mean sea level, k is the local upward unit normal, Ω is the Earth's angular velocity, λ and ϕ are longitude and latitude, respectively, e_r is a unit vector directed outward from the axis of rotation, and $\epsilon(\lambda, \phi, z)$ represents all anisotropic contributions. Even if the Earth were perfectly spherical, effective gravity would not act uniformly across the surface of the Earth. Nor would it act entirely in the vertical. According to (6.1), the centrifugal acceleration deflects g into the horizontal, introducing a component of gravity along surfaces of constant z. The centrifugal acceleration also causes the vertical component of g to vary with location, from a minimum of 9.78 m s^{-2} at the equator to a maximum of 9.83 m s^{-2} at the poles. Anisotropic contributions, although more complicated, have a similar effect.

These contributions introduce horizontal components of gravity. They must be balanced by other forces that act along surfaces of constant geometric altitude. Thus,

(1.16) represents but one of several component equations that are required to describe hydrostatic equilibrium. Horizontal forces that are introduced by gravity can overshadow those which actually control the motion of an air parcel. For instance, rotation introduces a horizontal component that must be balanced by a variation of pressure along surfaces of constant geometric altitude. That horizontal pressure gradient exists even under static conditions, wherein an observer in the reference frame of the Earth would anticipate Newton's second law to reduce to simple hydrostatic balance in the vertical. To avoid such complications, it is convenient to introduce a coordinate system that consolidates all components of gravity into a single vertical component.

6.2 GEOPOTENTIAL COORDINATES

Gravity is conservative. The specific work performed during a cyclic displacement of mass through the Earth's gravitational field therefore vanishes

$$\Delta w_g = - \oint \boldsymbol{g}(\lambda, \phi, z) \cdot d\boldsymbol{\xi} = 0, \tag{6.2}$$

where $d\boldsymbol{\xi}$ denotes the incremental displacement of an air parcel. According to Sec. 2.1.4, (6.2) is a necessary and sufficient condition for the existence of an exact differential

$$
\begin{aligned}
d\Phi &= \delta w_g \\
&= -\boldsymbol{g} \cdot d\boldsymbol{\xi},
\end{aligned}
\tag{6.3}
$$

which defines the gravitational potential or *geopotential* Φ. In (6.3), $d\Phi$ equals the specific work performed against gravity to complete the displacement $d\boldsymbol{\xi}$. Then, per discussion in Sec. 2.1.4, gravity is an irrotational vector field that may be expressed in terms of its potential function

$$\boldsymbol{g} = -\nabla\Phi. \tag{6.4}$$

Surfaces of constant geopotential are not spherical like surfaces of constant geometric altitude (Fig. 6.1). Centrifugal acceleration in the rotating reference frame of the Earth distorts geopotential surfaces into oblate spheroids. The component of $\nabla\Phi$ along surfaces of constant geometric altitude introduces a horizontal component of gravity. It must be balanced by a horizontal pressure gradient force.

Representing gravity is simplified by transforming from pure spherical coordinates, in which elevation is fixed along surfaces of constant geometric altitude, to geopotential coordinates, in which elevation is fixed along surfaces of constant geopotential. Geopotential surfaces may be used for coordinate surfaces because Φ increases monotonically with altitude (6.3), ensuring a one-to-one relationship between those variables. Introducing the aforementioned transformation is tantamount to measuring elevation z along the line of effective gravity. The latter will be termed *height* to distinguish it from geometric altitude. Defining z in this manner agrees with the usual notion of local vertical being plumb with the line of gravity. It will be adopted as the convention hereafter.

In geopotential coordinates, gravity has no horizontal component. Consequently, in terms of height, (6.4) reduces to simply

$$d\Phi = g dz, \tag{6.5}$$

where g denotes the magnitude of g. The latter acts in the direction of decreasing height - not geometric altitude. Using mean sea level as a reference elevation[1] yields an expression for the absolute geopotential

$$\Phi(z) = \int_0^z g \, dz'. \tag{6.6}$$

Having dimensions of specific energy, the geopotential represents the work that must be performed against gravity to raise a unit mass from mean sea level to a height z. Evaluating Φ from (6.6) requires measured values of effective gravity throughout. However, because it is a point function, Φ is independent of path by (6.2) and the Exact Differential Theorem (2.7). Therefore, Φ depends only on height z. For the same reason, surfaces of constant geopotential coincide with surfaces of constant height.

Using surfaces of constant geopotential for coordinate surfaces consolidates gravity into the vertical coordinate, height. However, g still varies with z through (6.5). To account for that variation, it is convenient to introduce yet another vertical coordinate in which the dependence on height is absorbed. *Geopotential height* is defined as

$$Z = \frac{1}{g_0} \int_0^z g \, dz$$
$$= \frac{1}{g_0} \Phi(z), \tag{6.7}$$

where $g_0 = 9.8 \, \mathrm{m^2 \, s^{-1}}$ is a reference value reflecting the average over the surface of the Earth. Then

$$dZ = \frac{1}{g_0} d\Phi$$
$$= \frac{g}{g_0} dz. \tag{6.8}$$

Because it accounts for the variation of gravity, geopotential height simplifies the expression for hydrostatic balance. In terms of Z, (1.16) reduces to

$$dp = -\rho g \, dz$$
$$= -\rho g_0 \, dZ. \tag{6.9}$$

When elevation is represented in terms of geopotential height, gravity is described by the constant value g_0.

While having these formal advantages, geopotential height is nearly identical to height because $g \cong g_0$ throughout the homosphere. In fact, the two measures of elevation differ by less that 1% below 60 km. Therefore, Z may be used interchangeably with z. This will be our convention hereafter. Then g is understood to refer to the constant reference value g_0.

6.3 HYDROSTATIC BALANCE

In terms of geopotential height, hydrostatic balance is expressed by

$$g \, dz = -v \, dp. \tag{6.10}$$

---

[1] Mean sea level is nearly coincident with a surface of constant geopotential.

The ideal gas law transforms this into

$$dz = -\frac{RT}{g}dlnp$$

$$= -Hdlnp,$$

(6.11)

where the scale height H is defined in (1.17). According to (6.11), the change of height is proportional to the local temperature and to the change of $-lnp$. As vertical changes of temperature and H are small compared with changes of pressure, the quantity $-lnp$ serves as a dimensionless measure of height.

6.3.1 Hypsometric equation

Consider a layer bounded by two isobaric surfaces $p = p_1(x, y, z)$ and $p = p_2(x, y, z)$, as illustrated in Fig. 6.2. Integrating (6.11) between those surfaces obtains

$$\Delta z = z_2 - z_1 = -\langle H \rangle ln\left(\frac{p_2}{p_1}\right),$$

(6.12.1)

where

$$\langle H \rangle = \frac{R}{g}\langle T \rangle$$

(6.12.2)

and

$$\langle T \rangle = \frac{\int_{p_1}^{p_2} T\,dlnp}{\int_{p_1}^{p_2} dlnp}$$

(6.12.3)

define the layer-mean scale height and temperature, respectively, and $z_1 = z(x, y, p_1)$ and $z_2 = z(x, y, p_2)$ denote the heights of the bounding isobaric surfaces. Thus

$$z_2 - z_1 = \frac{R}{g}\int_{p_2}^{p_1} T\,dlnp.$$

(6.12.4)

Known as the *hypsometric equation*, (6.12) asserts that the *thickness* of a layer bounded by two isobaric surfaces is proportional to the mean temperature of that layer and the pressure difference across it.

In regions of cold air, the e-folding scale H is short. Pressure then decreases upward sharply (1.17). The vertical separation of isobaric surfaces in such regions is therefore compressed (Fig. 6.2). In regions of warm air, H is tall. Pressure then decreases upward slowly. The separation of isobaric surfaces in those regions is therefore expanded.

6.3.2 Meteorological Analyses

If $z_1 = 0$, the hypsometric equation provides the height of the upper surface $z_2 = z(x, y, p_2)$. For example, the height of the 500-hPa surface in Fig. 1.9 follows as a vertical integral of temperature. It may be determined from measurements of temperature between that surface and mean sea level ($\sim$1000 hPa) or, alternatively, between that surface and a reference isobaric surface, if the height of the reference surface is known. Such temperature measurements are made routinely by rawinsondes (Sec. 1.6.1). Also measuring air motion and humidity, rawinsondes ascend via balloon twice-daily. They provide coverage that is dense over populated continent, but sparse over ocean and

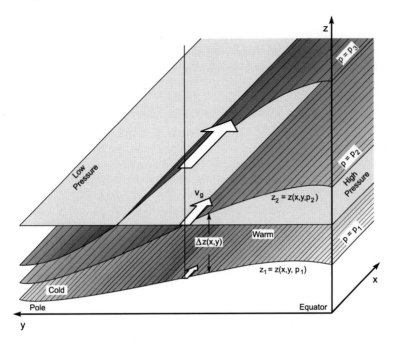

Figure 6.2 Isobaric surfaces (stippled) in the presence of a horizontal temperature gradient. Vertical spacing of isobaric surfaces is compressed in cold air and expanded in warm air, which makes the height of an individual isobaric surface low in the former and high in the latter. Because pressure decreases upward monotonically, a surface of constant height (shaded) then has low pressure where isobaric height is low and high pressure where isobaric height is high. Therefore, contours of height for an individual isobaric surface may be interpreted similarly to isobars on a constant height surface. Also indicated is the geostrophic velocity v_g (Sec. 12.1), which is directed parallel to contours of isobaric height and has magnitude proportional to its gradient.

in the tropics (Fig. 1.37). Temperature measurements are also made remotely by satellite. Satellite measurements of temperature provide continuous global coverage, albeit asynoptic.[2] Satellite retrievals also have coarser vertical resolution ($\sim$4-8 km) than that provided by rawinsondes ($\sim$100 m). Both are assimilated operationally in *synoptic analyses* of the instantaneous circulation, like those presented in Figs. 1.9 and 1.10.

Synoptic analyses fall into two categories:

(1) The *surface analysis* displays properties at the ground, but extrapolated to a single reference level: mean sea level. It thus displays sea level pressure (SLP) on a surface of constant height $z = 0$.

---

[2] The term *synoptic* refers to the simultaneous distribution of some property, for example, a snapshot over the Earth of the instantaneous height of the 500-hPa surface (Fig. 1.9a). A polar-orbiting satellite, however, observes different sites at different times. Termed *asynoptic*, such measurements sample the entire Earth only after 12–24 hours.

(2) An *upper air analysis* displays properties on a surface of constant pressure. Accordingly, an upper air analysis is an *isobaric analysis*: It represents a quasi-horizontal section of the circulation along a surface where $p = $ const.[3]

Figure 1.9 displays the 500-hPa analysis. On an isobaric surface, displaying pressure is uninformative; it's constant. Displayed instead is the height of the 500-hPa surface. This representation displays height on a surface of constant p. It thus interchanges the roles of p and z. Elevation is then represented by pressure, which becomes the vertical coordinate (e.g., $p = 700$ hPa, 500 hPa, 300 hPa, etc).

In isobaric coordinates, pressure becomes the independent variable. The height of an isobaric surface then becomes the dependent variable, for example, $z_{500} = z(x, y, 500$ hPa$)$. A theoretical basis for using pressure to describe elevation follows from the hypsometric equation. Because H is positive, (6.12) implies a one-to-one relationship between height and pressure: A particular height corresponds to a particular pressure. Pressure may therefore be used alternatively to height as the vertical coordinate. Isobaric coordinates, developed in Chap. 11, use surfaces of constant pressure for coordinate surfaces. Although they evolve with the circulation, isobaric coordinates afford several advantages.

The horizontal distribution of height on an isobaric surface, $z = z(x, y, p)$, may be interpreted analogously to the distribution of pressure on a surface of constant height, $p = p(x, y, z)$, for example, on the mean height of that isobaric surface. Pressure decreases with height monotonically. Consequently, low height of an isobaric surface corresponds to low pressure on a surface of constant height (Fig. 6.2). Just the reverse applies to high values. For this reason, contours of height on an isobaric surface mirror contours of pressure on a surface of constant height. From hydrostatic equilibrium, that pressure represents the weight of the overlying atmospheric column. The distribution of isobaric height therefore reflects the horizontal distribution of atmospheric mass.

According to Fig. 1.9, the height of the 500-hPa surface slopes downward toward the pole at mid-latitudes, where z_{500} decreases from 5900 m to 5000 m. The decrease of 500-hPa height mirrors the poleward decrease of layer-mean temperature between the surface and 500 hPa. $\langle T \rangle_{500}$ is plotted in Fig. 6.3 for the same time as in Fig. 1.9a. The wavy region where $\langle T \rangle_{500}$ and z_{500} decrease sharply delineates the *polar front*. It separates cold polar air from warmer air at lower latitude. As is apparent from Fig. 1.9a, that region of sharp temperature gradient coincides with strong westerlies of the jet stream, which are tangential to contours of isobaric height. The jet stream thus marks the interface between cold polar air and warmer air at lower latitude. A steepening of the meridional temperature gradient steepens the gradient of 500-hPa height (6.12). That, in turn, contracts the horizontal spacing of height contours in Fig. 1.9a. For reasons developed in Chap. 12, this contraction is accompanied by an intensification of the jet.

Meridional gradients of temperature and height are particularly steep east of Asia and North America. Those features coincide with the North Pacific and North Atlantic storm tracks. They are manifest in the time-mean circulation (Fig. 1.9b) as local

---

[3] This representation is motivated by observational considerations. Sounding measurements at different sites are returned at specified values of pressure, which are interpolated seamlessly along an isobaric surface.

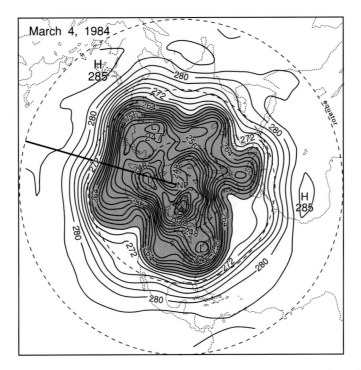

Figure 6.3 Mean temperature (°K) of the layer between 1000 hPa and 500 hPa for March 4, 1984. Bold solid line marks a meridional cross section displayed in Fig. 6.4.

intensifications of the jet stream. Anomalies of cold air that punctuate the polar front in Fig. 6.3 (e.g., the cold outbreak over the North Atlantic) introduce synoptic-scale depressions of the 500-hPa surface in Fig. 1.9a. By deflecting contours of isobaric height, those features disrupt the jet stream. In their absence, the jet would be nearly zonal (e.g., Fig. 1.9b).

Contours of 500-hPa height that delineate the polar front in Fig. 1.9a correspond to isobars on the constant-height surface: $z \sim 5.5$ km, with pressure decreasing toward the pole. Thus less atmospheric column lies above that elevation poleward of the front than equatorward of the front. Figure 6.4 presents a meridional cross section of thermal structure and motion in Figs. 6.3 and 1.9a. The horizontal distribution of mass just noted results from compression of isobaric surfaces poleward of the front, which is positioned near 40 N, and expansion of those surfaces equatorward of the front. This introduces a tilt to isobaric surfaces at midlatitudes, one which steepens with height. Found at the same latitudes is the jet stream. Like the tilt of isobaric surfaces, westerlies intensify upward, exceeding 75 m/s near the tropopause. Similar behavior is evident at high latitude. It marks the base of the polar-night jet, which intensifies in the stratosphere.

6.4 STRATIFICATION

The lapse rates Γ_d and Γ_s, (2.33) and (5.32), describe the evolution of a displaced air parcel under unsaturated and saturated conditions, respectively. They apply formally

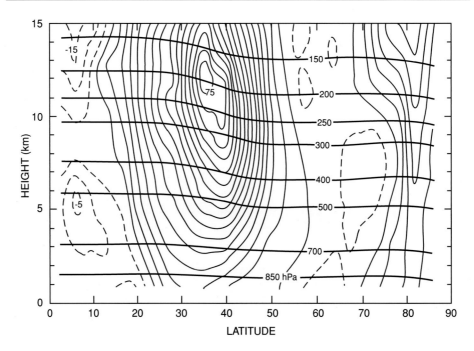

Figure 6.4 Meridional cross section of thermal structure and motion in Figs. 1.9a and 6.3 at the position indicated in the latter. Vertical spacing of isobaric surfaces (heavy solid curves) is expanded at low latitude and compressed at high latitude, introducing a slope at mid-latitudes that steepens with height. Zonal wind speed (contoured in m s^{-1}) delineates the subtropical jet, which coincides with the steep slope of isobaric surfaces. Both intensify upward to a maximum at the tropopause. Contour increment: 5 m s^{-1}.

with z representing geopotential height and g constant. Neither, however, has a direct relationship to the temperature of the parcel's environment – because, under adiabatic conditions, a displaced parcel is thermally isolated.

Thermal properties of the environment are dictated by the history of air at a given location, namely, by where that air has been and what thermodynamic influences it has experienced. The *environmental lapse rate* is defined as

$$\Gamma = -\frac{dT}{dz},$$ (6.13)

where T refers to the ambient temperature. Like parcel lapse rates, $\Gamma > 0$ corresponds to temperature decreasing with height. Conversely, $\Gamma < 0$ corresponds to temperature increasing with height. In the latter case, the profile of environmental temperature is said to be *inverted*.

The compressibility of air leads to atmospheric mass being stratified, as is reflected in the sharp upward decrease of density and pressure (Fig. 1.2). Because the environment is in hydrostatic equilibrium, the distribution of pressure is related to thermal structure through the hypsometric relation. Incorporating (6.11) transforms (6.13) into

$$-\frac{1}{H}\frac{dT}{dlnp} = \Gamma$$

or

$$\frac{d\ln T}{d\ln p} = \frac{R}{g}\Gamma$$
$$= \kappa \frac{\Gamma}{\Gamma_d}. \tag{6.14}$$

6.4.1 Idealized stratification

For certain classes of thermal structure, (6.14) can be integrated analytically to obtain the distributions of pressure and other thermal properties inside an atmospheric layer.

Layer of constant lapse rate

If $\Gamma = \text{const} \neq 0$, (6.14) yields

$$\frac{T}{T_s} = \left(\frac{p}{p_s}\right)^{\frac{\Gamma}{\Gamma_d}\kappa}, \tag{6.15}$$

where the subscript s refers to the base of the layer. Because

$$T(z) = T_s - \Gamma z, \tag{6.16}$$

(6.15) may be used to obtain the vertical distribution of pressure

$$\frac{p}{p_s} = \left[1 - \kappa\frac{\Gamma}{\Gamma_d}\left(\frac{z}{H_s}\right)\right]^{\frac{\Gamma_d}{\kappa\Gamma}}. \tag{6.17}$$

Then (2.31) implies the vertical profile of potential temperature

$$\theta(z) = (T_s - \Gamma z)\left[1 - \kappa\frac{\Gamma}{\Gamma_d}\left(\frac{z}{H_s}\right)\right]^{-\frac{\Gamma_d}{\Gamma}}, \tag{6.18}$$

where $p_s = p_0 = 1000$ hPa has been presumed for the base of the layer.

As shown in Fig. 6.5, the compressibility of air makes pressure decrease with height for all Γ. The same is true of density. However, potential temperature decreases with height only for $\Gamma > \Gamma_d$. It increases with height for $\Gamma < \Gamma_d$. In both cases, θ varies sharply with height because the pressure term in (6.18) dominates over the temperature term. As a result, environmental potential temperature can change by several hundred Kelvin in just a couple scale heights. Such thermal structure is typical in the stratosphere (cf. Fig. 18.19). In contrast, the potential temperature of a displaced air parcel is independent of height, at least under unsaturated conditions.

Thermal structure described by (6.16)–(6.18) can apply locally (e.g., inside a layer), even if temperature does not vary linearly throughout the atmosphere. Should it apply for the entire range of height, (6.16) has one further implication. For $\Gamma > 0$, positive temperature requires the atmosphere to have a finite upper bound

$$z_{top} = \frac{T_s}{\Gamma}, \tag{6.19}$$

where p vanishes. For $\Gamma \leq 0$, no upper bound exists.

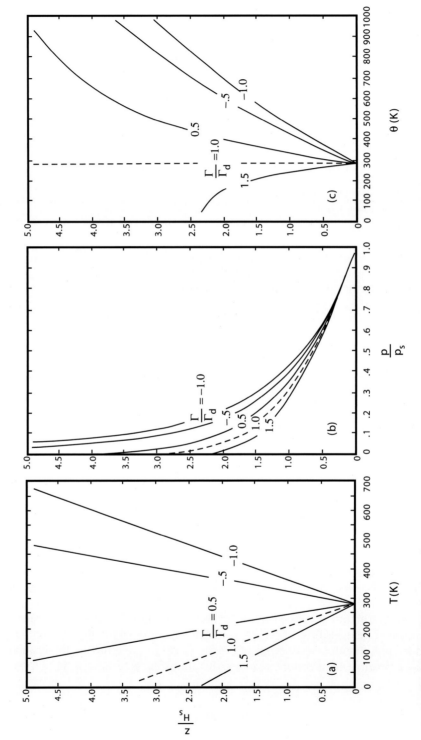

Figure 6.5 Vertical profiles of (a) temperature, (b) pressure, and (c) potential temperature inside layers of constant lapse rate Γ.

Isothermal layer

For the special case $\Gamma = 0$, (6.17) is indeterminate. Reverting to the hydrostatic relation (6.11) yields

$$z = -H ln \left(\frac{p}{p_s} \right)$$

or

$$\frac{p}{p_s} = e^{-\frac{z}{H}}, \tag{6.20}$$

which is identical to (1.17) under these circumstances. Then the vertical profile of potential temperature is simply

$$\theta(z) = T_s e^{\kappa \frac{z}{H}} \\ = \theta_s e^{\kappa \frac{z}{H}}. \tag{6.21}$$

Under isothermal conditions, environmental potential temperature increases by a factor of e every $\kappa^{-1} \cong 3$ scale heights.

Adiabatic layer

For the special case $\Gamma = \Gamma_d$, (6.15) reduces to

$$\frac{T}{T_s} = \left(\frac{p}{p_s} \right)^{\kappa}. \tag{6.22}$$

The distribution of pressure (6.17) then becomes

$$\frac{p}{p_s} = \left[1 - \kappa \frac{z}{H_s} \right]^{\frac{1}{\kappa}}. \tag{6.23}$$

Equation (6.22) is the same relationship between temperature and pressure implied by Poisson's equation (2.30.2) that defines potential temperature (2.31). Accordingly, the vertical profile of potential temperature (6.18) reduces to

$$\theta(z) = \theta_s = \text{const.} \tag{6.24}$$

6.5 LAGRANGIAN INTERPRETATION OF STRATIFICATION

As noted earlier, hydrostatic equilibrium applies in the presence of motion as well as under static conditions. Therefore, each of the stratifications in Sec. 6.4 is valid even if a circulation is present, as is invariably the case. Under those circumstances, vertical profiles of temperature, pressure, and potential temperature (6.16)–(6.18) correspond to horizontal-mean thermal structure. They may be interpreted as averages over many ascending and descending air parcels. Interpreting thermal structure in this manner provides some insight into the mechanisms that shape mean stratification.

For a layer of constant lapse rate, the relationship between temperature and pressure (6.15) resembles one implied by Poisson's equation (2.30.2), but for a polytropic process (Sec. 2.5.1) with $\frac{\Gamma}{\Gamma_d} \kappa$ in place of κ. Consequently, we may associate the thermal structure in (6.16)–(6.18) with a vertical rearrangement of air in which individual parcels evolve diabatically according to a polytropic process. In that description, air parcels moving vertically exchange heat with their surroundings in such proportion for their temperatures to vary linearly with height. For an individual air parcel, the foregoing process has a polytropic specific heat that satisfies

$$\frac{R}{(c_p - c)} = \frac{\Gamma}{\Gamma_d}\kappa$$

or

$$c = c_p \left(1 - \frac{\Gamma_d}{\Gamma}\right). \tag{6.25}$$

The corresponding heat transfer for the parcel is then described by (2.37). Its change of potential temperature follows from (2.40).

6.5.1 Adiabatic stratification: A paradigm of the troposphere

Consider a layer characterized by

$$\Gamma = \Gamma_d. \tag{6.26.1}$$

Such stratification is termed *adiabatic*. Then (6.25) implies

$$c = 0$$
$$\delta q = d\theta = 0. \tag{6.26.2}$$

Thus, individual air parcels evolve adiabatically. Under these circumstances, air parcels undergo vertical motion but do not interact thermally with their surroundings. Such behavior may be regarded as the limiting case of when vertical motion is fast compared with diabatic influences. It is typical of cumulus and sloping convection. Although the vertical motion of parcels is adiabatic, they may still exchange heat at the boundaries of the layer. In fact, in the absence of mechanical forcing, such heat transfer is necessary to maintain vertical motion.

The evolution of an individual parcel may be treated as a circuit that it follows between the upper and lower boundaries of the layer, where heat transfer is concentrated. Such a circuit is depicted in Fig. 6.6 for a layer representative of the troposphere. At the lower boundary, an individual parcel moves horizontally long enough for heat transfer to occur. Isobaric heat absorption (e.g., through absorption of LW radiation and transfer of sensible heat from the surface) increases the parcel's potential temperature (Fig. 6.7a). The accompanying increase of temperature (Fig. 6.7b) causes the parcel to become positively buoyant and rise. If the time scale of vertical motion is short compared with that of heat transfer, the parcel ascends along an adiabat in state space. Upon reaching the upper boundary, it again moves horizontally. There, isobaric heat rejection (e.g., through LW emission to space) decreases the potential temperature and temperature of the parcel, which therefore becomes negatively buoyant. The parcel then sinks along a different adiabat until it has returned to the surface and completed a thermodynamic cycle.

If it remains unsaturated (e.g., below the LCL), each air parcel comprising the layer preserves its potential temperature away from the boundaries. Consequently, as it traverses the layer, each parcel traces out a uniform profile of θ. The horizontal-mean potential temperature $\bar{\bar{\theta}}(z)$ is equivalent to an average over all such parcels at

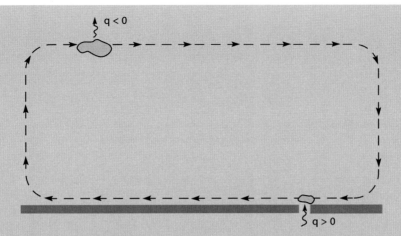

Figure 6.6 Idealized circuit followed by an air parcel during which it absorbs heat at the base of a layer and rejects heat at its top, with adiabatic vertical motion between.

a given elevation. Therefore, it too is independent of height:

$$\bar{\bar{\theta}}(z) = \text{const.} \tag{6.27.1}$$

As is true for an individual parcel, a uniform vertical distribution of potential temperature corresponds to the constant environmental lapse rate (Sec. 2.4.2)

$$\bar{\bar{\Gamma}} = \Gamma_d. \tag{6.27.2}$$

Consequently, the thermal structure (6.24) may be regarded as the horizontal mean of a layer in which air is being efficiently overturned.

This interpretation may be extended to saturated conditions (e.g., inside a layer of cloud). An air parcel's evolution is then pseudo-adiabatic, so its equivalent potential temperature is conserved. Efficient overturning of air will make the horizontal-mean equivalent potential temperature $\bar{\bar{\theta}}_e(z)$ independent of height

$$\bar{\bar{\theta}}_e(z) = \text{const.} \tag{6.28.1}$$

From Sec. 5.4.3, a uniform vertical distribution of $\bar{\bar{\theta}}_e(z)$ corresponds to a mean environmental lapse rate

$$\bar{\bar{\Gamma}} = \Gamma_s. \tag{6.28.2}$$

The corresponding stratification is termed *saturated adiabatic*.

Even though all parcels inside the layer evolve in like fashion, local behavior will differ from the horizontal mean – because conserved properties may still vary from one parcel to another (e.g., due to different histories experienced by those parcels at the layer's boundaries). An ascending parcel will have values

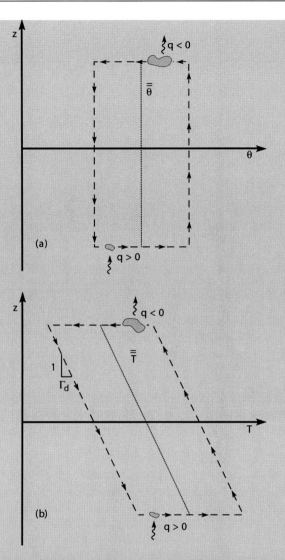

Figure 6.7 Thermodynamic cycle followed by the air parcel in Fig. 6.6 in terms of (a) potential temperature and (b) temperature. Horizontally averaged behavior for a layer comprised of many such parcels also indicated (dotted).

of θ and θ_e that are greater than horizontal-mean values because that parcel will have recently absorbed heat at the lower boundary. Conversely, a descending parcel will have values of θ and θ_e that are less than horizontal-mean values because that parcel will have recently rejected heat at the upper boundary. The actual stratification of the layer at any instant will therefore vary with position.

Were heat transfer at the boundaries eliminated, buoyantly driven motion would "spin down" through turbulent and molecular diffusion. Because diffusion destroys gradients between individual parcels, this limiting state is characterized by homogeneous distributions of θ and θ_e. It thus corresponds to the environmental lapse rate $\Gamma = \Gamma_d$ or Γ_s. The limiting homogeneous state therefore has the same stratification as the horizontal-mean stratification of the layer that is being efficiently overturned. The latter may therefore be regarded as "statistically well-mixed."

The stratification (6.28) is characteristic of the troposphere. Close to the saturated adiabatic lapse rate (Fig. 1.2), mean thermal structure in the troposphere reflects efficient overturning of moist air by cumulus and sloping convection. Air ascends at the saturated adiabatic lapse rate Γ_s ($\sim$5.5 K/km). After being dehydrated through condensation and precipitation, it descends at the dry adiabatic lapse rate Γ_d ($\sim$10 K/km). The observed global-mean lapse rate in the troposphere, $\overline{\overline{\Gamma}} \cong 6.5$ K/km, falls between the two, close to Γ_s. Air overturned by convection is continually re-humidified through contact with warm ocean surface. Because convective motion operates on time scales of a day or shorter, it makes the troposphere statistically well-mixed.

In traversing the circuit, the parcel in Fig. 6.7 absorbs heat at high temperature and rejects heat at low temperature. By the Second Law (Sec. 3.2), net heat is absorbed over a cycle. Then the First Law (2.13) implies that the parcel performs net work during its traversal of the circuit. An individual parcel in the above circulation therefore behaves as a heat engine. In fact, the thermodynamic cycle in Fig. 6.7 is analogous to the Carnot cycle pictured in Fig. 3.2, except that heat transfer occurs isobarically instead of isothermally. More expansion work is performed by the parcel during ascent than is performed on the parcel during descent. Net work performed during the cycle is reflected in the area circumscribed by the parcel's evolution in Fig. 6.7b.

Net heat absorption and work performed by individual air parcels make the general circulation of the troposphere behave as a heat engine, one that is driven thermally by heat transfer at its lower and upper boundaries. Work performed by individual parcels is associated with a redistribution of mass: Air that is effectively warmer and lighter at the lower boundary is exchanged with air that is effectively cooler and heavier at the upper boundary. This redistribution of mass represents a conversion of potential energy into kinetic energy. The conversion of energy maintains the general circulation against frictional dissipation.

6.5.2 Diabatic stratification: A paradigm of the stratosphere

The idealized behavior just described relies on heat transfer being confined to the lower and upper boundaries of the layer, where an air parcel resides long enough for diabatic influences to become important. Between the boundaries, the time scale of motion is short. For motion that operates on longer time scales, typical of the stratosphere, the evolution of an individual air parcel is not adiabatic.

Radiative transfer is the primary diabatic influence outside the boundary layer and cloud. It is characterized by cooling rates of order 1 K day^{-1} in the troposphere (see Fig. 8.24). Cooling rates as large as 10 K day^{-1} occur in the stratosphere and near cloud (Fig. 9.36).

For comparison, the cooling rate associated with adiabatic expansion follows from (2.33) as

$$\left. \frac{dT}{dt} \right|_{ad} = -\Gamma' \frac{dz}{dt}$$

$$= -\Gamma' w,$$

(6.29.1)

where T and z refer to an individual parcel, Γ' denotes its lapse rate, and

$$w = \frac{dz}{dt}$$

(2.29.2)

is the parcel's vertical velocity. Cumulus and sloping convection are characterized by a vertical velocity of 0.1 m s^{-1}. Under unsaturated conditions, this gives an adiabatic cooling rate of $\frac{dT}{dt}|_{ad} = 84$ K day^{-1}. Even under saturated conditions, the pseudo-adiabatic cooling rate is of order 50 K day^{-1}. Both are much greater than radiative cooling rates. Therefore, heat transfer in cumulus and sloping convection can, to good approximation, be neglected. On the other hand, motion in the stratosphere is characterized by a vertical velocity of 1 mm s^{-1}. This gives an adiabatic cooling rate of only $\frac{dT}{dt}|_{ad} = 0.84$ K day^{-1}. For such motion, heat transfer cannot be neglected. Radiative influences may even dominate temperature changes that are introduced by expansion and compression.

Under these circumstances, an air parcel moving vertically interacts thermally with its environment. Because θ is no longer conserved, diabatic motion implies different thermal structure. Consider a layer characterized by

$$\Gamma_d > \Gamma = \text{const.}$$

(6.30.1)

Such stratification is termed *subadiabatic*. Temperature decreases with height slower than dry adiabatic. By (6.18), θ increases with height (cf. Fig. 6.5). Then (6.25), together with (2.37), gives

$$c > 0$$

$$\delta q > 0 \qquad\qquad dT > 0$$

(6.30.2)

$$\delta q < 0 \qquad\qquad dT < 0$$

for $\Gamma < 0$ (temperature increasing with height). For $\Gamma > 0$ (temperature decreasing with height), c and δq satisfy reversed inequalities. Both imply that an air parcel absorbs heat during ascent, when it experiences expansional cooling, and rejects heat during descent, when it experiences compressional warming. The parcel's potential temperature therefore increases with height. As illustrated in Fig. 6.8, heat is rejected at high temperature at the upper boundary and absorbed at low temperature at the lower boundary. By the Second Law, net heat is rejected over a cycle. Then the First Law implies that net work must be performed on the parcel for it to traverse the circuit.

Opposite to the troposphere, this behavior is characteristic of the stratosphere. Net heat rejection and work performed on individual air parcels make the general circulation of the stratosphere behave as a refrigerator. It is driven mechanically by waves that propagate upward from the troposphere. By transferring momentum, planetary waves exert an influence on the stratosphere analogous to paddle work.

When dissipated, they drive air toward the winter pole, where it experiences radiative cooling (see Figs. 18.11; 8.27). Individual parcels then sink across isentropic surfaces to lower θ (Fig. 18.10). Conversely, air at low latitude experiences radiative warming. Parcels there rise across isentropic surfaces to higher θ.

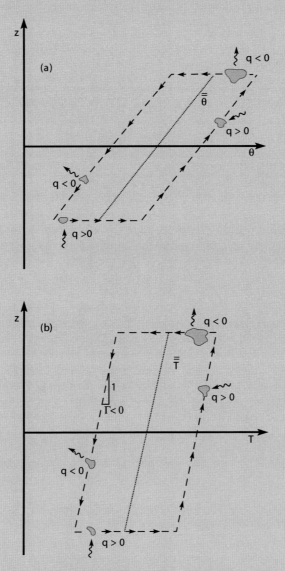

Figure 6.8 As in Fig. 6.7, but for an air parcel whose vertical motion is diabatic and whose temperature increases with height.

For both, vertical motion is slow enough to make diabatic influences important. It enables the temperature of individual parcels to increase with height. The gradual nature of vertical motion in the stratosphere reflects its opposition by buoyancy. Unlike the troposphere, buoyancy in the stratosphere opposes vertical motion because, invariably, warm (high-θ) air overlies cool (low-θ) air. To exchange effectively-heavier air at lower levels with effectively-lighter air at upper levels, work must be performed against the opposition of buoyancy. The rearrangement of mass represents a conversion of kinetic energy (that of the waves driving the motion) into potential energy. Manifest in temperature, the potential energy is dissipated thermally through LW emission to space.

SUGGESTED REFERENCE

Atmospheric Thermodynamics (1981) by Iribarne and Godson contains a thorough discussion of atmospheric statics.

PROBLEMS

1. Evaluate the discrepancy between geometric and geopotential height, exclusive of anisotropic contributions to gravity, up to 100 km.
2. Show that an atmosphere of uniform negative lapse rate need have no upper bound.
3. For reasons developed in Chap. 12, large-scale horizontal motion is nearly tangential to contours of geopotential height, approximated by the geostrophic velocity

$$v_g = \frac{1}{f}\left(-\frac{\partial \Phi}{\partial y}, \frac{\partial \Phi}{\partial x}\right),$$

   where $v = (u, v)$ denotes the horizontal component of motion in the eastward (x) and northward (y) directions and Φ is the geopotential along an isobaric surface. Suppose that satellite measurements provide the 3-dimensional distributions of temperature and of the mixing ratio of a long-lived chemical species. From those measurements, describe an algorithm to determine 3-dimensional air motion on large scales at heights great enough to ignore the variation of surface pressure.
4. An upper-level depression, typical of extratropical cyclones during amplification, affects isobaric surfaces aloft but leaves the surface pressure distribution undisturbed. Consider the depression of isobaric height associated with the temperature distribution

$$T(\lambda, \phi, \xi) = \overline{\overline{T}} + \overline{T}\cos(\phi) + T'e^{-\frac{\lambda^2 + (\phi - \phi_0)^2}{L^2}} \cdot f(\xi),$$

   where $\overline{\overline{T}}$, $\overline{T}$, and T' reflect global-mean, zonal-mean, and disturbance contributions to temperature, respectively, $\xi = -\ln(\frac{p}{p_0})$, with $p_0 = 1000$ hPa, is a measure of elevation,

$$f(\xi) = \begin{cases} -\cos\left(\pi \frac{\xi}{\xi_T}\right) & \xi \leq \xi_T \\ \\ 0 & \xi > \xi_T, \end{cases}$$

   with $\xi_T = -\ln(\frac{100}{1000})$ reflecting the tropopause elevation, the surface pressure $p_s = p_0$ remains constant, $\phi_0 = \frac{\pi}{4}$, $L = 0.25$ rads, and $\overline{\overline{T}} = 250$ K, $\overline{T} = 30$ K, and $T' = 5$ K.
   (a) Derive an expression for the 3-dimensional distribution of geopotential height.
   (b) Plot the vertical profiles of disturbance temperature and height in the center

of the anomaly. (c) Plot the horizontal distribution of geopotential height for the mid-tropospheric level $\xi = \frac{\xi_T}{2}$, as a function of λ and $\phi > 0$. (d) Plot a vertical section at latitude ϕ_0, showing the elevations ξ of isobaric surfaces as functions of λ.

5. Under the conditions of (4), how large must the anomalous temperature T' be for the disturbance to form a cut-off low at the mid-tropospheric level $\xi = \frac{\xi_T}{2}$?

6. Unlike mid-latitude depressions, hurricanes are warm-core cyclones that are invariably accompanied by a low in surface pressure. Consider a hurricane which is associated with the temperature distribution

$$T(\lambda, \phi, \xi) = \overline{T} + T'(\lambda, \phi)e^{-\alpha\xi},$$

where $\overline{T}$ and T' reflect zonal-mean and disturbance contributions to temperature, $\xi = -\ln\left(\frac{p}{p_0}\right)$, with $p_0 = 1000$ hPa, is a measure of elevation,

$$T'(\lambda, \phi) = \overline{T}\alpha\xi_s(\lambda, \phi)e^{\alpha\xi_s(\lambda,\phi)},$$

in which

$$\xi_s(\lambda, \phi) = 0.1 e^{-\frac{\lambda^2 + \phi^2}{L^2}},$$

with $\alpha^{-1} = \frac{\xi_T}{2}$, and $\xi_T = \ln\left(\frac{100}{1000}\right)$ reflect the surface and tropopause pressures, respectively, and $L = 0.1$ rads. (a) Derive an expression for the 3-dimensional distribution of geopotential height. (b) Plot the vertical profiles of disturbance temperature and height in the center of the anomaly. (c) Plot the horizontal distribution of geopotential height for the mid-tropospheric level $\xi = \frac{\xi_T}{2}$. (d) Plot a vertical section at the equator, showing the elevations ξ of isobaric surfaces as functions of λ.

7. A surface analysis plots pressure adjusted to a common reference level: mean sea level (MSL). Different locations can then be compared meaningfully. Derive an approximate expression for the pressure at MSL in terms of the local surface height z_s, surface pressure p_s, and the global-mean surface temperature $\overline{\overline{T}} = 288$ K. (b) A surface analysis shows a ridge with a maximum of 1040 hPa over Denver, which lies at an elevation of 5300' above MSL. What surface pressure was actually recorded there?

8. Deep convection inside a region of the tropics releases latent heat to achieve a column heating rate of 250 W m^{-2}. (a) If, uncompensated, that heating would produce a temperature perturbation that is invariant with height up to the tropopause, calculate the implied change of thickness for the 1000 hPa–100 hPa layer after 1 day (cf. Fig. 7.8). (b) What process, in reality, compensates the tendency of tropospheric thickness implied in (a)?

9. Show that the polytropic potential temperature in Prob. 2.16 is conserved under the diabatic conditions described in Sec. 6.5.2.

10. Use (2.27), with z replaced by $-\ln p$, to relate the area circumscribed in Fig. 6.7 to the work performed by the air parcel during one thermodynamic cycle.

11. Approximate the thermodynamic cycle in Fig. 6.7 by one comprised of two iso-baric legs and two adiabatic legs to show that (a) net work is performed during a complete cycle and (b) the work performed is proportional to the change of temperature associated with isobaric heat transfer.

12. Show that sub-adiabatic stratification requires an individual parcel moving verti-cally to, in general, absorb (reject) heat as it ascends (descends).

SEVEN

Static stability

The property responsible for the distinctly different character of the troposphere and stratosphere is vertical stability. Because it applies even under motionless conditions and because displacements of air remain in hydrostatic balance, vertical stability is also referred to as *static stability*. Although forces are never far from hydrostatic equilibrium, vertical motions are introduced by forced lifting over elevated terrain and through buoyancy. Buoyantly driven motion is related closely to the stability of the atmospheric mass distribution. The latter, in turn, is shaped by transfers of energy between the Earth's surface, the atmosphere, and deep space. By promoting convection in some regions and suppressing it in others, vertical stability controls a wide range of properties.

7.1 REACTION TO VERTICAL DISPLACEMENT

Because of hydrostatic equilibrium, together with the compressibility of air, density decreases upward regardless of temperature structure (Fig. 6.5b). Thus, mean stratification invariably has lighter air configured over heavier air. This suggests stability with respect to vertical displacements. Were air incompressible, this would indeed be the case.

Consider an arbitrary air parcel, inside the layer pictured in Fig. 7.1. In a linear stability analysis, this parcel will be used to establish how the layer reacts to infinitesimal disturbances from equilibrium. Although the analysis focuses on an individual parcel, stability actually refers to the layer as a whole.

Suppose the parcel is disturbed by a virtual displacement $\delta z' = z'$, where primes distinguish properties of the parcel from those of its environment. Through expansion or compression, the parcel automatically adjusts to the environmental pressure to

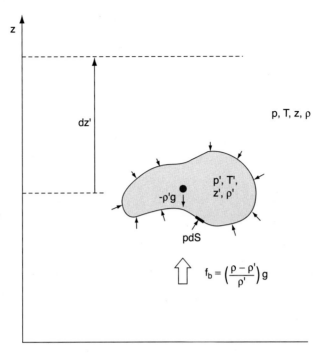

Figure 7.1 Schematic of an air parcel, the properties of which are distinguished by primes, inside a designated layer. The specific buoyancy force f_b results from an imbalance between the parcel's weight and the net pressure force acting over its surface.

preserve mechanical equilibrium. Thus

$$p' = p$$

throughout the parcel's displacement. Further, so long as the displacement occurs on a time scale short compared with that of heat transfer with the environment, the parcel's evolution may be regarded as adiabatic.

Per unit volume, Newton's second law for the parcel is

$$\rho' \frac{d^2 z'}{dt^2} = -\rho'g - \frac{\partial p'}{\partial z'}, \qquad (7.1.1)$$

which follows from a development analogous to the one leading to (1.16). For the environment, the momentum balance is simply hydrostatic equilibrium:

$$0 = \rho g - \frac{\partial p}{\partial z}. \qquad (7.1.2)$$

Subtracting (7.1) and incorporating mechanical equilibrium yields for the parcel

$$\rho' \frac{d^2 z'}{dt^2} = (\rho - \rho')g. \qquad (7.2.1)$$

This relationship also follows from Archimedes' principle (Prob. 7.3). Per unit mass, the parcel's momentum balance is then described by

$$\frac{d^2 z'}{dt^2} = \left(\frac{\rho - \rho'}{\rho'}\right) g$$

$$= f_b,$$

(7.2.2)

where f_b is the (specific) net buoyancy force experienced by the parcel. With (7.1), the buoyancy force is recognized as an imbalance between the parcel's weight and the vertical component of pressure force acting over its surface.

Consider how the displaced parcel's temperature varies in relation to that of its environment. With the gas law, (7.2.2) may be expressed

$$\frac{d^2 z'}{dt^2} = g\left(\frac{T' - T}{T}\right).$$

(7.3)

For small displacements from its undisturbed elevation, the parcel's temperature decreases with height at the parcel lapse rate: $\Gamma' = \Gamma_d$ (Γ_s) under unsaturated (saturated) conditions. Thus

$$T' = T_0 - \Gamma' z'$$

(7.4.1)

describes the first-order variation of the parcel's temperature from its undisturbed value T_0, which equals the environmental temperature at the parcel's undisturbed height. On the other hand, the temperature of the environment decreases with height at the local environmental lapse rate Γ, so the corresponding variation of environmental temperature is

$$T = T_0 - \Gamma z'.$$

(7.4.2)

The parcel's lapse rate is determined by its thermodynamics. Under adiabatic conditions, Γ' is independent of the environmental lapse rate Γ – because the parcel is then thermally isolated. The environmental lapse rate, on the other hand, is determined by the history of air residing in the layer (e.g., by where that air has been and what thermodynamic influences it has experienced).

Incorporating (7.4) transforms the vertical momentum balance (7.3) into

$$\frac{d^2 z'}{dt^2} = \frac{g}{T}(\Gamma - \Gamma')z'$$

$$= f_b.$$

(7.5)

Hence, the specific buoyancy force experienced by the parcel is proportional to its displacement z' and to the difference between lapse rates of the parcel and the environment.

7.2 STABILITY CATEGORIES

If the parcel is unsaturated,

$$\Gamma' = \Gamma_d.$$

(7.6.1)

Because Γ' is then constant, (7.4.1) is also valid for finite displacements. If the parcel is saturated,

$$\Gamma' = \Gamma_s,$$

(7.6.2)

where Γ_s implies its value at the undisturbed height.

STABILITY CATEGORIES

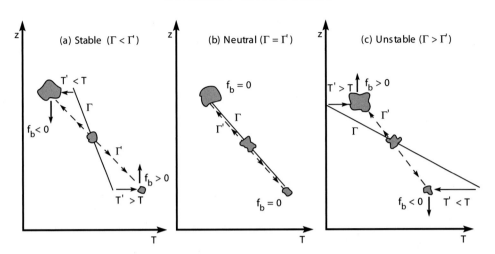

Figure 7.2 Buoyancy reaction experienced by a displaced air parcel, in terms of the environmental lapse rate Γ and the parcel lapse rate Γ', which equals Γ_d (Γ_s) under unsaturated (saturated) conditions. (a) $\Gamma < \Gamma'$, (b) $\Gamma = \Gamma'$, (c) $\Gamma > \Gamma'$.

7.2.1 Stability in terms of temperature

For either of the above conditions, three possibilities exist (Fig. 7.2):

(1) $\Gamma < \Gamma'$: Environmental temperature decreases with height slower than the displaced parcel's temperature. Γ is *subadiabatic*. Then (7.5) implies

$$\frac{1}{z'}\frac{d^2 z'}{dt^2} = \frac{f_b}{z'} < 0.$$

The parcel experiences a buoyancy force f_b that opposes its displacement z'. If it is displaced upward, the parcel becomes heavier than its environment and thus negatively buoyant. Conversely, if it is displaced downward, the parcel becomes lighter than its environment and positively buoyant. This buoyancy reaction constitutes a *positive restoring force*, one that acts to restore the system to its undisturbed height – irrespective of the sense of the displacement. The atmospheric layer is then said to be *statically stable*. The environmental temperature profile is said to possess *positive stability*.

(2) $\Gamma = \Gamma'$: Environmental temperature decreases with height at the same rate as the parcel's temperature. Γ is *adiabatic*. Under these circumstances,

$$\frac{1}{z'}\frac{d^2 z'}{dt^2} = \frac{f_b}{z'} = 0,$$

so the buoyancy force f_b vanishes and the displaced parcel experiences no restoring force. The atmospheric layer is then said to be *statically neutral*. The environmental temperature profile is said to possess *zero stability*.

(3) $\Gamma > \Gamma'$: Environmental temperature decreases with height faster than the parcel's temperature. Γ is *superadiabatic.* Then

$$\frac{1}{z'}\frac{d^2 z'}{dt^2} = \frac{f_b}{z'} > 0.$$

The parcel experiences a buoyancy force f_b that reinforces its displacement z'. If it is displaced upward, the parcel becomes lighter than its environment and thus positively buoyant. Conversely, if it is displaced downward, the parcel becomes heavier than its environment and negatively buoyant. This buoyancy reaction constitutes a *negative restoring force*, one that drives the system away from its undisturbed height – irrespective of the sense of the displacement. The atmospheric layer is then said to be *statically unstable*. The temperature profile is said to possess *negative stability*.

Even though density decreases upward in all three cases, through buoyancy, each leads to a very different response. The layer's stability is not determined by its vertical profile of density because the density of a displaced air parcel changes through expansion and compression. Depending on whether its density decreases with height faster or slower than environmental density, a displaced parcel will find itself heavier or lighter than its environment. It will, therefore, experience a restoring force that is either positive or negative.

The degree of stability or instability is reflected in the magnitude of the restoring force. The latter is proportional to the difference between the lapse rates of the environment and the parcel. According to (7.5), the parcel lapse rate Γ' serves as a reference against which the local environmental lapse rate Γ is measured to determine stability (Fig. 7.3). Under unsaturated conditions, $\Gamma' = \Gamma_d$. If environmental temperature decreases with height slower than Γ_d (i.e., Γ is subadiabatic), the layer is stable. If environmental temperature decreases with height at a rate equal to Γ_d (i.e., Γ is adiabatic), the layer is neutral. If environmental temperature decreases with height faster than Γ_d (i.e., Γ is superadiabatic), the layer is unstable.

Under saturated conditions (e.g., inside a layer of stratiform cloud), the same criteria hold, but with Γ_s in place of Γ_d. Because

$$\Gamma_s < \Gamma_d,$$

Fig. 7.3 implies that static stability is violated more easily under saturated conditions than under unsaturated conditions. That is, the range of stable lapse rate is narrower under saturated conditions than under unsaturated conditions. Reduced stability of saturated air follows from the release (absorption) of latent heat, which warms (cools) an ascending (descending) parcel. By adding positive (negative) buoyancy, the release (absorption) of latent heat reinforces vertical displacements, destabilizing the layer.

7.2.2 Stability in terms of potential temperature

Stability criteria can also be expressed in terms of potential temperature. For unsaturated conditions, logarithmic differentiation of (2.31) gives

$$\frac{1}{\theta}\frac{d\theta}{dz} = \frac{1}{T}\frac{dT}{dz} - \frac{\kappa}{p}\frac{dp}{dz}.$$

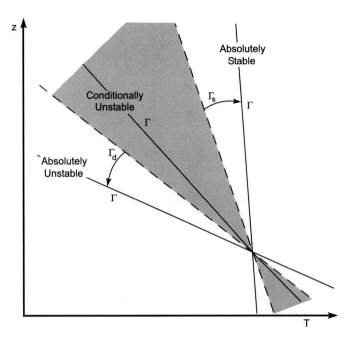

Figure 7.3 Vertical stability in terms of temperature, determined by the environmental lapse rate Γ in relation to the adiabatic lapse rate Γ'.

Hydrostatic equilibrium and the gas law transform this into

$$\frac{1}{\theta}\frac{d\theta}{dz} = \frac{1}{T}\frac{dT}{dz} + \frac{\kappa}{H}.$$

Then with (1.17), the vertical gradient of potential temperature may be expressed in terms of the lapse rate

$$\frac{d\theta}{dz} = \frac{\theta}{T}(\Gamma_d - \Gamma). \tag{7.7}$$

As it follows from the definitions of potential temperature and lapse rate, (7.7) also governs the environment. An environmental lapse rate $\Gamma < \Gamma_d$ (subadiabatic) corresponds to $\frac{d\theta}{dz} > 0$. Potential temperature then increases upward. An environmental lapse rate $\Gamma = \Gamma_d$ (adiabatic) corresponds to $\frac{d\theta}{dz} = 0$. The profile of potential temperature is then uniform. An environmental lapse rate $\Gamma > \Gamma_d$ (superadiabatic) corresponds to $\frac{d\theta}{dz} < 0$. Potential temperature then decreases upward.

Thus, for unsaturated conditions, the stability criteria established previously in terms of temperature translate into the simpler criteria in terms of potential temperature:

$$\frac{d\theta}{dz} > 0 \qquad \text{(stable)}$$

$$\frac{d\theta}{dz} = 0 \qquad \text{(neutral)} \tag{7.8}$$

$$\frac{d\theta}{dz} < 0 \qquad \text{(unstable)}.$$

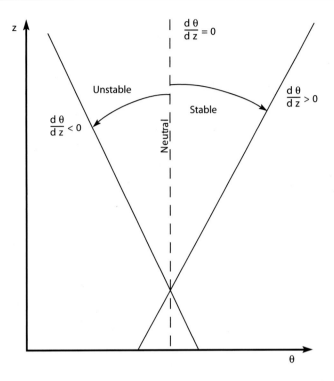

Figure 7.4 Vertical stability, under unsaturated conditions, in terms of potential temperature.

Illustrated in Fig. 7.4, these criteria have the same form as those expressed in terms of temperature. However, the reference profile corresponding to neutral stability is now just a uniform distribution of θ. This has a parallel in the vertical stability of water, which is incompressible. If potential temperature increases with height, air that is effectively warmer and lighter (with account of compressibility) overlies air that is effectively colder and heavier. The layer is then stable. If potential temperature is uniform with height, the layer is neutral. If θ decreases with height, air that is effectively colder and heavier overlies air that is effectively warmer and lighter. The layer is then unstable.

The simplified stability criteria (7.8) reflect the close relationship between θ and atmospheric motion. Recall that potential temperature is conserved for individual air parcels because it accounts for expansion work. Therefore, a displaced parcel traces out a uniform profile of θ; it defines the reference thermal structure corresponding to neutral stability. The same is true of temperature for an incompressible fluid. It is for this reason that potential temperature plays a role for air analogous to the role that temperature plays for water. Hence, the vertical distribution of θ describes the effective stratification of mass when compressibility is taken into account.

Under saturated conditions, θ is not conserved. However, θ_e is conserved. It bears a relationship to Γ_s analogous to the role that θ bears to Γ_d under unsaturated conditions. Using arguments similar to that above, it may be shown that the criteria for stability under saturated conditions are also given by (7.8), but with θ_e in place of θ

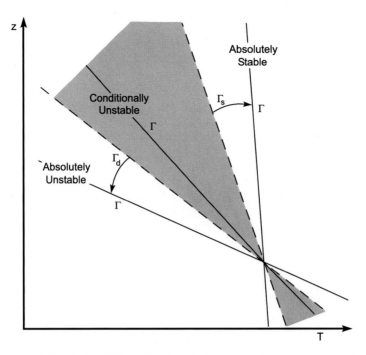

Figure 7.5 Vertical stability of moist air in terms of temperature and the environmental lapse rate Γ.

(Prob. 7.16). Thus, the vertical distribution of θ_e describes the effective stratification of mass, when both compressibility and latent heat release are taken into account.

7.2.3 Moisture dependence

Humidity complicates the description of vertical stability by introducing the need for two categories: One is valid under unsaturated conditions. The other is valid under saturated conditions, wherein the release of latent heat must be accounted for. Because $\Gamma_s < \Gamma_d$, three possibilities exist (Fig. 7.5). If

$$\Gamma < \Gamma_s, \tag{7.9.1}$$

a layer is *absolutely stable*. A small displacement of air will be met by a positive restoring force – irrespective of whether the air is saturated, because (7.9.1) ensures that a displaced parcel will cool (warm) faster than its environment. If

$$\Gamma > \Gamma_d, \tag{7.9.2}$$

the layer is *absolutely unstable*. A small displacement of air will result in a negative restoring force – irrespective of whether the air is saturated, because (7.9.2) ensures that a displaced parcel will cool (warm) slower than its environment. If

$$\Gamma_s < \Gamma < \Gamma_d, \tag{7.9.3}$$

the layer is *conditionally unstable*. A small displacement of air will then result in a positive restoring force if the layer is unsaturated, but a negative restoring force if it is saturated.

The first two possibilities apply irrespective of whether or not the parcel is saturated – because the restoring force does not change sign. The third possibility, however, hinges on the parcel's state. For finite displacements, an air parcel that is initially unsaturated can become saturated and vice versa. If the layer is conditionally unstable, the restoring force experienced by such a parcel will change sign. Such conditions are treated in Sec. 7.4.

7.3 IMPLICATIONS FOR VERTICAL MOTION

The stability of a layer determines its ability to support vertical motion and thus to support transfers of heat, momentum, and constituents. To conserve mass, vertical motion must be compensated by horizontal motion (cf. Fig. 5.4). Hydrostatic stability therefore also influences horizontal transport. Three-dimensional (3D) turbulence, which disperses atmospheric constituents, involves both vertical and horizontal motion. Consequently, suppressing vertical motion simultaneously suppresses the horizontal component of 3D eddy motion and hence turbulent dispersion.

A layer that is stably stratified inhibits vertical motion. Small vertical displacements, introduced mechanically by flow over elevated terrain or thermally through isolated heating, are then opposed by the positive restoring force of buoyancy (Prob. 7.5). Conversely, a layer that is unstably stratified promotes vertical motion through the negative restoring force of buoyancy. Work performed against buoyancy under stable conditions or by buoyancy under unstable conditions reflects a conversion between potential energy and kinetic energy. It controls the layer's evolution.

The vertical momentum balance for an unsaturated air parcel may be expressed

$$\frac{d^2 z'}{dt^2} + N^2 z' = 0, \tag{7.10.1}$$

where

$$N^2 = \frac{g}{T}(\Gamma_d - \Gamma)$$
$$= -\frac{f_b}{z'} \tag{7.10.2}$$

defines the *buoyancy* or *Brunt-Väisäillä frequency* N. Equation (7.10.1) describes a simple harmonic oscillator, with N^2 reflecting the "stiffness" of the buoyancy spring. From (7.7), the Brunt-Väisäillä frequency may be expressed simply

$$N^2 = g\frac{d\ln\theta}{dz}. \tag{7.10.3}$$

Proportional to the restoring force of buoyancy, N^2 is a measure of stability.

If a layer is stable, $N^2 > 0$. Then (7.10.1) has solutions of the form

$$z'(t) = Ae^{iNt} + Be^{-iNt}. \tag{7.11.1}$$

A displaced parcel oscillates about its undisturbed height, with kinetic energy repeatedly exchanged with potential energy. The stronger the stability (i.e., the stiffer the buoyancy spring), the more rapid is the parcel's oscillation, and the smaller is its displacement. Owing to the positive restoring force of buoyancy, a small imposed displacement remains small. By limiting vertical motion, positive stability also imposes a constraint on horizontal motion (e.g., suppressing eddy motion associated with 3D turbulence).

Under neutral conditions, $N^2 = 0$ and the restoring force vanishes. Imposed displacements are then met with no opposition. In that event, the solution of (7.10.1) grows linearly with time. It describes a displaced parcel moving inertially, with no conversion between potential and kinetic energy. Under these circumstances, a small imposed displacement ultimately evolves into finite displacements of air. That long-term behavior violates the linear analysis leading to (7.10.1), which is predicated on infinitesimal displacements. However, in the presence of friction, displacements remain bounded (Prob. 7.17).

Under positive and neutral stability, air displacements can remain small enough for the mean stratification of a layer to be preserved. A layer of negative stability evolves very differently. In that event, $N^2 = -\hat{N}^2 < 0$ and solutions of (7.10.1) have the form

$$z'(t) = Ae^{\hat{N}t} + Be^{-\hat{N}t}. \tag{7.11.2}$$

A parcel's displacement then grows exponentially with time through reinforcement by the negative restoring force of buoyancy. The first term in (7.11.2) dominates the long-term behavior, violating the linear analysis that led to (7.10.1). Unlike behavior under positive and neutral stability, air displacements need not remain bounded. Except for small $\hat{N}$, displacements amplify exponentially – even in the presence of friction (Prob. 7.18). A small initial disturbance then evolves into fully developed convection. Nonlinear effects eventually limit amplification by modifying the stratification of the layer. By rearranging mass, convective cells alter N^2 and hence the buoyancy force experienced by individual air parcels. The simple linear description (7.10), which relies on mean properties of the layer remaining constant, then breaks down.

Amplifying motion described by (7.11.2) is fueled by a conversion of potential energy, which is associated with the the vertical distribution of mass, into kinetic energy, which is associated with convective motion. By feeding off of the layer's potential energy, air motion modifies its stratification. Fully developed convection results in efficient vertical mixing. It rearranges the conserved property θ (θ_e) into a distribution that is statistically homogeneous (Sec. 6.5). According to (7.8), this limiting distribution corresponds to a state of neutral stability. Thus, small disturbances to an unstable layer amplify, eventually evolving into fully developed convection. Convective overturning neutralizes the instability by mixing θ (θ_e) into a uniform distribution. In that limiting state, no more potential energy is available for conversion to kinetic energy. Without a regeneration of instability, convective motion then decays through frictional dissipation.

7.4 FINITE DISPLACEMENTS

Stability criteria established in Sec. 7.2 hold for infinitesimal displacements. A parcel that is initially unsaturated then remains so and likewise for one that is initially saturated. For finite displacements, a disturbed air parcel can evolve between saturated and unsaturated conditions. Such displacements are particularly relevant to conditional instability (7.9.3). A displaced parcel may then experience a restoring force of

one sign initially, but of opposite sign thereafter. If the layer itself is displaced vertically (e.g., in sloping convection or through forced lifting over elevated terrain), finite
displacement enables the layer's stability itself to change.

7.4.1 Conditional instability

Consider an unsaturated parcel that is displaced upward inside the conditionally unstable layer shown in Fig. 7.6a. Depicted there is a simple representation of the tropical
troposphere, which is assigned a constant and conditionally unstable lapse rate, and
the lower stratosphere, which is assigned a lapse rate of zero. Below its LCL, the displaced parcel cools at the dry adiabatic lapse rate Γ_d. It therefore cools more rapidly
than its environment (7.9.3). The parcel thus finds itself cooler than its environment
and hence negatively buoyant. It experiences a positive restoring force, one that intensifies with height (7.5). Work must therefore be performed against buoyancy to complete
the displacement.

   If enough work is performed, the parcel will eventually cross its LCL. It then cools
slower, due to the release of latent heat, at the saturated adiabatic lapse rate Γ_s. Because
the layer is conditionally unstable, the parcel also cools slower than its environment
(7.9.3). Consequently, the temperature difference between the parcel and its environment narrows. The positive restoring force of buoyancy then weakens with height.

   After sufficient displacement, the temperature profile of the parcel crosses the
profile of environmental temperature. There, the parcel finds itself warmer than its
environment. Positively buoyant, it thereafter experiences a negative restoring force.
Buoyancy then performs work on the parcel, which accelerates upward autonomously.
The height where this occurs is the *Level of Free Convection* (LFC).

   Above the LFC, the parcel accelerates upward until, eventually, its temperature
crosses the profile of environmental temperature a second time. The height where
this occurs is the Level of Neutral Buoyancy (LNB). Above the LNB, the parcel again
finds itself cooler than its environment and thus negatively buoyant. Buoyancy then
opposes further ascent. Air that accelerated upward above the LFC therefore diverges
horizontally in a neighborhood of the LNB, forming extensive anvil.

   Energetics of the displaced parcel provide some insight into deep convective
motion. Under adiabatic conditions, the buoyancy force is conservative (e.g., the work
performed along a cyclic path vanishes). Therefore, the potential energy P of the displaced parcel may be introduced in a manner similar to that used in Sec. 6.2 to introduce the geopotential

$$dP = \delta w_b$$
$$= -f_b dz', \tag{7.12}$$

where δw_b is the incremental work performed against buoyancy. Defining a reference
value of zero potential energy at the undisturbed height z_0 and incorporating (7.2)
obtains

$$P = \int_{z_0}^{z} \left(\frac{v - v'}{v} \right) g dz'$$
$$= \int_{p_0}^{p} (v' - v) dp. \tag{7.13}$$

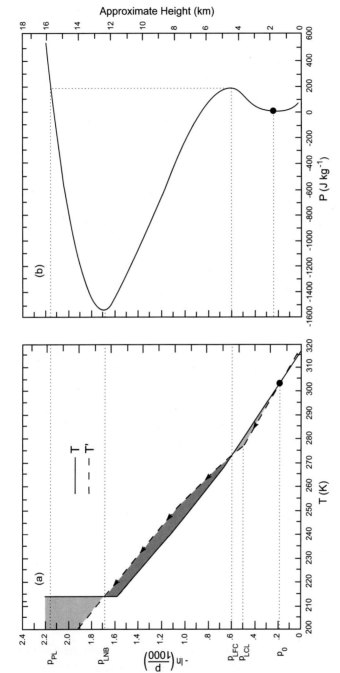

Figure 7.6 (a) Environmental temperature (solid), under conditionally unstable conditions symbolic of the tropical troposphere ($\Gamma \cong 8.5$ K km^{-1}) and stratosphere, and parcel temperature (dashed) displaced from the undisturbed level $p_0 = 800$ hPa. Parcel temperature decreases with height along a dry adiabat below the LCL and along a saturated adiabat above the LCL. It crosses the profile of environmental temperature at the *level of free convection* p_{LFC} and again at the Level of Neutral Buoyancy p_{LNB} above the tropopause. Not far overhead is the limiting height of convective overshoot, the Penetration Level p_{PL}. (b) Potential energy of the displaced parcel, which is proportional to minus the cumulative area $A(p)$ under the parcel's temperature profile and above the environmental temperature profile that is shaded in (a).

Then, with the gas law, the parcel's potential energy becomes

$$P(p) = -R \int_p^{p_0} (T' - T) d\ln p$$

$$= -RA(p),$$

(7.14)

where $A(p)$ is the cumulative area between the temperature profile of the parcel T' and that of the environment T to the level p (Fig. 7.6a).

Above the undisturbed level but below the LFC, $T' < T$. P then increases upward (Fig. 7.6b). Beneath the undisturbed level, T' and T are reversed. Hence P increases downward as well. Increasing P away from the undisturbed level represents a "potential well" in which the parcel is bound. Air beneath the LFC is thus trapped. It can be released from the potential well only if work is performed against the positive restoring force of buoyancy. Achieving that condition requires finite work and hence finite vertical displacement.

If sufficient work is performed to lift the parcel to its LFC, then P reaches a maximum. Thereafter, $T' > T$, so P decreases upward. Under conservative conditions,

$$\Delta K = -\Delta P,$$

(7.15.1)

which follows from (7.2). Decreasing P above the LFC represents a conversion of potential energy to kinetic energy. Buoyancy work accelerates the parcel upward, driving deep convection.

The total potential energy available for conversion to kinetic energy is termed the *Convective Available Potential Energy* (CAPE). Represented by the dark-shaded area in Fig. 7.6a, it equals the total work performed by buoyancy above the LFC. That, in turn, equals the large drop of potential energy in Fig. 7.6b. Because the parcel's temperature eventually crosses the environmental temperature profile a second time (at the LNB), CAPE is finite. So is the kinetic energy that can be acquired by the parcel. An upper bound on the parcel's kinetic energy follows from (7.15.1) as

$$\frac{w'^2}{2} = P(p_{LFC}) - P(p_{LNB})$$

$$= R \int_{p_{LNB}}^{p_{LFC}} (T' - T) d\ln p$$

(7.15.2)

$$= \text{CAPE}.$$

It describes adiabatic ascent under conservative conditions. In practice, however, mixing with the environment makes a cumulus updraft inherently nonconservative (cf. Fig. 5.4). The upward velocity in (7.15.2) is therefore realized only inside the core of a broad updraft, where air is isolated from entrainment and mixing that damp upward motion at the periphery of the updraft.

Above the LNB, T' and T are again reversed. P therefore increases upward above p_{LNB}. Negatively buoyant, the parcel is then bound in another potential well. Despite opposition by buoyancy, the parcel overshoots its new equilibrium level – due to the kinetic energy that it acquired above the LFC. However, P increases sharply above p_{LNB}, because environmental temperature above the tropopause diverges rapidly from the parcel's temperature. Consequently, penetration into the stable layer overhead is shallow compared with the depth traversed through the conditionally unstable layer below. Eventually, the parcel reaches a height where all of the kinetic energy that it

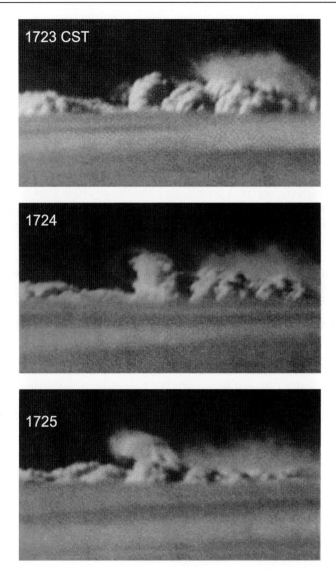

Figure 7.7 Convective domes overshooting the anvil of a cumulonimbus complex, at 2-minute intervals. Fibrous features visible in the second plate mark ice particles that nucleate spontaneously when stratospheric air is displaced by convection. After Fujita (1992).

acquired above the LFC has been re-converted into potential energy. This defines the *Penetration Level* (PL). As the motion is presumed to be conservative, the PL coincides with the height where P equals its previous maximum at the LFC.

The penetration level provides an estimate of the height of "convective overshoots", which accompany cumulus towers. However, like the maximum updraft in (7.15.2), this estimate is only an upper bound. In practice, some of the kinetic energy driving convective overshoots is siphoned off by mixing with the environment. Figure 7.7

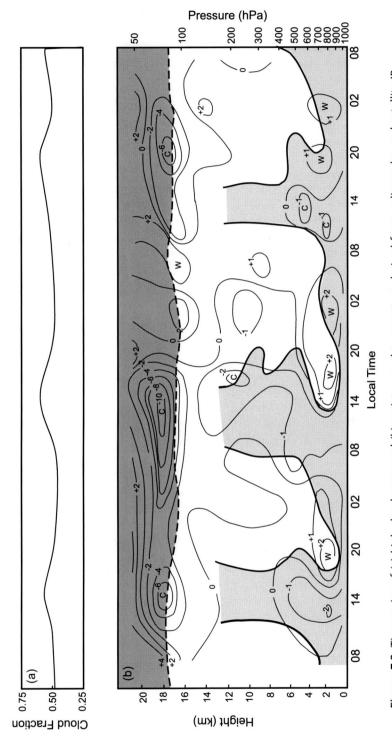

Figure 7.8 Time series of (a) high cloud cover and (b) moisture and temperature derived from radiosondes and satellite IR radiances during the winter MONEX observing campaign. Dark shading indicates the fractional coverage by cold convective cloud anvil. Relative humidity exceeding 80% is shaded light. Temperature departure from the local time-mean value is contoured. Adapted from Johnson and Kriete (1982).

185

shows an overshooting cumulonimbus complex at 2-minute intervals. Several convective domes have penetrated above the anvil, where air eventually diverges horizontally. Negatively buoyant, overshooting cloud splashes back into the anvil, but not before it has been sheared and mixed with stratospheric air. Thus, convective overshoots lead to an irreversible rearrangement and eventually mixing of tropospheric and stratospheric air.

This feature makes penetrative convection instrumental in processes that control the mean elevation of the tropopause. The tropopause is determined by a competition between vertical overturning, which prevails in the troposphere and drives stratification toward neutral stability, and radiative transfer, which prevails in the stratosphere and drives stratification toward positive stability. Figure 7.8 displays anomalous temperature, moisture, and cloud cover for an area in the tropics, each as a function of time. Towering moisture (light shading in Fig. 7.8b) and the amplification of cold high cloud (Fig. 7.8a) mark episodes of deep cumulus convection. They are contemporaneous with the tropopause (dashed) ascending to higher elevation and, simultaneously, with it becoming anomalously cold (contours) – by as much as 10 K. Jointly, these anomalies reflect an upward extension of positive lapse rate that characterizes the troposphere. It invades levels that previously were stable and in the stratosphere.

7.4.2 Entrainment

In practice, the vertical velocity and penetration level of a cumulus updraft are reduced from adiabatic values by nonconservative effects. Environmental air, which is cooler and drier (low θ_e), is entrained into and mixed with updraft air, which is anomalously warm and moist (high θ_e); cf. Fig. 5.4. Simultaneously, high-θ_e air is detrained from the updraft, mixed with low-θ_e air in the environment.[1] Such mixing depletes ascending parcels of positive buoyancy and kinetic energy. Only in the core of a broad updraft are adiabatic values ever realized. Mixing also modifies the environment of a cumulus updraft. This mediates gradients of temperature and moisture that control buoyancy.

For reasons established in Chap. 9, entrainment of environmental air is intrinsic to cumulus convection. It renders the evolution of ascending parcels diabatic. The impact this process exerts on convection is illustrated by regarding entrained environmental air to become uniformly mixed with ascending air inside a moist updraft. Following Holton (2004), we consider a parcel of mass m inside the updraft pictured in Fig. 5.4. Through mixing, additional mass is incorporated into that parcel. The parcel may then be identified with a certain percentage of the overall mass crossing a given level. If μ is a conserved property (e.g., $\mu = r$ beneath the LCL), then $m\mu'$ represents how much of that property is associated with the parcel under consideration. As before, primes distinguish properties of the parcel from those of the environment.

---

[1] Detrainment, which refers to the transfer of air to the environment, compensates entrainment that is introduced through mixing (e.g., along the wall of a cumulus updraft). Detrainment also applies to air that diverges from an updraft, for example, in the formation of anvil, which is likewise accompanied by mixing.

After mass dm of environmental air has been mixed into it, the parcel will have mass $m + dm$ and property $\mu' + d\mu'$. Conservation of μ then requires

$$(m + dm)(\mu' + d\mu') = m\mu' + dm\mu, \tag{7.16.1}$$

where $dm\mu$ represents how much of the conserved property has been incorporated into the parcel from the environment. If the above process occurs during time dt, expanding (7.16.1) and neglecting terms higher than first order leads to

$$\frac{d\mu'}{dt} = \frac{d(lnm)}{dt}(\mu - \mu'), \tag{7.16.2}$$

which governs the conserved property inside the entraining updraft. More generally, if μ is not conserved but rather is produced per unit mass at the rate S_μ, its evolution inside the aforementioned parcel is governed by

$$\frac{d\mu'}{dt} = \frac{d(lnm)}{dt}(\mu - \mu') + S_\mu. \tag{7.17}$$

The time rate of change following the parcel is related to its vertical velocity $w' = \frac{dz'}{dt}$ through the chain rule:

$$\frac{d}{dt} = w\frac{d}{dz}, \tag{7.18}$$

where $\frac{d}{dz}$ refers to a change with respect to the parcel's height. Then (7.17) becomes

$$w\frac{d\mu'}{dz} = \frac{w}{H_e}(\mu - \mu') + S_\mu, \tag{7.19}$$

in which $H_e^{-1} = \frac{d(lnm)}{dz}$ defines the mass *entrainment height*. H_e represents the height for upwelling mass to increase by a factor of e. It characterizes the rate at which the entraining thermal expands due to incorporation of environmental air (Fig. 5.4).[2] If $\mu = ln\theta_e$, the source term $S_{\theta_e} = 0$ because θ_e is conserved. Then, with (5.27), (7.19) reduces to

$$\frac{d(ln\theta'_e)}{dz} = \frac{1}{H_e}\left\{ln\left(\frac{\theta}{\theta'}\right) + \frac{l}{c_p}\left[\frac{r_c}{T} - \frac{r'_c}{T'}\right]\right\}. \tag{7.20}$$

Replacing saturation values by the actual mixing ratios and temperatures by those corresponding to the LCL (Sec. 5.4.2) and noting that differences of temperature are small compared with differences of moisture leads to

$$\frac{d(ln\theta'_e)}{dz} \cong \frac{1}{H_e}\left\{ln\left(\frac{\theta}{\theta'}\right) + \frac{l}{c_pT}(r - r')\right\}$$
$$= \frac{1}{H_e}\left\{ln\left(\frac{T}{T'}\right) + \frac{l}{c_pT}(r - r')\right\}. \tag{7.21}$$

Because $r' > r$ and $T' > T$, θ'_e decreases upward. This is to be contrasted with the parcel's behavior under conservative conditions (e.g., in the absence of entrainment).

---

[2] A cumulus cloud expands with height slower than the moist updraft supporting it because mixing with drier environmental air leads to evaporation of condensate and dissolution of cloud along its periphery (Sec. 9.3.3).

Its mass is then fixed and $\theta'_e = $ const. The two sinks of θ'_e on the right-hand side of (7.21) reflect transfers of sensible and latent heat to the environment. Those properties are mixed across the parcel's control surface (cf. Fig. 2.1).

Suppose now μ equals the specific momentum w. Then $w = 0$ for the environment. Equation (7.19) then reduces to

$$w \frac{dw}{dz} = -\frac{w^2}{H_e} + S_w$$

or

$$\frac{dK'}{dz} = -\frac{2}{H_e} K' + \frac{S_K}{w}, \tag{7.22}$$

where $K' = \frac{w^2}{2}$ is the specific kinetic energy of the parcel. The production of kinetic energy is related to the production of upward momentum: $S_K = w S_w$. Unlike the source term for θ_e, $S_K \neq 0$. K is not conserved, even in the absence of entrainment, because the parcel's kinetic energy changes through work that is performed on it by buoyancy.

Differentiating (7.12) along with (7.15) implies

$$\frac{dK'}{dz} = f_b \tag{7.23.1}$$

under conservative conditions (e.g., for a parcel of fixed mass, wherein $H_e \to \infty$). Hence we identify

$$S_K = w f_b$$
$$= w g \left(\frac{T' - T}{T} \right). \tag{7.23.2}$$

Then (7.22) becomes

$$\frac{dK'}{dz} = g \left(\frac{T' - T}{T} \right) - \frac{2}{H_e} K'. \tag{7.24}$$

The sink of K' on the right hand side represents turbulent drag. It is exerted on the ascending parcel through the incorporation of momentum from the environment. By diluting kinetic energy, entrainment reduces the parcel's acceleration from what it would be under the action of buoyancy alone. That, in turn, limits its penetration into the stable layer above the LNB, to about an entrainment height (Prob. 7.21).

7.4.3 Potential instability

Until now, we have considered displacements of individual air parcels within a fixed layer. Suppose the layer itself is displaced vertically (e.g., in sloping convection or forced lifting over elevated terrain). Then changes of the layer's thermodynamic properties alter its stability. Stratification of moisture plays a key role in this process because different levels need not achieve saturation simultaneously.

Consider an unsaturated layer in which potential temperature increases with height,

$$\frac{d\theta}{dz} > 0, \tag{7.25.1}$$

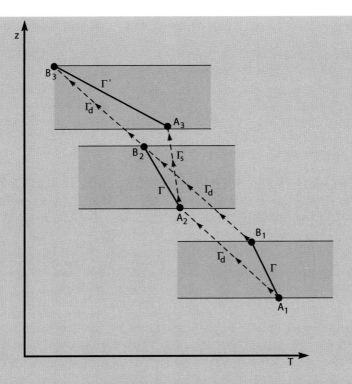

Figure 7.9 Successive positions and thermal structures of a potentially unstable layer (shaded), in which moisture decreases sharply with height, that is displaced vertically. Air is initially unsaturated, so all levels cool at the dry adiabatic lapse rate. However, displacement to position 2 leads to saturation of air at the layer's base (A_2). Thereafter, that air cools at the saturated adiabatic lapse rate, whereas air at the layer's top (B_2) continues to cool at the dry adiabatic lapse rate.

but in which equivalent potential temperature decreases with height,

$$\frac{d\theta_e}{dz} < 0. \tag{7.25.2}$$

By (7.8), the layer is stable. The difference $\theta_e - \theta$ for an individual parcel reflects the total latent heat available for release. Therefore, (7.25.2) represents a decrease with height of mixing ratio (e.g., as would develop over warm ocean).

Suppose that this layer is displaced vertically and that changes in its thickness can be ignored (Fig. 7.9). Because the layer is unsaturated, air parcels along a vertical section all cool at the dry adiabatic lapse rate Γ_d (e.g., between positions 1 and 2). The layer's profile of temperature is then preserved through the displacement. So is the layer's lapse rate and hence its stability.

Once some of the layer becomes saturated, this is no longer true. Lower levels, because they have greater mixing ratio, achieve saturation sooner than upper levels. Above their LCL, lower levels cool slower ($A_2 \rightarrow A_3$), at the saturated adiabatic lapse rate Γ_s. However, upper levels, which remain unsaturated, continue to cool at the dry adiabatic lapse rate Γ_d ($B_2 \to B_3$). Differential cooling between lower and upper levels then swings the temperature profile counter-clockwise, destabilizing the layer. This destabilization follows from the release of latent heat at lower levels, which heat the layer from below. Sufficient vertical displacement will render the layer unstable with respect to saturated conditions. This can occur through finite displacement, even if, initially, the layer was absolutely stable with respect to infinitesimal displacement. A layer for which this is possible is *potentially unstable*. The criterion for potential instability is (7.25.2). It implies an upward decrease of mixing ratio, as is often found above warm ocean.

A layer characterized by (7.25.2), but with the reversed inequality, is *potentially stable*. The criterion (7.25.2) then reflects an upward increase of mixing ratio. It accounts for the release of latent heat in upper levels, which, once they become saturated, stabilizes the layer. If a potentially stable or unstable layer becomes fully saturated, then it satisfies the usual criteria for instability – because all levels evolve in the same manner.

7.4.4 Modification of stability under unsaturated conditions

A layer's stability can also change through vertical compression and expansion, even under unsaturated conditions. Consider the layer in Fig. 7.10, which is displaced to a new height and thermodynamic state (distinguished by primed variables) from an initial height and state. Because θ is conserved under unsaturated conditions, the difference of θ across the layer is preserved:

$$d\theta' = d\theta.$$

Then the vertical gradient of potential temperature satisfies

$$\left(\frac{d\theta}{dz}\right)' dz' = \left(\frac{d\theta}{dz}\right) dz. \tag{7.26.1}$$

A horizontal section of layer having initial area dA must also satisfy conservation of mass:

$$\rho \, dA \, dz = \rho' \, dA' \, dz'. \tag{7.26.2}$$

Combining (7.26) obtains

$$\left(\frac{d\theta}{dz}\right)' = \left(\frac{d\theta}{dz}\right)\left(\frac{\rho'}{\rho}\right)\left(\frac{dA'}{dA}\right) \tag{7.27}$$

for the new vertical gradient of potential temperature inside the layer.

Sinking motion and horizontal divergence of air correspond to

$$\frac{\rho'}{\rho} > 1$$

$$\frac{dA'}{dA} > 1, \tag{7.28}$$

respectively. Each requires vertical compression of the layer: the former through a reduction of the layer's overall volume and the latter to compensate horizontal expansion of the layer. By (7.27), these motions steepen the vertical gradient of potential temperature. They increase the stability of a layer that is initially stable.

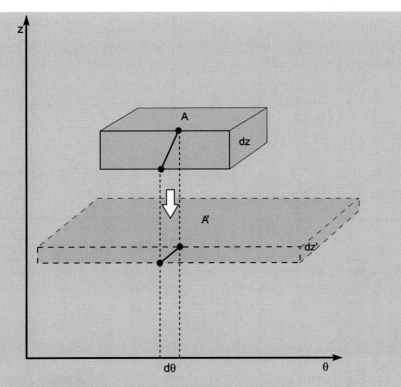

Figure 7.10 Successive positions and thermal structures of a layer that remains unsaturated and therefore preserves the potential temperature difference across it.

Similarly, they increase the instability of a layer that is initially unstable. For reasons developed in Chap. 12, stable air inside an anticyclone experiences sinking motion, or "subsidence." It is accompanied near the surface by horizontal divergence. Together, these features of motion inside an anticyclone intensify stability. They lead to the formation of a *subsidence inversion*. Found at heights as low as a kilometer, it caps vertical dispersion, trapping pollutants near the ground.

In a cyclone, just the reverse occurs: Air experiences ascending motion and horizontal convergence. These features of motion inside a cyclone imply (7.28) with the inequalities reversed. The direction in Fig. 7.10 is likewise reversed. Vertical expansion then weakens the vertical gradient of potential temperature. It drives a stable layer towards neutral stability. Ascending motion and horizontal convergence inside a cyclone can eliminate the stability of a layer, providing conditions favorable for convection.

7.5 STABILIZING AND DESTABILIZING INFLUENCES

The preceding development illustrates that stability can be modified by internal motion that rearranges air. More important to the overall stability of the atmosphere is external heat transfer, which shapes thermal structure. The vertical distribution of θ provides

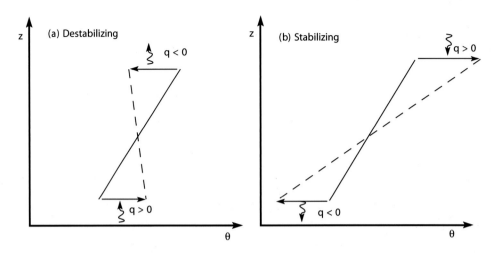

Figure 7.11 (a) Destabilizing influences: Heating from below and cooling from above act to swing the profile of potential temperature counterclockwise. (b) Stabilizing influences: Cooling from below and heating from above act to swing the profile of potential temperature clockwise.

insight into how stability is established because it reflects the effective stratification of mass.

In light of the First Law (2.36), the stability criteria (7.8) imply that a layer is desta-bilized by heating from below or cooling from aloft (Fig. 7.11a). Each rotates the ver-tical profile of θ counter-clockwise and thus toward instability. These are precisely the influences exerted on the troposphere by exchanges of energy with the Earth's surface and deep space. Absorption of LW radiation and transfers of latent and sen-sible heat from the Earth's surface warm the troposphere from below. LW emission at upper levels cools the troposphere from above. Both drive the vertical profile of θ counter-clockwise and the lapse rate toward superadiabatic values. By destabilizing the stratification, these forms of heat transfer maintain convective overturning. The latter, because it operates on short time scales, maintains the troposphere close to moist neutral stability. It makes the general circulation of the troposphere behave as a heat engine (Fig. 6.7), driven thermally by heat transfer from the Earth's surface.

Conversely, cooling from below or heating from above stabilize a layer (Fig. 7.11b). Each rotates the profile of θ clockwise and thus toward increased stability. These forms of heat transfer are symbolic of the stratosphere. There, ozone heating increases upward, stabilizing thermal structure. By inhibiting vertical motion, strong stability suppresses turbulent dispersion, forcing constituents to become stratified. Work must then be performed against buoyancy to drive vertical motion. This makes the general circulation of the stratosphere behave as a refrigerator (Fig. 6.8), driven mechanically by the absorption of wave activity that propagates up from below.

According to Secs. 7.2 and 7.4, moisture also figures in the stability of a layer. If the lapse rate is conditionally unstable (7.9.3), humidifying a layer lowers the LCL and hence the LFC of individual parcels. That, in turn, increases CAPE, which fuels deep convection (cf. Fig. 7.6). Likewise, humidifying lower portions of the layer (e.g., through contact with warm ocean) increases the equivalent potential temperature there. This

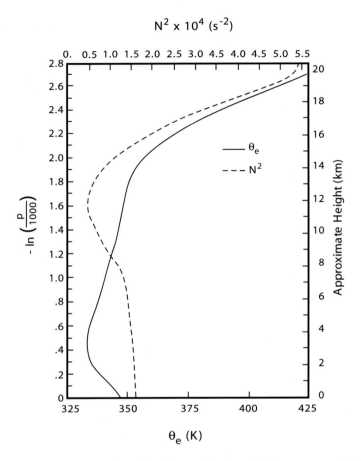

Figure 7.12 Mean vertical profiles of equivalent potential temperature (solid) and buoyancy frequency squared (dashed) over Indonesia, obtained by averaging radiosonde observations between longitudes of 90E and 105E and latitudes of 10S and 10N during March 1984 (cf. Fig. 1.30).

reflects a transfer of latent heat across the layer's lower boundary. It swings the profile of θ_e counter-clockwise, driving $\frac{d\theta_e}{dz}$ toward negative values. Thus, introducing moisture from below, as occurs over warm ocean, drives the troposphere toward potential instability.

For conditional and potential instability, finite displacements of air – whether they be of an individual air parcel or of a layer *in toto* – eventually result in buoyantly driven convection. However, for each, a finite potential well must be overcome before the instability can be released. Due to the source of water vapor at its base, lower portions of the tropical troposphere are potentially unstable. Figure 7.12 presents mean distributions of θ_e and N^2 over the equatorial western Pacific. $N^2 > 0$ implies that θ increases upward throughout (7.10.3). The layer is stable with respect to unsaturated conditions. However, in the lowest 3 km, θ_e decreases upward, making the layer potentially unstable. Despite such instability and, more generally, conditional instability, deep convection develops only in preferred regions, those where forced lifting is prevalent (Fig. 1.30). Over the Indian Ocean and western Pacific, equatorward-moving air is driven by thermal contrast between ocean and neighboring continent. It converges horizontally to produce forced ascent. The latter lifts air above the LFC, releasing it from the potential well that constrains air at lower levels. Similarly, diurnal heating of tropical landmasses imparts enough buoyancy for surface air over the maritime continent, South America, and tropical Africa to likewise overcome the potential well. Once released at the LFC, that air accelerates upward through work that is performed by buoyancy.

7.6 TURBULENT DISPERSION

7.6.1 Convective mixing

Vertical stability controls the development of vertical motion and hence mixing by 3D turbulence. Regions of weak or negative stability favor convective overturning, which results in efficient mixing of air. This situation is common near the ground because absorption of SW radiation destabilizes the surface layer. Introducing moisture from below has a similar effect. By increasing the equivalent potential temperature of surface air, it makes the overlying layer potentially unstable. When conditions are favorable, small disturbances evolve into fully developed convection. It, in turn, neutralizes the instability by rearranging mass vertically.

As noted earlier, the troposphere has a global-mean lapse rate of about 6.5 K km^{-1}. This is close to neutral stability with respect to saturated conditions ($\Gamma_s \cong 5.5$ °K/km). The mean stratification reflects two salient features of the troposphere: (1) Tropospheric air remains moisture laden, through contact with warm ocean. (2) It is efficiently overturned in cumulus and sloping convection. Efficient vertical exchange carries air between the surface and the upper troposphere in a matter of days, only hours inside cumulus towers. Because it operates on a time scale that is short compared with diabatic influences, convective mixing drives thermal structure of the troposphere toward $\theta_e = $ const and a mean state of moist neutral stability (Sec. 6.5).

Cumulus convection is favored in the tropical troposphere by high SST and strong insolation. Both support large transfers of radiative, latent, and sensible heat from the Earth's surface, which destabilize the troposphere. Opposing those destabilizing influences is cumulus convection. Reinforced by the release of latent heat, it maintains much of the tropical troposphere close to moist neutral stability. Cumulus convection also operates outside the tropics. However, there, it is organized by synoptic weather systems. Frontal motions associated with cyclones, like the one in the eastern Atlantic in Fig. 1.28, lift air through sloping convection. By displacing moist air that originated in the tropics, often conditionally or potentially unstable, those large-scale motions favor cumulus convection. Hence, they too drive stratification toward moist neutral stability.

7.6.2 Inversions

Contrary to the troposphere, the stratosphere is characterized by strong positive stability. Ozone heating makes temperature increase upward. Potential temperature then increases upward sharply (Fig. 6.5), from about 350 K near the tropopause to more than 1000 K near 10 hPa. The upward vertical gradient of θ produces a strong positive restoring force. It suppresses vertical motion and 3D turbulence, forcing chemical constituents to become stratified.

Unlike the troposphere, where air is efficiently overturned, the stratosphere has a characteristic time scale for vertical exchange of months to years. Its layered structure is occasionally illustrated by volcanic eruptions, which hurl debris above the tropopause. During the 1950s and 1960s, nuclear tests provided similar illustrations. Aerosol particles introduced by those sources have very slow settling rates. Consequently, they move with individual bodies of air, behaving as tracers. Observed by satellite, stratospheric aerosol exhibits little or no vertical motion. Instead, it fans out horizontally, forming cloud of global dimension that remains in the stratosphere for many months (see Fig. 9.6).

The long residence time of stratospheric aerosol enables it to alter SW absorption at the Earth's surface for extended durations. This feature makes volcanic aerosol a consideration for climate. The great eruption of Krakatoa destroyed the Indonesian island of the same name in 1883. Manifest in the color of sunsets, it modified radiative properties of the stratosphere for years afterwards. Pinatubo exerted a similar influence. As is apparent in Fig. 1.27, Pinatubo influenced global-mean surface temperature by reducing the transmission of SW radiation. Through cumulus convection, which was affected indirectly, it also influenced water vapor in the upper troposphere.

Tropospheric aerosol is introduced routinely by deforestation. It was magnified during the Kuwaiti oil fires, which followed the Persian Gulf War. In contrast to stratospheric aerosol, tropospheric aerosol is not long lived. It is removed on a time scale of only days, by rain out and other scavenging processes associated with convection (Chap. 9).

Convection does not penetrate appreciably above the tropopause, where N^2 and the restoring force of buoyancy increase sharply (Fig. 7.12). Upon encountering the sharp increase of stability, a cumulus updraft is quickly drained of positive buoyancy. It fans out horizontally to form cloud anvil, as typifies the mature stage of cumulonimbus thunderstorms (Fig 7.7). Similar behavior is occasionally observed at intermediate levels of the troposphere, when amplifying cumulus encounter a shallow inversion through which the updraft is able to penetrate (see Fig. 9.18).

Temperature structure inside an anticyclone is often inverted. A subsidence inversion can form less than a kilometer above the surface (Sec. 7.4). It confines pollutants to a shallow layer near the ground, where they are introduced. Because wind inside an anticyclone also tends to be light, pollutants are neither dispersed nor advected away from source regions. Their concentrations therefore increase during periods of high emission. The same mechanisms inhibit cloud formation by confining moisture near the surface. Consequently, anticyclones are typified by cloud-free but polluted conditions.

7.6.3 Life cycle of the nocturnal inversion

Heat transfer from the Earth's surface and convection compete for control of stratification. Their competition is illustrated by the formation and breakup of the *nocturnal inversion*, strong stability near the surface that often develops at night. Arid and cloud-free conditions permit efficient LW cooling to space. Surface temperature then decreases sharply after sunset. To preserve thermal equilibrium, heat is absorbed, radiatively and conductively, from the overlying air. This process cools the surface

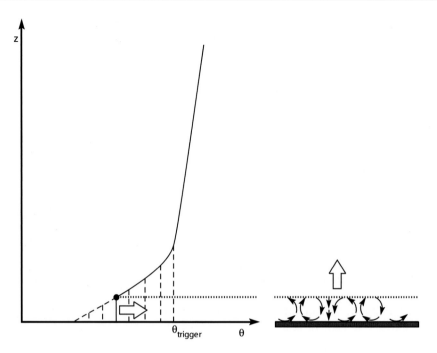

Figure 7.13 Schematic illustrating the breakdown of the nocturnal inversion, which forms during night. Following sunrise, surface heating and heat transfer to the overlying air destabilize a shallow layer adjacent to the ground. Convection then develops spontaneously and, through vertical mixing, restores the thermal structure inside that layer to neutral stability. Continued heat transfer from the ground requires θ to be mixed over a progressively deeper layer, which erodes the nocturnal inversion from below. When convection has advanced through the entire layer of strong stability, the nocturnal inversion has been destroyed. Deep convection can then penetrate into weakly stable air overhead, dispersing pollutants that were previously trapped near the surface in stable air.

layer from below, stabilizing it. Sufficient cooling will swing the temperature profile inverted (Fig. 7.13). Potential temperature near the ground then increases upward sharply.

Strongly stable, the nocturnal inversion suppresses vertical motion. It also suppresses 3D turbulence, which requires vertical eddy motion. Without turbulent dispersion, pollutants emitted at the surface are frozen in the stable air. Their concentrations therefore increase steadily near pollution sources. Such behavior is particularly noticeable in urban areas, during early morning periods of heavy traffic.

After sunrise, absorption of SW radiation warms the ground. It influences the surface layer in just the reverse manner. To preserve thermal equilibrium, heat is then transferred to the overlying air, initially through conduction and radiative transfer. These forms of heat transfer warm the surface layer from below. They increase potential temperature near the ground, swinging its vertical profile counter clockwise. In this manner, a shallow layer near the surface is destabilized. Thermal convection then develops spontaneously, neutralizing the instability. By rearranging air vertically, heat is transferred from the surface (high θ) and mixed vertically across a shallow layer of convective overturning. Inside that layer, θ is homogenized, neutralizing the layer's stability. The same process operates on other conserved properties, such as the mixing ratios of water vapor and pollutants. Like θ, each is driven toward a uniform vertical profile inside the layer of convective overturning (see Fig 13.6).

Absorption of SW radiation and upward heat transfer continue to destabilize the surface layer. That, in turn, maintains convective overturning. To preserve neutral stability, additional heat transferred to the atmosphere must be mixed vertically over a deepening layer. The layer driven toward neutral stability ($\theta \cong$ const) therefore expands upward. So do pollutants that were frozen near the surface in stable air. As surface heating continues, convective mixing systematically erodes the nocturnal inversion from below (Fig. 7.13). Simultaneously, it reduces pollution concentrations by diluting them over an ever deeper layer.

After sufficient surface heating, convection will have advanced through the entire nocturnal inversion. At that point, it reaches air of weak stability overhead (e.g., air that was unaffected by surface cooling during the night). The surface temperature at which this condition is achieved is the *break* or *trigger temperature*. At this temperature, deep convection can develop from the source of positive buoyancy at the ground. Pollutants are then dispersed vertically over a deep volume. They are also dispersed horizontally by 3D turbulence, which develops spontaneously. Together, these forms of transport result in a sharp reduction of pollution concentrations. The latter is often attended by an abrupt improvement in visibility. By permitting vertical transport of water vapor, the breakup of the nocturnal inversion also enables the formation of convective cloud. Cumulus appears soon thereafter.

7.7 RELATIONSHIP TO OBSERVED THERMAL STRUCTURE

Thermal structure is strongly influenced by convection, which operates on a time scale much shorter than radiative heating. In the tropics, cumulus convection is the dominant form of vertical motion. The height to which it prevails is dictated by CAPE (Sec. 7.4.1). Produced by transfers of heat from the Earth's surface, CAPE determines the LNB, which caps deep convection. It is reflected in the θ_e of surface air.

An air parcel initially at the surface will, if lifted above the LFC, accelerate upward. By the time it reaches the upper troposphere, nearly all of the parcel's water vapor will have been removed through condensation, attended by the release of its latent heat. The pseudo isentrope ($\theta'_e =$ const) along which the parcel evolves will then coincide with an isentrope ($\theta' =$ const); see Prob. 5.20. Because $p' = p$ (mechanical equilibrium), the condition of neutral buoyancy that defines the LNB (7.3)

$$T'_{LNB} = T_{LNB} \tag{7.29.1}$$

is then equivalent to

$$\theta'_{LNB} = \theta_{LNB}. \tag{7.29.2}$$

As the parcel is now devoid of water vapor, this reduces to the condition

$$\theta_{LNB} \cong \theta'_e. \tag{7.30}$$

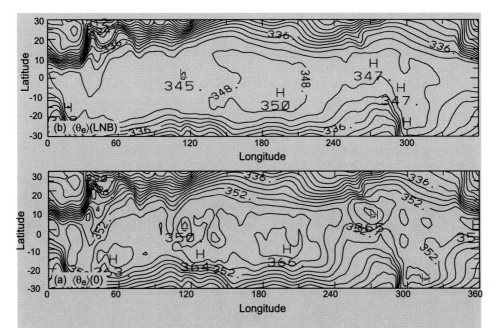

Figure 7.14 (a) Geographical distribution of equivalent potential temperature at the Earth's surface (mean plus 1 std deviation), $\theta_e(0)$. (b) As in (a),
but at the Level of Neutral Buoyancy (LNB), accounting for entrainment of
cumulus updrafts (7.21). After Salby et al. (2003).

The LNB corresponds to the height where environmental θ equals θ'_e of cumulus
updrafts.

Under adiabatic conditions (e.g., in the absence of entrainment), (7.30) simplifies
to

$$\theta_{LNB} = \theta_e(z = 0)$$

because θ'_e is then conserved, equal to its initial value. In the presence of entrainment, θ'_e decreases upward. Reflecting diabatic losses of sensible and latent heat,
this reduces the limiting values of θ', which is no longer constant even in the upper
troposphere. By (7.29.2), entrainment therefore lowers the LNB.

Plotted in Fig. 7.14a is the horizontal distribution of $\theta_e(z = 0)$ during November 1991 through February 1992. Values maximize in the equatorial Pacific, where
$\theta_e(z = 0)$ approaches 370 K. That value corresponds to θ_{LNB} under adiabatic conditions. The isentropic surface with $\theta = 370$ K is positioned near the tropical
tropopause, just above 100 hPa. Although deep cumulus form in the equatorial
Pacific during this period, few reach this high. Instead, cloud coverage decreases
sharply above 200 hPa (see Figs 9.24, 9.25).

Plotted in Fig. 7.14b is the horizontal distribution of θ_{LNB}, with entrainment
accounted for through (7.21). The pattern resembles that of $\theta_e(z = 0)$, but with
values reduced. θ_{LNB} has values of about 345 K over most of the tropics, 350 K in
the equatorial Pacific. $\theta = 345$ K positions the LNB near 200 hPa, several kilometers
lower than is possible under adiabatic conditions. It is at this level that cloud anvil
is most extensive and air diverges horizontally (Fig. 9.26a).

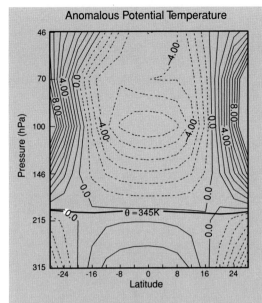

Figure 7.15 Departure of environmental θ from its NS mean. Superimposed is the zonal-mean isentrope $\theta \cong \theta_e = 345$ K (bold), which marks the LNB of cumulus updrafts (cf. Fig. 7.14b). Adapted from Salby et al. (2003).

The mechanism that lowers the LNB below its adiabatic level is entrainment. By incorporating sensible and latent heat from the environment, it dilutes the buoyancy of a cumulus updraft over an expanding volume (cf. Figs. 5.4, 9.16). Simultaneously, the environment is ventilated by cumulus detrainment. Transfers of heat and moisture from the updraft force environmental thermal structure, driving it toward the vertical structure inside cumulus updrafts.

Figure 7.15 plots, as a function of latitude and pressure in the upper troposphere, the departure of environmental θ from its NS mean. Contemporaneous with θ_{LNB} in Fig. 7.14b, the anomaly reveals layers of heating and cooling. In the tropics, anomalous θ is positive at levels below 200 hPa, the LNB of cumulus updrafts. From (2.36), it reflects heating of the environment, which increases θ at most levels of the tropical troposphere.[4] The layer of heating coincides with deep convection, as comprises the ITCZ (Fig. 1.30). It prevails at levels up to the LNB. At those levels, cumulus updrafts are positively buoyant. They are therefore warmer than their surroundings ($\theta' > \theta$). Cumulus detrainment of high-θ air thus warms the environment, serving as a heat source. The same process transfers total water from updrafts. It therefore humidifies the environment, serving as a moisture source (Sec. 9.3.4).

Above 200 hPa, anomalous θ reverses sign. The reversal is nearly coincident with the zonal-mean isentrope $\theta \cong \theta_e = 345$ K (bold), which marks the LNB of cumulus updrafts; cf. Fig. 7.14b. Anomalous θ becomes strongly negative near 100 hPa, weakening at higher levels. From (2.36), negative values reflect cooling of the environment. Appearing above the LNB, the layer of cooling coincides with convective overshoots (Fig. 7.7). It extends a couple of kilometers above the LNB. Convective overshoots at those levels are negatively buoyant. Hence they are colder than their

---

[4] Except at lower levels, areal coverage by cloud is small. It decreases in the upper troposphere to only a few percent (cf. Fig. 9.24). Consequently, large-scale thermal structure in Fig. 7.15 describes regions that are chiefly cloud-free.

surroundings ($\theta' < \theta$). Cumulus detrainment of low-θ air therefore cools the environment, serving as a heat sink. This process drives environmental thermal structure toward the nearly uniform vertical distribution of θ inside cumulus updrafts. By driving Γ toward Γ_d and neutralizing stability, it acts to raise the tropopause and decrease its temperature (cf. Fig. 7.8).

SUGGESTED REFERENCES

Vertical stability is treated in detail in *Atmospheric Thermodynamics* (1981) by Iribarne and Godson.

Environmental Aerodynamics (1978) by Scorer contains an advanced treatment of entrainment, including laboratory simulations.

PROBLEMS

A pseudo-adiabatic chart is provided in Appendix F.

1. Use Newton's 2nd law to derive the momentum balance (7.1.1) for an individual air parcel.

2. In early morning, the profile of potential temperature in a layer adjacent to the surface is described by

$$\frac{\theta}{\theta_0} = e^{-\frac{z}{h_1}} - \frac{z}{h_2},$$

   with $h_1 > h_2$. If condensation can be ignored, (a) characterize the static stability of this layer, (b) determine the profile of buoyancy frequency, (c) determine the limiting distribution of θ for $t \to \infty$, after sufficient surface heating for convection to develop.

3. Use Archimedes' principle to establish (7.2.2).

4. A simple harmonic oscillator comprised of a 1-kg weight and an elastic spring with constant k provides an analogue of buoyancy oscillations. Calculate the effective spring constant for unsaturated conditions representative of (a) the troposphere: $\Gamma = 6.5$ °K km^{-1} and $T = 260$ °K, (b) the stratosphere: $\Gamma = -4$ °K km^{-1} and $T = 250$ °K.

5. An impulsive disturbance imposes a vertical velocity w_0 on an individual air parcel. If the layer containing that parcel has nonnegative stability N, (a) determine the parcel's maximum vertical displacement, as a function of N, under linear conditions, (b) describe the layer's evolution under nonlinear conditions, in relation to N.

6. Upstream of isolated rough terrain, the surface layer has thermal structure

$$T(z) = T_0 \frac{a}{a+z},$$

   where $a > 0$ and $T_0 > a\Gamma_d$. (a) Determine the upstream profile of potential temperature. (b) Characterize the upstream vertical stability. (c) Describe the mean stratification far downstream of the rough terrain.

7. Derive the identity

$$\frac{1}{\theta}\frac{d\theta}{dz} = \frac{1}{T}\frac{dT}{dz} + \frac{\kappa}{H}$$

   which relates the vertical gradients of temperature and potential temperature.

8. An early morning sounding reveals the temperature profile

$$T(z) = \begin{cases} T_0 + \left(\frac{\Gamma_d}{2}\right) z & z < 1 \text{ km} \\ T_0 + \left(\frac{\Gamma_d}{2}\right) - \left(\frac{\Gamma_d}{4}\right)(z - 1) & z \geq 1 \text{ km}. \end{cases}$$

(a) Determine the *trigger temperature* for deep convection to develop. (b) If, following sunrise, the ground warms at $3°C$ per hour, estimate the time when cumulus clouds will appear.

9. Radiative heating leads to temperature in the stratosphere increasing up to the stratopause, above which it decreases in the mesosphere. If the temperature profile at a certain station is approximated by

$$T(z) = T_0 \frac{z}{b + \left(\frac{z}{h}\right)^2},$$

determine (a) the profile of potential temperature, (b) the profile of Brunt-Väisäillä frequency squared N^2. (c) For reasons developed in Chap. 14, gravity waves can propagate through regions of positive static stability, which provides the positive restoring force for their oscillations. In terms of the variable b, how small must h be to block propagation through the mesosphere? (d) By noting that gravity waves introduce vertical displacements, discuss the behavior of air in the mesosphere under the conditions in (c).

10. Analyze the stability and energetics for the stratification and parcel in Fig. 7.6, but under initially saturated conditions.

11. (a) To what height would the parcel in Prob. 7.10 penetrate under purely adiabatic conditions? (b) How/why does this differ from reality? (See Prob. 7.21)

12. Estimate the maximum height of convective towers under the conditions in Fig. 7.6, but with the isothermal stratosphere replaced by one with a lapse rate of $-3°K \text{ km}^{-1}$.

13. The approach of a warm front is often heralded by deteriorating visibility and air quality. Consider a frontal surface that slopes upward and to the right, separates warm air on the left from cold air on the right, and moves to the right. The intersection of this surface with the ground marks the front. (a) Characterize the vertical stability at a station ahead of the front. (b) Discuss the vertical dispersion of pollutants that are introduced at the ground. (c) Describe how the features in (a) and (b) and pollution concentrations evolve as the front approaches the station.

14. A morning temperature sounding over Florida reveals the profile

$$T = \begin{cases} 20 - 8(z - 1) \ (°C) & z \geq 1 \text{ km} \\ 20 \ (°C) & z < 1 \text{ km}. \end{cases}$$

If the mixing ratio at the surface is 20 g kg^{-1}, determine the LFC.

15. The potential temperature inside a layer varies as

$$\frac{\theta}{\theta_0} = -(z - a)^2 + a^2 + 1 \qquad z > 0.$$

(a) Characterize the stability of this layer. (b) How large a vertical displacement would a parcel originally at $z = 0$ have to undergo to be released from this layer?

16. Establish stability criteria analogous to (7.8) under saturated conditions.

17. Rayleigh friction approximates turbulent drag in proportion to an air parcel's velocity:

$$D = Kw,$$

where D is the specific drag experienced by the parcel, w is its velocity, and the Rayleigh friction coefficient K has dimensions of inverse time. Show that, in the presence of Rayleigh friction, vertical displacements remain bounded under conditions of neutral stability.

18. Show that, for given Rayleigh friction coefficient K, vertical displacements remain bounded for small instability but become unbounded for sufficiently large instability.

19. The lapse rate at a certain tropical station is constant and equal to $7\,^\circ\mathrm{K\,km^{-1}}$ from 1000 hPa to 200 hPa, above which conditions are isothermal. At the surface, the temperature is 30°C and the relative humidity is sufficiently great for vertical displacements to achieve immediate saturation. Approximate the saturated adiabat by the constant lapse rate: $\Gamma_s \cong 6.5\,^\circ\mathrm{K\,km^{-1}}$, to estimate the maximum vertical velocity attainable inside convective towers. (see Prob. 7.21)

20. The weather service forecasts the intensity of convection in terms of a *thermal index TI*, which is defined as the local ambient temperature minus the temperature anticipated inside a thermal under adiabatic conditions. The more negative is TI, the stronger are updrafts inside thermals. A morning sounding over an arid region reveals a lapse rate of $-4\ \mathrm{K\,km^{-1}}$ in the lowest kilometer and a constant lapse rate of $6\ \mathrm{K\,km^{-1}}$ above, with the surface temperature being 10°C. If air is sufficiently dry to remain unsaturated and if the surface temperature is forecast to reach a maximum of 40°C, (a) determine the thermal index then as a function of height for $z > 1$ km, (b) to what height will updrafts intensify? (c) Calculate the maximum updraft and maximum height reached by updrafts anticipated under adiabatic conditions if, at the time of maximum temperature, mean thermal structure has been driven adiabatic in the lowest kilometer.

21. In practice, vertical velocities inside a thermal differ significantly from adiabatic values. Entrainment limits the maximum updraft to that near the level of maximum buoyancy. Under the conditions in Prob. 7.20, but accounting for entrainment, calculate the profile of kinetic energy and note the height of maximum updraft inside a thermal for an entrainment height of (a) infinity, (b) 1 km, (c) 100 m.

22. Establish relationship (7.21) that governs behavior inside an entraining thermal.

23. Consider the conditions in Prob. 7.19, but in light of entrainment. If the ambient mixing ratio varies as $r = r_0 e^{-\frac{z}{h}}$, with $h = 2$ km, and the entrainment height is 1 km, calculate the profile of equivalent potential temperature inside a moist thermal.

24. Consider a constant tropospheric lapse rate of $\Gamma = 7\,^\circ\mathrm{K\,km^{-1}}$ and mixing ratio that varies as $r = r_0 e^{-\frac{z}{h}}$, with $h = 2$ km. If the troposphere can be regarded as a layer close to saturation, evaluate its stability to large-scale lifting (e.g., in sloping convection) in the presence of moisture that is representative of (a) mid-latitudes: $r_0 = 5\ \mathrm{g\,kg^{-1}}$, (b) the tropics: $r_0 = 25\ \mathrm{g\,kg^{-1}}$.

Radiative transfer

The thermal structure and stratification discussed in Chaps. 6 and 7 are shaped in large part by radiative transfer. According to the global-mean energy budget (Fig. 1.32), of some 542 W m^{-2} that is absorbed by the atmosphere, more than 80% is supplied through radiative transfer: 368 W m^{-2} through absorption of LW radiation from the Earth's surface and another 68 W m^{-2} through direct absorption of SW radiation. Similarly, the atmosphere loses energy through LW emission to the Earth's surface and to space.

To maintain thermal equilibrium, components of the atmospheric energy budget must, on average, balance. Vertical transfers of radiant energy involved in this balance are instrumental in determining thermal structure and motion that characterize individual layers. Likewise, the dependence of radiative transfer on latitude, temperature, and cloud introduces horizontal variations into the energy budget. Those variations lead to net radiative heating at low latitude and net radiative cooling at middle and high latitudes (Fig. 1.34c). To maintain thermal equilibrium locally, the latter must be compensated by a poleward transfer of heat. This heat transfer is accomplished by the general circulation. The general circulation, in turn, is driven by the horizontal distribution of radiative heating and cooling. Understanding these features of the Earth-atmosphere system therefore requires an understanding of how radiation interacts with the atmosphere.

8.1 SHORTWAVE AND LONGWAVE RADIATION

Energy transfer in the atmosphere involves radiation in two distinct bands of wavelength: shortwave radiation emitted by the sun and longwave radiation emitted by the Earth's surface and atmosphere. Owing to the disparate temperatures of the emitters, these two forms of radiation are concentrated at wavelengths λ that are widely separated. Figure 8.1a shows blackbody emission spectra for temperatures of 6000°K and 288°K, corresponding to the equivalent blackbody temperature of the sun and the global-mean surface temperature of the Earth.[1] The spectrum of SW radiation is concentrated at wavelengths shorter than 4 μm. It peaks in the visible near $\lambda = 0.5$ μm. Wings of the SW spectrum extend into the UV ($\lambda < 0.3$ μm) and into the near-infrared ($\lambda > 0.7\mu$m). On the other hand, the spectrum of LW radiation peaks well into the IR, at a wavelength of about 10 μm. Wings of the LW spectrum extend down to wavelengths of about 5 μm and out to the microwave region ($\lambda > 100$ μm). Overlap between the spectra of SW and LW radiation is negligible. They may therefore be treated independently by considering wavelengths shorter and longer than 4 μm.

Also shown as a function of wavelength is the fractional absorption of radiation passing from the top of the atmosphere (TOA) to a height of 11 km (Fig. 8.1b) and to the ground (Fig. 8.1c). Energetic radiation: $\lambda < 0.3$ μm, is absorbed at high levels through photodissociation and ionization of O_2 and O_3. As a result, very little of the UV incident on TOA reaches the tropopause. By contrast, most of the spectrum in the visible and near-IR arrives at the tropopause unattenuated. However, in passing from 11 km to the ground, the remaining SW spectrum is substantially absorbed in the IR by water vapor and carbon dioxide, absolute concentrations of which are large in the troposphere. Consequently, the spectrum of SW radiation reaching the surface is concentrated at visible wavelengths. To those wavelengths, the atmosphere is mostly transparent.

In contrast to SW radiation, LW radiation emitted by the Earth's surface is almost completely absorbed – by H_2O across a wide band centered at 6.3 μm and another in the far-infrared, by CO_2 in a band centered at 15 μm (near the peak of the LW emission spectrum), and by a variety of trace gases including O_3, CH_4, and N_2O. Therefore, most of the LW energy emitted by the ground is captured in the overlying layer of air. That energy must be re-emitted, half upward and half back downward. The upward component is then absorbed again in the next overlying layer. It must likewise be re-emitted, half upward and half back downward. In this fashion, energy emitted by the Earth's surface undergoes repeated absorption and re-emission in adjacent atmospheric layers until, eventually, it is rejected to space. This sequence of radiative exchanges traps LW radiation and, through mechanisms developed below, elevates the surface temperature of the Earth.

Comparing Figs. 8.1b and 8.1c indicates that most of the LW energy emitted by the Earth's surface is absorbed in the troposphere. Only in the *atmospheric window* at wavelengths of 8–12 μm is absorption weak enough for much of the LW radiation emitted by the surface to pass freely to space. The 9.6-μm band of ozone, which is positioned inside this window, is the only strong absorber at those wavelengths. Most of that absorption takes place in the stratosphere, where ozone is concentrated.

8.1.1 Spectra of observed SW and LW radiation

The most important property of the SW spectrum is the *solar constant*, which equals the flux of radiant energy that arrives at TOA (at the mean Earth–sun distance), integrated over wavelength. The solar constant is weakly variable, having a value of about

---

[1] The representation in Fig. 8.1a, logarithmic in wavelength versus wavelength times energy flux, is area preserving. Equal areas under a curve in different bands of wavelength represent equal power.

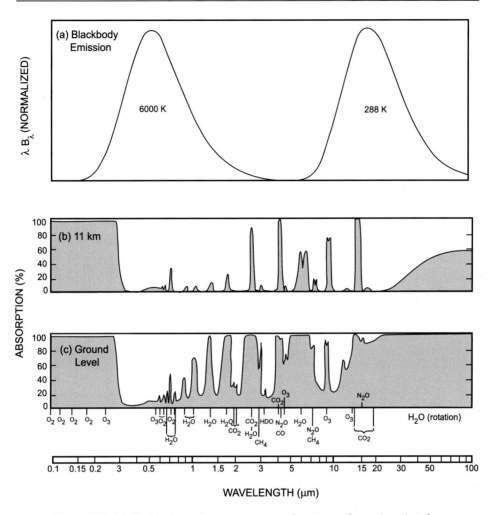

Figure 8.1 (a) Blackbody emission spectra as functions of wavelength λ for temperatures corresponding to the sun and the Earth's global-mean surface temperature. The representation: λ times energy flux vs $\log(\lambda)$, is area preserving – equal areas under a spectral curve represent equal power. (b) Absorption as a function of λ along a vertical path from the top of the atmosphere to a height of 11 km. (c) As in (b), but to the surface. Adapted from Goody and Yung (1989).

1370 W m^{-2}. When distributed over the Earth, this SW flux leads to the daily *insolation* shown in Fig. 1.33, which follows primarily from the length of day and solar inclination for particular latitudes and times of year.

Spectra of observed SW and LW radiation are shaped by the absorption characteristics described above. At TOA (Fig. 8.2), the solar spectrum resembles the emission spectrum of a blackbody at about 6000°K. That temperature is characteristic of the sun's *photosphere*, where most of the radiation is emitted. Wavelengths shorter than 300 nm deviate from this simple picture. Instead, they reflect temperatures of 4500–5000°K, which are found somewhat higher, near the base of the sun's *chromosphere*.

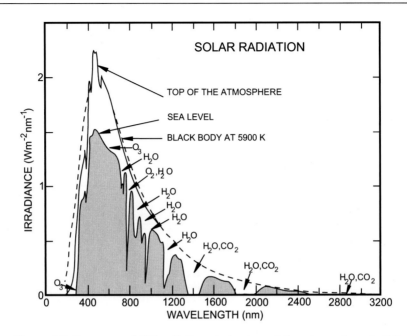

Figure 8.2 Spectrum of SW radiation at the top of the atmosphere (solid) and at the Earth's surface (stippled), compared against the emission spectrum of a blackbody at 6000 K (dashed). Individual absorbing species indicated. Adapted from Coulson (1975).

At the Earth's surface, the wings of the solar spectrum have been mostly clipped off, in the UV by ozone and molecular oxygen and in the IR by absorption bands of water vapor and carbon dioxide. SW radiation is then limited to wavelengths of 0.3–2.4 μm.

Wavelengths shorter than 300 nm account for a small fraction of the SW energy flux. They are absorbed fairly high in the Earth's atmosphere through photodissociation and ionization. As shown in Fig. 8.3, these energetic components penetrate to altitudes in inverse relation to their wavelengths. Wavelengths shorter than 200 nm do not penetrate appreciably below 50 km. Wavelengths shorter than 150 nm are confined even higher, to the thermosphere. Exceptional is the discrete *Lyman-α emission* at 121 nm. It corresponds to transitions from the gravest excited state of hydrogen. The Lyman-α line happens to coincide with an optical window in the banded absorption structure at these wavelengths. This feature enables Lyman-α radiation to penetrate into the lower thermosphere and upper mesosphere, where it is involved in photo-ionization and dissociation.

The wavelength of SW radiation is also inversely related to the altitude in the solar atmosphere from which it is emitted. As indicated in Fig. 8.4, wavelengths longer than 300 nm originate in the photosphere. Wavelengths shorter than 200 nm are emitted from the chromosphere. Very energetic components, with wavelengths shorter than 50 nm, originate even higher, in the corona. Solar variability increases with altitude in the solar atmosphere. For this reason, temporal variations in the SW spectrum are concentrated in the far UV. Those wavelengths make only a minor contribution to the solar constant.

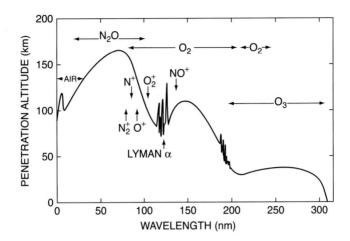

Figure 8.3 Penetration altitude, where SW flux at wavelength λ is attenuated by a factor of e^{-1}. Absorbing species and thresholds for photo-ionization indicated. After Herzberg (1965).

Two phenomena dominate solar variability. The 27-day rotation cycle of the sun leads to variations of several percent in the UV. However, it decreases to less than 1% at wavelengths longer than 250 nm (WMO, 1986). Solar activity also varies with the prevalence of dark regions, *sunspots*. The latter evolve through an 11-year cycle that modulates solar emission. It is noteworthy that, according to proxy evidence, solar activity experienced a distinct minimum during the seventeenth century. The so-called *Maunder minimum* coincided with the Little Ice Age (Sec. 1.6.2). Between extremes of the 11-year solar cycle, *solar-max* and *solar-min*, activity changes noticeably at outer reaches of the sun's atmosphere. Owing to the inverse relationship between solar altitude and the emitted wavelength, SW variability decreases sharply with increasing λ (Fig. 8.4). This sharply limits the altitudes in the Earth's atmosphere that are affected by such variations. At $\lambda = 160$ nm, the peak-to-peak variation in flux between solar-max and solar-min is about 10%. By 300 nm, however, it has decreased to less than 1%. In terms of the solar constant and the majority of energy reaching the Earth's surface, the variation between solar-max and solar-min is of order 0.1% (Willson et al., 1986).

The spectrum of LW radiation emitted to space by the Earth-atmosphere system is presented in Fig. 8.5. Measured by satellite, it too resembles the emission spectrum of a blackbody. However, it corresponds to a temperature of 288 K only at wavelengths in the atmospheric window: 8–12 μm (wavenumbers λ^{-1} of 800–1200 cm^{-1}). Only there does radiation emitted by the Earth's surface pass freely to space. Other wavenumbers are trapped, absorbed, and then re-emitted by overlying water vapor and cloud. Those atmospheric absorbers emit radiation to space at temperatures colder than the Earth's surface.

The strong absorption bands of CO_2 and O_3 sharply reduce emission to space at 15 μm and 9.6 μm, respectively. Outgoing radiation at those wavelengths corresponds to blackbody temperatures that are distinctly colder than that within the atmospheric window, where outgoing radiation emanates directly from the Earth's surface. The 15-μm absorption of CO_2 is positioned near the center of the LW emission spectrum. It takes a bite out of the spectrum that would otherwise be emitted to space.

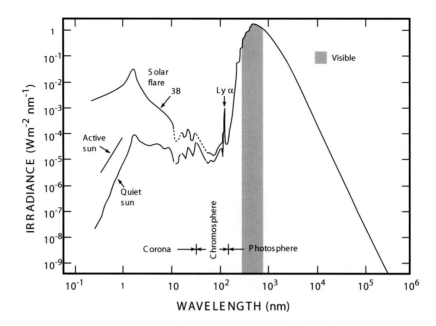

Figure 8.4 Spectrum of solar irradiance (flux) as a function of wavelength under normal and disturbed conditions. Visible band and emission levels in the solar atmosphere indicated. Adapted from Smith and Gottlieb (1974) with permission of Kluwer Academic Publishers.

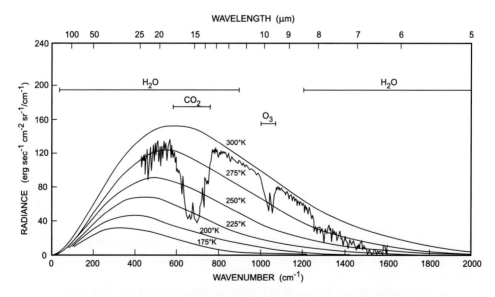

Figure 8.5 Spectrum of outgoing LW radiation over (215 W, 15 N) observed by Nimbus-4 IRIS, as a function of wavenumber λ^{-1}. Blackbody spectra for different temperatures and individual absorbing species indicated. Adapted from Liou (1980).

Radiation at those wavelengths corresponds to a blackbody temperature of about 220 K. That temperature reflects an altitude in the upper troposphere, above most of the column abundance of CO_2 (Fig. 1.7). The 9.6-μm absorption of ozone corresponds to a temperature of about 270 K. That temperature reflects an altitude in the upper stratosphere, above most of the column abundance of ozone. Along with water vapor, these minor constituents block LW radiation that is emitted by the Earth's surface. By absorbing that radiation and re-emitting part of it back downward, they increase surface temperature over what it would be in the absence of an atmosphere (namely, over the effective blackbody temperature of 255 K). Known as the *greenhouse effect*, this process follows from the atmosphere's selective transmissivity: It is transparent to SW radiation that arrives from the sun. However, it is opaque to LW radiation that is re-emitted by the Earth's surface.

8.2 DESCRIPTION OF RADIATIVE TRANSFER

8.2.1 Radiometric quantities

Understanding how radiation influences atmospheric properties requires a quantitative description of radiative transfer. The latter is complicated by the 3-dimensional nature of radiation and its dependence on wavelength and directionality. As illustrated in Fig. 8.6, radiant energy traversing a surface with unit normal n has contributions from all directions, which are characterized by the unit vector $\hat{\Omega}$. Such multidimensional radiation is termed *diffuse*. If it arrives along a single direction (like incoming solar), it is termed *parallel-beam radiation*. The monochromatic *intensity* or *radiance* is given by $I_\lambda = I_\lambda \hat{\Omega}$. It represents the rate that energy of wavelengths λ to $\lambda + d\lambda$ flows per unit area through an increment of solid angle $d\Omega$ in the direction $\hat{\Omega}$. Having dimensions of $\frac{power}{wavelength \cdot area \cdot steradian}$, $I_\lambda(x, \hat{\Omega})$ is a function of position x and direction $\hat{\Omega}$. It characterizes a pencil of radiation that traverses the surface at a *zenith angle θ* from its normal. Then the rate energy flows in the direction $\hat{\Omega}$, per unit cross-sectional area and wavelength interval, is $I d\Omega$.

The component of intensity normal to the surface is $I_\lambda \cdot n = I_\lambda (\hat{\Omega} \cdot n) = I_\lambda \cos\theta$. Integrating it over the half-space of solid angle in the positive n direction defines the monochromatic *flux* or *irradiance* crossing the surface

$$
\begin{aligned}
F_\lambda^+ &= \int_{2\pi} I_\lambda \cdot n \, d\Omega^+ \\
&= \int_{2\pi} I_\lambda \cos\theta \, d\Omega^+ \\
&= \int_0^{2\pi} \int_0^{\frac{\pi}{2}} I_\lambda(\phi, \theta) \cos\theta \sin\theta \, d\theta \, d\phi.
\end{aligned}
\tag{8.1}
$$

F_λ^+ has dimensions of $\frac{power}{area \cdot wavelength}$. The *total flux* then follows by integrating over wavelength

$$
F^+ = \int_0^\infty F_\lambda^+ d\lambda,
\tag{8.2}
$$

which has the dimensions of energy flux: $\frac{power}{area}$. The monochromatic flux F_λ^+ thus represents the spectral density at wavelength λ of the total flux F^+.

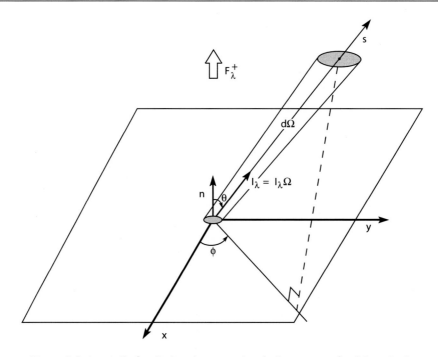

Figure 8.6 A pencil of radiation that occupies the increment of solid angle $d\Omega$ in the direction $\hat{\Omega}$ and traverses a surface with unit normal n. The monochromatic *intensity* or *radiance* passing through the pencil: $\boldsymbol{I}_\lambda = I_\lambda \hat{\boldsymbol{\Omega}}$, describes the rate at which energy inside the pencil crosses the surface per unit area, steradian, and wavelength. Integrating the component normal to the surface: $I_\lambda(\hat{\boldsymbol{\Omega}} \cdot \boldsymbol{n}) = I_\lambda \cos\theta$, over the half-space of 2π steradians in the positive n direction yields the monochromatic forward *flux* or *irradiance* F_λ^+.

F_λ^+ describes the forward energy flux in the n direction. The backward energy flux F_λ^- may be obtained by integrating I in the negative n direction over the opposite half-space. Both are nonnegative quantities. The monochromatic *net flux* in the n direction is then the difference between the forward and backward components

$$F_\lambda = F_\lambda^+ - F_\lambda^-. \tag{8.3}$$

Applying (8.1) and (8.3) to the three coordinate directions obtains the net flux vector $\boldsymbol{F}_\lambda$. In the special case of *isotropic* radiation, intensity is independent of direction: $I_\lambda \neq I_\lambda(\hat{\boldsymbol{\Omega}})$. Under these circumstances, (8.1) reduces to

$$F_\lambda^+ = \pi I_\lambda. \tag{8.4}$$

Because the forward and backward components are then equal, the net flux in any direction vanishes.

Electromagnetic radiation can interact with matter through three basic mechanisms: absorption, scattering, and emission. A pencil of radiation occupying the solid angle $d\Omega$ is attenuated in proportion to the density and absorption characteristics of the medium through which it passes. Energy can also be scattered out of the solid angle, which likewise attenuates energy flowing through $d\Omega$. Together, absorption and scattering account for the net *extinction* of energy passing through the pencil of radiation. Conversely, energy emitted into the solid angle or scattered into $d\Omega$ from other directions intensifies the flow of energy through that pencil of radiation. These interactions are governed by a series of laws that describe the transfer of radiation through matter.

8.2.2 Absorption

In the absence of scattering, absorption of energy from a pencil of radiation is expressed by the following principle.

Lambert's Law

The fractional energy absorbed from a pencil of radiation is proportional to the mass traversed by the radiation. For an incremental distance ds, this principle is expressed by

$$\frac{dI_\lambda}{I_\lambda} = -\rho\sigma_{a\lambda}ds, \tag{8.5.1}$$

where the constant of proportionality $\sigma_{a\lambda}$ is the specific *absorption cross section* (also referred to as the *mass absorption coefficient*), which has units of $\frac{\text{area}}{\text{mass}}$. The cross section symbolizes the area of the pencil lost by passing through an increment of mass $\rho dAds$ or, equivalently, the effective absorbing area of that mass – each for the wavelength λ. The foregoing principle may also be expressed in terms of the particle number density n (e.g., of molecules or aerosol) through which the radiation passes

$$\frac{dI_\lambda}{I_\lambda} = -n\hat{\sigma}_{a\lambda}ds, \tag{8.5.2}$$

where the absorption cross section $\hat{\sigma}_{a\lambda}$ is referenced to an individual particle and has dimensions of area. Thus, $\hat{\sigma}_{a\lambda}$ symbolizes the effective absorbing area posed by a particle in the path of the radiation. In either representation, Lambert's law for attenuation of energy is linear in the intensity I_λ.

The *absorption coefficient*

$$\begin{aligned} \beta_{a\lambda} &= \rho\sigma_{a\lambda} \\ &= n\hat{\sigma}_{a\lambda}, \end{aligned} \tag{8.6.1}$$

has dimensions of inverse length. It measures the characteristic distance over which energy is attenuated. The density in (8.6) corresponds to the mass of absorber. In a mixture, it may follow from several constituent gases. Collecting contributions from different constituents gives

$$\beta_{a\lambda} = \sum_i r_i\rho\sigma_{a\lambda i} \tag{8.6.2}$$

$$= \sum_i r_i n\hat{\sigma}_{a\lambda i}, \tag{8.6.3}$$

where r_i denotes the mass mixing ratio of the ith absorbing species.

Integrating Lambert's law along the path of radiation obtains

$$I_\lambda(s) = I_\lambda(0)e^{-\int_0^s \rho\sigma_{a\lambda}ds'}. \tag{8.7}$$

In the absence of scattering and emission, the intensity inside a pencil of radiation decreases exponentially with the *optical path length*

$$
\begin{aligned}
u(s) &= \int_0^s \rho \sigma_{a\lambda} \, ds' \\
     &= \int_0^s \beta_{a\lambda} \, ds'.
\end{aligned}
\tag{8.8}
$$

$u(s)$ is the dimensionless distance traversed by radiation, weighted according to the density and absorption cross section of the medium. Because Lambert's law involves no directionality, it applies to flux as well as intensity.

The monochromatic *transmissivity* describes the fraction of incident radiation that remains in the pencil at a given distance. It is given by

$$
\begin{aligned}
\mathcal{T}_\lambda(s) &= \frac{I_\lambda(s)}{I_\lambda(0)} \\
                       &= e^{-u(s)}.
\end{aligned}
\tag{8.9}
$$

Transmissivity decreases exponentially with optical path length through an absorbing medium. Conversely, the monochromatic *absorptivity* a_λ represents the fraction of incident radiation that has been absorbed from the pencil during the same traversal. Because

$$
\mathcal{T}_\lambda + a_\lambda = 1
\tag{8.10}
$$

in the absence of scattering and emission, the absorptivity follows as

$$
a_\lambda(s) = 1 - e^{-u(s)}.
\tag{8.11}
$$

With increasing path length through an absorbing medium, absorptivity increases exponentially, approaching unity. In that limit, the medium is said to be *optically thick*.

8.2.3 Emission

To maintain thermal equilibrium, a substance that absorbs radiant energy must also emit it. Like absorption, the emission of energy into a pencil of radiation is proportional to the mass involved. The basis for describing thermal emission is the theory of blackbody radiation. It was developed by Planck and contemporaries near the turn of the century.

Blackbody radiation is an idealization that corresponds to the energy emitted by an isolated cavity in thermal equilibrium. Such radiation is characterized by the following properties:

- The radiation is uniquely determined by the temperature of the emitter.
- For a given temperature, the radiant energy emitted is the maximum possible at all wavelengths.
- The radiation is isotropic.

In addition to being a perfect emitter, a blackbody is also a perfect absorber: Radiation incident upon a blackbody is completely absorbed at all wavelengths.

Planck's Law

To explain blackbody radiation, Planck postulated that the energy E of molecules is quantized and can undergo only discrete transitions that satisfy

$$\Delta E = \Delta n \cdot h\nu, \tag{8.12}$$

where n is integer, h is Planck's constant (Appendix C), and $\nu = \frac{c}{\lambda}$ is the frequency of electromagnetic radiation emitted or absorbed to accomplish the energy transition ΔE. Hence, the radiation emitted or absorbed by individual molecules is quantized in *photons*, which carry energy in integral multiples of $h\nu$. On this basis, Planck derived the following relationship between the spectrum of intensity B_λ emitted by a population of molecules and their absolute temperature

$$B_\lambda(T) = \frac{2hc^2}{\lambda^5 (e^{\frac{hc}{K\lambda T}} - 1)}, \tag{8.13}$$

where K is the Boltzmann constant. Plotted in Fig. 8.7, the *Planck* or *blackbody spectrum* (8.13) increases with temperature. It possesses a single maximum at wavelength λ_m. At wavelengths shorter than λ_m, B_λ decreases sharply. At wavelengths longer than λ_m, it involves a comparatively wide band. Because blackbody radiation is isotropic, the form of the Planck spectrum applies to flux (8.4) as well as to intensity.

Wien's Displacement Law

The wavelength of maximum intensity follows from (8.13) as

$$\lambda_m = \frac{2897}{T} (\mu m). \tag{8.14}$$

It decreases with increasing temperature of the emitter. This feature of the emission spectrum allows the *brightness temperature* of a body to be inferred from radiation emitted by it. For instance, SW radiation incident on TOA is concentrated in the visible. It peaks near a wavelength of 0.480 μm (Fig. 8.2), which corresponds to blue light. Wien's displacement law then implies an effective solar temperature of about 6000°K. On the other hand, the 288°K mean surface temperature of Earth corresponds to maximum emission at a wavelength of about 10 μm (Fig. 8.1).

Analogous reasoning can be applied to individual wavelengths via (8.13). For instance, minima in the observed spectrum of outgoing LW radiation (Fig. 8.5) appear at 15 μm and 9.6 μm. They can be identified with emission by CO_2 in the upper troposphere, at the top of its column abundance, and by O_3 in the upper stratosphere, at the top of its column abundance. Each corresponds to a blackbody temperature that is significantly colder than the Earth's surface temperature. In fact, surface temperature is manifest in the spectrum of outgoing LW radiation only in the atmospheric window at 8–12 μm.

The Stefan-Boltzmann Law

The total flux emitted by a blackbody follows by integrating over the electromagnetic spectrum. In combination with (8.4), this yields

$$\begin{aligned} F &= \pi B(T) \\ &= \sigma T^4, \end{aligned} \tag{8.15}$$

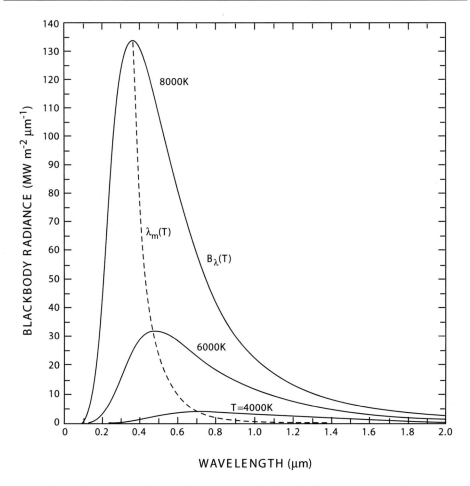

Figure 8.7 Spectra of emitted intensity $B_\lambda(T)$ for blackbodies at several temperatures, with wavelength of maximum emission $\lambda_m(T)$ indicated.

where σ is the Stefan-Boltzmann constant (Appendix C).[2]

The quartic temperature dependence in (8.15) implies LW emission that differs sharply between cold and warm objects. This feature of LW radiation is inherent in IR imagery like that presented in Figs. 1.28 and 1.30. The Sahara desert at 1200 UT is very dark, having temperature in excess of 315°K. It emits LW flux in excess of 550 W m^{-2}. (This strong surface emission is mediated by the atmosphere, which limits the outgoing LW flux at TOA to 350 W m^{-2} or less.) At the same time, deep convective tower over Indonesia has temperature as cold as 180°K. It emits LW flux of only $\sim$60 W m^{-2}. The disparate LW emission by these regions of different temperature symbolizes geographical variations in the Earth's energy budget. It is also symbolic of the diurnal variation

[2] The Stefan-Boltzmann law and Wien's displacement law predate Planck's postulate of the quantization of energy and his derivation of (8.13). While accounting for these earlier results, Planck's law resolved the failure of classical mechanics to explain the thermal emission spectrum at short wavelengths.

of emission, which follows from changes during the day of surface temperature and cloud cover.

Kirchoff's Law

The preceding laws determine the emission into a pencil of radiation during its passage through a medium that behaves as a blackbody. Real substances absorb and emit radiation at rates smaller than those of a perfect absorber and emitter. A *gray body* absorbs and emits with efficiencies that are independent of wavelength. The *absorptivity* a_λ of a substance is defined as the ratio of the intensity it absorbs to $B_\lambda(T)$. According to (8.11), the absorptivity of a layer approaches unity with increasing optical path length. At that point, it behaves as a blackbody. Similarly, the *emissivity* ϵ_λ is defined as the ratio of intensity emitted by a substance to $B_\lambda(T)$. According to the properties of blackbody radiation, it too must approach unity with increasing optical path length.

Consider two infinite plates that communicate with one another only radiatively: one a blackbody and the other a gray body with constant absorptivity a and emissivity ϵ. Suppose the plates are in thermal equilibrium, so they absorb and emit energy in equal proportion. Suppose further that the plates have different temperatures. Introducing a conducting medium between the plates would then drive the system out of thermal equilibrium by permitting heat to flow from the warmer to the colder plate. To restore thermal equilibrium, radiation would then have to transfer energy from the colder plate back to the warmer plate. This would violate the second law of thermodynamics. It follows that the *radiative equilibrium temperature* of the gray plate, that temperature at which it emits energy at the same rate as it absorbs energy from its surroundings, must be identical to that of the black plate.

For the gray plate to be in equilibrium, the flux of energy it emits: $\epsilon\sigma T^4$, must equal the flux of energy it absorbs from the black plate: $a\sigma T^4$. Applying this reasoning to substances of arbitrary monochromatic absorptivity and emissivity leads to the conclusion

$$\epsilon_\lambda = a_\lambda. \qquad (8.16)$$

Known as *Kirchoff's law*, (8.16) asserts that a substance emits radiation at each wavelength as efficiently as it absorbs it.

In general, the radiative efficiency of a substance varies with wavelength. As illustrated in Prob. 8.10, an illuminated surface with absorptivity a_{SW} in the SW and a_{LW} in the LW has a radiative-equilibrium temperature that varies according to the ratio $\frac{a_{SW}}{a_{LW}}$. A surface such as snow absorbs weakly in the visible but strongly in the IR. It thus has a lower radiative-equilibrium temperature than a gray surface, for which the absorptivity is independent of wavelength. In a similar manner, the selective transmissivity of the atmosphere increases the radiative-equilibrium temperature of the Earth's surface over what it would be in the absence of an atmosphere. To SW radiation, the atmosphere is transparent. Passing freely through the atmosphere, that radiation is absorbed at the Earth's surface. To LW radiation, the atmosphere is opaque. Re-emitted by the Earth's surface to preserve thermal equilibrium, that radiation is trapped by the overlying atmosphere, which absorbs at those wavelengths (Prob. 8.21). To offset reduced transmission, the surface must emit more LW radiation, which requires a higher temperature (8.15). Surface temperature must therefore increase, just enough for the LW radiation that is rejected to space at TOA to balance the SW radiation that is absorbed. The increased surface temperature represents the greenhouse effect. It is related directly to the difference of atmospheric opacity to SW and LW radiation (Prob. 8.22).

Kirchoff's law is predicated on a state of *local thermodynamic equilibrium* (LTE), wherein temperature is uniform and radiation is isotropic within some small volume. Those conditions are satisfied when energy transitions are dominated by molecular collision. For the most important radiatively active gases, they hold at pressures greater than 0.01 hPa (e.g., at altitudes below 60 km). At greater height, LTE breaks down because the interval between collisions is no longer short compared with the lifetime of excited states associated with absorption and emission. Kirchoff's law is then invalid.

With (8.16), the fractional energy emitted into a pencil of radiation along an incremental distance ds may be expressed

$$\frac{dI_\lambda}{B_\lambda} = d\epsilon_\lambda = da_\lambda$$

$$= \rho\sigma_{a\lambda}ds$$

by Lambert's law. Then

$$dI_\lambda = \rho\sigma_{a\lambda}J_\lambda ds, \tag{8.17}$$

where

$$J_\lambda = B_\lambda(T) \tag{8.18}$$

defines the monochromatic *source function* in the absence of scattering. Equation (8.18) holds under the conditions of Kirchoff's law, namely, under LTE. Otherwise, the contribution to the source function from emission is more complex.

8.2.4 Scattering

Beyond absorption and emission, energy passing through a pencil of radiation is also modified by scattering. The scattering process refers to the extraction and subsequent re-emission of energy by matter. A population of molecules possesses, in addition to its translational energy associated with temperature, electronic, vibrational, and rotational energies. Internal to individual molecules, those forms of energy can be excited by absorbing a photon. The absorbed energy can then be released in several ways. The simplest is when the excited internal energy is converted into translational energy through molecular collision, or "thermalized." This corresponds to thermal absorption of radiation. Thermal emission occurs through the reverse process. The excited energy can also be re-emitted by molecules – in wavelengths and directions different from those of the incident radiation (Fig. 8.8). This constitutes scattering. Analogous interactions occur between radiation and atmospheric aerosol.

The foregoing process is described in terms of the monochromatic *scattering cross section* $\sigma_{s\lambda}$, which symbolizes the fractional area removed from a pencil of radiation

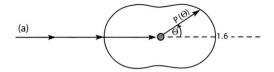

Figure 8.8 Angular distribution of radiation scattered from (a) small particles (of radius $a \ll \lambda$), which is representative of *Rayleigh scattering* of SW radiation by air molecules, and (b) large particles ($a \gg \lambda$), which is representative of *Mie scattering* of SW radiation by cloud droplets (Sec. 9.4.1). Phase function P is plotted in terms of the scattering angle Θ and in (b) for a scattering population with the refractive index of water and an effective size parameter $x_e = 2\pi \frac{a_e}{\lambda} = 5$. Note: The compressed scale in (b) implies that energy redirected by large particles is dominated by forward scattering. Larger particles produce even stronger forward scattering (cf. Fig. 9.30).

through scattering. As absorption and scattering are both linear in I_λ, their effects are additive. The monochromatic *extinction cross section* is defined by

$$k_\lambda = \sigma_{a\lambda} + \sigma_{s\lambda}. \tag{8.19.1}$$

The *extinction coefficient* then follows as

$$\beta_{e\lambda} = \rho k_\lambda, \tag{8.19.2}$$

where ρ is understood to refer to the optically active species. Lambert's law (8.5) and its consequences for attenuation then hold with k_λ in place of $\sigma_{a\lambda}$.

Scattering also modifies how energy is introduced into a pencil of radiation. This makes the source function more complex than the contribution from emission alone (8.18). While scattering removes radiation from one direction, it introduces it into other directions. Alternatively, photons can be scattered into a particular solid angle $d\Omega$ from all directions. If photons are introduced into the pencil of radiation through only one encounter with a particle, the process is termed *single scattering*. If more encounters are involved, the process is termed *multiple scattering*.

The *single scattering albedo*

$$\omega_\lambda = \frac{\sigma_{s\lambda}}{k_\lambda} \tag{8.20}$$

represents the fraction of radiation lost through extinction that is scattered out of a pencil of radiation. Then

$$1 - \omega_\lambda = \frac{\sigma_{a\lambda}}{k_\lambda} \tag{8.21}$$

represents the fraction lost through extinction that is absorbed from the pencil of radiation. The directionality of the scattered component is described in the *phase function* $P_\lambda(\hat{\Omega}, \hat{\Omega}')$. It represents the fraction of radiation scattered by an individual particle from the direction $\hat{\Omega}'$ into the direction $\hat{\Omega}$. If the phase function is normalized according to

$$\frac{1}{4\pi} \int_{4\pi} P_\lambda(\hat{\Omega}, \hat{\Omega}') d\Omega' = 1, \tag{8.22}$$

$\omega_\lambda \frac{P_\lambda(\hat{\Omega}, \hat{\Omega}')}{4\pi} d\Omega'$ represents the fraction of radiation lost through extinction from the pencil in the direction $\hat{\Omega}'$ that is scattered into another pencil in the direction $\hat{\Omega}$. The scattered contribution to the source function is then

$$J_{s\lambda} = \frac{\omega_\lambda}{4\pi} \int_{4\pi} I_\lambda(\hat{\Omega}') P_\lambda(\hat{\Omega}, \hat{\Omega}') d\Omega'. \tag{8.23}$$

Combining contributions from emission and scattering yields the total source function

$$J_\lambda = (1 - \omega_\lambda) B_\lambda(T) + \frac{\omega_\lambda}{4\pi} \int_{4\pi} I_\lambda(\hat{\Omega}') P_\lambda(\hat{\Omega}, \hat{\Omega}') d\Omega'. \tag{8.24}$$

8.2.5 The Equation of Radiative Transfer

Collecting the extinction of energy (8.19) and the introduction of energy (8.17) yields the net rate at which the intensity inside a pencil of radiation changes in the s direction

$$\frac{dI_\lambda}{\rho k_\lambda ds} = -I_\lambda + J_\lambda, \tag{8.25}$$

which is the *radiative transfer equation* in general form. In the absence of scattering, (8.25) reduces to

$$\frac{dI_\lambda}{\rho k_\lambda ds} = -I_\lambda + B_\lambda(T), \tag{8.26}$$

where k_λ then equals the absorption cross section $\sigma_{a\lambda}$. The *optical thickness* is defined by

$$\chi_\lambda(s) = \int_s^0 \rho k_\lambda ds' \tag{8.27.1}$$
$$= -u(s).$$

It increases in the negative s direction (i.e., opposite to $\hat{\Omega}$). Then

$$d\chi_\lambda = -\rho k_\lambda ds, \tag{8.27.2}$$

casts the energy budget for a pencil of radiation into the canonical form

$$\frac{dI_\lambda}{d\chi_\lambda} = I_\lambda - B_\lambda(T), \tag{8.28}$$

which is known as *Schwartzchild's equation*.

The solution of Schwartzchild's equation may be expressed formally with the aid of an integrating factor. Multiplying by $e^{-\chi_\lambda}$ allows (8.28) to be written

$$\frac{d}{d\chi_\lambda}\left[e^{-\chi_\lambda}I_\lambda\right] = -e^{-\chi_\lambda}B_\lambda(T).$$

Upon integrating from 0 to $\chi_\lambda(s)$, it yields

$$I_\lambda(s) = I_\lambda(0)e^{\chi_\lambda(s)} - \int_0^{\chi_\lambda(s)} B_\lambda[T(\chi_\lambda')]e^{\chi_\lambda(s)-\chi_\lambda'}d\chi_\lambda'. \tag{8.29}$$

The first term in (8.29) describes an exponential decrease with optical path length of the incident intensity $I_\lambda(0)$, as embodied in Lambert's law (8.7). The second term describes the cumulative emission and absorption between 0 and $\chi_\lambda(s)$. If the properties T, ρ, and k_λ are known as functions of s, (8.29) uniquely determines the intensity along the path of radiation.

8.3 ABSORPTION CHARACTERISTICS OF GASES

8.3.1 Interaction between radiation and molecules

The absorptivity of a medium approaches unity with increasing optical path length, irrespective of wavelength (Sec. 8.2.1). Along finite path lengths (e.g., through an atmosphere of bounded mass), absorption may be large at some wavelengths and small at others, according to the optical properties of the medium. For a gas, the absorption spectrum a_λ is concentrated in a complex array of lines that correspond to transitions between the discrete electronic, vibrational, and rotational energy levels of molecules. Electronic transitions are stimulated by radiation at UV and visible wavelengths. Vibrational and rotational transitions occur at IR wavelengths. At pressures greater than about 0.1 hPa (e.g., below 60 km), the internal energy acquired by absorbing a photon is quickly thermalized. In addition to discrete absorption characteristics, a continuum of absorption occurs at shorter wavelengths in connection with photodissociation and ionization of molecules. The latter occur at wavelengths of X-ray, UV, and, to a lesser degree, of visible radiation. They are possible for all λ shorter than the threshold to break molecular and electronic bonds.

Figure 8.9 displays absorption spectra in the IR for optically active gases corresponding to a vertical path through the atmosphere. All of the species are minor constituents of air. The most important are water vapor, carbon dioxide, and ozone. Each of these triatomic molecules is capable of undergoing simultaneous vibration-rotation transitions. They produce a clustering of absorption lines. At coarse resolution, like spectra in Figs. 8.1 and 8.9, those clusters appear as continuous bands of absorption. Water vapor absorbs strongly in a broad band centered near 6.3 μm and in another at 2.7 μm. Both bands correspond to transitions from vibrationally excited states: 6.3 μm absorption to the ν_2 vibrational mode (Fig. 8.10a), whereas the band at 2.7 μm involves both ν_1 and ν_3 modes of vibration. Rotational transitions of H_2O are less energetic. They lead to absorption at wavelengths longer than 12 μm (cf. Fig. 8.1). Carbon dioxide is excited vibrationally by wavelengths near 15 μm and also near 4.3 μm. The former corresponds to the transverse mode ν_2 (Fig. 8.10b), whereas the latter corresponds to the longitudinal vibration ν_3. Ozone absorbs strongly at wavelengths near 9.6 μm in connection with vibrational transitions. This absorption band coincides with

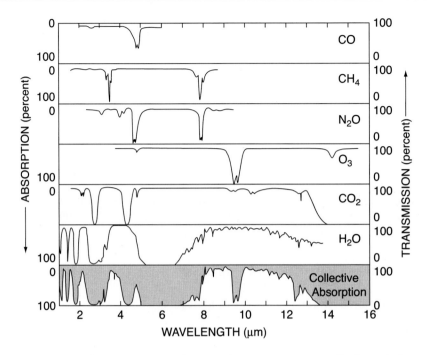

Figure 8.9 Absorption spectra for LW radiation passing vertically through the atmosphere as contributed by strong absorbing gases. Adapted from *Handbook of Geophysics and Space Environment* (1965).

the atmospheric window at 8–12 μm (Fig. 8.1). It therefore enables stratospheric ozone to interact radiatively with the troposphere and the Earth's surface. Except for ozone, most of the absorption by these species takes place in the troposphere, where absolute concentrations are large (cf. Figs. 8.1b,c). Nitrous oxide, methane, carbon monoxide, and CFC-11 and -12 also have absorption lines within the range of wavelength shown in Fig. 8.9. However, they are of secondary importance.

At higher resolution, IR absorption features in Fig. 8.9 actually comprise a complex array of lines. Figure 8.11 illustrates this in high-resolution spectra for the 15 μm band of CO_2 (Fig. 8.11a) and for the rotational band of H_2O at wavelengths of 27–31 μm (Fig. 8.11b). Absorption lines in the 15-μm band of CO_2 are spaced regularly in wavenumber. By contrast, the rotational band of water vapor involves a random distribution of absorption lines. In principle, all such lines must be accounted for in radiative calculations. However, their number and complexity make *line-by-line calculations* impractical for most applications. Instead, radiative calculations rely on *band models*. The latter represent the gross characteristics of absorption spectra in particular ranges of wavelength. Under cloud-free conditions, they provide an economical yet accurate alternative (e.g., Goody et al., 1989; Mlawer et al., 1997).

Similar features characterize the absorption spectrum at visible and UV wavelengths, where radiation interacts with molecular oxygen and ozone. In addition to discrete features, the more energetic radiation at these wavelengths produces continuous bands of absorption in connection with photodissociation and ionization of molecules. Figure 8.12 shows the absorption cross section $\hat{\sigma}_{a\lambda}$ for O_2 in the UV. The *Herzberg continuum* lies at wavelengths of 242–200 nm, where O_2 is dissociated into

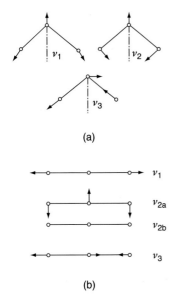

(a)

(b)

Figure 8.10 Normal modes of vibration for triatomic molecules corresponding to (a) H_2O and O_3, in which atoms are configured triangularly, and (b) CO_2, in which atoms are configured longitudinally. After Herzberg (1945).

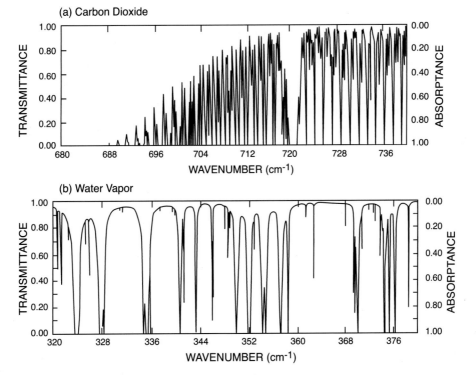

Figure 8.11 Absorption spectra in (a) the 15-μm band of CO_2 and (b) the rotational band of H_2O at 27–31 μm. Adapted from McClatchey and Selby (1972).

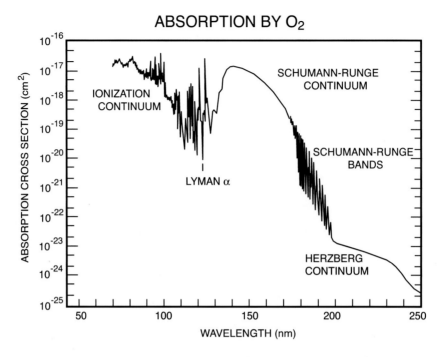

Figure 8.12 Absorption cross section as a function of wavelength for molecular oxygen. After Brasseur and Solomon (1986). Reprinted by permission of Kluwer Academic Publishers.

two ground-state oxygen atoms. At shorter wavelengths, the absorption spectrum is marked by the discrete *Schumann-Runge bands*, where O_2 is vibrationally excited. These excited bound states are unstable; they eventually produce two ground-state oxygen atoms.

Each of the vibrational bands in the Schumann-Runge system is actually comprised of many rotational lines, which are shown at high resolution in Fig. 8.13. The rotational line structure is fairly regular at long wavelength. However, it gradually degenerates to an almost random distribution at short wavelength. Wavelengths shorter than 175 nm are absorbed in the *Schumann-Runge continuum* (Fig. 8.12), in which O_2 is dissociated into two oxygen atoms (one electronically excited). At still shorter wavelengths, absorption exhibits an irregular banded structure that eventually merges into the ionization continuum at wavelengths shorter than 102 nm.

Actual absorption by individual bands in Fig. 8.12 varies strongly with altitude. Even though it has the smallest values of $\hat{\sigma}_{a\lambda}$, the Herzberg continuum at 200–242 nm dominates absorption by O_2 up to 60 km – because shorter wavelengths have already been removed at higher altitude (Fig. 8.3). Above 60 km, the Schumann-Runge bands are dominant. The Schumann-Runge continuum prevails in the thermosphere, where very energetic UV is present.

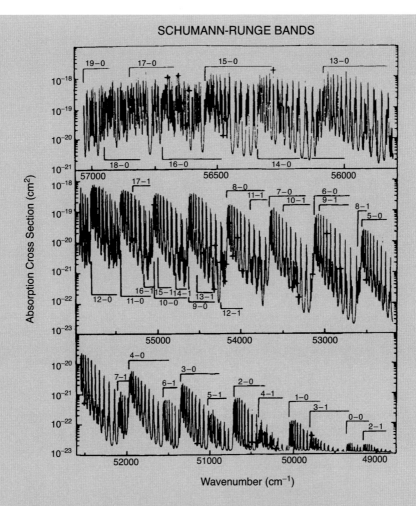

Figure 8.13 Absorption cross section in the Schummann-Runge absorption bands of molecular oxygen. Shown at high spectral resolution. After Kocharts (1971). Reprinted by permission of Kluwer Academic Publishers.

Ozone absorbs at longer wavelengths, which penetrate to lower altitudes. The primary absorption of UV by ozone is in the *Hartley band* at wavelengths of 200–310 nm (Fig. 8.14). At long wavelength, the Hartley band merges with the *Huggins bands* at 310–400 nm. Ozone also absorbs in the visible in the *Chappuis band* at wavelengths of 400–850 nm, which is important at altitudes below 25 km. In all of these bands, absorption follows from photodissociation of ozone. The Hartley and Chappuis bands are continuous. Conversely, the Huggins bands involve a spectrum of diffuse lines. The Chappuis band is weaker than the Hartley and Huggins bands. However, its importance is underscored by its overlap with the peak of the SW spectrum. For this reason, O_3 is rapidly photo-dissociated in sunlight – even at the Earth's surface. This property couples ozone in the stratosphere to snow and cloud cover in the troposphere, which magnify the reflected SW flux.

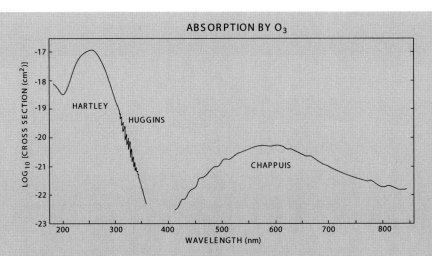

Figure 8.14 Absorption cross section of ozone, as a function of wavelength. *Sources:* Andrews et al. (1987) and WMO (1986).

Absorption and therefore radiative heating follow from the cross sections of individual species, weighted by their respective mixing ratios (Fig. 8.15). Absorption at wavelengths longer than 200 nm is dominated by photodissociation of O_3. Absorption at shorter wavelengths is dominated by photodissociation of O_2. Owing to the inverse relationship between wavelength and penetration altitude (Fig. 8.3), these bands influence different levels. Absorption by O_3 in the Hartley and Huggins bands at $\lambda > 200$ nm dominates in the stratosphere and mesosphere. There, it provides the primary source of heating. Ozone absorption in the Chappuis band at $\lambda > 400$ nm becomes important below 25 km. On the other hand, absorption by O_2 at $\lambda < 200$ nm prevails only only above 60 km. Despite its secondary contribution to absorption, photodissociation of O_2 plays a key role in the energetics of the stratosphere and mesosphere because it produces atomic oxygen, which supports ozone formation (1.27).

8.3.2 Line broadening

Absorption lines in Figs. 8.11 and 8.13 are not truly discrete. Rather, each occupies a finite band of wavelength, due to practical considerations surrounding molecular absorption and emission. The spectral width of an absorbing line is described in terms of a shape factor. The absorption cross section at frequency v is expressed

$$\sigma_{av} = S f(v - v_0),\tag{8.30.1}$$

where

$$S = \int \sigma_v \, dv\tag{8.30.2}$$

is the *line strength*, v_0 is the line center, and the shape factor f accounts for line broadening.

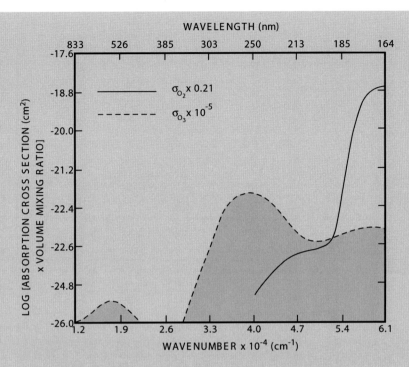

Figure 8.15 Relative contributions to absorption cross section from ozone and molecular oxygen. Adapted from Brasseur and Solomon (1984). Reprinted by permission of Kluwer Academic Publishers.

A fundamental source of line width is *natural broadening*, which results from the finite lifetime of excited states. Perturbations to the radiation stream then introduce a natural width that has the *Lorentz line shape*

$$f_L(\nu - \nu_0) = \frac{\alpha_L}{\pi[(\nu - \nu_0)^2 + \alpha_L^2]}, \tag{8.31.1}$$

where

$$\alpha_L = (2\pi\bar{t})^{-1} \tag{8.31.2}$$

is the half-width at half-power of the line (Fig. 8.16) and $\bar{t}$ is the mean lifetime of the excited state. For vibrational and rotational transitions in the IR, natural broadening is insignificant compared with other sources of line width. The two most important stem from the random motion of molecules and collisions between them.

Collisional or *pressure broadening* results from perturbations to the absorbing or emitting molecules. Introduced through encounters with other molecules, those perturbations destroy the phase coherence of radiation. The spectral width introduced by collision is modeled in the Lorentz line shape (8.31.1), but with the collisional half-width

$$\alpha_c = \alpha_0 \left(\frac{p}{p_0}\right)\left(\frac{T_0}{T}\right)^{\frac{1}{2}}, \tag{8.32}$$

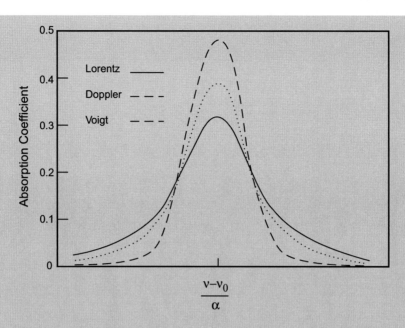

Figure 8.16 Lorentz, Doppler, and Voigt shape factors describing finite spectral line width, plotted as functions of $\frac{v-v_0}{\alpha}$, with α the half width of each line shape. Curves normalized to achieve the same area under each. Reconstructed from Andrews et al. (1987).

where $\alpha_0 \cong 0.1$ cm^{-1} is the half-width at standard temperature and pressure T_0, p_0. From kinetic theory, α_c is inversely proportional to the mean time between collisions.

Molecular motion v along the line of sight introduces another source of line width. *Doppler broadening* follows from the frequency shift

$$v = v_0 \left(1 \pm \frac{v}{c}\right) \tag{8.33.1}$$

of individual molecules undergoing emission or absorption. In concert with the Boltzmann probability distribution

$$P(v) = \sqrt{\frac{m}{2\pi KT}} e^{-\frac{mv^2}{2KT}}, \tag{8.33.2}$$

where m is the molecular mass, (8.33.1) leads to the shape factor

$$f_D(v - v_0) = \frac{1}{\alpha_D \sqrt{\pi}} e^{-\left(\frac{v-v_0}{\alpha_D}\right)^2}, \tag{8.34.2}$$

where

$$\alpha_D = \frac{v_0}{c} \sqrt{\frac{2KT}{m}} \tag{8.34.2}$$

is the Doppler half-width divided by a factor of $\sqrt{ln2}$. The Doppler line shape is illustrated in Fig. 8.16, along with the collisional line shape for $\alpha_D = \alpha_c$. Unlike the collisional half-width, α_D does not depend on pressure.

Below 30 km, the pressure dependence in α_c makes collisional broadening dominant for IR bands of CO_2 and H_2O. At greater height, Doppler and natural broadening become significant at wavelengths in the visible and UV. Natural broadening of the Schumann-Runge bands of O_2 becomes comparable to collisional broadening in the upper stratosphere and mesosphere. Those two sources of line width are treated jointly in the *Voigt line shape*, which is superposed in Fig. 8.16. When plotted as functions of $\frac{\nu - \nu_0}{\alpha}$, with α the half width of each line shape, the Doppler profile is dominant near the line center, whereas the Lorentz profile prevails in the wings of the line. The Voigt profile then gives a line shape between the two.

8.4 RADIATIVE TRANSFER IN A PLANE PARALLEL ATMOSPHERE

In a stratified atmosphere, absorbers vary sharply with height. It is therefore convenient to treat radiative transfer within the framework of a *plane parallel atmosphere*, wherein

- Curvature associated with sphericity of the Earth is ignored.
- The medium is regarded as horizontally homogeneous and the radiation field horizontally isotropic.

Then, along a slant path ds (Fig. 8.17), a pencil of radiation inclined from the vertical at a zenith angle θ traverses an atmospheric layer of thickness

$$dz = \mu ds, \tag{8.35.1}$$

where

$$\mu = cos\theta. \tag{8.35.2}$$

It is convenient to introduce the *optical depth*

$$\tau_\lambda = \int_z^\infty \rho k_\lambda dz'$$
$$= \frac{1}{g}\int_0^p k_\lambda dp' = \mu\chi_\lambda, \tag{8.36}$$

which is measured downward from the top of the atmosphere. The energy budget for a pencil of radiation (8.28) is then transformed into

$$\mu\frac{dI_\lambda}{d\tau_\lambda} = I_\lambda - J_\lambda, \tag{8.37}$$

where $I_\lambda = I_\lambda(\tau_\lambda, \mu)$ and $J_\lambda = J_\lambda(\tau_\lambda, \mu)$. Proceeding as in the treatment of (8.28) leads to a formal expression for upwelling radiation ($0 < \mu \leq 1$) in terms of that at the surface ($\tau_\lambda = \tau_{s\lambda}$)

$$I_\lambda(z, \mu) = I_\lambda(\tau_{s\lambda}, \mu)e^{\frac{\tau_\lambda(z)-\tau_{s\lambda}}{\mu}} + \int_{\tau_\lambda(z)}^{\tau_{s\lambda}} J_\lambda(\tau_\lambda', \mu)e^{\frac{\tau_\lambda(z)-\tau_\lambda'}{\mu}}\frac{d\tau_\lambda'}{\mu} \quad 0 < \mu \leq 1 \tag{8.38.1}$$

and for downwelling radiation ($-1 \leq \mu < 0$) in terms of that at TOA ($\tau_\lambda = 0$)

$$I_\lambda(z, \mu) = I_\lambda(0, \mu)e^{\frac{\tau_\lambda(z)}{\mu}} - \int_0^{\tau_\lambda(z)} J_\lambda(\tau_\lambda', \mu)e^{\frac{\tau_\lambda(z)-\tau_\lambda'}{\mu}}\frac{d\tau_\lambda'}{\mu} \quad -1 \leq \mu < 0. \tag{8.38.2}$$

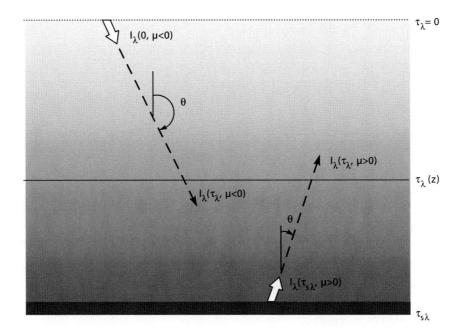

Figure 8.17 Plane parallel atmosphere, in which a pencil of radiation is inclined at the zenith angle $\theta = cos^{-1}\mu$. Elevation is measured by the optical depth for a given wavelength: τ_λ, which increases downward from zero at the top of the atmosphere to a surface value of $\tau_{s\lambda}$.

As before, the first terms on the right hand sides of (8.38) describe the extinction of radiation emanating from the boundaries. The second terms account for the cumulative emission, absorption, and scattering into the pencil between the boundaries and the height z. If I_λ is known at the upper and lower boundaries and if J_λ can be specified in terms of known properties, (8.38) determines the radiation field throughout the atmosphere. Alternatively, if J_λ, ρ, and k can all be specified in terms of temperature (e.g., in the absence of scattering (8.18) and under hydrostatic equilibrium (1.17)), (8.38) determines the thermal structure of the atmosphere. The expressions for the radiation field then represent integral equations for the temperature distribution $T(z)$. Their solution defines the *radiative-equilibrium thermal structure*, that for which radiative components of the energy budget are in balance.

The upwelling and downwelling fluxes are give by

$$F_\lambda^\uparrow(\tau_\lambda) = 2\pi \int_0^1 I_\lambda(\tau_\lambda, \mu)\mu d\mu$$

$$F_\lambda^\downarrow(\tau_\lambda) = -2\pi \int_{-1}^0 I_\lambda(\tau_\lambda, \mu)\mu d\mu.$$

(8.39)

Horizontal homogeneity and isotropy make these the essential descriptors of radiative transfer in a plane parallel atmosphere. Integrating (8.38) over zenith angle and incorporating (8.4) obtains

$$F_\lambda^\uparrow(\tau_\lambda) = 2\pi \int_0^1 I_\lambda(\tau_{s\lambda}, \mu) e^{\frac{\tau_\lambda - \tau_{s\lambda}}{\mu}} \mu d\mu + 2\pi \int_{\tau_\lambda}^{\tau_{s\lambda}} \int_0^1 J_\lambda(\tau_\lambda', \mu) e^{\frac{\tau_\lambda - \tau_\lambda'}{\mu}} d\mu d\tau_\lambda' \quad (8.40.1)$$

$$F_\lambda^\downarrow(\tau_\lambda) = -2\pi \int_{-1}^0 I_\lambda(0, \mu) e^{\frac{\tau_\lambda}{\mu}} \mu d\mu + 2\pi \int_0^{\tau_\lambda} \int_{-1}^0 J_\lambda(\tau_\lambda', \mu) e^{\frac{\tau_\lambda - \tau_\lambda'}{\mu}} d\mu d\tau_\lambda'. \quad (8.40.2)$$

For LW radiation and in the absence of scattering, the source function is just $B_\lambda(T)$. Upwelling radiation at the surface, which is treated as black, is then given by

$$I_\lambda(\tau_{s\lambda}, \mu) = B_\lambda(T_s) \qquad 0 < \mu \le 1. \quad (8.41.1)$$

Likewise, downwelling LW radiation at TOA must vanish, so

$$I_\lambda(0, \mu) = 0 \qquad -1 \le \mu < 0. \quad (8.41.2)$$

The upwelling and downwelling fluxes of LW radiation may then be written

$$F_\lambda^\uparrow(\tau_\lambda) = 2\pi B_\lambda(T_s) \int_0^1 e^{\frac{\tau_\lambda - \tau_{s\lambda}}{\mu}} \mu d\mu + 2 \int_{\tau_\lambda}^{\tau_{s\lambda}} \int_0^1 \pi B_\lambda[T(\tau_\lambda')] e^{\frac{\tau_\lambda - \tau_\lambda'}{\mu}} d\mu d\tau_\lambda' \quad (8.42.1)$$

$$F_\lambda^\downarrow(\tau_\lambda) = 2 \int_0^{\tau_\lambda} \int_0^1 \pi B_\lambda[T(\tau_\lambda')] e^{\frac{\tau_\lambda' - \tau_\lambda}{\mu}} d\mu d\tau_\lambda'. \quad (8.42.2)$$

The integrations over zenith angle may be expressed in terms of the exponential integral (Abramowitz and Stegun, 1972)

$$E_n(\tau) = \int_1^\infty \frac{e^{-x\tau}}{x^n} dx, \quad (8.43.1)$$

which satisfies

$$\frac{dE_n}{d\tau} = -E_{n-1}(\tau). \quad (8.43.2)$$

Letting $x = \mu^{-1}$ transforms (8.42) into

$$F_\lambda^\uparrow(\tau_\lambda) = 2\pi B_\lambda(T_s) E_3(\tau_{s\lambda} - \tau_\lambda) + 2 \int_{\tau_\lambda}^{\tau_{s\lambda}} \pi B_\lambda[T(\tau_\lambda')] E_2(\tau_\lambda' - \tau_\lambda) d\tau_\lambda' \quad (8.44.1)$$

$$F_\lambda^\downarrow(\tau_\lambda) = 2 \int_0^{\tau_\lambda} \pi B_\lambda[T(\tau_\lambda')] E_2(\tau_\lambda - \tau_\lambda') d\tau_\lambda'. \quad (8.44.2)$$

Then integrating over wavelength recovers the total upward and downward fluxes

$$F^\uparrow(\tau) = 2 \int_0^\infty \pi B_\lambda(T_s) E_3(\tau_{s\lambda} - \tau_\lambda) d\lambda + 2 \int_\tau^{\tau_s} \int_0^\infty \pi B_\lambda[T(\tau_\lambda')] E_2(\tau_\lambda' - \tau_\lambda) d\lambda d\tau' \quad (8.45.1)$$

$$F^\downarrow(\tau) = 2 \int_0^\tau \int_0^\infty \pi B_\lambda[T(\tau_\lambda')] E_2(\tau_\lambda - \tau_\lambda') d\lambda d\tau', \quad (8.45.2)$$

which describe LW radiative transfer in a non-scattering atmosphere.

8.4.1 Transmission function

Implementing (8.45) is complicated by the rapid variation with wavenumber of absorption cross section, which involves thousands of vibrational and rotational lines in the IR. In place of line-by-line calculations, radiative transfer is calculated with the aid of a band-averaged transmission function that embodies the gross characteristics of the absorption spectrum in particular ranges of wavelength.

Averaging (8.44) over a frequency interval $\Delta \nu$, where $\nu = c\lambda^{-1}$ is symbolic of wavenumber, yields the band-averaged fluxes

$$F_{\bar{\nu}}^{\uparrow}(\tau_{\bar{\nu}}) = 2\pi B_{\bar{\nu}}(T_s) \int_{\Delta \nu} E_3(\tau_{s\nu} - \tau_{\nu}) \frac{d\nu}{\Delta \nu} + 2 \int_{\tau_{\bar{\nu}}}^{\tau_{s\nu}} \pi B_{\bar{\nu}}[T(\tau_{\bar{\nu}}')] \int_{\Delta \nu} E_2(\tau_{\nu}' - \tau_{\nu}) \frac{d\nu}{\Delta \nu} d\tau' \quad (8.46.1)$$

$$F_{\bar{\nu}}^{\downarrow}(\tau_{\bar{\nu}}) = 2 \int_0^{\tau_{\bar{\nu}}} \pi B_{\bar{\nu}}[T(\tau_{\bar{\nu}}')] \int_{\Delta \nu} E_2(\tau_{\nu} - \tau_{\nu}') \frac{d\nu}{\Delta \nu} d\tau_{\nu}'. \quad (8.46.2)$$

In (8.46), $B_{\bar{\nu}}$ may be used for the band-averaged emission because the Planck function varies smoothly with frequency. The *band transmissivity* or *transmission function* is defined as

$$\mathcal{T}_{\bar{\nu}}(\tau_{\bar{\nu}}, \mu) = \frac{1}{\Delta \nu} \int_{\Delta \nu} e^{-\frac{\tau_{\nu}}{\mu}} d\nu. \quad (8.47)$$

Decreasing downward, $\mathcal{T}_{\bar{\nu}}$ represents the fractional intensity in the band and in direction μ that reaches optical depth $\tau_{\bar{\nu}}$. The *band absorptivity* is then

$$a_{\bar{\nu}} = 1 - \mathcal{T}_{\bar{\nu}}, \quad (8.48)$$

analogous to (8.11).

Averaging (8.47) over zenith angle defines the *diffuse flux transmission function* for the band

$$\mathcal{T}_{\bar{\nu}}(\tau_{\bar{\nu}}) = \frac{\int_0^1 \mathcal{T}_{\bar{\nu}}(\tau_{\bar{\nu}}, \mu) \mu d\mu}{\int_0^1 \mu d\mu}$$

$$= 2 \int_0^1 \mathcal{T}_{\bar{\nu}}(\tau_{\bar{\nu}}, \mu) \mu d\mu. \quad (8.49)$$

It may be expressed in terms of the exponential integral as

$$\mathcal{T}_{\bar{\nu}}(\tau_{\bar{\nu}}) = 2 \int_{\Delta \nu} E_3(\tau_{\nu}) \frac{d\nu}{\Delta \nu}. \quad (8.50)$$

With the aid of (8.43.2), the band-averaged fluxes may then be expressed in terms of the flux transmission function

$$F_{\bar{\nu}}^{\uparrow}(\tau_{\bar{\nu}}) = \pi B_{\bar{\nu}}(T_s) \mathcal{T}_{\bar{\nu}}(\tau_{s\bar{\nu}} - \tau_{\bar{\nu}}) - \int_{\tau_{\bar{\nu}}}^{\tau_{s\bar{\nu}}} \pi B_{\bar{\nu}}[T(\tau_{\nu}')] \frac{d\mathcal{T}_{\bar{\nu}}(\tau_{\nu}' - \tau_{\bar{\nu}})}{d\tau_{\nu}'} d\tau_{\nu}' \quad (8.51.1)$$

$$F_{\bar{\nu}}^{\downarrow}(\tau_{\bar{\nu}}) = \int_0^{\tau_{\bar{\nu}}} \pi B_{\bar{\nu}}[T(\tau_{\nu}')] \frac{d\mathcal{T}_{\bar{\nu}}(\tau_{\bar{\nu}} - \tau_{\nu}')}{d\tau_{\nu}'} d\tau_{\nu}'. \quad (8.51.2)$$

In principle, absorption characteristics of the band $\Delta \nu$ determine the corresponding transmission function and the vertical fluxes through (8.51). Collecting contributions from all absorbing bands then obtains the total upwelling and downwelling fluxes. However, in practice, even individual absorption bands are so complex as to make direct calculations impractical (cf. Figs. 8.11, 8.13). Instead, the transmission function is evaluated with the aid of band models that capture the salient features of the absorption spectrum in terms of properties such as mean line strength, line spacing, and line width. The regular band model of Elsasser (1938) treats a series of evenly spaced Lorentz lines, such as those comprising the 15-μm

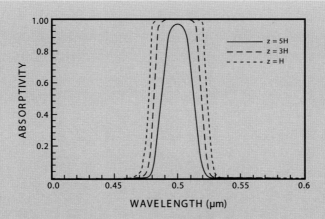

Figure 8.18 Absorptivity contributed by an individual absorption line at 0.5 μm for an absorber with a uniform mixing ratio and for the layer extending from the top of the atmosphere to a height z of 5 scale heights, 3 scale heights, and 1 scale height. As the layer extends downward, where absorber concentration increases exponentially, the line becomes saturated, first at its center but eventually over a widening range of wavelength.

band of CO_2 (Fig. 8.11a). The Goody (1952) random model treats a spectrum of lines that are randomly spaced, as is typical of the 6.3-μm band of water vapor (Fig. 8.11b).

The transmission function, although more complex than the exponential attenuation of monochromatic radiation (8.9), is far simpler than a line-by-line calculation over the band. It is instructive to consider how the transmission function, or alternatively the absorptivity (8.48), behaves with optical depth for an individual absorption line. Below a certain height, the absorptivity of a line may be represented in terms of the Lorentz profile (8.31.1). As τ_ν increases, so does a_ν (Fig. 8.18). Eventually, the absorptivity reaches unity near the line center. At that point, the corresponding frequencies are fully absorbed. The line is then *saturated*. This represents the limit of *strong absorption*.

Subsequent absorption can occur only in the wings of the line, which are pushed out by the ever widening region of saturation. In a band containing many lines, saturation leads to the absorption spectrum filling in between discrete features. This occurs first for the narrowly separated vibration-rotation transitions (e.g., Figs. 8.11, 8.13). Over sufficient optical depth, those clusters merge to form continuous bands of absorption, as appear in Figs. 8.1 and 8.9. The strong-absorption limit applies to bands that dominate vertical energy transfer in the atmosphere. It is the basis for a powerful approximation that reduces the 3-dimensional description of diffuse LW radiation to a 1-dimensional description.

8.4.2 Two-stream approximation

Embodied in the flux transmission function (8.50) is an integral over zenith angle, which collects contributions from diffuse radiation, and another integral over

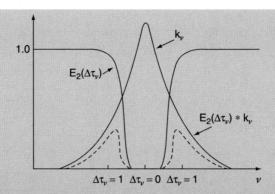

Figure 8.19 Factors $E_2(\Delta\tau_\nu)$ and k_ν appearing in the integral over zenith angle that is implicit to the band-averaged upwelling and downwelling fluxes (8.51). Due to their reciprocal nature, the factors' product is concentrated at frequencies corresponding to an optical depth of $\Delta\tau_\nu \cong 1$, which is associated with an effective zenith angle of $\overline{\mu}^{-1} \cong \frac{5}{3}$. Adapted from Andrews et al. (1987).

frequency. Because $E_3(\tau_\nu)$ depends on frequency, the two integrations are related. In the strong-absorption limit, the integral over zenith angle can be approximated fairly accurately by incorporating the spectral character of $T_{\bar{\nu}}$.

Integrals on the right hand sides of (8.51) involve the term

$$\frac{dT_{\bar{\nu}}}{d\tau_\nu'}(\Delta\tau_\nu)d\tau_\nu' = 2\int_{\Delta\nu} E_2(\Delta\tau_\nu)\rho k_\nu dz' \frac{d\nu}{\Delta\nu}, \tag{8.52.1}$$

where

$$\Delta\tau_\nu = |\tau_\nu' - \tau_\nu|. \tag{8.52.2}$$

If the band described by $T_{\bar{\nu}}$ is saturated, the factors k_ν and $E_2(\Delta\tau_\nu)$ have reciprocal behavior in frequency. As depicted in Fig. 8.19, $E_2(\Delta\tau_\nu)$ vanishes near the center of the band, for which the medium is optically thick. Conversely, it approaches unity in the wings of the band, for which the medium is optically thin. A transition between these extremes occurs at frequencies corresponding to an optical thickness of $\Delta\tau_\nu \cong 1$. The behavior of k_ν is just reversed. k_ν is large near the center of the band, but vanishes in the wings.

Due to the reciprocal nature of $E_2(\Delta\tau_\nu)$ and k_ν, their product in (8.52.1) is concentrated about an optical thickness of $\Delta\tau_\nu \cong 1$. Consequently, most of the energy exchange between two levels in (8.51) occurs along paths that have an optical thickness of unity. Along paths of optical thickness much greater than 1, the medium is optically thick. Radiation is then absorbed before it can traverse the levels, which therefore experience little interaction. Along paths of optical thickness much smaller than 1, the medium is optically thin. Radiation then traverses the levels with little attenuation, so the levels again experience little interaction.

For any specified $\Delta\tau_\nu$, an effective inclination $\bar{\mu}$ may be found such that the exponential integral over zenith angle is given by

$$2E_3(\Delta\tau_\nu) = e^{-\frac{\Delta\tau_\nu}{\bar{\mu}}}. \tag{8.53.1}$$

The parameter $\bar{\mu}^{-1}$ is called the *diffusivity factor*. As contributions to (8.51) are restricted to $\Delta\tau_\nu \cong 1$, it too is restricted to a narrow range of values. For $\Delta\tau_\nu = 1$, the diffusivity factor assumes the value

$$\bar{\mu}^{-1} \cong \frac{5}{3}. \tag{8.53.2}$$

The implication of this analysis is that the zenith angle dependence in (8.51) may be eliminated in favor of an effective inclination $\bar{\mu}$. The latter reduces the multidimensional description to integrals over optical depth alone. Diffuse transmission which is embodied in $\mathcal{T}_{\bar{\nu}}$ is then equivalent to that of a collimated beam inclined at a zenith angle of $53°$

$$\mathcal{T}_{\bar{\nu}}(\tau_{\bar{\nu}}, \mu) \cong \mathcal{T}_{\bar{\nu}}(\tau_{\bar{\nu}}, \bar{\mu}). \tag{8.53.3}$$

Alternatively, it is equivalent to a beam inclined at zero zenith angle but through an optical thickness that is expanded by a factor of $\frac{5}{3}$.

Known as the *exponential kernel approximation*, this simplification follows from the fact that radiative exchange in strong bands is dominated by spectral intervals in which $\Delta\tau_\nu \cong 1$. For this reason, most of the LW radiation emitted by the Earth's surface is captured in the lower troposphere. The latter is rendered optically thick by strong absorption bands of water vapor and carbon dioxide. Likewise, incident SW radiation in particular wavelengths of UV is absorbed in the stratosphere over a limited range of altitude (Fig. 8.3). The latter is rendered optically thick by photodissociation at those wavelengths. The development leading to (8.53) is one of several so-called *two stream approximations* that eliminate the zenith angle dependence in $F^\uparrow$ and $F^\downarrow$. Applicable under fairly general circumstances (even in the presence of scattering), this formalism leads to diffusivity factors in the range $\frac{3}{2} \leq \bar{\mu}^{-1} \leq 2$, depending on the particular approximation adopted.

The full description of radiative transfer in a plane parallel atmosphere may be reduced to a vertical description by taking $\mu = \pm\bar{\mu}$ for upwelling and downwelling radiation and introducing the transformation

$$\tau_\lambda^* = \bar{\mu}^{-1}\tau_\lambda \tag{8.54.1}$$

$$F_\lambda^{\uparrow\downarrow}(\tau_\lambda^*) = \pi I_\lambda(\tau_\lambda, \pm\bar{\mu}) \tag{8.54.2}$$

$$J_\lambda^*(\tau_\lambda^*) = \pi J_\lambda(\tau_\lambda, \pm\bar{\mu}), \tag{8.54.3}$$

which relies on hemispheric isotropy. Then integrating the radiative transfer equation (8.37) over upward and downward half-spaces yields for the budgets of upwelling and downwelling radiation

$$\frac{dF_\lambda^\uparrow}{d\tau_\lambda^*} = F_\lambda^\uparrow - J_\lambda^* \tag{8.55.1}$$

$$-\frac{dF_\lambda^\downarrow}{d\tau_\lambda^*} = F_\lambda^\downarrow - J_\lambda^*. \tag{8.55.2}$$

The foregoing development applies to LW radiation, which is inherently diffuse. SW radiation, which is parallel-beam in the absence of scattering, involves only a single direction μ.

8.5 THERMAL EQUILIBRIUM

8.5.1 Radiative equilibrium in a gray atmosphere

We are now in a position to evaluate the thermal structure toward which radiative transfer drives the atmosphere. To do so, we consider a simple model of the Earth's atmosphere: a *gray atmosphere* is transparent to SW radiation but absorbs LW radiation with a constant absorption cross section that is independent of wavelength, temperature, and pressure.

Consider a plane parallel gray atmosphere that is nonscattering, motionless, and in thermal equilibrium with incoming SW radiation and with a black underlying surface. Individual layers of the atmosphere then interact only through LW radiation.

For an incremental layer of thickness dz, the First Law (2.22.2) implies

$$\rho c_p \frac{dT}{dt} - \frac{dp}{dt} = \rho \dot{q}$$

$$= -\frac{dF}{dz},$$

(8.56.1)

where

$$F = F^\uparrow - F^\downarrow$$

(8.56.2)

is the net LW flux integrated over wavelength and $\dot{q}$ is the local rate heat is absorbed per unit mass or the *specific heating rate*. The convergence of LW flux on the right hand side of (8.56.1) represents the local radiative heating rate per unit volume. It forces the atmosphere's thermal structure. In equilibrium, the left hand side vanishes. Then the net flux must be independent of height

$$F = \text{const.}$$

(8.57)

This implies that radiative energy does not accumulate within individual layers.

The upwelling and downwelling components of F are governed by (8.55). Adding and subtracting yields the equivalent system

$$\frac{dF}{d\tau^*} = \bar{F} - 2B^*$$

(8.58.1)

$$\frac{d\bar{F}}{d\tau^*} = F,$$

(8.58.2)

where

$$\bar{F} = F^\uparrow + F^\downarrow$$

(8.58.3)

represents the total flux emanating from an incremental layer and all quantities have been integrated over wavelength. With (8.57), (8.58.2) gives

$$\bar{F} = F\tau^* + c,$$

(8.59.1)

whereas (8.58.1) reduces to

$$\bar{F} = 2B^*.$$

(8.59.2)

Emission is then described by

$$B^*(\tau^*) = \frac{F}{2}\tau^* + B_0^*.$$

(8.60)

It increases linearly with optical depth, from its value B_0^* at TOA (Fig. 8.20).

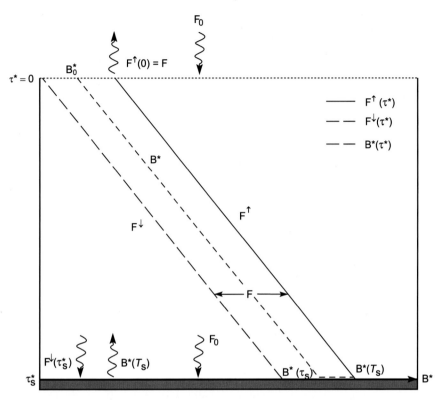

Figure 8.20 Upwelling and downwelling LW fluxes and LW emission in a gray atmosphere that is in radiative equilibrium with an incident SW flux F_0 and a black underlying surface. Note: the emission profile is discontinuous at the surface.

For thermal equilibrium of the Earth-atmosphere system, the incident SW flux

$$F_0 = (1 - \mathcal{A})\bar{F}_s \qquad (8.61)$$

(Sec. 1.4.1) must be balanced by the outgoing LW flux at the top of the atmosphere

$$F^\uparrow(0) = F_0. \qquad (8.62)$$

Because the downwelling LW flux at TOA vanishes, $F(0)$ and $\bar{F}(0)$ each reduce to $F^\uparrow(0)$. Then, by (8.57),

$$F = F_0.$$
$$= \text{const.} \qquad (8.63)$$

The net LW flux at any level equals the SW flux at TOA. Incorporating (8.59.2) then yields

$$B^*(\tau^*) = \frac{F_0}{2}(\tau^* + 1), \qquad (8.64)$$

which is plotted in Fig. 8.20 along with $F^\uparrow$ and $F^\downarrow$. Under radiative equilibrium, the upwelling and downwelling fluxes differ by a constant. Net heating at any level therefore vanishes (8.56.1).

Consider now thermal equilibrium of the surface. Because it is black, the surface fully absorbs the incident SW flux. It also absorbs the downwelling LW flux emitted by the atmosphere $F^\downarrow(\tau_s^*)$. These must be balanced by the upwelling LW flux emitted by the surface:

$$B^*(T_s) = F_0 + F^\downarrow(\tau_s^*). \qquad (8.65)$$

Subtracting (8.59.2) and (8.63) gives the downwelling LW flux emitted by the atmosphere

$$F^\downarrow(\tau_s^*) = B^*(\tau_s^*) - \frac{F_0}{2}. \qquad (8.66)$$

Combining it with (8.65) yields

$$B^*(T_s) = B^*(\tau_s^*) + \frac{F_0}{2}. \qquad (8.67)$$

According to (8.67), the temperature under radiative equilibrium is discontinuous at the surface: The ground, with emission $B^*(T_s)$, is warmer than the overlying air, with emission $B^*(\tau_s^*)$.

The emission profile (8.64) determines the radiative-equilibrium temperature through

$$\begin{aligned} B^*(\tau^*) &= \pi B[T(\tau^*)] \\ &= \sigma T^4. \end{aligned} \qquad (8.68)$$

Because the atmosphere is hydrostatic, density decreases with height exponentially (1.17). By (8.36), so does τ^*. Further, most of the atmosphere's opacity follows from water vapor and cloud (Fig. 8.1), which are concentrated in the lowest levels. Therefore, the linear variation with optical depth of B^* (8.64) translates into a steep decrease with height of temperature. At the surface, the lapse rate is infinite because, there, the radiative-equilibrium temperature is discontinuous.

Optical depth is taken to vary with height exponentially

$$\tau = \tau_s e^{-\frac{z}{h}}, \qquad (8.69)$$

where $h = 2$ km is symbolic of the absolute concentration of water vapor (Fig. 1.17). Then $F_0 = 240$ W m^{-2} obtains the temperature profiles shown in Fig. 8.21 (solid). For given $\tau_s > 0$, the surface temperature exceeds that in the absence of an atmosphere ($\tau_s = 0$). It increases with the atmosphere's optical depth. This dependence on τ_s is a manifestation of the greenhouse effect. Radiative equilibrium temperature is maximum at the ground, decreasing upward steeply. It jumps discontinuously to the surface temperature T_s. As $z \to \infty$, the radiative-equilibrium temperature approaches a finite limiting value $\left(\frac{F_0}{2\sigma}\right)^{\frac{1}{4}} \cong 215°$K, called the *skin temperature*.

In a neighborhood of the surface, temperature decreases upward sharply – more sharply, in fact, than the saturated adiabatic lapse rate $\Gamma_s \cong 5.5°$K km^{-1} (dotted). The latter corresponds to moist neutral stability. Thus, in the lowest levels, each of the radiative-equilibrium temperature profiles in Fig. 8.21 is statically unstable. For an optical depth of $\tau_s = 4$, the radiative-equilibrium profile is conditionally unstable below

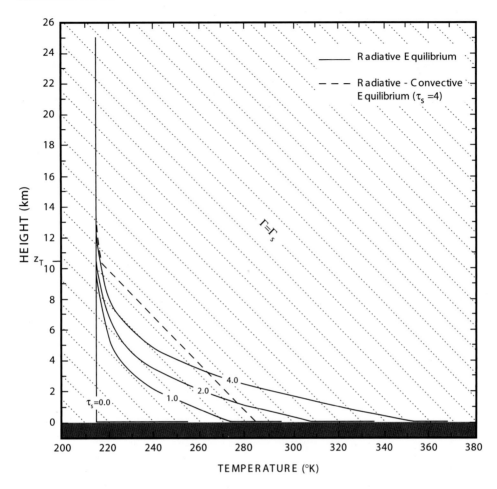

Figure 8.21 Radiative equilibrium temperature (solid) for the gray atmosphere in Fig. 8.20, with a profile of optical depth representative of water vapor (8.69). Presented for several atmospheric optical depths τ_s. Saturated adiabatic lapse rate (dotted) and radiative-convective equilibrium temperature for $\tau_s = 4$ (dashed) superposed.

6 km, and absolutely unstable below about 5 km. At the ground, it has a lapse rate of $\sim 36°$K km^{-1}.

8.5.2 Radiative-convective equilibrium

If air motion is accounted for, the conditions produced under radiative equilibrium cannot be maintained. Convection develops spontaneously, neutralizing the unstable stratification that was introduced by radiative transfer. It leads to *radiative-convective equilibrium.* Superposed in Fig. 8.21 for $\tau_s = 4$ (dashed), radiative-convective equilibrium produces two layers of distinctly different structure. Below a height z_T, thermal structure is controlled by convective overturning that is driven by radiative heating and its destabilization of the stratification. This layer constitutes the troposphere in this framework. Because convective overturning operates on a time scale much shorter than radiative transfer, it drives stratification below z_T to neutral stability. Under saturated conditions, this corresponds to a lapse rate of Γ_s (Sec. 7.6.1). Heat supplied at the Earth's surface is transferred upward and mixed vertically over the convective layer. This process maintains a uniform profile of equivalent potential temperature, $\theta_e \cong$ const. In the layer above z_T, thermal structure remains close to radiative equilibrium because, at those levels, radiative transfer stabilizes the stratification. This layer constitutes the stratosphere in this framework. Tropospheric air displaced upward by convection cools along the saturated adiabat. It therefore penetrates little above the tropopause height z_T, where the saturated-adiabatic and radiative-equilibrium profiles cross.

The particular form of radiative-convective equilibrium assumed depends on the parameterization of cloud and convective heat flux at the surface, which dictate the height of the tropopause. For instance, taking the height of the convective layer to equal the maximum height of instability under radiative equilibrium predicts one tropopause height. Taking the air temperature at $z = 0$ to equal the surface temperature under radiative equilibrium leads to a different tropopause height.

An alternative that circumvents ambiguities surrounding convection requires the convective layer to supply the same upward radiation flux at the tropopause as would be supplied under radiative equilibrium (Goody and Yung, 1995). This formalism is tantamount to presuming that the stratosphere is unaffected by vertical motion below. It provides a self-consistent rationale for determining radiative-convective equilibrium. Integrating (8.55) as in the development of (8.29) leads to expressions valid under arbitrary conditions for the upward and downward radiation fluxes:

$$F^{\uparrow}(\tau^*) = B(T_s)e^{\tau^* - \tau_s^*} - \int_{\tau_s^*}^{\tau^*} e^{\tau^* - \tau'} B^*[T(\tau')]d\tau' \tag{8.70.1}$$

$$F^{\downarrow}(\tau^*) = \int_0^{\tau^*} e^{\tau' - \tau^*} B(\tau')d\tau', \tag{8.70.2}$$

where convection has been presumed efficient enough to maintain the surface at the same temperature as the overlying air. Then equating (8.70.1) to the upward flux under radiative equilibrium gives

$$F^{\uparrow}(\tau^*) = \frac{F_0}{2}(\tau^* + 2), \tag{8.71}$$

which follows in the same manner as (8.66). For $\tau_s = 4$, this treatment yields a surface temperature of $T_s \cong 285°K$, close to the observed global-mean surface temperature of the Earth. Superposed in Fig. 8.21 is the radiative-convective equilibrium temperature that results (dashed). Temperature under radiative-convective equilibrium is cooler in the lower troposphere and warmer in the upper troposphere than results under radiative equilibrium. Stability in the lowest levels has been neutralized. Radiative-convective equilibrium achieves a tropopause at about 10.5 km, in qualitative agreement with observed thermal structure. Relative to the characteristic extinction scale, $h = 2$ km, z_T lies several optical depths above the Earth's surface. Temperature in the stratosphere is therefore close to the atmosphere's skin temperature.

The corresponding upward and downward LW fluxes are plotted as functions of height in Fig. 8.22a (dashed), along with those under radiative equilibrium (solid). Reduced temperature in the lower troposphere leads to smaller upward and downward

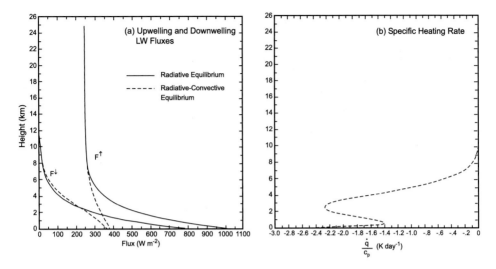

Figure 8.22 (a) Upwelling and downwelling fluxes as functions of height in the gray atmosphere in Fig. 8.20 for radiative equilibrium (solid) and radiative-convective equilibrium (dashed). (b) Specific heating rate under radiative-convective equilibrium.

fluxes. They merge with the radiative-equilibrium profiles above z_T. To satisfy overall equilibrium, $F^\downarrow$ approaches 0, whereas $F^\uparrow$ approaches 240 W m^{-2} as $z \to \infty$. Although the radiative-equilibrium fluxes preserve a constant difference, the net flux under radiative-convective equilibrium varies with altitude below z_T. Implied is nonzero radiative heating inside the troposphere (8.56).

The specific heating rate $\frac{\dot{q}}{c_p}$, which has dimensions of degrees per day, is plotted in Fig. 8.22b. It is negative throughout the troposphere, which experiences net radiative cooling (Fig. 1.32). Radiative cooling approaches zero inside the stratosphere, where thermal structure comes under radiative equilibrium. Inside the troposphere, radiative cooling must, at each level, be balanced by convective heating. The latter is supplied by upward heat transfer from the Earth's surface. This is consistent with observed thermal structure inside the tropical troposphere, where deep convection prevails (Sec. 7.7). Notice that radiative cooling is greatest at (1) the surface, where temperature is a maximum, and (2) the top of the extinction layer ($z \cong 2$–4 km), corresponding to water vapor, above which LW emission passes freely to space. Reaching a maximum of about $2°$K day^{-1}, the cooling rate in Fig. 8.22b is in qualitative agreement with detailed calculations of radiative cooling based on observed behavior (cf. Fig. 8.24a).

More sophisticated calculations of radiative-convective equilibrium incorporate distributions of optically active species, mean cloud cover, and surface absorption of SW radiation. Figure 8.23 shows the results of a calculation that accounts for radiative transfer by water vapor, carbon dioxide, and ozone. The radiative-equilibrium temperature (solid) decreases steeply from about $340°$K at the surface to a tropopause value of about $185°$K near 10 km. Most of the troposphere is absolutely unstable. In the stratosphere, temperature increases upward due to absorption of SW radiation by ozone. Radiative-convective equilibrium (dashed) produces thermal structure which resembles that of the gray atmosphere. Relative to thermal structure under radiative

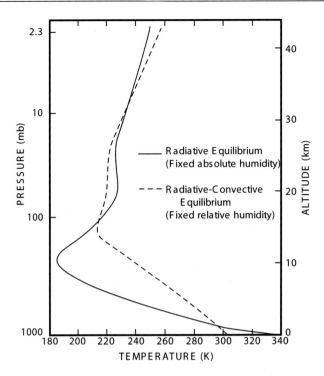

Figure 8.23 Temperature under radiative equilibrium (solid) and radiative-convective equilibrium (dashed) from calculations that include mean distributions of water vapor, carbon dioxide, and ozone. Adapted from Manabe and Wetherald (1967).

equilibrium, temperature is cooler in the lower troposphere and warmer in the upper troposphere.

8.5.3 Radiative heating

Knowledge of the mean distributions of temperature and optically active constituents enables LW heating to be inferred via (8.56) from the upwelling and downwelling fluxes. For an individual band, the specific heating rate may be expressed

$$
\begin{aligned}
\frac{\dot{q}}{c_p} &= -\frac{1}{\rho c_p}\frac{dF}{dz^*}\\
&= \frac{k}{c_p}\frac{dF}{d\tau^*},
\end{aligned}
\tag{8.72}
$$

where (8.53) has been incorporated into height and the frequency dependence is implicit. Differentiating and subtracting (8.51) gives

$$
\begin{aligned}
\frac{dF}{d\tau^*} = {}& B^*(T_s)\frac{dT(\tau_s^* - \tau^*)}{d\tau^*} - \int_{\tau^*}^{\tau_s^*} B^*[T(\tau')]\frac{d^2 T(\tau' - \tau^*)}{d\tau' d\tau^*}\,d\tau'\\
&- \int_0^{\tau^*} B^*[T(\tau')]\frac{d^2 T(\tau^* - \tau')}{d\tau' d\tau^*}\,d\tau' + B^*[T(\tau^*)]\left.\frac{dT(|\tau' - \tau^*|)}{d\tau'}\right|_{\tau'=\tau_-^*}^{\tau'=\tau_+^*}.
\end{aligned}
$$

The specific heating rate can then be arranged into the form

$$\frac{\dot{q}(\tau^*)}{c_p} = \frac{k(\tau^*)}{c_p} \left\{ B^*[T(\tau^*)] \frac{dT(\tau^*-0)}{d\tau^*} + \left[B^*(T_s) - B^*(\tau^*) \right] \frac{dT(\tau_s^*-\tau^*)}{d\tau^*} \right.$$

$$- \int_{\tau^*}^{\tau_s^*} \left[B^*(\tau') - B^*(\tau^*) \right] \frac{d^2T(\tau'-\tau^*)}{d\tau^* d\tau'} d\tau' \qquad (8.73)$$

$$\left. - \int_0^{\tau^*} \left[B^*(\tau') - B^*(\tau^*) \right] \frac{d^2T(\tau^*-\tau')}{d\tau^* d\tau'} d\tau' \right\}.$$

Equation (8.73) collects contributions to $\frac{\dot{q}}{c_p}$ from LW interactions between the level τ^* and its environment. $\frac{dT(\tau^*-\tau')}{d\tau'}$ then serves as an influence function between that level and another τ'. The first term on the right hand side describes interaction with space. Because $\frac{dT}{d\tau} < 0$ (8.47), this term is always negative. It represents cooling to space. The second term, which describes interaction with the surface, is negative when the level τ^* is warmer than the ground. The last two terms are *exchange integrals* that collect contributions from levels τ' above and below τ^*. Because $\frac{d^2T}{d\tau d\tau'} < 0$, those contributions represent cooling when the temperature at level τ^* exceeds that at level τ'.

Figure 8.24a illustrates the dominant contributions to LW heating. Except for ozone, the primary LW absorbers cool the atmosphere. Water vapor dominates LW cooling in the troposphere. Accounting for 80–90% of overall cooling, it leads to a globally averaged cooling rate of about $2°K$ day^{-1}. In the stratosphere, the 15-μm band of CO_2 dominates LW cooling. Together with the 9.6-μm band of O_3, it produces a maximum cooling rate near the stratopause of $\sim 12°K$ day^{-1}. Consistent with the dominance of CO_2 cooling in the stratosphere is a cooling trend observed at those altitudes, which accompanies increasing CO_2 and a warming trend near the Earth's surface (Fig. 1.42). Because the 9.6-μm band of ozone lies within the atmospheric window, it also produces heating in the lower stratosphere, by absorbing upwelling LW radiation from below.[3]

In contrast to LW radiation, SW radiation produces only heating, as the atmosphere does not emit at those wavelengths. For a particular wavelength, the specific heating rate is given by

$$\dot{q} = -k\mu_s \frac{dF}{d\tau},$$

where $\mu_s = -\mu > 0$ refers to the *solar zenith angle* and, as above, the wavelength dependence is implicit. Incorporating (8.7) transforms this into

$$\dot{q} = -k\mu_s \frac{d}{d\tau} \left[I_0 e^{-\frac{\tau}{\mu_s}} \right]$$

$$= kI_0 e^{-\frac{\tau}{\mu_s}}. \qquad (8.74)$$

---

[3] Even though specific heating rates in the stratosphere are larger than those in the troposphere, volume heating rates $\rho\dot{q}$ are smaller by two orders of magnitude.

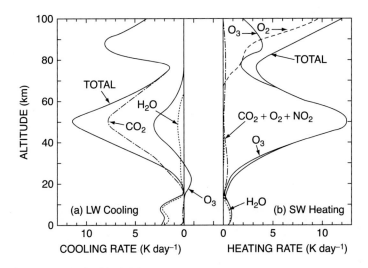

Figure 8.24 Global-mean profiles of (a) LW cooling and (b) SW heating. Contributions from individual radiatively active constituents also shown. After London (1980).

Then the volume heating rate follows as

$$\rho \dot{q} = \rho k I_0 e^{-\frac{\tau}{\mu_s}}$$
$$= n \hat{k} I_0 e^{-\frac{\tau}{\mu_s}},$$

(8.75)

where ρ and n refer to absorbers at the wavelength under consideration.

Heating varies strongly with the absolute concentration n of absorber. Dominated by the variation of air density, the absorber concentration may be modeled as

$$\frac{n}{n_0} = e^{-\frac{z}{H}}.$$

Then the optical depth also varies exponentially and (8.75) becomes

$$\rho \dot{q} = n_0 \hat{k} I_0 e^{-\left[\tau_0 e^{-\frac{z}{H}} + \frac{z}{H}\right]},$$

(8.76.1)

where

$$\tau_0 = \frac{n_0 \hat{k} H}{\mu_s}.$$

(8.76.2)

The volume heating rate (8.76) possesses a single maximum at the height

$$\frac{z_0}{H} = ln \tau_0.$$

(8.77)

It corresponds to a slant optical path from TOA of unity. Above that level, heating decreases exponentially with air density. Below, it decreases even faster due to attenuation in the first exponential term of (8.76.1). Therefore, SW heating is concentrated within about a scale height of the level z_0. It defines the *Chapman layer* for this wavelength. Consequently, the penetration altitude in Fig. 8.3 for a particular wavelength also describes where most of the SW absorption and heating occur. As shown in Fig. 8.25, increased solar zenith angle reduces SW heating, because the vertical flux is

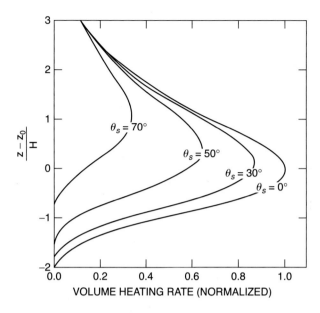

Figure 8.25 Volume heating rate due to absorption of SW radiation at solar zenith angle θ_s. Concentration of heating within a depth of order H defines the *Chapman layer* for a particular wavelength. After Banks and Kocharts (1973).

diminished according to μ_s. It also increases the altitude z_0, because the optical path length traversed by solar radiation is elongated.

The extinction cross section k for different wavelengths results from different species (8.6). They, in turn, have different vertical profiles. Consequently, the full spectrum of SW radiation produces a series of Chapman layers, which follow from the vertical distributions of optically active species. According to Fig. 8.3, energetic radiation at wavelengths shorter than 200 nm is absorbed in the upper mesosphere and thermosphere by O_2 in the Schumann-Runge bands and continuum. Wavelengths of 200–300 nm are absorbed in the mesosphere and stratosphere by O_3 in the Hartley band. Longer wavelengths penetrate to the surface, with partial absorption in the visible by ozone and in the near IR by water vapor and carbon dioxide.

Ozone accounts for most of the SW absorption in the stratosphere and mesosphere (Fig. 8.24b). It produces maximum heating near the stratopause, even though the absolute concentration of ozone maximizes considerably lower (Fig. 1.20). In the global-mean, SW heating and LW cooling at these heights nearly cancel. The middle atmosphere is therefore close to radiative equilibrium. In the troposphere, water vapor is the dominant SW absorber. Achieving a global-mean heating rate of about 1°K day^{-1}, SW absorption does not completely offset LW cooling. Thus, radiative transfer leads to net cooling of the troposphere of order 1°K day^{-1}. It is compensated by convective heating, which derives from the upward transfer of heat from the Earth's surface (Fig. 1.32).

Heating rates depicted in Fig. 8.24 apply to a clear atmosphere under global-mean conditions. Relative to those, local conditions can deviate dramatically. Figure 8.26 shows that, under conditions typical of the tropical troposphere, SW heating rates can

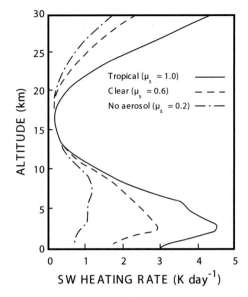

Figure 8.26 SW heating rate under different atmospheric conditions relating to solar elevation and the presence of aerosol. After Liou (1980).

be substantially greater (see also Reynolds et al., 1975). Likewise, the radiative balance in the stratosphere suggested by Fig. 8.24 for the global mean does not apply locally. As shown in Fig. 8.27, the winter stratosphere experiences strong net cooling – more than $8°K day^{-1}$ inside polar night. It implies this region is warmer than radiative equilibrium. As in the troposphere, net radiative cooling must be balanced by a mechanical transfer of energy. That energy transfer occurs through planetary waves

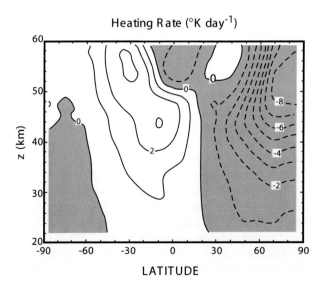

Figure 8.27 Net radiative heating in the middle atmosphere during January, as derived from Nimbus-7 LIMS. After Kiehl and Solomon (1986).

(Fig. 1.10). By driving the circulation out of radiative equilibrium, they perform work on the stratosphere. Manifest in temperature warmer than radiative equilibrium, that work is converted into heat and eventually rejected to space via LW cooling.

8.6 THERMAL RELAXATION

If a layer is in radiative equilibrium with its surroundings, LW and SW contributions to its radiation budget balance. Net heating then vanishes. If disturbed from radiative equilibrium (e.g., by motions that displace air into different radiative environments), a layer will experience net heating or cooling (8.73), which tends to restore the layer to radiative equilibrium. For example, polar air drawn equatorward by a cyclone will find itself colder than the radiative-equilibrium temperature of its new environment (Fig. 1.18). Net radiative heating then acts to destroy the temperature anomaly by driving that air toward the local radiative-equilibrium temperature. Because the circulation is related to thermal structure, damping anomalous temperature also damps anomalous motion.

In the absence of other forms of heat transfer, the temperature field is described by (8.56), with the heating rate given by (8.73). The first term in (8.73) describes cooling to space. It often dominates radiative heating. Figure 8.28 compares the total heating for water vapor, carbon dioxide, and ozone against the contributions from cooling to space alone. Over much of the troposphere and stratosphere, cooling to space provides an accurate approximation to net radiative heating. Exceptional is ozone in the lower stratosphere, which absorbs upwelling 9.6-μm radiation from below.

The agreement in Fig. 8.28 under widely varying circumstances is the basis for the *cool-to-space approximation*. For a particular frequency band, it is expressed by

$$\frac{dT}{dt} \cong -\frac{1}{\rho c_p} B^*[T(z^*)] \frac{dT_{\bar{\nu}}(z^* - \infty)}{dz^*}, \tag{8.78}$$

where vertical motion (e.g., $\frac{dp}{dt}$) is ignored, z^* includes the diffusivity factor (8.54.1), and summation over different absorbers is understood. Away from the surface and in the absence of strong curvature of the emission profile, (8.78) accurately represents cooling due to both water vapor and carbon dioxide. These conditions hold for temperature disturbances of large vertical scale. Interactions with underlying and overlying layers are then negligible. Under these conditions, emission to space of monochromatic LW radiation takes place from a Chapman layer, analogous to absorption of SW radiation. Emission over a finite spectral band can thus be interpreted in terms of a series of Chapman layers.

Consider a disturbance $T'(z, t)$ to an equilibrium temperature profile $T_0(z)$.[4] If $\frac{T'}{T_0} \ll 1$ and if the cool-to-space approximation applies, then emission in (8.78)

---

[4] The equilibrium thermal structure may be maintained by heat transfer other than radiation alone. However, we will take T_0 to represent the radiative-equilibrium temperature.

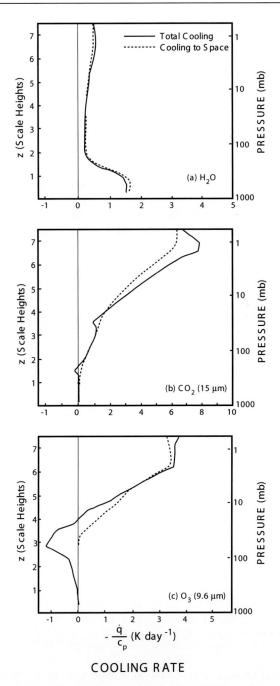

Figure 8.28 Comparison of the total cooling rate (solid) vs the contribution from cooling to space (dashed) for (a) water vapor, (b) carbon dioxide, and (c) ozone. Adapted from Rodgers and Walshaw (1966).

may be linearized in T', yielding

$$\frac{\partial T'}{\partial t} = -\frac{1}{\rho_0 c_p}\frac{dB^*}{dT}[T_0(z^*)]\frac{dT_{\bar{v}}(z^* - \infty)}{dz^*}\cdot T' \tag{8.79}$$

$$= -\alpha(z^*)T'.$$

Known as the *Newtonian cooling approximation*, (8.79) governs the evolution of anomalous temperature. The Newtonian cooling coefficient $\alpha(z^*)$ varies spatially through the equilibrium thermal structure $T_0(z^*)$. This approximation is of great practical importance because it eliminates the rather cumbersome interactions in (8.73) in favor of a simple expression that depends only on the local temperature. It is also linear in disturbance temperature. Even though it breaks down for CO_2 in the mesosphere (Andrews et al., 1987), the Newtonian cooling approximation is widely adopted.

Under the influence of Newtonian cooling, a temperature disturbance relaxes exponentially toward radiative equilibrium. Owing to its linearity, (8.79) may be cast into the form of an e-folding time

$$-\frac{1}{T'}\frac{\partial T'}{\partial t} = \frac{4\sigma T_0^3}{\rho_0 c_p}\frac{dT_{\bar{v}}(z^* - \infty)}{dz^*} \tag{8.80}$$

$$= t_{rad}^{-1}.$$

t_{rad} measures the efficiency of radiative transfer in relation to other factors that influence thermal structure. Taking values representative of the troposphere (Prob. 8.17) yields a radiative time scale of order 10 days.

The time scale $t_{rad} = \alpha^{-1}$ is an order of magnitude longer than the characteristic time scale of air motion. Radiative transfer is therefore inefficient compared with dynamical influences. For this reason, air motion controls tropospheric properties like thermal structure and vertical stability. In the stratosphere, where water vapor and cloud are rare, cooling to space is stronger. There, t_{rad} decreases to only 3–5 days. Although shorter, this is still long enough to permit air motion to influence many stratospheric properties.

More accurate formulations of heat transfer apply to temperature disturbances whose vertical scales are not large enough to neglect exchange between neighboring layers. Plotted in Fig. 8.29 is the thermal damping rate for sinusoidal vertical structure with several vertical wavelengths λ_z. Thermal dissipation minimizes in the limit $\lambda_z \to \infty$. This limit corresponds to deep temperature structure that is governed by Newtonian cooling (8.79). Under those circumstances, t_{rad} decreases from 10–15 days in the troposphere to about 5 days near the stratopause. Smaller vertical scales experience significantly faster damping due to exchange with neighboring layers. Relaxation times of only a couple of days are then experienced in the upper stratosphere and mesosphere.

8.7 THE GREENHOUSE EFFECT

The radiative-equilibrium surface temperature T_s is significantly warmer than that in the absence of an atmosphere. Combining (8.67) and (8.64) for a gray atmosphere yields

$$B^*(T_s) = \frac{F_0}{2}(\tau_s^* + 2). \tag{8.81}$$

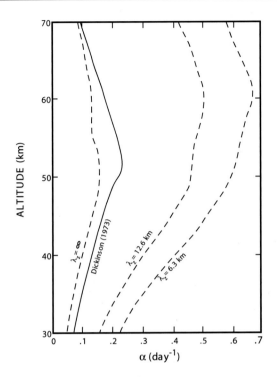

Figure 8.29 Thermal damping rate for temperature disturbances with sinusoidal vertical structure of vertical wavelength λ_z (dashed). After Fels (1982). Newtonian cooling rate (solid), from a calculation by Dickinson (1973), is superposed.

Through (8.68), (8.81) determines T_s under radiative equilibrium. Surface temperature increases with the optical depth of the atmosphere, a manifestation of the greenhouse effect. The surface temperature under radiative-convective equilibrium, although smaller, is likewise controlled by τ_s, for example, through (8.70) and (8.71).

The atmosphere's optical depth depends on its composition through radiatively active species. As described in Sec. 1.2.4, several have increased steadily during the last two centuries, inclusive of anthropogenic contributions. Presented in Fig. 8.30 is an estimate of the respective increase in greenhouse warming of the Earth's surface during the last two centuries, inferred from proxy records of composition (Sec. 1.6.2). Carbon dioxide is greatest, increasing downwelling LW radiation by $\sim$1.5 Wm^{-2}. Methane, nitrous oxide, halocarbons, and ozone are comparatively rare. Yet, together, they introduce almost as much additional warming as CO_2. Collectively, these anthropogenic gases represent additional warming of the Earth's surface of about 3 Wm^{-2}. For reference, the additional warming introduced by a doubling of CO_2 is, from Figs. 8.30 and 1.14, seen to be $\sim$4 Wm^{-2}.

Offsetting the increased greenhouse warming is increased cooling of the Earth's surface by contemporaneous changes of aerosol. Entering the energy budget through albedo, those changes have likewise been inferred, somehow, from proxy records. Supported by surface changes, the inferred changes of aerosol represent an offset of about -1.5 Wm^{-2}.

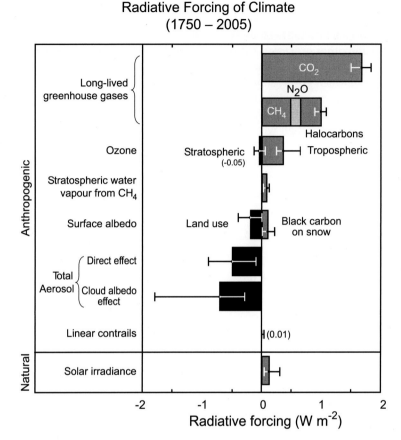

Figure 8.30 Inferred increase in radiative warming of the Earth's surface during the last two centuries from several influences. Not accounted for are changes of water vapor and cloud. Adapted from IPCC (2007).

The residual, $+1.5$ Wm^{-2}, represents net warming. It is about 0.5% of the 327 Wm^{-2} of overall downwelling LW radiation that warms the Earth's surface (Fig. 1.32). The vast majority of that warming is contributed by water vapor. Together with cloud, it accounts for 98% of the greenhouse effect. How water vapor has changed in relation to changes of the comparatively minor anthropogenic species (Fig. 8.30) is not known.

The additional surface warming introduced by anthropogenic increases in greenhouse gases amounts to about 75% of that which would be introduced by a doubling of CO_2. Arrhenius' estimate of 5–6°K for the accompanying increase of surface temperature (Sec. 1.2.4) then translates into ~4°K. Yet, the observed change of global-mean temperature since the mid nineteenth century is only about 1°K (Sec. 1.6.1). The discrepancy points to changes of the Earth-atmosphere system (notably, involving the major absorbers, water vapor and cloud) that develop in response to imposed perturbations, like anthropogenic emission of CO_2.

8.7.1 Feedback in the climate system

Surface temperature depends on the atmosphere's optical depth. The latter, in turn, depends on atmospheric composition through radiatively active species. Those constituents are produced and destroyed by surface processes. Water vapor is produced at ocean surfaces through evaporation. Carbon dioxide is produced by decomposition of organic matter. Ozone is destroyed at the Earth's surface through oxidation. These and other processes that control radiatively active species are temperature dependent. This dependence introduces the possibility of feedback between the two sides of (8.81), which governs surface temperature through optical depth.

Because climate properties are interdependent, a perturbation of one will introduce changes in others. Secondary changes modify the direct response of surface temperature. The process through which this additional change is introduced is a *feedback mechanism*. If it reinforces the direct response of surface temperature, the process represents positive feedback. If it opposes the direct response of surface temperature, the process represents negative feedback. Climate feedback mechanisms are numerous. Some of the more important are highlighted.

Consider the energy balance at TOA. In the framework of Sec. 8.5, the net radiation (Sec. 1.4.2), averaged globally, is described by

$$\mathcal{R} = F_0 - F^\uparrow(\tau_s = 0), \tag{8.82.1}$$

where, with (1.30), (8.61) and (8.68),

$$F_0 = \frac{F_s}{4}(1 - \mathcal{A}) \tag{8.82.2}$$

$$F^\uparrow(0) = \epsilon \sigma T_s^4, \tag{8.82.3}$$

and ϵ represents an effective emissivity for the Earth-atmosphere system. Under equilibrium, $\mathcal{R} = 0$.

Suppose now that the system is perturbed, for example, by an increase of CO_2 that upsets the energy balance by dF Wm^{-2}. This is equivalent to an increase in SW heating of dF: $F_0 \rightarrow F_0 + dF$. All components of the system will then undergo an adjustment to restore equilibrium. In terms of surface temperature, the change of net radiation is thus

$$d\mathcal{R} = 0 = \left(\frac{\partial \mathcal{R}}{\partial F} + \frac{\partial \mathcal{R}}{\partial T_s} \frac{dT_s}{dF} \right) dF, \tag{8.83}$$

where dF is the *direct radiative forcing*. As $\partial \mathcal{R} = \partial F$, (8.83) reduces to

$$\frac{dT_s}{dF} = -\left(\frac{\partial \mathcal{R}}{\partial T_s} \right)^{-1}$$

$$= \Lambda. \tag{8.84}$$

The factor Λ defines the *climate sensitivity* with respect to a radiative perturbation dF. It measures the response of surface temperature to a perturbation dF in the energy balance: $dT_s = \Lambda dF$. With (8.82), the climate sensitivity becomes

$$\Lambda = \left(\frac{F_s}{4} \frac{\partial \mathcal{A}}{\partial T_s} + \frac{\partial F^\uparrow(0)}{\partial T_s} \right)^{-1}. \tag{8.85}$$

Albedo and upwelling LW flux in (8.85) depend on other properties, such as cloud cover and humidity. The contribution to Λ from the change of an individual property can be evaluated by holding others fixed.

Emission Feedback

An increase of T_s introduced through increased heating dF will produce an increase of LW emission at TOA. By cooling the Earth-atmosphere system, the increase of emission opposes the increase of T_s. It thus constitutes negative feedback. If albedo and effective emissivity in (8.82) are fixed, then emission is the only feedback present. The climate sensitivity then reduces to

$$\Lambda = \left(4\epsilon\sigma T_s^3\right)^{-1} = \left(4\sigma T_e^3\right)^{-1}$$

$$= \left[4\frac{F^\uparrow(0)}{T_e}\right]^{-1} \tag{8.86}$$

$$\cong 0.27°K\left(Wm^{-2}\right)^{-1}.$$

A perturbation to the energy balance of 1 Wm^{-2} requires a response of surface temperature of $0.27°K$. The former is equivalent to an increase in solar constant of $\sim 0.5\%$.

Temperature – Water Vapor Feedback

Owing to the strong temperature dependence in the Claussius-Clapeyron relation (4.38), saturation vapor pressure e_w increases sharply with T_s. An increase of surface temperature therefore increases the saturation vapor pressure e_w, which enables the vapor pressure e to increase. This increases τ_s and, from (8.81), further increases T_s. The response of water vapor thus reinforces the temperature increase that is introduced directly through dF. It constitutes positive feedback.

Such changes enter (8.82) through $F^\uparrow(0)$, implicitly through the effective emissivity ϵ. Representing them analytically is not feasible. Instead, the dependence on T_s of $F^\uparrow(0)$ has been evaluated empirically. Satellite observations indicate that $F^\uparrow(0)$ increases with T_s approximately linearly, $\frac{\partial F^\uparrow(0)}{\partial T_s}$ being in the range 1.55–1.85 $Wm^{-2}°K^{-1}$ (North, 1975; Cess, 1976; Warren and Schneider, 1979). Supporting evidence comes from numerical simulations under radiative-convective equilibrium. Observations indicate that, despite large regional changes during its annual cycle, the relative humidity of the atmosphere as a whole varies little. Simulations with RH fixed reveal that $F^\uparrow(0)$ does not increase with T_s quartically, as in (8.82.3). Rather, it increases almost linearly, with $\frac{\partial F^\uparrow(0)}{\partial T_s} \cong 2.2$ $Wm^{-2}°K^{-1}$ (Manabe and Wetherald, 1967). If albedo remains fixed, then, inclusive of such feedback, (8.85) yields a climate sensitivity of

$$\Lambda \cong 0.55°K\left(Wm^{-2}\right)^{-1}. \tag{8.87}$$

As it rests on an observed relationship between $F^\uparrow(0)$ and T_s, (8.87) does not discriminate to water vapor; it also includes feedbacks from emission and cloud.[5] Together, these feedbacks amplify the climate sensitivity. The response of surface temperature is now twice as large as would be required in the presence of only emission feedback (8.86).

A doubling of CO_2 over preindustrial levels would perturb the energy budget by 4 Wm^{-2}. From (8.87), it would thus be expected to produce warming of surface temperature of $\sim 2°K$. The present increase of CO_2 over preindustrial levels corresponds

---

[5] An increase of cloud acts in two opposing directions: It sharply increases optical depth, which increases T_s (8.81). Simultaneously, it sharply increases albedo, which decreases T_s (8.82.2). Although these influences enter at leading order, the complex dependence of cloud on temperature, humidity, and other factors leaves cloud feedback poorly understood.

to a LW perturbation of ~1.7 Wm^{-2} (Fig. 8.30). Via (8.87), it translates into an increase in surface temperature of ~1 K. This value is broadly consistent with the observed increase of Global Mean Temperature over the last century (Sec. 1.6). It is not consistent with the increase of GMT during closing decades of the twentieth century. The latter implied a trend twice as great (Fig. 1.39). Indeed, after the turn of the century, GMT underwent a general decline for several years. Its increase during closing decades of the twentieth century is therefore unrepresentative of the secular variation of GMT. Because it pertains to a comparatively short interval, that warming is strongly contributed to by climate variability, as prevailed during earlier decades of the twentieth century.

Temperature – CO_2 feedback

Like other trace species, CO_2 is produced and destroyed at the Earth's surface. Involving a number of reservoirs and processes that are difficult to document, individual sources and sinks are poorly quantified. Net emission of CO_2, however, is determined unambiguously in the instrumental record (Fig. 1.43). As discussed in Sec. 1.6.2, the net emission rate of CO_2 varies coherently with T_s. An increase of T_s introduced by a radiative perturbation dF thus leads to an increase in the emission rate of CO_2, and, hence, cumulatively in CO_2. The latter, in turn, increases τ_s, reinforcing the increase of T_s introduced directly by dF. The dependence of CO_2 emission on temperature thus constitutes positive feedback. It is analogous to the positive feedback between temperature and water vapor. However, feedback between temperature and CO_2 is weaker – by two orders of magnitude (e.g., 4 Wm^{-2} vs. 327 Wm^{-2}; cf. Fig. 1.32). Similar considerations apply to methane, which, like CO_2 is produced by decomposition of organic matter.

Such feedback can be treated in a manner analogous to water vapor. It is, in fact, already represented in the empirical relationship between T_s and $F^{\uparrow}(0)$. Derived from observed changes, that relationship accounts for feedback, not only with water vapor, but with all temperature-dependent species.

Increased temperature increases net emission of CO_2. Decreased temperature has the reverse effect. Each represents an imbalance between sources and sinks of CO_2. It is noteworthy that the positive sensitivity to temperature, $d\dot{r}_{CO_2}/dT_s$, is not restricted to small perturbations. As is evident in Fig. 1.43, the dependence on temperature applies to changes of $\dot{r}_{CO_2}$ as large as 100%. Also noteworthy is that the correspondence applies to changes of temperature that are clearly of different origin. Following the eruption of Pinatubo, when SW heating decreased, $\dot{r}_{CO_2}$ decreased by more than 50% (Fig. 1.27). During the 1997–1998 El Nino, when SST increased, $\dot{r}_{CO_2}$ increased by more than 100%.[6] To maintain stability, there must exist a negative feedback on CO_2, one that is sufficiently strong to bridle the enhancement of CO_2 emission by positive feedback from temperature. That negative feedback involves sinks of CO_2 at the Earth's surface (Sec 1.2.4). It is analogous to the negative feedback that bridles the enhancement of water vapor emission by positive feedback from temperature. Unlike CO_2, the negative feedback on water vapor involves sinks in the atmosphere.

The satellite record of GMT, in concert with the instrumental record of CO_2 is long enough to provide a population of climate perturbations, wherein the

---

[6] The large increase of CO_2 emission then is consistent with the oceanic source of CO_2, which dominates transfers to the atmosphere (see Fig. 17.11)

Plate 0 Visible imagery during August 2005, revealing twin cyclones that formed in the equatorial Pacific. Being entrained into them is haze from Indonesian forest fires, which developed during prolonged drought under El Nino conditions (negative phase of the Southern Oscillation). While it prevailed over SE Asia, the dense aerosol that was mantained by those fires led to widespread health restrictions and public closures. Courtesy of Australian Bureau of Meteorology and Japan Meterological Agency.

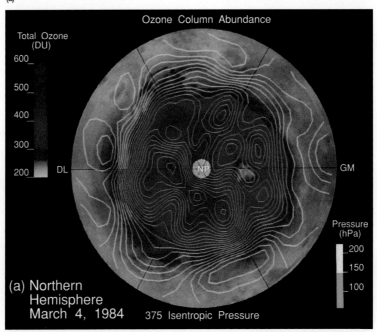

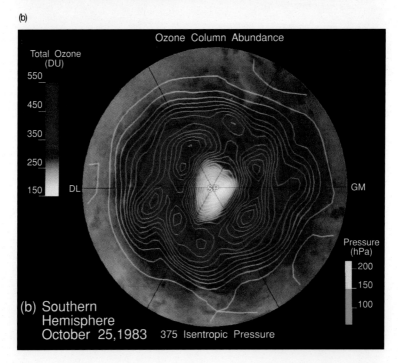

Plate 1 Distribution of total ozone (color) from the Total Ozone Mapping Spectrometer (TOMS) on board Nimbus-7 and pressure (contours) on the 375°K isentropic surface (see Sec. 2.4.1) over (a) the Northern Hemisphere on March 4, 1984 and (b) the Southern Hemisphere on October 25, 1983.

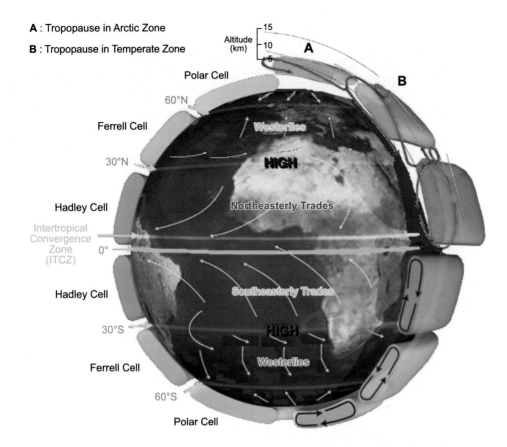

A : Tropopause in Arctic Zone

B : Tropopause in Temperate Zone

Altitude (km)

15
10
5

A

B

Polar Cell

60°N

Ferrell Cell

30°N

Westerlies

HIGH

Hadley Cell

Northeasterly Trades

Intertropical Convergence Zone (ITCZ)

0°

Hadley Cell

Southeasterly Trades

30°S

HIGH

Ferrell Cell

Westerlies

60°S

Polar Cell

Plate 2 Schematic of the mean circulation, which is comprised of Hadley cells equatorward of 30°, Ferrell cells at mid-latitudes, and polar cells at higher latitude. The lower branch of the Hadley cell involves northeasterly and southeasterly Trade Winds, which meet in the InterTropical Convergence Zone (ITCZ), where air rises to form deep cumulus convection and heavy rainfall. *Source*: sealevel.jpl.nasa.gov.

Walker Circulation

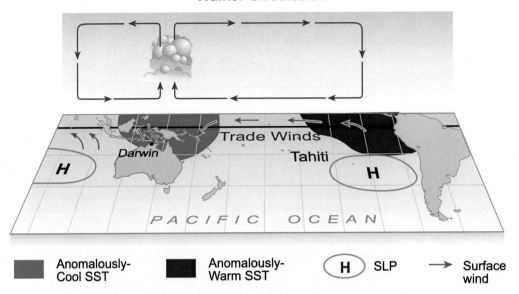

Plate 3 Schematic of the Pacific Walker circulation. *Source*: http://www.bom.gov.au/lam/climate/levelthree/analclim.

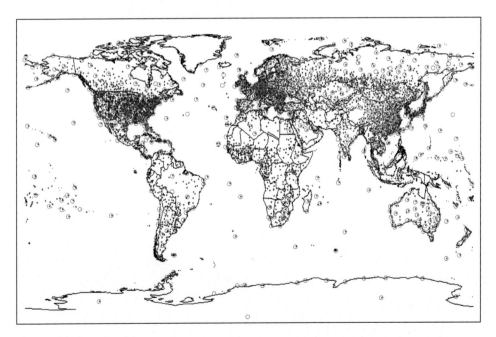

Plate 4 Stations comprising the global radiosonde network, where vertical profiles of atmospheric structure are measured twice daily at 0000 and 1200 UT (open circles), and the ground network of surface temperature measurements (dots). *Sources:* Gruber and Haimberger (2008), Peterson and Vose (1997).

Temperature Trend

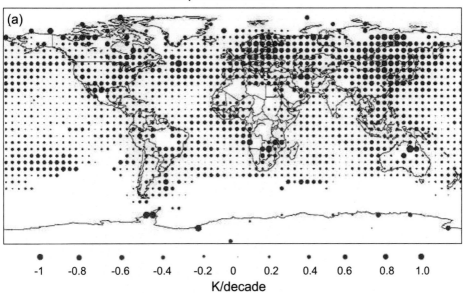

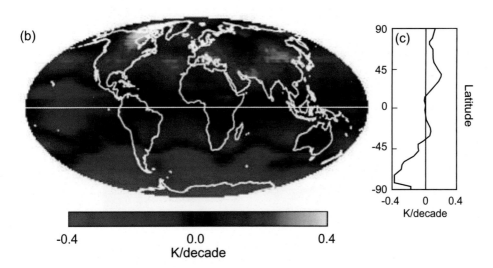

Plate 5 Temperature trend during the satellite era, as function of geographical position, derived from (a) the surface record and (b) the satellite record.
(c) Zonal-mean trend in the sattellite record, which has global coverage.
Sources: IPCC (2007), Mears et al. (2003).

Plate 6 (a) Schematic of the mountain wave that develops leeward of an elongated range. *Lenticular clouds* are fixed with respect to orography, while air flows through them. The positions of *rotor clouds* are also fixed, but those fragmented features are highly unsteady, with turbulent ascent (descent) along their leading (trailing) edges. Westward tilt with height (e.g., of wave crests) corresponds to upward propagation of energy (Chap. 14), a feature that is often visible in the structure of lenticular clouds. (b) Longitudinal view through the mountain wave. Smooth lenticular clouds form in a stack, separated by clear regions, due to stratification of moisture. These laminar features contrast sharply with turbulent rotor clouds below. (c) Transverse view from the ground. Vorticity and severe turbulence are made visible by looping cloud matter that comprises rotor. Courtesy of V. Haynes and L. Feierabend (Soaring Society of Boulder).

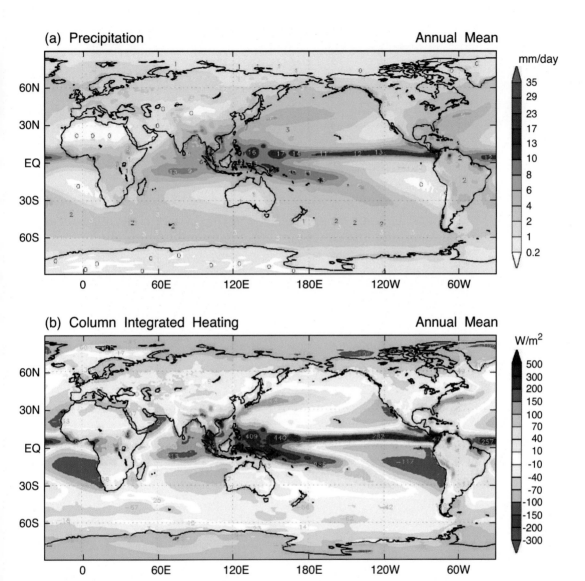

Plate 7 (a) Annual-mean precipitation rate (mm/day). (b) Column-integrated atmospheric heating (W m^{-2}). Both are concentrated inside the ITCZ, where precipitation and atmospheric heating exceed 10 mm/day and 250 W m^{-2}, respectively. Support comes from the South Pacific Convergence Zone (SPCZ), the South Atlantic Convergence Zone (SACZ), and the North Atlantic and North Pacific storm tracks. *Source:* Japanese 25-yr Climatology, cooperative project of Japan Meteorological Agency (JMA) and Central Research Institute of Electric Power Industry (CRIEPI) (http://ds.data.jma.go.jp).

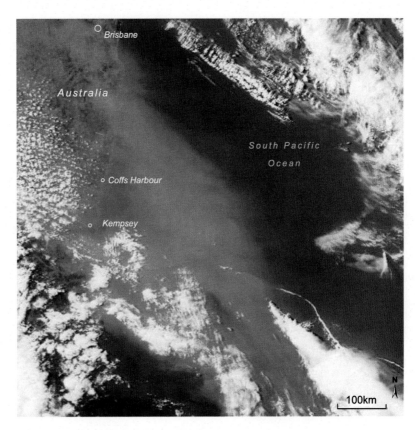

Plate 8 Visible imagery during September 2009, when the eastern seaboard of Australia was blanketed by red dust that originated 1000 km to the west. Notice the change of albedo relative to neighboring ocean.

Plate 9 Field of view across Sydney Harbor (a) under normal conditions and (b) during the dust storm of September 2009. With permission, copyright 2010 Fairfax Digital.

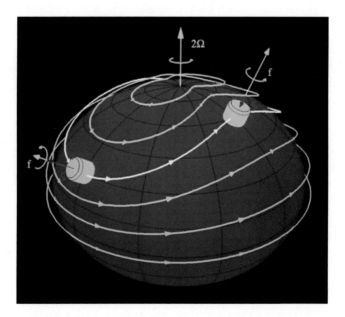

Plate 10 Schematic illustrating the reaction of an air parcel to meridional displacement. Displaced equatorward, an eastward-moving parcel spins up cyclonically to conserve absolute vorticity. Northward motion induced ahead of it then deflects the parcel's trajectory poleward back toward its undisturbed latitude. The reverse process occurs when the parcel overshoots and is displaced poleward of its undisturbed latitude.

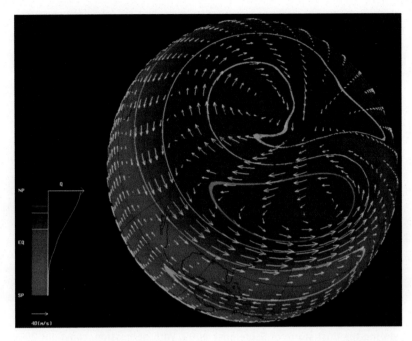

Plate 11 Distribution of potential vorticity Q and horizontal motion in a 2-dimensional calculation representative of the stratospheric circulation at 10 hPa under disturbed conditions. Color and velocity scales shown at left. High-Q polar air (blue), which marks the polar-night vortex, has been displaced well off the pole and distorted by an amplified planetary wave. Replacing it is low-Q air from equatorward (red), which has been advected into the polar cap. It spins up anticyclonically, forming a reversed circulation with easterly circumpolar flow at high latitude. These features are characteristic of a *stratospheric sudden warming* (cf. Fig. 18.13).

Solstitial Swing of Temperature

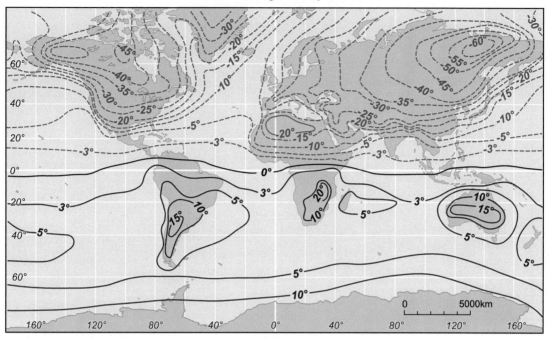

Plate 12 Annual swing of surface temperature: January–July.

(a) January

Surface Pressure and Motion

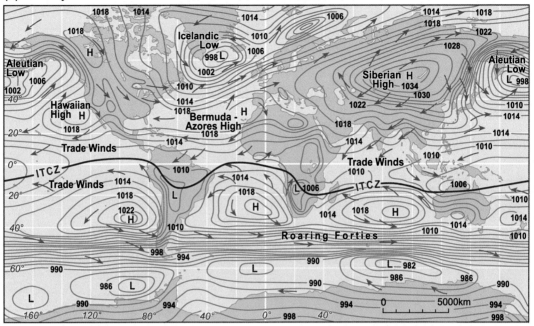

(b) July

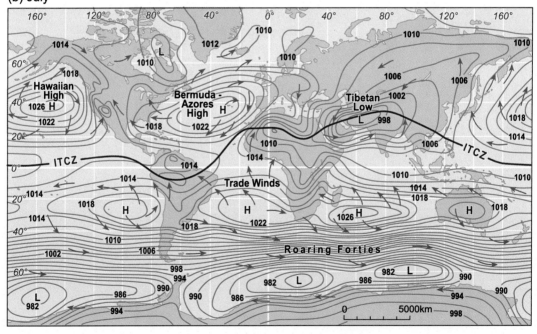

Plate 13 Climatological-mean SLP and surface motion for (a) January and (b) July.

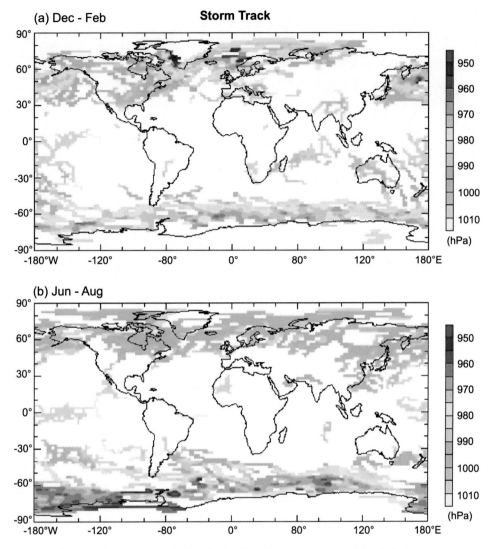

Plate 14 Mean intensity of cyclonic disturbances, measured by their depression of SLP, during (a) December–February 1984 and (b) June–August 1984. *Source:* http://data.giss.nasa.gov (10.07.10).

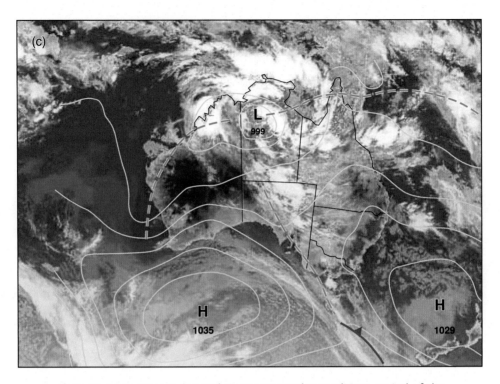

Plate 15 Instantaneous SLP and IR imagery under conditions typical of the Australian monsoon. Deep convection is organized by the monsoon low, as well as trough axes that emanate from it. Cirrus blow off at upper levels spirals outward anticyclonically.

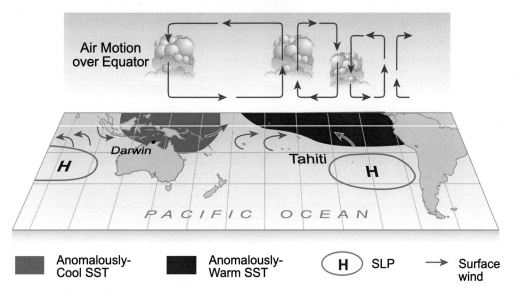

Plate 16 Disturbed Walker circulation during El Nino. Adapted from http://www.bom.gov.au (11.07.10).

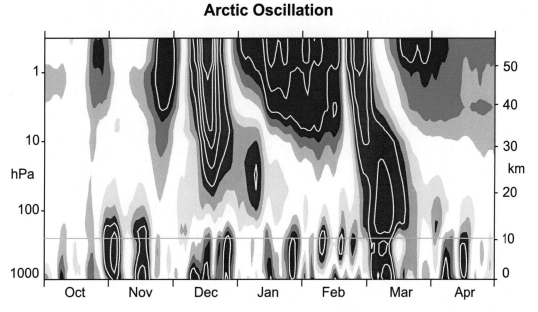

Plate 17 Time-height section of dimensionless AO index during 1998–1999. AO index represents the intensity of the leading EOF of geopotential height. After Baldwin and Dunkerton (2001).

(a)

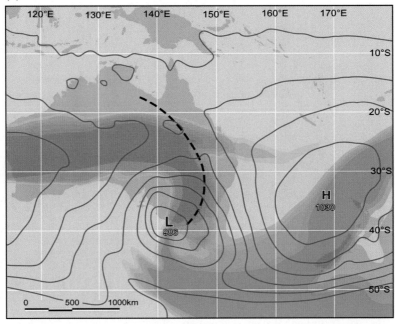

(b)

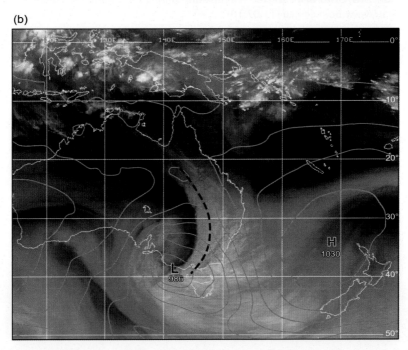

Plate 18 Conditions accompanying an amplifying baroclinic system that has reached maturity. (a) Distribution of SLP (contoured) and 300-hPa isotachs stronger than 20 m/s, which delineate the jet stream (shaded). (b) Contemporaneous water vapor image from MTSAT, representing humid air (light) and dry air (dark).

Surface Salinity

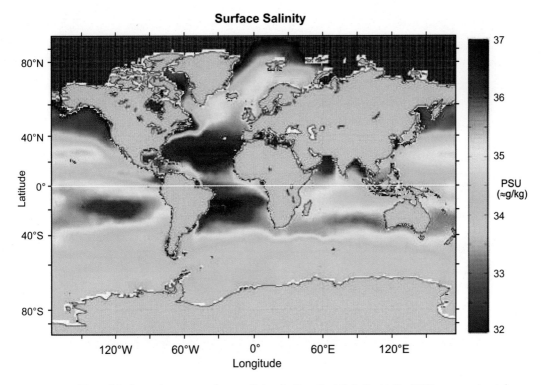

Plate 19 Annual-mean surface salinity, in Practical Salinity Units (PSU), approximately equal to g/kg. *Source:* Antonov et al. (2006).

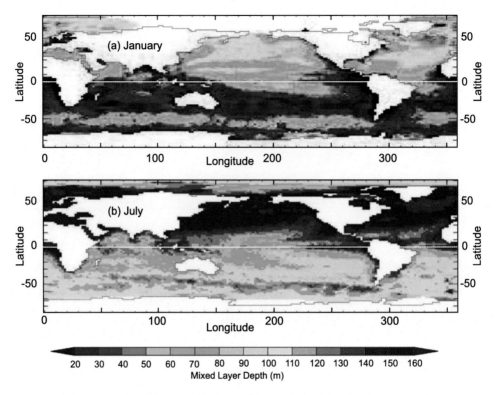

Plate 20 Depth of the mixed layer, as function of geographical position, during (a) January and (b) July. *Source:* http:/www./locean-ipsl.upmc.fr.

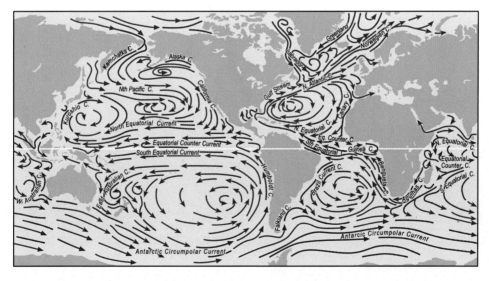

Plate 21 Surface circulation during January, illustrating water that is comparatively warm (red) and cold (blue). *Source:* US Navy Oceanographic Office.

Thermohaline Circulation

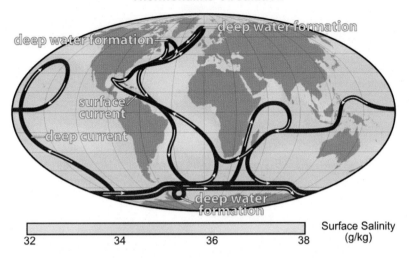

Surface Salinity (g/kg)

32   34   36   38

Plate 22 Schematic of the thermohaline circulation, along with surface salinity (shaded). Deep water (blue) forms in the far north Atlantic, near Greenland, and in the Weddell Sea, near Antarctica. After meandering through the global ocean, it eventually percolates upward, forming surface water (red). The time for the entire circuit to be completed is 1000–2000 years. *Source:* http://earthobservatory.nasa.gov.

SW Heating

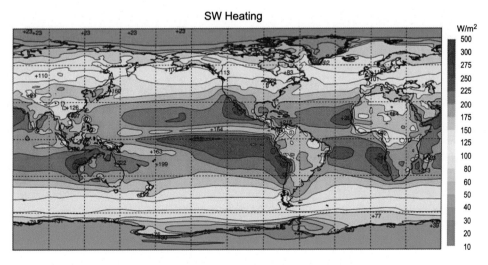

Plate 23 Annual-mean absorption of SW radiation at the Earth's surface. SW heating maximizes in the subtropics, especially in the eastern oceans, which are comparatively free of cloud and have large albedo. Derived from ERA-40 reanalysis. *Source:* Kallberg et al. (2005).

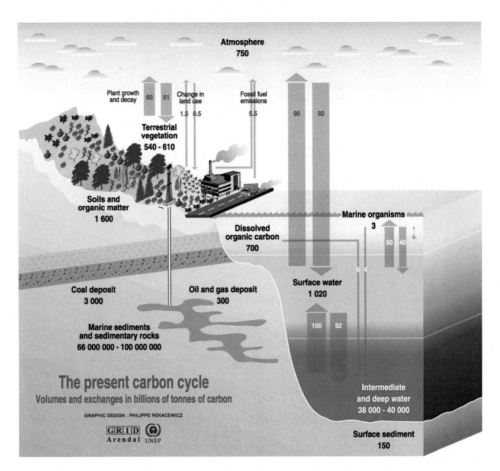

Plate 24 Estimated global carbon cycle, illustrating stores of carbon, in GtC, and transfers in GtC/yr, where 1 GtC $= 10^9$ tons of carbon. *Source:* http://maps.grida.no/go/graphic/the-carbon-cycle, design by Philippe Rekacewicz, UNEP/GRID-Arendal (11.07.10).

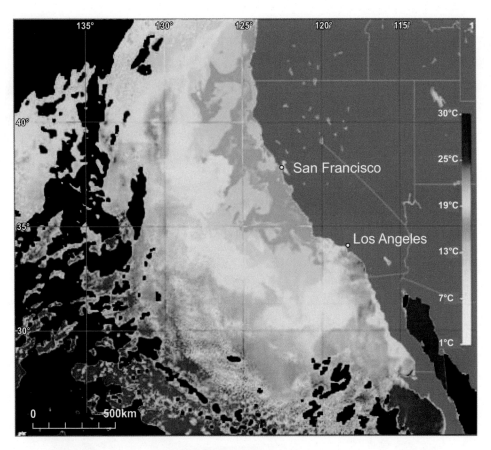

Plate 25 Surface temperature off the California coast. Temperature decreases shoreward, attaining values 10–15°C colder than in the open ocean. Black shading represents sites that are overcast, where data is unavailable. *Source:* http://oceanmotion.org.

(a) Dec – Feb 1999: La Nina Conditions

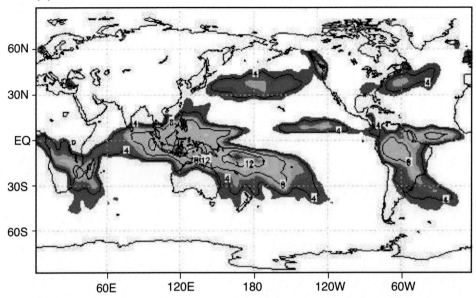

(b) Dec – Feb 1998: El Nino Conditions

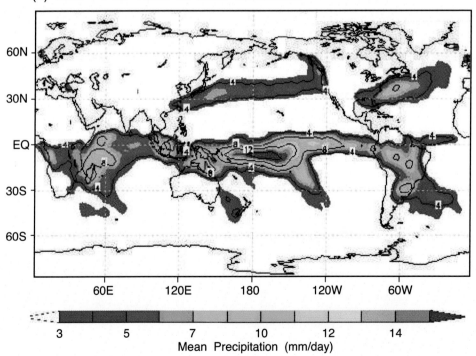

Mean Precipitation (mm/day)

Plate 26 Global distribution of precipitation (a) during Dec–Feb 1999, under La Nina conditions, when rainfall was concentrated in the western Pacific, and (b) during Dec–Feb 1998, under El Nino conditions, when rainfall was displaced eastward into the central Pacific. Also evident during northern winter are the storm tracks over the north Atlantic and north Pacific, where precipitation is organized by amplifying synoptic weather systems. *Source:* http://www.esrl.noaa.gov (15.07.10).

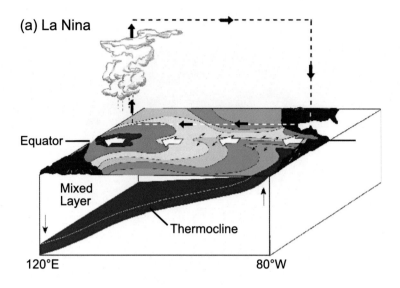

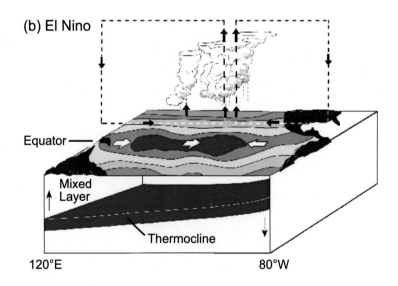

Plate 27 Ocean thermal structure, surface wind, SST (contoured, with red warmest), and Walker circulation under (a) La Nina conditions and (b) El Nino conditions. *Source:* http://www.pmel.noaa.gov.

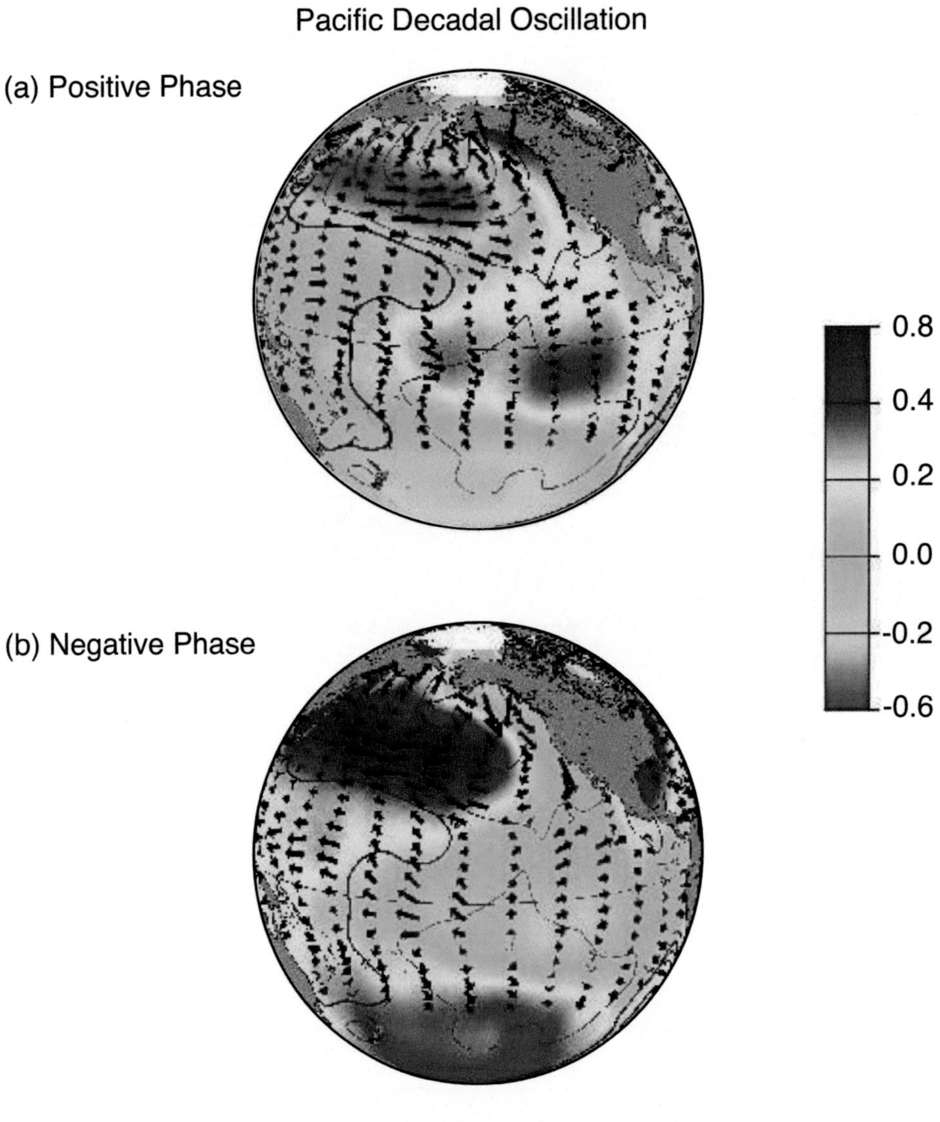

Pacific Decadal Oscillation

(a) Positive Phase

(b) Negative Phase

Plate 28 Anomalous SST (shaded), SLP (contoured), and surface wind associated with the Pacific Decadal Oscillation (PDO) during its (a) positive phase, when SST in the equatorial eastern Pacific is anomalously warm, and (b) negative phase, when SST in the equatorial eastern Pacific is anomalously cool. *Source:* http://jisao.washington.edu.

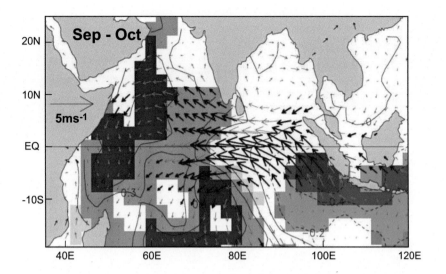

Plate 29 Anomalous SST (contoured), regions of >90% significance shaded with red significant and warm and blue significant and cold, and surface wind associated with Indian Ocean dipole (IOD). After Saji et al. (1999). Reprinted by permission of Macmillan Publishers Ltd: Nature (Saji et al., 1999), Copyright (1999).

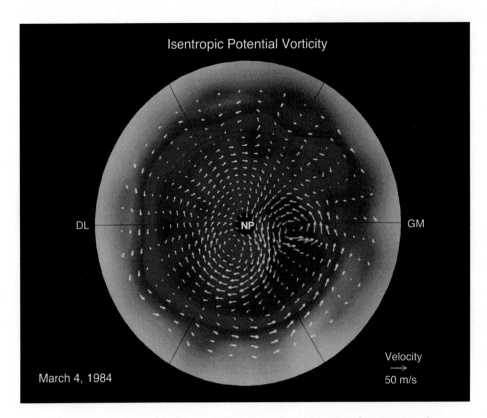

Plate 30 Distributions of potential vorticity Q and horizontal motion on the 850°K isentropic surface (near 10 hPa) for March 4, 1984. High-Q polar air (blue), which marks the polar-night vortex, has been displaced off the pole and distorted by a large amplitude planetary wave, with secondary eddies appearing along its tail. Low-Q air from equatorward (red) that is advected into the polar cap spins up anticyclonically, forming a reversed circulation with easterly circumpolar flow at high latitude (cf. 14.26).

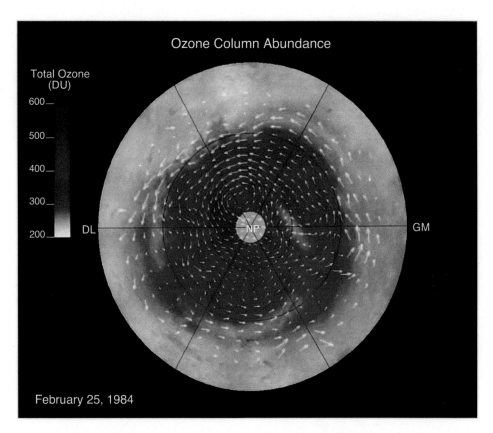

Plate 31 Total ozone Σ_{O_3} and horizontal motion on the 400°K isentropic surface in the Northern Hemisphere on February 25, 1984, during a stratospheric warming when zonal flow was disturbed down to 70 hPa. A tongue of enhanced Σ_{O_3} coincides with cross-polar flow between longitudes of 150 E and 30 W, where air descends along isentropic surfaces and undergoes compression. Air displaced equatorward in the Eastern Hemisphere ascends isentropically and undergoes expansion, introducing anomalously low Σ_{O_3}.

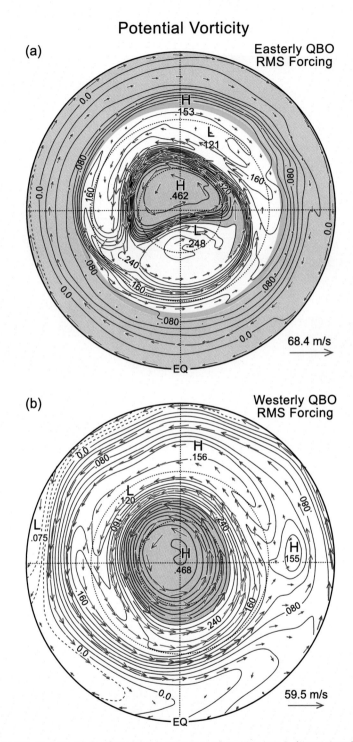

Plate 32 Distributions of potential vorticity and motion in a 2-dimensional calculation representative of the polar-night vortex (a) during the QBO easterlies
(red), when the zero-wind line (green), which is the critical line for stationary
planetary waves, invades the winter subtropics, and (b) during the QBO westerlies, when the critical line for stationary planetary waves recedes into the
summer subtropics. Adapted from O'Sullivan and Salby (1990).

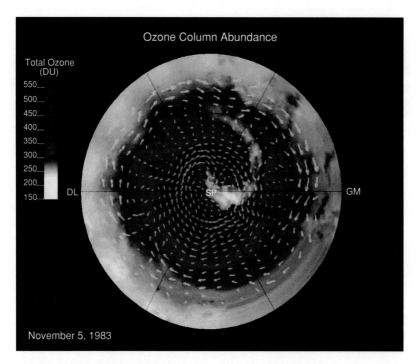

Plate 33 Distributions of total ozone and horizontal motion on the 325°K isentropic surface over the Southern Hemisphere on November 5, 1983, during the final warming. Ozone-depleted air (white), which had previously been confined inside the polar-night vortex, escapes in a tongue of anomalously-lean values (white) that spirals anticyclonically into mid-latitudes to conserve potential vorticity.

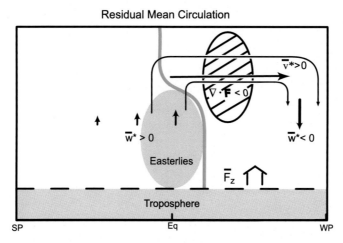

Plate 34 Schematic representation of the residual mean circulation $(\bar{v}^*, \bar{w}^*)$. Driven by absorption of EP flux $(\nabla \cdot \boldsymbol{F})$ that is transmitted upward by planetary waves (18.27). The latter is measured by the upward EP flux at 100 hPa, integrated over latitude $\bar{F}_z$ (18.32). During QBO easterlies (red), the zero-wind line (green), which is the critical line for stationary planetary waves, invades the winter hemisphere (cf. Fig. 18.17a).

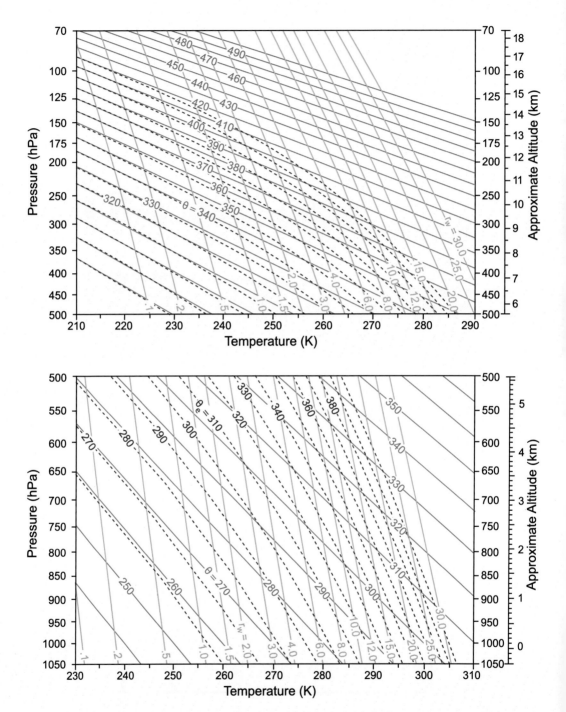

Plate 35 Pseudo-Adiabatic Chart: Adiabats (blue-solid), which are characterized by constant values of potential temperature θ; pseudo-adiabats (red-dashed), which are characterized by constant values of equivalent potential temperature θ_e; and isopleths of saturation mixing ratio with respect to water r_w (green), all as functions of temperature and pressure with the ordinate proportional to $\left(\frac{p_0}{p}\right)^{\kappa}$.

Earth-atmosphere system was disturbed from equilibrium. Those perturbations reveal the sensitivity of CO_2 emission.

The climate sensitivity of CO_2 emission with respect to a change of temperature, $\frac{d\dot{r}_{CO_2}}{dT_s}$, is described in the same fashion as the climate sensitivity of temperature with respect to a change of radiative heating, $\frac{dT_s}{dF}$. Coherent changes of $\dot{r}_{CO_2}$ and T_s in Fig. 1.43 imply a mean sensitivity

$$\frac{d\dot{r}_{CO_2}}{dT_s} = c[\dot{r}_{CO_2}|T_s]\frac{\langle \dot{r}^2_{CO_2}\rangle^{\frac{1}{2}}}{\langle T^2_s\rangle^{\frac{1}{2}}}$$

$$\approx 3.5 \frac{\text{ppmv/yr}}{\text{K}},$$

where $c[\dot{r}_{CO_2}|T_s]$ is the respective correlation and $\langle\ \rangle$ denotes time mean. Revealed by natural perturbations to the Earth-atmosphere system, the sensitivity accounts for much of the observed variation of CO_2 emission on interannual time scales (Fig. 1.43). It establishes that GMT cannot increase without simultaneously increasing CO_2 emission – from natural sources. Further, the coherent but out-of-phase changes in ^{12}C and ^{13}C (Fig. 1.43a, b) indicate that sources involved in interannual changes of CO_2 emission have isotopic composition similar to those involved in longer-term changes of CO_2 (Fig. 1.14). Their close correspondence to surface temperature begs the question: Does this dependence on T_s also contribute to changes of CO_2 emission on longer time scales?

During the satellite era, GMT in the record from MSU increased between 1979 and 2009 with a mean trend of $\frac{dT_s}{dt} \approx 0.125°\text{K/decade}$. This represents warming over 3 decades of $\Delta T_s \approx 0.38°\text{K}$. In concert with the above sensitivity, it implies an increase in CO_2 emission rate of $\Delta\dot{r}_{CO_2} \approx 1.3$ ppmv/yr. The observed trend in $\dot{r}_{CO_2}$ during the satellite era is $d\dot{r}_{CO_2}/dt \approx 0.26$ ppmv/yr · decade (Fig 1.43). It represents an effective increase in emission rate over 3 decades of $\Delta\dot{r}_{CO_2} \approx 0.8$ ppmv/yr. Hence the increase in CO_2 emission rate anticipated from the increase in T_s alone is comparable to the observed increase in $\dot{r}_{CO_2}$. In fact, it exceeds the observed increase. Including increased emission from sources that are independent of temperature (e.g., anthropogenic sources) would magnify the anticipated increase even further.

During the twentieth century, GMT increased with a mean trend of ~0.075 K/decade (Sec. 1.6.1). With the above sensitivity, this translates into an increase in CO_2 emission rate of $\Delta\dot{r}_{CO_2} \approx 2.6$ ppmv/yr. The observed increase in emission rate follows from Fig. 1.43, wherein $\dot{r}_{CO_2}$ currently approaches 2.2 ppmv/yr, and from the ice core record (Fig. 1.14), which gives at the opening of the twentieth century $\dot{r}_{CO_2} \approx 0.2$ ppmv/yr. Differencing yields an observed increase over the twentieth century of $\Delta\dot{r}_{CO_2} \approx 2.0$ ppmv/yr. The increase in CO_2 emission anticipated from the increase in T_s alone is again comparable to the observed increase in $\dot{r}_{CO_2}$. If anything, it too exceeds the observed increase.[7]

The results for the two periods are in broad agreement. Together with the strong dependence of CO_2 emission on temperature (Fig. 1.43), they imply that a significant portion of the observed increase in $\dot{r}_{CO_2}$ derives from a gradual increase in surface

---

[7] The modest overestimate of the increase in emission rate (2.6 vs 2.0 ppmv/yr) is to be expected. On very-long time scales, mediating influences will eventually limit the response of natural sources to a change of temperature, invalidating the sensitivity that is apparent on interannual time scales. Yet, the comparatively-minor discrepancy with the observed increase in emission rate indicates that, even on the time scale of a century, the sensitivity observed on interannual time scales remains approximately valid.

temperature. Some of that warming is consistent with a rebound from the Little Ice Age (Moberg et al., 2005); cf. Fig. 1.45. Through integrated emission, the same implication must apply to CO_2 itself.

An estimate of this contribution follows from the temperature reconstruction in Fig. 1.45, which, as reconstructions go, is a conservative description of centennial changes during the last millennium. The rebound from the Little Ice Age involved systematic warming between the years 1600 and 1900 with a mean trend of $\sim 0.02°$K/decade. This is about half of the mean trend over the entire instrumental record, which dates back to about 1850: $\sim 0.045°$K/decade (IPCC, 2007). With the above sensitivity of CO_2 emission, it implies an increase during the twentieth century of $\Delta r_{CO_2} \approx 55$ ppmv. The observed increase is ~ 80 ppmv. Such changes occur on time scales that are long compared to the length of instrumental records. Nevertheless, they are short compared to the time scales of heat transfer with the deep ocean (Chap. 17).

The preceding analysis follows from the observed sensitivity of CO_2 in a population of climate perturbations. It accounts for a significant component of observed changes during the satellite era and, more generally, during the twentieth century as a whole. The resemblance between observed changes of CO_2 and those anticipated from increased surface temperature also points to a major inconsistency between proxy records of previous climate. Proxy CO_2 from the ice core record (Fig 1.13) indicates a sharp increase after the nineteenth century. At earlier times, proxy CO_2 becomes amorphous: Nearly homogeneous on time scales shorter than millennial, the ice core record implies virtually no change of atmospheric CO_2. According to the above sensitivity, it therefore implies a global-mean climate that is "static," largely devoid of changes in GMT and CO_2. Proxy temperature (Fig. 1.45), on the other hand, exhibits centennial changes of GMT during the last millennium, as large as 0.5–1.0°K. In counterpart reconstructions, those changes are even greater (Section 1.6.2). It is noteworthy that, unlike proxy CO_2 from the ice core record, proxy temperature in Fig. 1.45 rests on a variety of independent properties. In light of the observed sensitivity, those centennial changes of GMT must be attended by significant changes of CO_2 during the last millennium. They reflect a global-mean climate that is "dynamic," wherein GMT and CO_2 change on a wide range of time scales. The two proxies of previous climate are incompatible. They cannot both be correct.

Ice – Albedo Feedback

Coverage by ice is a significant contributor to the Earth's albedo. It decreases with an increase of temperature. By (8.82.2), this increases F_0. From (8.81), it further increases T_s. The response of ice cover thus reinforces the temperature increase that is introduced directly by dF. Like water vapor, it constitutes positive feedback.

We consider this influence in a simple model of the Earth-atmosphere system, one that includes interaction between radiative transfer and ice. Albedo is prescribed to vary with the idealized temperature dependence

$$\mathcal{A} = \begin{cases} \mathcal{A}_0 = \text{const} & T_s < 230 \text{ K} \\ \mathcal{A}_0 - \eta\,(T_s - 230) & 230 \leq T_s \leq 270 \text{ K} \\ \mathcal{A}_0 - \eta\,(270 - 230) = \text{const} & T_s > 270 \text{ K}. \end{cases} \tag{8.88}$$

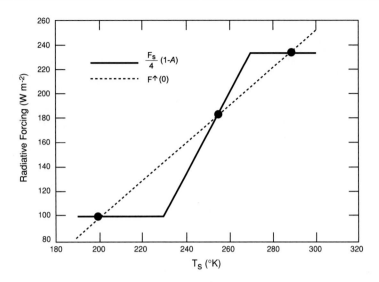

Figure 8.31 SW heating (solid) and LW cooling (dashed) at TOA, as functions of surface temperature T_s, in a simple climate system that includes ice-albedo feedback. With simplified dependence of albedo (8.88) and OLR (8.90), the values $A_0 = 0.70$, $\eta = 0.01$, $F_0^\uparrow(0) = -212\,\mathrm{Wm^{-2}}$, and $\gamma = 1.55\,\mathrm{Wm^{-2}/^\circ K}$ make $\frac{F_s}{4}\eta > \gamma$ and $\Lambda < 0$. The system then possesses three equilibrium states (solid circles), at which SW heating equals LW cooling. Adapted from Kiehl (1992).

For T_s warmer than 270°K, the system is unglaciated; albedo then reduces to that of cloud and other fixed contributors. For T_s colder than 230°K, it is fully glaciated. Between those extreme states, ice cover decreases linearly with increasing T_s. Equilibrium requires $\mathcal{R} = 0$ or

$$\frac{F_s}{4}(1 - \mathcal{A}) = F^\uparrow(0). \tag{8.89}$$

Adopting for $F^\uparrow(0)$ the linear dependence on T_s

$$F^\uparrow(0) = F_0^\uparrow(0) + \gamma\,T_s\,, \tag{8.90}$$

together with (8.88) and (8.89), then defines a system of three simultaneous equations in the unknowns $\mathcal{A}$, $F^\uparrow(0)$, and T_s. With $\mathcal{A}_0$, η, $F_0^\uparrow(0)$, and γ specified, solutions correspond to those values of T_s at which curves for the left- and right-hand sides of (8.89) intersect. In those states, SW heating equals LW cooling and the glaciated system is in equilibrium. The behavior is illustrated in Fig. 8.31, which plots the left- and right-hand sides of (8.89) for values that make $\frac{F_s}{4}\eta > \gamma$. The idealized climate system then possesses three equilibrium states. The warmest ($T_s = 288^\circ$K) represents a state of no ice. ($\mathcal{A}$ is then invariant under small changes of T_s.) The coldest ($T_s = 202^\circ$K) represents a state that is fully glaciated. ($\mathcal{A}$ is then likewise invariant under small changes of T_s.) Between those extreme states is the intermediate equilibrium state ($T_s = 254^\circ$K). It corresponds to partial glaciation. $\mathcal{A}$ then decreases with small increases of T_s.

The climate system governed by (8.88)–(8.90) involves just a couple of variables. Even though highly simplified, it is capable of complex and wide-ranging behavior. The system's response to perturbation falls into three categories (Fig. 8.32), according to the sensitivities to increased T_s of increased SW heating (via reduced $\mathcal{A}$) and increased LW cooling (via increased $F^\uparrow(0)$):

(1) $\frac{F_s}{4}\eta < \gamma$ (Fig. 8.32a): SW heating (solid) increases with increasing T_s *slower* than does LW cooling (dashed). The system then possesses only one equilibrium state, at which the curves for SW heating and LW cooling intersect. Under these conditions, (8.85) gives

$$\Lambda > 0.$$

Perturbing the TOA energy balance by $dF > 0$ is equivalent to introducing an infinitesimal upward displacement to the curve for SW heating: $\frac{F_s}{4}(1 - \mathcal{A}) \to \frac{F_s}{4}(1 - \mathcal{A}) + dF$ (dotted). The system evolves to its perturbed state along the curve for LW cooling. Upon reaching the displaced curve for perturbed heating, equilibrium is reestablished. As is apparent from Fig. 8.32a, this involves surface warming: $dT_s = \Lambda dF > 0$. The latter is required to offset the positive perturbation dF (heating) via an opposing perturbation $d[\frac{F_s}{4}(1 - \mathcal{A}) - F^\uparrow(0)]$, which, under these conditions, is negative (cooling).

(2) $\frac{F_s}{4}\eta = \gamma$ (Fig. 8.32b): SW heating increases with increasing T_s *at the same rate* as does LW cooling. The system then possesses a continuum of equilibrium states, at which the curves for SW heating and LW cooling coincide. Under these conditions, (8.85) gives

$$\Lambda = \infty.$$

Perturbing the TOA energy balance by $dF > 0$ is again equivalent to an infinitesimal upward displacement of the curve for SW heating. However, because it is parallel to the curve for LW cooling, reestablishing equilibrium now requires a finite change of temperature. The system must evolve along the curve for LW cooling to a temperature warmer than the extremal temperature: $T_s > 270$ K, where it reaches the curve for perturbed heating and equilibrium is reestablished. Although stimulated by an infinitesimal perturbation in radiative heating, the temperature change $dT_s = \Lambda dF$ required to offset it and restore equilibrium is finite.

(3) $\frac{F_s}{4}\eta > \gamma$ (Fig. 8.32c): SW heating increases with increasing T_s *faster* than does LW cooling. The system then possesses three equilibrium states, at which the curves for SW heating and LW cooling intersect (Fig. 8.31). Under these conditions, (8.85) gives

$$\Lambda < 0.$$

From the intermediate equilibrium state, the nearest perturbed equilibrium state is reached via surface cooling: $dT_s = \Lambda dF < 0$. The latter is required to offset the positive perturbation dF (heating) via an opposing perturbation $d[\frac{F_s}{4}(1 - \mathcal{A}) - F^\uparrow(0)]$, which, under these conditions, is negative (cooling) only if $dT_s < 0$.

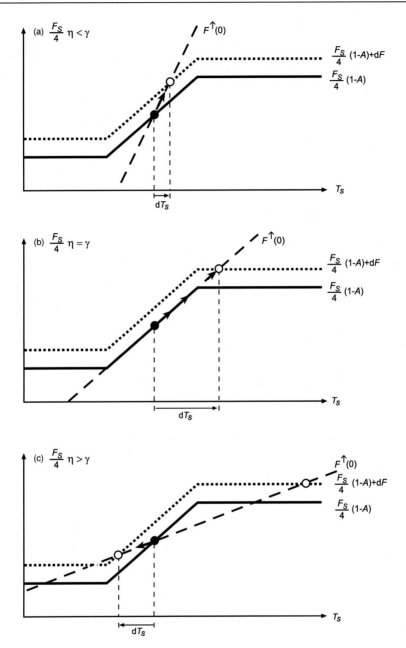

Figure 8.32 SW heating (solid) and LW cooling (dashed) at TOA as functions of surface temperature T_s in a simple climate system, inclusive of ice-albedo feedback, that is governed by (8.88)–(8.90). Superimposed are states of equilibrium, where SW heating equals LW cooling (solid circles) and the SW heating when perturbed infinitesimally by $dF > 0$ (dotted). (a) $\frac{F_s}{4}\eta < \gamma$: SW heating increases with increasing T_s *slower* than does LW cooling. When perturbed by infinitesimal heating, $dF > 0$, the system evolves to a new equilibrium state (open circle) through surface warming, $dT_s > 0$. (b) $\frac{F_s}{4}\eta = \gamma$: SW heating increases with increasing T_s *at the same rate* as does LW cooling. When perturbed by infinitesimal heating, $dF > 0$, the system evolves to a new equilibrium state through surface warming, $dT_s > 0$, that is finite. (c) $\frac{F_s}{4}\eta > \gamma$: SW heating increases with increasing T_s *faster* than does LW cooling. When perturbed by infinitesimal heating, $dF > 0$, the system evolves to the nearest new equilibrium state through surface cooling, $dT_s < 0$.

More sophisticated models of this form can possess one, two, or three equilibria, depending on the solar constant; see Hartmann (1994) for an overview. Under certain conditions, a partially glaciated state can be in equilibrium, but unstable. The system will then jump to another equilibrium state, one that is fully glaciated. Developing spontaneously (e.g., from an infinitesimal perturbation), this finite shift to the fully glaciated state occurs through positive feedback that goes unchecked (Sec. 8.7.2). The reverse change has been proposed in relation to methane that is sequestered beneath Arctic ice. Receding ice would release that methane, which would then reinforce greenhouse warming of the Earth's surface (Sec. 1.2.4). That, in turn, would accelerate the recession of ice, and so forth. How this hypothesis fits with the seasonality of CH_4 emission from Arctic tundra is unclear. During the freeze-in months of autumn, methane emission is as large as it is during unfrozen months of the entire summer (Mastepanov, 2008).

8.7.2 Unchecked feedback

Positive feedback such as that between temperature and water vapor reinforces small perturbations to the Earth-atmosphere system. Were it to continue without opposition, such feedback would produce a large shift of T_s, along with shifts of other properties that characterize climate. A paradigm of such changes is the so-called *runaway greenhouse effect*. It is used to explain the present state of the Venusian atmosphere.

The evolution of a planet's atmosphere occurs through slow discharge of gases from the planet's surface. In the Earth's atmosphere, water vapor operates in this manner. Atmospheric uptake of water vapor is limited by its saturation vapor pressure. The latter depends only on temperature (e.g., on surface temperature T_s). Because water vapor is responsible for atmospheric opacity, its vapor pressure e translates into optical depth τ_s. At saturation, $e = e_w(T_s)$. Plotting T_s against $log(e_w)$ produces the saturation curve in Fig. 8.33 (dotted). It separates heterogeneous states of water from vapor phase alone.

Now the surface temperature of a planet is determined by the optical depth of its atmosphere τ_s and the incident solar flux F_0, e.g., via (8.81) under radiative equilibrium or (8.70) under radiative-convective equilibrium. Thus the surface temperature T_s is a function of e, which is symbolic of τ_s. $T_s(e)$ defines a family of radiative-convective equilibrium curves, each one for a different value of F_0. Two are plotted in Fig. 8.33. The curve for a particular F_0 can be thought to represent the evolution of a given planet. For small e, $T_s \cong (\frac{F_0}{\sigma})^{\frac{1}{4}}$. T_s then reflects the initial state of the planet, when all of its water resided at the surface in condensed phase (solid circle).

The situation for Earth is represented in the curve for Planet 1 (solid). Initially unsaturated, the atmosphere absorbs water vapor from the surface. This increases e, its LW opacity τ_s, and hence its surface temperature T_s. Continued absorption of water vapor and its positive feedback onto temperature then drive the state of the atmosphere to the right along the radiative-convective equilibrium curve. Eventually, the atmosphere's state encounters the saturation curve, where $e = e_w$ (open circle). Feedback is then checked by condensation, which returns H_2O to the planet's surface as fast as it is evaporated. Further increases of e are possible only through increases of $e_w(T_s)$. For the conditions of Planet 1, saturation is achieved at a surface temperature of about $287°K$. These conditions describe a saturated system that contains multiple phases of water. They are characteristic of the present state of the Earth's atmosphere.

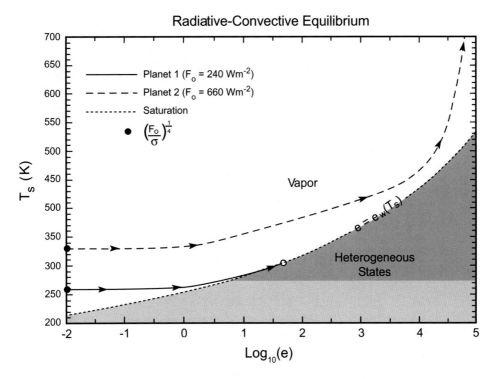

Figure 8.33 Surface temperature under radiative-convective equilibrium, as a function of vapor pressure for water (symbolic of atmospheric optical depth to LW radiation τ_s). Radiative-convective equilibrium curves shown for conditions representative of Earth (Planet 1) and Venus (Planet 2), which characterize atmospheric evolution from an initial state when all water resided at the surface in condensed phase. Positive feedback between temperature and water vapor drives the state of a planet's atmosphere to greater temperature and humidity. For Planet 1 (solid), the radiative-equilibrium curve eventually encounters the saturation curve (dotted), where no more water is absorbed by the atmosphere and positive feedback ceases. However, the radiative equilibrium curve for Planet 2 (dashed) never encounters the saturation curve. It therefore continues to evolve through positive feedback until all water has been incorporated into the atmosphere. The calculation of radiative-convective equilibrium presumes a constant albedo and relates optical depth to vapor pressure as $\tau_s = 4 \frac{e(T)}{e(288K)}$.

Saturation prevents H_2O from reaching great heights, where it would be photodissociated and ultimately destroyed when the resulting atomic hydrogen was lost to space (Sec. 1.2.2). Saturation of the atmosphere has thus maintained water on the Earth.

The curve for Planet 2 corresponds to Venus (dashed). It describes a very different evolution. Closer to the sun, Venus has a larger F_0 and hence a higher initial surface temperature. As before, the atmosphere's state evolves to the right along the radiative-convective equilibrium curve. However, because of its higher initial temperature,

Planet 2 never encounters the saturation curve. Feedback of water vapor onto T_s then goes unchecked. Water evaporates from the planet's surface without bound. Eventually, all of the planet's H_2O has been absorbed by the atmosphere. Planet 2 has then achieved an optical depth and temperature that vastly exceed those present initially. This evolution is symbolic of Venus, which has mean surface temperature of 750°K – some 400°K warmer than would exist in the absence of its atmosphere.

Atmospheric composition implied by the evolution of Planet 2 also differs fundamentally from that of Planet 1. For initial water abundance comparable to that of Earth, the atmosphere of Planet 2 would achieve a state in which H_2O was its primary constituent. Without condensation, water vapor would then occupy all levels of the atmosphere. It would therefore suffer photodissociation by UV, allowing atomic hydrogen to escape to space. The free oxygen released could then recombine only at the surface through oxidation processes, which produce CO_2. In this fashion, water would be systematically eliminated, replaced by carbon dioxide. This scenario is consistent with the present-day atmosphere of Venus, which is rich in CO_2.

8.7.3 Simulation of climate

The feedbacks developed in Sec. 8.7.1 can amplify the response of the Earth-atmosphere system over its direct radiative response. In principle, those feedbacks are represented in global climate models. GCMs are, increasingly, coupled atmosphere-ocean models. The large heat capacity of the ocean, supported by transfers of latent heat, make climate simulations inherently sensitive to the ocean simulation (Sec. 17.2) Yet, limited observations of the deep-ocean circulation leave those simulations largely unvalidated.

GCMs predict that doubling CO_2 over pre-industrial levels will lead to warming of GMT by 3–7 K (IPCC, 2007). Those predictions are considerably greater than the warming anticipated by the observed feedback between temperature and water vapor ($\sim$2 K), which in turn is consistent with observed warming during the twentieth century. The magnified response represents an amplification of warming through feedback mechanisms inherent in GCMs. As climate projections rest on those internal feedbacks, which are poorly understood, the accuracy with which GCMs reproduce observed changes is pivotal.

Global climate models are sophisticated extensions of the idealized models considered above. Treatments of climate properties in different GCMs are as varied as they are complex. For some properties, like cloud cover, ice, and vegetation, they must resort to empirical relationships or simply ad hoc parameterization. For others, the governing equations cannot even be defined. Together with the ocean simulation, these limitations introduce errors, which can be substantial. Along with discrepancies between GCMs, they leave in question how faithfully climate feedbacks are represented (see, e.g., Tsushima and Manabe, 2001; Lindzen and Choi, 2009).

Discrepancies from observations and between models are smallest for global-mean properties, such as those represented in idealized models of climate (Sec. 8.7.1). Typified by GMT, global-mean properties benefit from a cancellation of error that is spatially incoherent: Fluctuations in one region operate independently of fluctuations in others. Regional properties, on the other hand, have error that is intrinsically greater, reflected in their greater variability (cf. Figs. 1.38, 1.39). For those, discrepancies with observations and between models can be substantial (IPCC, 2007).

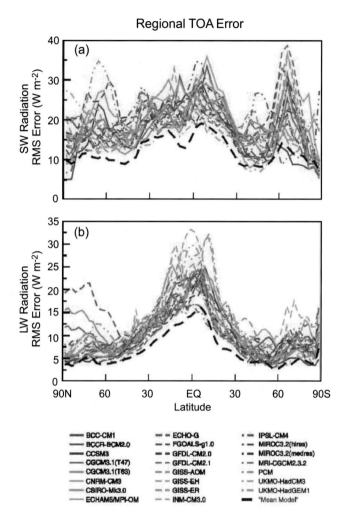

Figure 8.34 RMS error in time-mean regional fluxes at TOA simulated by several GCMs, relative to coincident fluxes observed by satellites in ERBE. (a) Error in reflected SW flux. (b) Error in outgoing LW flux. After IPCC (2007).

The accuracy of GCMs is reflected in the skill with which they simulate the TOA energy budget: the driver of climate. By construction, GCMs achieve global-mean energy balance. How faithfully the energy budget is represented locally, however, is another matter. The local energy budget forces regional climate, along with the gamut of weather phenomena that derive from it. This driver of regional conditions is determined internally – through the simulation of local heat flux, water vapor, and cloud. Symbolizing the local energy budget is net radiation (Fig. 1.34c), which represents the local imbalance between the SW and LW fluxes F_0 and $F^{\uparrow}(0)$ in the TOA energy budget (8.82). Local values of those fluxes have been measured around the Earth by the three satellites of ERBE. The observed fluxes, averaged over time, have then been compared against coincident fluxes from climate simulations, likewise averaged. Figure 8.34 plots, for several GCMs, the rms error in simulated fluxes, which have been referenced against those observed by ERBE. Values represent the regional error in the (time-mean) TOA energy budget. The error in reflected SW flux, $\frac{F_s}{4} - F_0$ in the global mean (8.82), is of order 20 Wm^{-2} (Fig. 8.34a). Such error prevails at most latitudes. Differences in error between models (an indication of intermodel discrepancies) are almost as large, 10-20 Wm^{-2}. The picture is much the same for outgoing LW flux (Fig. 8.34b). For $F^\uparrow(0)$, the rms error is of order 10-15 Wm^{-2}. It is larger for all models in the tropics, where the error exceeds 20 Wm^{-2}.

The significance of these discrepancies depends on application. Overall fluxes at TOA are controlled by water vapor and cloud (Fig. 1.32) – the major absorbers that account for the preponderance of downwelling LW flux to the Earth's surface. Relative to those fluxes, the errors in Fig. 8.34 are manageable: Of order 10% for outgoing LW and 20% for reflected SW. Relative to minor absorbers, however, this is not the case. The entire contribution to the energy budget from CO_2 is about 4 Wm^{-2}. Errors in Fig. 8.34 are an order of magnitude greater. Consequently, the simulated change introduced by increased CO_2 (2-4 Wm^{-2}), even inclusive of feedback, is overshadowed by error in the simulated change of major absorbers.

Discrepancies between GCMs arise from inaccuracies in climate properties and from differences in how those properties are represented. Much of the discrepancy surrounds the representation convection and its influence on water vapor and cloud, the absorbers that account for most of the downwelling LW flux to the Earth's surface. The involvement of convection is strongly suggested by models of radiative-convective equilibrium. Those simulations are inherently sensitive to how convection and cloud are prescribed. Cloud is especially significant to radiative considerations because it sharply modifies the atmosphere's scattering characteristics, which determine albedo, and its absorption characteristics, which determine optical depth.

SUGGESTED REFERENCES

Fundamentals of Modern Physics (1967) by Eisberg provides an excellent overview of Planck's theory and the emergence of modern physics.

An Introduction to Atmospheric Radiation (1980) by Liou includes discussions of solar variability and band models. An advanced treatment of radiative transfer is presented in *Atmospheric Radiation* (1995) by Goody and Yung, which includes a comparison of two-stream approximations. Applications of radiative transfer to remote sensing are discussed in *Atmospheric Physics* (2000) by Andrews.

Aeronomy of the Middle Atmosphere (1986) by Brasseur and Solomon and *Middle Atmosphere Dynamics* (1987) by Andrews et al. treat radiative transfer in the stratosphere and mesosphere.

Global Physical Climatology (1994) by Hartmann provides a detailed treatment of the energy budget, including climate feedbacks.

Modeling the Atmospheric Component of the Climate System (1992) by Kiehl contains a nice overview of the hierarchy of climate models. An overview of current models is presented in Bader et al. (2008).

PROBLEMS

1. Calculate the brightness temperature of the Earth based on (a) observed broadband OLR of 235 W m^{-2}, (b) the outgoing narrow-band flux (Fig. 8.5) at $\lambda = 11$ μm, (c) the outgoing narrow-band flux at $\lambda = 9.6$ μm, (d) the outgoing narrow-band flux at $\lambda = 15$ μm.

2. As the sun cools, its spectrum will shift toward longer wavelengths. Estimate the change in the Earth's equivalent blackbody temperature if the peak in the SW spectrum is displaced from its current position of 0.48 μm to a yellower wavelength of 0.55 μm.

3. Orbital ellipticity brings the Earth some 3.5% nearer the sun during January than during July. This is responsible for the slight asymmetry of insolation apparent in Fig. 1.33. (a) Calculate the corresponding change in the Earth's equivalent blackbody temperature. (b) Discuss this change in relation to other changes that are introduced by hemispheric asymmetries of the Earth's surface and atmospheric circulation.

4. Consider a 250°K isothermal atmosphere that contains a single absorbing gas of uniform mixing ratio 0.05. The specific absorption cross section of that gas, as a function of wavelength, is given by

$$\sigma_{a\lambda} = 1.0e^{-\left(\frac{\lambda-\lambda_0}{\alpha}\right)^2} \, \text{m}^2\text{kg}^{-1},$$

where $\lambda_0 = 0.5$ μm and $\alpha = .01$ μm. (a) Calculate, as a function of λ, the optical path length for downward-traveling radiation from the top of the atmosphere to the height z. (b) Plot the absorptivity of this layer, as a function of λ, for $z = 5, 3$, and 1 scale heights.

5. Show that a layer having optical thickness $\Delta u << 1$ has absorptivity $\Delta a = \Delta u$.

6. The top of the photosphere has a radius of $7 \, 10^8$ m. If the mean Earth-sun distance is $1.5 \, 10^{11}$ m, calculate the equivalent blackbody temperature of the photosphere.

7. Venus has an albedo of 0.77 and a mean distance from the sun of $1.1 \, 10^{11}$ m. (a) Calculate the equivalent blackbody temperature of Venus. (b) Discuss this value in relation to the observed surface temperature of 750°K.

8. Consider an isothermal atmosphere that is optically thick to LW radiation. (a) How much of the LW radiance emitted to space at zero zenith angle originates in its uppermost 3 optical depths? (b) Determine the altitude range, in scale heights, occupied by the layer between optical depths of 0.1 and 3.0 under isothermal conditions.

9. A flat plate sensor on board a satellite behaves as a gray body with constant absorptivity a. Calculate its radiative-equilibrium temperature when the sensor faces the sun if (a) $a = 0.8$, (b) $a = 0.2$.

10. As for Prob. 8.9, but if the sensor has different absorptivities to SW and LW radiation of (a) a_{SW} and a_{LW}, respectively, (b) $a_{SW} = 0.2$ and $a_{LW} = 0.8$.

11. Under the conditions of Prob. 8.10, calculate the sensor's temperature as a function of time after the satellite has been rotated to face the sensor away from the sun. Presume that the sensor has an area of 0.1 m^2 and a specific heat of 2 J K^{-1}.

12. The hood of an automobile may be modeled as a gray flat plate with LW absorptivity 0.9 and the atmosphere as a gray layer with LW absorptivity 0.8. In this framework, estimate the hood's temperature for a solar zenith angle of 25° if (a) the automobile's color is light with a SW absorptivity of 0.2, (b) the automobile's color is dark with a SW absorptivity of 0.6. (c) What physical process would, in practice, mediate the values in (a) and (b)?

13. A simple model approximates the Earth's surface as a gray body and the atmosphere as a gray layer with LW absorptivities of 1.0 and 0.8, respectively. Use

this model to estimate the radiative-equilibrium temperature of the ground in the presence of 50% cloud cover (fully reflective) and under (a) perpetual summertime conditions over vegetated terrain: a solar zenith angle of $25°$ and the surface having a SW absorptivity of 0.6, (b) perpetual wintertime conditions over snow-covered terrain: a solar zenith angle of $60°$ and the surface having a SW absorptivity of 0.2.

14. An 80% solar eclipse occurs during morning, when the temperature would normally be $20°C$. Use a gray atmosphere in radiative equilibrium with a black surface to estimate the change of surface temperature that would result were the eclipse to persist indefinitely (cf. Prob. 8.18).

15. Discuss how the competing influences of radiative transfer and convective mixing control stratification, in light of the diurnal variation of insolation.

16. The CO_2 absorption band is already saturated (Fig. 8.1). How then can increased levels of CO_2 produce global warming?

17. (a) Calculate the characteristic time scale of thermal damping in the troposphere based on $\rho_0 = 1.2$ kg m^{-3}, the equivalent blackbody temperature $T_e = 255°K$ (which is symbolic of where most OLR is emitted), and a characteristic depth of 5 km (which corresponds to the preceding temperature under radiative-convective equilibrium: Fig. 8.21). (b) What is the heat capacity of the surface under this approximation?

18. Use Newtonian cooling in Prob. 8.17 to estimate the amplitude ΔT_s of the diurnal cycle of surface temperature, where ΔT_s is half the nocturnal depression of temperature from its daytime maximum. Under equinoctial conditions and for a surface temperature maximum of $300°K$, estimate ΔT_s for an optical depth (a) $\tau_s = 4$ that is representative of the global-mean conditions and (b) $\tau_s = 1$ that is representative of arid conditions.

19. Under radiative equilibrium, evaluate the change of solar constant that would increase the Earth's surface temperature by $1°K$.

20. Use Figs. 1.13 and 1.24, in concert with Fig. 8.30, to determine an upper bound on total LW warming of the Earth's surface by CO_2 and CH_4 (the dominant greenhouse gases that have anthropogenic contributions). Apply the result, in concert with Fig. 1.32, to determine a lower bound on the respective LW warming by water vapor and cloud.

21. Consider a simple model of the Earth-atmosphere system, comprised of an isothermal gray layer with SW and LW absorptivities a_{SW} and a_{LW}, respectively, and a black underlying surface. Both are illuminated by a SW flux F_0 and are in thermal equilibrium. The energy budget may be formulated by tracing energy through repeated absorptions and reemissions by the surface and atmosphere. (a) Develop an arithmetic progression for the fractional energy absorbed by the atmosphere following successive transmissions from the surface to construct a series representation for the net energy absorbed by the atmosphere. (b) Use the series representation to calculate the radiative-equilibrium temperature of the atmosphere for $F_0 = (1 - A)\frac{F_s}{4} = 240$ W m^{-2} (Sec. 1.3), $a_{SW} = 0.20$, and $a_{LW} = 0.94$, which are representative of values in Fig. 1.32.

22. Use the results of Prob. 8.21 to determine the radiative-equilibrium temperature of the surface (a) for the conditions given, (b) in the absence of an atmosphere, (c) as in (a), but if the atmosphere's LW absorptivity is increased to $a_{LW} = 0.98$.

23. Estimate the level where the collisional half-width equals the Doppler half-width for (a) the water vapor absorption line at $\lambda^{-1} = 352$ cm^{-1} (Fig. 8.11), (b) the CO_2 absorption line at $\lambda^{-1} = 712$ cm^{-1}.

24. Consider a discretely stratified atmosphere comprised of N isothermal layers, each of which is transparent to SW radiation but has an LW absorptivity a. Collectively, the layers are in radiative equilibrium with one another, an incident SW flux F_0, and with a black underlying surface. (a) Derive expressions for the flux F_s emitted by the surface for $N = 1$, $N = 2$, and then generalize those expressions to arbitrary N. (b) Let $a = \frac{\tau_s}{N}$ and take the limit $N \to \infty$ to recover expression (8.81) for a continuously stratified atmosphere. (c) Verify that, even though they have different optical characteristics, the atmospheres for different N all have the same equivalent blackbody temperature. Explain why.

25. Derive expressions (8.38) for the upwelling and downwelling intensities in a plane parallel atmosphere as functions of optical depth.

26. Consider a gray atmosphere with a LW specific absorption coefficient k and an underlying surface with SW and LW absorptivities a_{SW} and a_{LW}, respectively. (a) Determine the distribution of radiative-equilibrium temperature as a function of optical depth, a_{SW}, and a_{LW}. (b) For $a_{LW} = 1$, discuss the limiting behavior: $a_{SW} \to 0$. (c) For $a_{SW} = 1$, discuss the limiting behavior: $a_{LW} \to 0$.

27. Show that the upwelling flux inside a gray atmosphere which is in radiative equilibrium with a black underlying surface is given by (8.71).

28. The Venusian atmosphere is composed chiefly of carbon dioxide, with $g = 8.8$ m s^{-2}, $c_p = 8.44 \ 10^2$ J kg^{-1}°K^{-1} for CO_2, a mean distance from the sun of 0.70 that for Earth, and an albedo of $\mathcal{A} = 0.77$. (a) Use radiative equilibrium and the observed surface temperature $T_s = 750$°K to estimate the optical depth of the Venusian atmosphere. (b) Calculate the temperature distribution under radiative-convective equilibrium based on (1) the profile of optical depth (8.69), with $h = \frac{RT_s}{g}$, (2) the presumption that the convective layer is warm enough to prevent saturation, and (3) in place of the constraint used in Sec. 8.5.2 to determine the tropopause height, requiring the temperature at the mid-level of the hydrostatically unstable layer to equal that under radiative equilibrium. (c) Plot the profiles of upwelling radiation, downwelling radiation, and radiative heating.

CHAPTER NINE

Aerosol and cloud

Radiative transfer is modified importantly by cloud. Owing to its high reflectivity in the visible, cloud shields the Earth-atmosphere system from solar radiation. It therefore introduces cooling in the SW energy budget of the Earth's surface, offsetting the greenhouse effect. Conversely, the strong absorptivity in the IR of water and ice sharply increases the optical depth of the atmosphere. Cloud thus introduces warming in the LW energy budget of the Earth's surface, reinforcing the greenhouse effect.

We develop cloud processes from a morphological description of atmospheric aerosol, without which cloud would not form. The microphysics controlling cloud formation is then examined. Macrophysical properties of cloud are developed in terms of environmental conditions that control the formation of particular cloud types. These fundamental considerations culminate in descriptions of radiative and chemical processes that involve cloud.

9.1 MORPHOLOGY OF ATMOSPHERIC AEROSOL

Small particulates suspended in air are produced and removed through a variety of processes. Those processes make the composition, size, and distribution of atmospheric aerosol widely variable (Table 9.1). Aerosol concentrations are smallest over ocean (Fig. 1.26), where a particle number density of $n = 10^3$ cm^{-3} is representative. They are greatest over industrialized continental areas, where $n > 10^5$ cm^{-3} is observed. These and other distinctions lead to two broad classes of tropospheric aerosol: *continental* and *marine*.

9.1.1 Continental aerosol

Continental aerosol includes (1) crustal species that are produced by erosion of the Earth's surface, (2) combustion and secondary components related to anthropogenic

Table 9.1. Properties of atmospheric aerosol and cloud

Type of Particulate	Altitude (km)	Horizontal Scale (km)	Frequency of Occurrence	Composition	Mass Loading (mg/m^3)	Optical Depth (at 0.55 μm)	Mean Particle Radius (μm)	Principal Size Range (μm)
Stratus, cumulus, nimbus cloud	1–18	10–1,000	0.5	Water, ice	1,000–10,000	~1–100	10–1,000	Variable
Cirrus cloud	7–16	10–1,000	0.3	Ice	10–100	~1	~10–100	Variable
Fog	0–1	10–100	Sporadic	Water	10–100	1–10	~10	10–50
Tropospheric aerosol	0–10	1,000–10,0000 (ubiquitous)	1	Sulfate, nitrate, minerals	0.01–0.1	~0.1	0.1–1	~0.3
Ocean haze	0–1	100–1,000	0.3	Sea salt sulfate	0.1–1	0.1–1	0.5	~0.3
Dust storm	0–3	10–1,000	Sporadic	Silicates, clays	< 1 – > 100	1–10	1–10	10–100
Volcanic cloud	5–35	100–10,000	Sporadic	Mineral ash, sulfates	< 1 – > 1,000	0.1–10	0.1–10	1–10
Smoke	0–10	1–100	Sporadic (from fires)	Soot, ash, tars	0.1–1	~0.1–10	0.1–1	~0.3
Stratospheric aerosol	10–30	1,000–10,000 (ubiquitous)	1	Sulfate	0.001–0.01	~0.01	0.1	~0.1
Polar stratospheric cloud	15–25	10–1,000	0.1 (winter only)	HNO$_3$/H$_2$0 ice	0.001–0.01	~0.01–1	1–10	~1
Polar mesospheric cloud	80–85	~200 (polar regions above 50°)	0.1 (summer only)	Ice	~0.0001	~0.0001–0.01	~0.05	~0.02
Meteoric dust	50–90	10–1,000	0.5–1	Minerals, carbon	~0.00001	~0.00001	≤ 0.01	Wide range incl. micro-meteors

After WMO (1988).

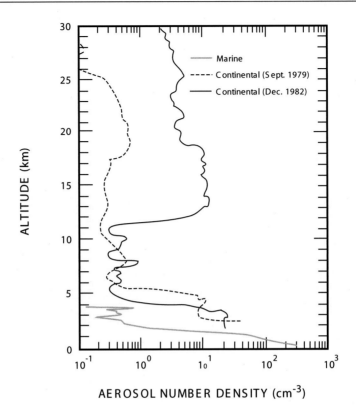

Figure 9.1 Aerosol number density n as a function of height in continental and marine environments. Profile of marine aerosol corresponds to particle radii $a > 0.5$ μm. The steep decrease of n above the marine boundary layer mirrors a similar decrease in dew point temperature (not shown). Profiles of continental aerosol correspond to $a > 0.15$ μm and are shown before and after the eruption of El Chichon. *Sources:* Patterson (1982), WMO (1988).

activities, and (3) carbonaceous components consisting of hydrocarbons and elemental carbon. Crustal aerosol is produced in subtropical deserts like the Sahara, southwestern United States, and southeast Asia (see Fig. 9.43). Combustion and secondary aerosols originate chiefly in the industrialized regions of North America, Europe, and Asia. These continental source regions give the Northern Hemisphere greater number densities than the Southern Hemisphere. Carbonaceous aerosol has large sources in tropical regions due to agricultural burning and, through reaction with other species, to vegetative emissions.

Aerosol number density decreases upward from the surface (Fig. 9.1), which serves as a source for particles and gaseous precursors. In the troposphere, n has a characteristic height of only 2 km. This shallow depth reflects the absolute concentration of water vapor (Fig. 1.17) and the removal of aerosol by precipitation processes. In the stratosphere, n decreases upward slower.

Aerosol is characterized by its *size distribution:* $n(a)$, which denotes the number density of particles with radii smaller than a. Size distribution is a monotonically increasing function of a, with $n(a)$ approaching n as $a \to \infty$. The *size spectrum:* $\frac{dn}{da}$

AEROSOL SIZE SPECTRUM

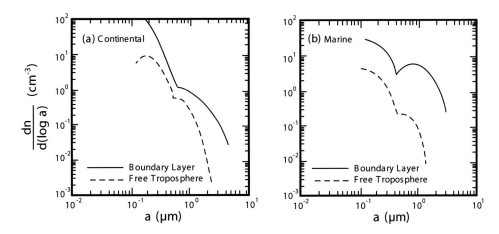

Figure 9.2 Aerosol size spectra, as functions of particle radius a, inside the planetary boundary layer and aloft for (a) continental environment and (b) marine environment. From Patterson (1982). Copyright by Spectrum Press (A. Deepak Publishing).

(Fig. 9.2) represents the contribution to $n = \int \frac{dn}{da} da$ from particles with radii between a and $a + da$. As for the spectrum of radiance (Fig. 8.1), plotting $\frac{dn}{d(\log a)}$ versus $\log a$ is area-preserving: Equal areas under the spectrum represent equal contributions to n from different size ranges. Crustal aerosol is comprised chiefly of windblown silicates. It involves particles with radii $a > 1$ μm. Secondary and combustion aerosols, which form through condensation, contain submicron-scale particles. These different production mechanisms give continental aerosol a bimodal size spectrum (Fig. 9.2a). Particles are concentrated in two distinct ranges of a. The size spectrum diminishes sharply with increasing a. The greatest contribution to n arrives from particles with $a < 0.1$ μm – the so-called *Aitken nuclei*. Despite their relative abundance, Aitken nuclei constitute less than half of the surface area of aerosol (Fig. 1.26), which figures centrally in thermodynamic and chemical processes. They constitute only about 10% of its mass.

Combustion and secondary aerosols form through several mechanisms: (1) gas-to-particle conversion, wherein gaseous precursors undergo physical and chemical changes that result in particle formation, (2) direct condensation of combustion products, and (3) direct emission of liquid and solid particles. Gas-to-particle conversion is responsible for much of the sulfate and nitrate that dominate urban aerosol. That process is exemplified by a series of reactions that begins with the gaseous pollutant sulfur dioxide

$$OH + SO_2 \rightarrow HSO_3, \tag{9.1}$$

followed by oxidation of HSO_3 to form H_2SO_4 vapor. Sulfuric acid vapor has a low saturation pressure. Therefore, it readily condenses, or *nucleates*, into liquid particles – especially in the presence of water vapor. Once formed, nuclei grow rapidly to radii of 0.01–0.1 μm. Larger aerosol can then develop through coagulation of individual nuclei

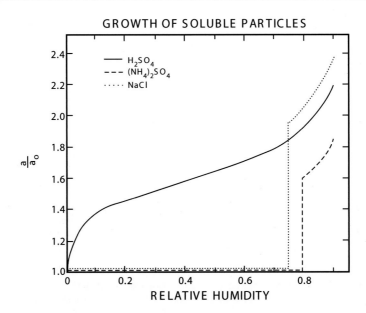

Figure 9.3 Aerosol particle radius as a function of relative humidity for liquid droplets and hygroscopic solids. Adapted from Patterson (1982). Copyright by Spectrum Press (A. Deepak Publishing).

to form *accumulation particles*. Sulfate aerosol is also produced through reaction of sulfur dioxide with ammonia inside cloud droplets. Upon evaporating, those droplets leave behind sulfate particles. In the troposphere, $(NH_4)_2SO_4$, NH_4HSO_4, and H_2SO_4 are all present. In the stratosphere, supercooled H_2SO_4 droplets are the prevalent form of sulfate. Nitrogen emissions from combustion and natural sources lead to analogous products. Sulfate and nitrate are both important in the troposphere. However, nitrate is especially prevalent in urban areas because the prevalence of N_2 in air makes it an inevitable byproduct of combustion.

Carbonaceous aerosols such as soot consist of submicron-size particles. Natural sources include pollen and spores, as well as complex hydrocarbon vapors, like isoprene. Emitted via plant transpiration, those vapors react with oxides of nitrogen and subsequently nucleate.

Once formed, aerosol nuclei interact strongly with water vapor. Through condensation, liquid particles such as H_2SO_4 enlarge steadily with increasing relative humidity (Fig. 9.3). Conversely, hygroscopic solids such as NaCl and $(NH_4)_2SO_4$ remain dry below a threshold of $RH \cong 80\%$. Particles then dissolve, undergo a discontinuous enlargement, and thereafter enlarge with increasing RH similar to liquid aerosol.

Anthropogenic sources account for about 30% of aerosol production. The anthropogenic component of sulfate plays a role in cloud formation. Over urban areas, it exceeds 60% of production. The distribution of continental aerosol is strongly affected by sedimentation, especially for large crustal species. Fall speeds range from 50 cm s^{-1} for large silicates ($a \cong 50$ μm) to 0.03 cm s^{-1} for smaller ($a \cong 1$ μm) particles. As a result, large particles tend to be confined to a neighborhood of their source regions.

PRODUCTION OF MARINE AEROSOL

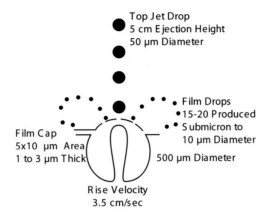

Figure 9.4 Formation of marine aerosol particles from a bursting ocean bubble. Adapted from Patterson (1982). Copyright by Spectrum Press (A. Deepak Publishing).

9.1.2 Marine aerosol

Composed primarily of sea salt, marine aerosol has a smaller overall concentration. Its number density decreases sharply above the boundary layer (Fig. 9.1), resembling the distribution of moisture. Like continental aerosol, marine aerosol has a size spectrum that is bimodal (Fig. 9.2b). It reflects two classes of droplets that form when ocean bubbles burst (Fig. 9.4). Water entrained into a bubble is ejected vertically in a stream of drops ($a \cong 25 \ \mu$m). Upon evaporating, they leave behind large ($a > 1 \ \mu$m) particles of sea salt. The thin film comprising the bubble's surface shatters, releasing droplets of $1 \ \mu$m and smaller. Upon evaporating, they produce smaller particles of sea salt.

9.1.3 Stratospheric aerosol

Aerosol is introduced into the stratosphere through penetrative convection and volcanic eruptions. It also forms in situ through gas-to-particle conversion from precursors like SO_2. That process maintains a background level of H_2SO_4 droplets, known as the *Junge Layer*. Its long residence time makes stratospheric aerosol fundamentally different from tropospheric aerosol. Small particulates with slow fall speeds are dynamically isolated by strong static stability. Sequestered from removal processes associated with precipitation, aerosol can survive in the stratosphere long after it is introduced. This feature enables stratospheric aerosol to alter SW absorption at the Earth's surface for extended durations. The great eruption of Krakatoa in 1883, which destroyed the Indonesian island of the same name, altered SW transmission and sunsets for 3 years afterwards (e.g., Humphreys, 1964). The eruption of Pinatubo had a similar influence, albeit weaker (Fig. 1.27).

Aerosol number density decreases upward in the stratosphere slower than in the troposphere (Fig. 9.1). Following major volcanic eruptions, aerosol number density actually increases into the stratosphere. This was evident following the eruption of El Chichon in 1982. A clear signature of El Chichon was registered in the transmission

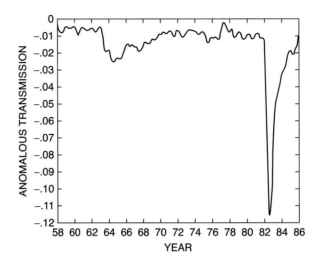

Figure 9.5 Change of direct solar transmission through the atmosphere, as observed over Hawaii by ground-based instruments. Note: Reduction of direct transmission is partially offset by an increase of diffuse transmission (e.g., Hoffman, 1988). The remainder is backscattered to space or absorbed. After WMO (1988).

of direct SW (Fig. 9.5), which decreased sharply in 1982. Through scattering and absorption, reduced transmission weakens SW heating of the Earth's surface. Such behavior was observed in temperature in the lower troposphere following Pinatubo (Fig. 1.27). Accompanying it was reduced water vapor in the upper troposphere, which is humidified by deep convection (Sec. 9.3.4). Increased extinction in volcanic debris can also increase radiative heating. It in turn can produce ascent of air in which the debris is suspended. Such was the case following the eruption of Pinatubo (WMO, 1995).

In the stratosphere, n has a characteristic vertical scale of 7 km. Resembling the scale height of air density, it suggests that relative concentration of aerosol (e.g., $\frac{n}{n_d}$) is approximately conserved for individual air parcels. Owing to the absence of removal processes, stratospheric aerosol behaves as a tracer. Figure 9.6 shows the cloud of volcanic debris introduced by the eruption of Mount St. Helens. Two months after the eruption, the cloud had dispersed zonally but not meridionally. Its confinement to high latitudes of the summer hemisphere follows from dynamical characteristics of the stratospheric circulation (Chap. 18).

9.2 MICROPHYSICS OF CLOUD

Large particles ($a > 1\ \mu$m) are vulnerable to sedimentation. This process accounts for 10–20% of the overall removal of atmospheric aerosol. A more efficient removal mechanism is *washout*. Cloud droplets that form on aerosol particles carry them back to the Earth's surface. This removal mechanism is closely related to cloud formation.

9.2.1 Droplet growth by condensation

The simplest means of forming cloud is through *homogeneous nucleation*, wherein pure vapor condenses to form droplets. Suppose a small embryonic droplet forms

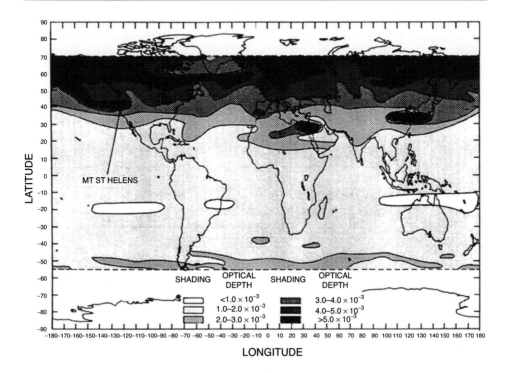

Figure 9.6 Mean aerosol optical depth at 1 μm 2 months after the eruption of Mt. St. Helens. Constructed by averaging satellite measurements from the Stratospheric Aerosol and Gas Experiment (SAGE) between July and August 1980. After Kent and McCormick (1984), copyright by the American Geophysical Union.

through chance collision of vapor molecules. The survival of that droplet is determined by a balance between condensation and evaporation. The equilibrium between the droplet and surrounding vapor is described by the Gibbs free energy (3.16.2), but with one modification. In addition to expansion work, a spherical droplet also performs work in association with its surface tension σ, which has dimensions of $\frac{\text{energy}}{\text{area}}$. Thus, $\sigma\,dA$ represents the work to form the incremental area dA of interface between vapor and liquid.

For a heterogeneous system that is comprised of the droplet and surrounding vapor, the fundamental relation for the Gibbs free energy (4.13) becomes

$$dG = -S\,dT + V\,dp + (\mu_v - \mu_w)\,dm_v + \sigma\,dA, \tag{9.2.1}$$

where

$$
\begin{aligned}
dm_v &= -dm_w \\
&= -n_w\,dV_w,
\end{aligned}
\tag{9.2.2}
$$

and n_w and V_w denote the number density and volume of the droplet.[1]

---

[1] In phase transformations away from saturation, the system is out of chemical equilibrium. The chemical potentials in (9.2) therefore need not be equal.

The difference of chemical potential between the vapor and liquid phases may be expressed in terms of the vapor and saturation vapor pressures by appealing to the fundamental relation (3.16) for the individual phases.[2] Under an isothermal and reversible change of pressure de,

$$d\mu_v = v_v de \qquad (9.3.1)$$

for the vapor and

$$d\mu_w = v_w de \qquad (9.3.2)$$

for the liquid. Subtracting gives

$$d(\mu_v - \mu_w) = (v_v - v_w)de$$
$$\cong v_v de. \qquad (9.4)$$

The gas law for an individual molecule of vapor is

$$ev_v = KT \qquad (9.5)$$

(e.g., Lee, Sears, and Turcotte, 1973), where K is the Boltzmann constant. Introducing it into (9.4) gives

$$d(\mu_v - \mu_w) = KT\, dln\, e.$$

Integrating from the saturation pressure (at which $\mu_v = \mu_w$) then obtains

$$\mu_v - \mu_w = KT ln\left(\frac{e}{e_w}\right). \qquad (9.6)$$

Incorporating (9.6) into (9.2) and integrating from a reference state of pure vapor (with the vapor pressure maintained) yields the change of Gibbs free energy

$$\Delta G = -V_w n_w KT ln\left(\frac{e}{e_w}\right) + \sigma A.$$

Thus, forming a spherical droplet of radius a corresponds to the change of Gibbs free energy

$$\Delta G = 4\pi a^2 \sigma - \frac{4}{3}\pi a^3 n_w KT ln\left(\frac{e}{e_w}\right). \qquad (9.7)$$

Figure 9.7 shows the free energy as a function of droplet radius for subsaturation ($\frac{e}{e_w} < 1$), saturation ($\frac{e}{e_w} = 1$), and supersaturation ($\frac{e}{e_w} > 1$). For subsaturation and saturation, G increases monotonically with droplet radius a. Now under isothermal isobaric conditions, a system approaches thermodynamic equilibrium by reducing its free energy (3.23). Consequently, droplet formation is not favored for $\frac{e}{e_w} \leq 1$. Those conditions characterize stable equilibrium and $G(a)$ an unbounded potential well. A droplet formed through the chance collision of vapor molecules will spontaneously evaporate, restoring the system to its original state. Under supersaturated conditions, $G(a)$ possesses a maximum at a critical radius a_c. Beyond that radius,

---

[2] Surface tension introduces a pressure difference between the droplet and surrounding vapor. It makes the pressure during a phase transformation variable. Strictly, that process should be treated as isothermal and isochoric, for which F is the appropriate free energy. However, discrepancies with G under isothermal and isobaric conditions are small enough to be ignored.

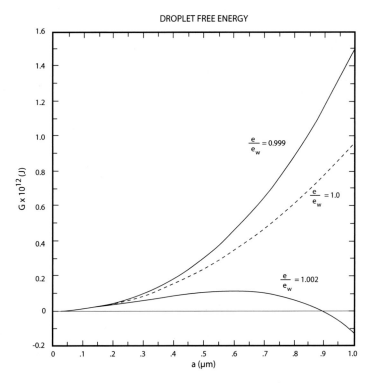

Figure 9.7 Gibbs free energy of a pure-water droplet produced through homogeneous nucleation, as a function of droplet radius. Shown for subsaturation ($\frac{e}{e_w} < 1$), saturation ($\frac{e}{e_w} = 1$), and supersaturation ($\frac{e}{e_w} > 1$).

free energy decreases. Condensation then becomes more efficient than evaporation. Under these circumstances, $G(a)$ characterizes conditional equilibrium and a bounded potential well. A droplet formed with $a < a_c$ evaporates back to the initial state ($G = 0$). But one formed with $a > a_c$ grows spontaneously through condensation of vapor. The system then diverges from its original state and the droplet is said to be *activated.*

The critical radius for a given temperature and vapor pressure follows by differentiating (9.7)

$$a_c = \frac{2\sigma}{n_w KT \ln\left(\frac{e}{e_w}\right)}. \tag{9.8}$$

Known as *Kelvin's formula*, (9.8) may be rearranged for the *critical* or *equilibrium supersaturation*

$$\epsilon_c = \left(\frac{e}{e_w}\right)_c - 1. \tag{9.9}$$

For supersaturation greater than ϵ_c, a droplet of radius a grows spontaneously through condensation. As shown in Fig. 9.8, small droplets have higher ϵ_c than large droplets. The larger is the droplet the smaller is the supersaturation necessary to maintain it. As $a \to \infty$, $\epsilon_c \to 0$. This behavior reflects the so-called *curvature effect* of surface tension

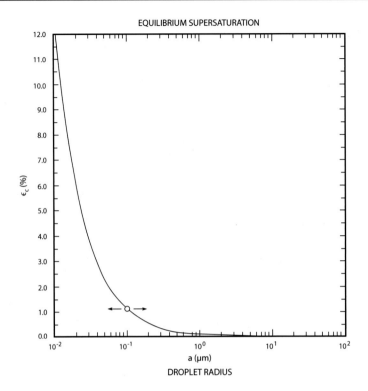

Figure 9.8 Critical supersaturation $\epsilon_c = (\frac{e}{e_w})_c - 1$ at 278 K necessary to sustain a droplet of radius a. The critical value of ϵ describes an unstable equilibrium, wherein perturbations in a are reinforced. A perturbed droplet then either evaporates or grows through condensation away from the equilibrium state.

(9.2). It makes the equilibrium vapor pressure over a spherical droplet (a finite) greater than that over a plane surface ($a = \infty$).

The curve for ϵ_c describes an unstable equilibrium, which therefore cannot be maintained. Adding a vapor molecule drives a droplet of radius a above the equilibrium curve. The droplet then has a supersaturation ϵ above its equilibrium value ϵ_c. Therefore, the droplet continues to grow through condensation. Removing a water molecule has the reverse effect. The droplet then has a supersaturation below its equilibrium value ϵ_c. Therefore, the droplet continues to evaporate. In either case, a perturbed droplet evolves away from the equilibrium state.

Values of ϵ_c represent the barrier to droplet formation via the chance collection of vapor molecules. This is associated with the work required to overcome surface tension. Only droplets large enough for ϵ to exceed ϵ_c will survive. Even then, an embryonic droplet as large as 0.01 μm (large by molecular standards) still requires a supersaturation of 12% to survive. Yet, supersaturations exceeding 1% are rarely observed.[3] Over

---

[3] Near the tropical tropopause, where temperature can decrease below 190°K (Figs. 1.7; 7.8), supersaturation with respect to ice of several tens of percent have been reported (Spichtinger

most of the troposphere, $\epsilon = 0.1\%$ is typical. Consequently, the formation of most cloud cannot be explained by homogeneous nucleation.

Instead, cloud droplets form through *heterogeneous nucleation*, wherein water vapor condenses onto existing particles of atmospheric aerosol. Termed *cloud condensation nuclei* (CCN), such particles support condensation at supersaturations well below those required for homogeneous nucleation. Particles that are *wettable* allow water to spread over their surfaces. Such particles provide ideal sites for condensation because they then resemble a droplet of pure water, one which, at observed supersaturations, could not attain that size through homogeneous nucleation. According to Fig. 9.8, the larger such a nucleus is, the lower is its equilibrium supersaturation and hence the more it favors droplet growth through condensation. If other conditions are equal, large nuclei therefore activate sooner than small nuclei.

Even more effective are hygroscopic particles, such as sodium chloride and ammonium sulfate. In the presence of moisture, NaCl and $(NH_4)_2SO_4$ absorb vapor and readily dissolve (Fig. 9.3). The resulting solution has a saturation vapor pressure below that of pure water – because e_w is proportional to the absolute concentration of water molecules on the surface of the droplet. Consequently, a droplet containing dissolved salt favors condensation more than would a droplet of pure water of the same size. The so-called *Köhler curves* (Fig. 9.9) describe the equilibrium supersaturations for solutions that contain specified amounts of solute. A droplet that develops on a soluble nucleus therefore evolves along the curve corresponding to the fixed mass of that solute. The presence of NaCl sharply reduces the equilibrium supersaturation below that for a pure-water droplet (homogeneous nucleation), which is superimposed on Fig. 9.9 (dashed). For fixed a, ϵ_c decreases with increasing solute. Eventually, it becomes negative; that is, equilibrium is achieved for $RH < 100\%$. Nuclei that are more soluble activate sooner and are more favorable to droplet growth.

The equilibrium described by a given Köhler curve differs according to whether the droplet radius is larger or smaller than a at the maximum of the curve. The latter defines the threshold radius for growth, a_t. A droplet with $a < a_t$ is in stable equilibrium: Adding a vapor molecule drives the droplet along its Köhler curve to higher ϵ_c. The droplet's actual supersaturation ϵ then lies beneath its equilibrium value. The droplet therefore evaporates back to its original size. Removing a water molecule has the reverse effect. In either case, a perturbed droplet is restored to its original state. Droplets with $a < a_t$ cannot grow through condensation; they comprise *haze*. Conversely, a droplet with $a > a_t$ is in unstable equilibrium: Adding a vapor molecule drives that droplet along its Köhler curve to lower ϵ_c. The actual supersaturation then exceeds the droplet's equilibrium value. The droplet therefore continues to grow through condensation. Removing a water molecule has the reverse effect. In either case, a perturbed droplet evolves away from its initial state. The behavior is analogous to that of a pure-water droplet at its critical supersaturation (Fig. 9.8). Droplets with $a > a_t$ can enlarge into cloud drops.

Only a small fraction of tropospheric aerosol actually serves as CCN. Many particles are too small, not wettable, or insoluble. Continental aerosol is the dominant source of CCN because of its greater number density. It also has a high concentration of soluble sulfate, which is found in many cloud droplets. By contrast, sea salt is not common

et al., 2006; Jensen et al., 2004; Kramer et al., 2009). However, those values are unrepresentative of conditions that prevail elsewhere in the troposphere.

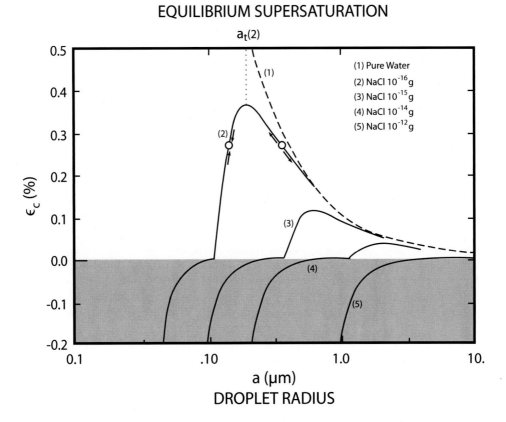

Figure 9.9 Critical supersaturation necessary to sustain a droplet of radius a and containing specified amounts of solute (solid). Critical supersaturation for a pure-water droplet (dashed) is superposed. For a given amount of solute, the corresponding *Köhler curve* describes different forms of equilibrium, depending on whether a is smaller or larger than the threshold radius at the maximum of the curve, a_t. For $a < a_t$, a droplet is in stable equilibrium: Perturbations in a are opposed by condensation and evaporation, which restore the droplet to its original state. For $a > a_t$, a droplet is in unstable equilibrium: Perturbations in a are reinforced, so the droplet either evaporates toward a_t or grows through condensation away from its original state.

in cloud droplets – not even over maritime regions. Instead, CCN in maritime cloud are composed chiefly of sulfate particles. Through anthropogenic emission, they have nearly doubled globally (IPCC, 1990). Sulfate aerosol also has a natural source: It forms through condensation of dimethylsulphide (DMS), which is emitted by phytoplankton at the surface of ocean.

Cumulus cloud has droplet number densities of order 10^2 cm^{-3} over maritime regions and 10^3 cm^{-3} over continental regions (Fig. 9.10). The greater number density of continental cloud mirrors the greater number density of continental aerosol. It also implies a smaller mean droplet size for continental cumulus. Nearly invariant between these cloud categories is the density of condensate, or *liquid water content* ρ_l (g m^{-3}). With droplets that are more numerous, continental cumulus therefore has a smaller

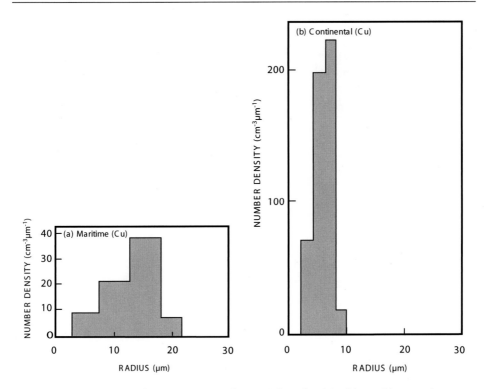

Figure 9.10 Droplet size spectra for cumulus cloud in (a) maritime environment and (b) continental environment. Note compressed scale in (b). Area under each curve reflects the overall number density n, which is an order of magnitude greater in continental cumulus than in maritime cumulus. *Source:* Pruppacher (1981).

mean droplet size. Droplets inside continental cumulus are also more homogeneous in size than droplets inside maritime cumulus: their size spectrum is narrow, concentrated about a preferred radius of $\sim$5 μm. Maritime cloud droplets have a wider size spectrum. It reflects their formation on soluble sulfates, which activate over a wide range of sizes.

Liquid water content varies greatly across a cumulus complex (Fig. 9.11). Nevertheless, it remains strongly correlated with upward motion. ρ_l increases vertically to a maximum in the upper half of a cloud. Liquid water content then decreases sharply, vanishing at cloud top. It is noteworthy that ρ_l observed in nonprecipitating cloud seldom approaches adiabatic values ρ_l^{ad}, namely, values that would be attained by air parcels rising in isolation from environmental air. In fact, $\frac{\rho_l}{\rho_l^{ad}}$ rarely exceeds 0.5. At cloud top it vanishes. Implied is vigorous entrainment of drier environmental air at the walls and top of a cloud system (Sec. 7.4.2). For similar reasons, the greatest supersaturation is found in the core of a broad cumulus. Only there is updraft air isolated and hence undiluted by environmental air.

Before precipitation can occur, cloud droplets must enlarge by several orders of magnitude. After passing over the peak of its Köhler curve, a droplet can grow through condensation. The droplet's mass m then increases at the rate vapor diffuses across

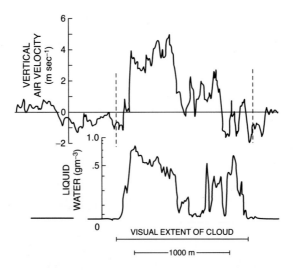

Figure 9.11 Variation of liquid water content ρ_l and vertical motion across a cumulus complex. After Warner (1969).

its surface

$$\frac{dm}{dt} = 4\pi r^2 v \frac{d\rho_v}{dr}, \tag{9.10}$$

where r denotes radial distance, v is the diffusion coefficient for water vapor, and ρ_v is its density. Integrating from the droplet radius $r = a$ to infinity and expressing the droplet's mass in terms of the density of liquid water ρ_w obtains

$$\frac{da}{dt} = \frac{v}{a\rho_w}[\rho_v(\infty) - \rho_v(a)].$$

With the gas law, this can be written

$$a\frac{da}{dt} = v\left[\frac{\rho_v(\infty)}{\rho_w}\right]\frac{e(\infty) - e(a)}{e(\infty)}. \tag{9.11}$$

For droplets larger than 1 μm, the curvature and solute effects (Figs. 9.8, 9.9) are small enough to take $e(a) \cong e_w$. Then (9.11) may be expressed

$$a\frac{da}{dt} \cong v\left[\frac{\rho_v(\infty)}{\rho_w}\right]\epsilon. \tag{9.12.1}$$

If the right-hand-side is approximately constant, this gives

$$a(t) = \left\{2v\left[\frac{\rho_v(\infty)}{\rho_w}\right]\epsilon t\right\}^{\frac{1}{2}}. \tag{9.12.2}$$

Droplet growth through condensation (Fig. 9.12) is rapid at small radius, but slows as a droplet enlarges. This implies that a droplet population will become increasingly homogeneous in size as it matures. Although explaining the preliminary stage of droplet growth, condensation is too slow at large a to account for observed droplets in precipitating cloud.

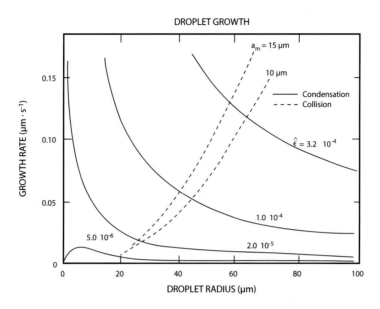

Figure 9.12 Droplet growth rate $\frac{da}{dt}$ due to condensation (solid) and collision (dashed) inside a cloud with liquid water content of $\rho_l = 1.0$ g m^{-3}, as functions of droplet radius a. Growth rate due to condensation shown for different supersaturations $\hat{\epsilon} = \left[\frac{\rho_v(\infty)}{\rho_w}\right]\epsilon$ (see text). Growth rate due to collision shown for a droplet population whose size spectrum has a maximum at radius a_m. Adapted from Matveev (1967).

9.2.2 Droplet growth by collision

Cloud that lies entirely beneath the freezing level is called *warm cloud*. Droplets inside warm cloud also grow through coalescence, when they collide with other droplets. Large droplets have faster fall speeds than small droplets. A large *collector drop* then sweeps through a volume of smaller droplets. The collection efficiency E is a dimensionless quantity that reflects the effective collecting area posed to other droplets. It is small for collector radii less than 20 μm. However, it increases sharply with collector size. For this reason, the wing of the droplet spectrum at large a is instrumental in forming precipitation.[4]

The rate a collector drop grows through collision with a homogeneous population of smaller droplets may be expressed

$$\frac{dm}{dt} = \pi a^2 E \rho_l \Delta w, \qquad (9.13)$$

where a refers to the collector drop and Δw is the difference in fall speed between it and the population of smaller droplets. Rearranging as in the derivation of (9.12) yields

$$\frac{da}{dt} = \frac{Ew}{4}\frac{\rho_l}{\rho_w}, \qquad (9.14)$$

---

[4] The collection efficiency also increases with the radius of the droplet population that is encountered by the collector drop (small droplets being swept around the collector drop without impact). Beyond a limiting droplet radius, however, coalescence is not favored.

Table 9.2. Fall speed of water drops

	Diameter (μm)	Fall Speed (cm/s)	Evaporation Distance (m)
Cloud Droplets	1	0.003	
	5	0.076	
	10	0.30	
	20	1.0	<1
	50	7.6	
	100	27	
Raindrops	200	72	150
	500	206	
	1000	403	
	2000	649	4200
	3000	806	
	5000	909	

Fall distance valid at 900 mb, 278 K, and *RH* = 90%. *Source: Smithsonian Meteorological Tables* (1958), Smithsonian Institution, Washington DC; Battan (1984).

where $\Delta w \cong w$ has been presumed for the collector drop. Because w and E both increase with the radius of the collector drop, (9.14) predicts growth that accelerates with its size. For this reason, growth due to collision eventually dominates that due to condensation. According to Fig. 9.12, droplets smaller than 20 μm grow primarily through condensation (solid) – because fall speeds are small enough for such droplets to move in unison with the surrounding air, making collisions infrequent. However, droplets larger than 20 μm grow primarily through collision (dashed). Because it favors large drops, collision tends to broaden the size spectrum from the homogeneous population that is favored by condensation. It thus provides a mechanism for a small fraction of the drops, those of large radius, to grow much faster than the rest of the population.

Radii with $a > 100$ μm dominate precipitation that reaches the ground. As seen in Table 9.2, they achieve fall speeds faster than 1 m/s. During their descent to warmer levels, cloud droplets and precipitation drops experience evaporation. The distance for complete evaporation increases with radius. Cloud droplets ($a < 100$ μm) are completely evaporated after falling less than a meter. Hence, they are too small to comprise precipitation. Rain drops ($a > 100$ μm) can fall several hundred meters before evaporating. Yet, even the largest fall only ~5 km before evaporating. For this reason, cloud with base higher than this value, prevalent over desert, produce no rainfall – irrespective of the column abundance of water. Marine cumulus have larger droplet sizes due to comparatively fewer CCN. Consequently, they are more likely to precipitate than continental cumulus of similar dimension. Collision also tends to limit the size of cloud droplets to a couple of millimeters because, during impact, larger drops fragment into smaller ones.

9.2.3 Growth of ice particles

Cloud that extends above the freezing level is called *cold cloud*. Cold cloud is seldom composed entirely of ice. In fact, nearly 50% of cloud warmer than $-10°$C and nearly

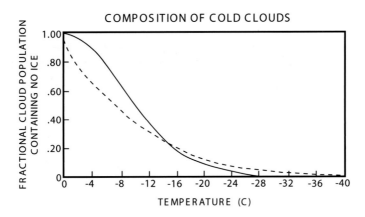

Figure 9.13 Fractional occurrence of unglaciated cloud, as a function of temperature, observed over Germany (solid) and Minnesota (dashed). *Source:* Pruppacher (1981).

all cloud warmer than $-4°C$ contains no ice at all (Fig. 9.13). Droplets in such cloud remain in the metastable state of supercooled water (Prob. 3.2). However, 90% of cloud colder than $-20°C$ contains some ice.

The reason why cold cloud is not entirely *glaciated* is that ice can form only under special circumstances. Homogeneous nucleation of ice involves the formation of an ice embryo through chance collection of water molecules inside a droplet. This process is not favored at temperatures above $-36°C$. At those temperatures, the free energy of formation increases for embryos smaller than a critical dimension – analogous to homogeneous nucleation of droplets (Fig. 9.7). The free energy then characterizes conditional equilibrium and a bounded potential well. An ice embryo smaller than the critical dimension for formation by chance collection of water molecules will then disperse. At temperatures above $-36°C$, the critical dimension for homogeneous nucleation is greater than 20 μm. Consequently, homogeneous nucleation of ice occurs only in very high cloud (see Fig. 7.7).

Heterogeneous nucleation occurs when water molecules collect onto an existing particle of a special type. Known as a *freezing nucleus*, it has molecular structure similar to ice. An ice embryo formed in this manner assumes the size of the freezing nucleus. It can therefore grow at temperatures much warmer than that permitting homogeneous nucleation. Maritime cloud, whose droplet size spectrum is broad with larger droplets (Fig. 9.10), glaciates more readily than does continental cloud. Crustal aerosol composed of clay is a prevalent form of freezing nucleus. But even at $-20°C$, fewer than one out of every hundred million aerosol particles serves as an ice nucleus. This leaves many cloud droplets supercooled. Aircraft encountering such cloud are therefore prone to icing.

Once formed, ice particles can grow through condensation of vapor, or *deposition*. The saturation vapor pressure with respect to ice is lower than that with respect to water (4.40). A cloud that is nearly saturated with respect to water may therefore be supersaturated with respect to ice.[5] A particle's growth is therefore accelerated if it

---

[5] Saturation mixing ratios with respect to water and ice typically differ by 10% or less. However, at very cold temperatures representative of the upper troposphere, the differences can be substantial (see Probs. 5.19, 18.4).

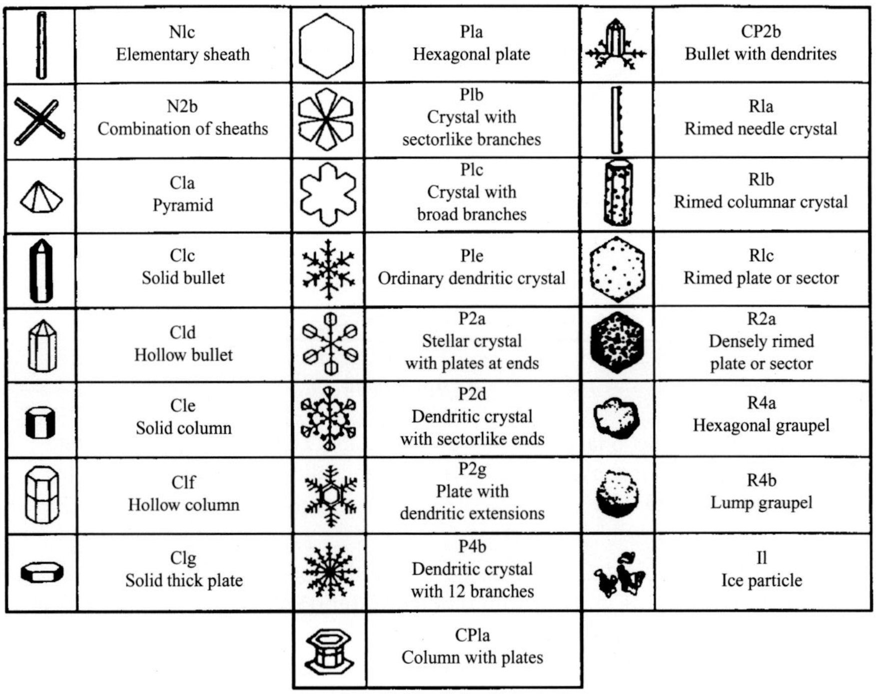

	Nlc Elementary sheath		Pla Hexagonal plate		CP2b Bullet with dendrites
	N2b Combination of sheaths		Plb Crystal with sectorlike branches		Rla Rimed needle crystal
	Cla Pyramid		Plc Crystal with broad branches		Rlb Rimed columnar crystal
	Clc Solid bullet		Ple Ordinary dendritic crystal		Rlc Rimed plate or sector
	Cld Hollow bullet		P2a Stellar crystal with plates at ends		R2a Densely rimed plate or sector
	Cle Solid column		P2d Dendritic crystal with sectorlike ends		R4a Hexagonal graupel
	Clf Hollow column		P2g Plate with dendritic extensions		R4b Lump graupel
	Clg Solid thick plate		P4b Dendritic crystal with 12 branches		Il Ice particle
			CPla Column with plates		

Figure 9.14 Ice crystals observed in cold cloud. Adapted from Pruppacher and Klett (1978). Reprinted by permission of Kluwer Academic Publishers.

freezes. Similarly, in a mixed population, ice particles grow faster than droplets. Ice particles also grow through collision with supercooled droplets, a process known as *riming*. This produces a layered structure about the original ice particle. Beyond a certain dimension, the original structure is no longer discernible. The resulting particle is then referred to as *graupel*. Such particles assume irregular shapes, often proliferated with cavities. Lastly, ice particles can grow through coagulation, which is favored at temperatures above $-5°C$. These mechanisms lead to a wide array of shapes and sizes. Included are bullets, plates, slender needles, and complex configurations of hexagonal crystals (Fig. 9.14).

9.3 MACROSCOPIC CHARACTERISTICS OF CLOUD

9.3.1 Formation and classification of cloud

Most cloud develops through vertical motion, when moist air is lifted above its LCL and becomes supersaturated. Cloud falls into three broad categories:

(1) *Stratiform* (layered) *cloud* develops from large-scale lifting of a stable layer. Characteristic of sloping convection, this process is exemplified by warm moist air that overrides cold heavier air along a warm front. Vertical motion that forms stratus cloud is of order 1 cm s^{-1}. The characteristic lifetime of stratus is of order a day.

Figure 9.15 Hook-shaped *cirrus uncinus*. After WMO (1969).

(2) *Cumuliform* (piled) *cloud* develops from isolated updrafts. Associated with cellular convection, such cloud grows through positive buoyancy. The latter derives from sensible heat transfer from the Earth's surface and latent heat that is released during condensation. Both sources of buoyancy make cumulus cloud dynamic. Developing cumulus evolves on a time scale of minutes. Supporting updrafts are of order 1 m s^{-1}. However, in organized mature cells like *cumulus congestus* and *cumulonimbus*, they can be several tens of m s^{-1}. The characteristic lifetime of cumulus ranges from a few minutes to hours.

(3) *Cirriform* (fibrous) *cloud* develops through either of the preceding forms of lifting. Found at great height, cirrus cloud is composed chiefly of ice particles. The larger particles descend in fall streaks or *mares' tails*. They give cirrus its characteristic fibrous appearance (Fig. 9.15).

Additional terms are used to specify particular cloud types. For example, *nimbo* and *alto* denote precipitation-bearing and mid-level, respectively. A complete tabulation is provided in WMO (1969).

Cumulus cloud develops in association with a buoyant updraft (Fig. 5.4). Known as a *thermal*, the updraft may be regarded as a succession of buoyant air parcels. Laboratory simulations reveal that the development of a thermal is closely related to its dissipation (Fig. 9.16a). A mass of buoyant fluid is transformed into a vortex

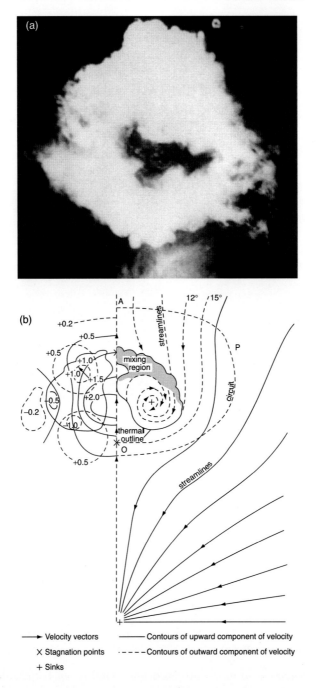

Figure 9.16 (a) Laboratory simulation of buoyant fluid inside a thermal. After advancing 1–2 diameters, a mass of buoyant fluid has been turned inside out. Environmental fluid entrained into its core produces a hole in the center of the buoyant mass. (b) Motion and streamlines about the moving body of fluid. Mixing with surrounding fluid causes the advancing mass to expand at an angle from the vertical of about 15°. Adapted from Scorer (1978).

ring (analogous to a smoke ring). The latter is characterized by a hole in its center. Punctuating its advancing surface is a complex array of protrusions or *turrets*. A mass of buoyant fluid penetrates into the surroundings only by entraining environmental fluid into its center (Fig. 9.16b). This process dilutes the buoyancy of a thermal. It also causes it to expand – at an angle of about $15°$ from the vertical. In the process, the buoyant mass is turned completely inside out after advancing only about $1\frac{1}{2}$ diameters. At that stage, the buoyant mass has become thoroughly mixed with surrounding fluid. Its buoyancy has then been destroyed. The direct involvement of vorticity is illustrated in the trajectory of individual fluid parcels. Some complete several circuits about the advancing vortex ring before falling out of its influence.

Expansion is less evident in cumulus cloud because the latter undergoes dissolution. Mixing with drier environmental air along the cloud's periphery leads to its evaporation. As observed, liquid water content is well below adiabatic values, mixing occurs throughout a cumulus cloud. Except for very broad cloud, some environmental air is communicated even into the core of cumulus. Consequently, a cumulus can grow to appreciable height only by expanding horizontally – enough to isolate saturated air in its core from drier environmental air that is mixed across its walls. Rising motion inside the updraft is compensated by gradual sinking motion in its environment. That subsidence is distributed over a broad area. It is promoted by evaporative cooling when saturated updraft air mixes with unsaturated environmental air and dissolves. By inhibiting ascent, subsiding motion in the environment organizes new parcels into the existing updraft, a process that favors cloud growth. Likewise, buoyant parcels are most likely to produce cloud growth if the region into which they advance has already been humidified and its stability reduced by preceding parcels.

A developing cumulus is marked by sharply defined turrets along its boundary, especially at its top. The turrets dissolve through evaporation as they penetrate into unsaturated air, replaced by new ones. The base of a developing cumulus is also sharply defined. It is level or even concave down – because warmer temperature in the thermal's core elevates the LCL (5.24). Conversely, a decaying cumulus is diffuse, rendered amorphous by turbulent mixing. That process leads to evaporation, quickly blurring sharp features that were created during the cloud's development. The cloud base may then become concave up, reflecting a reversal of vertical motion and the horizontal temperature gradient. Cloud matter that descends beneath the cloud base is termed *virga*. It is referred to as precipitation only if it reaches the ground (corresponding to drop radii $a > 100 \ \mu\text{m}$).

Thermal growth is inhibited by strong vertical stability. Figure 9.17 illustrates a thermal that encounters and partially penetrates through a layer of strong stability. After penetrating the stable layer, buoyant fluid continues to advance, eventually diffusing via detrainment and turbulent mixing with the surroundings. Fluid deflected by the stable layer fans out, forming anvil. Such behavior appears occasionally at intermediate levels in the troposphere (Fig. 9.18). It forms invariably when a cumulonimbus tower encounters stable air at the tropopause. Penetrative overshoots at the tropopause are accompanied by turbulent entrainment (Fig. 7.7). That process mixes tropospheric and stratospheric air, which eventually splash back to the LNB to occupy the resulting anvil.

The stratiform anvil of a mature cumulonimbus occasionally evolves into small cellular convection. LW emission at its top and absorption at its base destabilize the anvil layer, introducing convective overturning (Sec. 9.4.2). Domes of sinking virga then define *mammatus* (breast-shaped) *cloud* (Fig. 9.19). Their descent is promoted

(a)   (b)   (c)   (d)

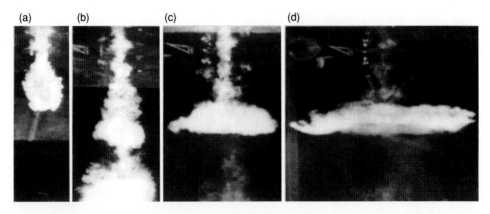

Figure 9.17 Laboratory simulation of a thermal encountering a layer of strong static stability. Upon reaching the stable layer, the negatively buoyant plume is detrained horizontally into an anvil. Fluid penetrating the stable layer continues to advance and eventually diffuses through turbulent mixing. After Scorer (1978).

by evaporative cooling. Spaced quasi-regularly, mammatus features resemble cells in Rayleigh-Bénard convection (Fig. 9.20), structure that develops in a shallow fluid that is heated differentially at its top and bottom. The horizontal dimension of Rayleigh-Bénard cells reflects the depth of the layer that is convectively overturned.[6]

Behaving similarly is stratiform cloud that is destabilized by absorption of LW radiation from the surface, producing radiative warming of its base, and emission to space, producing radiative cooling of its top. Instability introduced by such radiative heating and cooling can transform a continuous stratiform layer into an array of cumulus cells with a preferred horizontal dimension. *Stratocumulus* and *fair weather cumulus* that are found behind a cold front develop in this fashion (e.g., Fig. 9.21). Cold air that is advected over warmer surface is then heated at its base and cooled radiatively at its top.

Cloud also forms in the absence of buoyancy through forced lifting over elevated terrain. *Orographic cloud*, unlike other forms, does not mark a particular body of air. Rather, it marks a control volume through which air flows. *Lenticular* (lens-shaped) *cloud* forms inside an organized wave pattern (Fig. 9.22a). This smooth cloud, often forming in a stack of lenses, develops when the air stream is deflected over a mountain range. Each lens in Fig. 9.22a continually forms along its leading edge, where air is displaced above its LCL. It then dissolves along its trailing edge, where air returns below its LCL. Also known as *mountain wave cloud*, lenticular is often found at great height. On occasion, it is observed even in the stratosphere, where it is termed *nacreous cloud*. Vertical motion that accompanies wave cloud has been harnessed by sailplanes to achieve record-breaking altitudes – as high as 50,000′. The vertical structure of lenticular cloud is controlled by environmental conditions. When moisture is stratified

---

[6] According to Rayleigh-Bénard theory, conversion of potential energy into kinetic energy becomes increasingly efficient with decreasing horizontal dimension of the cells. Energy conversion is eventually offset by frictional dissipation (e.g., mixing of momentum across the walls of the cells), which becomes important at horizontal dimensions comparable to the depth of the fluid. It is no accident that cumulonimbus cells have horizontal dimensions comparable to the depth of the troposphere.

Figure 9.18 A developing cumulus that encounters and penetrates through a shallow layer of strong static stability. Courtesy of Ronald L. Holle (Holle Photography).

in layers, lenticular cloud forms in a stack (Fig. 9.22b). Such vertical structure can result from towers of anomalous moisture that have been left upstream by prior convection. When sheared by the circulation, they fold to form layered structure.

The smooth form of lenticular characterizes air flow that is turbulence-free or *laminar*. That steady motion is often replaced just downstream by severe turbulence. It is

Figure 9.19 *Mammatus cloud* that develops from a cumulonimbus anvil. Supplied by the National Center for Atmospheric Research, University Corporation for Atmospheric Research, National Science Foundation.

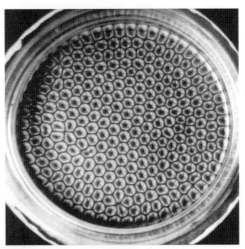

Figure 9.20 Laboratory simulation of Rayleigh-Bénard convection. A fluid layer heated at its base evolves into a regular array of hexagonal cells with a preferred horizontal scale. Fluid ascends in the center of each cell and descends along its edges. After Koschmieder and Pallas (1974).

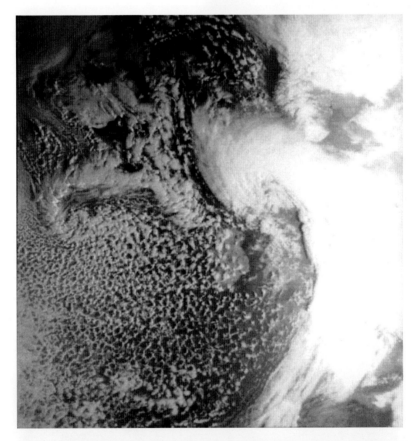

Figure 9.21 Visible image revealing cellular cloud structure behind a cold front. The cloud pattern is organized into a quasi-regular array of cells with a preferred horizontal scale. The cold front is part of a mature cyclone that advances across the British isles. Cloud cover ahead of the cold front marks the warm sector of the cyclone, which is just occluding north of Ireland. After Scorer (1986).

Figure 9.22 (a) Schematic of the mountain wave that develops leeward of an elongated range. *Lenticular clouds* are fixed with respect to orography, while air flows through them. The positions of *rotor clouds* are also fixed, but those fragmented features are highly unsteady, with turbulent ascent (descent) along their leading (trailing) edges. Westward tilt with height (e.g., of wave crests) corresponds to upward propagation of energy (Chap. 14), a feature that is often visible in the structure of lenticular clouds. (b) Longitudinal view through the mountain wave. Smooth lenticular clouds form in a stack, separated by clear regions, due to stratification of moisture. These laminar features contrast sharply with turbulent rotor clouds below. (c) Transverse view from the ground. Vorticity and severe turbulence are made visible by looping cloud matter that comprises rotor. Courtesy of V. Haynes and L. Feierabend (Soaring Society of Boulder). See color plate section: Plate 6.

made partially visible by *fractus* (fragmented) *cloud*. So-called *rotor cloud* is evident at levels beneath the lenticular in Figs. 9.22a,b. It is "dynamic." Looping cloud matter in Fig. 9.22c makes visible strong updrafts and downdrafts. Severe turbulence accompanying that vertical motion has been responsible for numerous aircraft disasters over mountainous terrain.

Because orographic cloud remains fixed, it makes cloud behavior unreliable as a tracer of air motion. More faithfully tracking the movement of individual bodies of air is cumulus cloud. However, even it is complicated by nonconservative behavior (e.g., condensation and evaporation), which limits its usefulness for diagnosing air motion.

The foregoing cloud forms develop through upward motion. Ascent reduces the saturation mixing ratio of air through adiabatic cooling. Cloud can also develop isobarically through heat rejection, which cools air below its dew point. Conductive and radiative cooling lead to the formation of fog and marine stratus over cold ocean.

The stratosphere is isolated from convective motion and therefore from the source of water vapor at the Earth's surface. Consequently, cloud rarely forms above the tropopause. Nacreous cloud is an exception. It develops through the mountain wave, which can propagate into the stratosphere (Chap 14). Nacreous cloud is a subset of more ubiquitous Polar Stratospheric Cloud (PSC). Because the stratosphere is very dry, PSC forms only under very cold conditions. Temperature over the Antarctic decreases below 190 K during Austral winter, more than 20K colder than temperature over the wintertime Arctic (Chap. 18). For this reason, PSC occurs over the Antarctic with much greater frequency and depth than over the Arctic.

Even over Antarctica, PSC is tenuous by comparison with tropospheric cloud. Some 3–5 km thick, PSC forms at heights of 15–25 km. It is often layered. Formation of PSC is strongly correlated with temperature. PSC is seldom sighted at temperatures warmer than 195 K. Conversely, its likelihood approaches 100% at temperatures below 185 K (WMO, 1988). Two distinct classes of PSC are observed. Type I PSC is composed of frozen droplets of nitric acid trihydrate (NAT), which has a low saturation vapor pressure. Submicron in scale, PSC I particles form at temperatures of 195–190 K. They have the appearance of haze. Type II PSC may be nucleated by Type I particles or by background aerosol. It contains much larger crystals of water ice. PSC II particles are several tens of microns and larger. Forming at temperatures of 185–190 K, they have features similar to cirrus. Their larger sizes give type II particles significant fall speeds. This feature makes them important in chemical considerations (Chap. 18).

9.3.2 Microphysical properties of cloud

Microphysical properties vary with cloud type (Table 9.3). Stratus has number densities $n \sim 300 \text{ cm}^{-3}$ and droplet radii of $a \sim 4$ μm. Liquid water content is more variable, but of order 0.5 g m^{-3}. Similar numbers apply for cumulus. However, cumulonimbus is distinguished by much smaller number density and much greater liquid water content. Implied are droplets that are significantly larger. Compared with water cloud, cirrus has much smaller number densities; they are observed over a wide range: 10^{-7} – 10 cm^{-3} (Dowling and Radke, 1990; Knollenberg et al., 1993). Its ice water content ρ_i is likewise smaller (e.g., $\rho_i < 0.1$ g m^{-3}). The characteristic dimension of cirrus ice crystals ranges from several tens of microns to as large as 1 mm. PSC has number densities similar to cirrus. It forms under analogous environmental conditions.

Table 9.3. *Microphysical properties of cloud*[%]

Cloud Type	n (cm^{-3})	$\langle a \rangle$ (μm)	ρ_l (g m^{-3})	$\langle \hat{\sigma}_e \rangle$ @0.5 μm (μm^2)	$\langle \beta_e \rangle$ @0.5 μm (km^{-1})
Stratus (St)	300.	3.	0.15	450.	135.
Stratocumulus (Sc)	250.	5.	0.3	120.	30.
Nimbostratus (Ns)	300.	4.	0.4	400.	120.
Cumulus (Cu)	300.	4.	0.5	200.	60.
Cumulonimbus (Cb)	75.	5.	2.5	500.	38.
Cirrus (Ci)	0.03	250.	0.025[*]	1.10^4	0.5
Tropical Cirrus (Cs)	0.10	800.	0.20[*]	4.10^4	0.4
PSC I	≤ 1	0.5	10^{-6*}	10	0.005
PSC II	≤ 0.10	50.	0.05[*]	100	0.015

[%] Representative values of number density n, mean droplet radius $\langle a \rangle$, liquid water content ρ_l, mean extinction cross section $\langle \hat{\sigma}_e \rangle$, and mean extinction coefficient $\langle \beta_e \rangle$ for different cloud types. Typical values are subject to a wide range of variability, especially for cirrus. *Sources:* Carrier et al. (1967), WMO (1988), Dowling and Radke (1990), Liou (1990), Knollenberg et al. (1993), Turco et al. (1993).
[*] Ice water content.

Figure 9.23 presents size spectra representative of several cloud types. Stratus and cumulus have the simplest spectra, which possess a single maximum. Spectra of nimbostratus and cumulonimbus exhibit secondary maxima, at radii larger than the mean: $\langle a \rangle \cong 5$ μm. The wings of these size spectra ($a > 20$ μm), are comparatively unpopulated. Nevertheless, they describe particles that grow most rapidly through collision and consequently favor precipitation (Sec. 9.2.2). Spectra of cirrus can also possess a secondary maximum, but at much larger size (e.g., at $100 - 1000$ μm). Those larger particles have rapid fall speeds. They produce mares' tails, characteristic of hook-shaped *cirrus uncinus* (Fig. 9.15).

9.3.3 Cloud dissipation

Cloud dissolves through evaporation when unsaturated environmental air is entrained into and mixed with saturated air. By diluting high θ_e and momentum inside an updraft with lower values from the environment, entrainment siphons off positive buoyancy and kinetic energy. Affected likewise is remaining water vapor, along with the attendant latent heat that would otherwise be available to reinforce upward motion (Sec. 7.4.2). These features make cumulus entrainment the principal dissipation mechanism for cumulus. Cloud growth can continue only if it is offset by a resupply of sensible and latent heat (high θ_e) from below.

When particles become so large that they can no longer be supported by the updraft (e.g., $a > 100$ μm), cloud is dissipated by precipitation or, in the case of cirrus, sedimentation. Falling cloud particles exert downward drag on the air through which they descend. If they dissolve, they also subject that air to evaporative cooling. Both influences oppose the updraft. In doing so, they inhibit the supply of sensible and latent heat that is necessary to maintain the cloud against entrainment (Prob. 9.17).

Entrainment is less effective in stratiform cloud because of its horizontal extent and weaker vertical velocities. However, radiative heating at its base and cooling at

CLOUD PARTICLE
SIZE SPECTRA

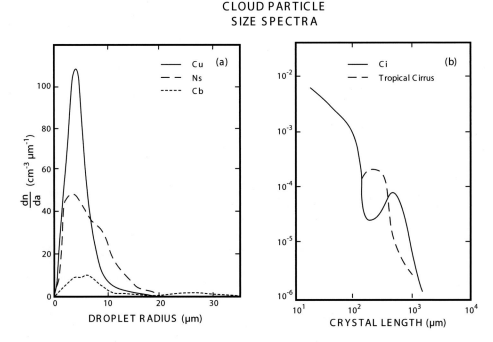

Figure 9.23 Cloud particle size spectra for (a) droplets inside water clouds and (b) crystals inside ice clouds. Adapted from Liou (1990).

its top destabilize stratus. This stimulates entrainment of drier air overhead, which acts to dissolve layered cloud. Radiative heating can also dissolve cloud by elevating its temperature above the dew point. Such heating leads to daily burn off of fog and marine stratus. This dissipation mechanism, together with formation mechanisms associated with surface heating, make diurnal variations an important feature of the cloud field (cf. Fig. 9.39).

9.3.4 Cumulus detrainment: Influence on the environment

At the wall of a cumulus updraft is entrainment of environmental air (Fig. 5.4). It is accompanied by detrainment of updraft air, which ventilates the environment. Thus, while it dilutes the θ_e of cumulus updrafts, this process simultaneously drives the environment toward the vertical structure inside updrafts (Sec. 7.7). The impact of cumulus detrainment is amplified in the upper troposphere. There, updrafts decelerate and diverge horizontally, forming extensive cirrostratus anvil.

Figure 9.24 presents a histogram of the fractional coverage by cloud, or the *cloud fraction η*. It is plotted as a function of brightness temperature, which translates into height. Representative of the tropics, η decreases upward exponentially, to less than 10% in the upper troposphere. Its sharp decrease with height reflects the histogram of updraft widths: Only the broadest updrafts escape damping by entrainment and, even then, only inside their cores. η levels off at temperatures colder than 235 K,

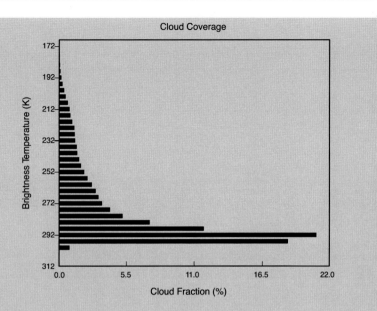

Figure 9.24 Histogram of fractional coverage by cloud, as a function of brightness temperature, which translates into altitude. After Salby et al. (2004a).

corresponding to ~250 hPa. It then decreases sharply at temperatures colder than 220 K, corresponding to levels above 200 hPa.[7]

Even though cloud fraction decreases upward monotonically, the coverage by individual anvils does not. Plotted in Fig. 9.25, also as a function of brightness temperature, is the mean areal extent of contiguous anvil during the same period as that presented in Fig. 7.14. Over South America (Fig. 9.25a), $\theta_{LNB} \cong 345$ K. This positions the LNB near 200 hPa, which is representative of much of the tropics (Sec 7.7). The extent of anvil increases upward, maximizing at brightness temperatures of 230–210 K. Corresponding to the layer 250–150 hPa, the maximum coincides with the LNB. It also coincides with the plateau in η, above which cloud fraction decreases sharply (Fig. 9.24). Over the equatorial Pacific (Fig. 9.25b), θ_{LNB} is 5–10 K warmer. This positions the LNB some 3 km higher than elsewhere in the tropics. There too, the extent of anvil increases upward. However, it now maximizes at a brightness temperature of 200 K, some 3 km higher.

In each region, the areal extent of anvil maximizes at levels neighboring the local LNB. The maximum marks extensive cirrostratus anvil. That cloud structure forms through detrainment of cumulus updrafts in the ITCZ. A small fraction of such air, inside the cores of broad updrafts, overshoots the LNB. It does so through kinetic energy that it acquired above the LFC (Fig. 7.6). After mixing with air overhead, it splashes back into an extensive anvil deck at the LNB, where updraft air diverges horizontally (Fig. 7.7).

---

[7] Although rare, some cloud is observed at brightness temperatures as cold as 180°K. It reflects overshoots above the tropical tropopause, where environmental temperature is ~200°K (Fig. 1.7).

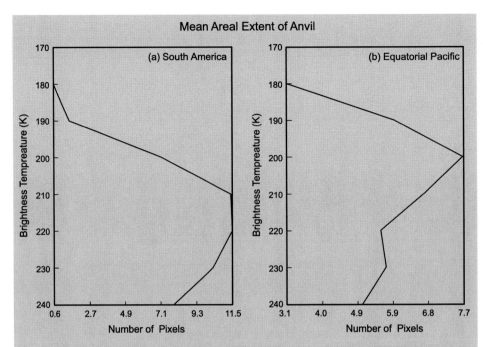

Figure 9.25 Mean areal extent of contiguous anvil (in number of 0.5°
pixels), as a function of brightness temperature, for the same period as
Fig. 7.14. After Salby et al. (2003).

In addition to cloud structure, such behavior is also manifest in the vertical
structure of horizontal motion. Velocity divergence measures the collective outflow
from deep cumulus towers (Sec. 10.5). Plotted in Fig. 9.26a for the equatorial Pacific,
horizontal divergence maximizes at the same levels as extensive anvil. Air that
diverges from cumulus updrafts in the ITCZ forms poleward motion. The latter
comprises the upper branch of the Hadley circulation (Fig. 1.35). By supplying total
water to an expansive area, that air also humidifies the layer of strong detrainment.
Above the tropical tropopause, horizontal divergence reverses sign. Representing
convergence, it implies weak downwelling of environmental air at lower levels, just
above the layer of strong detrainment. The implied vertical motion is consistent
with cumulus detrainment above the LNB, where it serves as a heat sink for the
environment (Sec. 7.7).

Plotted in Fig. 9.26b for the same location and period is anomalous environmental mixing ratio (deviation from its NS mean). As η at these levels is only a couple
of percent (Fig. 9.24), anomalous r represents the humidity of regions that are
chiefly cloud-free. In the upper troposphere, positive anomalous r reflects a source
of environmental water vapor. It is consistent with detrainment of total water from
cumulus updrafts, a process that humidifies the environment. The implied source
maximizes near 200 hPa, within the layer of large anvil extent and magnified detrainment (Figs 9.25b, 9.26a).

Overlying the layer where anomalous mixing ratio maximizes is a layer in which
environmental RH exceeds 80% (shaded). Coincident with strong outflow, it appears
within the layer of anomalously cold temperature (Fig. 7.15). The latter, which is

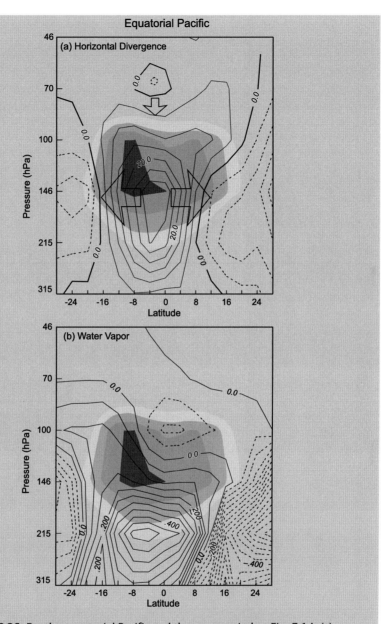

Figure 9.26 For the equatorial Pacific and the same period as Fig. 7.14, (a) horizontal velocity divergence, which measures the collective outflow from cumulus towers, and (b) anomalous humidity, departure from the NS mean of the logarithm of environmental mixing ratio. Shaded is RH exceeding 60%, 70%, and 80%. After Salby et al. (2003).

forced by cumulus overshoots, sharply reduces r_c. It thus drives environmental air toward saturation. These features of humidity structure are consistent with ice cloud being found preferentially at sites where RH is high (Wu et al, 2005). Inside cumulus overshoots, ice water content can be substantial. However, it involves large particles, which are removed efficiently by precipitation to lower levels. Much the same applies to neighboring anvil.

Above the layer that is nearly saturated, anomalous mixing ratio reverses sign. Negative values at those levels reflect a sink of environmental water vapor. Their vertical position suggests a close relationship to the layer of nearly saturated conditions just below. Implied is efficient removal of total water as cumulus updrafts cross the layer of high RH. Because environmental air at those levels is nearly saturated, entrainment becomes ineffective at limiting condensation inside updrafts. Particle growth can then proceed unhindered, achieving large sizes that precipitate out (Danielsen, 1982). Depleted of total water, cumulus updrafts experiencing detrainment at higher levels would then dehydrate the environment.

9.4 RADIATIVE TRANSFER IN AEROSOL AND CLOUD

In the presence of liquid and solid particles, photons experience repeated reflection and diffraction. The path traveled by SW radiation is therefore greatly elongated. Supported by absorption inside particles, this sharply increases the optical depth posed to incident radiation.

9.4.1 Scattering by molecules and particles

Rayleigh scattering

The simplest treatment of scattering is due to Rayleigh (1871). It describes the interaction of sunlight with molecules. *Rayleigh scattering* applies to particles much smaller than the wavelength of radiation. It considers a molecule that is exposed to electromagnetic radiation as an oscillating dipole. The strength of this oscillation is measured by the *dipole moment p*. It is related to the electric field E_0 of incident radiation as

$$p = \alpha E_0, \tag{9.15}$$

where α is the *polarizability* of the scatterer. Established through interaction with the wave's electromagnetic field, the dipole in turn radiates a scattered wave with electric field E. At large distance r from the dipole, radiation emitted at an angle γ from p behaves as a plane wave whose amplitude satisfies

$$E = \frac{1}{c^2} \frac{sin\gamma}{r} \frac{\partial^2 p}{\partial t^2}, \tag{9.16.1}$$

where c is the speed of light (e.g., Jackson, 1975). Incorporating (9.15) and plane wave radiation of the form $e^{ik(r-ct)}$ then obtains

$$E = -E_0 k^2 \alpha \frac{sin\gamma}{r} e^{ik(r-ct)}. \tag{9.16.2}$$

Consider sunlight at a *scattering angle* Θ from the path of incident radiation (Fig. 9.27). Sunlight is unpolarized: Its electric field E_0 is distributed isotropically over directions orthogonal to the path of propagation. It may therefore by represented in independent components E_{0x} and E_{0y} of equal magnitude. These incident electric fields induce dipole moments p_x and p_y, which in turn radiate scattered waves. Radiation scattered at the angle Θ makes an angle $\gamma_x = \frac{\pi}{2}$ from p_x and an angle $\gamma_y = \frac{\pi}{2} - \Theta$

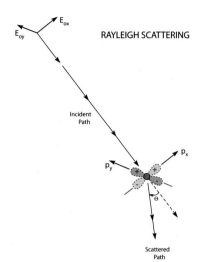

RAYLEIGH SCATTERING

Figure 9.27 Rayleigh scattering of SW radiation by air molecules. Unpolarized sunlight is characterized by radiation with equal and independent electric fields $\boldsymbol{E}_{0x}$ and $\boldsymbol{E}_{0y}$, which induce dipole moments $\boldsymbol{p}_x$ and $\boldsymbol{p}_y$ at the scatterer. Those in turn emit radiation at angles Θ from the path of incident radiation with electric fields $\boldsymbol{E}_x$ and $\boldsymbol{E}_y$ that are determined by $\boldsymbol{p}_x$ and $\boldsymbol{p}_y$ (see text). The orthogonal electric fields $\boldsymbol{E}_{0x}$ and $\boldsymbol{E}_{0y}$ have been chosen to position the scattering angle Θ in the plane formed by the path of incident radiation and $\boldsymbol{E}_{0y}$.

from $\boldsymbol{p}_y$. Then (9.16.2) gives the corresponding electric fields of scattered radiation

$$
\begin{aligned}
\boldsymbol{E}_x &= -\boldsymbol{E}_{0x}k^2\alpha\frac{1}{r}e^{ik(r-ct)} \\
\boldsymbol{E}_y &= -\boldsymbol{E}_{0y}k^2\alpha\frac{cos\Theta}{r}e^{ik(r-ct)}.
\end{aligned}
\tag{9.17}
$$

The intensity of radiation is proportional to the average of $|\boldsymbol{E}|^2$. Incident sunlight of intensity I_0 is then scattered into the angle Θ with intensity

$$
I(\Theta, r) = I_0 k^4 \frac{\alpha^2}{r^2}\frac{(1 + cos^2\Theta)}{2}.
\tag{9.18}
$$

In terms of the phase function (8.22) and wavelength $\lambda = \frac{2\pi}{k}$, (9.18) may be cast into the canonical form

$$
I(\Theta, r) = I_0 \frac{\alpha^2}{r^2}\frac{32\pi^4}{3\lambda^4}P(\Theta),
\tag{9.19.1}
$$

with

$$
P(\Theta) = \frac{3}{4}(1 + cos^2\Theta).
\tag{9.19.2}
$$

The scattered intensity (Fig. 8.8a) has maxima in the forward ($\Theta = 0°$) and backward ($\Theta = 180°$) directions, with equal energy directed into each half-space.

The scattered flux follows as in (9.19), but with the incident flux F_0 in place of I_0. Integrating the scattered flux over a sphere of radius r obtains the scattered power

$$
\mathcal{P} = F_0\alpha^2\frac{128\pi^5}{3\lambda^4},
\tag{9.20}
$$

which has dimensions of $\frac{energy}{time}$. The scattering cross section for an individual molecule is then

$$
\begin{aligned}
\hat{\sigma}_s &= \frac{\mathcal{P}}{F_0} \\
&= \alpha^2\frac{128\pi^5}{3\lambda^4}.
\end{aligned}
\tag{9.21}
$$

Finally, the scattered intensity at distance r may be expressed

$$I(\Theta, r) = I_0 \frac{\hat{\sigma}_s}{4\pi r^2} P(\Theta). \tag{9.22}$$

Electromagnetic field theory relates the polarizability α to the dimensionless refractive index

$$m = m_r - m_i. \tag{9.23}$$

The real and imaginary parts of m refer to the phase speed and absorption of electromagnetic radiation in a medium, relative to those in a vacuum. An ensemble of scatterers with number density n has polarizability

$$\alpha = \frac{3}{4\pi n} \left(\frac{m^2 - 1}{m^2 + 2} \right). \tag{9.24}$$

At wavelengths of visible radiation, absorption by air molecules is small enough for m_i to be ignored, whereas m_r is close to unity. Then (9.24) reduces to

$$\alpha \cong \frac{m_r - 1}{2\pi n}. \tag{9.25}$$

The scattering cross section for an individual molecule is thus given by

$$\hat{\sigma}_s = \frac{32\pi^3 (m_r - 1)^2}{3n^2 \lambda^4}. \tag{9.26}$$

Scattering then yields an optical depth for the atmosphere of

$$\tau = \hat{\sigma}_s(\lambda) \int_0^\infty n(z)dz. \tag{9.27}$$

Rayleigh scattering explains why the sky appears blue. The λ^{-4} dependence of $\hat{\sigma}_s$ implies that short wavelengths are scattered by air molecules much more effectively than long wavelengths. Blue light ($\lambda \sim 0.42\ \mu$m) is scattered five times more than red light ($\lambda \sim 0.65\ \mu$m). It therefore arrives at the Earth's surface as diffuse radiation, emanating from all directions. When the sun is near the horizon, sunlight passes obliquely through the atmosphere and therefore along an extended path. Shorter wavelengths are then scattered out of the incident beam, leaving only longer (red) wavelengths to illuminate the sky. On average, about 40% of the SW flux is scattered out of the incident beam in the near-UV. Conversely, less than 1% is removed in the near-IR. Overall, about 10% of SW radiation incident on the atmosphere is scattered by molecules. Half of it is returned to space, contributing to the Earth's albedo (Fig. 1.32). Most of that scattering takes place in the lowest 10 km, where $n(z)$ is large. For this reason, increasing elevation witnesses a darkening of the sky, along with an intensification and whitening of direct sunlight. Rayleigh scattering influences ozone photochemistry and stratospheric heating. It enhances the SW flux available to the Chappuis bands of O_3 (Sec. 8.3.1).

Owing to its relationship to dipole radiation, Rayleigh scattering has certain polarization properties. In the forward and backward directions ($\Theta = 0, 180°$), scattered radiation is unpolarized – like incident sunlight. However, scattered sunlight becomes increasingly polarized at intermediate angles. It is fully polarized at directions orthogonal to the path of incident radiation ($\Theta = 90°$). For this reason, the sky assumes a

INDEX OF REFRACTION

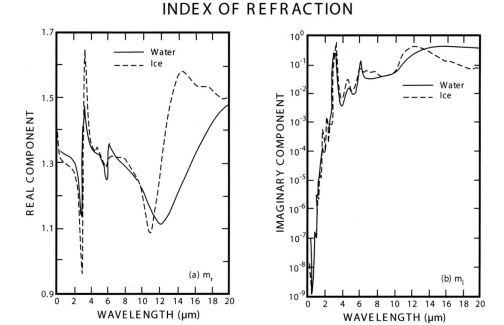

Figure 9.28 Index of refraction for water (solid) and ice (dashed) as functions of wavelength. Adapted from Liou (1990).

deeper blue when viewed through a polarizer at directions orthogonal to incident sunlight.

Mie scattering

Scattering from spherical particles of arbitrary dimension was first treated by Mie (1908). *Mie scattering* applies to the interaction of radiation with aerosol and cloud droplets. Figure 9.28 presents the refractive indices of water and ice, each as a function of wavelength. m_r is of order unity, but varies with λ. This feature makes diffraction and reflection by spheres of condensate wavelength-dependent. m_i is proportional to the absorption coefficient $\beta_{a\lambda}$:

$$m_i = \beta_{a\lambda} \frac{\lambda}{4\pi}. \qquad (9.28)$$

It is small in the visible. However, m_i increases sharply at $\lambda > 1 \ \mu$m for both condensed phases. Consequently all but tenuous cloud is optically thick in the IR.

Application of Mie scattering to atmospheric aerosol relies on particles being sufficiently separated for their interactions with the radiation field to be treated independently. Mie theory solves Maxwell's equations for the electromagnetic field about a dielectric sphere of radius a in terms of an expansion in spherical harmonics and Bessel functions; see Liou (1980) for a detailed treatment. Properties of the scattered wave field emerge in terms of the dimensionless size parameter

$$x = 2\pi \frac{a}{\lambda}. \qquad (9.29)$$

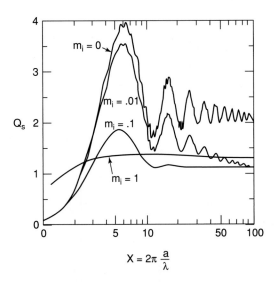

Figure 9.29 Scattering efficiency from a dielectric sphere having real component of refractive index $m_r = 1.33$ for several imaginary components m_i, as functions of the dimensionless size parameter $x = 2\pi \frac{a}{\lambda}$. After Hansen and Travis (1974). Reprinted by permission of Kluwer Academic Publishers.

The dimensionless *scattering* and *extinction efficiencies*

$$Q_s = \frac{\hat{\sigma}_s}{\pi a^2}$$

$$Q_e = \frac{\hat{k}}{\pi a^2}$$

(9.30)

represent the fractional area of the incident beam that is removed through interaction with the sphere by scattering and *in toto*. The scattering efficiency is shown in Fig. 9.29 for a sphere with $m_r = 1.33$ and for several values of m_i. For conservative scattering ($m_i = 0$), Q_s increases quartically for $x \ll 1$ corresponding to Rayleigh scattering. It attains a maximum near $x \cong 2\pi$ (i.e., at $\lambda \cong a$). The scattering efficiency then passes through a series of oscillations. They result from interference between light transmitted through and diffracted about the sphere.

In the limit of large radius, the oscillations in scattering efficiency damp out, yielding the limiting behavior

$$Q_s \sim 2 = \text{const} \qquad x \to \infty$$

(9.31)

for conservative scattering ($m_i = 0$). Hence a sphere much larger than the wavelength of incident radiation scatters twice as much energy as it intercepts. Scattered radiation then includes contributions from energy that is diffracted about the sphere as well as energy that is redirected by reflection inside the sphere. This limiting behavior applies to the interaction of SW radiation with cloud droplets (cf. Fig. 9.23). Notice that, in this limit, Q_s depends only weakly on wavelength. Further, droplet size spectra span several of the weak oscillations that do remain. As a result, those oscillations experience cancellation. The scattering efficiency is therefore indiscriminate over wavelength. This feature of Mie scattering explains why cloud appears white.

For nonconservative scattering ($m_i > 0$), Q_s exhibits similar behavior. However, the oscillations due to interference are damped out with increasing absorption (because radiation emerges from the sphere progressively weaker). By $m_i = 1$, only the maximum near $x \cong 2\pi$ remains. Increased attenuation inside the sphere also reduces the limiting value of Q_s for $x \to \infty$, by absorbing energy that would otherwise contribute to the scattered wave field. Far from the sphere, radiation exhibits strong forward scattering (see Fig. 8.8b). However, for an individual sphere, it is complicated by rapid fluctuations with Θ that result from interference.

Under the conditions of independent scattering, properties of a population of spheres follow by collecting contributions from the individual particles. The scattering and extinction coefficients are then described by

$$\beta_s = \int \hat{\sigma}_s dn(a)$$
$$\beta_e = \int \hat{k} dn(a),$$

(9.32)

where $dn(a) = \frac{dn}{da} da$ represents the droplet size spectrum. The single-scattering albedo for the population (8.20) is characterized by

$$\omega = \frac{\beta_s}{\beta_e}.$$

(9.33)

Similarly, the optical depth of a cloud of thickness Δz_c may be estimated as

$$\tau_c = \beta_e \Delta z_c,$$

(9.34)

where β_e is representative of the cloud as a whole.

Figure 9.30 shows, for several wavelengths, the phase function (solid) corresponding to a droplet size spectrum representative of cumulus (Fig. 9.23). Compared with Rayleigh scattering (dashed), Mie scattering possesses much stronger directionality. All λ shown exhibit strong forward scattering. This is especially true of short wavelengths, which are sharply peaked about $\Theta = 0$ and experience little absorption (Fig. 9.28b). A weaker maximum appears at backward scattering for wavelengths in the near-IR and visible.

The mean extinction cross section for the population is given by $\langle \sigma_e \rangle = \frac{\beta_e}{n}$ (Fig. 9.31a). It is nearly constant across the visible, for which cloud droplets are large (9.31). Weak absorption at those wavelengths leads to a single scattering albedo (Fig. 9.31b) of near unity. Hence $\sigma_e \cong \sigma_s$ and scattering is likewise wavelength independent. Individual components of sunlight are thus scattered with equal efficiency, so cloud appears white. The extinction cross section attains a maximum near 5 μm, at a wavelength comparable to the mean radius of cloud droplets (Fig. 9.23). It decreases at longer wavelength. Because $\omega \sim 1$ for $\lambda < 10$ μm, most of the extinction results from scattering. At wavelengths of SW radiation, the latter dominates over absorption. Exceptional are narrow bands at $\lambda \cong 3$ μm and 6 μm, where absorption spectra of water and ice are peaked (Fig. 9.28b). At $\lambda > 10$ μm, ω decreases due to diminishing Q_s and increasing absorption. Similar optical properties follow from a droplet size spectrum representative of stratus.

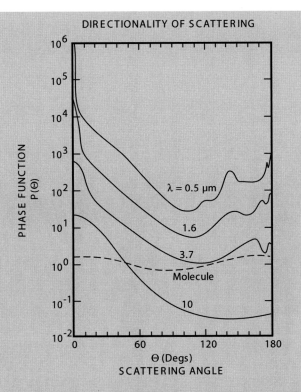

DIRECTIONALITY OF SCATTERING

Figure 9.30 Phase function for scattering from the cumulus droplet population in Fig. 9.23, for several wavelengths (solid). Phase function for Rayleigh scattering (dashed) superposed. *Source:* Liou (1990).

CLOUD OPTICAL PROPERTIES

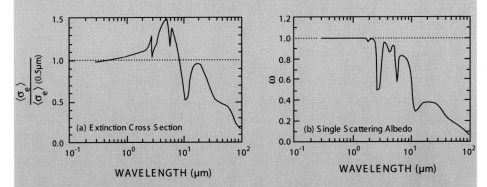

Figure 9.31 (a) Extinction cross section normalized by $\langle \sigma_e \rangle (0.5 \ \mu m) = 166 \ \mu m^2$ and (b) single scattering albedo for the cumulus droplet population in Fig. 9.23. Adapted from Liou (1990).

Extinction by cloud droplets greatly increases the SW opacity of the atmosphere. For cumulus 1 km thick and a droplet number density of 300 cm^{-3}, (9.34) gives an optical depth at $\lambda = 0.5$ μm of $\tau_c = 50$. The cloud is optically thick to SW radiation. Organized cumulus of the same dimensions can have $\tau_c > 100$. Even shallow stratus has optical depths in the visible well in excess of 10. In the IR, τ_c is smaller. However, it is still large enough to render all but tenuous cloud optically thick to LW radiation. At 10 μm, the foregoing conditions for cumulus lead to an optical depth of about 20.

Scattering by nonspherical ice particles is complicated by their irregular structure and anisotropy. Even though their rudimentary geometry is hexagonal, ice crystals assume a wide range of shapes (Fig. 9.14). They can produce a myriad of optical effects associated with orientation, directionality, and polarization. The familiar halo of thin cirrus results from prismatic crystals with a preferred orientation.

9.4.2 Radiative transfer in a cloudy atmosphere

How cloud and aerosol influence atmospheric thermal structure is now evaluated through the radiative transfer equation. Consider a homogeneous scattering layer that is illuminated by a solar flux F_s, which is inclined at the zenith angle $\theta_s = cos^{-1}\mu_s$ (Fig. 9.32). The source function for diffuse SW radiation (8.24) then becomes

$$J(\phi, \mu, \tau) = \frac{\omega}{4\pi} \int_0^{2\pi} \int_{-1}^1 I(\phi', \mu', \tau)P(\phi, \mu; \phi', \mu')d\phi'd\mu' + \frac{\omega}{4\pi}F_s P(\phi, \mu; \phi', -\mu_s)e^{-\frac{\tau}{\mu_s}},$$

(9.35)

where the wavelength dependence is implicit and emission is ignored. The first term on the right-hand side represents contributions to diffuse radiation from multiple scattering of the diffuse intensity I. The second term represents the contribution from single scattering of direct solar radiation. Transmission of diffuse SW radiation through a cloudy atmosphere is then governed by the radiative transfer equation.

Scattering transforms (8.25) into an integro-differential equation. It is treated by expanding the phase function $P(\cos \Theta)$, where Θ is the 3-dimensional scattering angle, in a series of spherical harmonics and the diffuse intensity I in a like manner; see Liou (1980) for a formal treatment. Expressing $P(cos\Theta)$ in terms of the spherical coordinates (ϕ, θ) and integrating over ϕ then yields the transfer equation governing the azimuthal-mean component $I(\mu, \tau)$ of the diffuse radiation field

$$\mu\frac{dI}{d\tau} = I - \frac{\omega}{2}\int_{-1}^1 I(\mu', \tau)P(\mu; \mu')d\mu' - \frac{\omega}{4\pi}F_s P(\mu; -\mu_s)e^{-\frac{\tau}{\mu_s}},$$

(9.36.1)

where

$$P(\mu; \mu') = \sum_j c_j P_j(\mu)P_j(\mu'),$$

(9.36.2)

P_j is the Legendre polynomial of degree j, and the expansion coefficients c_j follow from orthogonality properties of the $P_j(\mu)$. For single scattering, the convolution integral is omitted. The diffuse intensity is then directly proportional to the phase function P. Approximate solutions for multiple scattering can be obtained by numerical techniques. A simple but enlightening solution follows from the two-stream approximation.

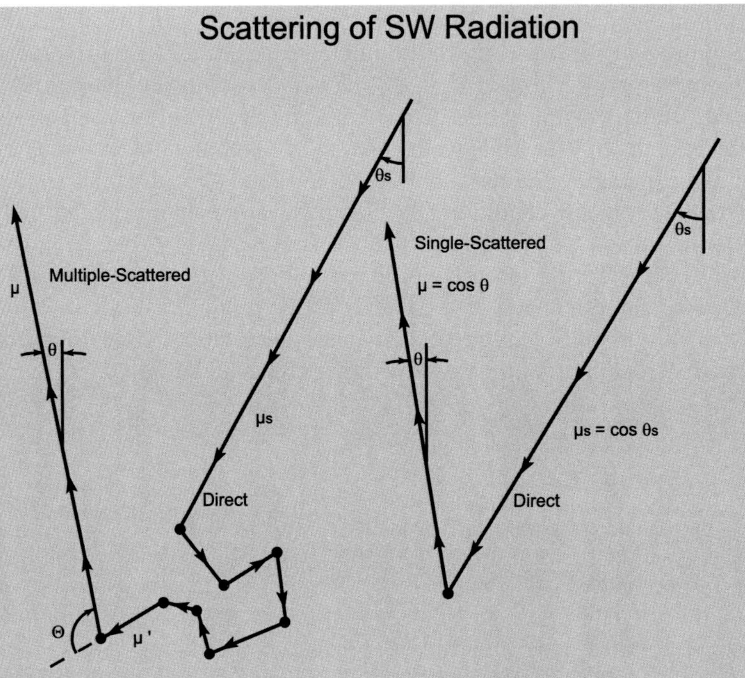

Scattering of SW Radiation

Figure 9.32 Transmission of diffuse SW radiation through a scattering layer.

Owing to the involvement of Legendre polynomials, the integral in (9.36) can be evaluated efficiently with Gaussian quadrature at a finite number of zenith angles μ_j (e.g., Rektorys, 1969). Each describes a different stream of radiation. The two-stream approximation then follows by taking just two angles: $j = \pm 1$, which correspond to the diffusivity factor $\bar{\mu}^{-1} = \sqrt{3}$, i.e., to the upwelling and downwelling streams $I^{\uparrow\downarrow} = I(\pm\bar{\mu}, \tau)$. Incorporating the orthogonality properties of Legendre polynomials (Abramowitz and Stegun, 1972) and the transformation (8.54) leads to a system of two ordinary differential equations for the downwelling and upwelling fluxes of diffuse SW radiation

$$-\frac{dF^\downarrow}{d\tau^*} = F^\downarrow - \omega\left[fF^\downarrow + (1-f)F^\uparrow\right] - S^+ e^{-\frac{\bar{\mu}}{\mu_s}\tau^*} \tag{9.37.1}$$

$$\frac{dF^\uparrow}{d\tau^*} = F^\uparrow - \omega\left[fF^\uparrow + (1-f)F^\downarrow\right] - S^- e^{-\frac{\bar{\mu}}{\mu_s}\tau^*}, \tag{9.37.2}$$

where τ^* is given by (8.54),

$$f = \frac{1+g}{2} \tag{9.37.3}$$

and $(1-f)$ represent the fractional energy that is forward and backward scattered, respectively,

$$g = \frac{1}{2}\int_{-1}^{1} P(cos\Theta)cos\Theta d(cos\Theta) \tag{9.37.4}$$

is an *asymmetry factor*, and

$$S^\pm = \omega \frac{F_s}{4}(1 \pm 3g\bar{\mu}\mu_s). \qquad (9.37.5)$$

The asymmetry factor g equals the first moment of the phase function. It measures the difference in scattering between the forward and backward half spaces. For isotropic scattering, $g = 0$ and $f = \frac{1}{2}$. The same holds for Rayleigh scattering, which is symmetric. However, the strong forward lobe of Mie scattering (Figs. 8.8b, 9.30) leads to $f > \frac{1}{2}$, with $g \cong 0.85$ representative of water cloud.

Subtracting and adding (9.36), introducing the net and total fluxes

$$F = F^\uparrow - F^\downarrow \qquad (9.38.1)$$

$$\bar{F} = F^\uparrow + F^\downarrow, \qquad (9.38.2)$$

and differentiating with respect to τ^* results in the equivalent 2nd order system

$$\frac{d^2 F}{d\tau^{*2}} - \gamma^2 F = Z e^{-\frac{\bar{\mu}}{\mu_s}\tau^*} \qquad (9.39.1)$$

$$\frac{d^2 \bar{F}}{d\tau^{*2}} - \gamma^2 \bar{F} = \bar{Z} e^{-\frac{\bar{\mu}}{\mu_s}\tau^*}, \qquad (9.39.2)$$

where

$$\gamma^2 = (1 - \omega)(1 + \omega - 2\omega f) \qquad (9.39.3)$$

$$Z = \frac{\bar{\mu}}{\mu_s} \bar{S} + (1 - \omega)S$$

$$\bar{Z} = -\frac{\bar{\mu}}{\mu_s} S - (1 + \omega - 2\omega f)\bar{S} \qquad (9.39.4)$$

and

$$S = S^+ - S^-$$

$$\bar{S} = S^+ + S^-. \qquad (9.39.5)$$

The system (9.39) governs the transmission of diffuse SW radiation. It has general solution

$$F^\uparrow = C\alpha_+ e^{\gamma\tau^*} + D\alpha_- e^{-\gamma\tau^*} + E_+ e^{-\frac{\bar{\mu}}{\mu_s}\tau^*}$$

$$F^\downarrow = C\alpha_- e^{\gamma\tau^*} + D\alpha_+ e^{-\gamma\tau^*} + E_- e^{-\frac{\bar{\mu}}{\mu_s}\tau^*}, \qquad (9.40.1)$$

where

$$\alpha_\pm = \frac{1 \pm \beta}{2}$$

$$\beta^2 = \frac{1 - \omega}{1 + \omega - 2\omega f} \qquad (9.40.2)$$

$$E_\pm = \frac{\bar{\delta} \pm \delta}{2}$$

$$\bar{\delta} = \frac{\mu_s^2}{\bar{\mu}^2 - \mu_s^2 \gamma^2} \bar{Z} \qquad \delta = \frac{\mu_s^2}{\bar{\mu}^2 - \mu_s^2 \gamma^2} Z.$$

The integration constants C and D are determined by boundary conditions. τ^* is measured from the top of the cloud layer, which has an effective optical depth τ_c^*. Likewise, the underlying surface is black. Then the downwelling diffuse flux vanishes at the top of the scattering layer, whereas the upwelling diffuse flux vanishes at the bottom:

$$F^\downarrow(0) = 0$$
$$F^\uparrow(\tau_c^*) = 0.$$

(9.41)

The integration constants then become

$$C = \frac{E_-\alpha_- e^{-\gamma\tau_c^*} - E_+\alpha_+ e^{-\frac{\beta}{\mu_s}\tau_c^*}}{\alpha_+^2 e^{\gamma\tau_c^*} - \alpha_-^2 e^{-\gamma\tau_c^*}}$$

$$D = \frac{E_+\alpha_- e^{-\frac{\beta}{\mu_s}\tau_c^*} - E_-\alpha_+ e^{\gamma\tau_c^*}}{\alpha_+^2 e^{\gamma\tau_c^*} - \alpha_-^2 e^{-\gamma\tau_c^*}}.$$

(9.42)

The solution can be used to evaluate the cloud albedo and transmissivity

$$\mathcal{A}_c = \frac{F^\uparrow(0)}{\mu_s F_s}$$

(9.43.1)

$$\mathcal{T}_c = \frac{F^\downarrow(\tau_c^*) + \mu_s F_s e^{-\frac{\beta}{\mu_s}\tau_c^*}}{\mu_s F_s}.$$

(9.43.2)

In (9.43.2), the direct-solar contribution becomes negligible for cloud that is optically thick. The cloud absorptivity follows as $1 - \mathcal{A}_c - \mathcal{T}_c$.

Figure 9.33 plots the cloud albedo and transmissivity, each as a function of optical depth and solar zenith angle; values are for microphysical properties representative of cumulus (Fig. 9.23) at a wavelength of 0.5 μm. Cloud albedo increases sharply at small optical depth (Fig. 9.33a). It exceeds 50% for $\tau_c > 20$, typical of water cloud. Hence, even shallow stratus are highly reflective. Transmissivity decreases with increasing τ_c – nearly complementary to cloud albedo. Thus, most incident SW energy is either reflected or transmitted. The absorptivity, $1 - \mathcal{A}_c - \mathcal{T}_c$, is of order 10% across much of the range of optical depth shown. Absorption by droplets and surrounding vapor is comparable. Because the cloud layer is optically thick, most of the absorption occurs near its top. There, the heating rate $\frac{\dot{q}}{c_p}$ reaches 10 Kday^{-1} and greater. Cloud albedo also increases with increasing solar zenith angle (Fig. 9.33b), which elongates the slant optical path. For $\tau_c = 20$, $\mathcal{A}_c$ increases from 53% for overhead sun to 75% when the sun is on the horizon. The behavior of $\mathcal{T}_c$ is nearly complementary. It decreases from 32% for overhead sun to only 10% when the sun is on the horizon. Shallower cloud depends even more strongly on solar zenith angle because the slant optical path then varies between optically thin and optically thick conditions.

The factor that influences cloud optical properties most strongly is liquid water content. It determines τ_c. The optical depth of a cloud may be expressed in terms of the column abundance of liquid water,

$$\Sigma_l = \int_0^\infty \rho_l dz$$
$$= \Delta z_c \cdot \frac{4\pi \rho_w}{3} \int a^3 dn(a),$$

(9.44)

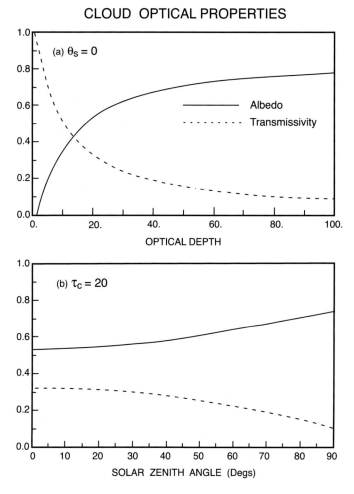

Figure 9.33 Albedo and transmissivity at 0.5 μm for a cloud layer with micro-physical properties of cumulus in Fig. 9.23 and nearly conservative scattering (a) for overhead sun, as functions of cloud optical depth and (b) for a cloud optical depth of 20, as functions of solar zenith angle.

which is called the *liquid water path*. In the limit of large particles, which applies to SW radiation, the extinction cross section asymptotically approaches $\hat{\sigma}_e = Q_e \pi a^2 \cong 2\pi a^2$ (Fig. 9.29). The optical depth may then be expressed

$$\tau_c = \frac{3\Sigma_l}{2\rho_w a_e},$$ (9.45.1)

where

$$a_e = \frac{\int a(\pi a^2)dn(a)}{\int (\pi a^2)dn(a)}$$ (9.45.2)

is a scattering-equivalent mean droplet radius. Most water cloud has liquid water content exceeding 0.1 g m^{-3} (Table 9.3). Therefore, cloud that is more than a kilometer thick has liquid water paths Σ_l of 100 g m^{-2} and greater. Such cloud then has optical depth in the visible of several tens and greater. As displayed in Fig. 9.34, this leads to

CLOUD OPTICAL PROPERTIES

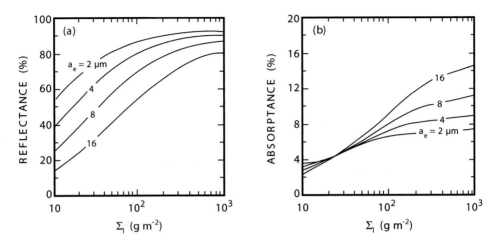

Figure 9.34 SW reflectance (a) and absorptance (b) as functions of liquid water path, for a solar zenith angle of 60° and for different scattering-equivalent droplet radii a_e. Adapted from Slingo (1989).

albedo of 50% and greater. Deep cloud with $\Sigma_l > 1000$ g m^{-2} has albedo that exceeds 80% (cf. bright features in Fig. 1.29).

For given Σ_l, cloud with small droplets has a higher albedo and smaller absorptivity than cloud with large droplets (Fig. 9.34). The increased reflectivity of cloud containing small droplets follows from the increased number densities in such cloud and the reduced absorption of small droplets. Its smaller droplets and greater number densities make continental cloud more reflective than maritime cloud of the same type and liquid water path (e.g., Twomey, 1977). For the same reason, precipitation ($a > 100$ μm) sharply increases the absorptivity of cloud. It is for this reason that precipitating cloud appears dark. Ice cloud, although more complex, is likewise highly reflective across the visible. Its dependence on microphysical properties resembles that of water cloud. Thin cirrus has ice water paths of $\Sigma_i < 5$ g m^{-2}. It has albedo of order 10% and absorptivity of only a couple of percent (Paltridge and Platt, 1981).

According to Fig. 9.31b, scattering by cloud particles falls off at $\lambda > 4$ μm, where water absorbs strongly (Fig. 9.28b). Figure 9.35 illustrates that water cloud 3 km thick, which is highly reflective at $\lambda < 1$ μm, becomes fully absorbing at $\lambda > 4$ μm.

The foregoing development governs the transmission through cloud of SW radiation. Transmission of LW radiation can be treated in similar fashion if the scattering of direct solar in (9.37) is replaced by the emission of terrestrial radiation: $(1 - \omega)B^*[T]$. Inside a cloud layer of temperature T_c, upwelling and downwelling LW fluxes are then described by

$$\frac{dF^\uparrow}{d\tau^*} = F^\uparrow - \omega\left[fF^\uparrow + (1 - f)F^\downarrow\right] - (1 - \omega)B^*[T_c] \qquad (9.46.1)$$

$$-\frac{dF^\downarrow}{d\tau^*} = F^\downarrow - \omega\left[fF^\downarrow + (1 - f)F^\uparrow\right] - (1 - \omega)B^*[T_c], \qquad (9.46.2)$$

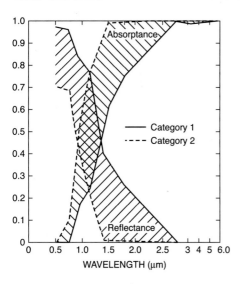

Figure 9.35 Reflectance and absorptance of 3-km thick stratus, as functions of wavelength for overhead sun and for a range of microphysical properties bounded by categories 1 and 2. Category 1 is based on a droplet distribution of characteristic radius 6 μm, $n = 10^2$ cm^{-3}, and $\rho_l = 0.30$ g m^{-3}. Category 2 includes rain with a droplet distribution of characteristic radius 600 μm, $n = 0.001$ cm^{-3}, and $\rho_l = 2.1$ g m^{-3}. *Source:* Cox (1981).

where the cloud is sufficiently thin for T_c to be treated as constant. A particular solution of (9.46) is the blackbody spectrum B^*. Proceeding as in the treatment of SW radiation leads to the general solution

$$F^\uparrow = C\alpha_+ e^{\gamma \tau^*} + D\alpha_- e^{-\gamma \tau^*} + (1-\omega)B^*[T_c]$$
$$F^\downarrow = C\alpha_- e^{\gamma \tau^*} + D\alpha_+ e^{-\gamma \tau^*} + (1-\omega)B^*[T_c],$$

(9.47)

which applies inside the scattering layer.

To illustrate the influence of cloud on the LW energy budget, we consider a statistically homogeneous scattering layer within the framework of radiative-convective equilibrium (Sec. 8.5.2). Microphysical properties incorporated in the treatment of SW radiation (Fig. 9.23) are used to define a cloud layer of thickness $\Delta z_c = 1$ km and fractional coverage η that is situated at a height z_c in the troposphere – all under the conditions of radiative-convective equilibrium (Fig. 8.21). In the spirit of a gray atmosphere, radiative characteristics at 10 μm are used to define the broadband transmission properties for LW radiation. The fractional cloud cover is adjusted to give an albedo of 30%. $F^\uparrow$ and $F^\downarrow$ are described by (8.70) outside the cloud layer and by (9.47) inside the cloud layer. Matching fluxes across the scattering layer then leads to behavior analogous to that under simple radiative-convective equilibrium (Fig. 8.22), except for the addition of the cloud layer.

Figure 9.36a shows upwelling and downwelling fluxes in the presence of a cloud layer of optical depth $\tau_c = 20$ that is situated at $z_c = 8$ km. The equilibrium temperature structure (not shown) is identical to that in Fig. 8.21, except that temperatures beneath the cloud layer are uniformly warmer. The surface temperature T_s has increased from its cloud-free value of 285 K to 312 K. Below the cloud layer, upwelling and downwelling fluxes resemble those under cloud-free conditions (Fig. 8.22a), except likewise displaced to higher values. Inside the cloud layer, the profiles of $F^\uparrow$ and $F^\downarrow$ undergo

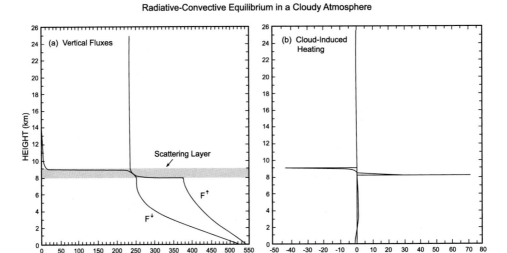

Figure 9.36 Energetics inside a cloudy atmosphere that is in radiative-convective equilibrium and contains a 1 km thick scattering layer with microphysical properties of cumulus in Fig. 9.23. The fractional cloud cover is adjusted to give an albedo of 0.30. (a) Upwelling and downwelling LW fluxes. (b) Cloud-induced heating: $\frac{\dot{q}}{c_p} - (\frac{\dot{q}}{c_p})^{CS}$, where the superscript refers to radiative-convective equilibrium under clear-sky conditions (Fig. 8.22).

a sharp adjustment that drives them into coincidence. This behavior follows from the cloud being optically thick. The LW flux is then isotropic, equal to σT^4. Above the cloud layer, $F^\uparrow$ and $F^\downarrow$ recover the cloud-free forms in Fig. 8.22a to satisfy boundary conditions at TOA (8.63).

Sharp changes of $F^\uparrow$ and $F^\downarrow$ in Fig. 9.36a introduce strong LW warming near the base of the cloud layer and strong LW cooling near its top. Figure 9.36b plots the cloud-induced heating: $\frac{\dot{q}}{c_p} - (\frac{\dot{q}}{c_p})^{CS}$, where $(\frac{\dot{q}}{c_p})^{CS}$ is the clear-sky heating profile under radiative-convective equilibrium (Fig. 8.22b). LW absorption (heating) is concentrated within an optical depth of the cloud base. LW emission (cooling) is concentrated within an optical depth of the cloud top. These regions of net absorption and emission are shallow, analogous to Chapman layers (Sec. 8.5.3). They introduce substantial heating and cooling locally (8.72). LW heating approaches 70 K day^{-1} near the cloud's base. LW cooling approaches 50 K day^{-1} near its top. By re-emitting energy downward, the cloud also introduces heating beneath its base. However, it is much weaker than the heating and cooling at the cloud's boundaries.

LW cooling in upper portions of a cloud is offset by SW heating, but only partially. Anvil cirrus experiences net radiative cooling at its top of several tens of K day^{-1} and heating at its base as large as 100 K day^{-1} (Platt et al., 1984a). Radiative cooling at its top and heating at its base destabilize a cloud layer. That, in turn, introduces convection, which entrains drier environmental air that tends to dissolve cloud (Fig. 9.19). Radiative heating also tends to dissolve cloud by elevating its saturation mixing ratio. Such behavior is typified by stratus burn off.

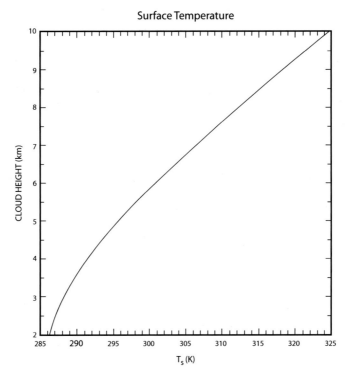

Figure 9.37 Surface temperature for the cloudy atmosphere in Fig. 9.36, as a function of the cloud layer's height.

The cloud layer in Fig. 9.36 is optically thick. Therefore, it influences the TOA LW energy budget chiefly through cloud-top emission. The latter decreases with the cloud's temperature and elevation. Figure 9.37 shows the dependence on cloud height of surface temperature under radiative-convective equilibrium. Situating the cloud below 2 km recovers nearly the same surface temperature as under cloud-free conditions. The scattering layer is then embedded inside the optically-thick layer of water vapor. It therefore has little influence on the effective emission temperature of the atmosphere. However, $z_c > 2$ km leads to warmer surface temperature. T_s increases steadily with the cloud's height. By $z_c = 8$ km, T_s has increased $27°$K to maintain $F^\uparrow(\tau^* = 0) = (1 - \mathcal{A})\overline{F}_s$ and hence maintain the equivalent blackbody temperature of the Earth (Prob. 9.26).

Water cloud becomes optically thick to LW radiation for liquid water paths Σ_l greater than about 20 g m^{-2}. It then behaves as a blackbody. The LW characteristics of ice cloud are more complex, at least for cloud that is optically thin. The emissivity (LW absorptivity) of thin cirrus with ice water paths less than 10 g m^{-2} (typical of cirrus 1 km thick) is 0.5 or smaller. It can vary diurnally with convection (Platt et al., 1984b). However, even cirrus emits approximately as a blackbody for Δz_c sufficiently large. Figure 9.38 shows spectra of OLR in the tropics over a clear-sky region and several neighboring cirrus-covered regions of different optical depth. Cirrus reduces the radiation emitted to space throughout the LW, except near 15 μm. There, radiation

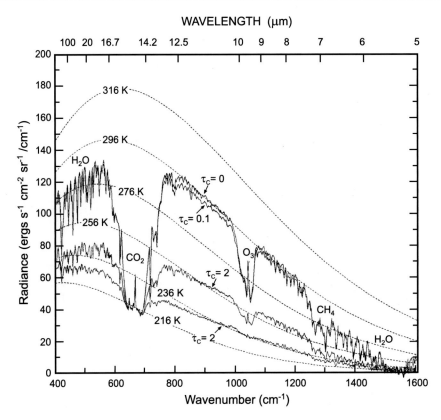

Figure 9.38 Spectrum of outgoing LW radiation observed by Nimbus-4 IRIS for a clear sky region ($\tau_c = 0$) at (134E,12N) and neighboring cirrus-covered regions ($\tau_c > 0$). Courtesy of B. Carlson (NASA/Goddard Institute for Space Studies).

emitted by the surface has already been absorbed by underlying CO_2. By an optical depth of $\tau_c = 5$, OLR has approached the blackbody spectrum with a temperature equal to the cirrus temperature of 230 K. Hence, the effect of cirrus is to remove the layer emitting to space to a higher level and colder temperature. Outgoing radiation is most reduced in the atmospheric window between 8 and 12 μm, where, under cloud-free conditions, radiation emitted by the surface passes freely to space.

Contrary to the preceding treatment, real cloud has limited horizontal extent. A finite cloud scatters energy out its sides. This enables neighboring clouds to interact. Energy lost through the sides of a cloud exceeds SW absorption for aspect ratios as small as $\frac{1}{20}$ (Cox, 1981). Finite extent is therefore important for all but extensive stratiform. This dependence makes optical characteristics of cloud sensitive to the vertical distribution of microphysical properties. A cloud which has small droplets near its base is less reflective than one which has small droplets near its top because, in the former, much of the SW energy has already been scattered out the sides before the highly reflective droplets are encountered.

9.5 ROLES OF CLOUD AND AEROSOL IN CLIMATE

Cloud modifies the global energy balance by altering the absorption and scattering characteristics of the atmosphere. The largest variations of albedo and OLR in Fig. 1.34 are associated with cloud. Convection also supports large transfers of sensible and latent heat from the Earth's surface (Fig. 1.32). They represent another important heat source for the atmosphere. Radiative effects similar to cloud are introduced by aerosol, which controls cloud formation through microphysical processes (Sec. 9.2).

9.5.1 Involvement in the global energy budget

Influence of cloud cover

The role of cloud in global energetics may be interpreted as a forcing in the TOA radiative energy balance. For LW radiation, the large opacity of cloud increases the optical depth of the atmosphere. Enhancing the greenhouse effect (8.81), this favors increased surface temperature. Thus, cloud introduces warming in the LW energy budget. This warming varies with cloud-top temperature and height (Fig. 9.37). Cold cirrus also cools the lower stratosphere via exchange in the 9.6 μm band of ozone (Sec. 8.5.3). For SW radiation, the high reflectivity of cloud decreases the incoming solar flux. This favors reduced surface temperature (8.81). Thus, cloud introduces cooling in the SW energy budget. Unlike the LW effect, cloud albedo is insensitive to cloud height. Low cloud, like that in visible imagery (Fig. 1.29), is therefore almost as bright as high cloud. The zenith angle dependence of scattering (Fig. 9.33b) enhances albedo at high latitude. Cloud also introduces heating by absorbing SW radiation.

Cloud routinely covers about 50% of the Earth. It accounts for about half of the Earth's albedo (eg, for ~0.15). Marine stratocumulus (e.g., over the southeastern Atlantic in Fig. 1.29) is particularly important. Because of its extensive coverage, it reflects much of the incident SW radiation. However, because its cloud-top temperature does not differ substantially from that of the surface, marine stratocumulus does not appreciably modify OLR and hence the LW energy budget (cf. Fig. 1.28). In contrast, deep cumulus over Africa, South America, and the maritime continent has very cold cloud top temperature, as much as 100 K colder than the underlying surface. Consequently it sharply reduces OLR from that under cloud-free conditions.

The SW and LW effects of cloud are each variable. Owing to the zenith angle dependence of scattering (Fig. 9.33b) and to systematic variations in cloud cover and type, the most important component of cloud variability is diurnal. Figure 9.39 shows the daily variation of cloud properties over the southeastern Pacific, which is populated by marine stratocumulus, and over the Amazon basin, the site of deep continental cumulus. Marine stratocumulus (Fig. 9.39a) is present throughout the day. However, it undergoes a 50% variation in albedo. This resembles the variation of fractional cloud cover, which results from SW heating during daylight hours. The corresponding variation of OLR is small. Continental cumulus (Fig. 9.39b) undergoes a 25% variation of OLR. It resembles the daily life cycle of cumulonimbus tower and cirrostratus anvil. The corresponding variation of fractional cloud cover is small. Cirrostratus anvil is glaciated, markedly colder than the surface. It persists well into the night, several hours after convection has dissipated (Platt et al., 1984a).

A quantitative description of how cloud figures in the global energy budget is complicated by its dependence on microphysical properties and interactions with the

DIURNAL VARIATION OF CLOUD PROPERTIES

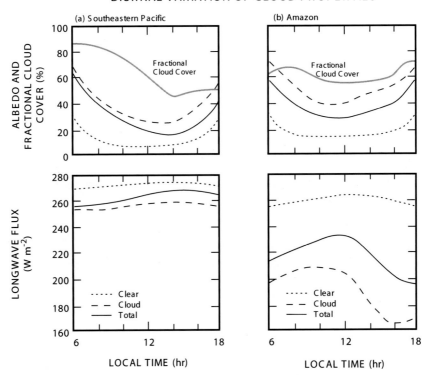

Figure 9.39 Top-of-the-atmosphere radiative properties, as functions of local time, over (a) the southeastern Pacific, which is populated by marine stratocumulus, and (b) the Amazon basin, which is populated by organized cumulonimbus and anvil cirrus. Adapted from Minnis and Harrison (1984).

surface. These complications are circumvented by comparing radiative fluxes at TOA under cloudy vs clear-sky conditions. Over a given region, the column-integrated radiative heating rate must equal the difference between the energy flux absorbed and that emitted to space

$$\dot{Q} = \int_0^\infty \rho \dot{q} dz$$
$$= (1 - \mathcal{A})F_s - F_{LW}(z = \infty),$$
(9.48)

where $\mathcal{A}$ is the albedo of the region and F_s and $F_{LW}(\infty)$ denote the downward SW and upward LW fluxes at TOA. The *cloud radiative forcing* is defined by

$$C = \dot{Q} - \dot{Q}^{CS},$$
(9.49)

where $\dot{Q}^{CS}$ is the heating rate under clear-sky conditions. It represents the net radiative influence of cloud on the column energy budget of the region. Incorporating (9.49) into (9.48) yields

$$C = (\mathcal{A}^{CS} - \mathcal{A})F_s + \left(F_{LW}^{CS} - F_{LW}\right).$$
(9.50)

From (9.50), the SW and LW components of cloud forcing are identified as

$$C_{SW} = (A^{CS} - A)F_s$$
$$= F_{SW}^{CS} - F_{SW} \tag{9.51.1}$$

$$C_{LW} = F_{LW}^{CS} - F_{LW}. \tag{9.51.2}$$

If the region has fractional cloud cover η, the outgoing SW and LW fluxes are given by

$$F_{SW} = (1 - \eta)F_{SW}^{CS} + \eta F_{SW}^{OC}$$
$$F_{LW} = (1 - \eta)F_{LW}^{CS} + \eta F_{LW}^{OC}, \tag{9.52}$$

where the superscripts refer to clear-sky and overcast subregions. Then the SW and LW components of cloud radiative forcing may be written

$$C_{SW} = \eta(F_{SW}^{CS} - F_{SW}^{OC})$$
$$= \eta F_s(A^{CS} - A^{OC}) \tag{9.53.1}$$

$$C_{LW} = \eta(F_{LW}^{CS} - F_{LW}^{OC}). \tag{9.53.2}$$

The components of cloud forcing (9.53) can be evaluated directly from broadband fluxes of outgoing LW and SW radiation that are measured by satellite. Figure 9.40 shows time-averaged distributions of C_{SW}, C_{LW}, and C. Longwave forcing (Fig. 9.40a) is large in centers of deep convection over tropical Africa, South America, and the maritime continent, where C_{LW} approaches 100 W m^{-2} (cf. Fig. 1.30b). Secondary maxima appear in the maritime ITCZ and in the North Pacific and North Atlantic storm tracks (Sec. 1.2.5). Shortwave forcing (Fig. 9.40b) is strong in the same regions, where $C_{SW} < -100$ W m^{-2}. Negative SW forcing is also strong over extensive marine stratocumulus in the eastern oceans and over the Southern Ocean, coincident with the storm track of the Southern Hemisphere. Inside the centers of deep tropical convection, SW and LW cloud forcing nearly cancel. They leave small values of C throughout the tropics (Fig. 9.40c). Negative C_{SW} in the storm tracks and over marine stratocumulus then dominates positive C_{LW}, especially over the Southern Ocean. It prevails in the global-mean cloud forcing.

Globally averaged values of C_{LW} and C_{SW} are about 30 and -45 W m^{-2}, respectively. Net cloud forcing is then -15 W m^{-2}. It represents radiative cooling of the Earth-atmosphere system. This is four times as great as the additional warming of the Earth's surface that would be introduced by a doubling of CO_2. Latent heat transfer to the atmosphere (Fig. 1.32) is 90 W m^{-2}. It is an order of magnitude greater. Consequently, the direct radiative effect of increased CO_2 would be overshadowed by even a small adjustment of convection (Sec. 8.7).

While circumventing many of the complications surrounding cloud behavior, cloud forcing is limited in several respects. Foremost is the fact that C gives only the column-integrated effect of cloud. It provides no information on the vertical distribution of that radiative heating and cooling. Shortwave cloud forcing represents cooling. It is concentrated near the Earth's surface, because the principal effect of increased albedo is to shield the ground from incident SW. Longwave cloud forcing represents warming. It is manifest in heating near the base of cloud and cooling near its top (Fig. 9.36b).

(a) Longwave Cloud Forcing

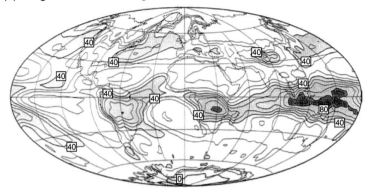

(b) Shortwave Cloud Forcing

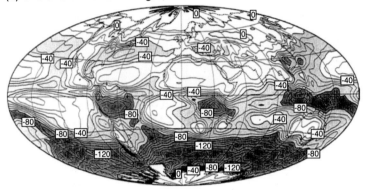

(c) Net Radiative Cloud Forcing

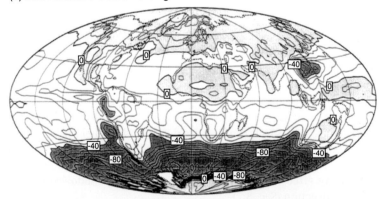

Figure 9.40 Cloud radiative forcing during northern winter derived from ERBE measurements on board the satellites ERBS and NOAA-9 for the (a) LW energy budget, (b) SW energy budget, and (c) net radiative energy budget. Courtesy of D. Hartmann (U. Washington).

That radiative forcing depends intrinsically on the vertical distribution of cloud. For instance, deep cumulonimbus and comparatively shallow cirrostratus can have identical cloud-top temperature, yielding the same LW forcing of the TOA energy budget. However, they have very different optical depths, producing very different vertical distributions of radiative heating. The strong correlation between water vapor and cloud cover introduces another source of uncertainty. Absorption by water vapor biases C_{LW} toward higher values (Hartmann and Doelling, 1991).

Cloud also introduces heating through the release of latent heat. Precipitation leads to a net transfer of heat from ocean to the atmosphere. Latent heating is particularly important for organized deep cumulus, which produces copious rainfall in the tropics. The column-integrated latent heating rate follows from the specific latent heat l and the precipitation rate $\dot{P}$ as

$$Q = l\rho_w \dot{P}. \tag{9.54}$$

$\dot{Q}$ is a thermodynamic counterpart of TOA radiative fluxes in (9.48). According to (9.54), each mm day^{-1} of precipitation translates into $\sim$25 W m^{-2} in the column energy budget of the atmosphere.

Figure 9.41a illustrates the geographical distribution of precipitation. Large precipitation is found inside the ITCZ, where $\dot{P}$ exceeds 10 mm day^{-1}. There, latent heating introduces more than 250 W m^{-2} of warming into the column energy budget of the atmosphere, which is plotted in Fig. 9.41b. Acting in the opposite sense is radiative cooling over the eastern oceans. Characterized by cold SST and extensive marine stratocumulus, those regions introduce 50–150 W m^{-2} of cooling into the column energy budget (cf. Fig. 1.29). The striking resemblance between latent heating and column-integrated heating in Fig. 9.41 makes the former a major energy source for the atmosphere. It also makes the atmosphere intrinsically sensitive to SST (Sec. 17.6).

The region of strong latent heating is the same one implied by cloud forcing to experience strong radiative heating and cooling. However, latent heating of the atmospheric column is three times greater than SW or LW forcing individually. Further, those components of radiative forcing experience strong cancellation inside convective centers, making latent heating an order of magnitude stronger than net radiative forcing by cloud. Latent heating is also distinguished by vertical structure. Whereas cloud radiative forcing is sharply concentrated in the vertical, latent heating inside deep cumulus influences much of the tropical troposphere. Latent heat release inside tropical convection is plotted in Fig. 9.42. It leads to a specific heating rate $\frac{\dot{q}}{c_p}$ of order 10 K day^{-1}. Through detrainment, that heat is transferred to the environment of deep cumulus (Sec. 7.7). Its deep distribution makes latent heat release a major source of energy for the tropical troposphere and the atmosphere as a whole (Fig. 9.41).

Influence of aerosol

Atmospheric aerosol introduces similar effects, albeit weaker. Like cloud, aerosol is an efficient scatterer of SW radiation. It alters optical properties following major volcanic eruptions (Fig. 9.5). Another source of aerosol is fire. Displayed on the cover is visible imagery from the GMS satellite (see color plate section). It reveals haze from Indonesian forest fires, which were supported by prolonged El Nino conditions (Sec. 17.6). Being entrained into twin tropical cyclones, such haze led to health restrictions and public closures. Anthropogenic sources of aerosol exert a similar influence. Near industrial centers in China, aerosol can become optically thick. There, reductions of SW flux at the surface of 10 Wm^{-2} per decade have been reported (Norris and Wild, 2009). Yet,

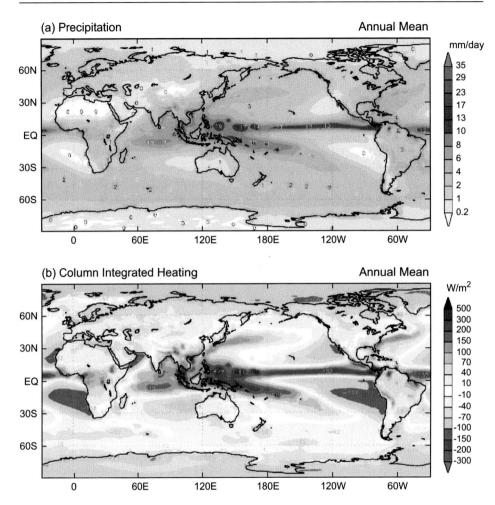

Figure 9.41 (a) Annual-mean precipitation rate (mm/day). (b) Column-integrated atmospheric heating (W m^{-2}). Both are concentrated inside the ITCZ, where precipitation and atmospheric heating exceed 10 mm/day and 250 W m^{-2}, respectively. Support comes from the South Pacific Convergence Zone (SPCZ), the South Atlantic Convergence Zone (SACZ), and the North Atlantic and North Pacific storm tracks. *Source:* Japanese 25-yr Climatology, cooperative project of Japan Meteorological Agency (JMA) and Central Research Institute of Electric Power Industry (CRIEPI) (http://ds.data.jma.go.jp). See color plate section: Plate 7.

over Japan, the same period witnessed no significant change. Generally, such effects are confined to a neighborhood of aerosol source regions, by efficient scavenging processes associated with convection.

Depending on its composition and the underlying surface, aerosol can either enhance albedo or diminish it. Over dark ocean, aerosol increases albedo, introducing cooling into the SW energy budget. Conversely, over bright polar ice, arctic haze (involving anthropogenic pollutants), reduces albedo (Ackerman, 1988). Aerosol also

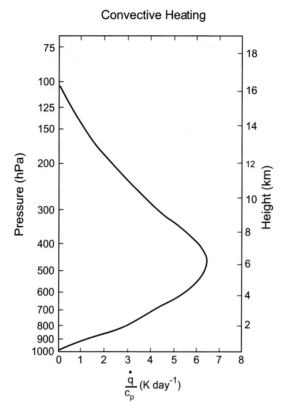

Figure 9.42 Vertical profile of the heating rate inside tropical convection, normalized by a precipitation rate of 10 mm day^{-1}. *Source:* Yanai et al. (1973).

contributes significantly to SW absorption inside the tropical troposphere (Fig. 8.26). To illustrate how these factors determine the net effect of aerosol, we consider a natural species: desert dust.

During disturbed periods, desert dust becomes optically thick. Once raised, it can be lifted inside vigorous desert convection as high as 500 hPa. Exposed to strong wind, it is then carried hundreds, sometimes thousands, of kilometers from its source. Figure 9.43 displays visible imagery during September 2009, when the eastern seaboard of Australia was blanketed by red dust that originated a thousand kilometers to the west. At some sites, the Earth's surface is entirely masked; the dust cloud is optically thick. It streams well off the continent, into the Pacific. Notice the change of albedo relative to neighboring ocean. Conditions inside the layer that is optically thick are displayed in Fig. 9.44, compared against the same field of view under normal conditions. Inside the layer that is optically thick, scattering is isotropic. Visibility is limited to a few tens of meters.

The influence of dust on the energy budget depends on its composition in relation to optical properties of the underlying surface. Dust absorbs SW radiation, reducing albedo over bright surface, like shallow stratus. It then introduces warming into the SW energy budget of the column (9.48). However, over dark surface like ocean, dust increases the albedo (Fig. 9.43). It then introduces cooling into the SW energy budget.

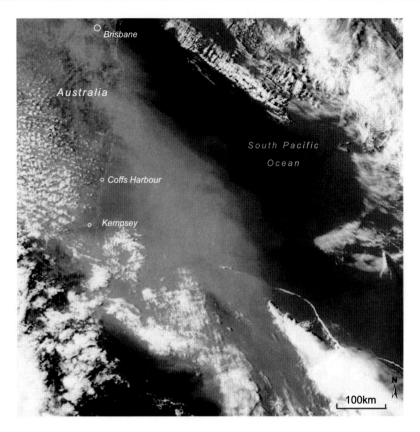

Figure 9.43 Visible imagery during September 2009, when the eastern seaboard of Australia was blanketed by red dust that originated 1000 km to the west. Notice the change of albedo relative to neighboring ocean. See color plate section: Plate 8.

The LW energy budget is influenced by desert aerosol in a manner similar to cloud. When convected to great height, desert aerosol can reduce the effective emission temperature, with commensurate warming in the column energy budget.[8]

The energy budget of desert aerosol is dominated by SW effects. Over surfaces with appreciable albedo, like the Sahara, those SW effects introduce net warming of the column. The same holds for the atmosphere alone. The vertical distribution of radiative heating, however, is determined by both LW and SW components. Figure 9.45 plots the net heating rate for Saharan dust that is distributed between the Earth's surface and 500 hPa. Over cloud-free ocean (Fig. 9.45a), absorption of upwelling LW radiation introduces significant heating near the surface, where $\frac{\dot{q}}{c_p} > 2$ K day^{-1}. For optical depths greater than 1, SW absorption introduces a second heating maximum near the middle of the dust layer. Over cloud-free desert (Fig. 9.45b), greater surface albedo backscatters much of the transmitted SW radiation, which is then absorbed inside

---

[8] The reduction of effective emission temperature can exceed that of cloud at the same height because, unlike cloud, temperature at the top of the dust layer decreases with height dry adiabatically.

Figure 9.44 Field of view across Sydney Harbor (a) under normal conditions and (b) during the dust storm of September 2009. With permission, copyright 2010 Fairfax Digital. See color plate section: Plate 9.

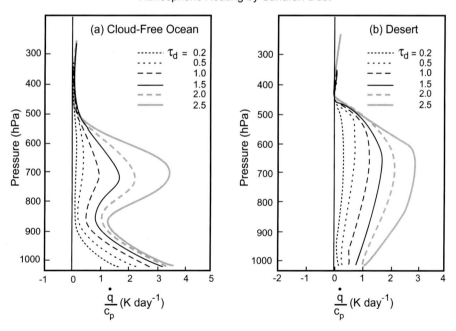

Figure 9.45 Vertical profiles of heating rate for Saharan dust that is distributed between 1000 hPa and 500 hPa and has optical depths τ_d over (a) cloud-free ocean and (b) desert. Adapted from Carlson and Benjamin (1982). Copyright by Spectrum Press (A. Deepak Publishing).

the dust layer. Net heating is thus dominated by SW absorption. It leads to $\frac{\dot{q}}{c_p}$ of order 1–2 K day^{-1} across much of the dust layer. Because it heats the atmosphere and cools the underlying surface, the net effect in both cases is to stabilize the stratification.

Beyond its direct radiative effect, aerosol influences cloud formation. This role of aerosol is potentially more significant than its direct radiative effect. Aerosol underpins the microphysical properties of cloud, like condensation, droplet size distribution, and precipitation (Rosenfeld, 2006). As these features shape the evolution of cloud, indirectly, aerosol influences macrophysical properties such as cloud growth and latent heating. An outbreak of desert dust alters the coverage and type of cloud, favoring stratiform. Through such dependence, a change in the concentration, size, or composition of atmospheric aerosol can introduce changes in cloud cover and in the radiative energy budget.

9.5.2 Involvement in chemical processes

Cloud also supports important interactions with chemical species. Precipitation is the primary removal mechanism for atmospheric pollutants. By serving as CCN, aerosol particles are incorporated into cloud droplets and precipitated to the Earth's surface. This process scavenges air of most nuclei activated and others absorbed through collision with cloud droplets. The efficiency of washout accounts for the short residence time of tropospheric aerosol (of order a day). Washout also operates on gaseous pollutants that are water soluble.

Aerosol particles support heterogeneous chemistry, which requires the presence of multiple phases. Cloud particles that comprise PSC are involved in the formation of the Antarctic ozone hole (Chap. 18). PSC also introduces an important transport mechanism, which can likewise influence ozone depletion. If they attain sufficient size, PSC particles undergo sedimentation – like cirrus ice particles. Water vapor and nitrogen are then removed from the lower stratosphere.

Chemical reactions similar to those occurring inside PSC take place on natural aerosol. The concentration of natural aerosol increases sharply following major volcanic eruptions (Fig. 9.1). Enhanced aerosol surface area then becomes comparable to that normally observed over the Antarctic in connection with PSC. This mechanism may have figured in diminished ozone that was observed following the eruption of El Chichon (Hoffmon and Solomon, 1989). More generally, however, ozone, even over the Antarctic, undergoes large changes from one year to the next. They are introduced through changes of the circulation (Chap. 18). Influencing transport and temperature, interpreting such changes requires an understanding of air motion.

SUGGESTED REFERENCES

Atmospheric aerosol is discussed in the monograph *Atmospheric Aerosols: Their Formation, Optical Properties, and Effects* (1982), edited by Deepak.

An overview of cloud microphysics is presented in *Atmospheric Science: An Introductory Survey* (2006) by Wallace and Hobbs. Advanced treatments can be found in *Microphysical Processes in Clouds* (1993) by Young and in *Microphysics of Clouds and Precipitation* (1978) by Pruppacher and Klett.

International Cloud Atlas (WMO, 1969) presents a complete classification of cloud.

Cloud Dynamics (1993) by Houze is a detailed treatment of the subject. An advanced treatment, including laboratory simulations, is presented in *Environmental Aerodynamics* (1978) by Scorer.

Classical Electrodynamics (1975) by Jackson develops Rayleigh scattering. A comprehensive treatment of cloud radiative processes is provided in *Radiation and Cloud Processes in the Atmosphere* (1990) by Liou.

Hartmann et al. (1986) contains an overview of how cloud and surface influence the Earth's energy budget.

PROBLEMS

1. In terms of the size distribution $n(a)$, derive an expression for the size spectrum of (a) aerosol surface area $S(a)$ and (b) aerosol mass $M(a)$, if particles have uniform density ρ.

2. A population of cloud droplets has the log-normal size spectrum

$$\frac{dn}{(loga)} = \frac{n_0}{\sigma\sqrt{2\pi}} e^{-\frac{ln^2\left(\frac{a}{a_m}\right)}{2\sigma_2}},$$

   where $n_0 = 300$ cm^{-3}, $a_m = 5$ μm is the median droplet radius, and $\sigma = 4$ μm is the rms deviation of $ln(a)$. (a) Plot the size spectrum $\frac{dn}{d(loga)}$, the surface area spectrum $\frac{dS}{d(loga)}$, and the mass spectrum $\frac{dM}{d(loga)}$. (b) Contrast the distributions of those properties over droplet radius a.

3. Provide an expression for the potential barrier that must be overcome for a water droplet to be activated.

4. (a) Determine the critical radius at 0.2% supersaturation for a droplet of pure water. (b) How large must a droplet containing 10^{-16} g of sodium chloride become to grow spontaneously?

5. A cumulus cloud forms inside a moist thermal. The base of the thermal is at 1000 hPa, where $T_0 = 30°C$ and $r_0 = 10$ g kg^{-1}. (a) If, at all elevations, the thermal is 2°C cooler near its periphery than in its core (other factors being equivalent), determine the height difference between the cloud base at its center versus at its periphery. (b) How would mixing of moisture with the environment modify this result?

6. The flow about a falling sphere of radius a and velocity w is *laminar* if the dimensionless Reynolds number

$$Re = \frac{\rho a w}{\mu}$$

   is less than 5000 (Chap. 13), where $\mu = 1.7 \ 10^{-5}$ kg m^{-1}s^{-1} is the coefficient of viscosity for air. Under these circumstances, the sphere experiences a viscous force

$$D = -6\pi \mu a w$$

   known as *Stokes drag*. (a) Calculate the terminal velocity w_t of a spherical cloud droplet of radius a. (b) For what range of a does the preceding result apply if the flow becomes turbulent for $Re > 5000$? (c) If the flow about them remains laminar, how large must droplets grow before they can no longer be supported by an updraft of 1 m s^{-1}?

7. Consider a population of cloud droplets having size distribution $n(a)$ under the conditions of Prob. 9.6. (a) Derive an expression for the specific drag force exerted on air by the falling droplet population. (b) What is required to maintain an updraft in the presence of those droplets?

8. Apply the results of Prob. 9.6 to estimate the time for volcanic dust particles of density $2.6 \; 10^3$ kg m^{-3} to reach the Earth's surface from 50 hPa if the tropopause is at 200 hPa and the particles are spherical with a mean radius of (a) 1.0 μm, (b) 0.1 μm.

9. When the ground is snow covered, the sky appears brighter. Why?

10. Rayleigh scattering intensifies inversely with wavelength. (a) Why is the sky not violet? (b) Use the Planck spectrum for an equivalent blackbody temperature of 6000 K to show that scattered SW radiation maximizes at wavelengths of blue light.

11. Calculate the atmospheric albedo due to Rayleigh scattering for overhead sun and (a) $\lambda = 0.7 \; \mu$m, (b) $\lambda = 0.5 \; \mu$m, (c) $\lambda = 0.3 \; \mu$m. The refractive index of air at 1000 hPa and 273°K is approximated by

$$(m-1)10^6 = 6.4328 \; 10^1 + \frac{2.94981 \; 10^4}{146 - \lambda^{-2}} + \frac{2.554 \; 10^2}{41 - \lambda^{-2}},$$

with λ in μm and $m-1$ proportional to density.

12. Following the eruption of Krakatoa, sunsets were modified for several years. How small must particles have been?

13. (a) Use the mean latent heat flux to the atmosphere: 90 W m^{-2} (Fig. 1.32), to estimate the global-mean precipitation rate. (b) Use the result in (a) to estimate the fraction of the water vapor column in the tropics that is processed by convection during one day. Reconcile this fraction with the coverage by and time scale of deep convection.

14. Use the mean annual precipitation in Fig. 9.41 and a characteristic tropopause height of 16 km to estimate the latent heating rate inside the ITCZ.

15. Estimate the entrainment height (Sec. 7.4.2) for a thermal of diameter (a) 100 m, (b) 1 km.

16. A stack of lenticular forms from moisture anomalies upstream that are sheared and advected through the mountain wave by the prevailing flow. Describe their evolution.

17. The milky appearance of the sky, often observed from the surface, disappears at an altitude of a kilometer or two. Why?

18. Nimbus cloud generally appears darker than other cloud. Why?

19. Use the scattered intensity (9.19) to obtain the total scattered power (9.20) for Rayleigh scattering.

20. Transmission of diffuse SW radiation through a scattering layer is described in terms of the phase function $P(cos\Theta)$, where Θ is the 3-dimensional scattering angle. Express $cos\Theta$ in terms of the zenith angles θ and θ' and the azimuthal angles ϕ and ϕ' of scattered and incident radiation, respectively.

21. Scattering of microwave radiation by cloud droplets and precipitation enables cloud properties to be measured by radar. (a) Verify that Rayleigh scattering is a valid description of such behavior. (b) The backscattering coefficient (8.19), which measures the reflected power per unit length of scattering medium, follows from

the scattering cross section as

$$\beta_s(\pi) = n\hat{\sigma}_s P(\pi),$$

where n is the particle number density and $P(\pi)$ is the phase function for backward scattering. Derive an expression for $\beta_s(\pi)$ in terms of the radius a of scatterers and the fractional volume $\eta = n \cdot \frac{4}{3}\pi a^3$ occupied by them to show that, for fixed η (eg, fixed liquid water content), the reflectivity is proportional to a^6.

22. Show that the blackbody function $B^*[T]$ is a particular solution of (9.46).

23. For the cloud droplet population in Prob. 9.2, compute (a) the scattering-equivalent mean droplet radius, (b) the liquid water path for a cloud 4 km deep, (c) the cloud's optical depth, (d) the cloud's absorptivity for $\lambda = 0.5$ μm and a solar zenith angle of 30°.

24. Calculate the optical depth posed to SW radiation by a 5-km cumulus congestus (an organization of cumulus), which has an average liquid water content of 0.66 g m^{-3} and a scattering-equivalent mean droplet radius of 12 μm.

25. A homogeneous cloud layer of optical depth τ_c is illuminated by the solar flux F_s at a zenith angle $\theta_s = cos^{-1}\mu_s$. If the surface is gray with absorptivity a, (a) derive expressions for the upwelling and downwelling diffuse fluxes inside the scattering layer, (b) plot the albedo and transmissivity, as functions of a, for overhead sun and $\tau_c = 20$.

26. Consider the radiative-convective equilibrium for the cloudy atmosphere in Fig. 9.36. (a) Explain why the surface temperature in Fig. 9.37 increases linearly with cloud height once the cloud is elevated sufficiently. (b) Nearly identical results follow for other cloud optical depths. Why?

27. Satellite observations reveal the following properties for a region of maritime convection in the tropics: fractional cloud cover of 0.5, average clear-sky and overcast albedos of 0.2 and 0.8, respectively, and average clear-sky and overcast LW fluxes of 350 W m^{-2} and 100 W m^{-2}, respectively. Calculate the average SW, LW, and net cloud forcing for the region.

28. Use the directionality of scattering to assess the change in overall SW transmission accompanying the reduction of direct transmission following the eruption of El Chichon in Fig. 9.5.

29. Satellite measurements of SW radiation enable the albedo of cloud to be determined. From it, cloud optical depth can be inferred. (a) Plot cloud albedo as a function of optical depth for a solar zenith angle of 30°, presuming a homogeneous scattering layer and a black underlying surface. (b) Discuss the sensitivity of cloud albedo to changes in optical depth for $\tau_c > 50$. What implications does this have for inferring optical depth?

30. A satellite measuring backscattered SW radiation and LW radiation emitted at 11 μm observes a tropical cloud with an albedo of 0.9 and a brightness temperature of 220 K. If the solar zenith angle is 30° and a scattering-equivalent radius of 5 μm applies, use the result of Prob. 9.29 to estimate (a) the liquid water path of the cloud, (b) the cloud's average liquid water content if it is 1 km thick.

CHAPTER TEN

Atmospheric motion

Under radiative equilibrium, the First Law reduces to a balance between radiative transfers of energy. This simple energy balance determines the radiative-equilibrium thermal structure. It is valid for a resting atmosphere. However, radiative equilibrium breaks down for an atmosphere in motion. Heat is then also transferred mechanically, by air motion. The radiative-equilibrium thermal structure (Fig. 8.21) differs conspicuously from the observed global-mean temperature (Fig. 1.2). Under radiative equilibrium, temperature is much too warm at the surface. It decreases upward too steeply. The discrepancy from observed temperature points to the importance of mechanical heat transfer. Under radiative-convective equilibrium, mechanical heat transfer is accounted for – by convection. It reconciles the equilibrium thermal structure with that observed, but only vertical structure that is characteristic of the global-mean.

Horizontal thermal structure differs significantly from that observed (Fig. 10.1). Under radiative-convective equilibrium (Fig. 10.1a), temperature decreases poleward at all latitudes, from ~315 K over the equator to colder than 180 K over the winter pole. Observed zonal-mean temperature (Fig. 10.1b) decreases poleward more gradually. It is some 20 K cooler over the equator and 40 K warmer over the winter pole. By comparison, radiative-convective equilibrium produces a meridional temperature gradient that is too steep. It accounts for mechanical transfer of heat, but only in the vertical.

At observed temperatures, individual latitudes are not in radiative equilibrium, not even in radiative-convective equilibrium. According to the distribution of net radiation (Fig. 1.34c), low latitudes experience radiative heating. Conversely, middle and high latitudes experience radiative cooling. To maintain thermal equilibrium locally, those radiative imbalances must be compensated by a poleward transfer of heat: from the tropics, where the radiative energy budget is in surplus, to middle and high

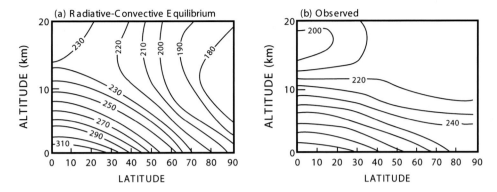

Figure 10.1 Zonal-mean temperature as a function of latitude and height (a) under radiative-convective equilibrium and (b) observed during northern winter. Without horizontal heat transfer, radiative-convective equilibrium establishes a meridional temperature gradient that is much steeper than observed. *Sources:* Liou (1990) and Fleming et al. (1988).

latitudes, where it is in deficit. This horizontal heat transfer is accomplished by the general circulation of the atmosphere and ocean, most of it by the atmospheric circulation. Understanding how observed thermal structure is maintained therefore requires an understanding of atmospheric motion and horizontal heat transfer that it accomplishes.

10.1 DESCRIPTION OF ATMOSPHERIC MOTION

Until now, the development has focused on an individual air parcel, a material element of infinitesimal dimension. Thermodynamic and hydrostatic influences acting on that discrete system must now be complemented by dynamical considerations that control its movement. Together, these properties constitute the individual or *Lagrangian description* of fluid motion. The latter represents atmospheric behavior in terms of the collective properties of material elements that comprise the atmosphere. Because the basic laws of physics apply to a discrete bounded system, they are developed most simply within the Lagrangian framework. Despite this conceptual advantage, in practice, the Lagrangian description of atmospheric behavior can be cumbersome, especially in numerical applications. It requires representing not only the thermal, mechanical, and chemical histories of individual air parcels, but also tracking their positions – not to mention their distortion as they move through the circulation.

Like any fluid system, the atmosphere is a continuum. It is therefore composed of infinitely many such discrete systems. In principle, all of them must be represented in the Lagrangian description of atmospheric behavior. For this reason, it is more convenient to describe atmospheric behavior in terms of field variables, which represent the distributions of properties at a particular instant. The distribution of temperature $T(\mathbf{x}, t)$, where $\mathbf{x} = (x, y, z)$ denotes 3-dimensional position, is a scalar field variable. So is the distribution of mixing ratio $r(\mathbf{x}, t)$. Likewise for the component of velocity in the ith coordinate direction $v_i(\mathbf{x}, t)$. Collecting the scalar components of velocity in the three coordinate directions gives the 3-dimensional velocity field $\mathbf{v}(\mathbf{x}, t) = v_i(\mathbf{x}, t)$; $i = 1, 2, 3$. Velocity is a vector field variable. At each location, it has both magnitude and direction.

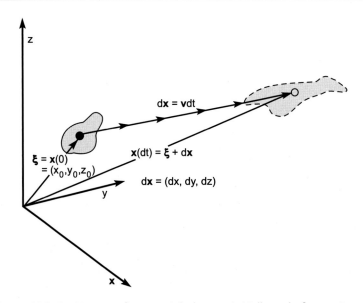

Figure 10.2 Positions $\boldsymbol{x}$ of a material element initially and after an interval dt. The element's initial position $\boldsymbol{x}_0 = (x_0, y_0, z_0)$ defines its *material coordinate* $\boldsymbol{\xi}$. During the interval dt, the element traces out an increment of *parcel trajectory* $d\boldsymbol{x} = \boldsymbol{v} dt$, where $\boldsymbol{v}$ is the element's instantaneous velocity.

Like thermodynamic and chemical variables, the velocity field $\boldsymbol{v}(\boldsymbol{x}, t)$ is a property of the fluid system. It describes the circulation. Because air motion transfers heat and chemical constituents, $\boldsymbol{v}(\boldsymbol{x}, t)$ is coupled to other field properties. Collectively, the distributions of such properties constitute the *field* or *Eulerian description* of fluid motion. The latter is governed by the equations of continuum mechanics.

The Eulerian description simplifies the representation of atmospheric behavior. However, physical laws governing atmospheric behavior, like the first and second laws of thermodynamics and Newton's laws of motion, apply directly to a fixed collection of matter. For this reason, the equations governing atmospheric behavior are developed most intuitively in the Lagrangian framework. In the Eulerian framework, the field property at a specified location involves different material elements at different times. Despite this complication, the Eulerian description is related to the Lagrangian description by a fundamental kinematic constraint: The field property at a given location and time must equal the property possessed by the material element which occupies that position at that time.

10.2 KINEMATICS OF FLUID MOTION

A material element (Fig. 10.2) can be identified by its initial position

$$\boldsymbol{\xi} = \boldsymbol{x}_0$$
$$= \left(x_0, y_0, z_0 \right), \tag{10.1.1}$$

which is referred to as the *material coordinate* of that element. The locus of points

$$\boldsymbol{x}(t) = \left(x(t), y(t), z(t) \right) \tag{10.1.2}$$

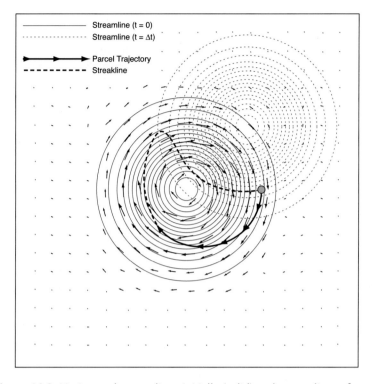

Figure 10.3 Motion and streamlines initially (solid) and streamlines after an interval Δt (dashed) for a vortex translating uniformly. Superposed is (1) the *parcel trajectory* at time Δt (bold solid) of a material element that initially coincides with the stippled circle and (2) the *streakline* at time Δt (bold dashed) that originates from the same position.

traced out by the element defines a *material* or *parcel trajectory*. It is uniquely determined by the material coordinate $\boldsymbol{\xi} = \boldsymbol{x}_0$ and the velocity field $\boldsymbol{v}(\boldsymbol{x}, t)$. With $\boldsymbol{\xi}$ held fixed, the vector $d\boldsymbol{x} = (dx, dy, dz)$ describes the incremental displacement of the material element during the interval dt. The element's velocity is then

$$\boldsymbol{v} = \frac{d\boldsymbol{x}}{dt}, \tag{10.2.1}$$

with components

$$u = \frac{dx}{dt}$$
$$v = \frac{dy}{dt} \tag{10.2.2}$$
$$w = \frac{dz}{dt}.$$

Because it is evaluated for fixed $\boldsymbol{\xi}$, the time derivative in (10.2) corresponds to an individual material element. It therefore represents the Lagrangian time rate of change, the rate of change on which the development has focused until now. By the kinematic constraint that relates Eulerian and Lagrangian descriptions, the velocity in (10.2) must equal the field value $\boldsymbol{v}(\boldsymbol{x}, t)$ at the material element's position $\boldsymbol{x}(t)$ at time t.

The velocity field defines a family of *streamlines* (Fig. 10.3), which, at any instant, are everywhere tangential to $v(x, t)$. If local velocities do not change with time, the motion field is *steady*. Parcel trajectories then coincide with streamlines. This feature of the motion follows from (10.2) because a parcel is then always displaced along the same streamline, which remains fixed. Under unsteady conditions, streamlines move to new positions. Parcel trajectories then no longer coincide with streamlines. Initially tangential to one streamline, the velocity of a parcel will displace it to a different streamline, with which its velocity is again temporarily tangential.

The motion is also characterized by a *streakline*. The latter is the locus of points traced out by a dye that is released into the flow at a particular location. As illustrated in Fig. 10.3, a streakline represents the positions at a given time of contiguous material elements that have passed through the location where the tracer is released. Like parcel trajectories, streaklines coincide with streamlines only if the velocity field is steady.

In addition to translation, a material element can undergo rotation and deformation as it moves through the circulation. These effects are embodied in the velocity gradient

$$\nabla v = \frac{\partial v_i}{\partial x_j}, \qquad i, j = 1, 2, 3 \tag{10.3}$$

which is a 2-dimensional tensor with the subscripts i and j referring to three Cartesian coordinate directions. The local velocity gradient $\nabla v(x)$ determines the relative motion between material coordinates. It therefore describes the distortion experienced by the material element located at x. Two material coordinates that are separated initially by dx_j experience relative motion

$$\begin{aligned}
dv_i &= \frac{\partial v_i}{\partial x_j} dx_j, \\
&= \frac{\partial v_i}{\partial x_1} dx_1 + \frac{\partial v_i}{\partial x_2} dx_2 + \frac{\partial v_i}{\partial x_3} dx_3,
\end{aligned} \tag{10.4}$$

where repeated indices in the same term imply summation and the vector product – the so-called *Einstein notation*. How the material element evolves through relative motion is elucidated by separating the velocity gradient into symmetric and antisymmetric components:

$$\frac{\partial v_i}{\partial x_j} = e_{ij} + \omega_{ij}, \tag{10.5.1}$$

where

$$e_{ij} = \frac{1}{2}\left(\frac{\partial v_i}{\partial x_j} + \frac{\partial v_j}{\partial x_i}\right) \tag{10.5.2}$$

$$\omega_{ij} = \frac{1}{2}\left(\frac{\partial v_i}{\partial x_j} - \frac{\partial v_j}{\partial x_i}\right). \tag{10.5.3}$$

The symmetric component $e(x, t)$ of the velocity gradient defines the *rate of strain* or *deformation tensor* acting on the material element at x. Diagonal components of e describe variations of velocity "in the direction of motion." They characterize the rate of longitudinal deformation, or "stretching," in the three coordinate directions. For example, e_{11} describes the rate of elongation of a material segment

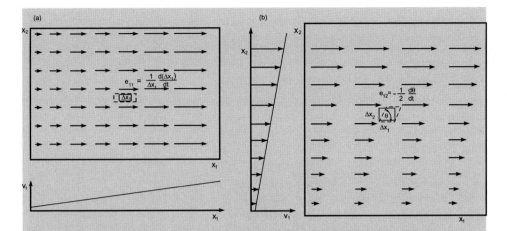

Figure 10.4 (a) Stretching of a material element in a longitudinally varying flow. The element's rate of elongation is reflected in the *normal strain rate* e_{11}. (b) Shearing of a material element in a transverse-varying flow. The rate at which the material angle θ decreases is reflected in the *shear strain rate* e_{12}.

aligned with coordinate 1 (Fig. 10.4a). Collecting all three components of stretching gives the *trace* of e, which describes the divergence of the velocity field

$$\frac{\partial v_i}{\partial x_i} = \nabla \cdot \boldsymbol{v}. \tag{10.6}$$

It represents the rate of increase in volume or *dilatation* of the material element. Under pure shear, the field $\boldsymbol{v}(\boldsymbol{x}, t)$ varies only in directions transverse to the motion. Diagonal elements of $e(\boldsymbol{x}, t)$ then vanish, so $\nabla \cdot \boldsymbol{v} = 0$: The material element may experience distortion, but no change of volume.

Off-diagonal components of $e(\boldsymbol{x}, t)$ describe variations of velocity "orthogonal to the motion." They characterize the rate of transverse deformation, or "shear," experienced by the material element at $\boldsymbol{x}$. As pictured in Fig. 10.4b, $2e_{12}$ represents the rate of decrease of the angle between two material segments aligned initially with coordinates 1 and 2. In the case of rectilinear motion that varies only longitudinally, off-diagonal components of $e(\boldsymbol{x}, t)$ vanish. The material element at $\boldsymbol{x}$ may then experience stretching, but no shear.

The antisymmetric component of $\nabla \boldsymbol{v}(\boldsymbol{x}, t)$ also describes relative motion, but not deformation. Because it contains only three independent components, the tensor $\boldsymbol{\omega}$ defines a vector

$$\hat{\boldsymbol{\omega}} = \begin{bmatrix} \omega_{23} \\ -\omega_{13} \\ \omega_{12} \end{bmatrix}. \tag{10.7}$$

The vector $\hat{\boldsymbol{\omega}}(\boldsymbol{x}, t)$ is related to the curl of the motion field or *vorticity*

$$2\hat{\boldsymbol{\omega}} = \nabla \times \boldsymbol{v}, \tag{10.8}$$

which represents twice the local angular velocity. Thus, $\hat{\boldsymbol{\omega}}(\boldsymbol{x}, t)$ represents the angular velocity of the material element at position $\boldsymbol{x}$. If symmetric components of

$\nabla \boldsymbol{v}(\boldsymbol{x}, t)$ vanish, so does the local deformation tensor $e(\boldsymbol{x}, t)$. The material element's motion then reduces to translation plus rigid body rotation.

10.3 THE MATERIAL DERIVATIVE

To transform to the Eulerian description, the Lagrangian derivative that appears in conservation laws must be expressed in terms of field properties. Consider a field variable $\psi = \psi(x, y, z, t)$. The incremental change of property ψ is described by the total differential

$$
\begin{aligned}
d\psi &= \frac{\partial \psi}{\partial t} dt + \frac{\partial \psi}{\partial x} dx + \frac{\partial \psi}{\partial y} dy + \frac{\partial \psi}{\partial z} dz \\
&= \frac{\partial \psi}{\partial t} dt + \nabla \psi \cdot d\boldsymbol{x},
\end{aligned}
\tag{10.9}
$$

where dx, dy, dz, and dt are suitably defined increments in space and time. Let $d\boldsymbol{x} = (dx, dy, dz)$ and dt in (10.9) denote increments of space and time with the material coordinate $\boldsymbol{\xi}$ fixed. Then $d\boldsymbol{x}$ represents the displacement of the material element $\boldsymbol{\xi}$ during the time increment dt. It describes the increment of parcel trajectory shown in Fig. 10.2. With position and time related in this manner, the total differential describes the incremental change of property ψ observed "in a frame moving with the material element." Differentiating with respect to time (with $\boldsymbol{\xi}$ fixed) gives the time rate of change of ψ "following a material element":

$$
\begin{aligned}
\frac{d\psi}{dt} &= \frac{\partial \psi}{\partial t} + u \frac{\partial \psi}{\partial x} + v \frac{\partial \psi}{\partial y} + w \frac{\partial \psi}{\partial z} \\
&= \frac{\partial \psi}{\partial t} + \boldsymbol{v} \cdot \nabla \psi.
\end{aligned}
\tag{10.10}
$$

It defines the *Lagrangian* or *material derivative* of the field variable $\psi(\boldsymbol{x}, t)$.

According to (10.10), the material derivative includes two contributions. The first, $\frac{\partial \psi}{\partial t}$, is the *Eulerian* or *local derivative*. It represents the rate of change of the material property ψ introduced by temporal changes in the field variable at the position $\boldsymbol{x}$ where the material element is located. The second contribution, $\boldsymbol{v} \cdot \nabla \psi$, is the *advective contribution*. It represents the change of the material property ψ due to motion of the material element to positions of different field values. Even if the field is steady, namely if $\psi(\boldsymbol{x}, t) = \psi(\boldsymbol{x})$ so that local values do not change with time, the property of a material element does change if that element moves across contours of $\psi(\boldsymbol{x})$. The rate that ψ changes for the material element is then given by the component of its velocity in the direction of the gradient of ψ times that gradient, namely by $\boldsymbol{v} \cdot \nabla \psi$.

10.4 REYNOLDS' TRANSPORT THEOREM

Owing to the kinematic relationship between Lagrangian and Eulerian descriptions, the material derivative emerges naturally in the laws governing field properties. Changes of ψ described by (10.10) apply to an infinitesimal material element. Consider now changes for a finite material volume $V(t)$ (i.e., one containing a fixed collection of matter). Because $V(t)$ has finite dimension, we must account for

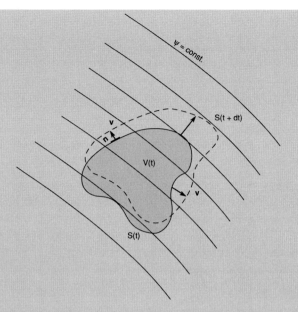

Figure 10.5 Finite material volume $V(t)$, containing a fixed collection of matter, that is displaced across a field property ψ.

variations of velocity v and property ψ across the material volume. Consider the integral property

$$\int_{V(t)} \psi(x, y, z, t)\,dV'$$

over the finite system. The time rate of change of this property follows by differentiating the volume integral. However, the time derivative $\frac{d}{dt}$ cannot be commuted inside the integral because the limits of integration (viz. the position and form of the material volume) are themselves variable.

The time rate of change of the integral material property has two contributions (Fig. 10.5), analogous to those treated above for an infinitesimal material element:

(1) Values of $\psi(x, t)$ within the instantaneous material volume change temporally due to unsteadiness of the field.

(2) The material volume moves to regions of different field values. Relative to a frame moving with $V(t)$, such motion introduces a flux of property ψ across the material volume's surface $S(t)$.

The first contribution is just the collective time rate of change of $\psi(x, t)$ within the material volume:

$$\int_{V(t)} \frac{\partial \psi}{\partial t}\,dV'.$$

The second is the net transfer of ψ across the boundary of $V(t)$. If $S(t)$ has the local outward normal n and local velocity v, the local flux of ψ relative to that section of material surface is $-\psi v$. Then the local flux of ψ into the material volume is given by $-\psi \boldsymbol{v} \cdot -\boldsymbol{n} = \psi \boldsymbol{v} \cdot \boldsymbol{n}$. Integrating over $S(t)$ gives the net rate that ψ is transferred into the material volume $V(t)$ due to its motion across contours of ψ:

$$\int_{S(t)} \psi \boldsymbol{v} \cdot \boldsymbol{n} dS'.$$

Collecting the two contributions and applying Gauss' theorem obtains

$$\frac{d}{dt} \int_{V(t)} \psi dV' = \int_{V(t)} \left\{ \frac{\partial \psi}{\partial t} + \nabla \cdot (\boldsymbol{v}\psi) \right\} dV'$$

$$= \int_{V(t)} \left\{ \frac{d\psi}{dt} + \psi \nabla \cdot \boldsymbol{v} \right\} dV', \qquad (10.11)$$

for the time rate of change of the integral material property. Known as *Reynolds' transport theorem*, (10.11) relates the time rate of change of some property of a finite body of fluid to the corresponding field variable and to the velocity field $\boldsymbol{v}(\boldsymbol{x}, t)$. As such, it constitutes a transformation between the Lagrangian and Eulerian descriptions of fluid motion.[1] To develop the equations that govern field variables in the Eulerian description of atmospheric motion, we apply Reynolds' theorem to properties of an arbitrary material volume.

10.5 CONSERVATION OF MASS

Let $\psi = \rho(\boldsymbol{x}, t)$. An arbitrary material volume $V(t)$ then has mass

$$\int_{V(t)} \rho(\boldsymbol{x}, t) dV'.$$

Because that system is comprised of a fixed collection of matter, the time rate of change of its mass must vanish

$$\frac{d}{dt} \int_{V(t)} \rho(\boldsymbol{x}, t) dV' = 0. \qquad (10.12)$$

Applying Reynolds' transport theorem transforms (10.12) into

$$\int_{V(t)} \left\{ \frac{d\rho}{dt} + \rho \nabla \cdot \boldsymbol{v} \right\} dV' = 0. \qquad (10.13)$$

This relation must hold for "arbitrary" material volume $V(t)$. It follows that the integrand must vanish identically.

Conservation of mass for individual bodies of air thus requires

$$\frac{d\rho}{dt} + \rho \nabla \cdot \boldsymbol{v} = 0 \qquad (10.14.1)$$

or, in flux form,

$$\frac{\partial \rho}{\partial t} + \nabla \cdot (\rho \boldsymbol{v}) = 0. \qquad (10.14.2)$$

Known as the *continuity equation*, (10.14) provides a constraint on the field variables $\rho(\boldsymbol{x}, t)$ and $\boldsymbol{v}(\boldsymbol{x}, t)$. It must be satisfied throughout the domain and continuously in time.

---

[1] Mathematically, the transport theorem is a generalization to a moving body of fluid of Leibniz' rule for differentiating an integral with variable limits. It enables the time derivative of the integral property to be expressed in terms of an integral of time and space derivatives.

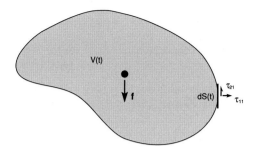

Figure 10.6 Finite material volume $V(t)$ that experiences a specific body force f internally and a stress tensor τ on its surface.

If material volume does not change for individual bodies of air, the motion is said to be *incompressible*. The specific volume $\alpha = 1/\rho$ for an arbitrary material element is then constant. Under these conditions, (10.14.1) reduces to

$$\nabla \cdot \boldsymbol{v} = 0,$$

which provides a simpler constraint on the motion.

Consider a specific field property f (viz. one referenced to a unit mass). Then $\psi = \rho f$ represents the absolute concentration of that property. Reynolds' transport theorem implies

$$\frac{d}{dt} \int_{V(t)} \rho f \, dV' = \int_{V(t)} \left\{ \frac{d}{dt}(\rho f) + \rho f (\nabla \cdot \boldsymbol{v}) \right\} dV'$$
$$= \int_{V(t)} \left\{ \rho \frac{df}{dt} + \left[\frac{d\rho}{dt} + \rho \nabla \cdot \boldsymbol{v} \right] \right\} dV'. \qquad (10.15)$$

By the continuity equation, the term in square brackets vanishes. Thus (10.15) reduces to the identity

$$\frac{d}{dt} \int_{V(t)} \rho f \, dV' = \int_{V(t)} \rho \frac{df}{dt} dV'. \qquad (10.16)$$

For an absolute concentration, the time rate of change of an integral material property assumes this simpler form.

10.6 THE MOMENTUM BUDGET

10.6.1 Cauchy's Equations of Motion

In an inertial reference frame, Newton's second law of motion applied to the material volume $V(t)$ may be expressed

$$\frac{d}{dt} \int_{V(t)} \rho \boldsymbol{v} \, dV' = \int_{V(t)} \rho f \, dV' + \int_{S(t)} \boldsymbol{\tau} \cdot \boldsymbol{n} \, dS', \qquad (10.17)$$

where $\rho \boldsymbol{v}$ is the absolute concentration of momentum, f is the specific body force acting internal to the material volume, and τ is the *stress tensor* acting on its surface (Fig. 10.6). The stress tensor τ is a counterpart of the deformation tensor e. It represents the vector force per unit area exerted on surfaces normal to the three coordinate directions. Then $\boldsymbol{\tau} \cdot \boldsymbol{n}$ is the vector force per unit area exerted on the section of material surface with unit normal $\boldsymbol{n}$.

For common fluids such as air and water, the stress tensor is symmetric

$$\tau_{ji} = \tau_{ij}. \tag{10.18}$$

Like $e(\mathbf{x}, t)$, the local stress tensor contains two basic contributions. Diagonal components of $\boldsymbol{\tau}(\mathbf{x}, t)$ define *normal stresses*. They act orthogonal to surfaces with normals in the three coordinate directions. The component $\tau_{11}(\mathbf{x}, t)$ describes the force per unit area acting in the direction of coordinate 1 on an element of surface with normal in the same direction (Fig. 10.6). Off-diagonal components of $\boldsymbol{\tau}(\mathbf{x}, t)$ define *shear stresses* that act tangential to those surfaces. The component τ_{21} describes the force per unit area acting in the direction of coordinate 2 on an element of surface with normal in the direction of coordinate 1.

Each of the stresses in $\boldsymbol{\tau}$ also represents a flux of momentum. The normal stress τ_{ii} (summation suspended) represents the flux of i momentum in the i direction. The shear stress τ_{ji} represents the flux of j momentum in the i direction. Formally, these fluxes are accomplished by molecular diffusion of momentum, which gives fluid viscosity. However, in large-scale atmospheric motion, similar fluxes of momentum are accomplished through mixing by small-scale turbulence. Much faster than molecular diffusion, turbulent fluxes are treated in analogous fashion. Both render the behavior of an individual material element diabatic because they transfer heat and momentum across its boundary (see Fig. 2.1).

In the atmosphere, the body force and stress tensor are either prescribed or are determined by the velocity field. Incorporating Reynolds' transport theorem for an absolute concentration (10.16) and Gauss' theorem transforms (10.17) into

$$\int_{V(t)} \rho \frac{d\mathbf{v}}{dt} dV' = \int_{V(t)} \{\rho \mathbf{f} + \nabla \cdot \boldsymbol{\tau}\} dV'$$

or

$$\int_{V(t)} \left\{ \rho \frac{d\mathbf{v}}{dt} - \rho \mathbf{f} - \nabla \cdot \boldsymbol{\tau} \right\} dV' = 0. \tag{10.19}$$

As before, (10.19) must hold for arbitrary material volume. The integrand must then vanish identically. Thus Newton's second law for individual bodies of air requires

$$\rho \frac{d\mathbf{v}}{dt} = \rho \mathbf{f} + \nabla \cdot \boldsymbol{\tau}. \tag{10.20}$$

Known as *Cauchy's equations*, (10.20) provide constraints on the field properties $\rho(\mathbf{x}, t)$ and $\mathbf{v}(\mathbf{x}, t)$. They must hold pointwise throughout the domain and continuously in time.[2]

The body force relevant to the atmosphere is gravity

$$\mathbf{f} = \mathbf{g}, \tag{10.21}$$

which is prescribed. On the other hand, internal stresses are determined autonomously by the motion. A *Newtonian fluid* like air has a stress tensor that is linearly proportional to the local rate of strain $e(\mathbf{x}, t)$. In the absence of motion, $\boldsymbol{\tau}$ reduces to the normal stresses exerted by pressure. For our purposes, it suffices

[2] In the absence of friction, these are called *Euler's equations*.

to define the stress tensor as

$$\tau = \begin{bmatrix} -p & & & \\ & & 2\mu e_{ij} & \\ & -p & & \\ & 2\mu e_{ji} & & \\ & & & -p \end{bmatrix}, \tag{10.22}$$

where μ is the *coefficient of viscosity*.[3] Equation (10.22) is the form assumed by the stress tensor for an incompressible fluid. Shear stresses internal to the fluid then arise from shear in the motion field $\frac{\partial v_i}{\partial x_j}$. When expressed in terms of specific momentum, the governing equations involve the *kinematic viscosity* $v = \frac{\mu}{\rho}$. Also called the molecular *diffusivity*, v has dimensions $\frac{length}{time^2}$. Turbulent momentum transfer is frequently modeled as diffusion. An eddy diffusivity is then used in place of v (Chap. 13).

With the stress tensor τ defined by (10.22), Cauchy's equations of motion reduce to

$$\frac{d\boldsymbol{v}}{dt} = \boldsymbol{g} - \frac{1}{\rho}\nabla p - \boldsymbol{D}, \tag{10.23.1}$$

where

$$\begin{aligned} \boldsymbol{D} &= -\frac{1}{\rho}\nabla \cdot \boldsymbol{\tau} \\ &= -\frac{1}{\rho}\nabla \cdot (\mu\nabla\boldsymbol{v}) = -\frac{1}{\rho}\frac{\partial}{\partial x_j}\mu\frac{\partial v_i}{\partial x_j} \end{aligned} \tag{10.23.2}$$

denotes the specific *drag force* exerted on the material element at location $\boldsymbol{x}$. Known as the *momentum equations*, (10.23) are a simplified form of the *Navier Stokes equations*. The latter embody the full representation of τ in terms of e and constitute the formal description of fluid motion. Conceptually, (10.23) asserts that the momentum of a material element changes according to the resultant force exerted on it by gravity, pressure gradient, and frictional drag.

10.6.2 Momentum equations in a rotating reference frame

Because they follow from Newton's laws of motion, the momentum equations apply in an inertial reference frame. However, the reference frame of the Earth (in which atmospheric motion is observed) is rotating and therefore noninertial. To apply in that reference frame, the momentum equations must be modified. Scalar quantities like $\rho(\boldsymbol{x}, t)$ appear the same in inertial and noninertial reference frames. So do their Lagrangian derivatives. However, vector quantities differ between those reference frames. Vector variables that describe a material element's motion (e.g., $\boldsymbol{x}$, $\boldsymbol{v} = \frac{d\boldsymbol{x}}{dt}$, and $\boldsymbol{a} = \frac{d\boldsymbol{v}}{dt}$) must therefore be corrected to account for acceleration of the Earth's reference frame.

Consider a reference frame rotating with angular velocity $\boldsymbol{\Omega}$ (Fig. 10.7). A vector $\boldsymbol{A}$ which is constant in that frame rotates when viewed in an inertial reference frame.

---

[3] Strictly, normal stresses also include a contribution from viscosity, but that effect is small enough to be neglected for most applications (e.g., Aris, 1962).

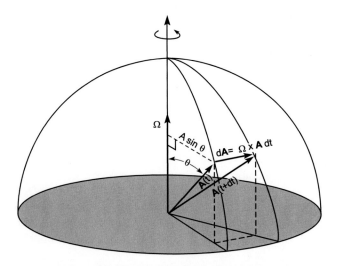

Figure 10.7 A vector A that is fixed in a rotating reference frame changes in an inertial reference frame during an interval dt by the increment $|dA| = A\sin\theta \cdot \Omega dt$ in a direction orthogonal to the plane of A and Ω, or by the vector increment $dA = \Omega \times A dt$.

During an interval dt, A changes by a vector increment dA which is perpendicular to the plane of A and Ω and which has magnitude

$$|dA| = A\sin\theta \cdot \Omega dt,$$

where θ is the angle between A and Ω. Hence, from an inertial reference frame, the vector A changes at the rate

$$\left|\frac{dA}{dt}\right| = A\Omega\sin\theta$$

and in a direction perpendicular to the plane of A and Ω. It follows that the time rate of change of A apparent in an inertial reference frame is described by

$$\left(\frac{dA}{dt}\right)_i = \Omega \times A. \tag{10.24}$$

More generally, a vector A that has the time rate of change $\frac{dA}{dt}$ in a rotating reference frame has the time rate of change

$$\left(\frac{dA}{dt}\right)_i = \frac{dA}{dt} + \Omega \times A \tag{10.25}$$

in an inertial reference frame.

Consider the position x of a material element. By (10.25), the element's velocity $v = \frac{dx}{dt}$ apparent in an inertial reference frame is

$$v_i = v + \Omega \times x. \tag{10.26}$$

Similarly, the acceleration that is apparent in the inertial frame is given by

$$\left(\frac{dv_i}{dt}\right)_i = \frac{dv_i}{dt} + \Omega \times v_i.$$

Incorporating the velocity apparent in the inertial frame (10.26) obtains

$$\left(\frac{d\boldsymbol{v}_i}{dt}\right)_i = \left(\frac{d\boldsymbol{v}}{dt} + \boldsymbol{\Omega} \times \boldsymbol{v}\right) + \boldsymbol{\Omega} \times (\boldsymbol{v} + \boldsymbol{\Omega} \times \boldsymbol{x})$$

$$= \frac{d\boldsymbol{v}}{dt} + 2\boldsymbol{\Omega} \times \boldsymbol{v} + \boldsymbol{\Omega} \times (\boldsymbol{\Omega} \times \boldsymbol{x})$$

(10.27)

for the acceleration apparent in the inertial frame.

According to (10.27), rotation of the reference frame introduces two corrections to the material acceleration. The first, $2\boldsymbol{\Omega} \times \boldsymbol{v}$, is the *Coriolis acceleration*. Perpendicular to an air parcel's velocity and to the *planetary vorticity* $2\boldsymbol{\Omega}$, it represents an apparent force in the rotating frame of the Earth. The Coriolis acceleration is important for motions that have time scales comparable to that of the Earth's rotation (Chap. 12). The second correction, $\boldsymbol{\Omega} \times (\boldsymbol{\Omega} \times \boldsymbol{x})$, is the *centrifugal acceleration* of an air parcel. Likewise an apparent force, it was treated in Chap. 6. When geopotential coordinates are used, this apparent force is absorbed into effective gravity.

Incorporating the material acceleration apparent in an inertial reference frame (10.27) transforms the momentum equations (10.23) into a form valid in the rotating frame of the Earth:

$$\frac{d\boldsymbol{v}}{dt} + 2\boldsymbol{\Omega} \times \boldsymbol{v} = -\frac{1}{\rho}\nabla p - g\boldsymbol{k} - \boldsymbol{D},$$

(10.28)

where g is understood to denote effective gravity and $\boldsymbol{k}$ the upward normal in the direction of increasing geopotential (i.e., $g\boldsymbol{k} = \nabla\Phi$). The correction $2\boldsymbol{\Omega} \times \boldsymbol{v}$ represents the "Coriolis acceleration" when it appears on the left-hand side of the momentum balance. If moved to the right-hand side, it represents the "Coriolis force" $-2\boldsymbol{\Omega} \times \boldsymbol{v}$, an apparent force that acts on an air parcel in the rotating frame of the Earth. Because it acts orthogonal to the parcel's displacement, the Coriolis force performs no work.

10.7 THE FIRST LAW OF THERMODYNAMICS

Applied to the material volume $V(t)$, the First Law may be expressed

$$\frac{d}{dt}\int_{V(t)} \rho c_v T dV' = -\int_{S(t)} \boldsymbol{q} \cdot \boldsymbol{n} dS' - \int_{V(t)} \rho p \frac{d\alpha}{dt} dV' + \int_{V(t)} \rho \dot{q} dV',$$

(10.29)

where $c_v T$ represents the specific internal energy. Forcing it on the right-hand side are three terms: (1) $\boldsymbol{q}$ is the local heat flux, so $-\boldsymbol{q} \cdot \boldsymbol{n}$ represents the heat flux "into" the material volume. (2) $\alpha = \frac{1}{\rho}$ is the specific volume, so $p\frac{d\alpha}{dt}$ represents the specific work rate. (3) $\dot{q}$ denotes the specific rate of internal heating (e.g., associated with the latent heat release and frictional dissipation of motion).

The local rate of expansion work is related to the dilatation of the material element occupying that position. Because

$$\frac{1}{\rho}\frac{d\rho}{dt} = \alpha\frac{d\left(\frac{1}{\alpha}\right)}{dt}$$

$$= -\frac{1}{\alpha}\frac{d\alpha}{dt},$$

(10.30)

the continuity equation (10.14) may be expressed

$$\frac{1}{\alpha}\frac{d\alpha}{dt} = \nabla \cdot \boldsymbol{v}. \tag{10.31}$$

Incorporating (10.31), along with Reynolds' transport theorem (10.16) and Gauss' theorem, transforms (10.29) into

$$\int_{V(t)} \left\{ \rho c_v \frac{dT}{dt} + \nabla \cdot \boldsymbol{q} + p\nabla \cdot \boldsymbol{v} - \rho\dot{q} \right\} dV' = 0.$$

Again, because $V(t)$ is arbitrary, the integrand must vanish identically.

Therefore, the First Law applied to individual bodies of air requires

$$\rho c_v \frac{dT}{dt} = -\nabla \cdot \boldsymbol{q} - p\nabla \cdot \boldsymbol{v} + \rho\dot{q}, \tag{10.32}$$

which provides a constraint on the field variables involved. Of the three thermodynamic properties represented, only two are independent because air behaves as a pure substance (Sec. 2.1.4); they must also satisfy the gas law. Similarly, the heat flux $\boldsymbol{q}$ and the internal heating rate $\dot{q}$ are determined autonomously by properties already represented. Therefore, they introduce no additional unknowns.

It is convenient to separate the heat flux into radiative and diffusive components:

$$\begin{aligned} \boldsymbol{q} &= \boldsymbol{q}_R + \boldsymbol{q}_T \\ &= \boldsymbol{F} - k\nabla T, \end{aligned} \tag{10.33}$$

where $\boldsymbol{F}$ is the net radiative flux (Sec. 8.2) and k denotes the thermal conductivity in Fourier's law of heat conduction. The first law then becomes

$$\rho c_v \frac{dT}{dt} + p\nabla \cdot \boldsymbol{v} = -\nabla \cdot \boldsymbol{F} + \nabla \cdot (k\nabla T) + \rho\dot{q}. \tag{10.34}$$

Known as the *thermodynamic equation*, (10.34) expresses the rate that a material element's internal energy changes in terms of the rate that work is performed on it and the net rate it absorbs heat through convergence of radiative and diffusive energy fluxes.

The thermodynamic equation can be expressed more compactly in terms of potential temperature. For an individual air parcel, the fundamental relation

$$du = Tds - pd\alpha \tag{10.35.1}$$

relates the change of entropy to other material properties. Incorporating the identity between entropy and potential temperature (3.25) transforms this into

$$c_p T d\ln\theta = du + pd\alpha. \tag{10.35.2}$$

Differentiating with respect to time, with the material coordinate $\boldsymbol{\xi}$ fixed, obtains

$$c_p T \frac{d\ln\theta}{dt} = \frac{du}{dt} + p\frac{d\alpha}{dt}, \tag{10.36}$$

where $\frac{d}{dt}$ represents the Lagrangian derivative (10.10). Multiplying by ρ and introducing the continuity equation (10.4) transforms this into

$$\frac{\rho c_p T}{\theta}\frac{d\theta}{dt} = \rho c_v \frac{dT}{dt} + p\nabla \cdot \boldsymbol{v}. \tag{10.37}$$

Then incorporating (10.37) into (10.34) absorbs the compression work into the time rate of change of θ, yielding the thermodynamic equation

$$\rho \frac{c_p T}{\theta} \frac{d\theta}{dt} = -\nabla \cdot \boldsymbol{F} + \nabla \cdot (k\nabla T) + \rho \dot{q}. \tag{10.38}$$

Collectively, the continuity, momentum, and thermodynamic equations represent five partial differential equations in five independent field variables: three components of motion and two independent thermodynamic properties. Advective contributions to the material derivative, like advection of momentum $\boldsymbol{v} \cdot \nabla \boldsymbol{v}$ and of temperature $\boldsymbol{v} \cdot \nabla T$, make the governing equations nonlinear – quadratically. Their solution requires initial conditions that specify the preliminary state of the atmosphere and boundary conditions that specify its properties along physical borders. Referred to as *the equations of motion*, these equations govern the behavior of a compressible, stratified atmosphere in a rotating reference frame. As such, they constitute the starting point for dynamical investigations, as well as for investigations of chemistry and radiation in the presence of motion.

SUGGESTED REFERENCES

Vectors, Tensors, and the Basic Equations of Fluid Mechanics (1962) by Aris develops the kinematics of fluid motion, along with the governing equations from a Lagrangian perspective.

An Introduction to Dynamic Meteorology (2004) by Holton contains alternate derivations of the continuity and thermodynamic equations.

PROBLEMS

1. Consider the 2-dimensional motion

$$u(x, y, t) = \frac{\partial \psi}{\partial y}$$

$$v(x, y, t) = -\frac{\partial \psi}{\partial x},$$

where

$$\psi(x, y, t) = 4e^{-\left\{ \left(x - x_0(t)\right)^2 + \left(y - y_0(t)\right)^2 \right\}}$$

defines a family of streamlines and

$$x_0(t) = y_0(t) = t.$$

Plot the parcel trajectory and streakline from $t = 0$ to 4 beginning at (a) $(x, y) = (0, 1)$, (b) $(x, y) = (-1, 0)$, (c) $(x, y) = (0, 0)$.

2. For the motion in Prob. 10.1, plot at $t = 1, 2, 3, 4$ the material volume which, at $t = 0$, is defined by the radial coordinates:

$$1 - \Delta r < r < 1 + \Delta r$$

$$-\Delta\phi < \phi < \Delta\phi$$

for (a) $\Delta r = 0.5$ and $\Delta\phi = \frac{\pi}{4}$, (b) $\Delta r = 0.1$ and $\Delta\phi = 0.1$. (c) Contrast the deformations experienced by the material volumes in (a) and (b) and use them to infer the limiting behavior for Δr and $\Delta\phi \to 0$.

3. Simplify the equations of motion for the special case of (a) incompressible motion, wherein the volume of a material element is conserved, (b) adiabatic motion.

4. Describe the circumstances under which a property ψ is conserved, yet varies spatially in a steady flow.

5. A free surface is one that moves to alleviate any stress and maintain $\tau = 0$. If the rate of strain tensor is dominated by the vertical shears $\frac{\partial u}{\partial z}$ and $\frac{\partial v}{\partial z}$, describe how the horizontal velocity varies adjacent to a free surface.

6. A geostationary satellite is positioned over latitude $\phi_0 = 30°$ and longitude $\lambda_0 = 0°$. From it, a projectile is fired northward at a speed v_0. By ignoring the satellite's altitude, calculate the longitude where the projectile crosses 45° latitude if (a) $v_0 = 1000$ m s^{-1}, (b) $v_0 = 100$ m s^{-1}, (c) $v_0 = 10$ m s^{-1}. (d) For each case, evaluate the dimensionless time for the traversal, scaled by the rotation period of the Earth.

7. Show that the antisymmetric component of the velocity gradient tensor may be expressed in terms of the angular velocity vector $\hat{\omega}$ in (10.7).

8. Show that the vorticity equals twice the local angular velocity (10.8).

9. Use the vector identity (D.14) in Appendix D to show that the material acceleration may be expressed

$$\frac{d\boldsymbol{v}}{dt} = \frac{\partial \boldsymbol{v}}{\partial t} + \nabla \frac{|\boldsymbol{v}|^2}{2} + \boldsymbol{\zeta} \times \boldsymbol{v},$$

where $\boldsymbol{\zeta} = \nabla \times \boldsymbol{v}$.

10. Demonstrate that the Coriolis force does not enter the budget of specific kinetic energy $\frac{|\boldsymbol{v}|^2}{2}$.

11. Show that, for 2-dimensional nondivergent motion in an inertial reference frame, the vorticity $\zeta = \boldsymbol{k} \cdot \nabla \times \boldsymbol{v}$ is a conserved property.

12. Consider a parcel with local speed v, the natural coordinate s measured along the parcel's trajectory, and a unit vector $\boldsymbol{s}$ that is everywhere tangential to the trajectory. (a) Express the material derivative in terms of v, s, and $\boldsymbol{s}$. (b) Show that the material acceleration described by

$$\frac{d\boldsymbol{v}}{dt} = \frac{dv}{dt}\boldsymbol{s} + v\frac{d\boldsymbol{s}}{dt}.$$

(c) Interpret the two accelerations appearing on the right.

CHAPTER ELEVEN

Atmospheric equations of motion

In vector form, the equations of motion are valid in any coordinate system. However, those equations do not lend themselves to application and standard methods of solution. To be useful, the governing equations must be expressed in scalar form. They then depend on the coordinate system.

We develop the scalar equations of motion within the general framework of curvilinear coordinates. In addition to accounting for geometric distortions inherent to spherical coordinates, that framework allows a straightforward development of the equations in thermodynamic coordinates, which afford a number of simplifications.

11.1 CURVILINEAR COORDINATES

Consider the *Cartesian coordinates*

$$x = (x_1, x_2, x_3).$$ (11.1.1)

They are measured from planar coordinate surfaces: $x_i = $ const. If those planes are perpendicular, (x_1, x_2, x_3) are referred to as *rectangular Cartesian coordinates.* More generally, consider the *curvilinear coordinates*

$$\hat{x} = (\hat{x}_1, \hat{x}_2, \hat{x}_3).$$ (11.1.2)

They are measured from coordinate surfaces that need not be planar nor mutually orthogonal (Fig. 11.1). Insofar as x and $\hat{x}$ are both valid coordinate systems, there exists a transformation

$$\hat{x} = \hat{x}(x)$$ (11.2)

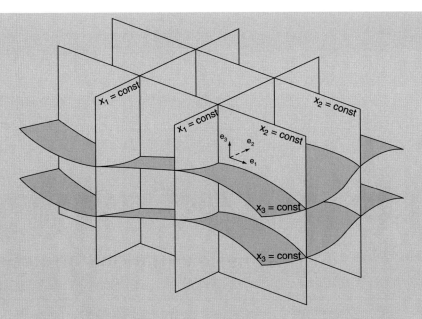

Figure 11.1 Coordinate surfaces $x_i = $ const for curvilinear coordinates. Coordinate vectors e_i point in the directions of increasing x_i.

from one representation to the other. It constitutes a mapping between all coordinates $\boldsymbol{x}$ in the original system and coordinates $\hat{\boldsymbol{x}}$ in the curvilinear coordinate system.

The mapping (11.2) is unique and therefore invertible, provided that the *Jacobian* of the transformation

$$J(\boldsymbol{x}, \hat{\boldsymbol{x}}) = \left| \frac{\partial \hat{x}_i}{\partial x_j} \right| \tag{11.3}$$

is nonzero for all $\boldsymbol{x}$. The Jacobian applies irrespective of whether the original coordinate system is Cartesian. It has the reciprocal property

$$J(\boldsymbol{x}, \hat{\boldsymbol{x}}) \cdot J(\hat{\boldsymbol{x}}, \boldsymbol{x}) = 1. \tag{11.4}$$

Unlike Cartesian coordinates, the curvilinear coordinates $\hat{x}_i$ need not represent length. For example, the spherical coordinates, longitude λ and latitude ϕ, represent angular displacements. The physical length ds_i corresponding to an increment of the curvilinear coordinate $\hat{x}_i$ is accounted for in the *metric scale factor h_i*:

$$ds_i = h_i d\hat{x}_i \tag{11.5.1}$$

$$h_i^{-1} = |\nabla \hat{x}_i|, \tag{11.5.2}$$

where the summation convention is suspended and ∇ refers to differentiation with respect to the original coordinates x_i. Similarly, the physical volume corresponding to the incremental element $dV_{\hat{x}} = d\hat{x}_1 d\hat{x}_2 d\hat{x}_3$ in the curvilinear coordinate system is described by

$$dV_x = J(\hat{x}, x)dV_{\hat{x}}$$
$$= \frac{1}{J(x, \hat{x})} dV_{\hat{x}}.$$

(11.6)

The curvilinear coordinate system is *orthogonal* if coordinate surfaces are mutually perpendicular. Coordinate vectors $\hat{e}_i$, which point in the directions of increasing $\hat{x}_i$ (Fig. 11.1), are then likewise mutually perpendicular

$$\hat{e}_i \cdot \hat{e}_j = \delta_{ij}.$$

Under those circumstances, the Jacobian reduces to

$$J(\hat{x}, x) = h_1 h_2 h_3$$

(11.7)

(Prob. 11.10). Similarly, the physical volume corresponding to the element $d\hat{x}_1 d\hat{x}_2 d\hat{x}_3$ in the curvilinear system is just the product of the corresponding physical lengths (11.5).

The Jacobian accounts for distortions of physical length in all three curvilinear coordinates. A special case arise when only the third coordinate is transformed

$$\hat{x}(x) = (x_1, x_2, \hat{x}_3).$$

(11.8)

Such is the case when surfaces of constant height are replaced by isobaric surfaces. The Jacobian is then given by

$$J(x, \hat{x}) = \begin{vmatrix} 1 & 0 & 0 \\ 0 & 1 & 0 \\ \frac{\partial \hat{x}_3}{\partial x_1} & \frac{\partial \hat{x}_3}{\partial x_2} & \frac{\partial \hat{x}_3}{\partial x_3} \end{vmatrix}$$

$$= \frac{\partial \hat{x}_3}{\partial x_3}.$$

(11.9)

For this special class of transformations, (11.4) reduces to the simple reciprocal property

$$\frac{\partial \hat{x}_3}{\partial x_3} = \left(\frac{\partial x_3}{\partial \hat{x}_3}\right)^{-1},$$

(11.10)

which is not true in general.

Vector operations in arbitrary curvilinear coordinates can now be expressed in terms of the corresponding metric scale factors:

$$\nabla \psi = \frac{1}{h_1} \frac{\partial \psi}{\partial \hat{x}_1} \hat{e}_1 + \frac{1}{h_2} \frac{\partial \psi}{\partial \hat{x}_2} \hat{e}_2 = \frac{1}{h_3} \frac{\partial \psi}{\partial \hat{x}_3} \hat{e}_3$$

(11.11.1)

$$\nabla \cdot A = \frac{1}{h_1 h_2 h_3} \left[\frac{\partial}{\partial \hat{x}_1}(h_2 h_3 A_1) + \frac{\partial}{\partial \hat{x}_2}(h_1 h_3 A_2) + \frac{\partial}{\partial \hat{x}_3}(h_1 h_2 A_3) \right]$$

(11.11.2)

$$\nabla \times A = \begin{vmatrix} h_1 \hat{e}_1 & h_2 \hat{e}_2 & h_3 \hat{e}_3 \\ \frac{\partial}{\partial \hat{x}_1} & \frac{\partial}{\partial \hat{x}_2} & \frac{\partial}{\partial \hat{x}_3} \\ h_1 A_1 & h_2 A_2 & h_3 A_3 \end{vmatrix}$$

(11.11.3)

$$\nabla^2 \psi = \frac{1}{h_1 h_2 h_3} \left[\frac{\partial}{\partial \hat{x}_1}\left(\frac{h_2 h_3}{h_1} \frac{\partial \psi}{\partial \hat{x}_1}\right) + \frac{\partial}{\partial \hat{x}_2}\left(\frac{h_1 h_3}{h_2} \frac{\partial \psi}{\partial \hat{x}_2}\right) \frac{\partial}{\partial \hat{x}_3}\left(\frac{h_1 h_2}{h_3} \frac{\partial \psi}{\partial \hat{x}_3}\right) \right].$$

(11.11.4)

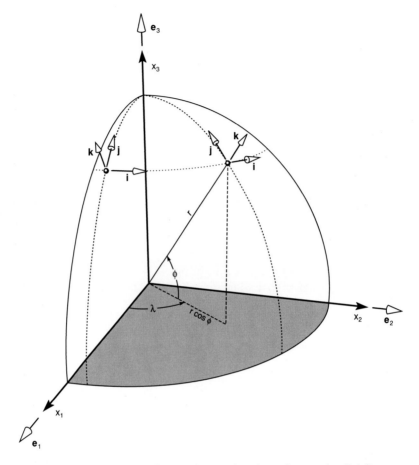

Figure 11.2 Spherical coordinates: longitude λ, latitude ϕ, and radial distance r. Coordinate vectors $e_\lambda = i$, $e_\phi = j$, and $e_r = k$ change with position (e.g., relative to fixed coordinate vectors e_1, e_2, and e_3 of rectangular Cartesian coordinates).

11.2 SPHERICAL COORDINATES

Consider the rectangular Cartesian coordinates $x = (x_1, x_2, x_3)$ having origin at the center of the Earth and the spherical coordinates

$$\hat{x} = (\lambda, \phi, r),$$

both fixed with respect to the Earth (Fig. 11.2). Spherical coordinate surfaces, $\lambda = $ const, $\phi = $ const, and $r = $ const, are then: (1) vertical planes intersecting the Cartesian x_3 axis, (2) conical shells with apex at the origin $x = 0$, and (3) concentric spherical shells, respectively. The corresponding unit vectors

$$\begin{aligned} e_\lambda &= i \\ e_\phi &= j \\ e_r &= k \end{aligned} \tag{11.12}$$

are everywhere perpendicular to those coordinate surfaces. Because coordinate surfaces are mutually perpendicular, so are the unit vectors. The spherical coordinates (λ, ϕ, r) constitute an orthogonal coordinate system.

The rectangular Cartesian coordinates can be expressed in terms of the spherical coordinates as

$$x_1 = r \ cos\phi \ cos\lambda$$
$$x_2 = r \ cos\phi \ sin\lambda \tag{11.13.1}$$
$$x_3 = r \ sin\phi.$$

The latter may be inverted for the spherical coordinates:

$$\lambda = tan^{-1}\left(\frac{x_2}{x_1}\right)$$
$$\phi = tan^{-1}\left[\frac{x_3}{x_1^2 + x_2^2}\right] \tag{11.13.2}$$
$$r = \sqrt{x_1^2 + x_2^2 + x_3^2}.$$

Metric scale factors for the spherical coordinates then follow as

$$h_1 = r cos\phi$$
$$h_2 = r \tag{11.14}$$
$$h_3 = 1$$

(Prob. 11.2). Because $\hat{x}$ constitutes an orthogonal coordinate system, the Jacobian of the transformation follows from (11.7) as

$$J(\hat{x}, x) = r^2 cos\phi, \tag{11.15}$$

which is nonzero except at the poles. At $\phi = \pm 90°$, coordinate surfaces $\phi = $ const converge onto the x_3 axis. The incremental volume encased by two such surfaces vanishes there, so the transformation is singular. Vector operations follow from (11.11) as

$$\nabla\psi = \frac{1}{r cos\phi}\frac{\partial\psi}{\partial\lambda}i + \frac{1}{r}\frac{\partial\psi}{\partial\phi}j + \frac{\partial\psi}{\partial r}k \tag{11.16.1}$$

$$\nabla \cdot A = \frac{1}{r cos\phi}\frac{\partial A_\lambda}{\partial\phi} + \frac{1}{r cos\phi}\frac{\partial}{\partial\phi}\left(cos\phi A_\phi\right) + \frac{1}{r^2}\frac{\partial}{\partial r}\left(r^2 A_r\right) \tag{11.16.2}$$

$$\nabla \times A = \frac{1}{r^2 \ cos\phi}\begin{vmatrix} r cos\phi i & rj & k \\ \frac{\partial}{\partial\lambda} & \frac{\partial}{\partial\phi} & \frac{\partial}{\partial r} \\ r cos\phi A_\lambda & r A_\phi & A_r \end{vmatrix} \tag{11.16.3}$$

$$\nabla^2\psi = \frac{1}{r^2 \ cos^2\phi}\frac{\partial^2\psi}{\partial\lambda^2} + \frac{1}{r^2 \ cos\phi}\left[cos\phi\frac{\partial\psi}{\partial\phi}\right] + \frac{1}{r^2}\frac{\partial}{\partial r}\left(r^2\frac{\partial\psi}{\partial r}\right). \tag{11.16.4}$$

From (11.14), physical displacements in the directions of increasing longitude, latitude, and radial distance are described by

$$dx = r cos\phi \, d\lambda$$
$$dy = r d\phi \qquad\qquad (11.17.1)$$
$$dz = dr,$$

in which vertical distance is measured by height

$$z = r - a, \qquad\qquad (11.17.2)$$

where a denotes the mean radius of the Earth.[1] Physical velocities in the spherical coordinate system are then expressed by

$$u = \frac{dx}{dt} = r cos\phi \frac{d\lambda}{dt}$$
$$v = \frac{dy}{dt} = r \frac{d\phi}{dt} \qquad\qquad (11.18)$$
$$w = \frac{dz}{dt} = \frac{dr}{dt}.$$

The vector equations of motion may now be cast in terms of spherical coordinates. With the above expressions, derivatives of scalar quantities transform directly. However, Lagrangian derivatives of vector quantities are complicated by the dependence on position of the coordinate vectors i, j, and k. Each rotates in physical space under a displacement of longitude or latitude. For example, an air parcel moving along a latitude circle at a constant speed u has velocity $v = ui$. The latter appears constant in the spherical coordinate representation. However, in physical space, it actually rotates (Fig. 11.2). The parcel therefore experiences an acceleration, which must be accounted for in the equations of motion.

Consider the velocity

$$v = ui + vj + wk.$$

The spherical coordinate vectors i, j, and k are functions of position $\hat{x}$. Therefore, the material acceleration is actually

$$\frac{dv}{dt} = \frac{du}{dt}i + \frac{dv}{dt}j + \frac{dw}{dt}k + u\frac{di}{dt} + v\frac{dj}{dt} + w\frac{dk}{dt}$$
$$= \left(\frac{dv}{dt}\right)_C + u\frac{di}{dt} + v\frac{dj}{dt} + w\frac{dk}{dt}, \qquad\qquad (11.19.1)$$

where the subscript refers to the basic form of the material derivative in Cartesian geometry

$$\left(\frac{d}{dt}\right)_C = \frac{\partial}{\partial t} + v \cdot \nabla. \qquad\qquad (11.19.2)$$

To evaluate corrections on the right hand side of (11.19), the spherical coordinate vectors are expressed in terms of the fixed rectangular Cartesian coordinate

---

[1] This representation is only approximate, because it ignores departures from sphericity of height surfaces (Sec. 6.2).

vectors e_1, e_2, e_3

$$i = -\sin\lambda\, e_1 + \cos\lambda\, e_2$$
$$j = -\sin\phi\cos\lambda\, e_1 - \sin\phi\sin\lambda\, e_2 + \cos\phi\, e_3 \qquad (11.20)$$
$$k = \cos\phi\cos\lambda\, e_1 + \cos\phi\sin\lambda\, e_2 + \sin\phi\, e_3,$$

demonstration of which is left as an exercise. From these and reciprocal expressions for e_1, e_2, and e_3 in terms of i, j, and k, Lagrangian derivatives of the spherical coordinate vectors follow as

$$\frac{di}{dt} = u\left(\frac{\tan\phi}{r}j - \frac{1}{r}k\right)$$
$$\frac{dj}{dt} = -u\frac{\tan\phi}{r}i - \frac{v}{r}k \qquad (11.21)$$
$$\frac{dk}{dt} = \frac{u}{r}i + \frac{v}{r}j$$

(see Prob. 11.11). Then material accelerations in the spherical coordinate directions become

$$\frac{du}{dt} = \left(\frac{du}{dt}\right)_C - \frac{uv\tan\phi}{r} + \frac{uw}{r} \qquad (11.22.1)$$

$$\frac{dv}{dt} = \left(\frac{dv}{dt}\right)_C + \frac{u^2\tan\phi}{r} + \frac{vw}{r} \qquad (11.22.2)$$

$$\frac{dw}{dt} = \left(\frac{dw}{dt}\right)_C - \left(\frac{u^2 + v^2}{r}\right). \qquad (11.22.3)$$

Corrections appearing on the right-hand sides of (11.22) are referred to as *metric terms*. They describe accelerations that result from curvature of the coordinate system.

Because the atmosphere occupies a thin shell, it is customary to simplify the radial dependence by neglecting small fractional changes of r in (11.16)–(11.22). The *shallow atmosphere approximation* makes use of the fact that $z \ll a$. It takes $r = a$ and ignores the geometric divergence associated with vertical displacements. The vector operations (11.16) then reduce to

$$\nabla\psi = \frac{1}{a\cos\phi}\frac{\partial\psi}{\partial\lambda}i + \frac{1}{a}\frac{\partial\psi}{\partial\phi}j = \frac{\partial\psi}{\partial z}k \qquad (11.23.1)$$

$$\nabla \cdot A = \frac{1}{a\cos\phi}\frac{\partial A_\lambda}{\partial\phi} + \frac{1}{a\cos\phi}\frac{\partial}{\partial\phi}\left(\cos\phi A_\phi\right) + \frac{\partial A_z}{\partial z} \qquad (11.23.2)$$

$$\nabla \times A = \frac{1}{a^2\cos\phi}\begin{vmatrix} a\cos\phi\, i & a\, j & k \\ \frac{\partial}{\partial\lambda} & \frac{\partial}{\partial\phi} & \frac{\partial}{\partial z} \\ a\cos\phi A_\lambda & aA_\phi & A_z \end{vmatrix} \qquad (11.23.3)$$

$$\nabla^2\psi = \frac{1}{a^2\cos^2\phi}\frac{\partial^2\psi}{\partial\lambda^2} + \frac{1}{a^2\cos\phi}\left[\cos\phi\frac{\partial\psi}{\partial\phi}\right] + \frac{\partial^2\psi}{\partial z^2}, \qquad (11.23.4)$$

in which height is formally adopted as the vertical coordinate.

With this approximation and material accelerations (11.22), the equations of motion can be cast into component form. The planetary vorticity (Sec. 10.6.2) can be expressed

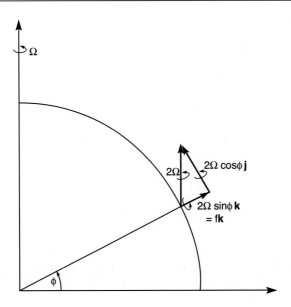

Figure 11.3 Planetary vorticity $2\boldsymbol{\Omega}$ decomposed into horizontal and vertical components.

in terms of horizontal and vertical components of the Earth's rotation

$$2\boldsymbol{\Omega} = 2\Omega(cos\phi\,\boldsymbol{j} + sin\phi\,\boldsymbol{k}) \tag{11.24}$$

(Fig. 11.3). The scalar equations of motion in spherical coordinates then become:

$$\frac{du}{dt} - 2\Omega(v\,sin\phi - w\,cos\phi) = -\frac{1}{\rho\,a\,cos\phi}\frac{\partial p}{\partial\lambda} + uv\frac{tan\phi}{a} - \frac{uw}{a} - D_\lambda \tag{11.25.1}$$

$$\frac{dv}{dt} + 2\Omega u\,sin\phi = -\frac{1}{\rho a}\frac{\partial p}{\partial\phi} - \frac{u^2\,tan\phi}{a} - \frac{uw}{a} - D_\phi \tag{11.25.2}$$

$$\frac{dw}{dt} - 2\Omega u\,cos\phi = -\frac{1}{\rho}\frac{\partial p}{\partial z} - g + \frac{u^2 + v^2}{a} - D_z \tag{11.25.3}$$

$$\frac{d\rho}{dt} + \rho\nabla\cdot\boldsymbol{v} = 0 \tag{11.25.4}$$

$$\rho c_v\frac{dT}{dt} + p\nabla\cdot\boldsymbol{v} = \dot{q}_{net}, \tag{11.25.5}$$

where

$$\frac{d}{dt} = \frac{\partial}{\partial t} + \frac{u}{a\,cos\phi}\frac{\partial}{\partial\lambda} + \frac{v}{a}\frac{\partial}{\partial\phi} + w\frac{\partial}{\partial z} \tag{11.25.6}$$

and

$$\dot{q}_{net} = -\nabla\cdot\boldsymbol{F} + \nabla\cdot(k\nabla T) + \rho\dot{q} \tag{11.25.7}$$

denotes the net heating rate from all diabatic sources.

11.2.1 The traditional approximation

Although it simplifies the mathematics, using height as the vertical coordinate with r replaced by the constant value a has an important drawback: the resulting equations do not possess an angular momentum principle, like the one satisfied by the equations in full spherical geometry (Prob. 11.12). On conceptual grounds alone, this is an important deficiency because much of atmospheric dynamics concerns how angular momentum is concentrated into strong jets that characterize the general circulation. The failure of (11.25) to properly represent the angular momentum of individual air parcels follows from an inconsistency in the treatment of horizontal and vertical displacements. Geometric variations in the radial direction are neglected. However, Coriolis and metric terms proportional to $cos\phi$, which accompany radial displacements, are retained.

Terms proportional to $2\Omega cos\phi$ in (11.25) are associated with the horizontal component of planetary vorticity (11.24) and vertical motion. They are much smaller than other terms, so they are often neglected on the basis of scaling arguments (Sec. 11.4). Known as the *traditional approximation* (Eckart, 1960), the neglect of terms proportional to $2\Omega cos\phi$ is formally valid in the limit of strong stratification, wherein

$$\frac{N^2}{\Omega^2} \to \infty. \tag{11.26}$$

Air is then constrained to move quasi-horizontally. This makes the vertical component of planetary vorticity $2\Omega sin\phi$ dominant in the Coriolis acceleration $2\boldsymbol{\Omega} \times \boldsymbol{v}$. For a lapse rate of $\Gamma = 6.5$ K km^{-1}, (7.10) gives N^2 of order 10^{-4} s^{-2} and $\Omega^2 \cong 5 \ 10^{-9}$ s^{-2}. Hence $\frac{N^2}{\Omega^2} \cong 2 \ 10^4$ is in good agreement with (11.26). On the other hand, weak or unstable stratification is controlled by convection. The latter operates on time scales short enough to render both components of the Coriolis acceleration unimportant.

An alternate derivation of the equations of motion (Phillips, 1966) provides a rationale for neglecting the aforementioned terms. The component equations assume a simpler form if derived from the momentum equations in vector-invariant form. In an inertial reference frame, the vector equations of motion can be expressed

$$\left(\frac{\partial \boldsymbol{v}_i}{\partial t}\right)_i + \nabla\left(\frac{\boldsymbol{v}_i \cdot \boldsymbol{v}_i}{2}\right) + (\nabla \times \boldsymbol{v}_i) \times \boldsymbol{v}_i = -\frac{1}{\rho}\nabla p - g\boldsymbol{k} - \boldsymbol{D}, \tag{11.27.1}$$

where the subscript refers to the velocity and local time rate of change apparent in the inertial frame (Prob. 11.13). Evaluating metric scale factors at $r = a$ before derivatives are applied recovers the vector operations (11.23). Expressing the velocity and time derivative in terms of those apparent in the rotating frame of the Earth gives

$$\boldsymbol{v}_i = \boldsymbol{v} + \Omega a cos\phi \boldsymbol{i} \tag{11.27.2}$$

$$\left(\frac{\partial}{\partial t}\right)_i = \frac{\partial}{\partial t} - \Omega\frac{\partial}{\partial \lambda}. \tag{11.27.3}$$

Then incorporating (11.23) yields the simplified momentum equations

$$\frac{du}{dt} - \left(f + \frac{utan\phi}{a}\right)v = -\frac{1}{\rho a cos\phi}\frac{\partial p}{\partial \lambda} - D_\lambda \tag{11.28.1}$$

$$\frac{dv}{dt} + \left(f + \frac{utan\phi}{a}\right)u = -\frac{1}{\rho a}\frac{\partial p}{\partial \phi} - D_\phi \tag{11.28.2}$$

$$\frac{dw}{dt} = -\frac{1}{\rho}\frac{\partial p}{\partial z} - g - D_z, \tag{11.28.3}$$

where the *Coriolis parameter*

$$f = 2\Omega sin\phi \tag{11.28.4}$$

represents the vertical component of planetary vorticity (Fig. 11.3).

In (11.28), the horizontal momentum equations are free of metric and Coriolis terms proportional to w. The vertical momentum equation is free of all such terms. The preceding system possesses the conservation principle

$$\frac{d}{dt}[(u + \Omega acos\phi)acos\phi] = -acos\phi \left(\frac{1}{\rho acos\phi} \frac{\partial p}{\partial \lambda} + D_\lambda \right). \tag{11.29}$$

According to (11.29), the angular momentum of an air parcel changes through the torque about the axis of rotation exerted on it by longitudinal forces (Prob. 11.14). Consistent with the shallow atmosphere approximation, angular momentum is treated as though the air parcel remains at $r = a$.

11.3 SPECIAL FORMS OF MOTION

Certain conditions simplify the governing equations. For *incompressible motion*, the specific volume of an individual air parcel is conserved, so

$$\frac{d\rho}{dt} = 0.$$

Then the continuity equation (11.25.4) reduces to

$$\nabla \cdot \boldsymbol{v} = 0. \tag{11.30}$$

Because ρ is conserved, an air parcel that coincides initially with a particular isochoric surface: $\rho = $ const, remains on that surface – despite movement of that surface. Thus isochoric surfaces are material surfaces (viz. they are comprised of a fixed collection of matter). If the motion is steady, ρ surfaces are also stream surfaces, to which the motion is tangential. These conditions hold automatically for an incompressible fluid like water. They also hold for a compressible fluid like air if the velocity is steady and everywhere orthogonal to the gradient of density. Because large-scale motion is quasi-horizontal, the latter condition is approximately satisfied by the atmospheric circulation.

For *adiabatic motion*, individual air parcels experience no heat transfer with their surroundings (Sec. 3.6.1). The thermodynamic equation (10.38) then reduces to

$$\frac{d\theta}{dt} = 0, \tag{11.31}$$

which asserts that the potential temperature of an air parcel is conserved. Thus an air parcel that coincides initially with an isentropic surface, $\theta = $ const, remains on that surface (cf. Fig. 2.9). Isentropic surfaces are material surfaces.

The above are particular examples of a conserved property, which behaves as a material tracer. More generally, a property r that is conserved for individual air parcels obeys the continuity equation

$$\frac{dr}{dt} = 0. \tag{11.32}$$

Mixing ratios of long-lived chemical species approximately satisfy (11.32). The rates of production and destruction of such species formally appear on the right-hand side of (11.32). However, those processes are much slower than advective changes in the Lagrangian derivative, which are represented on the left-hand side. Water vapor, which is conserved away from cloud and the Earth's surface, behaves in this manner (cf. Fig. 1.18). So does ozone, which, in the lower stratosphere, has a photochemical lifetime of several weeks. Hence, to leading approximation, particular values of mixing ratio track the movement of individual bodies of air. To the same degree of approximation, surfaces $r = $ const are material surfaces.

11.4 PREVAILING BALANCES

The preceding equations govern motion in a compressible, stratified, and rotating atmosphere. Consequently, they describe a wide range of phenomena. While describing planetary-scale circulations that involve times scales of days and longer, the equations of motion also describe small-scale acoustic waves that have time scales of only fractions of a second. This generality needlessly complicates the description of phenomena that comprise the general circulation. To elucidate essential balances that control large-scale motion, it is useful to examine the relative sizes of various terms.

11.4.1 Motion-related stratification

Stratification is represented in the distributions of pressure and density. It involves two components: (1) a basic component associated with static conditions and (2) a small departure from it that is related exclusively to motion:

$$p_{tot} = p_0(z) + p(\mathbf{x}, t)$$
$$\rho_{tot} = \rho_0(z) + \rho(\mathbf{x}, t),$$

(11.33)

where $\mathbf{x} = (x, y, z)$ follows from (11.17). The static components p_0 and ρ_0 are functions of height alone. They are symbolized by global-mean pressure and density. Those components overshadow vertical variations associated with motion. It is therefore useful to eliminate them from the governing equations. What remains is then discriminated to variations that are related directly to motion.

Incorporating (11.33) into the horizontal momentum equation transforms the pressure gradient force into

$$-\frac{1}{(\rho_0 + \rho)} \nabla_h (p_0 + p) \cong -\frac{1}{\rho_0} \left(1 - \frac{\rho}{\rho_0}\right) \nabla_h p$$
$$\cong -\frac{1}{\rho_0} \nabla_h p,$$

(11.34)

where ∇_h denotes the horizontal gradient. In (11.34), $\frac{\rho}{\rho_0} \ll 1$ and $\frac{p}{p_0} \ll 1$ have been used, in combination with the binomial expansion, to ignore higher order terms. Because $p_0(z)$ depends on height alone, it drops out of (11.34). The horizontal momentum equations are therefore left in their original form (11.28).

In the vertical momentum equation, the basic components of stratification introduce vertical gradients that cannot be ignored. The right-hand side of (11.28.3) contains the net buoyancy force acting on an air parcel (7.2). Incorporating (11.33) transforms the buoyancy force into

$$f_b = -\frac{1}{(\rho_0 + \rho)}\frac{\partial}{\partial z}(p_0 + p) - g \cong -\frac{1}{\rho_0}\left(1 - \frac{\rho}{\rho_0}\right)\left(\frac{\partial p_0}{\partial z} + \frac{\partial p}{\partial z}\right) - g$$

$$\cong -\frac{1}{\rho_0}\left(\frac{\partial p}{\partial z} + \rho g\right), \tag{11.35}$$

wherein the basic stratification automatically satisfies hydrostatic equilibrium. The vertical momentum equation then becomes

$$\frac{dw}{dt} = -\frac{1}{\rho_0}\frac{\partial p}{\partial z} - \frac{\rho}{\rho_0}g - D_z. \tag{11.36}$$

11.4.2 Scale analysis

With dependent variables discriminated to motion, terms in the momentum equations may now be evaluated for relative importance. Away from the Earth's surface and regions of organized convection, the frictional drag D is small enough to be ignored. Large-scale atmospheric motion is then characterized by the scales

$$U = 10 \text{ m s}^{-1} \qquad L = 10^3 \text{ km} \qquad P = 10 \text{ mb} = 10^3 \text{ Pa}$$
$$W = 10^{-2} \text{ m s}^{-1} \quad H = 10 \text{ km} \qquad f_0 = 10^{-4} \text{ s}^{-1},$$

where U and W refer to horizontal and vertical motion, respectively, L and H are horizontal and vertical length scales that characterize the motion field, and $f_0 = 2\Omega sin\phi_0$ is the planetary vorticity at a representative latitude ϕ_0. The pressure scale P characterizes the departure from static conditions, namely the pressure variation associated directly with motion. If the Lagrangian derivative is dominated by advective changes (e.g., by $v \cdot \nabla$), then $\frac{L}{U}$ represents a time scale for advection that may be used to characterize $\frac{d}{dt}$.

Scaling velocities, lengths, and the Lagrangian derivative by the above (e.g., $u \to Uu$, $x \to Lx$, $\frac{d}{dt} \to \frac{U}{L}\frac{d}{dt}$, with u, x, and $\frac{d}{dt}$ then nondimensional) transforms the horizontal momentum equations (11.28.1) and (11.28.2) into the dimensionless forms

$$Ro\frac{du}{dt} - sin\phi v - Ro\left(\frac{L}{a}\right)tan\phi uv = -\frac{P}{f_0\rho_{00}UL} \cdot \frac{1}{\rho_0}\frac{\partial p}{\partial x} \tag{11.37.1}$$

$$Ro\frac{dv}{dt} + sin\phi u + Ro\left(\frac{L}{a}\right)tan\phi u^2 = -\frac{P}{f_0\rho_{00}UL} \cdot \frac{1}{\rho_0}\frac{\partial p}{\partial y}, \tag{11.37.2}$$

where $\rho_{00} = \rho_0(0)$ and $Ro = U/f_0L$. In (11.37), all variables are nondimensional, of order unity, and are related to their dimensional counterparts through multiplication by the preceding scale factors. The *Rossby number*

$$Ro = \frac{U}{f_0L} \tag{11.38}$$

is a dimensionless parameter that represents the ratio of the rotational time scale f_0^{-1} to the advective time scale $\frac{L}{U}$.

With the preceding scales, dimensionless factors in (11.37) are characterized by the following orders of magnitude:

$$Ro = 10^{-1} \qquad Ro\left(\frac{L}{a}\right) = 10^{-2} \qquad \frac{P}{f_0\rho_{00}UL} = 1.$$

The horizontal momentum equations are seen to be dominated by a balance between the Coriolis acceleration associated with the vertical component of planetary vorticity and the horizontal pressure-gradient force. Both are of order unity. The material acceleration, which is of order Ro, is comparatively small – because advection is slow compared with rotation. It follows that the horizontal motion of air is strongly influenced by rotation. Accelerations associated with metric terms and with vertical advection of momentum in $\frac{d}{dt}$ are even smaller. Those proportional to w that were eliminated in the derivation of (11.28) are much smaller.

Similar treatment transforms the vertical momentum equation (11.36) into the dimensionless form

$$Ro\left(\frac{W}{U}\right)\frac{dw}{dt} = -\left(\frac{L}{H}\right)\frac{P}{f_0\rho_{00}UL}\cdot\frac{1}{\rho_0}\frac{\partial p}{\partial z} - \left(\frac{P}{p_{00}}\right)\frac{g}{f_0U}\cdot\frac{\rho}{\rho_0},\qquad(11.39)$$

where $p_{00} = p_0(0)$. The preceding scales imply the orders of magnitude

$$Ro\left(\frac{W}{U}\right) = 10^{-3} \qquad \left(\frac{L}{H}\right)\frac{P}{f_0\rho_{00}UL} = 10^2 \qquad \left(\frac{P}{p_{00}}\right)\frac{g}{f_0U} = 10^2.$$

For large-scale motion, the vertical momentum equation is dominated by a balance between the vertical pressure-gradient and gravitational forces. Hence, like basic fields, the motion-related component of stratification is in hydrostatic equilibrium. Other vertical forces are much smaller. Vertical forces that balance in hydrostatic equilibrium are also two orders of magnitude stronger than those appearing in the horizontal momentum equations. The disparate magnitudes of those forces reflect the comparative influences of gravity and rotation (11.26).

The preceding analysis allows the equations of motion to be simplified. Inclusive of the basic stratification, the dimensional equations governing large-scale atmospheric motion in spherical coordinates then become

$$\frac{du}{dt} - \left(f + \frac{utan\phi}{a}\right)v = -\frac{1}{\rho a cos\phi}\frac{\partial p}{\partial \lambda} - D_\lambda \qquad(11.40.1)$$

$$\frac{dv}{dt} + \left(f + \frac{utan\phi}{a}\right)u = -\frac{1}{\rho a}\frac{\partial p}{\partial \phi} - D_\phi \qquad(11.40.2)$$

$$\frac{1}{\rho}\frac{\partial p}{\partial z} = -g, \qquad(11.40.3)$$

$$\frac{d\rho}{dt} + \rho\nabla\cdot\boldsymbol{v} = 0 \qquad(11.40.4)$$

$$\rho c_v \frac{dT}{dt} + p\nabla\cdot\boldsymbol{v} = \dot{q}_{net}. \qquad(11.40.5)$$

Known as the *primitive equations*, (11.24) represent the starting point for descriptions of large-scale atmospheric motion. Metric terms proportional to $tan\phi$ and horizontal drag have been retained in (11.40). However, although often included in numerical integrations, they are small enough to be ignored for many applications. In chemical considerations, the primitive equations are augmented by continuity equations of the form

$$\frac{dr_i}{dt} = \dot{P}_i - \dot{D}_i \qquad(11.40.6)$$

for the mixing ratio of the ith species, where $\dot{P}_i$ and $\dot{D}_i$ denote the local rates of production and destruction of that species.

The governing equations must be closed with boundary conditions. Spherical geometry is periodic, so cyclic continuity suffices for horizontal boundary conditions. Vertical boundary conditions constrain the vertical velocity. At the ground, air motion must be tangential to the Earth's surface, which has elevation $z_s(\lambda, \phi, t)$.[2] This is equivalent to requiring an air parcel that is initially in contact with the Earth's surface to track along that elevation:

$$
\begin{aligned}
w &= \frac{dz}{dt} \\
&= \frac{dz_s}{dt},
\end{aligned}
$$
(11.41)

where $\frac{d}{dt}$ is given by (11.25.6).[3] In the presence of viscosity, air motion must also satisfy the *no-slip condition*, which requires all components of the velocity at the surface to vanish. Upper boundary conditions are more difficult to apply because the atmosphere is unbounded. Nevertheless, similar constraints are often imposed at a finite height.

11.5 THERMODYNAMIC COORDINATES

11.5.1 Isobaric coordinates

Because they involve hydrostatic equilibrium, the primitive equations simplify when formulated with pressure as the vertical coordinate. Hence we consider a transformation from the standard spherical coordinates $x = (x, y, z)$ given by (11.17) to the modified coordinates $x_p = (x, y, p)$. Isobaric surfaces $p =$ const then replace constant height surfaces $z =$ const as coordinate surfaces. Horizontal derivatives must therefore be evaluated along isobaric surfaces. Using p as a vertical coordinate is possible because hydrostatic equilibrium (11.40.3) ensures that pressure decreases upward monotonically. p is therefore a single-valued function of z.

In casting the equations into isobaric coordinates, the variables p and z are interchanged. Pressure then becomes the independent variable and z becomes the dependent variable. For a specified pressure, $z(p)$ represents the height of an isobaric surface. It is contoured in Fig. 1.9. Isobaric surfaces are not perpendicular to other coordinate surfaces. Therefore, this representation is not an orthogonal coordinate system.

Consider a scalar quantity

$$
\psi(x, y, z, t) = \hat{\psi}\left(x, y, p(x, y, z, t), t\right).
$$
(11.42)

Hereafter, the caret will be omitted with the pressure dependence understood. By the chain rule, the vertical derivative can be expressed

$$
\left(\frac{\partial \psi}{\partial z}\right)_{xyt} = \left(\frac{\partial \psi}{\partial p}\right)_{xyt} \left(\frac{\partial p}{\partial z}\right)_{xyt},
$$
(11.43)

---

[2] The lower boundary condition may be applied on a surface internal to the atmosphere, in which case the elevation is time-dependent.

[3] Even though an air parcel at the ground must remain in contact with that boundary, neighboring parcels can undergo large vertical displacements through deformation of finite bodies of air (e.g., in convection; cf. Fig. 5.4).

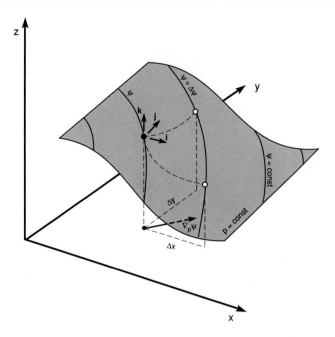

Figure 11.4 Variation of property ψ along an isobaric surface. Horizontal derivatives $(\frac{\partial \psi}{\partial x})_{ypt} = \underset{\Delta x \to 0}{lim} \frac{\Delta \psi}{\Delta x}$ and $(\frac{\partial \psi}{\partial y})_{xpt} = \underset{\Delta y \to 0}{lim} \frac{\Delta \psi}{\Delta y}$ are evaluated with changes of ψ along the isobaric surface $p = $ const. (For clarity, Δx and Δy have been chosen to place the incremented values of ψ on the same contour.) Those derivatives define the horizontal gradient evaluated along the isobaric surface $\nabla_p \psi = (\frac{\partial \psi}{\partial x})_{ypt} \, \boldsymbol{i} + (\frac{\partial \psi}{\partial y})_{xpt} \boldsymbol{j}$, which lies in the horizontal plane but reflects how ψ varies along the isobaric surface.

where subscripts denote variables that are held fixed. Likewise, differentiation with respect to x becomes

$$\left(\frac{\partial \psi}{\partial x}\right)_{yzt} = \left(\frac{\partial \psi}{\partial x}\right)_{ypt} + \left(\frac{\partial \psi}{\partial p}\right)_{xyt} \left(\frac{\partial p}{\partial x}\right)_{yzt}$$

and similarly for differentiation with respect to y. Thus, the horizontal gradient evaluated on surfaces of constant height, ∇_z, translates into

$$\nabla_z \psi = \nabla_p \psi + \left(\frac{\partial \psi}{\partial p}\right)_{xyt} \nabla_z p, \tag{11.44.1}$$

where

$$\nabla_p = \left(\frac{\partial}{\partial x}\right)_{ypt} \boldsymbol{i} + \left(\frac{\partial}{\partial y}\right)_{xpt} \boldsymbol{j} \tag{11.44.2}$$

denotes the horizontal gradient evaluated on an isobaric surface (Fig. 11.4). The local time derivative becomes

$$\left(\frac{\partial \psi}{\partial t}\right)_{xyz} = \left(\frac{\partial \psi}{\partial t}\right)_{xyp} + \left(\frac{\partial \psi}{\partial p}\right)_{xyt} \left(\frac{\partial p}{\partial t}\right)_{xyz}. \tag{11.45}$$

Incorporating the above expressions transforms the Lagrangian derivative into

$$\frac{d\psi}{dt} = \left(\frac{\partial\psi}{\partial t}\right)_{xyp} + \boldsymbol{v}_h \cdot \nabla_p\psi + \left(\frac{\partial\psi}{\partial p}\right)_{xyt}\left[\left(\frac{\partial p}{\partial t}\right)_{xyz} + \boldsymbol{v}_h \cdot \nabla_z p + w\left(\frac{\partial p}{\partial z}\right)_{xyt}\right]$$

$$= \left(\frac{\partial\psi}{\partial t}\right)_{xyp} + \boldsymbol{v}_h \cdot \nabla_p\psi + \frac{dp}{dt}\left(\frac{\partial\psi}{\partial p}\right)_{xyt},$$

(11.46)

where

$$\boldsymbol{v}_h = u\boldsymbol{i} + v\boldsymbol{j} \qquad (11.47)$$

denotes horizontal velocity. From (11.44.2), the second term in (11.46) represents horizontal advective changes evaluated on an isobaric surface.[4] The third term has the form of vertical advection if

$$\omega = \frac{dp}{dt} \qquad (11.48)$$

is identified as the vertical velocity in pressure coordinates. With p as the vertical coordinate, ω denotes the Lagrangian derivative of vertical position. It is positive in the direction of increasing p (i.e., *downward*). In isolation, ω does not have dimensions of velocity. Nonetheless, the product $\omega\frac{\partial\psi}{\partial p}$ has dimensions of vertical advection of the property ψ. The material derivative may then be expressed

$$\frac{d\psi}{dt} = \frac{\partial\psi}{\partial t} + (\boldsymbol{v} \cdot \nabla\psi)_p$$

$$= \frac{\partial\psi}{\partial t} + \boldsymbol{v}_h \cdot \nabla_p\psi + \omega\frac{\partial\psi}{\partial p},$$

(11.49)

in which p is held fixed unless otherwise noted and $(\boldsymbol{v} \cdot \nabla\psi)_p$ denotes 3-dimensional advection of ψ in pressure coordinates.

Letting $\psi = z(x, y, p, t)$ in (11.43) and (11.44), which for fixed p represents the height of an isobaric surface, obtains

$$\frac{\partial z}{\partial z} = 1$$

$$\nabla_z z = \left(\frac{\partial z}{\partial x}\right)_{yzt}\boldsymbol{i} + \left(\frac{\partial z}{\partial y}\right)_{xzt}\boldsymbol{j} = 0.$$

Then (11.43) reduces to the simple reciprocal property

$$\left(\frac{\partial z}{\partial p}\right)^{-1} = \frac{\partial p}{\partial z}$$

$$= -\rho g.$$

(11.50.1)

This relationship was foreshadowed earlier by the Jacobian (11.10) for transformations that involve the vertical coordinate alone. Similarly, (11.44) reduces to

$$\nabla_z p = -\left(\frac{\partial z}{\partial p}\right)^{-1}\nabla_p z$$

$$= \rho g\nabla_p z.$$

(11.50.2)

---

[4] Not to be confused with advective changes introduced by motion along that surface.

Incorporating (11.50.2) into the horizontal momentum equation transforms the pressure gradient force into

$$-\frac{1}{\rho}\nabla_z p = -g\nabla_p z$$

$$= -\nabla_p \Phi,$$

(11.51)

where the geopotential Φ is given by (6.7). Then, with (11.49), the horizontal momentum equations become

$$\frac{\partial \boldsymbol{v}_h}{\partial t} + (\boldsymbol{v} \cdot \nabla \boldsymbol{v}_h)_p + \left(f + u\frac{tan\phi}{a}\right)\boldsymbol{k} \times \boldsymbol{v}_h = -\nabla_p \Phi - \boldsymbol{D}_h,$$

(11.52)

where $\boldsymbol{D}_h$ denotes the horizontal component of drag. Simple substitution transforms the hydrostatic equation into

$$\frac{\partial \Phi}{\partial p} = -\alpha$$

$$= -\frac{RT}{p}.$$

(11.53)

The continuity equation follows from its expression in terms of specific volume. In isobaric coordinates, (10.31) becomes

$$\frac{d\alpha_p}{dt} = \alpha_p (\nabla \cdot \boldsymbol{v})_p,$$

(11.54.1)

where α_p refers to the incremental material volume between isobaric surfaces, dV_p, and

$$(\nabla \cdot \boldsymbol{v})_p = \nabla_p \cdot \boldsymbol{v}_h + \frac{\partial \omega}{\partial p}.$$

(11.54.2)

Then (11.6) implies

$$\alpha_p = J(\boldsymbol{x}, \boldsymbol{x}_p)\alpha.$$

(11.55)

According to (11.9),

$$J(\boldsymbol{x}, \boldsymbol{x}_p) = \left|\frac{\partial p}{\partial z}\right|$$

$$= \rho g.$$

(11.56)

Hence,

$$\alpha_p = \rho\alpha \cdot g$$

$$= g.$$

(11.57)

It follows that

$$\frac{d\alpha_p}{dt} = 0,$$

(11.58)

which asserts that the material volume in isobaric coordinates is conserved. The continuity equation (11.54) then reduces to

$$\nabla_p \cdot \boldsymbol{v}_h + \frac{\partial \omega}{\partial p} = 0.$$

(11.59)

In isobaric coordinates, the continuity equation has the same form as that for incompressible motion (11.30).

The thermodynamic equation can be developed from the First Law for an individual air parcel (2.22), which implies

$$c_p \frac{dT}{dt} - \alpha \frac{dp}{dt} = \dot{q}_{net}. \tag{11.60}$$

With (11.49), this becomes

$$\frac{\partial T}{\partial t} + (\boldsymbol{v}_h \cdot \nabla T)_p - \frac{\alpha \omega}{c_p} = \frac{\dot{q}_{net}}{c_p}. \tag{11.61.1}$$

Expansion work is seen to be proportional to the velocity across isobaric surfaces. As before, the thermodynamic equation assumes a more compact form when expressed in terms of potential temperature:

$$\frac{\partial \theta}{\partial t} + (\boldsymbol{v}_h \cdot \nabla \theta)_p = \frac{\theta}{c_p T} \dot{q}_{net}, \tag{11.61.2}$$

derivation of which is left as an exercise. Expansion work is thus absorbed into θ.

Collectively, the equations of motion in isobaric coordinates are then given by

$$\frac{d\boldsymbol{v}_h}{dt} + \left(f + u \frac{\tan\phi}{a} \right) \boldsymbol{k} \times \boldsymbol{v}_h = -\nabla_p \Phi - \boldsymbol{D}_h \tag{11.62.1}$$

$$-\frac{\partial \Phi}{\partial \ln p} = RT \tag{11.62.2}$$

$$(\nabla \cdot \boldsymbol{v})_p = 0 \tag{11.62.3}$$

$$\frac{d\theta}{dt} = \frac{\theta}{c_p T} \dot{q}_{net}, \tag{11.62.4}$$

with

$$\frac{d}{dt} = \frac{\partial}{\partial t} + \boldsymbol{v}_h \cdot \nabla_p + \omega \frac{\partial}{\partial p}. \tag{11.62.5}$$

The lower boundary condition requires air to maintain the surface elevation z_s, so

$$\frac{d\Phi}{dt} = g \frac{dz_s}{dt} \tag{11.63}$$

at $p = p_s(x, y, t)$.

Transforming the equations into this coordinate system yields the following simplifications:

(1) The pressure gradient force has been linearized through elimination of ρ.

(2) The continuity equation has reduced to a statement of 3-dimensional nondivergence. In addition to being linear, it is now "diagnostic" (i.e., it involves no time derivatives).

These simplifications are not acquired without a price. Coordinate surfaces $p = \text{const}$ are now time dependent. Their positions evolve with the circulation. Further, isobaric surfaces need not coincide with the Earth's surface. The lower boundary condition

(11.63) then involves different values of the vertical coordinate $p = p_s(x, y, t)$, complicating its application.

Complications in the lower boundary condition can be averted by introducing the modified pressure coordinate

$$\sigma = \frac{p}{p_s}. \tag{11.64}$$

Unlike p, σ preserves a constant value at the Earth's surface. Transforming to σ-coordinates passes the complication of the lower boundary condition into the governing equations. Nevertheless, σ-coordinates offer computational advantages. They are relied upon in GCMs; see Haltiner and Williams (1980) for a formal development.

11.5.2 Log-pressure coordinates

Combining some of the benefits of height and pressure is the modified vertical coordinate

$$z^* = -H ln\left(\frac{p}{p_{00}}\right), \tag{11.65.1}$$

where

$$H = \frac{RT_0}{g} \tag{11.65.2}$$

is treated as constant. At this level of approximation, the basic pressure and density that correspond to static conditions reduce to

$$p_0(z^*) = p_0(0)e^{-\frac{z^*}{H}}$$
$$\rho_0(z^*) = \rho_0(0)e^{-\frac{z^*}{H}}. \tag{11.66}$$

Log-pressure height is formally constant on isobaric surfaces. It is based on the stratification under static conditions, namely on global-mean properties that vary only with height. For this reason, log-pressure height is an approximation to geopotential height. Because variations of H are comparatively small, so are discrepancies between z^* and z. Under isothermal conditions, the two measures of elevation are identical.

In terms of z^*, the hydrostatic equation becomes

$$\frac{\partial \Phi}{\partial z^*} = \frac{RT}{H}. \tag{11.67}$$

The vertical velocity in log-pressure coordinates is given by

$$w^* = \frac{dz^*}{dt}$$
$$= -\frac{H}{p}\frac{dp}{dt} = -\frac{H}{p}\omega. \tag{11.68}$$

Then the Lagrangian derivative translates into

$$\frac{d}{dt} = \frac{\partial}{\partial t} + \boldsymbol{v}_h \cdot \nabla_{z^*} + w^*\frac{\partial}{\partial z^*}, \tag{11.69}$$

where ∇_{z^*} reflects the horizontal gradient evaluated on an isobaric surface. Likewise, vertical divergence in pressure coordinates becomes

$$\frac{\partial \omega}{\partial p} = \frac{\partial w^*}{\partial z^*} - \frac{w^*}{H}$$
$$= \frac{1}{\rho_0} \frac{\partial}{\partial z^*} \left(\rho_0 w^* \right). \tag{11.70}$$

It transforms the continuity equation into

$$\nabla_{z^*} \cdot \boldsymbol{v}_h + \frac{1}{\rho_0} \frac{\partial}{\partial z^*} \left(\rho_0 w^* \right) = 0. \tag{11.71}$$

The thermodynamic equation is transformed in similar fashion and is left as an exercise.

With the foregoing expressions, the equations of motion in log-pressure coordinates become

$$\frac{d\boldsymbol{v}_h}{dt} + \left(f + u\frac{\tan\phi}{a} \right) \boldsymbol{k} \times \boldsymbol{v}_h = -\nabla_{z^*}\Phi - \boldsymbol{D}_h \tag{11.72.1}$$

$$\frac{\partial \Phi}{\partial z^*} = \frac{RT}{H} \tag{11.72.2}$$

$$\nabla_{z^*} \cdot \boldsymbol{v}_h + \frac{1}{\rho_0} \frac{\partial}{\partial z^*} \left(\rho_0 w^* \right) = 0 \tag{11.72.3}$$

$$\left(\frac{\partial}{\partial t} + \boldsymbol{v}_h \cdot \nabla_{z^*} \right) \left(\frac{\partial \Phi}{\partial z^*} \right) + N^{*^2} w^* = \frac{\kappa}{H} \dot{q}_{net}, \tag{11.72.4}$$

where

$$\frac{d}{dt} = \frac{\partial}{\partial t} + \boldsymbol{v}_h \cdot \nabla_{z^*} + w^* \frac{\partial}{\partial z^*} \tag{11.72.5}$$

and

$$N^{*^2} = \frac{R}{H} \left(\frac{\partial T}{\partial z^*} + \frac{\kappa}{H} T \right) \tag{11.72.6}$$

represents the static stability in log-pressure coordinates. In the troposphere, N^{*^2} varies with height only weakly. To leading order, it can therefore be treated as constant. The lower boundary condition requires geometric vertical velocity to be specified (11.63). It becomes

$$\frac{\partial \Phi}{\partial t} + \boldsymbol{v}_h \cdot \nabla_{z^*}\Phi + \frac{RT}{H} w^* = g\frac{dz_s}{dt}. \tag{11.73}$$

At the level of approximation inherent to log-pressure coordinates, this may be evaluated at the constant elevation $z^* = z_s^*$, but with variations of z_s^* accounted for in the right-hand side of (11.73).

The equations in log-pressure coordinates have several advantages. Variables are analogous to those in physical coordinates, so they are easily interpreted. Yet, the pressure-gradient force, hydrostatic equation, and continuity equation retain nearly the same simplified forms as in isobaric coordinates. In addition, mathematical complications surrounding comparatively small variations of temperature are ignored.

11.5.3 Isentropic coordinates

The nearly adiabatic nature of air motion simplifies the governing equations when θ is treated as the vertical coordinate. Isentropic surfaces are then coordinate surfaces. Large-scale air motion is nearly tangential to them (Fig. 2.9).

We consider a transformation from the standard spherical coordinates $\mathbf{x} = (x, y, z)$ to the modified coordinates $\mathbf{x}_\theta = (x, y, \theta)$. For it to serve as a vertical coordinate, potential temperature must vary monotonically with altitude. Hydrostatic stability requires $\frac{\partial \theta}{\partial z} > 0$. The relationship between potential temperature and height is then single valued – so long as the stratification remains stable. As is true for isobaric coordinates, using isentropic surfaces as coordinate surfaces leads to a coordinate system that is nonorthogonal.

Consider the scalar variable

$$\psi = \hat{\psi}\left(x, y, \theta(x, y, z, t), t\right). \tag{11.74}$$

Proceeding as in the development of (11.43) and (11.44) transforms vertical and horizontal derivatives into

$$\frac{\partial \psi}{\partial z} = \frac{\partial \psi}{\partial \theta} \frac{\partial \theta}{\partial z} \tag{11.75.1}$$

$$\nabla_z \psi = \nabla_\theta \psi + \frac{\partial \psi}{\partial \theta} \nabla_z \theta, \tag{11.75.2}$$

where θ is held fixed unless otherwise noted and

$$\nabla_\theta = \left(\frac{\partial}{\partial x}\right)_{y\theta t} \mathbf{i} + \left(\frac{\partial}{\partial y}\right)_{x\theta t} \mathbf{j} \tag{11.75.3}$$

represents the horizontal gradient evaluated on an isentropic surface. Then the Lagrangian derivative becomes

$$\begin{aligned}
\frac{d\psi}{dt} &= \frac{\partial \psi}{\partial t} + (\mathbf{v} \cdot \nabla \psi)_\theta \\
&= \frac{\partial \psi}{\partial t} + \mathbf{v}_h \cdot \nabla_\theta \psi + \frac{d\theta}{dt} \frac{\partial \psi}{\partial \theta}.
\end{aligned} \tag{11.76}$$

As before, we identify

$$\omega_\theta = \frac{d\theta}{dt} \tag{11.77}$$

as the vertical velocity in potential temperature coordinates. Positive upwards, ω_θ represents the Lagrangian derivative of vertical position in this coordinate system.

Letting $\psi = z(x, y, \theta, t)$ in (11.75), which for fixed θ represents the height of an isentropic surface, leads to the identities

$$\frac{\partial \theta}{\partial z} = \left(\frac{\partial z}{\partial \theta}\right)^{-1} \tag{11.78.1}$$

$$\nabla_\theta z = -\frac{\partial z}{\partial \theta} \nabla_z \theta. \tag{11.78.2}$$

Then substituting (11.78.2) transforms (11.75.2) into

$$\nabla_z \psi = \nabla_\theta \psi - \frac{\partial \psi}{\partial z} \nabla_\theta z. \tag{11.79}$$

Taking $\psi = p$ and incorporating (11.79) transforms the pressure-gradient force in the horizontal momentum equations into

$$-\frac{1}{\rho}\nabla_z p = -\frac{1}{\rho}\nabla_\theta p + \frac{1}{\rho}\frac{\partial p}{\partial z}\nabla_\theta z$$
$$= -\frac{1}{\rho}\nabla_\theta p - g\nabla_\theta z. \tag{11.80}$$

Now Poisson's relationship for potential temperature (2.31) implies the identity

$$ln\theta = lnT - \kappa(lnp - lnp_0).$$

Applying the horizontal gradient evaluated on an isentropic surface obtains

$$c_p\nabla_\theta T = \frac{RT}{p}\nabla_\theta p$$
$$= \frac{1}{\rho}\nabla_\theta p. \tag{11.81}$$

Then the pressure gradient force (11.80) reduces to

$$-\frac{1}{\rho}\nabla_z p = -c_p\nabla_\theta T - g\nabla_\theta z$$
$$= -\nabla_\theta \Psi. \tag{11.82}$$

The *Montgomery streamfunction*

$$\Psi = c_p T + gz$$
$$= c_p T + \Phi \tag{11.83}$$

plays a role in isentropic coordinates analogous to the role in isobaric coordinates played by geopotential. With (11.82) and (11.76), the horizontal momentum equations become

$$\frac{\partial \boldsymbol{v}_h}{\partial t} + (\boldsymbol{v}\cdot\nabla\boldsymbol{v}_h)_\theta + \left(f + u\frac{tan\phi}{a}\right)\boldsymbol{k}\times\boldsymbol{v}_h = -\nabla_\theta\Psi - \boldsymbol{D}_h. \tag{11.84}$$

To transform the hydrostatic equation, we consider (11.75.1) with $\psi = p$, which gives

$$-\rho g\frac{\partial z}{\partial \theta} = \frac{\partial p}{\partial \theta}. \tag{11.85}$$

Poisson's relationship implies

$$\frac{1}{\theta} = \frac{1}{T}\frac{\partial T}{\partial \theta} - \frac{\kappa}{p}\frac{\partial p}{\partial \theta}.$$

Then substituting into (11.85) obtains

$$c_p\frac{\partial T}{\partial \theta} + g\frac{\partial z}{\partial \theta} = \frac{c_p T}{\theta}$$

or

$$\frac{\partial \Psi}{\partial \theta} = c_p\left(\frac{p}{p_0}\right)^\kappa, \tag{11.86}$$

which serves as the hydrostatic equation in isentropic coordinates.

The continuity equation follows from its expression in terms of specific volume. In isentropic coordinates, this becomes

$$\frac{d\alpha_\theta}{dt} = \alpha_\theta (\nabla \cdot \boldsymbol{v})_\theta,$$

(11.87.1)

where α_θ refers to the incremental material volume between isentropic surfaces, dV_θ, and

$$(\nabla \cdot \boldsymbol{v})_\theta = \nabla_\theta \cdot \boldsymbol{v}_h + \frac{\partial \omega_\theta}{\partial \theta}.$$

(11.87.2)

According to (11.6),

$$\alpha_\theta = J(\boldsymbol{x}, \boldsymbol{x}_\theta)\alpha,$$

(11.88)

where

$$J(\boldsymbol{x}, \boldsymbol{x}_\theta) = \left| \frac{\partial \theta}{\partial z} \right|.$$

(11.89)

Incorporating (11.85), with (11.78.1), obtains

$$\alpha_\theta = \rho\alpha \cdot g \left(\frac{\partial p}{\partial \theta} \right)^{-1}$$

$$= g \left(\frac{\partial p}{\partial \theta} \right)^{-1}.$$

(11.90)

Then the continuity equation (11.87) reduces to

$$\frac{d}{dt} \left(\frac{\partial p}{\partial \theta} \right)^{-1} = \left(\frac{\partial p}{\partial \theta} \right)^{-1} (\nabla \cdot \boldsymbol{v})_\theta,$$

or

$$\frac{d}{dt} \left(\frac{\partial p}{\partial \theta} \right) + \left(\frac{\partial p}{\partial \theta} \right) (\nabla \cdot \boldsymbol{v})_\theta = 0.$$

(11.91)

If $(\frac{\partial p}{\partial \theta})$ is identified with density, then (11.91) has the same form as the continuity equation in physical coordinates (11.25). The thermodynamic equation also has the same form as earlier.

Collectively, the equations of motion in isentropic coordinates are then given by

$$\frac{d\boldsymbol{v}_h}{dt} + \left(f + u\frac{\tan\phi}{a} \right) \boldsymbol{k} \times \boldsymbol{v}_h = -\nabla_\theta \Psi - \boldsymbol{D}_h$$

(11.92.1)

$$\frac{\partial \Psi}{\partial \theta} = c_p \left(\frac{p}{p_0} \right)^\kappa$$

(11.92.2)

$$\frac{d}{dt} \left(\frac{\partial p}{\partial \theta} \right) + \left(\frac{\partial p}{\partial \theta} \right) (\nabla \cdot \boldsymbol{v})_\theta = 0$$

(11.92.3)

$$\omega_\theta = \frac{\theta}{c_p T} \dot{q}_{net},$$

(11.92.4)

with

$$\frac{d}{dt} = \frac{\partial}{\partial t} + \boldsymbol{v}_h \cdot \nabla_\theta + \omega_\theta \frac{\partial}{\partial \theta}.$$

(11.92.5)

The lower boundary condition becomes

$$\frac{d\Psi}{dt} - \frac{d}{dt}\left(\theta\frac{\partial\Psi}{\partial\theta}\right) = g\frac{dz_s}{dt}.$$

(11.93)

These equations simplify the description of vertical motion, which is related directly to the rate at which heat is absorbed by an air parcel (11.92.4). Under adiabatic conditions, air motion is tangential to coordinate surfaces. ω_θ and vertical advection then vanish. Under diabatic conditions, the system is still advantageous because it relates vertical motion (difficult to measure because of its smallness) to quantities that are determined more reliably.

Another advantage of isentropic coordinates is enhanced resolution in regions of strong temperature gradient, as typify frontal zones. Figure 11.5 displays a frontal zone, represented in isobaric and isentropic coordinates. In the isobaric representation (Fig. 11.5a), motion and potential temperature vary sharply across the frontal zone. Representing those properties numerically therefore requires high resolution. The fine computational mesh, in turn, requires a short time step. Together, these requirements sharply magnify the computational burden necessary to treat such behavior. The sharp gradients in Fig. 11.5 develop through flow deformation (Sec. 10.2). During the amplification of synoptic weather systems (Chap. 16), it concentrates θ surfaces into a narrow zone. Such deformation, however, proceeds under nearly adiabatic conditions. It is therefore absorbed into the representation in isentropic coordinates. In that representation, θ surfaces maintain a fixed separation. This feature of isentropic coordinates magnifies the region of sharp behavior (Fig. 11.5b). Properties therefore vary smoothly, limiting the computational burden necessary to treat such behavior. Although offering these advantages, isentropic coordinates make the continuity equation more complex. They also suffer from variations of θ along the Earth's surface, which complicate the lower boundary condition.

Vertical Coordinate Representations

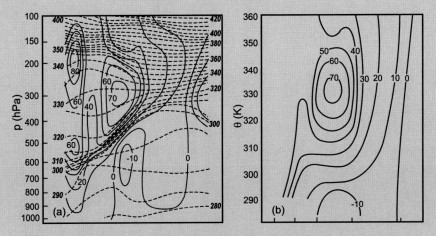

Figure 11.5 Cross section of isotachs (solid) and isentropes (dashed) through a frontal zone represented in (a) isobaric coordinates and (b) isentropic coordinates. *Sources*: Shapiro and Hastings (1973) and R. Bleck (U. Miami), personal communication.

SUGGESTED REFERENCES

Vectors, Tensors, and the Basic Equations of Fluid Mechanics (1962) by Aris provides a complete treatment of curvilinear coordinates.

A treatment of generalized vertical coordinates is given in *The Ceaseless Wind* (1986) by Dutton. Their advantages in numerical applications are developed in Bleck (1978a,b).

Numerical Prediction and Dynamic Meteorology (1980) by Haltiner and Williams describes σ coordinates and their application in large-scale simulations.

PROBLEMS

1. Show that spherical coordinates are expressed in terms of the rectangular Cartesian coordinates by (11.13).

2. Show that metric scale factors for spherical coordinates are given by (11.14).

3. In isobaric coordinates, (a) derive the thermodynamic equation in terms of potential temperature (11.61.2), (b) express the vertical velocity ω in terms of the geometric vertical velocity w, and (c) for $\omega = 100$ hPa day^{-1} at 500 hPa, estimate w.

4. Show that isochoric surfaces are material if the motion is steady and v is everywhere orthogonal to $\nabla \rho$.

5. (a) For a characteristic velocity of 10 m s^{-1}, at what horizontal scale does the Earth's rotation become important? (b) How will the Earth's rotation be manifested in the streamlines of steady flow?

6. The *geostrophic wind* v_g is the horizontal velocity that follows from a balance between the Coriolis and pressure-gradient forces. Provide expressions for u_g and v_g in (a) isobaric coordinates and (b) isentropic coordinates. (c) How does v_g behave as the equator is approached?

7. Use the result in Prob. 11.6, along with the hydrostatic equation, to derive an expression in isobaric coordinates for the vertical variation of geostrophic wind.

8. Isentropic surfaces slope meridionally more steeply than isobaric surfaces. Use the surface temperature in Fig. 10.1b and a uniform lapse rate of 6.5 K km^{-1} to estimate the heights over the equator and pole of (a) the 300 K isentropic surface and (b) the 700 mb isobaric surface.

9. Derive the thermodynamic equation in log-pressure coordinates (11.72.4).

10. Show that the Jacobian reduces to the product of the metric scale factors (11.7) if the curvilinear coordinate system is orthogonal.

11. (a) Show that the spherical coordinate vectors are expressed in terms of the Cartesian coordinate vectors by (11.20). (b) Then show that the spherical coordinate vectors have spatial derivatives

$$\frac{\partial i}{\partial \lambda} = \sin\phi j - \cos\phi k \qquad \frac{\partial i}{\partial \phi} = 0 \qquad \frac{\partial i}{\partial z} = 0$$

$$\frac{\partial j}{\partial \lambda} = -\sin\phi i \qquad \frac{\partial j}{\partial \phi} = -k \qquad \frac{\partial j}{\partial z} = 0$$

$$\frac{\partial k}{\partial \lambda} = \cos\phi i \qquad \frac{\partial k}{\partial \phi} = j \qquad \frac{\partial k}{\partial z} = 0.$$

(Hint: Obtain reciprocal identities for e_1, e_2, and e_3 in terms of i, j, and k.) (c) Evaluate the Lagrangian derivatives of the spherical coordinate vectors to obtain (11.21).

12. Derive the angular momentum principle implied by the vector equations of motion in full spherical geometry.

13. Show that the equations of motion in an inertial reference frame are expressed by (11.27). (Hint: See Prob. 10.9).

14. Show that the approximate system (11.28) has the angular momentum principle (11.29).

15. The equations of motion are expressed conveniently in terms of the *Exner function* $\pi = c_p(\frac{p}{p_0})^\kappa$ as the vertical coordinate, where $p_0 = 1000$ hPa. Transform the equations from isobaric coordinates to Exner coordinates.

16. Frictional drag beneath a cyclone produces horizontal convergence, which varies with height inside the planetary boundary layer as

$$-\nabla \cdot \boldsymbol{v}_h = \zeta e^{-\frac{z^*}{h}},$$

where $\zeta = (\nabla \times \boldsymbol{v}_g) \cdot \boldsymbol{k}$ is the (constant) vorticity of the geostrophic wind above the boundary layer and z^* is log-pressure height (Prob 11.6). (a) Determine the vertical motion as a function of height if $\frac{h}{H} \ll 1$. (b) Describe the divergent component of motion which must exist above the boundary layer ($z^* \gg h$) if, at sufficient height, vertical motion eventually vanishes.

CHAPTER TWELVE

Large-scale motion

Scale analysis indicates that large-scale motion is dominated by a balance between the Coriolis acceleration and the pressure-gradient force (11.37). Material acceleration is of order Rossby number, denoted $O(Ro)$. It is an order of magnitude smaller than the dominant terms. Metric terms are even smaller. The horizontal momentum equations can therefore be expressed in terms of the horizontal velocity v_h

$$\frac{dv_h}{dt} + fk \times v_h = -\nabla_p\Phi - D. \tag{12.1}$$

We use the prevalence of certain terms in (12.1) to illustrate the essential balances that control large-scale motion. In the framework of asymptotic series representation, dependent variables in (12.1) can be expanded in a power series in the small parameter Ro. The momentum equations can then be balanced with the expansion in Ro truncated to the gravest terms. In principle, this procedure can be carried out recursively: The momentum balance can be obtained at successively higher order in Ro, based on the balance at lower order. However, $Ro \ll 1$ makes terms omitted much smaller than those retained in the approximate momentum balance. Retaining only the gravest terms therefore often provides sufficient accuracy.

12.1 GEOSTROPHIC EQUILIBRIUM

To zero order in Ro, the momentum equations reduce to

$$fk \times v_g = -\nabla_p\Phi \tag{12.2.1}$$

where frictional drag D is presumed to be $O(Ro)$ or smaller. Reflecting a balance between the Coriolis acceleration and the horizontal pressure-gradient force, (12.2.1) defines *geostrophic equilibrium* (*geo* referring to Earth and *strophic* to turning). The

GEOSTROPHIC EQUILIBRIUM

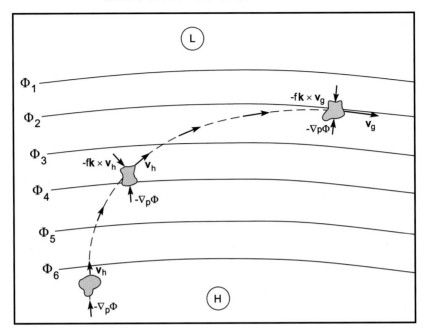

Figure 12.1 Evolution of a hypothetical air parcel that is initially motionless in the Northern Hemisphere and in stratification wherein contours of isobaric height $z = \frac{1}{g}\Phi$ are nearly straight, uniformly spaced, and, for the sake of illustration, fixed.

horizontal velocity that satisfies (12.2.1) is the *geostrophic velocity*

$$
\begin{aligned}
\boldsymbol{v}_g &= \frac{1}{f}\boldsymbol{k} \times \nabla_p\Phi \\
&= \frac{1}{f}\left(-\frac{\partial\Phi}{\partial y}, \frac{\partial\Phi}{\partial x}\right).
\end{aligned}
\tag{12.2.2}
$$

Similar expressions follow in isentropic coordinates with the Montgomery streamfunction.

According to (12.2.2), the geostrophic velocity is tangential to contours of height, $z = \frac{1}{g}\Phi$, with low height on the left (right) in the Northern (Southern) Hemisphere. This is equivalent to motion being along isobars on a constant height surface (Sec. 6.3). Such motion is apparent in the observed circulation (Fig. 1.9). It is just perpendicular to air motion that is normally observed in an inertial reference frame, namely, from high toward low pressure and, hence, perpendicular to isobars. This peculiarity of large-scale atmospheric motion (one of several related features) is referred to as the "geostrophic paradox." Derived from the Coriolis force, which acts orthogonal to motion, it was not resolved until the Earth's rotation was accounted for.

To illustrate how geostrophic equilibrium is established, consider a hypothetical initial value problem: An atmosphere, initially at rest, is characterized by contours of isobaric height that are nearly straight, uniformly spaced, and, for the sake of illustration, fixed (Fig. 12.1). At $t = 0$, an air parcel has $\boldsymbol{v}_h = 0$. The Coriolis force acting

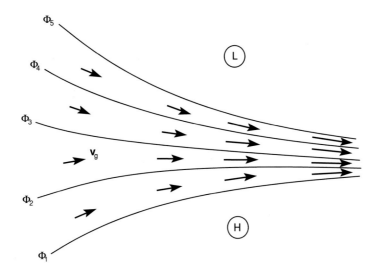

Figure 12.2 Variation of geostrophic velocity with changes of isobaric height.

on it therefore vanishes. In response to the pressure gradient force, the parcel accelerates in the direction of $-\nabla_p\Phi$, toward low height. As soon as motion develops, the Coriolis force $-f\mathbf{k} \times \mathbf{v}_h$ acts perpendicular to the horizontal velocity. In the Northern Hemisphere, where $f = 2\Omega sin\phi$ is positive, the Coriolis force deflects the parcel's motion toward the right. Were it in the Southern Hemisphere, the parcel would be deflected to the left. In either event, the Coriolis force performs no work on the parcel – because it acts orthogonal to the instantaneous motion. While the parcel's trajectory veers to the right, the pressure gradient force remains unchanged because, under the foregoing conditions, $\nabla_p\Phi$ is invariant of position. However, the Coriolis force changes continually, remaining proportional and orthogonal to $\mathbf{v}_h$. As the parcel accelerates under the pressure-gradient force, its trajectory veers increasingly to the right. Eventually, $\mathbf{v}_h$ has been deflected tangential to contours of Φ. At that point, the Coriolis force acts in direct opposition to the pressure gradient force. Both are perpendicular to the parcel's velocity. If $\mathbf{v}_h$ satisfies (12.2), the forces are in balance. The motion is then in geostrophic equilibrium. Consequently, the parcel experiences no further acceleration.[1]

Once this mechanical equilibrium is established, the parcel's motion is steady – except for gradual adjustments allowed for in contours of Φ. According to (12.2.2), geostrophic wind speed v_g is proportional to $\nabla_p\Phi$. It is therefore inversely proportional to the spacing of height contours (Fig. 12.2). Geostrophic wind speed increases into a region where height contours converge. It decreases into a region where they diverge. For either case, $\mathbf{v}_g$ remains tangential to height contours. In the absence of vertical motion, such behavior automatically satisfies conservation of mass: It is nondivergent, $\nabla \cdot \mathbf{v} = 0$. This feature of geostrophic motion makes height contours streamlines.

---

[1] If the Coriolis and pressure-gradient forces do not balance exactly, the motion will undergo an oscillation about contours of Φ, which, upon being damped out, leaves the parcel in geostrophic equilibrium.

The balance (12.2) is valid for curved motion, so long as an anomaly's horizontal scale is large enough and its velocity is slow enough for the material acceleration to be negligible relative to the pressure-gradient and Coriolis forces. (These conditions maintain $Ro \ll 1$). Geostrophic equilibrium then implies circular motion about a closed center of height. Low height lies to the left (right) in the Northern (Southern) Hemisphere. Therefore, geostrophic flow about a closed low is counterclockwise in the Northern Hemisphere. It is clockwise in the Southern Hemisphere. In each hemisphere, motion about a closed low has the same sense of rotation as the planetary vorticity $f\mathbf{k}$. Such motion is termed *cyclonic*. About a closed high, geostrophic motion is clockwise in the Northern Hemisphere. It is counterclockwise in the Southern Hemisphere. In each hemisphere, such motion has a sense of rotation opposite to the planetary vorticity. Motion about a closed high is termed *anticyclonic*.

Geostrophic motion is apparent in the 500-hPa circulation (Fig. 1.9a). Punctuating the jet stream are anomalies of low height. Motion about them is counterclockwise. In fact, the circumpolar flow itself may be regarded as a cyclonic vortex. It is implied by the hypsometric relationship (6.12). Poleward-decreasing temperature in the troposphere produces low isobaric height over the pole (Fig. 6.2). Geostrophic equilibrium then establishes a cyclonic circulation about the pole, manifest in the subtropical jet (Fig. 1.9b). The circulation at 10 hPa is also cyclonic, but intensified (Fig. 1.10b). Ozone heating at low latitude and LW cooling at high latitude (Fig. 8.27) produce poleward-decreasing temperature across the winter stratosphere. Through the hypsometric relationship, this establishes the polar-night vortex. Outside the vortex is the *Aleutian high*. It is accompanied by anticyclonic motion. Even though the instantaneous circulation can be highly disturbed from zonal symmetry (Figs. 1.9a, 1.10a), motion remains nearly tangential to contours of height.

12.1.1 Motion on an f plane

Consider motion in which meridional displacements of air are sufficiently narrow to ignore variations of f. Expanding f in a Taylor series about the reference latitude ϕ_0 and truncating to zeroth order gives the constant Coriolis parameter

$$f_0 = 2\Omega sin\phi_0. \tag{12.3.1}$$

In the same framework, it is convenient to neglect spherical curvature in favor of simple Cartesian geometry. A Cartesian plane tangent to the Earth at latitude ϕ_0 (Fig. 12.3) describes horizontal position in terms of the distances

$$x = acos\phi_0 \cdot \lambda$$
$$y = a \cdot (\phi - \phi_0), \tag{12.3.2}$$

but with the Earth's sphericity ignored. Together, these simplifications comprise the *f-plane* description of atmospheric motion.

On an f plane, the geostrophic velocity defines a nondivergent vector field. Tangential paths of v_g, which coincide with height contours, then serve as streamlines. These and other implications of nondivergence follow from the general representation of a vector field.

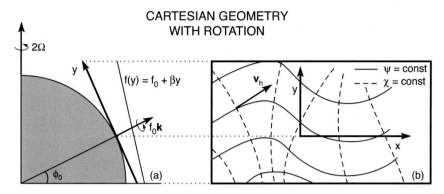

Figure 12.3 Approximation of the rotating spherical Earth by Cartesian geometry. On a plane tangent to the Earth at latitude ϕ_0, distance is measured by the coordinates $x = a\cos\phi_0 \cdot \lambda$ and $y = a \cdot (\phi - \phi_0)$, which increase to the east and north, respectively. (a) Approximating the Coriolis parameter by its zeroth order variation: $f(y) = f_0 = 2\Omega\sin\phi_0$, yields the *f-plane* description of atmospheric motion. Approximating it to first order: $f(y) = f_0 + \beta y$, where $\beta = \frac{df}{dy}|_{y=0}$, yields the *β-plane* description of atmospheric motion. (b) The horizontal motion field at any instant $v_h(x, y, t)$ can be expressed in terms of a *streamfunction* $\psi(x, y, t)$ and a *velocity potential* $\chi(x, y, t)$: $v_h = k \times \nabla\psi + \nabla\chi$, which represent the rotational and divergent components of v_h, respectively.

The Helmholtz Theorem

Any vector field v can be represented in terms of a *divergent* or irrotational component and a *solenoidal* or rotational component:

$$v = \nabla\chi + \nabla \times \psi, \tag{12.4}$$

where χ is a *scalar potential* (analogous to the potential function in Chap. 2) and ψ is a *vector potential*. The divergent component

$$v_d = \nabla\chi \tag{12.5.1}$$

possesses zero vorticity

$$\nabla \times v_d = 0. \tag{12.5.2}$$

The solenoidal component

$$v_s = \nabla \times \psi \tag{12.6.1}$$

possesses zero divergence

$$\nabla \cdot v_s = 0. \tag{12.6.2}$$

For the 2-dimensional field of horizontal velocity, (12.4) reduces to

$$v_h = \nabla\chi + k \times \nabla\psi, \tag{12.7.1}$$

where χ is the *velocity potential* and ψ is the *streamfunction*. The horizontal velocity field then has divergence

$$\nabla \cdot v_h = \nabla^2\chi \tag{12.7.2}$$

and vorticity

$$\nabla \times \boldsymbol{v}_h = \nabla^2 \psi \boldsymbol{k}. \tag{12.7.3}$$

The divergent component of motion $\boldsymbol{v}_d$ is orthogonal to contours of χ. The solenoidal component $\boldsymbol{v}_s$ is tangential to contours of ψ.

According to (12.2.2), geostrophic motion on an f plane has the form of a solenoidal vector field. It is characterized by the *geostrophic streamfunction*

$$\psi = \frac{1}{f_0}\Phi. \tag{12.8}$$

Because the divergence of $\boldsymbol{v}_g$ vanishes, the continuity equation implies little or no vertical motion.[2] If pressure variations along the Earth's surface can be ignored, integrating (11.54.2) upward to some isobaric surface obtains

$$\omega(x, y, p, t) = -\int_p^{p_s} \nabla \cdot \boldsymbol{v}_h dp. \tag{12.9}$$

As $\nabla \cdot \boldsymbol{v}_g = 0$ (12.6.2), geostrophic equilibrium implies zero vertical motion. Air parcels then simply exchange horizontal positions with no vertical rearrangement. The solenoidal character of large-scale motion follows from the Earth's rotation, which maintains the circulation close to geostrophic equilibrium. Vertical motion does occur in the large-scale circulation. However, it is higher order in Ro and therefore small. Such motion enters through the *ageostrophic* or divergent component of horizontal velocity $\boldsymbol{v}_d$.

Owing to its rotational character, geostrophic motion involves substantial deformation. Horizontal shear associated with vorticity (10.5) distorts finite bodies of air into complex forms. Figure 12.4 displays the evolution of a material volume as it is advected though a 2-dimensional cyclone. Shear strains the body into an elongated shape. Through this distortion, the transverse separation of boundaries collapses to small scales. Accompanying gradients therefore steepen. On those scales, turbulence and nonconservative processes act efficiently (Chap. 13). They homogenize sharp contrasts that have developed through shear strain. This consequence of rotational motion is tantamount to mixing. It is analogous to cream being stirred into a cup of coffee. Shear strains exaggerate fluid gradients until, eventually, they are dissolved by small-scale turbulence and molecular diffusion. In the atmosphere, such mixing is accomplished by horizontal eddy motions and by vertical motions in convection (cf. Figs. 1.18, 1.28).

12.2 VERTICAL SHEAR OF THE GEOSTROPHIC WIND

Geostrophic balance determines the horizontal structure of motion. Together with hydrostatic balance, it also determines the vertical structure. A fundamental principle of rotating fluid mechanics is the *Taylor-Proudman theorem*. It asserts that the motion

---

[2] Hereafter, the term *divergence* will be understood to refer to the divergence of the horizontal velocity.

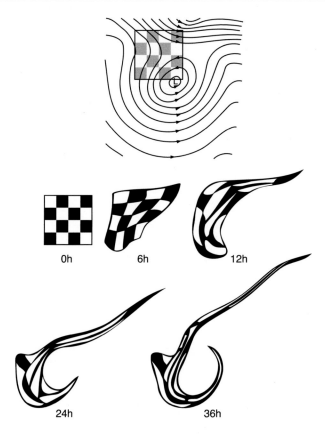

Figure 12.4 Deformation of a material volume at successive times inside a 2-dimensional cyclone, obtained by integrating the barotropic nondivergent vorticity equation. Adapted from Welander (1955). Copyright (1955) Munksgaard International Publishers Ltd., Copenhagen Denmark.

of a homogeneous incompressible fluid cannot vary along the axis of rotation. Motion then occurs in so-called *Taylor-Proudman columns*.[3] The atmosphere is not homogeneous nor incompressible. Nevertheless, it obeys an analogue of the Taylor-Proudman theorem, which relates horizontal motion to stratification.

12.2.1 Classes of stratification

Stratification is represented in the distributions of thermodynamic properties. For dry air, only two such properties are independent (Sec. 2.1.4). Therefore, any two families of thermodynamic surfaces uniquely describe the stratification (Fig. 12.5). Mathematically, this is expressed by $\theta = \theta(p, T)$, where p and T are functions of space and time.

---

[3] See Greenspan (1968) for a laboratory demonstration.

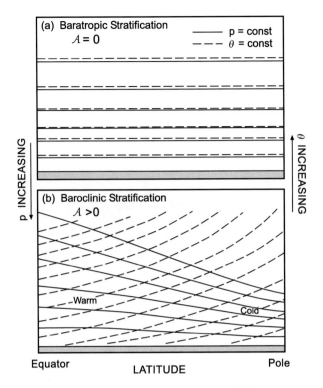

Figure 12.5 Thermal structure corresponding to (a) barotropic stratification, wherein isentropic surfaces coincide with isobaric surfaces and available potential energy $\mathcal{A}$ is zero (Sec. 15.1.3), and (b) baroclinic stratification,
wherein isentropic surfaces do not coincide with isobaric surfaces and $\mathcal{A}$ is
positive. The rotation of isobaric and isentropic surfaces from their positions
under barotropic stratification is symbolic of atmospheric heating at low latitude and cooling at middle and high latitudes (cf. Figs. 1.34c and 9.41b).

Should isentropic surfaces coincide with isobaric surfaces (Fig. 12.5a), $\theta = \theta(p)$.
The stratification is then said to be *barotropic*. Under barotropic stratification, the circulation possesses only one thermodynamic degree of freedom. It is represented in the
single independent family of thermodynamic surfaces. Because other thermodynamic
surfaces coincide with that family, specifying p uniquely determines θ. Jointly, they
determine the thermodynamic state and all other thermodynamic properties. Under
these circumstances, the geostrophic velocity will be seen to be invariant with height.
Air motion then occurs in columns.

More generally, two families of thermodynamic surfaces do not coincide (Fig.
12.5b), so $\theta = \theta(p, T)$. The stratification is then said to be *baroclinic*. Under baroclinic
stratification, the circulation possesses two thermodynamic degrees of freedom. Consequently, along any thermodynamic surface, other thermodynamic properties vary.
The geostrophic velocity then varies with height. However, it does so in direct proportion to the variation of temperature along isobaric surfaces.

12.2.2 Thermal wind balance

Under barotropic stratification, temperature does not vary along isobaric surfaces. Integrating the hypsometric relationship (6.12) upward from the ground then gives Φ. The latter has a contribution from the lower boundary, which may vary horizontally, plus a function of pressure alone. The horizontal distribution of height therefore does not change from one isobaric surface to another. It follows that the geostrophic velocity (12.1.2) is independent of elevation. The motion is columnar. Under barotropic stratification (wherein density is uniform along isobaric surfaces), geostrophic motion is invariant in the direction of $f\mathbf{k}$. The behavior is analogous to Taylor-Proudman flow for a homogeneous incompressible fluid.

Under baroclinic stratification, T varies along isobaric surfaces. Differentiating (12.1.2) with respect to p obtains

$$\frac{\partial \boldsymbol{v}_g}{\partial p} = \frac{1}{f}\boldsymbol{k} \times \nabla_p \frac{\partial \Phi}{\partial p}. \tag{12.10}$$

Incorporating the hydrostatic equation (11.62.2) then yields the vertical gradient of geostrophic velocity

$$\frac{\partial \boldsymbol{v}_g}{\partial \ln p} = \frac{R}{f}\boldsymbol{k} \times \nabla_p T \tag{12.11.1}$$

or, in terms of log-pressure height z,

$$\frac{\partial \boldsymbol{v}_g}{\partial z} = \frac{R}{Hf}\boldsymbol{k} \times \nabla_z T. \tag{12.11.2}$$

Known as *thermal wind balance*, (12.11) asserts that vertical shear of the geostrophic velocity is proportional to the horizontal temperature gradient along isobaric surfaces. The term *thermal wind* describes the incremental velocity $\Delta \boldsymbol{v}_g$ across a layer bounded by two isobaric surfaces and of thickness $-H\Delta \ln p$. According to (12.11), thermal wind is proportional to the horizontal gradient of temperature in that layer. Vertical shear is thus a measure of the departure from barotropic stratification.

From (12.11), an equatorward temperature gradient is accompanied by positive (westerly) shear of the zonal wind. Westerlies then intensify upward. Examples are found in the troposphere and winter stratosphere (Figs. 1.7, 1.8). Radiative heating at low latitude, supported by transfer of latent heat, and radiative cooling at middle and high latitudes establish an equatorward temperature gradient in the troposphere (Figs. 1.34c; 9.41b). The latter prevails in both hemispheres. Thermal wind balance then implies westerlies that intensify upward. They form the subtropical jets, which maximize near the tropopause (Fig. 1.8). In the stratosphere, ozone heating prevails at low latitude, whereas LW cooling to space prevails at high latitude of the winter hemisphere, e.g., inside polar night (Fig. 8.27). They establish an equatorward temperature gradient over a deep layer. Westerlies therefore intensify upward, forming the polar-night jet. Opposite behavior occurs in the summer stratosphere. There, solar insolation is greatest at high latitude (Fig. 1.33), producing a poleward temperature gradient. Zonal wind, which is westerly in the troposphere, therefore experiences easterly shear. It reverses not far above the tropopause. Easterlies then intensify upward, forming the summer easterly jet.

These features can also be inferred directly from geostrophic equilibrium and the hypsometric relationship (a consequence of hydrostatic balance). In the presence of a horizontal temperature gradient, the vertical spacing of isobaric surfaces is expanded in warm air (e.g., at tropical latitudes) but compressed in cold air (e.g., at polar latitudes); cf. Fig. 6.2. An equatorward gradient of isobaric height is produced. The gradient of height steepens with increasing elevation. Geostrophic equilibrium then requires westerlies to intensify upward. They form the jet stream, which coincides with the steep temperature gradient at mid-latitude. Separating warm tropical air from cold polar air (Fig. 6.3), that gradient marks the polar front. The steeper the temperature gradient, the sharper is the polar front and the stronger is the jet stream.

The distribution of net radiation (Fig. 1.34c), in concert with geostrophic equilibrium, thus favors strong westerlies at mid-latitude. Such motion transfers no heat poleward, as is required to offset radiative heating at low latitude and radiative cooling at middle and high latitudes. The same applies to water vapor and chemical constituents that have sources in the tropics. Consequently, the Earth's rotation tends to stratify properties meridionally, just as gravity tends to stratify them vertically.

Thermal wind balance also applies to zonally asymmetric motion. The polar front and jet stream are disturbed by synoptic weather systems, which prevail at mid-latitude (Fig. 1.9a). Those disturbances are typified by a cold-core low (Prob. 6.4). According to (12.11), such thermal structure produces cyclonic vertical shear. It reinforces cyclonic motion. A cold-core low therefore intensifies upward, achieving maximum wind near the tropopause. The reverse occurs inside a warm-core low, as typifies tropical cyclones (Prob. 6.6). That thermal structure produces anticyclonic shear. It opposes cyclonic motion. A warm-core low therefore weakens upward, achieving maximum wind at the surface.

Geostrophic equilibrium applies to motion that is sufficiently steady, slow, and weakly curved for material acceleration to be negligible relative to the Coriolis and pressure-gradient forces (i.e., for it to satisfy $Ro \ll 1$). Frictional drag must also be negligible. These conditions are well-satisfied by large-scale extratropical motion away from the Earth's surface. v_h is then approximately nondivergent. However, the conditions for geostrophic equilibrium break down near the equator. They are also invalidated if the horizontal scale of advection becomes small. Either renders terms in (12.1) that are $O(Ro)$ nonnegligible. These effects introduce an ageostrophic component to v_h. The latter is accompanied by divergence and hence by vertical motion (12.9).

12.3 FRICTIONAL GEOSTROPHIC MOTION

Inside the planetary boundary layer, frictional drag is no longer negligible. It is large enough to invalidate geostrophic equilibrium. Turbulent eddy motion within a kilometer of the surface is produced by strong vertical shear (see Fig. 13.3). It mixes momentum between bodies of air. The accompanying momentum flux exerts a shear stress on individual air parcels, making D in (12.1) of the same order as the Coriolis and pressure-gradient forces (Sec. 10.6).

Drag can be represented as *Rayleigh friction*, with linear drag coefficient K:

$$D = Kv_h. \tag{12.12}$$

Mechanical equilibrium then requires

$$f\mathbf{k} \times v_h = -\nabla_p \Phi - Kv_h. \tag{12.13}$$

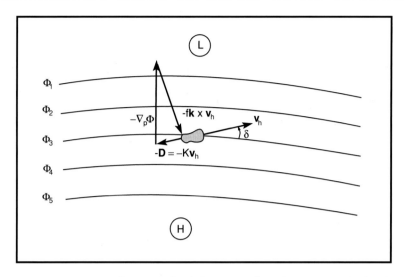

Figure 12.6 Frictional geostrophic balance in a flow characterized by nearly-straight contours of height and in the presence of Rayleigh friction.

Defining *frictional geostrophic equilibrium*, (12.13) may be solved for the components of horizontal motion (Prob. 12.11). Friction modifies both the magnitude and the direction of v_h. The ensuing mechanical equilibrium can be motivated by introducing drag into the geostrophic balance in Fig. 12.1. D reduces the the velocity of the air parcel, which reduces the Coriolis force $-f\boldsymbol{k} \times v_h$ acting on it. That, in turn, allows the pressure-gradient force to drive the parcel's motion across isobars toward low pressure (Fig. 12.6). Eventually, the motion achieves an angle δ from isobars such that the resultant of the three forces acting on the parcel vanishes. The air parcel is then in mechanical equilibrium.

Unlike simple geostrophic motion, frictional geostrophic motion is not solenoidal. The divergent component of v_h introduces vertical motion via continuity (12.9). Friction produces convergence into a center of low surface pressure (Fig. 12.7a). The convergence must be compensated overhead by rising motion. Frictional convergence into a cyclone organizes moisture near the surface, while forcing upward motion. Simultaneously, it reduces vertical stability (Sec. 7.4.4). These features of cyclonic motion encourage convection and cloud formation. Conversely, friction produces divergence out of a center of high surface pressure (Fig. 12.7b). It must be compensated overhead by subsidence. Frictional divergence out of an anticyclone leads to surface air being replaced by dry air from overhead. The accompanying subsidence opposes upward motion. It also produces a subsidence inversion, which traps water vapor and pollutants near the surface. These features of anticyclonic motion inhibit convection and cloud formation, while favoring the accumulation of pollutants.

The foregoing behavior is a consequence of vertical motion, which develops through the small departure from geostrophic equilibrium. The divergent component of motion under frictional geostrophic equilibrium is inversely proportional to f (Prob. 12.12). For this reason, vertical motion is favored in the tropics, where small f allows large departures from geostrophic equilibrium. The latter support thermally-direct

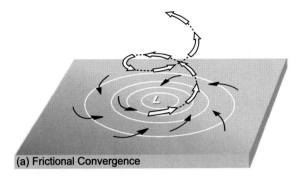

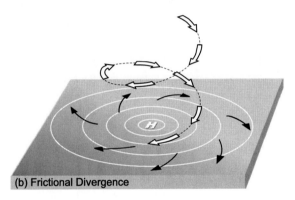

Figure 12.7 Vertical motion introduced by (a) frictional convergence into surface low pressure and (b) frictional divergence out of surface high pressure.

circulations, wherein air ascends over low pressure and descends over high pressure (Figs. 1.35, 1.36).

12.4 CURVILINEAR MOTION

The other factor driving the circulation out of geostrophic equilibrium is material acceleration. If the trajectory of an air parcel is sufficiently curved, it experiences a centripetal acceleration. Embodied in the advective contribution, $\boldsymbol{v}_h \cdot \nabla_p \boldsymbol{v}_h$, that acceleration makes Ro nonnegligible. Unsteadiness can have the same effect through the local time derivative $\frac{\partial \boldsymbol{v}_h}{\partial t}$. However, this contribution is typically less important than advective acceleration.

Consider steady motion in which an air parcel moves along a curved trajectory, characterized by the local radius of curvature R. The parcel trajectory then coincides with a streamline. It is convenient to introduce the *trajectory coordinates s and $\boldsymbol{n}$*, which increase along and orthogonal to the path of an individual parcel (Fig. 12.8). Coordinate vectors s and $\boldsymbol{n}$ are tangential and orthogonal to the local velocity $\boldsymbol{v}_h$, with $\boldsymbol{n}$ pointing to the left of s. R is defined to be positive if the center of curvature lies in the positive $\boldsymbol{n}$ direction (i.e., to the left of s). Curvature is then cyclonic if R is of the same sign as f ($fR > 0$) and anticyclonic if it is of opposite sign ($fR < 0$).

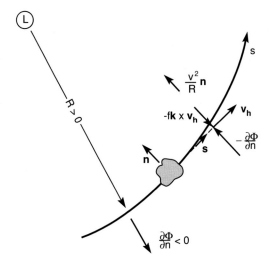

Figure 12.8 Trajectory coordinates s and n, which increase along and orthogonal to the path of an individual parcel. The coordinate vectors s and n are then tangential and orthogonal to the local velocity $\boldsymbol{v}_h$, with n pointing to the left of s to form a right-handed coordinate system: $\boldsymbol{s} \times \boldsymbol{n} = \boldsymbol{k}$. The trajectory's curvature is represented in the radius of curvature R, which is defined to be positive (negative) if the center of curvature lies in the positive (negative) n direction. The local velocity is then described by $\boldsymbol{v}_h = v\boldsymbol{s}$, with speed $v = \frac{ds}{dt}$. Centripetal acceleration $\frac{v^2}{R}\boldsymbol{n}$ follows from an imbalance between the pressure gradient force $-\frac{\partial \Phi}{\partial n}$ and the Coriolis force $-f\boldsymbol{k} \times \boldsymbol{v}_h$.

In this coordinate system, the horizontal velocity is described by

$$\boldsymbol{v}_h = v\boldsymbol{s}, \tag{12.14.1}$$

where the speed is simply

$$v = \frac{ds}{dt}. \tag{12.14.2}$$

The material acceleration of an individual parcel follows as

$$\frac{d\boldsymbol{v}_h}{dt} = \frac{dv}{dt}\boldsymbol{s} + v\frac{d\boldsymbol{s}}{dt}. \tag{12.15}$$

The first term on the right hand side represents longitudinal acceleration along the parcel's trajectory. The second represents centripetal acceleration. It acts transverse to the parcel's motion, following from curvature of its trajectory.

By the chain rule,

$$\begin{aligned}
\frac{d\boldsymbol{s}}{dt} &= \frac{d\boldsymbol{s}}{ds}\frac{ds}{dt} \\
&= v\frac{d\boldsymbol{s}}{ds}.
\end{aligned} \tag{12.16}$$

Analysis similar to that in Sec. 10.6.2 shows that the unit vector $\boldsymbol{s}$ changes at the rate

$$\frac{d\boldsymbol{s}}{dt} = \frac{v}{R}\boldsymbol{n}. \tag{12.17}$$

Substituting (12.16) then obtains

$$\frac{d\boldsymbol{v}_h}{dt} = \frac{dv}{dt}\boldsymbol{s} + \frac{v^2}{R}\boldsymbol{n}, \tag{12.18}$$

which describes the material acceleration in terms of components longitudinal and transverse to the motion. As $R \rightarrow \infty$, the centripetal acceleration approaches zero and the parcel's motion becomes rectilinear.

Incorporating (12.18) transforms the horizontal momentum equation (12.1) into

$$\frac{dv}{dt}\boldsymbol{s} + \frac{v^2}{R}\boldsymbol{n} = -fv\boldsymbol{n} - \frac{\partial \Phi}{\partial s}\boldsymbol{s} - \frac{\partial \Phi}{\partial n}\boldsymbol{n}. \tag{12.19}$$

The component equations in the $\boldsymbol{s}$ and $\boldsymbol{n}$ directions are thus:

$$\frac{dv}{dt} = -\frac{\partial \Phi}{\partial s} \tag{12.20.1}$$

$$\frac{v^2}{R} + fv = -\frac{\partial \Phi}{\partial n}. \tag{12.20.2}$$

Consider now motion that is tangential to contours of height, which serve as streamlines. Then $\frac{\partial \Phi}{\partial s}$ vanishes and (12.20.1) implies that $v = \text{const}$ along an individual streamline. Accordingly, we focus on the momentum budget in the normal direction. The centripetal acceleration $\frac{v^2}{R}$ represents the material acceleration in (11.37). It is $O(Ro)$, where the Rossby number

$$Ro = \frac{U}{fR}$$

is based on the length scale R. According to (12.20.2), centripetal acceleration follows from an imbalance between the Coriolis and pressure-gradient forces.

12.4.1 Inertial motion

If the centripetal and Coriolis accelerations are in balance,

$$\frac{v^2}{R} + fv = 0. \tag{12.21.1}$$

This balance applies in the absence of pressure gradient. The radius of curvature follows as

$$R = -\frac{v}{f}. \tag{12.21.2}$$

On an f plane, $R = \text{const}$. Streamlines then form circular orbits that are anticyclonic ($fR < 0$). The period of revolution is given by

$$\tau = \frac{\pi}{\Omega sin\phi}. \tag{12.22}$$

τ equals half a *sidereal day*: the time for a Foucalt pendulum to sweep through 180° of azimuth. The motion (12.21) describes an *inertial oscillation*. Parcels revolve opposite to the vertical component of planetary vorticity. Inertial oscillations have been observed in the oceans. However, they do not play a major role in the atmospheric circulation.

12.4.2 Cyclostrophic motion

If either R or f approaches zero, $Ro \to \infty$. These circumstances apply for small horizontal scale or near the equator. The material acceleration then dominates the Coriolis acceleration. Under these conditions, (12.20.2) reduces to the balance

$$\frac{v^2}{R} = -\frac{\partial \Phi}{\partial n}, \tag{12.23.1}$$

which defines *cyclostrophic equilibrium*. The *cyclostrophic wind speed*

$$v_c = \sqrt{-R\frac{\partial \Phi}{\partial n}} \tag{12.23.2}$$

may be either cyclonic or anticyclonic. However, it is always about low pressure (e.g., $\frac{\partial \Phi}{\partial n}$ has sign opposite to R).

   Cyclostrophic equilibrium applies to motion (1) with strong curvature, wherein the advective acceleration is large, or (2) in the presence of slow planetary rotation f, wherein the Coriolis force is weak. The former is typical of small-scale vortices like tornadoes. The latter is important in tropical cyclones. Slow planetary rotation is also a feature of the Venusian atmosphere. Unlike the Earth's atmosphere, the circulation of Venus is governed by cyclostrophic equilibrium.

12.4.3 Gradient motion

To first order in Ro, all three terms in (12.20.2) are retained. The resulting balance defines *gradient equilibrium*. Gradient wind balance describes horizontal motion when no two of the terms dominate. This often applies to tropical motions, for which fR is small.

   Equation (12.20.2) is quadratic in v. It can be solved for the *gradient wind speed*

$$v_{gr} = -\frac{fR}{2} \pm \sqrt{\frac{f^2 R^2}{4} - R\frac{\partial \Phi}{\partial n}}. \tag{12.24}$$

Not all roots of (12.24) are physically meaningful. Real wind speed requires

$$\frac{\partial \Phi}{\partial n} \begin{cases} < \frac{f^2 R}{4} & (R > 0) \\ \\ > \frac{f^2 R}{4} & (R < 0). \end{cases} \tag{12.25}$$

Subsequent analysis is restricted to regular cyclonic or anticyclonic motion in the Northern Hemisphere, for which $\frac{\partial \Phi}{\partial n} < 0$ (Fig. 12.8). This corresponds to cyclonic motion about a low and anticyclonic motion about a high, as prevails under geostrophic equilibrium (i.e., to zero order in Ro). The second inequality in (12.25) limits the pressure gradient that can be sustained by anticyclonic motion. Under these conditions, $\frac{\partial \Phi}{\partial n} < 0$ and $R < 0$. $\frac{\partial \Phi}{\partial n}$ is then sandwiched between $\frac{f^2 R}{4}$ and 0. Near the center of an anticyclone, $R \to 0$. The height distribution must therefore become flat and the accompanying motion weak (12.24). For cyclonic motion, no corresponding limitation applies. With $R > 0$, $\frac{\partial \Phi}{\partial n} < 0$ automatically satisfies the first inequality in (12.25). Hence, cyclones are free to intensify, whereas the amplification of anticyclones is limited. This distinction explains why wind inside an anticyclone is comparatively weak. It follows from the fact that the pressure-gradient force $(-\frac{\partial \Phi}{\partial n})$ reinforces the centripetal acceleration in a cyclone but opposes it in an anticyclone; cf. Fig. 12.8.

Letting $R \to \infty$ $(Ro \to 0)$ in (12.20.2) recovers geostrophic equilibrium. Conversely, letting $R \to 0$ $(Ro \to \infty)$ recovers cyclostrophic equilibrium. Intermediate to those extremes is the gradient wind. It is only approximated by the geostrophic wind. Eliminating the pressure gradient in favor of the geostrophic wind speed

$$v_g = -\frac{1}{f}\frac{\partial \Phi}{\partial n}$$

transforms (12.20.2) into

$$\frac{v_g}{v_{gr}} = 1 + \frac{v_{gr}}{fR}. \tag{12.26}$$

The geostrophic wind speed overestimates the gradient wind speed in cyclonic motion ($fR > 0$). It underestimates it in anticyclonic motion ($fR < 0$). The discrepancy from gradient wind (e.g., the error) is proportional to Ro. For large-scale extratropical motion, the discrepancy is of order 10–20%. However, discrepancies as large as 50–100% can occur inside intense cyclones and in tropical systems characterized by small fR.

12.5 WEAKLY DIVERGENT MOTION

The idealized balances considered in the preceding section are useful for describing the structure of large-scale motion. However, they provide no information on how the circulation evolves. Free of time derivatives, such equations are termed *diagnostic*. Unsteadiness and divergence are comparatively small. Nevertheless, they enable the circulation to evolve from one state to another. Equations that describe such changes are termed *prognostic*. A simple framework for describing how the circulation evolves may be developed in the absence of vertical motion.

12.5.1 Barotropic nondivergent motion

Consider motion that is horizontally nondivergent and under barotropic stratification. Continuity then implies that vertical motion vanishes (12.9). Thermal wind balance (12.11) implies that v_h is independent of height. Consequently, the motion is 2-dimensional or columnar. The governing equations in physical coordinates then reduce to

$$\frac{dv_h}{dt} + f\mathbf{k} \times v_h = -\alpha\nabla p \tag{12.27.1}$$

$$\nabla \cdot v_h = 0, \tag{12.27.2}$$

where friction is ignored and ∇ is understood to refer to the horizontal gradient. Vector identity (D.14) in Appendix D enables (12.27.1) to be written

$$\frac{\partial v_h}{\partial t} + \frac{1}{2}\nabla(v_h \cdot v_h) + (\zeta + f)\mathbf{k} \times v_h = -\alpha\nabla p, \tag{12.28.1}$$

where

$$\zeta = (\nabla \times v_h) \cdot \mathbf{k}$$
$$= \frac{\partial v}{\partial x} - \frac{\partial u}{\partial y} \tag{12.28.2}$$

is the vertical component of *relative vorticity*. Consistent with prior convention, relative vorticity is cyclonic if it has the same sign as the planetary vorticity f and anticyclonic otherwise. Applying the curl to (12.28.1) and making use of other vector identities in Appendix D yields the vorticity budget for an air parcel

$$\frac{\partial \zeta}{\partial t} + \boldsymbol{v}_h \cdot \nabla \zeta + \boldsymbol{v}_h \cdot \nabla f = (\nabla p \times \nabla \alpha) \cdot \boldsymbol{k}.$$

Because $\alpha = \alpha(p)$ under barotropic stratification and $f = f(y)$, this reduces to

$$\frac{d\zeta}{dt} + v\frac{\partial f}{\partial y} = 0. \tag{12.29}$$

Because

$$v\frac{\partial f}{\partial y} = \frac{df}{dt},$$

the vorticity budget can be expressed

$$\frac{d(\zeta + f)}{dt} = 0. \tag{12.30}$$

$(\zeta + f)$ represents the *absolute vorticity* of air, namely, that apparent to an observer in an inertial reference frame. According to (12.30), $(\zeta + f)$ is conserved for an individual air parcel. Thus, under barotropic nondivergent conditions, absolute vorticity behaves as a tracer of horizontal air motion. It is simply rearranged by the circulation. Air moving equatorward experiences decreasing f. It must compensate by spinning up cyclonically, increasing its relative vorticity ζ (cf. the outbreak of cold polar air in the cyclone west of Africa in Figs. 1.28 and 1.29). Air moving poleward must evolve in the opposite manner, decreasing its relative vorticity. On large scales, the circulation tends to remain close to barotropic stratification. For this reason, absolute vorticity is approximately conserved.

Because it is nondivergent, the preceding motion may be represented in terms of a streamfunction (12.6), which automatically satisfies the continuity equation (12.27.2). Incorporating the vorticity (12.7.3) then transforms (12.30) into

$$\left(\frac{\partial}{\partial t} - \frac{\partial \psi}{\partial y}\frac{\partial}{\partial x} + \frac{\partial \psi}{\partial x}\frac{\partial}{\partial y}\right)\left\{\nabla^2 \psi + f\right\} = 0, \tag{12.31}$$

where ψ is the geostrophic streamfunction, e.g., (12.8). Known as the *barotropic nondivergent vorticity equation*, (12.31) provides a closed prognostic system for the single unknown ψ. It governs the structure of motion and how it evolves. Advective terms in the material derivative make (12.31) quadratically nonlinear in the dependent variable ψ. With suitable initial conditions and boundary conditions, the barotropic nondivergent vorticity equation can be integrated for the motion at some later time. It was applied to perform seminal experiments in numerical weather prediction (e.g., Charney et al., 1950).

12.5.2 Vorticity budget under baroclinic stratification

Under more general circumstances, absolute vorticity is not conserved. However, the governing equations still possess a conservation principle, albeit more complex. Baroclinic stratification renders the motion 3-dimensional. Vorticity is then a

3-dimensional vector quantity. Despite this complication, the quasi-horizontal nature of large-scale motion makes the vertical component of vorticity dominant.

Exclusive of friction, the horizontal momentum equations in physical coordinates are

$$\frac{\partial u}{\partial t} + \boldsymbol{v} \cdot \nabla u - f v = -\alpha \frac{\partial p}{\partial x} \tag{12.32.1}$$

$$\frac{\partial v}{\partial t} + \boldsymbol{v} \cdot \nabla v + f u = -\alpha \frac{\partial p}{\partial y}, \tag{12.32.2}$$

where $\boldsymbol{v}$ and ∇ denote the full 3-dimensional velocity and gradient. Cross differentiating (12.32) with respect to x and y and subtracting obtains

$$\left(\frac{\partial}{\partial t} + \boldsymbol{v} \cdot \nabla\right)\left(\frac{\partial v}{\partial x} - \frac{\partial u}{\partial y}\right) + \frac{\partial v}{\partial x} \cdot \nabla v - \frac{\partial v}{\partial y} \cdot \nabla u + f\left(\frac{\partial u}{\partial x} + \frac{\partial v}{\partial y}\right) + v\frac{\partial f}{\partial y} = \frac{\partial p}{\partial x}\frac{\partial \alpha}{\partial y} - \frac{\partial p}{\partial y}\frac{\partial \alpha}{\partial x}.$$

This may be consolidated into

$$\frac{d(\zeta + f)}{dt} + f\nabla_z \cdot \boldsymbol{v}_h + \left(\frac{\partial \boldsymbol{v}}{\partial x} \cdot \nabla v - \frac{\partial \boldsymbol{v}}{\partial y} \cdot \nabla u\right) = (\nabla_z p \times \nabla_z \alpha) \cdot \boldsymbol{k}. \tag{12.33}$$

The third term in (12.33) may be written

$$\frac{\partial \boldsymbol{v}_h}{\partial x} \cdot \nabla v - \frac{\partial \boldsymbol{v}}{\partial y} \cdot \nabla u = \frac{\partial u}{\partial x}\left(\frac{\partial v}{\partial x} - \frac{\partial u}{\partial y}\right) + \frac{\partial v}{\partial y}\left(\frac{\partial v}{\partial x} - \frac{\partial u}{\partial y}\right) + \frac{\partial w}{\partial x}\frac{\partial v}{\partial z} - \frac{\partial w}{\partial y}\frac{\partial u}{\partial z}$$

$$= \zeta\nabla_z \cdot \boldsymbol{v}_h + \frac{\partial w}{\partial x}\frac{\partial v}{\partial z} - \frac{\partial w}{\partial y}\frac{\partial u}{\partial z}. \tag{12.34}$$

Similarly, the last term in (12.34) may be expressed

$$\frac{\partial w}{\partial x}\frac{\partial v}{\partial z} - \frac{\partial w}{\partial y}\frac{\partial u}{\partial z} = -\left(\xi\frac{\partial w}{\partial x} + \eta\frac{\partial w}{\partial y}\right), \tag{12.35.1}$$

where

$$\xi = \frac{\partial w}{\partial y} - \frac{\partial v}{\partial z} \tag{12.35.2}$$

$$\eta = \frac{\partial u}{\partial z} - \frac{\partial w}{\partial x} \tag{12.35.3}$$

represent the horizontal components of vector vorticity $\boldsymbol{\zeta} = (\xi, \eta, \zeta)$. Then the budget for the vertical component of vorticity becomes

$$\frac{d(\zeta + f)}{dt} = -(\zeta + f)\nabla_z \cdot \boldsymbol{v}_h + \left(\xi\frac{\partial w}{\partial x} + \eta\frac{\partial w}{\partial y}\right) + (\nabla_z p \times \nabla_z \alpha) \cdot \boldsymbol{k}, \tag{12.36}$$

wherein the Lagrangian derivative includes full 3-dimensional advection.

According to (12.36), the absolute vorticity of an air parcel can change through three mechanisms (Fig. 12.9):

<u>Vertical stretching:</u> By continuity, horizontal convergence $-\nabla_z \cdot \boldsymbol{v}_h$ is compensated by vertical stretching an air parcel (Fig. 12.9a). The reduced moment of inertia then leads to that material element spinning up to conserve angular momentum (in the same sense as its absolute vorticity). Horizontal divergence leads to the material element spinning down to conserve angular momentum.

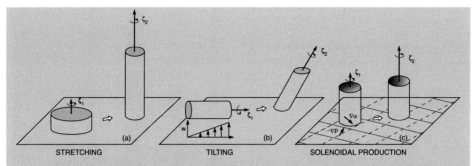

Figure 12.9 Forcing of the absolute vorticity $(f + \zeta)$ of an air parcel by (a) vertical stretching, which is compensated by horizontal convergence, (b) tilting, which exchanges horizontal and vertical vorticity, and (c) solenoidal production, which results from variation of density (stipled) across the pressure gradient force. The latter exerts a torque on the air parcel that is located inside a *solenoid* defined by two intersecting isobars (solid) and isochores (dashed).

Tilting: Horizontal shear of the vertical velocity $\nabla_z w$ deflects an air parcel, along with its vector angular momentum (Fig. 12.9b). This process transfers vorticity from horizontal components into the vertical component. A sharp horizontal temperature gradient is accompanied by strong vertical shear of the horizontal velocity (12.11). That shear magnifies the horizontal components of vorticity ξ and η, making tilting an important source of vertical vorticity.

Solenoidal production: Under baroclinic stratification, isobars on a surface of constant height do not coincide with isochores. Two sets of intersecting isobars and isochores then define a *solenoid* (Fig. 12.9c). Across it, pressure varies in one direction, but density varies in another. Variation of density across the pressure-gradient force then introduces a torque. The latter changes the angular momentum of the air parcel occupying a solenoid. Under barotropic stratification, density is uniform across the pressure gradient. The torque and solenoidal production of vorticity $\nabla_z p \times \nabla_z \alpha$ then vanish.

These forcing mechanisms reflect rearrangements of angular momentum – not its dissipation, which has been excluded by considering adiabatic and inviscid conditions. Scale analysis indicates that, in large-scale motion, vertical stretching dominates the forcing of absolute vorticity. Other mechanisms, which are generally smaller, derive from baroclinic stratification. The dominance of vertical stretching is a reflection of the nearly barotropic nature of large-scale motion.

If the vorticity budget is expressed in isentropic coordinates, tilting and solenoidal production disappear under adiabatic conditions. Vector identities in Appendix D enable the horizontal momentum equation (11.92.1) to be expressed

$$\frac{\partial \boldsymbol{v}_h}{\partial t} + \frac{1}{2}\nabla_\theta \left(\boldsymbol{v}_h \cdot \boldsymbol{v}_h\right) + \left(\zeta_\theta + f\right)\boldsymbol{k} \times \boldsymbol{v}_h + \omega_\theta \frac{\partial \boldsymbol{v}_h}{\partial \theta} = -\nabla_\theta \Psi, \qquad (12.37)$$

where $\zeta_\theta = (\nabla_\theta \times \boldsymbol{v}_h) \cdot \boldsymbol{k}$ is the relative vorticity evaluated on an isentropic surface. Under adiabatic conditions, the term representing vertical advection disappears.

Applying $k \cdot \nabla_\theta \times$ to (12.37) then obtains the vorticity budget

$$\frac{\partial(\zeta_\theta + f)}{\partial t} + v_h \cdot \nabla_\theta(\zeta_\theta + f) = -(\zeta_\theta + f)\nabla_\theta \cdot v_h$$

or

$$\frac{d(\zeta_\theta + f)}{dt} = -(\zeta_\theta + f)\nabla_\theta \cdot v_h. \tag{12.38}$$

The Lagrangian derivative now involves only horizontal advection. Because vertical motion vanishes identically, so does the tilting that appears in physical coordinates (12.36). Likewise, solenoidal production vanishes because the vertical vorticity ζ_θ is evaluated on an isentropic surface. Absolute vorticity then changes only through vertical stretching. Because of the simple balance in (12.38), horizontal velocity can be treated through the scalar properties, vorticity and divergence. This enables motion to be integrated in numerical frameworks more effectively than can velocity directly (Bourke, 1974). For this reason, such treatment is widely adopted in GCMs (see, e.g., Williamson et al., 1987).

Under adiabatic conditions, the continuity equation in isentropic coordinates (11.92.3) reduces to

$$\frac{d}{dt}\left(\frac{\partial p}{\partial \theta}\right) + \left(\frac{\partial p}{\partial \theta}\right)\nabla_\theta \cdot v_h = 0. \tag{12.39}$$

Combining this with (12.38) yields

$$\frac{1}{(\zeta_\theta + f)}\frac{d(\zeta_\theta + f)}{dt} = \frac{1}{\left(\frac{\partial p}{\partial \theta}\right)}\frac{d}{dt}\left(\frac{\partial p}{\partial \theta}\right).$$

The vorticity budget may then be expressed

$$\frac{dQ}{dt} = 0, \tag{12.40.1}$$

where

$$Q = \frac{\zeta_\theta + f}{-\frac{1}{g}\left(\frac{\partial p}{\partial \theta}\right)} \tag{12.40.2}$$

is one of several equivalent definitions of the *potential vorticity*, also called the *Ertel potential vorticity* (Ertel, 1942) and the *isentropic potential vorticity*.

Equation (12.40) is a generalization of the conservation principle for barotropic nondivergent motion. It asserts that, under inviscid adiabatic conditions, the potential vorticity of an air parcel is conserved. The parcel's absolute vorticity $(\zeta_\theta + f)$ can change. However, it does so in direct proportion to the change in vertical spacing $-\frac{1}{g}\left(\frac{\partial p}{\partial \theta}\right)$ of isentropic surfaces (11.85), which, under adiabatic conditions, are material surfaces. Despite its name, the conserved property Q does not have dimensions of vorticity; the denominator in (12.40.2) introduces other dimensions.

Potential vorticity is a dynamical tracer of horizontal motion. Under the same conditions that Q is conserved, vertical motion is determined by variations in the elevation of an isentropic surface, on which Q is evaluated. Unlike chemical tracers, potential vorticity is not a passive tracer: It is not advected "passively." Rather, the vorticity at a point induces motion about it. This property makes Q and the motion

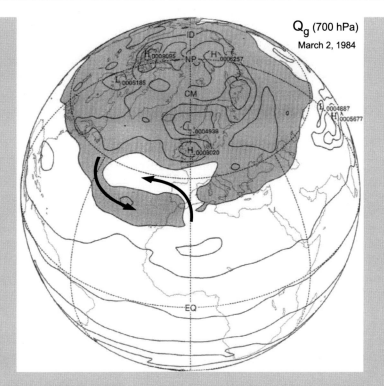

Figure 12.10 Distribution of quasi-geostrophic potential vorticity at 700 hPa on March 2, 1984, two days before the structure in Figs. 1.9a, 1.29. A cyclone off the coast of Africa has rearranged air, producing a dipole pattern in which high Q_g (shaded) has been folded south of low Q_g (cf. Fig. 16.6). A similar pattern appears over Europe, where Q_g has been folded by another cyclone.

field completely dependent on one another. According to the so-called *invertibility principle* (Hoskins et al., 1985), the distribution of Q at some instant uniquely determines the circulation. The conserved property Q may therefore be regarded as being self-advected.

In practice, friction and diabatic effects are slow enough away from the surface and outside convection for Q to be conserved over many advection times. Synoptic weather systems are marked by anomalies of potential vorticity (see Fig. 12.10). Forecasting such systems amounts to predicting the distribution of Q, which is controlled primarily by advection. Over time scales much longer than a day, nonconservative processes lead to production and destruction of Q (e.g., inside an individual air parcel). Turbulence, radiative transfer, and condensation eventually render the behavior of an air parcel diabatic. That, in turn, forces motion across isentropic surfaces. Such irreversibility leads to a vertical drift of air, motion that is ultimately manifest in the mean meridional circulation of the atmosphere (Sec. 3.6.1).

Unlike barotropic nondivergent motion (12.31), behavior described by (12.40) cannot be represented exclusively in terms of a streamfunction. Consequently, the vorticity budget alone does not provide sufficient information to describe the motion. Divergence, although comparatively small, interacts with vorticity in (12.38). In large-scale numerical integrations, a closed prognostic system is formed by augmenting the vorticity budget with the budget of divergence, which follows in similar fashion (see, e.g., Haltiner and Williams, 1980).

12.5.3 Quasi-geostrophic motion

Under baroclinic and diabatic conditions, the governing equations are complex. An approximate prognostic system governing large-scale motion is therefore useful. Such a system can be derived via an expansion in Ro – as was used to introduce the diagnostic description of geostrophic motion. For the resulting system to remain prognostic, the expansion must retain terms higher order than those in (12.2). In particular, it must retain horizontal divergence, which forces changes of absolute vorticity through vertical stretching (12.38).

We consider motion in which meridional displacements of air are $O(Ro)$ and hence sufficiently narrow to ignore the Earth's curvature. Analogous to the f-plane development, spherical geometry is approximated in a Cartesian system tangent to the Earth at latitude ϕ_0 (Fig. 12.3). For consistency with other terms in the equations, the variation with latitude of the Coriolis force must now be retained to order Ro. Expanding f in a Taylor series in y (which reflects the meridional displacement of an air parcel) and truncating to first order yields the β-plane approximation

$$f = f_0 + \beta y \qquad (12.41.1)$$

$$\beta = \left.\frac{df}{dy}\right|_{y=0}, \qquad (12.41.2)$$

where f_0 and y are given by (12.3). The motion is also presumed to satisfy the so-called *Boussinesq approximation*: Variations of density are ignored, except where they accompany gravity in the buoyancy force. The Boussinesq approximation neglects changes in the density of an air parcel that are introduced by changes of its pressure. It retains only those density changes that are introduced thermally. Vertical displacements of air must then be shallow compared with H.

Following the development in Sec. 11.4, all variables are nondimensionalized by characteristic scale factors. Dependent variables are then expanded in a power series in Rossby number. The dimensionless horizontal velocity becomes

$$\boldsymbol{v}_h = \boldsymbol{v}_g + Ro\boldsymbol{v}_a + \ldots, \qquad (12.42)$$

where $\boldsymbol{v}_g$ satisfies geostrophic balance with $f = f_0$ and the *ageostrophic velocity* $\boldsymbol{v}_a$ represents an $O(Ro)$ correction to it. Both are $O(1)$, as are all dependent variables in a formal scale analysis that consolidates the magnitudes of terms into dimensionless parameters like Ro (e.g., Charney, 1973). To streamline the development, we depart from the formal procedure by absorbing dimensionless parameters other than Ro into the dependent variables. Their magnitudes can then be inferred from the ensuing equations. This treatment enables the dimensionless equations to be expressed in forms nearly identical to their dimensional counterparts.

In log-pressure coordinates, the horizontal momentum equation, exclusive of friction, becomes

$$Ro\left\{\frac{\partial}{\partial t} + (\boldsymbol{v}_g + Ro\boldsymbol{v}_a + \ldots) \cdot \nabla_z\right\} (\boldsymbol{v}_g + Ro\boldsymbol{v}_a + \ldots) + Row\frac{\partial}{\partial z}(\boldsymbol{v}_g + Ro\boldsymbol{v}_a + \ldots)$$
$$+ (f_0 + Ro\beta y)\boldsymbol{k} \times (\boldsymbol{v}_g + Ro\boldsymbol{v}_a + \ldots) \tag{12.43}$$
$$= -\nabla_z(\Phi_g + Ro\Phi_a + \ldots),$$

where z denotes log-pressure height (11.65) and factors like β are understood to be dimensionless and $O(1)$. Because $\boldsymbol{v}_g$ satisfies (12.2), the Ro^0 balance drops out of (12.43). Under the Boussinesq approximation, w is smaller than $O(Ro)$, denoted $o(Ro)$. Thus vertical advection of momentum also drops out to $O(Ro)$. The momentum balance at $O(Ro)$ is then given by

$$\frac{d_g\boldsymbol{v}_g}{dt} + f_0\boldsymbol{k} \times \boldsymbol{v}_a + \beta y\boldsymbol{k} \times \boldsymbol{v}_g = -\nabla_z\Phi_a, \tag{12.44.1}$$

where

$$\frac{d_g}{dt} = \frac{\partial}{\partial t} + \boldsymbol{v}_g \cdot \nabla_z \tag{12.44.2}$$

is the time rate of change moving horizontally with the geostrophic velocity. According to (12.44), first-order departures from simple geostrophic equilibrium introduce a time rate of change or *tendency* into $\boldsymbol{v}_g$, behavior that is absent at $O(Ro^0)$.

The continuity equation (11.72.3) becomes

$$\frac{1}{\rho_0}\frac{\partial}{\partial z}(\rho_0 w) = -\nabla_z \cdot (\boldsymbol{v}_g + Ro\boldsymbol{v}_a + \ldots)$$
$$= -Ro\nabla_z \cdot \boldsymbol{v}_a. \tag{12.45}$$

Thus, w follows directly from $\boldsymbol{v}_a$, through the associated divergence. The ageostrophic velocity introduces a *secondary circulation*. Unlike the $O(Ro^0)$ geostrophic motion, the secondary circulation involves vertical motion. Under adiabatic conditions, the thermodynamic equation (11.72.4) becomes

$$\left\{\frac{\partial}{\partial t} + (\boldsymbol{v}_g + Ro\boldsymbol{v}_a + \ldots) \cdot \nabla_z\right\}\frac{\partial}{\partial z}\left(\Phi_g + Ro\Phi_a + \ldots\right) + N^2 w = 0$$

or

$$\frac{d_g}{dt}\left(\frac{\partial\Phi_g}{\partial z}\right) + N^2 w = 0, \tag{12.46}$$

where N^2 is evaluated with the basic stratification T_0. The Boussinesq approximation requires N^2 to be large enough to render vertical displacements small. Therefore, the second term in (12.46) must be retained to leading order.

Equations (12.44), (12.45), and (12.46) describe *quasi-geostrophic motion* on a β-plane. Accurate to $O(Ro)$, they have dimensional counterparts that are nearly identical. Expressing the momentum budget as in (12.28) and applying $\boldsymbol{k} \cdot \nabla_z \times$ obtains the vorticity equation

$$\frac{d_g\zeta_g}{dt} + v\beta + f_0\nabla_z \cdot \boldsymbol{v}_a = 0.$$

The budget of vorticity then becomes

$$\frac{d_g(\zeta_g + f)}{dt} = -f_0 \nabla_z \cdot \boldsymbol{v}_a,$$ (12.47.1)

where

$$\begin{aligned}\zeta_g &= \nabla_z^2 \psi \\ &= \frac{1}{f_0} \nabla_z^2 \Phi_g\end{aligned}$$ (12.47.2)

is the geostrophic vorticity. This is similar to the full vorticity equation in isentropic coordinates (12.38). However, the conditions of quasi-geostrophy make the relative vorticity ζ_g negligible compared with the planetary vorticity f_0 in the divergence term. With the continuity equation, (12.47) may be written

$$\frac{d_g(\zeta_g + f)}{dt} = f_0 \frac{1}{\rho_0} \frac{\partial}{\partial z}(\rho_0 w).^4$$

Then applying the operator $\frac{1}{\rho_0} \frac{\partial}{\partial z}(\rho_0)$ to the thermodynamic equation and eliminating w yields the *quasi-geostrophic potential vorticity equation*

$$\frac{d_g Q_g}{dt} = 0,$$ (12.48.1)

where

$$Q_g = \left\{ \nabla_z^2 \psi + f_0 + \beta y + \frac{1}{\rho_0} \frac{\partial}{\partial z}\left(\frac{f_0^2}{N^2} \rho_0 \frac{\partial \psi}{\partial z}\right)\right\}$$ (12.48.2)

is the *quasi-geostrophic potential vorticity*. Under the prevailing conditions, Q_g is a conserved property: It is invariant for individual air parcels. Because $\frac{d_g}{dt}$ involves only $\boldsymbol{v}_g$, (12.48) may be expressed in terms of the single dependent variable ψ

$$\left(\frac{\partial}{\partial t} - \frac{\partial \psi}{\partial y}\frac{\partial}{\partial x} + \frac{\partial \psi}{\partial x}\frac{\partial}{\partial y}\right)\left\{\nabla_z^2 \psi + f_0 + \beta y + \frac{1}{\rho_0}\frac{\partial}{\partial z}\left(\frac{f_0^2}{N^2}\rho_0\frac{\partial \psi}{\partial z}\right)\right\} = 0.$$ (12.49)

This is a generalization of the barotropic nondivergent vorticity equation (12.31) to weakly divergent motion.

Equation (12.49) constitutes a closed prognostic system for the streamfunction ψ. Like (12.31), it is quadratically nonlinear. With suitable initial and boundary conditions, it can be integrated for the structure and evolution of the motion. The conserved property Q_g is an analogue of isentropic potential vorticity (12.40.2). However, it is invariant following $\boldsymbol{v}_g$ rather than the actual horizontal motion.

Figure 12.10 shows the distribution of Q_g at 700 hPa 2 days prior to the analysis in Fig. 1.9. In the eastern Atlantic, cold high-Q_g air (shaded) is being drawn southeastward about a cyclone which matures into the one evident in Figs. 1.28 and 1.29. That cold air is being exchanged with warm low-Q_g air which is likewise being drawn about the cyclone, but in the opposite sense. These features mark a poleward transfer of warm air and an equatorward transfer of cold air. Together, they represent a poleward

---

[4] Under fully Boussinesq conditions, the stretching term simplifies to $f_0 \frac{\partial w}{\partial z}$ – the same as under incompressible conditions. With the density weighting retained, the approximation is referred to as *quasi-Boussinesq*. It accounts for vertical structure associated with the amplification of atmospheric waves (Chap. 14).

transfer of heat. The rearrangement of air introduces a dipole pattern, with high Q_g folded south of low Q_g. Similar structure appears over northern Europe, where the distribution of Q_g has been folded by a cyclone over the North Sea. Such systems are a ubiquitous feature of the tropospheric circulation. They continually develop along the jet stream, which coincides with the sharp temperature gradient that marks the polar front (Figs. 6.3, 6.4). The jet stream is thereby distorted, as warm air is swept poleward and cold air is swept equatorward (cf. Fig. 16.7). By exchanging high- and low-Q_g air, these disturbances to the jet stream transfer heat poleward. This dynamical heat transfer offsets radiative heating at low latitudes and radiative cooling at middle and high latitudes, maintaining thermal equilibrium locally (Sec. 1.5).

Observed structure of Q_g, like that in Fig 12.10, must be derived indirectly from temperature observations (12.48.2). Because that treatment requires repeated differentiation, the resulting structure tends to be noisy, contaminated by error. However, the basic signature emerges clearly in satellite imagery of water vapor, which, because it is conserved outside of cloud, registers large-scale air motion directly (cf. Figs. 16.6, 16.7).

The quasi-geostrophic equations describe motion in which the rotational component is dominant, only weakly coupled to the divergent component. As the equator is approached, f becomes small enough to invalidate those conditions. The divergent component of motion is then comparable to the rotational component. Tropical circulations must therefore be treated with the full primitive equations or with forms discriminated to the tropics, in which divergence and vorticity are both retained to leading order.

SUGGESTED REFERENCES

An Introduction to Dynamic Meteorology (2004) by Holton gives a complete treatment of gradient wind and an introduction to the concepts of vorticity and circulation.

Advanced treatments are presented in *Geophysical Fluid Dynamics* (1979) by Pedlosky and *Atmospheric and Oceanic Fluid Dynamics* (2006) by Vallis.

Hoskins et al. (1985) provides an overview of isentropic potential vorticity. Comparisons with quasi-geostrophic potential vorticity are given in *Atmosphere-Ocean Dynamics* (1982) by Gill.

PROBLEMS

1. The bullet train moves over straight and level track at $40°$ latitude. If the train's mass is 10^5 kg, how fast must it move to experience a transverse force equal to 0.1% of its weight?

2. Surface motion beneath a cyclone is described by the streamfunction and velocity potential

$$\psi(x, y) = \Psi \left[1 - e^{-\frac{(x^2+y^2)}{L^2}} \right]$$

$$\chi(x, y) = X e^{-\frac{(x^2+y^2)}{L^2}},$$

where Ψ and X are constants. Determine (a) the horizontal velocity, (b) the horizontal divergence, and (c) the vertical vorticity. (d) Sketch $\boldsymbol{v}_h(x, y)$ and discuss it in relation to $\psi(x, y)$ and $\chi(x, y)$. (e) Indicate the horizontal trajectory of an air parcel and discuss it in relation to 3-dimensional air motion.

3. The 500-hPa surface has height

$$
z_{500} = \begin{cases} \overline{\overline{z}}_{500} - \overline{z}_{500}\left(\frac{y}{a}\right) - z'_{500}\left[1 + cos\left(\pi \frac{x^2+y^2}{L^2}\right)\right] & (x^2+y^2)^{\frac{1}{2}} \le L \\ \overline{\overline{z}}_{500} - \overline{z}_{500}\left(\frac{y}{a}\right) & (x^2+y^2)^{\frac{1}{2}} > L, \end{cases}
$$

where the constants $\overline{\overline{z}}_{500}$, $\overline{z}_{500}$, and z'_{500} reflect global-mean, zonal-mean, and perturbation contributions to the height field; a is the radius of the Earth; and y is measured on a midlatitude f plane centered at latitude ϕ_0. (a) Provide an expression for the 500-hPa geostrophic flow $\boldsymbol{v}_{500}$. (b) Sketch $z_{500}(x, y)$ and $\boldsymbol{v}_{500}(x, y)$, presuming $z'_{500} \ll \overline{z}_{500}$. Discuss the contributions to $\boldsymbol{v}_{500}$, noting how the flow is influenced by the thermal perturbation. (c) Provide an expression for the absolute vorticity at 500 hPa and discuss its contributions.

4. A midlatitude cyclone centered at 45 N produces a radial height gradient at 500 hPa of 0.16 m km^{-1} at a characteristic radial distance of $R = a\Delta\theta$, with $\Delta\theta = 10°$. If the motion is sufficiently steady, calculate the corresponding (a) geostrophic wind and (b) gradient wind.

5. As in Prob. 12.4, but for a tropical cyclone centered at 10 N, which produces a radial height gradient at 850 hPa of 1.88 m km^{-1} at a characteristic radial distance corresponding to $\Delta\theta = 1°$.

6. A vertical sounding at 45 N reveals isothermal conditions with $T = 0°C$ below 700 hPa, but vertical shear characterized by the following wind profile:

$$
v = \begin{cases} 10 \text{ m s}^{-1} & \text{S} & 950 \text{ hPa} \\ 14.14 \text{ m s}^{-1} & \text{SW} & 850 \text{ hPa} \\ 10 \text{ m s}^{-1} & \text{S} & 750 \text{ hPa,} \end{cases}
$$

where the wind blows from direction indicated. (a) Calculate the mean horizontal temperature gradient in the layers between 950 hPa and 850 hPa and between 850 hPa and 750 hPa. (b) Estimate the rate at which the preceding layers would warm/cool locally through temperature advection (e.g., were temperature approximately conserved). (c) Use the result of (b) to estimate the lapse rate created between those layers after one day, noting the physical implications.

7. Consider the core of a tornado that is isothermal and in rigid-body rotation with angular velocity Ω. Derive an expression for the radial profile of pressure inside the tornado's core in terms of the pressure p_0 at its center.

8. Isobaric height in the Venusian atmosphere varies locally as

$$
\Phi(x, y) = \overline{\Phi}\left[1 + \epsilon \frac{x^2 + y^2}{L^2}\right].
$$

(a) If the motion is steady and frictionless, provide an expression for the horizontal velocity. (b) Can a similar result be obtained if the flow is transient? Why?

9. A stratospheric sudden warming is accompanied by a reversal of westerlies that comprise the polar-night vortex (Fig. 1.10). Occurring in just a couple of days, this flow reversal follows from a rearrangement of air by planetary waves (see Fig. 14.26). Suppose that the initial motion at middle and high latitudes is approximated by rigid body rotation, with angular velocity 0.25Ω, and that the reversed motion at high latitude is likewise approximated by rigid body rotation, but with zonal easterlies at 75 N of 20 m s^{-1}. If, during the reversal, motion remains approximately nondivergent, estimate from where air that eventually prevails over the polar cap must have originated.

10. The zonal circulation is disrupted by a planetary wave that advects air northward from 10 N to 50 N. If wind at 10 N is easterly with speed of 5 m s^{-1} and relative vorticity of 1.2 10^{-5} s^{-1}, estimate the column-averaged vorticity at 50 N if (a) air motion is barotropic nondivergent, (b) air motion is confined between an isentropic surface near 1000 hPa and another near the tropopause, which slopes downward toward the pole from 100 hPa in the tropics to 300 hPa at 50 N.

11. (a) Determine the components of frictional geostrophic motion tangential and orthogonal to contours of height, in terms of the corresponding geostrophic velocity. (b) Construct an expression for the angle δ by which v_h deviates from height contours. Estimate δ for a Rayleigh-friction coefficient of $K = 4$ days^{-1} at a latitude of (c) 45° and (d) 15°.

12. By applying the results of Prob. 12.11, (a) derive an expression for the divergence of frictional geostrophic motion on an f plane to show that, in the presence of Rayleigh friction, $\nabla \cdot v_h$ is proportional to the $\nabla^2 \Phi$. (b) From this expression, infer how vertical motion over a surface low depends on its latitude.

13. The Kelvin wave figures in tropical circulations. Representing a vertical overturning in the equatorial plane, it has the form of a traveling Walker circulation (Fig. 1.36). Horizontal motion in the Kelvin wave is zonal and described by

$$v_h(x, y) = e^{-\left(\frac{y}{Y}\right)^2} e^{i(kx - \sigma t)} i,$$

where y is measured on an equatorial betaplane ($\phi_0 = 0$) and Y corresponds to 10° of latitude. (a) Sketch the velocity field over one wavelength in the x direction. (b) Calculate the vorticity and divergence of the wave, sketching each on the equatorial beta-plane. (c) Derive a system of equations that determines the streamfunction and velocity potential of the wave. Outline the solution of this system.

14. Consider steady, barotropic, nondivergent motion on a midlatitude beta plane. The motion is zonal and uniform with $v = \overline{u}i$ upstream of an extended NS ridge at $x = 0$, which deflects streamlines meridionally from their undisturbed latitudes. If the disturbed motion downstream deviates from the upstream motion by a small meridional perturbation v', (a) construct a differential equation governing the trajectory $y(x)$ of an air parcel initially at $y = 0$, (b) obtain the general solution downstream of the obstacle, (c) determine the trajectory if the parcel is deflected at $(0, 0)$ with a slope of $tan(\alpha)$, and (d) interpret the motion downstream of the obstacle in terms of a restoring force.

15. (a) Explain why solenoidal production of vorticity vanishes in isentropic coordinates (12.38). (Hint: How many thermodynamic degrees of freedom are present in general? along an isentropic surface?) (b) Apply this argument to the budget of vorticity in isobaric coordinates.

16. Consider a steady vortex with circular streamlines and azimuthal velocity v. (a) Determine the form of the velocity profile $v(r)$ that possesses no vorticity for $r > 0$. (b) Calculate the *circulation* $\Gamma(r) = \oint v(r) \cdot ds = \int_0^{2\pi} v(r) r d\phi$ about the vortex. (c) Use Stokes' theorem to interpret the distribution of vorticity.

THIRTEEN

The planetary boundary layer

Previous development treats motion as a smoothly varying field property. However, atmospheric motion is always partially turbulent. Laboratory experiments provide a criterion for the onset of turbulence, in terms of the dimensionless *Reynolds number*

$$Re = \frac{LU}{\nu}, \tag{13.1}$$

where L and U are scales characterizing a flow and $\nu = \frac{\mu}{\rho}$ is the kinematic viscosity (Sec. 10.6). For Re greater than a critical value of about 5000, smooth laminar motion undergoes a transition to turbulent motion, which is inherently unsteady, 3-dimensional, and involves a spectrum of space and time scales. Even for a length scale as short as 1 m, the critical Reynolds number with $\nu = 10^{-5}$ m^2 s^{-1} is exceeded for velocities of only $U > 0.05$ m s^{-1}. Consequently, atmospheric motion inevitably contains some turbulence.[1]

Until now, turbulent transfers have been presumed small enough to treat an individual air parcel as a closed system. This approximation is valid away from the Earth's surface and outside convection. In the *free atmosphere*, the time scale for turbulent exchange is long, much longer than the 1-day time scale that characterizes advective changes of an air parcel. Budgets of momentum, internal energy, and constituents for an individual parcel may therefore be treated exclusive of turbulent exchanges with the parcel's environment – at least over time scales comparable to advection. This feature of the atmospheric circulation is reflected in the spectrum of kinetic energy (Fig. 13.1). In the free atmosphere, kinetic energy is concentrated at periods longer

---

[1] Stratification of mass leads to ν increasing exponentially with height. Therefore, a height is eventually reached where ν is sufficiently large to drive Re below the critical value and suppress turbulence. This transition occurs at the turbopause, see Fig. 1.4 (Prob. 13.3).

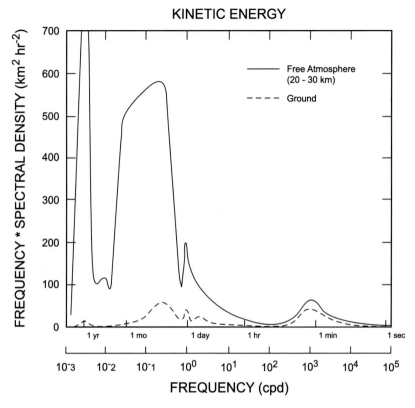

Figure 13.1 Power spectrum of atmospheric kinetic energy. Adapted from Vinnichenko (1970). Copyright (1970) Munksgaard International Publishers Ltd., Copenhagean, Denmark.

than a day, where it is associated with large-scale disturbances and seasonal variations. However, in a neighborhood of the surface, as much as half of the kinetic energy lies at periods of only minutes. Inside the *planetary boundary layer*, turbulent eddies are generated mechanically by strong shear as the flow adjusts sharply to satisfy the no-slip condition at the Earth's surface (see Fig. 13.3). Turbulence is also generated thermally through buoyancy, when the stratification is destabilized by heat transfer from the Earth's surface.

Inside the boundary layer, mixing between air parcels cannot be ignored. There, the time scale for turbulent exchange is comparable to that for advection. Turbulent exchanges of mass, heat, and momentum then make an air parcel an open system, rendering its behavior diabatic. Turbulent mixing transfers heat and moisture between the surface and the atmosphere. Because it destroys gradients inherent to large-scale motion, it also damps the circulation.

13.1 DESCRIPTION OF TURBULENCE

Turbulent fluctuations need not be hydrostatic. Therefore, the equations of motion are expressed most conveniently in physical coordinates. The limited vertical scale of turbulent eddies makes compressibility nonessential. However, variations of density associated with temperature fluctuations play a key role in stratified turbulence because, through buoyancy, they couple the motion to gravity. These features are embodied in the Boussinesq approximation (Sec. 12.5). Equivalent to incompressibility, the Boussinesq approximation allows density to be treated as constant, except where accompanied by gravity in the buoyancy force.

Consider the motion-related stratification (Sec. 11.4), which is regarded as a small departure from static conditions. In terms of the static density $\rho_0(z)$, the specific buoyancy force experienced by an air parcel (11.35) is

$$f_b = -\frac{1}{\rho_0}\frac{\partial p}{\partial z} - \frac{\rho}{\rho_0}g. \tag{13.2}$$

The gas law and Poisson's relation for potential temperature, along with the neglect of compressibility, then imply

$$\frac{\rho}{\rho_0} = -\frac{T}{T_0} + \frac{p}{p_0}$$
$$\cong -\frac{\theta}{\theta_0}. \tag{13.3}$$

Hence, the vertical momentum budget becomes

$$\frac{dw}{dt} - \frac{1}{\rho_0}\frac{\partial p}{\partial z} + g\frac{\theta}{\theta_0} - D_z. \tag{13.4}$$

Similarly, the thermodynamic equation becomes

$$\frac{d\theta}{dt} + w\frac{d\theta_0}{dz} = \frac{\theta_0}{c_p T_0}\dot{q}_{net}, \tag{13.5}$$

where $\theta_0(z)$ describes the basic stratification and θ the small departure from it. Under the Boussinesq approximation, the continuity equation reduces to a statement of 3-dimensional nondivergence. Then the equations governing turbulent motion become

$$\frac{du}{dt} - fv = -\frac{1}{\rho_0}\frac{\partial p}{\partial x} - D_x \tag{13.6.1}$$

$$\frac{dv}{dt} + fu = -\frac{1}{\rho_0}\frac{\partial p}{\partial y} - D_y \tag{13.6.2}$$

$$\frac{dw}{dt} = -\frac{1}{\rho_0}\frac{\partial p}{\partial z} + g\frac{\theta}{\theta_0} - D_z \tag{13.6.3}$$

$$\frac{\partial u}{\partial z} + \frac{\partial v}{\partial y} + \frac{\partial w}{\partial z} = 0 \tag{13.6.4}$$

$$\frac{d\theta}{dt} + w\frac{d\theta_0}{dz} = \frac{\theta_0}{c_p T_0}\dot{q}_{net}, \tag{13.6.5}$$

where

$$\frac{d}{dt} = \frac{\partial}{\partial t} + \boldsymbol{v} \cdot \nabla$$
$$= \frac{\partial}{\partial t} + u\frac{\partial}{\partial x} + v\frac{\partial}{\partial y} + w\frac{\partial}{\partial z} \tag{13.6.6}$$

includes 3-dimensional advection. The frictional drag, $\boldsymbol{D} = -\frac{1}{\rho}\nabla \cdot \boldsymbol{\tau}$, follows from the deformation tensor $\boldsymbol{e}$ (10.23). Under the Boussinesq approximation, it reduces to

$$\boldsymbol{D} = -\nu\nabla^2\boldsymbol{v}. \tag{13.6.7}$$

Compared with the time scale of turbulent eddies, radiative heating in $\dot{q}_{net}$ is slow enough to be ignored. However, thermal diffusion (e.g., conduction) operates efficiently on small scales that are created through turbulent deformation (cf. Fig. 12.4). Diffusion of momentum and heat are both involved in dissipation of turbulence, which takes place on the smallest scales of motion. Sharp gradients produced when anomalies are strained down to small dimensions are acted upon efficiently by molecular diffusion.[2] When those gradients are destroyed, so too is the energy that has been transferred to small scales. The foregoing process reflects a cascade of energy from large scales of organized motion to small scales, where it is dissipated by molecular diffusion.

13.1.1 Reynolds decomposition

Of relevance to the large-scale circulation is the time-averaged motion and how it is influenced by turbulent fluctuations. To describe such interactions, each field variable is separated into a slowly-varying mean component, which is denoted by an overbar, and a fluctuating component, which is denoted by a prime, e.g.,

$$\boldsymbol{v} = \bar{\boldsymbol{v}} + \boldsymbol{v}', \tag{13.7.1}$$

with

$$\overline{\boldsymbol{v}'} = 0. \tag{13.7.2}$$

For this decomposition to be meaningful, time scales of the two components must be widely separated. Only then can averaging be performed over an interval long enough for a stable mean to be recovered from fluctuating properties yet short enough for mean properties to be regarded as steady.

The continuity equation (13.6.4) enables the material acceleration to be expressed in terms of the divergence of a momentum flux

$$\frac{du}{dt} = \frac{\partial u}{\partial t} + \nabla \cdot (\boldsymbol{v}u) \tag{13.8.1}$$

$$\frac{dv}{dt} = \frac{\partial v}{\partial t} + \nabla \cdot (\boldsymbol{v}v) \tag{13.8.2}$$

$$\frac{dw}{dt} = \frac{\partial w}{\partial t} + \nabla \cdot (\boldsymbol{v}w), \tag{13.8.3}$$

[2] The scale dependence of diffusion can be inferred from the viscous drag (13.6.7), which increases quadratically with the inverse scale of motion anomalies.

where terms on the far right describe the 3-dimensional momentum carried by the velocity $\boldsymbol{v}$. Likewise, the material rate of change of temperature in (13.6.7) can be expressed in terms of the divergence of a heat flux

$$\frac{d\theta}{dt} = \frac{\partial \theta}{\partial t} + \nabla \cdot (\boldsymbol{v}\theta). \tag{13.9}$$

Expanding as in (13.7) and averaging then obtains the equations governing the time-mean motion

$$\frac{\overline{d}\,\overline{u}}{dt} - f\overline{v} = -\frac{1}{\rho_0}\frac{\partial \overline{p}}{\partial x} - \left[\frac{\partial \overline{u'u'}}{\partial x} + \frac{\partial \overline{u'v'}}{\partial y} + \frac{\partial \overline{u'w'}}{\partial z}\right] \tag{13.10.1}$$

$$\frac{\overline{d}\,\overline{v}}{dt} + f\overline{u} = -\frac{1}{\rho_0}\frac{\partial \overline{p}}{\partial y} - \left[\frac{\partial \overline{u'v'}}{\partial x} + \frac{\partial \overline{v'v'}}{\partial y} + \frac{\partial \overline{v'w'}}{\partial z}\right] \tag{13.10.2}$$

$$\frac{\overline{d}\,\overline{w}}{dt} = -\frac{1}{\rho_0}\frac{\partial \overline{p}}{\partial z} + g\frac{\overline{\theta}}{\theta_0} - \left[\frac{\partial \overline{u'w'}}{\partial x} + \frac{\partial \overline{v'w'}}{\partial y} + \frac{\partial \overline{w'w'}}{\partial z}\right] \tag{13.10.3}$$

$$\frac{\partial \overline{u}}{\partial x} + \frac{\partial \overline{v}}{\partial y} + \frac{\partial \overline{w}}{\partial z} = 0. \tag{13.10.4}$$

$$\frac{\overline{d}\,\overline{\theta}}{dt} + \overline{w}\frac{d\theta_0}{dz} = -\left[\frac{\partial \overline{u'\theta'}}{\partial x} + \frac{\partial \overline{v'\theta'}}{\partial y} + \frac{\partial \overline{w'\theta'}}{\partial z}\right], \tag{13.10.5}$$

where

$$\frac{\overline{d}}{dt} = \frac{\partial}{\partial t} + \overline{\boldsymbol{v}} \cdot \nabla \tag{13.10.6}$$

is the time rate of change following the mean motion and diffusive transfers of momentum and heat on scales of the mean flow are slow enough to be ignored.

Terms in (13.10.1)–(13.10.3) involve time-averaged products of fluctuating velocities. They represent the convergence of an eddy momentum flux (e.g., of eddy momentum carried by the eddy velocity). Mean x momentum (13.10.1) is forced by the convergence of eddy x momentum flux and so forth. Then, according to the discussion in Sec. 10.6, fluxes of x momentum carried by the three components of fluctuating velocity

$$\tau_{xx} = -\rho_0 \overline{u'u'} \tag{13.11.1}$$

$$\tau_{xy} = -\rho_0 \overline{u'v'} \tag{13.11.2}$$

$$\tau_{xz} = -\rho_0 \overline{u'w'} \tag{13.11.3}$$

represent stresses in the x direction that are exerted on the mean flow by turbulent motions. Referred to as *Reynolds stresses*, their convergence exerts a turbulent drag on the mean motion of an air parcel. Collectively, the Reynolds stresses define a turbulent stress tensor $\boldsymbol{\tau} = -\rho_0 \overline{\boldsymbol{v}'\boldsymbol{v}'}$. $\boldsymbol{\tau}$ is a counterpart of the stress tensor in (10.22). Its convergence represents the absorption of eddy momentum by the mean flow, which, in turn, forces $\overline{\boldsymbol{v}}$ in (13.10.1) (13.10.3). A similar interpretation applies to the mean potential temperature. The latter is forced by the convergence of eddy heat flux in (13.10.5)

$$\frac{q}{c_p} = \rho_0 \overline{\boldsymbol{v}'\theta'}. \tag{13.12}$$

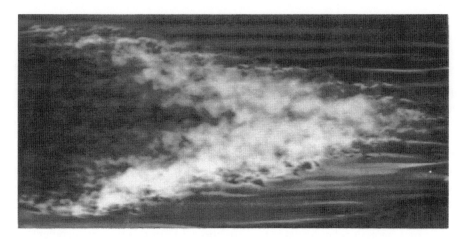

Figure 13.2 Turbulent spot introduced into a laminar flow (from right to left) at $Re = 4.0 \ 10^5$. After Van Dyke (1982).

13.1.2 Turbulent diffusion

Reynolds decomposition provides a framework for describing how turbulent motions interact with the mean flow. However, it affords little insight into how the turbulent fluxes of momentum and heat that force mean motion can be determined. The description of mean motion (13.10) can be closed only by resorting to empirical or *ad hoc* prescriptions that relate eddy fluxes to the mean fields.

The closure scheme adopted most widely is inspired by molecular diffusion, which smoothes out gradients. Although more complex, turbulence has a similar effect on mean properties. Figure 13.2 shows the evolution of a turbulent spot that has been introduced into a uniform flow at large Re. Streaklines of smoke captured by the spot are dispersed across the turbulent region, which spreads laterally with distance downstream. Therefore, mean concentration inside the turbulent region decreases steadily, assuming progressively weaker gradient as smoke is diluted over an ever-expanding volume. Turbulent dispersion is responsible for mean concentrations of pollutants diminishing following the breakdown of the nocturnal inversion (Sec. 7.6.3). It is also responsible for the radial expansion of buoyant thermals (Sec. 9.3).

Turbulent dispersion acts on all conserved properties. It is analogous to molecular diffusion, at least qualitatively. In this spirit, the eddy momentum flux can be expressed in terms of the gradient of mean momentum (13.6.7)

$$\boldsymbol{\tau} = \rho_0 K_M \cdot \nabla \overline{\boldsymbol{v}}, \qquad (13.13.1)$$

where the *eddy diffusivity* of momentum K_M is an analogue of the kinematic viscosity ν for molecular diffusion.[3] For the boundary layer, it suffices to consider mean motion that is horizontally homogeneous. Horizontal diffusion of momentum then vanishes and K_M pertains to the vertical flux alone. The vertical eddy flux of horizontal momentum

$$\overline{w'\boldsymbol{v}'} = -K_M \frac{\partial \overline{\boldsymbol{v}}}{\partial z} \qquad (13.13.2)$$

---

[3] For anisotropic turbulence, wherein statistical properties vary with direction, K_M is a tensor. The components of K_M then refer to turbulent dispersion in the horizontal and vertical directions. Their application in 2-dimensional models is described in Andrews et al. (1987).

is then proportional to the vertical gradient of mean momentum. Likewise, the vertical heat flux becomes

$$\overline{w'\theta'} = -K_H \frac{\partial \overline{\theta}}{\partial z},$$
(13.13.3)

where K_H is the eddy diffusivity of heat. It is proportional to the vertical gradient of mean θ.

Expressions (13.13) comprise the so-called *flux-gradient relationship*. Turbulent transfer in (13.10) then assumes the form of diffusion of mean properties (e.g., $K_M \nabla^2 \overline{v}$). In principle, eddy diffusivities can be incorporated as above. In practice, however, K_M and K_H depend on fluctuating field properties, which are unknown. They must therefore be evaluated empirically, for example, in terms of the eddy fluxes $\overline{w'u'}$ and $\overline{w'\theta'}$. Inside the planetary boundary layer, measured eddy diffusivities of momentum range between 1 and 10^2 m^2 s^{-1}. These values of K_M are orders of magnitude greater than its counterpart for molecular diffusion ($\nu = 10^{-5}$ m^2 s^{-1}). For this reason, turbulent diffusion is the chief source of drag and mechanical heating experienced by the mean flow.

13.2 STRUCTURE OF THE BOUNDARY LAYER

13.2.1 The Ekman Layer

Consider motion inside a boundary layer in which mean properties are steady and horizontally homogeneous. In terms of eddy diffusivity, the horizontal momentum equations become

$$-f\overline{v} = -\frac{1}{\rho_0} \frac{\partial \overline{p}}{\partial x} + K \frac{\partial^2 \overline{u}}{\partial z^2}$$
(13.14.1)

$$f\overline{u} = -\frac{1}{\rho_0} \frac{\partial \overline{p}}{\partial y} + K \frac{\partial^2 \overline{v}}{\partial z^2},$$
(13.14.2)

where $K = K_M$ is regarded as constant and the Boussinesq approximation has been invoked to ignore vertical advection of mean momentum. Eliminating the pressure gradient in favor of the geostrophic velocity, $\boldsymbol{v}_g = \frac{1}{\rho_0 f} \boldsymbol{k} \times \nabla p$, transforms (13.14) into

$$K \frac{\partial^2 \overline{u}}{\partial z^2} + f(\overline{v} - v_g) = 0$$
(13.15.1)

$$K \frac{\partial^2 \overline{v}}{\partial z^2} - f(\overline{u} - u_g) = 0.$$
(13.15.2)

Multiplying (13.15.1) by $i = \sqrt{-1}$ and adding it to (13.15.2) obtains the single equation

$$K \frac{\partial^2 \chi}{\partial z^2} - if\chi = -if\chi_g$$
(13.16.1)

in terms of the consolidated variables

$$\chi = \overline{u} + i\overline{v}$$
(13.16.2)

$$\chi_g = u_g + iv_g.$$
(13.16.3)

If $\boldsymbol{v}_g$ does not vary through the boundary layer and is oriented parallel to the x axis, (13.16) has the general solution

$$\chi(z) = A e^{(1+i)\gamma z} + B e^{-(1+i)\gamma z} + u_g$$
(13.17.1)

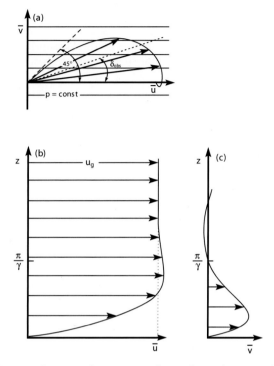

Figure 13.3 Mean horizontal motion inside an Ekman layer in which isobars are oriented parallel to the x axis. (a) Planform view of the Ekman spiral, with $\overline{\boldsymbol{v}} = (\overline{u}, \overline{v})$ shown at different levels. Vertical profiles of (b) $\overline{u}$ and (c) $\overline{v}$ through the boundary layer.

where

$$\gamma = \sqrt{\frac{f}{2K}}. \tag{13.17.2}$$

At the Earth's surface, the no-slip condition requires the horizontal velocity to vanish. Above the boundary layer, the motion must asymptotically approach the geostrophic velocity:

$$\overline{\boldsymbol{v}} = 0 \qquad z = 0$$
$$\overline{\boldsymbol{v}} \sim u_g \boldsymbol{i} \qquad z \to \infty. \tag{13.17.3}$$

Incorporating these boundary conditions and collecting real and imaginary parts then yield the component velocities

$$\overline{u}(z) = u_g \left(1 - e^{-\gamma z} \cos\gamma z\right) \tag{13.18.1}$$
$$\overline{v}(z) = v_g e^{-\gamma z} \sin\gamma z, \tag{13.18.2}$$

which define an *Ekman layer*.

The mean velocity in (13.18) describes an *Ekman spiral*. It is plotted in Fig. 13.3 as a function of height. Moving downward through the boundary layer witnesses a deflection of the mean velocity from its value in the free atmosphere, $\boldsymbol{v}_g = u_g \boldsymbol{i}$, across isobars: $\overline{\boldsymbol{v}}$ is deflected to the left in the Northern Hemisphere, but to the right in the Southern Hemisphere – always toward low pressure (Sec. 12.3). At the ground, $\overline{\boldsymbol{v}}$ attains a limiting deflection of $\delta = 45°$.

The deflection from isobars measures the turbulent drag that is exerted on the mean flow. Such drag drives the motion out of geostrophic equilibrium. It increases with mean vertical shear. The effective depth of the Ekman layer is given by $z = \frac{\pi}{\gamma}$, where $\bar{v}$ reverses sign. Above this height, $\bar{v}$ remains close to v_g (Fig. 13.3). Values of $K = 10$ m^2 s^{-1} and $f = 10^{-4}$ s^{-1} give $\frac{\pi}{\gamma} \cong 1$ km. In practice, a pure Ekman spiral is seldom observed.[4] Nonetheless, it captures the salient structure of the planetary boundary layer. A deflection from surface isobars $\delta_{obs} \cong 25°$ is typical.

13.2.2 The surface layer

Within the lowest 15% of the boundary layer, the assumption that $K \cong$ const breaks down. Instead, K, which is determined by the characteristic velocity and length scales of turbulent eddies, varies with height linearly. Within the surface layer, the no-slip condition limits eddies to a characteristic length scale that is proportional to the distance z from the ground. It gives them a characteristic velocity scale

$$u_* = \left(-\overline{u'w'}\right)^{\frac{1}{2}}. \tag{13.19}$$

Known as the *friction velocity*, u_* reflects the turbulent shear stress near the ground. The flux-gradient relationship (13.13) then requires

$$u_*^2 = kzu_*\frac{\partial \bar{u}}{\partial z}, \tag{13.20}$$

where $\bar{v} = \bar{u}\boldsymbol{i}$ near the surface has been presumed parallel to the x axis and $k = 0.4$ is the so-called *von Karman constant*. Integrating (13.20) gives the logarithmic vertical structure inside the surface layer

$$\left(\frac{\bar{u}}{u_*}\right) = \frac{1}{k}ln\left(\frac{z}{z_0}\right), \tag{13.21}$$

in which z_0 is a *roughness length* that characterizes the Earth's surface. The depth of the surface layer varies in relation to its roughness through z_0. For level vegetated terrain, z_0 has values less than 5 cm.

The structure (13.21) is a counterpart of the *viscous sublayer* that exists in turbulent flow over a flat plate (e.g., Schlicting, 1968). Inside it, fluxes of momentum and heat are independent of height. For neutral stability, the logarithmic behavior predicted by (13.21) is well-obeyed over a wide range of level terrain (Fig. 13.4). The same is true for stable stratification, within some distance of the surface.

13.3 INFLUENCE OF STRATIFICATION

Static stability modifies the character of the boundary layer by influencing the production and destruction of turbulence. Vertical shear of the mean flow leads to mechanical production of turbulent kinetic energy. Under neutral stability, that energy cascades equally into all three components of motion. The turbulence produced is isotropic.

---

[4] High Re characteristic of the atmosphere makes simple Ekman flow dynamically unstable.

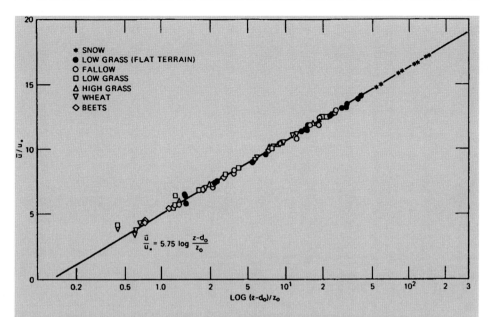

Figure 13.4 Observations of horizontal velocity versus height in terms of the friction velocity u_* and roughness length z_0 for different terrain characterized by the height d_0. After Paeschke (1937).

Turbulent kinetic energy is also produced through buoyancy. Unlike mechanical production, buoyancy can represent either a source or a sink of turbulent kinetic energy. Under unstable stratification, turbulent vertical velocities w' are reinforced by buoyancy. Under stable stratification, they are inhibited. In either event, buoyancy acts selectively on the vertical component of motion. Even though some of that energy cascades into the horizontal components, buoyancy makes turbulent motion anisotropic. In the presence of strong static stability, the vertical component may contain very little of the turbulent kinetic energy.

The degree of anisotropy is reflected in the dimensionless *flux Richardson number*

$$R_f = \frac{\frac{g}{\theta}\overline{w'\theta'}}{\overline{w'u'}\frac{\partial \overline{u}}{\partial z}}.$$ (13.22)

It represents the ratio of thermal to mechanical production of turbulent kinetic energy. Unlike other dimensionless parameters, R_f is a function of position. Inside the surface layer, downward flux of positive x momentum exerts a positive drag on the ground. It makes the denominator of (13.22) negative. Under unstable stratification, the vertical heat flux must be positive to transfer heat from lower to upper levels and hence drive the stratification towards neutral stability (Sec. 7.5). These conditions lead to $R_f < 0$. Large negative values of R_f occur in the presence of weak shear. Turbulence is then driven chiefly by buoyancy, termed *free convection*. Alternatively, if the heat flux vanishes, $R_f = 0$. Turbulence is then driven exclusively by shear. Under stable stratification, the vertical heat flux must be negative (Prob. 13.11). It follows that $R_f > 0$. Buoyancy then opposes vertical eddy motion, damping turbulent kinetic energy. Turbulence can be maintained under these conditions only if production of turbulent energy by shear exceeds its damping by buoyancy.

Analogous, but more readily evaluated, is the *gradient Richardson number*

$$Ri = \frac{\frac{g}{\theta}\frac{\partial \bar{\theta}}{\partial z}}{\left(\frac{\partial \bar{u}}{\partial z}\right)^2}$$

$$= \frac{N^2}{\left(\frac{\partial \bar{u}}{\partial z}\right)^2}. \tag{13.23.1}$$

With (13.13), the gradient Richardson number is related to the flux Richardson number as

$$R_f = \frac{K_H}{K_M} Ri. \tag{13.23.2}$$

If $K_m = K_H$, the two are equivalent. The gradient Richardson number reflects the dynamic stability of the mean flow. It therefore provides a criterion for the onset of turbulence. For $N^2 > 0$, Ri is positive. Buoyancy then stabilizes the motion by opposing vertical displacements. If the denominator in (13.13.1) is sufficiently small, so is the destabilizing influence of shear. Disturbances involving vertical motion are then damped out. Corresponding to Ri that is sufficiently large, these conditions describe sheared flow that is dynamically stable. Stability analysis reveals how large (e.g., Thorpe, 1969): For $Ri > \frac{1}{4}$, small disturbances are damped out by buoyancy. Sheared flow is then stable and turbulence is inhibited. Conversely, sheared flow becomes dynamically unstable if

$$Ri < \frac{1}{4}. \tag{13.24}$$

This criterion for dynamical instability is supported by laboratory experiments (Thorpe, 1971). It corresponds to shear production of turbulent kinetic energy that exceeds buoyancy damping by a factor of 4. Small disturbances then amplify into fully developed turbulence.[5] Sheared flow is also unstable for $Ri < 0$ (e.g., for $N^2 < 0$). Turbulence is then driven by both shear and buoyancy.

Stable stratification inhibits turbulence inside the boundary layer. But increasing shear near the Earth's surface eventually makes mechanical production of turbulent kinetic energy large enough to overcome damping by buoyancy. Consequently, a turbulent layer inevitably exists close to the ground. There, $\bar{u}$ varies with z logarithmically and fluxes of momentum and heat are independent of height (13.21). The depth of this layer is controlled by surface roughness and static stability. Strong stability limits the layer to a few tens of meters above the ground. Conversely, weak stability allows a deeper layer with smaller $\frac{\partial \bar{u}}{\partial z}$.

Static stability inside the boundary layer is controlled by heating and cooling of the surface layer. Under cloud-free conditions and over warm SST, surface heating

---

[5] Outside the boundary layer, the criterion for the onset of turbulence can be achieved locally (e.g., in association with fractus rotor cloud (Fig. 9.22) and severe turbulence found in the mountain wave; see Fig. 14.24).

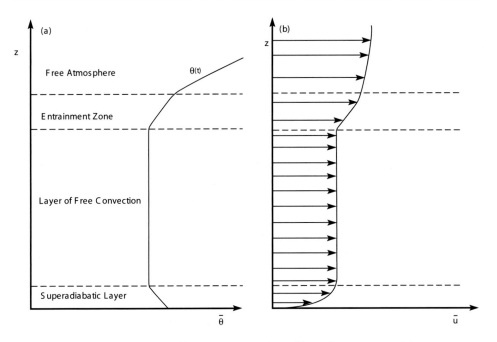

Figure 13.5 Schematic illustrating vertical profiles of mean potential temperature and horizontal motion inside a buoyantly-driven boundary layer. Air parcels emerging from the superadiabatic layer near the surface ascend through the layer of free convection, which extends to the stable layer aloft and in which $\bar{\theta}$ and $\bar{u}$ are homogenized. Continued heat supply at the surface enables the convective layer to advance upward by eroding the stable layer from below.

is strong. It favors buoyantly driven turbulence by destabilizing the stratification. These conditions lead to the formation of a superadiabatic layer immediately above the Earth's surface (Fig. 13.5a). Buoyant parcels emerge from the superadiabatic layer, entering a deeper layer of free convection. That layer is characterized by strong overturning and neutral stability ($\bar{\theta}$ = const). Vertical mixing of momentum makes the profile of mean horizontal motion likewise uniform (Fig. 13.5b).[6] The height of convective plumes is eventually limited by stable air overhead, which magnifies detrainment and mixing with the environment (Sec. 7.4). Nonetheless, continued heat supply at the surface enables the convective layer to expand upward, eroding the stable layer from below.

The foregoing process is responsible for the diurnal cycle of the boundary layer (Sec. 7.6.3). It is revealed by vertical soundings (Fig. 13.6). During early morning, θ increases upward in the lowest 300 m (Fig. 13.6a). Marking the nocturnal inversion, strong static stability has been neutralized only in a shallow layer near the surface. Specific humidity (Fig. 13.6b) decreases upward from the surface. By midday, surface

---

[6] The flux-gradient relationship (13.13.2) breaks down in the well-mixed layer. Turbulent drag can then be represented as Rayleigh friction (Sec. 12.3), the coefficient of which is determined by $\bar{\boldsymbol{v}}$; see Holton (2004).

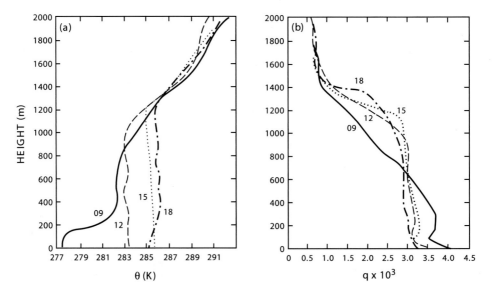

Figure 13.6 Observed profiles during the day of (a) potential temperature and (b) specific humidity. Adapted from Sun and Chang (1986).

convection has broken through the inversion and thoroughly mixed potential temperature and moisture across the lowest kilometer. Mean horizontal motion inside the layer of convective overturning (not shown) is likewise independent of height. Later during the day, the convective layer expands upward further, by a few hundred meters. By early evening, surface heating subsides, replaced by surface cooling. The nocturnal inversion then begins to be re-established.

13.4 EKMAN PUMPING

Turbulent drag inside the boundary layer drives the large-scale motion out of geostrophic equilibrium. By doing so, it introduces a divergent component of motion (Sec. 12.3). To satisfy continuity, divergence must be compensated by vertical motion. The latter imposes a secondary circulation on the $O(Ro^0)$ geostrophic flow in the free atmosphere. Frictional convergence inside the boundary layer leads to ascending motion above surface low pressure (Fig. 12.7). Frictional divergence leads to reversed motion above surface high pressure.

The secondary circulation imposed on the free atmosphere is forced by vertical motion at the top of the boundary layer. By continuity, the latter is related to the convergence of horizontal mass flux inside the Ekman layer. That, in turn, follows from the cross-isobaric component $\bar{v}$. Integrating vertically over the boundary layer obtains the column-integrated mass flux across isobars

$$
\begin{aligned}
M &= \rho_0(0) \int_0^\infty \bar{v}\,dz \\
&= \rho_0(0) \int_0^\infty u_g e^{-\gamma z} \sin\gamma z\,dz,
\end{aligned}
\tag{13.25.1}
$$

which reduces to

$$M = \frac{\rho_0(0)}{2\gamma} u_g. \tag{13.25.2}$$

Because $\bar{v}j$ is perpendicular to v_g, directed toward low pressure, the column-integrated mass flux across isobars may be expressed

$$M = \frac{\rho_0(0)}{2\gamma} k \times v_g. \tag{13.26}$$

Integrating the continuity equation (13.10.4) in similar fashion obtains

$$\nabla_z \cdot M + \rho_0(0)w_0 = 0, \tag{13.27}$$

where w_0 denotes the vertical velocity at the top of the boundary layer, or the *Ekman pumping*. Incorporating (13.26) then yields

$$w_0 = \frac{1}{2\gamma}\zeta_g(0), \tag{13.28.1}$$

where

$$\zeta_g(0) = \nabla_z^2 \psi(0) \tag{13.28.2}$$

is the geostrophic vorticity at the base of the free atmosphere. According to (13.28), Ekman pumping is directly proportional to the large-scale vorticity above the boundary layer. Positive vorticity (cyclonic motion) induces upward motion out of the boundary layer: Ekman pumping. It is compensated by frictional convergence inside the boundary layer (e.g., into surface low pressure). Negative vorticity (anticyclonic motion) induces subsidence into the boundary layer: *Ekman suction*. It is compensated by frictional divergence (e.g., out of surface high pressure). Values used previously to evaluate γ and $\zeta_g = 10^{-5}$ s^{-1}, typical of quasi-geostrophic motion, give w_0 of order 5 mm s^{-1}.

Frictional convergence and divergence inside the boundary layer, by redistributing mass, act to relieve imbalances of pressure. In doing so, they destroy pressure gradients that support large-scale motion. Divergent motion inside the boundary layer also has an indirect but more important influence. Upward motion out of the boundary layer must be compensated overhead by horizontal divergence in the free atmosphere (Fig. 13.7). According to the vorticity budget (12.36), such divergence weakens the absolute vorticity of air in the free atmosphere. To conserve angular momentum, a ring of air that expands radially must spin down. This loss of vorticity damps organized motion about a cyclone.

Under barotropic stratification, the horizontal velocity in the free atmosphere is independent of height (12.11). Integrating the continuity equation (13.10.4) from the top of the boundary layer to a height h where w has decreased to zero then yields

$$-w_0 = h\nabla_z \cdot v$$
$$= h\nabla_z \cdot v_a, \tag{13.29}$$

where v_a is the ageostrophic velocity that comprises the secondary circulation. The vorticity equation for quasi-geostrophic motion (12.47) on an f plane reduces to

$$\frac{d_g\zeta_g}{dt} = -f_0\nabla_z \cdot v_a. \tag{13.30}$$

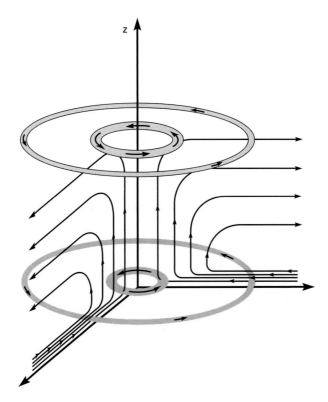

Figure 13.7 Schematic of the secondary circulation imposed on the $O(Ro^0)$ geostrophic motion above the boundary layer. Frictional convergence in the boundary layer beneath a cyclone is compensated by ascending motion and divergence in the free atmosphere. A ring of air diverging from the center of the cyclone must then spin down to conserve angular momentum, which decelerates organized motion above the boundary layer.

With (13.29) and (13.28), this becomes

$$\frac{d_g \zeta_g}{dt} = \frac{f_0}{2\gamma h} \zeta_g$$

or

$$\frac{1}{\zeta_g} \frac{d_g \zeta_g}{dt} = \left(\frac{f_0 K}{2h^2} \right)^{\frac{1}{2}} . \tag{13.31}$$

A vorticity anomaly, characteristic of a mid-latitude cyclone, thus decays exponentially as

$$\zeta_g = e^{-\frac{t}{\tau_E}}, \tag{13.32.1}$$

where

$$\tau_E = \left(\frac{f_0 K}{2h^2} \right)^{-\frac{1}{2}} \tag{13.32.2}$$

is an Ekman e-folding time. Hence, the secondary circulation imposed by Ekman pumping causes the primary circulation to spin down.[7] This process involves no actual dissipation. Rather, it reflects a redistribution of angular momentum, when absolute vorticity that is concentrated in a vorticity anomaly is diluted over a wider area. By Stokes theorem, the circulation along an expanding material contour $s(x, y)$

$$
\Gamma = \oint v_g \cdot ds
$$
$$
= \int_A \zeta_g dA
$$

(13.33)

remains unchanged. With values representative of the troposphere, (13.32) gives an e-folding time for spin down of order days. This is an order of magnitude faster than frictional dissipation in the free atmosphere.

SUGGESTED REFERENCES

A First Course in Turbulence (1972) by Tennekes and Lumley presents an introduction to the subject at the graduate level. *Statistical Fluid Mechanics* (1973) by Monin and Yaglom is an advanced treatment that develops the Lagrangian description of turbulent dispersion.

A meteorological treatment at the graduate level is given in *An Introduction to Boundary Layer Meteorology* (1988) by Stull. Relevant observations may be found in *Structure of the Atmospheric Boundary Layer* (1989) by Sorbjan.

PROBLEMS

1. Evaluate the Reynolds number for the following flows: (a) Stokes flow about a spherical cloud droplet (Prob. 9.6) of radius 10 μm and fall speed of 3 mm s^{-1}, (b) cumulus convection of characteristic dimension 1 km and velocity 1 m s^{-1}, (c) a midlatitude cyclone of characteristic dimension 10^3 km and velocity 5 m s^{-1}.

2. Estimate how strong vertical shear must become before turbulence forms under stability representative of (a) mean conditions in the troposphere: $N^2 = 1.10^{-4}$ s^{-2} and (b) mean conditions in the stratosphere: $N^2 = 6.10^{-4}$ s^{-2}. (c) How would these values be affected if, locally, isentropic surfaces are steeply deflected from the horizontal?

3. Estimate the height of the turbopause based on characteristic length and velocity scales of 1 km and 10 m s^{-1}, respectively.

4. Frictional dissipation is an essential feature of turbulence because it cascades energy from large scales of organized motion to small scales, on which molecular diffusion operates efficiently. The specific energy dissipation rate for homogeneous isotropic turbulence is described by

$$
\epsilon \cong \frac{U^3}{L},
$$

where U and L are the characteristic velocity and length scales of the large-scale motion that drives the cascade. Hence the rate of energy dissipation, which is concentrated at small scales, is dictated by the motion at large scales, which is

---

[7] Should a disturbance's vertical structure make $v_g(0) = 0$, Ekman pumping and divergence in the free atmosphere vanish. Ekman spin down is then eliminated.

nearly inviscid. Use this result to estimate the rate at which energy is dissipated inside a cumulonimbus cloud of characteristic dimension 10 km and velocity 10 m s^{-1}. Compare this value against the rate at which energy is generated by a large urban power plant (~1000 M W).

5. For the cumulonimbus cell in Prob. 13.4, estimate the precipitation rate necessary to sustain the motion.

6. A mid-latitude cyclone is associated with the disturbance height in the lower troposphere

$$z' = -Ze^{-\frac{x^2+y^2}{2L^2}},$$

where $Z = 200$ m, L $= 1000$ km, y is measured from $45°$, and the boundary layer has an eddy diffusivity of 10 m^2 s^{-1}. Calculate the maximum vertical velocity atop the boundary layer.

7. Observations reveal a nearly logarithmic variation with height of mean wind below 100 m above prairie grass land that has a roughness length of 4 cm. If the mean wind is 5 m s^{-1} at the top of this layer, calculate the stress exerted on the ground.

8. Calculate the characteristic spin-down time due to Ekman pumping for a cyclone at $45°$, vertical motion extending to the tropopause at 10 km, and an eddy diffusivity inside the boundary layer of 10 m^{2}s^{-2}. Compare this spin-down time with the time scale of radiative damping.

9. Derive the flux form of the momentum equations (13.8).

10. Recover (13.20) governing mean motion inside the surface layer.

11. Demonstrate that the turbulent vertical heat flux must be positive (negative) under unstable (stable) stratification.

12. Show that divergence associated with the secondary circulation above the boundary layer does not alter the circulation $\Gamma = \oint \boldsymbol{v} \cdot d\boldsymbol{s}$ along a material contour.

13. Inside a *well-mixed boundary layer* of depth h, convection makes mean horizontal motion and potential temperature invariant with height (Fig. 13.5), rendering the flux-gradient relationship inapplicable. The vertical momentum flux then decreases linearly with height from a surface maximum described by

$$\overline{w'\boldsymbol{v}'}|_s = -c_d|\overline{\boldsymbol{v}}|\overline{\boldsymbol{v}},$$

where c_d is a dimensionless *drag coefficient*. (a) Construct a counterpart of the Ekman equations (13.15) governing mean motion inside the well-mixed layer to show that the effect of turbulence is then equivalent to Rayleigh friction (Sec. 12.3). (b) Determine $\overline{\boldsymbol{v}}$ if $\boldsymbol{v}_g = u_g\boldsymbol{i}$ above the well-mixed boundary layer. (c) Calculate the time scale of Rayleigh friction for $h = 1$ km, $c_d = 5$ 10^{-3}, and a mean wind speed of 5 m s^{-1}.

14. The contrail formed from the exhaust of a commercial jet flying at 200 hPa contains liquid water. At a distance x from the jet, the contrail has a temperature $T(x)$ and a mixing ratio for total water $r_t(x)$. The contrail dissolves some distance x_d downstream of the jet through entrainment with ambient air, of temperature T_a and mixing ratio r_a. If, immediately aft of the jet, the exhaust plume has temperature $T(0) = T_0$ and total water mixing ratio $r_t(0) = r_0$, (a) formulate a differential equation governing $r_t(x)$, in terms of the entrainment length $L^{-1} = \frac{d(lnm)}{dx}$, which characterizes the rate at which the contrail expands through mixing with ambient air (Sec. 7.4.2), (b) derive expressions for $r_t(x)$ and $T(x)$, (c) derive an expression

for the saturation mixing ratio of the contrail $r_i(x)$, (d) determine the distance x_d at which the contrail dissolves for $L = 100$ m, $r_0 = 4$ g kg^{-1}, $r_a = 0.05$ g kg^{-1}, $T_0 = -20$ C, and $T_a = -50$ C, (e) as in (d), but under the warm moist conditions: $T_a = -30$ C, $r_a = 1.25$ g kg^{-1}, $T_0 = 0$ C, symbolic of conditions inside the warm sector of an advancing cyclone.

FOURTEEN

Wave propagation

The governing equations support several forms of wave motion. Atmospheric waves are excited when air is disturbed from equilibrium (e.g., mechanically when air is displaced over elevated terrain or thermally when air is heated inside convection). By transferring momentum, wave motions convey the influence of one region to another. This mechanism of interaction enables tropical convection to influence the extratropical circulation. It also enables the troposphere to perturb the stratosphere, driving it out of radiative equilibrium (Fig. 8.27).

14.1 DESCRIPTION OF WAVE PROPAGATION

Wave motion is possible in the presence of a positive restoring force. By opposing disturbances from equilibrium, the latter supports local oscillations in field properties. Under stable stratification, buoyancy provides such a restoring force (Sec. 7.3). The compressibility of air provides another. The variation with latitude of the Coriolis force provides yet another restoring force. It will be seen to support large-scale atmospheric disturbances.

14.1.1 Surface water waves

The description of wave motion is illustrated with an example under nonrotating conditions, which will serve as a model of buoyancy oscillations in the atmosphere. Consider disturbances to a layer of incompressible fluid of uniform density ρ and depth H (Fig. 14.1). The layer is bounded below by a rigid surface. It is bounded above by a *free surface*, namely, one that adjusts position to relieve any stress. Motion

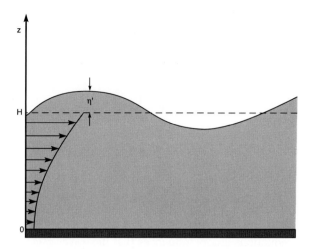

Figure 14.1 A layer of incompressible fluid of uniform density ρ and depth H. The layer is bounded below by a rigid surface and above by a free surface that has displacement η' from its mean elevation. The profile of horizontal motion for surface water waves is indicated.

inside this layer is governed by

$$\rho \frac{d\boldsymbol{v}}{dt} = -\nabla p - g\boldsymbol{k} \tag{14.1.1}$$

$$\nabla \cdot \boldsymbol{v} = 0, \tag{14.1.2}$$

with $\boldsymbol{v}$ and ∇ referring to 3-dimensional vector quantities.

We are interested in how the fluid layer responds to disturbances that are imposed through a deflection of its free surface. It is convenient to decompose the behavior into a zonal-mean basic state, which represents undisturbed conditions and is denoted by an overbar, and perturbations to it, which are denoted by a prime. For example,

$$\boldsymbol{v} = \overline{\boldsymbol{v}} + \boldsymbol{v}' \tag{14.2.1}$$

with

$$\overline{\boldsymbol{v}'} = 0. \tag{14.2.2}$$

This decomposition is analogous to the one applied in the treatment of turbulence (Chap. 13). For simplicity, the undisturbed zonal motion $\overline{\boldsymbol{v}} = \overline{u}\boldsymbol{i}$ is taken to be independent of x, analogous to motion in Fig. 1.8.

Incorporating (14.2) into equations (14.1) produces terms involving mean properties, perturbation properties, and products thereof. Perturbations are characterized by the wave amplitude ϵ. The latter symbolizes small departures from equilibrium, for instance, u', v', $p' = O(\epsilon)$. In the spirit of the asymptotic analysis for small Rossby number (Chap. 12), the resulting system can then be solved recursively to increasing order in ϵ. To zero order in ϵ (i.e., in the absence of waves), the governing equations reduce to a statement of hydrostatic equilibrium

$$\frac{\partial \overline{p}}{\partial z} = -\rho g, \tag{14.3}$$

which must be satisfied by the mean state. To first order in ϵ, with the basic state balance (14.3) eliminated, the governing equations reduce to

$$\frac{D\boldsymbol{v}'}{Dt} = -\nabla p' \tag{14.4.1}$$

$$\nabla \cdot \boldsymbol{v}' = 0, \tag{14.4.2}$$

where

$$\frac{D}{Dt} = \frac{\partial}{\partial t} + \bar{u}\frac{\partial}{\partial x} \tag{14.4.3}$$

defines the material derivative following the mean motion. Linear in perturbation quantities, equations (14.4) govern the wave field. They neglect products of perturbation quantities, specifically, the advective acceleration $\boldsymbol{v}' \cdot \nabla \boldsymbol{v}'$, which are $O(\epsilon^2)$.

The first-order equations ignore influences that the waves exert on the mean state. Such feedback is accounted for in the equations accurate to second order in ϵ. Upon eliminating the zero and first-order balances and averaging zonally, the second-order balance reduces to

$$\frac{\partial \bar{u}}{\partial t} = -\overline{(\boldsymbol{v}' \cdot \nabla \boldsymbol{v}')}$$
$$= -\frac{\partial(\overline{v'u'})}{\partial y} - \frac{\partial(\overline{w'u'})}{\partial z}. \tag{14.5}$$

Terms on the right-hand side of (14.5) are quadratic in perturbation velocity. They represent the convergence of wave momentum flux, $-\nabla \cdot \overline{\boldsymbol{v}'u'}$. This absorption of wave momentum exerts a drag on the mean flow. It forces zonal-mean motion – just as the convergence of turbulent momentum flux forces mean motion in (13.10). Corresponding changes in $\bar{u}$ then reflect a second-order correction to the zeroth-order steady motion.

The development is now restricted to the limit of small wave amplitude. Second-order terms in (14.5) are then small, so the basic state may be regarded as steady. Under these conditions, motion in the fluid layer is described by the equations accurate to first order in wave amplitude, which are said to have been *linearized*. If the basic state is in uniform motion, $\bar{u}$ is independent of position. Applying $\nabla \cdot$ to (14.4.1) and incorporating (14.4.2) consolidates the system into a single equation for the perturbation pressure

$$\nabla^2 p' = 0. \tag{14.6}$$

With fixed boundaries, Laplace's equation cannot describe wave propagation because it is diagnostic (Sec. 12.5). With boundaries that are time dependent, it can. Boundary conditions that govern the free surface introduce time dependence, making the complete system prognostic.

The free surface is an interface between fluids. It must satisfy two conditions. The *dynamic condition* requires the stress to vanish. As the surface is free, it adjusts to relieve any imbalance of forces. If the air pressure overhead is negligible, then the total pressure $(p = \bar{p} + p')$ must vanish. With hydrostatic equilibrium for the mean state (14.3), the total pressure on the free surface is

$$p = -\rho g \eta' + p' \qquad z = H + \eta', \tag{14.7.1}$$

where η' is the deflection of the free surface from its undisturbed elevation. Then the perturbation pressure at the top of the layer is given by

$$p' = \rho g \eta' \qquad z = H, \qquad (14.7.2)$$

which, to the same degree of approximation in (14.6), is evaluated at the mean elevation of the free surface. The *kinematic condition* requires the free surface to be a material surface. Its vertical displacement must therefore satisfy

$$\frac{d\eta'}{dt} = w \qquad z = H + \eta'. \qquad (14.8.1)$$

In the limit of small amplitude, this reduces to

$$\frac{D\eta'}{Dt} = w' \qquad z = H. \qquad (14.8.2)$$

Combining the dynamic condition (14.7.2) and the kinematic condition (14.8.2) obtains

$$\frac{1}{\rho g}\frac{Dp'}{Dt} = w' \qquad z = H. \qquad (14.9.1)$$

Incorporating (14.9.1) into the vertical momentum equation then gives the upper boundary condition

$$\frac{1}{g}\frac{D^2 p'}{Dt^2} = -\frac{\partial p'}{\partial z} \qquad z = H. \qquad (14.9.2)$$

Another boundary condition applies at the base of the layer, where vertical motion must vanish identically. There, the vertical momentum equation requires

$$\frac{\partial p'}{\partial z} = 0 \qquad z = 0. \qquad (14.10)$$

14.1.2 Fourier synthesis

Consider a small disturbance to the free surface. The system is positively stratified because the density of the liquid is much greater than that of the overlying air, which is ignored. Buoyancy then provides a positive restoring force. It drives fluid back toward its undisturbed position, setting up oscillations. A compact initial disturbance therefore radiates away in the form of waves.

Wave activity radiating away from the initial disturbance may assume a complex form, for example, because the structure of the disturbance imposes that form initially or because individual components of the wave field propagate differently. Equations governing the wave field are linear in perturbation quantities. They also have coefficients that are independent of position and time.

These features enable the wave field to be constructed from a superposition of plane waves, each of the form $e^{i(kx+ly-\sigma t)}$. They are synthesized in the Fourier integral

$$p'(\mathbf{x}, t) = \frac{1}{(2\pi)^3} \int \int \int_{-\infty}^{\infty} P_{kl}^{\sigma}(z) e^{i(kx+ly-\sigma t)} dk dl d\sigma$$
$$= \frac{1}{(2\pi)^3} \int \int \int_{-\infty}^{\infty} P_{\mathbf{k}}^{\sigma}(z) e^{i(\mathbf{k}\cdot\mathbf{x}-\sigma t)} dk d\sigma. \qquad (14.11.1)$$

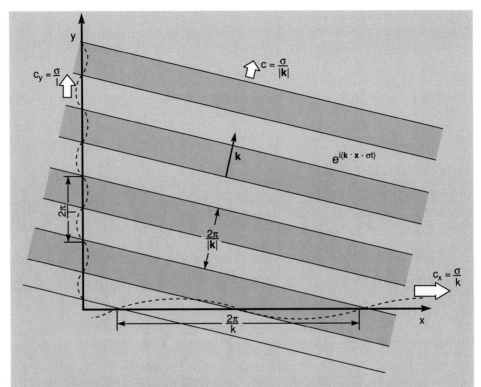

Figure 14.2 An individual plane wave component having wavenumber vector $\boldsymbol{k} = (k, l)$ and frequency σ. Lines of constant phase: $kx + ly - \sigma t = $ const, propagate parallel to $\boldsymbol{k}$ ($\sigma > 0$) or antiparallel to $\boldsymbol{k}$ ($\sigma < 0$), with *phase speed* $c = \frac{\sigma}{|\boldsymbol{k}|}$. Phase propagation in the x and y directions is described by the *trace speeds* $c_x = \frac{\sigma}{k}$ and $c_y = \frac{\sigma}{l}$. The latter equal or exceed c and are related to it through $\frac{1}{c_x^2} + \frac{1}{c_y^2} = \frac{1}{c^2}$, so they do not represent its components. As $\boldsymbol{k}$ becomes orthogonal to a direction, the trace speed in that direction becomes infinite.

Equation (14.11.1) represents an *inverse Fourier transform*. It describes p' at level z as a spectrum of monochromatic plane waves. Individual components of the spectrum have angular frequency σ and wavenumber vector $\boldsymbol{k} = (k, l)$. The complex amplitude $P_{kl}^{\sigma}(z)$ contains the magnitude and phase of wave component (k, l, σ). It must satisfy the governing equations, as well as boundary conditions. The corresponding wave component has lines of constant phase, $kx + ly - \sigma t = $ const (Fig. 14.2). The latter propagate in the direction of $\boldsymbol{k}$ with the *phase speed*

$$c = \frac{\sigma}{|\boldsymbol{k}|}, \tag{14.11.2}$$

where $|\boldsymbol{k}|^2 = k^2 + l^2$. Phase propagation in the x and y directions is described by the *trace speeds*

$$c_x = \frac{\sigma}{k}$$
$$c_y = \frac{\sigma}{l}. \tag{14.11.3}$$

Note, $c_x^2 + c_y^2 \neq c^2$. According to Fig. 14.2, the trace speed in a given direction equals or exceeds the phase speed. As k becomes orthogonal to that direction, it approaches infinity.

By the Fourier integral theorem (e.g., Rektorys, 1969), the complex amplitudes in (14.11.1) follow as

$$P_{kl}^{\sigma}(z) = \int \int \int_{-\infty}^{\infty} p'(x, y, z, t) e^{-i(kx+ly-\sigma t)} dx\, dy\, dt. \tag{14.12.1}$$

Equation (14.12.1) is the *Fourier transform* of the variable p' at level z. It describes a spectrum of complex wave amplitude over the space-time scales k, l, and σ. Real p' requires the amplitude spectrum to satisfy the Hermitian property

$$P_{-k-l}^{-\sigma} = P_{kl}^{\sigma*}. \tag{14.12.2}$$

Equations (14.11.1) and (14.12.1) constitute a one-to-one mapping between $p'(x, t)$ in physical space and $P_{kl}^{\sigma}(z)$ in Fourier space. The latter are therefore equivalent representations of the wave field. The *power spectrum*, $|P_{kl}^{\sigma}|^2$, then describes the distribution over wavenumber and frequency of variance $|p'|^2$. It characterizes the energy in the wave field. In general, a disturbance excites a continuum of space and time scales. Wave variance therefore involves an integral over wavenumber and frequency:

$$\int \int \int_{-\infty}^{\infty} |p'(x, t)|^2 dx\, dt = \frac{1}{(2\pi)^3} \int \int \int_{-\infty}^{\infty} |P_k^{\sigma}|^2 dk\, d\sigma. \tag{14.13}$$

The equivalence of wave variance in physical space and in Fourier space is known as *Parseval's theorem*.

Differentiating p' in physical space is equivalent to multiplying its transform P_{kl}^{σ} by the corresponding space and time scales:

$$\frac{\partial p'}{\partial x} \leftrightarrow ik P_{kl}^{\sigma}$$

$$\frac{\partial p'}{\partial y} \leftrightarrow il P_{kl}^{\sigma} \tag{14.14}$$

$$\frac{\partial p'}{\partial t} \leftrightarrow -i\sigma P_{kl}^{\sigma}.$$

Then applying (14.12.1) transforms the material derivative into

$$\frac{Dp'}{Dt} \to -i\omega P_{kl}^{\sigma}, \tag{14.15.1}$$

where

$$\omega = \sigma - k\overline{u}. \tag{14.15.2}$$

ω is the *intrinsic frequency*, that relative to the medium, for the component with frequency σ and zonal wavenumber k. It is the frequency observed from a frame moving with the mean motion $\overline{u}$. The term $-k\overline{u}$ represents a Doppler shift of the intrinsic frequency by background motion. A component propagating opposite to $\overline{u}$ is Doppler shifted to higher intrinsic frequency: $\omega > \sigma$. One propagating in the same direction as $\overline{u}$ is Doppler shifted to lower intrinsic frequency: $\omega < \sigma$.

Considering solutions of the form $e^{i(kx+ly-\sigma t)}$ is equivalent to applying the Fourier transform (14.12.1). The governing equations in physical space, (14.6), (14.9.2), and (14.10), are then transformed into their counterparts in Fourier space:

$$\frac{\partial^2 P_{kl}^{\sigma}}{\partial z^2} - |k|^2 P_{kl}^{\sigma} = 0 \tag{14.16.1}$$

$$\frac{\partial P_{kl}^{\sigma}}{\partial z} = 0 \qquad\qquad z = 0 \tag{14.16.2}$$

$$\frac{\partial P_{kl}^{\sigma}}{\partial z} - \frac{\omega^2}{g} P_{kl}^{\sigma} = 0 \qquad\qquad z = H. \tag{14.16.3}$$

Equations (14.16) govern the complex amplitude $P_{kl}^{\sigma}(z)$ of the plane wave component with wavenumber and frequency (k, l, σ). They depend on the intrinsic frequency ω, that relative to the mean flow $\bar{u}$. Differential dependence on x, y, and t has been reduced to algebraic dependence on the scales k, l, and σ for individual modes. The latter can be re-synthesized into $p'(x, t)$ via the inverse Fourier transform (14.11.1).

The general solution of (14.16.1) satisfying the lower boundary condition (14.16.2) is of the form

$$P_{kl}^{\sigma}(z) = A\cosh(|k|z) \tag{14.17}$$

(Fig. 14.1). Substituting into the upper boundary condition (14.16.3) then yields the algebraic identity

$$\omega^2 = g|k|\tanh(|k|H). \tag{14.18}$$

Equation (14.18) is called the *dispersion relation*. It must be satisfied by each component (k,l,σ) if the wave spectrum is to obey the governing equations. The dispersion relation expresses one of the space-time scales (e.g., frequency) in terms of the others. Therefore, the spatial scales of an individual wave component determine its frequency.

14.1.3 Limiting behavior

According to (14.18), different horizontal scales propagate relative to the mean flow at different intrinsic phase speeds

$$\hat{c} = \frac{\omega}{|k|}$$
$$= \left[gH \frac{\tanh(|k|H)}{|k|H} \right]^{\frac{1}{2}}. \tag{14.19}$$

Phase speed is plotted as a function of $|k|H$ in Fig. 14.3. It decreases with wavenumber monotonically. The longest waves ($|k|H \to 0$) therefore travel fastest.

In the limit of long wavelength,

$$\hat{c} \sim \sqrt{gH} \qquad\qquad |k|H \to 0. \tag{14.20}$$

The dispersion relation (14.20) describes *shallow water waves*. They have wavelengths $\lambda = \frac{2\pi}{|k|}$ long compared with the depth H of the fluid. Phase speed is then independent of wavelength. Consequently, different components of a shallow water disturbance propagate in unison. For this reason, shallow water waves are *nondispersive*: A disturbance retains its initial shape. Shallow water waves are also fast. In an ocean 5 km deep, $\hat{c} > 200$ m s^{-1} – comparable to the speed of sound. This feature of shallow water

PHASE SPEED

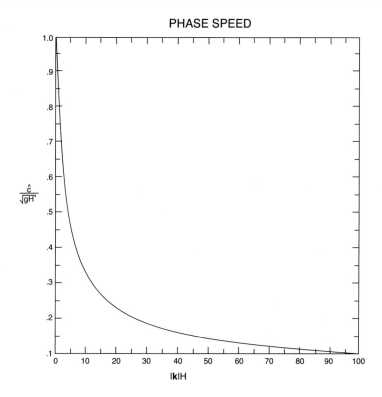

Figure 14.3 Intrinsic phase speed of surface water waves (equals phase speed in the absence of mean motion), as a function of horizontal wavenumber.

waves enables large-scale components of a tsunami to traverse an ocean basin in a matter of hours.[1]

In the longwave limit, amplitude (14.17) is independent of z (Fig. 14.4a). Then, by continuity (14.4.2) and the lower boundary condition, shallow water waves possess no vertical motion. Under these circumstances, the vertical momentum equation reduces to a statement of hydrostatic equilibrium. Hence, the longwave limit is equivalent to invoking hydrostatic balance. This feature of shallow water waves follows from the shallowness of vertical displacements in this limit, relative to horizontal displacements.

The opposite extreme is the limit of short wavelength. Then

$$\hat{c} \sim \sqrt{\frac{g}{|\boldsymbol{k}|}} \qquad |\boldsymbol{k}|H \to \infty. \tag{14.21}$$

The dispersion relation (14.21) describes *deep water waves*. They have wavelengths short compared with the depth of the fluid. In this limit, amplitude (14.17) decreases exponentially away from the free surface (Fig. 14.4b). It has the structure of an *edge wave*. Deep water waves do not sense the lower boundary because their energy there

---

[1] The tsunami excited by the magnitude-9 earthquake in Honshu Japan of March 11, 2011 reached California in just 10 hrs. http://www.youtube.com/user/NOAAPMEL?feature=mhum#p/c/3/PBZGH3yieLc (15/06/11)

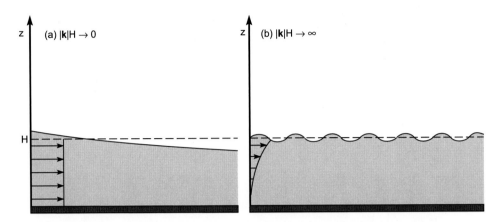

Figure 14.4 Surface water waves (a) in the longwave limit, where they assume the form of *shallow water waves*, with horizontal wavelengths long compared to the fluid depth and horizontal motion that is invariant with depth, and (b) in the shortwave limit, where they assume the form of *deep water waves*, with horizontal wavelengths short compared to the fluid depth and motion that decays exponentially away from the free surface as an edge wave.

is negligible. Unlike shallow water waves, they have phase speed that varies with wavelength. In a deep water disturbance, $\hat{c}$ increases with λ. Longer horizontal wavelengths in the disturbance are therefore faster than shorter wavelengths, which are left behind. For this reason, deep water waves are *dispersive*: A disturbance does not retain its initial shape. As it propagates away from its forcing, a deep water disturbance unravels into individual wave components.

14.1.4 Wave dispersion

Because a spectrum of waves is involved, interference among components can modify wave activity as it radiates away from the imposed disturbance. To illustrate such behavior, consider a *group* of plane waves defined by an incremental band of wavenumber that is centered at k, with l fixed (Fig. 14.5a). Wave activity in this band can be approximated by the rectangle shown. Variance is then shared equally by two discrete components with wavenumbers $k \pm dk$. Those components have frequencies $\sigma \pm d\sigma$, which are related to wavenumber through the dispersion relation (14.18). For unit amplitude, wave activity in the band is described by

$$\eta_k'(x, t) = \frac{1}{2}\left\{\cos\left[(k - dk)x - (\sigma - d\sigma)t\right] + \cos\left[(k + dk)x - (\sigma + d\sigma)t\right]\right\}$$

$$= \cos(dk \cdot x - d\sigma \cdot t) \cdot \cos(kx - \omega t), \tag{14.22.1}$$

which follows from (14.11.1) with the amplitude spectrum approximated as above and with the Hermitian property (14.12.2). Wave activity in the band may then be expressed

$$\eta_k'(x, t) = \cos\left[dk(x - \frac{\partial \sigma}{\partial k}t)\right] \cdot \cos(kx - \sigma t). \tag{14.22.2}$$

As illustrated in Fig. 14.5b, (14.22) describes a disturbance with wavenumber k and trace speed $c_x = \frac{\sigma}{k}$ that is modulated by an envelope of wavenumber dk. The envelope

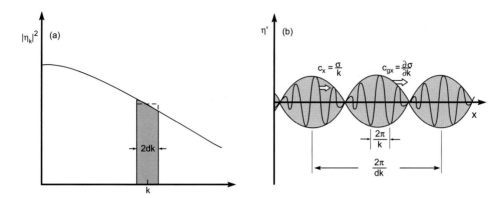

Figure 14.5 (a) Power spectrum of surface water waves as a function of zonal wavenumber. A *wave group* is defined by variance in the incremental band centered at k (shaded), which may be approximated by the rectangle shown. Wave activity is then carried by two discrete components at $k \pm dk$. (b) Wave activity in the band produces a disturbance of wavenumber k that is modulated by an envelope of wavenumber dk. Oscillations propagate with the trace speed $c_x = \frac{\sigma}{k}$. However, their envelope propagates with the *group speed* $c_{gx} = \frac{\partial \sigma}{\partial k}$.

propagates in the x direction with the *group speed*

$$
\begin{aligned}
c_{gx} &= \frac{\partial \sigma}{\partial k} \\
&= \frac{\partial \omega}{\partial k} + \overline{u}.
\end{aligned}
$$
(14.23.1)

Because it is concentrated inside the envelope, wave activity in the above band propagates, not with the phase speed, but with the group speed.

The two components in (14.22) approximate the spectrum inside a narrow band of wavenumber; they correspond to a discrete boxcar filter in k. Defining a group by a continuous Gaussian filter produces behavior analogous to that in Fig. 14.5b, but concentrated inside a Gaussian envelope. Of limited extent, it represents a packet of waves. Outside the envelope, oscillations vanish. Wave activity must therefore move with the envelope – regardless of the movement of individual crests and troughs inside it.

Similar reasoning applied to the wavenumber l, with k fixed, yields the group speed in the y direction:

$$
\begin{aligned}
c_{gy} &= \frac{\partial \sigma}{\partial l} \\
&= \frac{\partial \omega}{\partial l}.
\end{aligned}
$$
(14.23.2)

Wave activity thus propagates in the $x - y$ plane with the *group velocity*

$$
\begin{aligned}
\boldsymbol{c}_g &= \left(\frac{\partial \sigma}{\partial k}, \frac{\partial \sigma}{\partial l} \right) \\
&= \left(\frac{\partial \omega}{\partial k} + \overline{u}, \frac{\partial \omega}{\partial l} \right).
\end{aligned}
$$
(14.23.3)

A function of k and l, $\boldsymbol{c}_g$ describes the propagation of wave activity in individual bands of the spectrum (14.11.1). According to (14.23.3), background motion advects wave

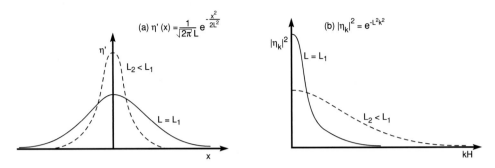

Figure 14.6 (a) A Gaussian disturbance imposed initially on the free surface. The length L characterizes the width of the disturbance. (b) The power spectrum of the disturbance, as a function of zonal wavenumber k. As L increases, variance becomes concentrated at small k in a *red* spectrum, which excites waves in the longwave limit. As L decreases, variance becomes distributed widely over k, approaching a *white* spectrum with most of the variance at large k, which excites waves in the shortwave limit.

activity with the basic flow. The intrinsic group velocity

$$\hat{c}_g = c_g - \overline{u}i \qquad (14.23.4)$$

gives the same information relative to the mean flow. According to (14.23.3), it is just the gradient of ω with respect to $\boldsymbol{k} = (k, l)$.

Let us return to the limiting forms of surface water waves described previously. They are considered in the absence of background motion, $\overline{u} = 0$. In the longwave limit, phase speed (14.20) is independent of $\boldsymbol{k}$. Because individual wave components all propagate with the same speed, a disturbance comprised of shallow water waves retains its initial shape. Such waves are nondispersive, because a disturbance that is initially compact remains so. The group speed of shallow water waves follows from (14.23) as

$$c_g = \sqrt{gH} \qquad (14.24)$$
$$= c.$$

Because all wavenumbers propagate with identical phase speed, a group of shallow water waves centered at wavenumber $\boldsymbol{k}$ propagates in the same direction and with the same speed.

An example is presented in Figs. 14.6 and 14.7. The free surface is initialized with a compact disturbance (Fig. 14.6a)

$$\eta'(x, y, 0) = \frac{1}{\sqrt{2\pi L}} e^{-\frac{x^2 + y^2}{2L^2}}, \qquad (14.25.1)$$

with characteristic horizontal scale L. Because it is not simple harmonic, that disturbance excites a continuum of waves. Transforming (14.25.1) via (14.12.1) with $t = 0$ gives its power spectrum (Fig. 14.6b)

$$|\eta_{kl}|^2 = e^{-L^2(k^2 + l^2)}. \qquad (14.25.2)$$

The excited wave field is then described by (14.11.1), with individual components subject to the dispersion relation (14.18). If the scale of the imposed disturbance is

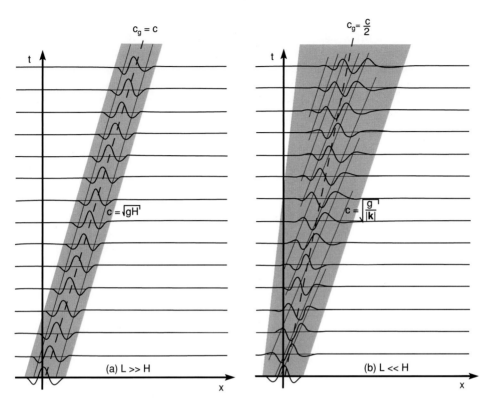

Figure 14.7 (a) Surface water waves produced by the initial disturbance in Fig. 14.6, with $L \gg H$. Shallow water waves radiate *nondispersively* away from the initial disturbance: Wave components with different k propagate at identical phase speed $c = \sqrt{gH}$ (solid). The envelope of wave activity then propagates at the same speed $c_g = c$ (dashed). Under these circumstances, the shape of the initial waveform is preserved, so wave activity (shaded) remains confined to the same range of x as initially. (b) As in (a) but for $L \ll H$ and a different x scale. Deep water waves radiate *dispersively* away from the initial disturbance: Wave components with different k propagate at different phase speeds $c = \sqrt{\frac{g}{|k|}}$ (solid), so the initial wave form unravels into a series of oscillations. Occupying a progressively wider range of x, the envelope of wave activity (shaded) propagates at half the median phase speed of individual components $c_g = \frac{c}{2}$ (dashed). Individual crests and troughs therefore overrun the envelope, disappearing at its leading edge, to be replaced by new ones at its trailing edge.

sufficiently long (e.g., $L = L_1 \gg H$), the power spectrum of η' (Fig. 14.6b) is very *red*: $|\eta_{kl}|^2$ decreases sharply with $(k^2 + l^2)$. Variance is then concentrated at small $|\mathbf{k}|H$, so the excited wave field is comprised chiefly of shallow water waves. Far from the initial disturbance, waves can be treated as planar. The disturbance (Fig. 14.7a) then translates outward jointly with its envelope at a uniform speed $c = c_g$, retaining its initial form. Therefore, wave activity (shaded) occupies the same range of x as initially, only displaced.

In the shortwave limit, the phase speed (14.21) depends on $|\mathbf{k}|$. Individual components (k, l) then propagate with different phase speeds. A disturbance comprised of deep water waves therefore cannot retain its initial shape. The difference in phase speed between components with small and large $|\mathbf{k}|$ causes a group to unravel with time. Such waves are said to be *dispersive* because a disturbance that is initially compact is eventually distributed over a wider domain. The group speed for deep water waves follows from (14.21) as

$$c_g = \frac{1}{2}\sqrt{\frac{g}{|\mathbf{k}|}}$$

$$= \frac{c}{2}. \tag{14.26}$$

A group of deep water waves thus propagates at exactly half their median phase speed. In this limit, c and c_g both depend on $|\mathbf{k}|$.

The behavior predicted by (14.26) is readily observed with the toss of a stone into a pond. It is recovered from (14.25) if the horizontal scale of the disturbance in Fig. 14.6a is sufficiently short (e.g., $L = L_2 \ll H$). The power spectrum of η' is then nearly *white* (Fig. 14.6b): $|\eta_{kl}|^2$ decreases with $|\mathbf{k}|$ slowly. Most of the excited variance therefore lies at large $|\mathbf{k}|H$. The disturbance is then comprised chiefly of deep water waves. Under these circumstances, the initial disturbance unravels into a series of oscillations (Fig. 14.7b), which propagate at different speeds. Wave activity therefore occupies an increasingly wider range of x. Contrary to shallow water waves (Fig. 14.7a), the envelope of wave activity translates outward slower than individual crests and troughs, at only half their speed (14.26). Waves inside it therefore overrun the envelope. They disappear at its leading edge, replaced by new ones that appear at its trailing edge.

14.2 ACOUSTIC WAVES

The simplest wave motions supported by the governing equations are sound waves. Compressibility provides the restoring force for simple acoustic waves, which have time scales short enough to ignore rotation, heat transfer, friction, and buoyancy. Acoustic waves are longitudinal disturbances: Fluid displacements are parallel to the propagation vector $\mathbf{k}$. We may, therefore, consider motion in the x direction. Under these circumstances, air motion is governed by

$$\frac{du}{dt} = -\frac{1}{\rho}\frac{\partial p}{\partial x} \tag{14.27.1}$$

$$\frac{d\rho}{dt} + \rho\frac{\partial u}{\partial x} = 0 \tag{14.27.2}$$

$$p\rho^{-\gamma} \text{ conserved}, \tag{14.27.3}$$

where Poisson's relation (2.30) serves as a statement of the First Law under adiabatic conditions. Applying the logarithm to (14.27.3) followed by the material derivative obtains

$$\frac{1}{\gamma}\frac{d\ln p}{dt} = \frac{1}{\rho}\frac{d\rho}{dt}.$$

Substituting into (14.27.2) yields

$$\frac{1}{\gamma}\frac{d\ln p}{dt} + \frac{\partial u}{\partial x} = 0. \tag{14.28}$$

For a homogeneous basic state (e.g., one that is isothermal and in uniform motion), the first order perturbation equations are then

$$\frac{Du'}{Dt} = -\frac{1}{\overline{\rho}}\frac{\partial p'}{\partial x}$$
$$\frac{Dp'}{Dt} + \gamma\overline{p}\frac{\partial u'}{\partial x} = 0.$$

(14.29)

These may be consolidated into a single second order equation

$$\frac{D^2 p'}{Dt^2} - \gamma R\overline{T}\frac{\partial^2 p'}{\partial x^2} = 0.$$

(14.30)

Equation (14.30) is a 1-dimensional wave equation for the perturbation pressure p'. From it, other field properties follow. The coefficients of (14.30) are independent of x and t. We may therefore consider solutions of the form $e^{i(kx-\sigma t)}$, tantamount to applying the Fourier transform (14.12.1). Substituting transforms (14.30) into the dispersion relation

$$\hat{c}^2 = \gamma R\overline{T}.$$

(14.31)

The right-hand side of (14.31) is the speed of sound. The intrinsic phase speed $\hat{c}$ is independent of k. Acoustic waves are thus nondispersive. Their group velocity equals their phase velocity.

14.3 BUOYANCY WAVES

More important to the atmosphere are disturbances that are supported by the positive restoring force of buoyancy. They exist under stable stratification. Known as *gravity waves*, these disturbances typically have time scales short enough to ignore rotation, heat transfer, and friction. Vertical motions induced by gravity waves need not be hydrostatic. The governing equations are therefore expressed most conveniently in physical coordinates. Gravity waves involve motion transverse to the propagation vector k. Consequently, they must be described in two dimensions. In the $x - z$ plane, air motion is described by

$$\frac{du}{dt} = -\frac{1}{\rho}\frac{\partial p}{\partial x}$$

(14.32.1)

$$\frac{dw}{dt} = -\frac{1}{\rho}\frac{\partial p}{\partial z} - g$$

(14.32.2)

$$\frac{1}{\rho}\frac{d\rho}{dt} + \frac{\partial u}{\partial x} + \frac{\partial w}{\partial z} = 0$$

(14.32.3)

$$\frac{d\ln p}{dt} - \gamma\frac{d\ln\rho}{dt} = 0.$$

(14.32.4)

As they account for compressibility, (14.32) also describe acoustic waves, which, at low frequency, are modified by buoyancy.

For a basic state that is isothermal, in uniform motion, and in hydrostatic equilibrium, the perturbation equations become

$$\frac{Du'}{Dt} = -gH\frac{\partial \hat{p}'}{\partial x}$$

(14.33.1)

$$\frac{Dw'}{Dt} = -gH\frac{\partial \hat{p}'}{\partial z} + g\hat{p}' - g\hat{\rho}'$$

(14.33.2)

$$\frac{D\hat{\rho}'}{Dt} + \frac{\partial u'}{\partial x} + \frac{\partial w'}{\partial z} - \frac{w'}{H} = 0 \qquad (14.33.3)$$

$$\frac{D}{Dt}\left(\hat{p}' - \gamma\hat{\rho}'\right) + \gamma\frac{N^2}{g}w' = 0, \qquad (14.33.4)$$

where

$$\hat{p}' = \frac{p'}{\overline{p}} \qquad\qquad \hat{\rho}' = \frac{\rho'}{\overline{\rho}} \qquad (14.33.5)$$

are scaled by the basic state stratification,

$$H = \frac{R\overline{T}}{g}, \qquad (14.33.6)$$

and

$$\frac{N^2}{g} = \frac{\partial \ln\overline{\theta}}{\partial z} \qquad (14.33.7)$$
$$= \frac{\kappa}{H}$$

corresponds to a buoyancy period of about 20 minutes.

Equations (14.33) constitute a closed system for the four unknowns: u', w', $\hat{p}'$, and $\hat{\rho}'$. Because the system has constant coefficients, solutions may be expressed in terms of plane waves (14.11.1). However, owing to the stratification of mass, a group of waves propagating vertically must adjust in amplitude to conserve energy. For quadratic quantities like $\overline{\rho}u'^2$ (kinetic energy density) to remain constant following the group, u', w', $\hat{p}'$, and $\hat{\rho}'$ must amplify vertically like $\overline{\rho}^{-\frac{1}{2}} = e^{\frac{z}{2H}}$. Considering solutions of the form $e^{\frac{z}{2H}+i(kx+mz-\sigma t)}$, where m denotes vertical wavenumber, then transforms the differential system (14.33) into an algebraic one. Consolidating the algebraic equations yields the dispersion relation

$$m^2 = k^2\left(\frac{N^2}{\omega^2} - 1\right) + \frac{\omega^2 - \omega_a^2}{c_s^2}, \qquad (14.34.1)$$

where

$$c_s^2 = \gamma R\overline{T} \qquad (14.34.2)$$

is the speed of sound. The term

$$\omega_a = \frac{c_s}{2H} \qquad (14.34.3)$$

is the *acoustic cutoff frequency* for vertical propagation. At frequencies $|\omega| < \omega_a$, acoustic waves do not propagate vertically.

Equation (14.34) describes *acoustic-gravity waves*, which involve buoyancy as well as compression. Their vertical propagation is controlled by the sign of m^2. For $m^2 > 0$, individual wave components oscillate in z. Vertically propagating, such waves are referred to as *internal waves* because their oscillatory structure occurs in the interior of a bounded domain.[2] Vertically propagating waves amplify upward like $e^{\frac{z}{2H}}$ (Fig. 14.8a). The upward amplification offsets the stratification of mass, making energy

---

[2] For freely propagating waves, which have form e^{imz}, this terminology is a misnomer. Internal modes, which have form $e^{imz} + e^{imz} = 2\cos mz$, are trapped vertically between boundaries, which lead to complete reflection.

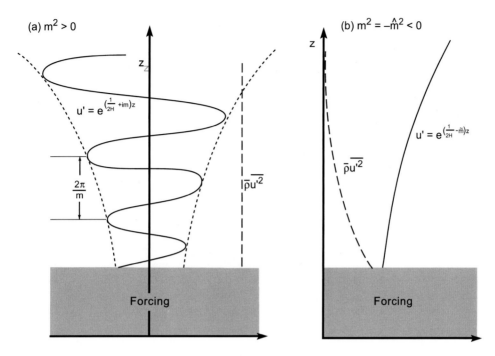

Figure 14.8 Vertical structure of (a) *internal waves* ($m^2 > 0$), which oscillate with z (solid) and amplify upward like $e^{\frac{z}{2H}}$ to make energy density (dashed) invariant with height, and (b) *external waves* ($m^2 < 0$), whose energy density (dashed) decays exponentially with height but whose amplitude (solid) can either amplify or decay depending on m.

density invariant with height. For $m^2 = -\hat{m}^2 < 0$, individual wave components vary exponentially in z (Fig. 14.8b). Oscillations then occur in-phase throughout the column. For the column energy to remain bounded, the energy density of these waves must decrease with height exponentially. Such waves are referred to as *evanescent* or *external* because they form along the exterior of a boundary. The edge-wave structure of surface water waves in the deep-water limit is an example (Fig. 14.4b). Even though their energy decreases upward, external waves in the range $\frac{1}{4H^2} < m^2 < 0$ amplify upward like $e^{(\frac{1}{2H} - \hat{m})z}$ due to stratification of mass.

The dispersion relation assumes two limiting forms associated with vertical propagation. They appear when m^2 is plotted as a function of k and ω (Fig. 14.9). For $\omega^2 \gg N^2$, (14.34) reduces to

$$k^2 + m^2 + \frac{1}{4H^2} \cong \frac{\omega^2}{c_s^2} \qquad \frac{\omega^2}{N^2} \to \infty. \qquad (14.35.1)$$

This limiting behavior describes high-frequency acoustic waves that are modified by stratification. It is also recovered if $N \to 0$, which eliminates the restoring force of buoyancy. Acoustic waves propagate vertically for ω greater than values on the upper curve for $m^2 = 0$ in Fig. 14.9 (shaded). For $k^2 + m^2 \to \infty$, their intrinsic phase speeds approach the speed of sound. In the longwave limit ($k \to 0$), the minimum ω for vertical propagation equals the acoustic cutoff ω_a. In the shortwave limit ($k \to \infty$), the

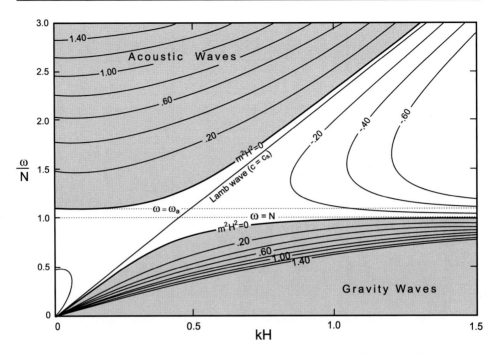

Figure 14.9 Vertical wavenumber squared contoured as a function of horizontal wavenumber and intrinsic frequency. Vertical propagation (shaded) occurs for acoustic waves at ω greater than the acoustic cutoff ω_a and for gravity waves at ω less than the buoyancy frequency N, where $\omega_a \cong 1.1 N$ under isothermal conditions. The Lamb wave connects the limiting behavior of acoustic waves for $m^2 = 0$ and $k \to \infty$ to the limiting behavior of gravity waves for $m^2 = 0$ and $k \to 0$. It propagates horizontally at the speed of sound.

minimum ω for vertical propagation tends to infinity. The corresponding waves have horizontal phase speeds that approach the speed of sound.

For $\omega^2 \ll \omega_a^2$, (14.34) reduces to

$$k^2 + m^2 + \frac{1}{4H^2} \cong \frac{N^2}{\omega^2} k^2 \qquad \frac{\omega^2}{\omega_a^2} \to 0, \qquad (14.35.2)$$

This limiting behavior describes low-frequency gravity waves that are modified by stratification of mass. It is also recovered if $c_s \to \infty$, which renders air motion incompressible. Gravity waves propagate vertically for ω less than values on the lower curve for $m^2 = 0$ in Fig. 14.9 (shaded). Unlike most vibrating systems, internal gravity waves have a high-frequency cutoff for vertical propagation: N. In the shortwave limit ($k \to \infty$), the maximum ω for vertical propagation approaches N. Horizontal phase speed then vanishes. The corresponding behavior describes column oscillations at the buoyancy frequency (7.11). In the longwave limit ($k \to 0$), the maximum ω for vertical propagation approaches zero. The corresponding waves propagate horizontally, with phase speeds that approach the speed of sound.

The regions of vertical propagation in Fig. 14.9 are well-separated. Between them, $m^2 < 0$ and waves are external (unshaded). Acoustic-gravity waves then propagate horizontally, with energy that decreases upward exponentially. They exert no net influence

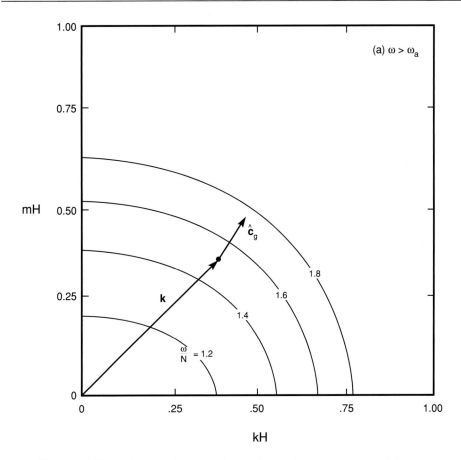

Figure 14.10 Intrinsic frequency of acoustic-gravity waves contoured as a function of wavenumber. Group velocity relative to the medium is directed orthogonal to contours of intrinsic frequency toward increasing ω. (a) Acoustic waves ($\omega > N$) have group velocity relative to the medium $\hat{c}_g$ that is nearly parallel to phase propagation $\boldsymbol{k}$. (b) Gravity waves ($\omega < N$) have $\hat{c}_g$ that is nearly orthogonal to $\boldsymbol{k}$, except close to $m = 0$, where both become horizontal.

in the vertical. Connecting the limiting behavior of internal acoustic waves for $k \to \infty$ and internal gravity waves for $k \to 0$ is the external *Lamb wave* (Lamb, 1910). It propagates horizontally at the speed of sound, for all k.

Dispersion characteristics of the acoustic and gravity branches of (14.34) are illustrated in Fig. 14.10, which plots ω as a function of $\boldsymbol{k} = (k, m)$. The intrinsic group velocity (14.23) is directed orthogonal to contours of intrinsic frequency toward increasing ω. On the other hand, phase propagates with the velocity $\hat{c}\frac{k}{|k|}$, parallel ($\omega > 0$) or anti-parallel ($\omega < 0$) to the wavenumber vector $\boldsymbol{k}$. For acoustic waves (Fig. 14.10a), contours of ω are elliptical. They are modified only slightly from their forms under unstratified conditions. The group velocity $\hat{c}_g$ is then close to the phase velocity $c\frac{k}{|k|}$ for all $\boldsymbol{k}$. Hence, acoustic waves are only weakly dispersive.

For gravity waves (Fig. 14.10b), contours of ω are hyperbolic. The situation is then quite different. In the limit of short wavelength ($k^2 + m^2 \to \infty$), contours of ω become

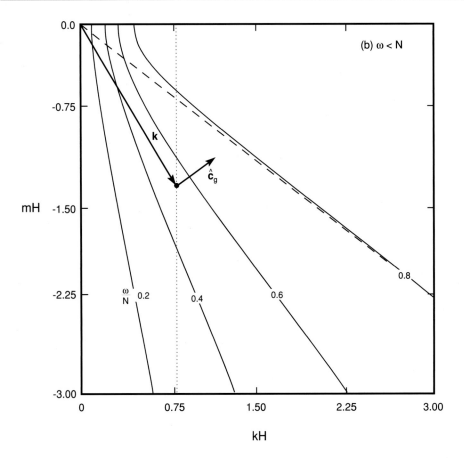

Figure 14.10 (*continued*)

straight lines that pass through the origin. The group velocity $\hat{c}_g$ is then orthogonal to the phase velocity $\hat{c}\frac{k}{|k|}$. Such waves are thus highly dispersive. A wave group defined by the vector k in Fig. 14.10b has phase that propagates downward and to the right. However, its envelope propagates upward and to the right. A packet of gravity waves (Fig. 14.11) propagates parallel to lines of constant phase (e.g. along crests and troughs), not perpendicular to them, as is observed of more familiar acoustic waves.[3] Consequently, downward propagation of phase translates into upward propagation of wave activity. This peculiarity follows from the dispersive nature of gravity waves. It applies to all waves whose vertical restoring force is buoyancy. In the limit of long wavelength ($k^2 + m^2 \rightarrow 0$), contours of ω become parallel to the m axis. Group velocity is then horizontal.

At very low frequency, rotation cannot be ignored. The Coriolis force then modifies the dispersion relation. It also makes parcel trajectories 3-dimensional. Under these conditions, air motion (u, w) associated with buoyancy oscillations experiences a Coriolis force fu. The latter drives parcel trajectories out of the $x - z$ plane,

---

[3] Lighthill (1978) presents a laboratory demonstration.

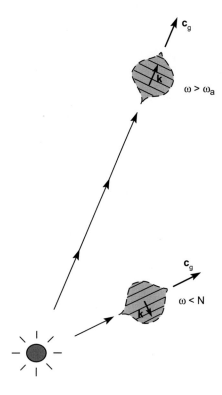

Figure 14.11 A packet of simple gravity waves ($\omega < N$) propagates along crests and troughs – just perpendicular to simple acoustic waves ($\omega > \omega_a$), which propagate orthogonal to crests and troughs.

introducing v. Disturbances in this range of frequency are referred to as *inertio-gravity waves*. They are treated in Gill (1982).

14.3.1 Shortwave limit

In the limit of short total wavelength ($k^2 + m^2 \to \infty$), the dispersion relation for gravity waves (14.35.2) reduces to

$$\frac{\omega^2}{N^2} = \frac{k^2}{k^2 + m^2}$$

$$= cos^2\alpha, \tag{14.36}$$

where α is the angle that $\boldsymbol{k}$ makes with the horizontal (Fig. 14.12). Contours of ω are then straight lines – the limiting behavior in Fig. 14.10b, wherein $\hat{c}_g$ becomes orthogonal to $\boldsymbol{k}$. Describing simple gravity waves, these disturbances have phase that propagates at progressively shallower angles as ω is increased. As ω approaches N, $\boldsymbol{k}$ becomes horizontal, with phase lines vertical.

The limiting behavior (14.36) applies to mountain waves (Fig. 9.22), which have horizontal wavelengths of about 10 km. Generated by forced ascent over elevated terrain, mountain waves are fixed with respect to the Earth. They have absolute frequencies $\sigma \cong 0$. Their intrinsic frequencies, however, do not vanish. Rather, they equal the Doppler shift $-k\overline{u}$. In westerly flow, contours of phase propagate westward relative to the medium. Recall that upward propagation of energy requires downward propagation of phase. Phase contours must therefore tilt westward with height above

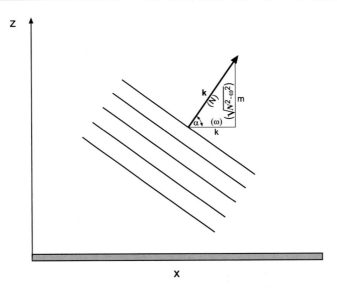

Figure 14.12 Schematic illustrating a simple gravity wave of wavenumber **k** that is inclined from the horizontal by an angle α.

their forcing. Such behavior is often observed in the leading edge of lenticular cloud (Fig. 9.22a).

Despite having zero phase velocity relative to the ground, orographically forced gravity waves have nonvanishing group velocity. Figure 14.13 displays the gravity wave pattern that is excited by steady westerly motion incident on a 2-dimensional ridge. Wave activity, manifest in oscillations of w', is transmitted upward and downstream. The dispersive nature of gravity waves leads to wave activity occupying a progressively wider range of x with increasing height above the source. Vertical motion induced by the oscillations can extend into the stratosphere (see Fig. 14.24). This feature of mountain waves has enabled sailplanes to reach altitudes as high as 50,000′ (17,000 m).

14.3.2 Propagation of gravity waves in an inhomogeneous medium

When the basic state is not isothermal and in uniform motion, wave propagation varies with position. Consider the propagation of gravity waves in a basic state that varies with height: $N^2 = N^2(z)$ and $\bar{u} = \bar{u}(z)$, and under the Boussinesq approximation (Sec. 12.5.3). The wave field is then governed by

$$\frac{Du'}{Dt} + w'\bar{u}_z = -gH\frac{\partial \hat{p}'}{\partial x} \tag{14.37.1}$$

$$\frac{Dw'}{Dt} = -gH\frac{\partial \hat{p}'}{\partial z} - g\hat{\rho}' \tag{14.37.2}$$

$$\frac{\partial u'}{\partial x} + \frac{\partial w'}{\partial z} = 0 \tag{14.37.3}$$

$$-\frac{D\hat{\rho}'}{Dt} + \frac{N^2}{g}w' = 0, \tag{14.37.4}$$

where (14.37.4) follows from (14.33.4) with the neglect of compressibility.

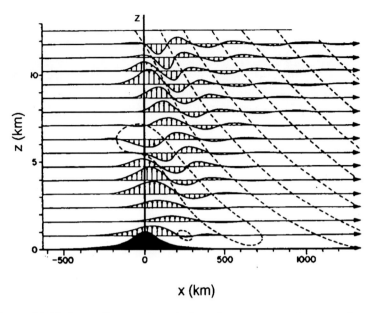

Figure 14.13 Streamlines and vertical motion accompanying a stationary gravity wave pattern that is excited by uniform flow over a two-dimensional ridge 100 km wide. After Queney (1948).

Applying the linearized material derivative to (14.37.2) gives

$$\frac{D^2 w'}{Dt^2} + gH\frac{D}{Dt}\left(\frac{\partial \hat{p}'}{\partial z}\right) = -g\frac{D\hat{\rho}'}{Dt}.$$

With (14.37.4), this becomes

$$\left[\frac{D^2}{Dt^2} + N^2\right]w' + gH\frac{D}{Dt}\left(\frac{\partial \hat{p}'}{\partial z}\right) = 0. \tag{14.38}$$

As coefficients are still independent of x and t, we consider solutions of the form $e^{i(kx-\sigma t)}$. Combining (14.37.1) with (14.37.3) obtains

$$gH\frac{\partial P_k^\sigma}{\partial z} = \frac{i}{k}\left\{(c_x - \overline{u})\frac{\partial^2 W_k^\sigma}{\partial z^2} + \overline{u}_{zz}W_k^\sigma\right\}, \tag{14.39}$$

where $P_k^\sigma(z)$ denotes the complex amplitude of $\hat{p}'$ (14.12.1). Then incorporating (14.39) into (14.38) yields

$$\frac{\partial^2 W_k^\sigma}{\partial z^2} + m^2(z)W_k^\sigma = 0, \tag{14.40.1}$$

in which

$$m^2(z) = \frac{N^2}{(c_x - \overline{u})^2} - k^2 + \frac{\overline{u}_{zz}}{c_x - \overline{u}} \tag{14.40.2}$$

serves as a 1-dimensional *index of refraction*.

Known as the *Taylor-Goldstein equation*, (14.40) describes the vertical propagation of wave activity in terms of local refractive properties of the medium. For $m^2 > 0$, $W_k^\sigma(z)$ is oscillatory in z. Wave activity then propagates vertically with a local wavelength $\frac{2\pi}{m(z)}$. Unlike propagation under homogeneous conditions (14.33), variations of m^2 can lead to reflection of wave activity. This effect couples upward and downward propagation. Oscillations governed by (14.40) do not amplify vertically, owing to the Boussinesq approximation that neglects compressibility. For $m^2 < 0$, vertical structure is evanescent. Vertical propagation is then prohibited. Wave activity incident on a region of $m^2 < 0$ must therefore be reflected.

The restoring force for buoyancy oscillations is controlled by static stability. In (14.40.2), increasing N^2 increases m^2, which supports vertical propagation. The greater N^2, the shorter is the vertical wavelength and the more vertical propagation is favored. On the other hand, increasing k^2 reduces m^2. Short horizontal wavelengths (large k) are therefore less able to propagate vertically than long horizontal wavelengths (small k). Vertical propagation is also influenced by the mean flow, chiefly through the intrinsic trace speed $c_x - \overline{u}$. For small flow curvature ($\overline{u}_{zz} \cong 0$), the first two terms on the right-hand side of (14.40.2) dominate. Increasing $|c_x - \overline{u}|$ corresponds to faster propagation relative to the mean flow or, for stationary waves, to stronger wind. It reduces m^2, inhibiting vertical propagation. The elongated vertical wavelength also makes wave activity increasingly vulnerable to reflection – because refractive properties do not vary gradually relative to the phase of oscillations. If m^2 is reduced to zero, the vertical wavelength becomes infinite (see Fig. 14.14b). Continued vertical propagation is then prohibited. Wave activity must therefore be fully reflected. The height where this occurs is called the *turning level*. Above the turning level, wave structure is external ($m^2 < 0$).

Decreasing $|c_x - \overline{u}|$ has the reverse effect. It favors vertical propagation by increasing m^2. The vertical wavelength is then shortened. This averts reflection by making variations in refractive properties comparatively gradual. Should $|c_x - \overline{u}|$ decrease to zero, (14.40) becomes singular. The index of refraction then becomes unbounded and the vertical wavelength collapses to zero (see Fig. 14.15b). The height where this occurs is called the *critical level*. Contrary to a turning level, which leads to reflection, the condition $m^2 \to \infty$ at the critical level leads to absorption of wave activity.

14.3.3 The WKB approximation

Analytical solutions to (14.40) can be obtained only for idealized profiles of $N^2(z)$ and $\overline{u}(z)$. However, an approximate class of solutions is valid when the basic state varies slowly compared with the phase of oscillations. It affords insight into how wave activity propagates – even under more general circumstances. The approximate solutions

$$W^{\pm}(z) = \frac{W_0^{\pm}}{\sqrt{m(z)}} e^{\pm i \int m(z) dz} \tag{14.41}$$

satisfy the related equation

$$\frac{\partial^2 W^{\pm}}{\partial z^2} + m^2(z) W^{\pm} = -R(z) m^2(z) W^{\pm}, \tag{14.42.1}$$

where

$$R(z) = \frac{1}{2m^3} \frac{\partial^2 m}{\partial z^2} - \frac{3}{4m^4} \left(\frac{\partial m}{\partial z} \right)^2. \tag{14.42.2}$$

The factor $R(z)$ accounts for the interdependence of W^+ and W^- and, hence, for wave reflection.

In the limit of $R \to 0$ everywhere, (14.42) reduces to the general form of (14.40). Reflection then vanishes and the upward- and downward-propagating waves may be treated independently. In the absence of dissipation, wave activity then propagates conservatively along *rays* that are smoothly refracted through the medium. The solution (14.41) defines the so-called *slowly varying* or *WKB approximation*.[4] Its validity (e.g., $R \ll 1$) rests on refractive properties changing on a scale long compared to the vertical wavelength. In practice, (14.41) often provides a qualitatively correct description of wave propagation, even when it cannot be justified formally. Should R become $O(1)$ somewhere, reflection ensues. It introduces the other of the components in (14.41). Clearly, $m^2 = 0$ violates the above condition. At the turning level, the local vertical wavelength is infinite, introducing strong reflection. There, the WKB approximation breaks down. Nevertheless, the wave field can be matched analytically across the turning level (see, e.g., Morse and Feschbach, 1953).

14.3.4 Method of geometric optics

Under conditions of the WKB approximation, wave activity can be traced from its source along rays that are defined by the local group velocity c_g. If

$$\bar{u}_{zz} \ll \frac{N^2}{|c_x - \bar{u}|}, \tag{14.43}$$

(14.40.2) has the same form as the dispersion relation for simple gravity waves in the limit of short wavelength (14.36). Components of c_g then follow from (14.23) as

$$c_{gx} = \frac{m^2 \omega}{k(k^2 + m^2)} + \bar{u}$$
$$c_{gz} = -\frac{m\omega}{(k^2 + m^2)}, \tag{14.44}$$

with $m = m(z)$, $\bar{u} = \bar{u}(z)$. A ray is everywhere tangent to c_g, with local slope in the $x - z$ plane

$$\frac{dz}{dx} = \frac{c_{gz}}{c_{gx}}. \tag{14.45}$$

Specifying its initial coordinates then determines the ray associated with an individual component (k, m) of the wave spectrum.

Turning Level

Orographically forced gravity waves involve a spectrum of horizontal wavenumbers k, all with zero frequency. Figure 14.14 presents rays of two components with $\sigma = 0$, under the conditions of (14.43), for $N^2 = $ const, and $\bar{u}(z)$ representative of the wintertime troposphere and stratosphere (Fig. 14.14a). Each ray emanates from

---

[4] The letters stand for Wentzel (1926), Kramer (1926), and Brillouin (1926), who rediscovered the procedure in different problems, after being introduced originally by Liouville (1837) and Green (1838).

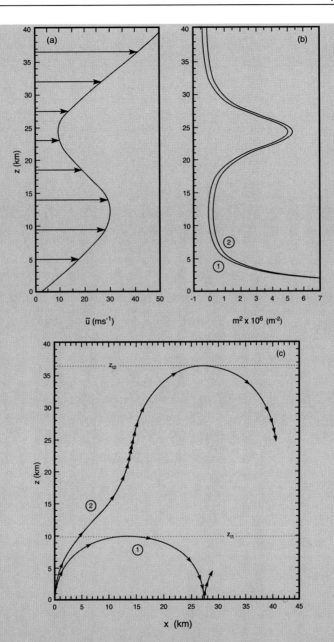

Figure 14.14 (a) Zonal motion representative of the extratropical troposphere and stratosphere during winter. (b) Vertical refractive index squared
following from mean motion in (a) and uniform static stability. (c) Rays for
two stationary wave components with a source at $(x, z) = (0, 0)$. Arrow
heads mark uniform increments of time moving along the ray at the group
velocity c_g. Component 1, which has short horizontal wavelength, encounters a *turning level* in the troposphere, where it is reflected downward.
Component 2, which has longer horizontal wavelength, propagates into
the stratosphere before encountering a turning level.

a point source located at the origin of the $x - z$ plane. Component 1 has short horizontal wavelength (large k). Westerly shear then drives ω^2 to N^2 and m^2 to zero near the tropopause (Fig. 14.14b). The ray for that component (Fig. 14.14c) is reflected at its turning level z_{t1}. Above z_{t1}, wave amplitude decays with height exponentially.

Because $m = 0$ at the turning level,

$$c_{gx} = \overline{u}$$
$$z = z_{t1}. \qquad (14.46)$$
$$c_{gz} = 0$$

Thus, relative to the medium, wave activity at $z = z_{t1}$ is motionless. Consider the time for wave activity to reach the turning level from some neighboring level below. The time for wave activity to propagate from a level z_0 to a level z is given by

$$\Delta t = \int_{z_0}^{z} \frac{dz'}{c_{gz}}. \qquad (14.47)$$

Near the turning level,

$$c_{gz} \sim -\frac{\omega m}{k^2} \qquad z \to z_{t1}. \qquad (14.48.1)$$

Because $\omega = -k\overline{u} < 0$ for $z < z_{t1}$ and $\overline{u}_z > 0$,

$$\omega \sim -N - k\overline{u}_z(z - z_{t1}) \qquad z \to z_{t1}. \qquad (14.48.2)$$

Substituting into the dispersion relation (14.36) obtains

$$m \sim k\left[2\frac{k}{N}\overline{u}_z(z_{t1} - z)\right]^{\frac{1}{2}} \qquad z \to z_{t1}. \qquad (14.48.3)$$

Then, near the turning level, the vertical group speed behaves as

$$c_{gz} \sim \left[2\frac{N}{k}\overline{u}_z(z_{t1} - z)\right]^{\frac{1}{2}} \qquad z \to z_{t1}. \qquad (14.49)$$

Incorporating (14.49) into (14.47) gives for the approach time

$$\Delta t = \left[\frac{2k}{N\overline{u}_z}(z_{t1} - z)\right]^{\frac{1}{2}}\Bigg|_{z}^{z_0} \qquad (14.50)$$
$$< \infty \qquad z \to z_{t1}.$$

The time for wave activity to reach the turning level is finite, even though c_{gz} vanishes there. So is the time for it to rebound and propagate away. It follows that wave activity can rebound from the turning level having suffered only finite dissipation. This is the premise for interpreting behavior near z_{t1} as reflection.

Wave activity trapped between its turning level and the surface is ducted horizontally inside the troposphere. Such trapping can produce oscillations far downstream of a wave source. It is revealed by extensive wave patterns in satellite cloud imagery (Fig. 1.11). In contrast, wave activity that propagates freely in the vertical (Fig. 14.13) produces only a few oscillations downstream. Cloud bands are therefore confined near the wave source (Fig. 9.22).

Component 2 of the wave spectrum has long horizontal wavelength (small k). Consequently, it does not encounter a turning level until wave activity has propagated well into the stratospheric jet (Fig. 14.14b). Although m^2 is reduced in strong westerlies of the tropospheric jet, it remains positive. This allows component 2 to propagate to higher levels. The region of small m^2 in the upper troposphere coincides with small vertical group speed. Propagation is then refracted into a shallower angle (Fig. 14.14c), only to be refracted more steeply again when wave activity emerges into weak westerlies of the lower stratosphere. At greater height, westerlies intensify again, with a commensurate decrease of m. Eventually, $\bar{u}$ in the stratospheric jet becomes strong enough to drive m^2 to zero. This occurs at the turning level z_{t2}, below which component 2 is vertically trapped.

Critical level

Consider now conditions representative of summer. Figure 14.15 shows the same features for component 2, but for easterly flow in the stratosphere. Having $c_x = 0$, component 2 is Doppler shifted to zero where $\bar{u}$ vanishes: the zero-wind line. There, the mean velocity matches the zonal trace speed: $\bar{u} = c_x = 0$. This condition occurs not far above the tropopause (Fig. 14.15a). At the critical level z_{c2}, $m^2 \to \infty$ and the vertical wavelength shrinks to zero (Fig. 14.15b).

Because $\omega = 0$ at the critical level,

$$
\begin{aligned}
c_{gx} = \bar{u} = 0 \\
c_{gz} = 0
\end{aligned}
\qquad z = z_{c2}.
\tag{14.51}
$$

Wave activity is again motionless relative to the medium. Consider the time for it to reach the critical level from some neighboring level z_0 below. For z near the critical level,

$$
c_{gz} \sim -\frac{\omega}{m} \qquad z \to z_{c2}.
\tag{14.52}
$$

Because $\omega = -k\bar{u} < 0$ for $z < z_{c2}$ and $\bar{u}_z < 0$,

$$
\omega \sim -k\bar{u}_z(z - z_{c2}) \qquad z \to z_{c2}.
\tag{14.53.1}
$$

Substituting into the dispersion relation gives

$$
m \sim \frac{N}{\bar{u}_z(z - z_{c2})} \qquad z \to z_{c2}.
\tag{14.53.2}
$$

Then, near the critical level, the vertical group speed behaves as

$$
c_{gz} \sim \frac{k}{N}\bar{u}_z^2(z - z_{c2})^2 \qquad z \to z_{c2}.
\tag{14.54}
$$

Incorporating (14.54) into (14.47) obtains for the approach time

$$
\Delta t = \frac{N}{k\bar{u}_z^2}(z - z_{c2})^{-1}\Big|_z^{z_0}
\tag{14.55}
$$

$$
\to \infty \qquad z \to z_{c2}.
$$

Unlike behavior at a turning level, the time for wave activity to reach the critical level is infinite. The critical level is, therefore, never actually encountered. Instead, wave activity becomes frozen relative to the medium, remaining in that state for infinite duration. Dissipation can then absorb it. The ray for component 2 (Fig. 14.15c) is refracted into the critical level, where it stalls and terminates.

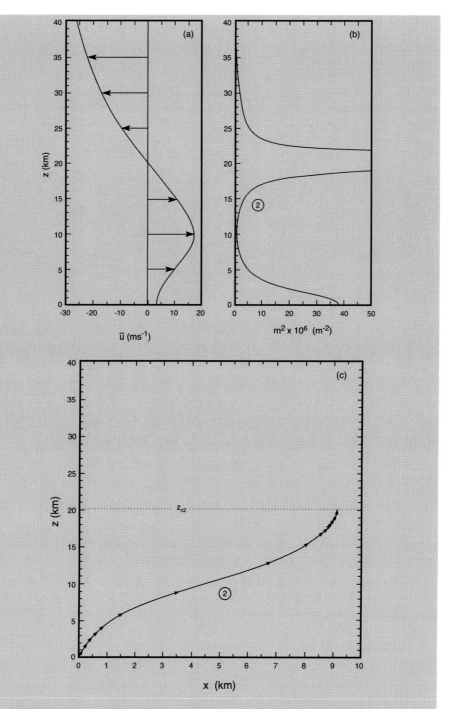

Figure 14.15 As in Fig. 14.14, except for conditions representative of summer and for component 2, the wave is Doppler-shifted to zero intrinsic phase speed: $c_x - \overline{u} = 0$, at the critical level, where $m^2 \to \infty$, group propagation stalls, and the ray terminates.

14.4 THE LAMB WAVE

Consider oscillations in an unbounded atmosphere and in the absence of forcing. These conditions define an unforced or eigenvalue problem. The latter determines the free modes or *normal modes* of the system. Their frequencies, at which waves can exist without forcing, are called *eignenfrequencies*. Should forcing be present at the one of the eigenfrequencies, the corresponding normal mode is greatly amplified relative to other wave components.

For oscillations that are hydrostatic, the perturbation equations (14.33) reduce to

$$\frac{Du'}{Dt} = -gH\frac{\partial \hat{p}'}{\partial x} \tag{14.56.1}$$

$$\left(H\frac{\partial}{\partial z} - 1\right)\hat{p}' + \hat{\rho}' = 0 \tag{14.56.2}$$

$$\frac{D\hat{\rho}'}{Dt} + \frac{\partial u'}{\partial x} + \left(\frac{\partial}{\partial z} - \frac{1}{H}\right)w' = 0 \tag{14.56.3}$$

$$\frac{D}{Dt}(\hat{p}' - \gamma\hat{\rho}') + \gamma\frac{\kappa}{H}w' = 0. \tag{14.56.4}$$

Considering solutions of the form $e^{i(kx-\sigma t)}$ transforms (14.56) into

$$-i\omega U_k^\sigma = -ikgHP_k^\sigma \tag{14.57.1}$$

$$\left(H\frac{\partial}{\partial z} - 1\right)P_k^\sigma + \rho_k^\sigma = 0 \tag{14.57.2}$$

$$-i\omega\rho_k^\sigma + ikU_k^\sigma + \left(\frac{\partial}{\partial z} - \frac{1}{H}\right)W_k^\sigma = 0 \tag{14.57.3}$$

$$-i\omega(P_k^\sigma - \gamma\rho_k^\sigma) + \gamma\frac{\kappa}{H}W_k^\sigma = 0. \tag{14.57.4}$$

Using (14.57.2) to eliminate density in (14.57.4) gives the vertical velocity

$$W_k^\sigma = \frac{i\omega H}{\kappa}\left[H\frac{\partial}{\partial z} - \kappa\right]P_k^\sigma. \tag{14.58}$$

Then, incorporating (14.58), (14.57.1), and (14.57.3) yields a 2nd-order differential equation for the perturbation amplitude $P_k^\sigma(z)$

$$\frac{\partial^2 P_k^\sigma}{\partial z^2} - \frac{1}{H}\frac{\partial P_k^\sigma}{\partial z} + \frac{\kappa g}{H}\frac{k^2}{\omega^2}P_k^\sigma = 0. \tag{14.59}$$

Solutions to (14.59) must satisfy boundary conditions. At the surface, w' must vanish. The upper boundary condition depends on whether the solution is vertically propagating or evanescent. If the solution is evanescent, it must have bounded column energy. If it is vertically propagating, far above the source, the solution must transmit wave activity upward – away from the region of excitation. These conditions define the so-called *finite-energy condition* and *radiation condition*, respectively. They apply at $z \to \infty$. Collectively, $P_k^\sigma(z)$ must then satisfy

$$\frac{\partial^2 P_k^\sigma}{\partial z^2} - \frac{1}{H}\frac{\partial P_k^\sigma}{\partial z} + \frac{\kappa g}{H}\frac{k^2}{\omega^2}P_k^\sigma = 0 \tag{14.60.1}$$

$$H\left[\frac{\partial}{\partial z} - \kappa\right]P_k^\sigma = 0 \qquad\qquad z = 0 \tag{14.60.2}$$

$$\text{finite energy/radiation} \qquad\qquad z \to \infty. \tag{14.60.3}$$

Equations (14.60) define a homogeneous boundary value problem for the vertical structure $P_k^\sigma(z)$, namely, one that has no forcing in the interior or on the boundaries. Nontrivial solutions exist only at particular frequencies.

Seeking solutions of the form $e^{(\frac{1}{2H}+im)z}$ transforms (14.60.1) into the dispersion relation

$$m^2 + \frac{1}{4H^2} = \frac{N^2}{\omega^2} k^2. \tag{14.61}$$

Equation (14.61) may be recognized as (14.35.2) in the conditional limit $k \to 0$. Horizontal wavelength is then long enough for vertical displacements to be hydrostatic. If $m^2 > 0$, solutions are vertically propagating. By (14.44), upward energy propagation (14.60.3) requires $m\omega < 0$, which corresponds to downward phase propagation. For $\omega > 0$, this requires $m < 0$. It is readily verified that these solutions do not satisfy the homogeneous lower boundary condition (14.60.2). Nor do their counterparts for $\omega < 0$. If $m^2 = -\hat{m}^2 < 0$, the solutions of (14.60.1) are external. The upper boundary condition, which requires bounded column energy, is satisfied by solutions of the form $e^{(\frac{1}{2H}-\hat{m})z}$. One of these also satisfies the lower boundary condition, namely, $\hat{m} = \frac{\kappa-\frac{1}{2}}{H}$. Representing a normal mode of the unbounded atmosphere, it has the vertical eigenstructure

$$P_k^\sigma(z) = e^{\kappa \frac{z}{H}}. \tag{14.62.1}$$

This vertical structure defines the *Lamb mode*. It makes w' vanish – not just at the surface, but everywhere (14.58). Buoyancy oscillations then vanish identically. The restoring force for Lamb waves is provided entirely by compressibility. For this external structure, (14.61) reduces to

$$\begin{aligned} \hat{c}^2 &= \gamma g H \\ &= c_s^2. \end{aligned} \tag{14.62.2}$$

Lamb waves propagate horizontally at the speed of sound. They bridge the acoustic and gravity manifolds in Fig. 14.9. Lamb waves are the normal modes of an unbounded, compressible, stratified atmosphere. Even though their energy decreases upward as an edge wave, Lamb waves amplify vertically. This feature makes them important in the upper atmosphere.

Because they are normal modes of the atmosphere, Lamb waves are excited preferentially by forcing that is indiscriminate over frequency. For example, an impulsive disturbance, such as a volcanic eruption, excites many wavenumbers and frequencies. The response spectrum to such forcing is unbounded at those wavenumbers and frequencies that satisfy (14.62.2). Removed from the source, it is therefore dominated by Lamb waves. Meteorological records include several impulsive disturbances to the atmosphere that were observed far from the source. Notable was the eruption of Krakatoa in 1883 (Chap. 9). Disturbances in surface pressure emanating from the eruption completed several circuits around the Earth before dissipating. Taylor (1929) used barometric records of the event to infer the vertical structure of atmospheric normal modes. By comparing arrival times at several stations, he showed that the compression wave emanating from Krakatoa propagated at the speed of sound. Then, with the aid of a system like (14.60), Taylor deduced the Lamb structure for atmospheric normal modes. A similar analysis was performed by Whipple (1930) for the impact of the great Siberian meteor in 1908. The Lamb vertical structure applies to atmospheric normal modes at all frequencies. It applies even at very low frequency, where another form of wave motion is supported by the Earth's rotation (Dikii, 1968).

14.5 ROSSBY WAVES

At frequencies below 2Ω, another class of wave motion exists, one that is possible only in a rotating medium. It is named for C.G. Rossby, who established its connection to weather phenomena (Rossby et al., 1939). These waves are also referred to as *rotational waves* and, on the gravest dimensions, *planetary waves*. The restoring force for Rossby waves is provided by the variation with latitude of the Coriolis force. It links them to the Earth's rotation.

14.5.1 Barotropic nondivergent Rossby waves

The simplest model of Rossby waves relies on the solenoidal character of large-scale motion (Sec. 12.1). Under nondivergent conditions, atmospheric motion is governed by conservation of absolute vorticity (12.30). On a beta plane, the latter is expressed by

$$\frac{d\zeta}{dt} + \beta v = 0.$$

Linearizing about a basic state that is barotropically stratified and in uniform motion yields the perturbation vorticity equation

$$\frac{D\zeta'}{Dt} + \beta v' = 0. \tag{14.63}$$

As the motion is nondivergent, it can be represented in terms of a streamfunction

$$\boldsymbol{v}'_h = \boldsymbol{k} \times \nabla \psi'. \tag{14.64}$$

Then the perturbation vorticity equation becomes

$$\frac{D}{Dt}\nabla^2 \psi' + \beta \frac{\partial \psi'}{\partial x} = 0. \tag{14.65}$$

Equation (14.65) is just the linearized form of the barotropic nondivergent vorticity equation (12.31), in which the absolute vorticity of an air parcel is conserved. Known as the *Rossby wave equation*, it reflects a balance between changes in the relative vorticity of an air parcel and changes in its planetary vorticity due to meridional displacement. The interaction of those influences is illustrated by an eastward-moving parcel that is deflected equatorward (Fig. 14.16). The parcel then experiences reduced planetary vorticity ($\delta f < 0$). To conserve absolute vorticity, it must spin up cyclonically ($\delta \zeta > 0$). Northward motion is therefore induced ahead of the parcel. It deflects the parcel's trajectory poleward – back toward its undisturbed latitude ϕ_0. Upon overshooting ϕ_0, the parcel spins up anticyclonically. Southward motion ahead of the parcel then deflects its trajectory equatorward – again back toward its undisturbed latitude. The variation with latitude of f thus exerts a torque on displaced air. It is analogous to the positive restoring force of buoyancy under stable stratification. This reaction enables air to cycle back and forth about its undisturbed latitude.

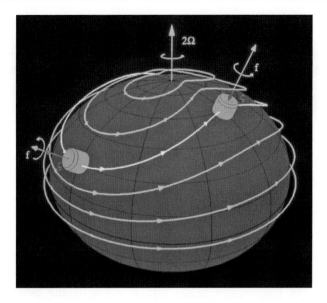

Figure 14.16 Schematic illustrating the reaction of an air parcel to meridional displacement. Displaced equatorward, an eastward-moving parcel spins up cyclonically to conserve absolute vorticity. Northward motion induced ahead of it then deflects the parcel's trajectory poleward, back toward its undisturbed latitude. The reverse process occurs when the parcel overshoots and is displaced poleward of its undisturbed latitude. See color plate section: Plate 10.

The coefficients of (14.65) are constant. Solutions may therefore be considered of the form $e^{i(kx+ly-\sigma t)}$. Substitution leads to the dispersion relation for Rossby waves

$$c_x - \overline{u} = -\frac{\beta}{k^2 + l^2}. \qquad (14.66)$$

Relative to the mean motion, Rossby waves propagate only westward.[5]

The intrinsic trace speed: $c_x - \overline{u} = \frac{\omega}{k}$, is proportional to the local gradient of planetary vorticity, β (12.41). It is inversely proportional to the horizontal wavenumber squared. The latter dependence makes Rossby waves dispersive. The gravest dimensions propagate westward fastest. Consequently, small scales have slow phase speeds. In westerly flow, they are swept eastward. Such behavior is typical of synoptic disturbances to the jet stream (Fig. 1.9). On the other hand, stationary Rossby waves, which are forced orographically, have $c_x = 0$. The dispersion relation can be satisfied only if $\overline{u} > 0$. Thus, stationary Rossby waves propagate away from forcing only if the mean flow is westerly.

The intrinsic frequency $|\omega|$ of Rossby waves is proportional to β. It has a high-frequency cutoff: 2Ω. This limiting frequency is analogous to the high-frequency cutoff N for the vertical propagation of gravity waves (Sec. 14.3).

The solenoidal character of Rossby waves distinguishes them from gravity waves, which, in contrast, are nearly irrotational. Because Rossby waves are almost nondivergent, the Helmholtz theorem (12.4) implies that, to leading order, horizontal motion

---

[5] The direction of Rossby wave propagation can be deduced from the vorticity pattern in Fig. 14.16 and its influence on the material contour shown. Southward motion behind the cyclonic anomaly displaces that segment of the material contour equatorward, shifting the wave trough westward. Northward motion ahead of it has the same effect on the wave crest.

is characterized by the vorticity field. To the same degree of approximation, vertical motion can be ignored. Conversely, gravity waves are determined chiefly by vertical motion, which interacts with buoyancy. To leading order, they are characterized by the divergence field. Although the essential properties of Rossby waves follow from the rotational component of motion, divergence nevertheless enters by forcing absolute vorticity (12.36).

14.5.2 Rossby wave propagation in three dimensions

To describe 3-dimensional wave propagation, divergence must be accounted for. Within the framework of quasi-geostrophy, air motion is governed by conservation of quasi-geostrophic potential vorticity. Linearizing (12.49) about an isothermal basic state in uniform motion recovers the perturbation potential vorticity equation for wave motion on a beta plane

$$\frac{D}{Dt}\left[\nabla^2\psi' + \left(\frac{f_0^2}{N^2}\right)\frac{1}{\bar{\rho}}\frac{\partial}{\partial z}\left(\bar{\rho}\frac{\partial\psi'}{\partial z}\right)\right] + \beta\frac{\partial\psi'}{\partial x} = 0, \tag{14.67}$$

where z refers to log-pressure height and $\psi' = \frac{1}{f_0}\Phi'$ is the geostrophic stream-function.

The coefficients are again constant. We therefore consider solutions of the form $e^{\frac{z}{2H}+i(kx+ly+mz-\sigma t)}$. Substitution into (14.67) recovers the dispersion relation for *quasi-geostrophic Rossby waves*

$$c_x - \bar{u} = -\frac{\beta}{k^2 + l^2 + \left(\frac{f_0^2}{N^2}\right)\left(m^2 + \frac{1}{4H^2}\right)}. \tag{14.68}$$

The denominator represents the effective total wavenumber squared. Equation (14.68) then has form analogous to (14.66), but modified by the stratification of mass. In the limit of strong stability ($\frac{N^2}{f_0^2} \to \infty$), (14.68) reduces to the dispersion relation for barotropic nondivergent Rossby waves. Much the same holds in the limit of long vertical wavelength ($m \to 0$).

Because their intrinsic phase speeds depend on wavenumber, Rossby waves are dispersive. Expressing (14.68) as

$$\omega = -\frac{\beta k}{k^2 + l^2 + \left(\frac{f_0^2}{N^2}\right)\left(m^2 + \frac{1}{4H^2}\right)} \tag{14.69}$$

leads to the components of group velocity

$$c_{gx} - \bar{u} = \beta\frac{k^2 - l^2 - \left(\frac{f_0^2}{N^2}\right)\left(m^2 + \frac{1}{4H^2}\right)}{\left[k^2 + l^2 + \left(\frac{f_0^2}{N^2}\right)\left(m^2 + \frac{1}{4H^2}\right)\right]^2} = \frac{\omega}{k}\left(1 + \frac{2\omega k}{\beta}\right) \tag{14.70.1}$$

$$c_{gy} = \frac{2\beta kl}{\left[k^2 + l^2 + \left(\frac{f_0^2}{N^2}\right)\left(m^2 + \frac{1}{4H^2}\right)\right]^2} = \frac{2l\omega^2}{\beta k} \tag{14.70.2}$$

$$c_{gz} = \frac{2\beta mk\left(\frac{f_0^2}{N^2}\right)}{\left[k^2 + l^2 + \left(\frac{f_0^2}{N^2}\right)\left(m^2 + \frac{1}{4H^2}\right)\right]^2} = \left(\frac{f_0^2}{N^2}\right)\frac{2m\omega^2}{\beta k}. \tag{14.70.3}$$

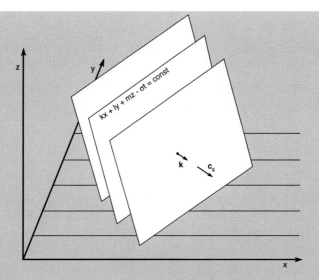

Figure 14.17 A stationary Rossby wave with group velocity that is upward and equatorward tilts westward in the same directions. Phase propagation relative to the medium is anti-parallel to **k**.

Wave activity propagates in the positive y direction if $kl > 0$. It propagates in the positive z direction if $km > 0$. Because $\omega < 0$, the exponential form adopted for ψ' then implies phase propagation in the opposite directions. Thus, equatorward phase propagation corresponds to energy propagation that is poleward. Downward phase propagation corresponds to energy propagation that is upward. For stationary waves in the presence of westerly flow, these conditions translate into phase surfaces that tilt westward in the direction of energy propagation. Under the foregoing circumstances, phase surfaces tilt westward in the upward and equatorward directions (Fig. 14.17).

Unlike c_{gy} and c_{gz}, the sign of $c_{gx} - \bar{u}$ depends on the magnitude of k. For $k < \frac{\beta}{2|\omega|}$ (long zonal wavelength), group propagation relative to the medium is westward. However, for $k > \frac{\beta}{2|\omega|}$ (short zonal wavelength), intrinsic group propagation is eastward. Separating westward and eastward group propagation is the locus of wavenumbers for which $c_{gx} - \bar{u}$ vanishes

$$k^2 = l^2 + \left(\frac{f_0^2}{N^2}\right)\left(m^2 + \frac{1}{4H^2}\right). \qquad (14.71)$$

These features of Rossby wave propagation can be inferred from the dispersion characteristics $\omega(k, l, m) = \text{const}$ (Fig. 14.18). Expressing (14.69) as

$$\left(k - \frac{\beta}{2|\omega|}\right)^2 + l^2 + \left(\frac{f_0^2}{N^2}\right)m^2 = \left(\frac{\beta}{2\omega}\right)^2 - \left(\frac{1}{2H}\right)^2 \qquad (14.72)$$

shows that surfaces of constant ω are nested ellipsoids centered on the points $(k, l, m) = (\frac{\beta}{2|\omega|}, 0, 0)$. The intrinsic group velocity (14.23) is directed orthogonal to those surfaces, toward increasing ω. Thus, $\hat{c}_g$ is upward for $m > 0$. It is equatorward

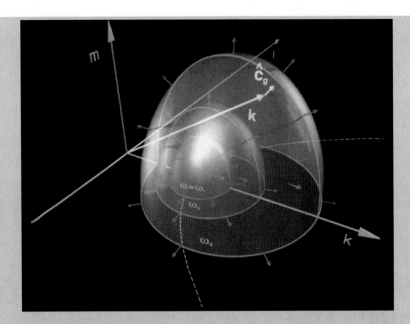

Figure 14.18 Intrinsic frequency for Rossby waves illustrated as a function of 3-dimensional wavenumber $\boldsymbol{k} = (k, l, m)$, with $\omega_3 > \omega_2 > \omega_1$. Group velocity relative to the medium is directed orthogonal to surfaces of intrinsic frequency toward increasing ω. For given l, zonal wavenumbers k greater than a critical value (dashed) have eastward $\hat{c}_g$, whereas smaller k have westward $\hat{c}_g$.

for $l < 0$. Each corresponds to phase propagation in the opposite direction. For k smaller than values defined by (14.71), indicated in Fig. 14.18 by the dashed curve, $\hat{c}_g$ is westward. Larger k propagate eastward.

The dispersion relation may be rearranged for the vertical wavenumber

$$m^2 = \left(\frac{N^2}{f_0^2} \right) \left[\frac{\beta}{\overline{u} - c_x} - |\boldsymbol{k}_h|^2 \right] - \frac{1}{4H^2}. \tag{14.73}$$

The gradient of planetary vorticity β provides the restoring force for horizontal displacements. It increases m^2 – analogous to the role played by N^2 for gravity waves (14.40). According to (14.73), Rossby waves propagate vertically only for a restricted range of zonal trace speed, specifically, for westward phase propagation relative to the mean flow ($c_x - \overline{u} < 0$). As for gravity waves, short horizontal wavelengths ($|\boldsymbol{k}_h|$ large) are less able to propagate upward than large horizontal wavelengths ($|\boldsymbol{k}_h|$ small).

For stationary waves ($c_x = 0$), (14.73) admits $m^2 > 0$ only for mean flow in the range

$$0 < \overline{u} < \frac{\beta}{|\boldsymbol{k}_h|^2 + \frac{f_0^2}{4N^2 H^2}}. \tag{14.74}$$

If the mean flow is easterly, Rossby waves are external. The same holds if the mean flow is westerly and exceeds a cutoff speed given by the right-hand side of (14.74). The influence of Rossby waves is then exponentially small far above their forcing. Notice that the cutoff wind speed decreases with increasing $|\boldsymbol{k}_h|^2$. Therefore, even when zonal wind is westerly, only the gravest horizontal wavelengths ($|\mathbf{k}_h|$ small) propagate vertically. Those wavelengths correspond to planetary waves (Fig. 1.10). Shorter wavelengths, characteristic of synoptic weather systems, have energy that is trapped near the surface.

These features of Rossby wave propagation were advanced by Charney and Drazin (1961) to explain observations from the International Geophysical Year (IGY), the first comprehensive campaign to observe the atmosphere. Collected during 1958, observations from the IGY revealed that the synoptic-scale disturbances which dominate the tropospheric circulation are conspicuously absent in the stratosphere (cf. Figs. 1.9, 1.10). In the summer stratosphere, eddies of all scales are absent. The time-mean circulation is then nearly circumpolar. The tropospheric circulation involves a spectrum of horizontal wavelengths. According to the preceding analysis, only planetary scale components of the wave spectrum can propagate into strong westerlies of the winter stratosphere (Fig. 1.8). Shorter components encounter turning levels not far above the tropopause (Fig. 14.14), below which they are trapped. During summer, stratospheric easterlies prevent all horizontal scales from propagating vertically. Rossby waves then encounter critical levels (Fig. 14.15), where $c_x = \overline{u}$ and wave activity is absorbed.

In the troposphere, kinetic energy is concentrated in synoptic-scale disturbances. Charney and Drazin argued that these features of Rossby wave propagation prevent kinetic energy from propagating to high levels. Were vertical propagation possible for energy-bearing eddies in the troposphere, the $e^{\frac{z}{2H}}$ amplification introduced by stratification would produce enormous temperatures at high altitude. The atmosphere would then possess a corona, similar to the sun. That, in turn, would result in the loss of nitrogen, oxygen, and, hence, of atmospheric mass (Figs. 1.5, 1.6).

Easterly flow in the tropical troposphere has a similar effect. It blocks horizontal wave propagation between the Northern and Southern Hemispheres. Large orographic features, such as the Alps and Himalayas, excite amplified planetary waves in the Northern Hemisphere. Owing to easterlies in the tropics, those planetary waves exert little influence on the circulation of the Southern Hemisphere, where orographic forcing of planetary waves is comparatively weak. These inter-hemispheric differences yield a broken storm track in the Northern Hemisphere, where amplified wave motions reinforce and weaken the storm track. However, they leave the storm track of the Southern Hemisphere comparatively undisturbed and hence zonally uniform (see Fig. 15.10).

14.5.3 Planetary wave propagation in sheared mean flow

Westerly mean flow during winter enables planetary-scale Rossby waves to radiate into the stratosphere. Due to stratification of mass, those waves amplify vertically. Where planetary wave activity propagates and is absorbed can be understood from a ray tracing analysis.

Consider zonal-mean flow that varies with latitude and height $\overline{u} = \overline{u}(y, z)$. Linearizing the quasi-geostrophic potential vorticity equation (12.49), with $-\frac{\partial \overline{\psi}}{\partial y} = \overline{u}$, yields the perturbation potential vorticity equation

$$\frac{D}{Dt}\left[\nabla^2 \psi' + \frac{1}{\overline{\rho}}\frac{\partial}{\partial z}\left(\frac{f_0^2}{N^2}\overline{\rho}\frac{\partial \psi'}{\partial z}\right)\right] + \beta_e \frac{\partial \psi'}{\partial x} = 0, \qquad (14.75.1)$$

where

$$\beta_e = \beta - \overline{u}_{yy} - \frac{1}{\rho} \frac{\partial}{\partial z} \left(\frac{f_0^2}{N^2} \rho \frac{\partial \overline{u}}{\partial z} \right)$$

$$= \frac{\partial \overline{Q}}{\partial y},$$

(14.75.2)

referred to as *beta effective*, represents the meridional gradient of zonal-mean vorticity. According to (14.75), positive curvature of $\overline{u}$ reduces β_e. It therefore weakens the restoring force introduced by meridional displacements of air. Conversely, negative curvature of $\overline{u}$ increases β_e. This dependence on mean shear is analogous to that of gravity wave propagation (14.40).

Considering solutions of the form $e^{\frac{z}{2H} + i(kx - \sigma t)}$ transforms (14.75) into a Helmholtz equation

$$\frac{\partial^2 \Psi_k^{\sigma}}{\partial y^2} + \frac{\partial}{\partial z} \left(\frac{f_0^2}{N^2} \frac{\partial \Psi_k^{\sigma}}{\partial z} \right) + \nu^2(y, z) \Psi_k^{\sigma} = 0,$$

(14.76.1)

where

$$\nu^2(y, z) = \frac{\beta_e}{\overline{u} - c_x} - k^2 - \frac{f_0^2}{4N^2 H^2}$$

(14.76.2)

represents a 2-dimensional index of refraction that describes propagation in the $y - z$ plane. In regions where $\nu^2(y, z) > 0$, solutions oscillate. In regions where $\nu^2(y, z) \leq 0$, they are evanescent. Therefore, propagation is excluded from regions where β_e is sufficiently small or where $\overline{u}$ is sufficiently strong. Should $\overline{u}(y, z) \to c_x$, ν^2 is unbounded. Equation (14.76) is then singular. The critical region defined by this condition differs from that of gravity waves, for which ν^2 is unbounded to either side of the critical line (cf. Fig. 14.15b). For Rossby waves, $\nu^2 \to +\infty$ on one side of the critical line, but $-\infty$ on the opposite side. Wave structure therefore oscillates rapidly on one side of the critical line but is sharply evanescent on the other.

Analytical solutions to (14.76) exist only under idealized circumstances. However, within the framework of the WKB approximation, wave propagation can be described in terms of monochromatic behavior along rays. The dispersion relation then assumes the same form as (14.69), but with β_e in place of β. Under these circumstances, propagation in the $y - z$ plane is controlled by

$$l^2 + \left(\frac{f_0^2}{N^2} \right) m^2 = \frac{\beta_e}{\overline{u} - c_x} - k^2 - \frac{f_0^2}{4N^2 H^2},$$

(14.77)

which is analogous to (14.40.2) for gravity waves. Figure 14.19 shows, for two stationary wave components, rays that are introduced into mean zonal flow that is representative of the wintertime troposphere and stratosphere. For component 1, planetary wave activity radiates upward and poleward into strengthening westerlies of the polar-night jet. It eventually encounters a turning line, where $l^2 + m^2$ vanishes and wave activity is reflected.

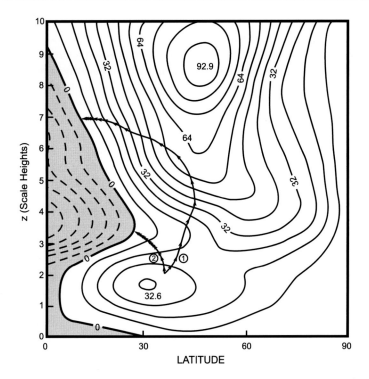

Figure 14.19 Rays of two stationary planetary wave components introduced into the lower stratosphere in zonal-mean winds (contoured in m s^{-1}) representative of northern winter on a mid-latitude beta plane. The ray for component 1, which initially propagates upward and poleward, encounters a turning line in strong westerlies of the polar-night jet, where wave activity is refracted equatorward. As the ray approaches easterlies, it next encounters a critical line ($\overline{u} = 0$), where c_g vanishes. Propagation then stalls, leaving wave activity to be absorbed, so the ray terminates. The ray for component 2 is initially directed upward and equatorward, so wave activity encounters the critical line and is absorbed even sooner. Thus wave activity introduced between tropical easterlies and strong polar westerlies can propagate vertically only a limited distance before being absorbed. Arrow heads mark uniform increments of time moving along a ray at the group velocity c_g.

Wave activity is then refracted equatorward, toward easterlies. The ray eventually encounters a critical line, where ω and the group velocity vanish. There, the ray stalls and terminates. Analysis similar to that in Sec. 14.3.4 for gravity waves shows that, ahead of the critical line, planetary wave activity is frozen in the medium. It is therefore absorbed. For component 2, planetary wave activity radiates initially upward and equatorward, toward easterlies. It encounters a critical line even sooner, so it too is absorbed. Dickinson (1968) used the foregoing analysis to explain the observed amplitudes of planetary waves. They are weaker than predicted by Charney and Drazin's (1961) 1-dimensional analysis, in which the longest horizontal wavelengths amplify vertically with little attenuation. The preceding analysis shows that planetary wave activity can propagate upward only a limited distance before it is refracted or reflected into easterlies, where it encounters a critical line and suffers strong absorption.

Horizontal wave propagation can be diagnosed in similar fashion. Figure 14.20a plots the anomalous planetary wave pattern observed in the troposphere during disturbed northern winters of El Nino. A series of positive and negative height anomalies marks a planetary wavetrain. Symbolic of wave activity along a ray, it radiates poleward from anomalous heating in the equatorial Pacific (stipled), where convection is amplified during El Nino. Along the wavetrain, β_e decreases poleward. At sufficiently high latitude, ν^2 has decreased to zero, representing a turning latitude. There, wave activity is reflected equatorward, influencing North America. According to (14.76), turning latitudes are encountered sooner by large k than small k. Hence, the gravest scales, those of planetary dimension, propagate poleward farthest. The resulting wavetrain is called a *teleconnection pattern*. It describes the influence of one region (e.g., the equatorial Pacific) on other regions.

The prominent ridge over western Canada and trough over the eastern United States characterize the so-called Pacific North America (PNA) pattern. The PNA pattern is observed in northern winters during El Nino, when convection and latent heating are amplified in the equatorial central Pacific. The anomalies in Fig. 14.20a disturb the extratropical planetary wave field, manifest in undulations of the jet stream (dashed). They upset the storm track and hence the development and propagation of synoptic weather systems.

Similar behavior is recovered in calculations in which anomalous convective heating is imposed in the equatorial central Pacific (Fig. 14.20b). As in observed structure, planetary wave activity propagates chiefly into the winter hemisphere. Propagation into the summer hemisphere is comparatively weak, blocked by easterlies in the summer subtropics that pose a critical line to planetary waves of low frequency. Such calculations capture the salient features of observed structure. However, there is considerable variance among them, as there is in observed structure; see Nigam (2003) for a review. These discrepancies reflect the involvement of many wavenumbers and frequencies, which are excited by unsteady convection. Those wave components disperse along different rays; Hoskins and Karoly (1981) present a ray-tracing analysis of stationary components. They also make the radiated response transient and sensitive to the distribution of zonal-mean wind (Jin and Hoskins, 1995; Branstator, 2002). The transient response is especially sensitive to tropical easterlies. The latter discriminate the wave response to westward components that propagate faster than easterlies neighboring the wave source (Garcia and Salby, 1987).

14.5.4 Transmission of planetary wave activity

Owing to the rotational nature of Rossby waves, the property propagated by them is related closely to vorticity. Under conservative conditions, the perturbation vorticity equation (14.75) may be written

$$\frac{DQ'}{Dt} + \frac{\partial \overline{Q}}{\partial y} v' = 0, \qquad (14.78.1)$$

where

$$Q' = \nabla^2 \psi' + \left(\frac{f_0^2}{N^2}\right) \frac{1}{\rho} \frac{\partial}{\partial z}\left(\overline{\rho} \frac{\partial \psi'}{\partial z}\right). \qquad (14.78.2)$$

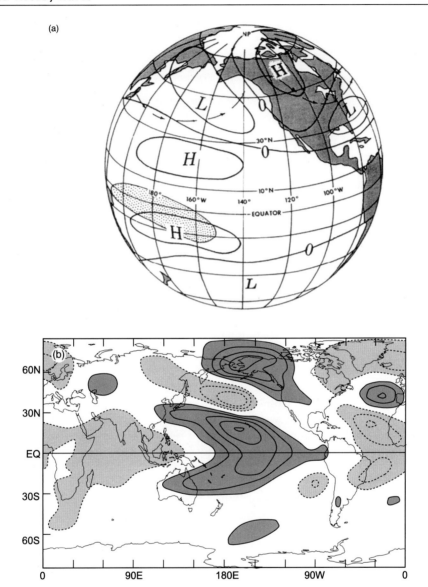

Figure 14.20 (a) Anomalous planetary wave field for northern winters during El Nino, when anomalous convection (stippled) is positioned in the tropical central Pacific. The anomalous ridge over western Canada and trough over the eastern United States characterize the so-called *Pacific North America (PNA) pattern* that upsets the normal track of the jet stream (wavy trajectory) and cyclone activity during El Nino winters. After Horel and Wallace (1981). (b) Anomalous height in the upper troposphere introduced by anomalous latent heating in the equatorial central Pacific. After de Weaver and Nigam (2004).

Multiplying by Q' and averaging zonally gives

$$\frac{\partial}{\partial t}\left(\frac{\overline{Q'^2}}{2}\right) + \frac{\partial \overline{Q}}{\partial y}\overline{v'Q'} = 0. \tag{14.79.1}$$

Incorporating (14.78.2) and (12.2.2) and integrating by parts, with field variables cyclic in x, then yields

$$\frac{\partial}{\partial t}\left(\frac{\overline{\rho}}{\frac{\partial \overline{Q}}{\partial y}}\frac{\overline{Q'^2}}{2}\right) - \frac{\partial}{\partial y}\left(\overline{\frac{\partial \psi'}{\partial x}\frac{\partial \psi'}{\partial y}}\right) + \left(\frac{f_0^2}{N^2}\right)\frac{1}{\overline{\rho}}\frac{\partial}{\partial z}\left(\overline{\rho}\,\overline{\frac{\partial \psi'}{\partial x}\frac{\partial \psi'}{\partial z}}\right) = 0. \tag{14.79.2}$$

With the hydrostatic relation (11.67), (14.79.2) may be cast into the canonical form

$$\frac{\partial A}{\partial t} + \nabla \cdot \boldsymbol{F} = 0, \tag{14.80.1}$$

where

$$A = \frac{\overline{\rho}}{\frac{\partial \overline{Q}}{\partial y}}\left(\frac{\overline{Q'^2}}{2}\right) \tag{14.80.2}$$

and

$$F_y = -\overline{\rho}\cdot\overline{u'v'} \tag{14.80.2}$$

$$F_z = \frac{f_0 R}{N^2 H}\overline{\rho}\cdot\overline{v'T'}. \tag{14.80.3}$$

Equation (14.80.1) has the form of a conservation relation. A represents the density of wave activity associated with Rossby waves. Proportional to the *potential enstrophy* $\frac{\overline{Q'^2}}{2}$, A is an analogue of wave action for gravity waves (Lighthill, 1978). The vector field $\boldsymbol{F}$ is the *Eliassen-Palm (EP) flux*. It represents the flux of wave activity. EP flux is therefore related to group velocity as

$$\boldsymbol{F} = \boldsymbol{c}_g A. \tag{14.81}$$

$\boldsymbol{F}$ is everywhere tangential to rays. Under conservative conditions, the property A is propagated along rays without loss. The divergence of EP flux is, from (14.79.1) and (14.80.1), seen to equal the northward flux of potential vorticity

$$\nabla \cdot \boldsymbol{F} = \overline{v'Q'}. \tag{14.82}$$

As Q is a conserved property, its northward flux describes meridional transport in general, namely, of all conserved properties. $\nabla \cdot \boldsymbol{F}$ is thus a direct measure of meridional transport by Rossby waves.

According to (14.80), $\boldsymbol{F}$ also represents the flux of westward momentum, the property transmitted by Rossby waves. $\nabla \cdot \boldsymbol{F}$ therefore equals the convergence of momentum flux. Describing wave absoption, it will be seen in Chap. 18 to represent the effective wave forcing of zonal-mean momentum. The basic state (e.g., $\overline{u}$ and $\overline{T}$) is, in general, influenced by wave activity through the convergence of wave momentum flux and of wave heat flux, as illustrated by (14.5). Equations (14.80) show that, for Rossby waves, those fluxes do not operate independently. The upward flux of westward momentum, F_z is determined by the northward flux of heat $\overline{v'T'}$. Together with (14.82), equations (14.80) show further that, if the wave field is conservative and steady,

$$\nabla \cdot F = \overline{v'Q'} = 0. \tag{14.83}$$

The convergence of EP flux then vanishes, as does the northward flux of potential vorticity. The waves then exert no net influence on the mean state. Under these conditions, meridional transport likewise vanishes. Rossby waves therefore influence the mean state and induce meridional transport only through nonvanishing $\nabla \cdot F$. Associated with absorption or emission of wave activity, $\nabla \cdot F$ is introduced by dissipation and transience.

14.6 WAVE ABSORPTION

In the absence of dissipation, wave activity is conserved following a wavepacket (e.g., moving along a ray with the velocity c_g).[6] When dissipation is present, this is no longer true. Wave activity is then absorbed from individual wavepackets. Deposited in the mean flow, it forces $\overline{u}$ through the convergence of momentum flux (14.5). To illustrate how absorption alters the wave field, we consider the propagation of planetary waves in the presence of radiative damping.

Planetary waves have vertical wavelengths much longer than a scale height. Radiative transfer then reduces to LW cooling to space (Sec. 8.6). For small departures from the equilibrium temperature, such radiative transfer is approximated by Newtonian cooling (8.79). The linearized form of the thermodynamic equation (12.46) then becomes

$$\frac{D}{Dt}\left(\frac{\partial \Phi'}{\partial z}\right) + N^2 w' = -\alpha\left(\frac{\partial \Phi'}{\partial z}\right), \tag{14.84.1}$$

where the Newtonian cooling rate α describes the damping of temperature perturbations (units of inverse time). Rearranging terms casts (14.84.1) into the form

$$\left(\frac{D}{Dt} + \alpha\right)\frac{\partial \psi'}{\partial z} + \frac{N^2}{f_0}w' = 0. \tag{14.84.2}$$

Equation (14.84.2) can be used to eliminate w' in the linearized vorticity equation (12.47),

$$\left[\frac{d(\zeta + f)}{dt}\right]' = f_0\frac{1}{\overline{\rho}}\frac{\partial}{\partial z}\left(\overline{\rho}w'\right), \tag{14.85}$$

where the left-hand side accounts for shear in the mean flow $\overline{u} = \overline{u}(y, z)$. The perturbation vorticity equation then becomes

$$\frac{D}{Dt}\left[\nabla^2\psi' + \frac{1}{\overline{\rho}}\frac{\partial}{\partial z}\left(\frac{f_0^2}{N^2}\overline{\rho}\frac{\partial \psi'}{\partial z}\right)\right] + \alpha\frac{1}{\overline{\rho}}\frac{\partial}{\partial z}\left(\frac{f_0^2}{N^2}\overline{\rho}\frac{\partial \psi'}{\partial z}\right) + \beta_e\frac{\partial \psi'}{\partial x} = 0. \tag{14.86}$$

Equation (14.86) is a generalization of (14.75) that accounts for thermal dissipation.

Within the framework of the WKB approximation, solutions are considered of the form $e^{\frac{z}{2H} + i(kx + ly + mz - \sigma t)}$. This transforms (14.86) into the dispersion relation

$$\left[1 + i\left(\frac{\alpha}{\omega}\right)\right]m^2 = m_0^2, \tag{14.87.1}$$

---

[6] Amplitude, however, varies along a ray (14.41); see Lighthill (1978).

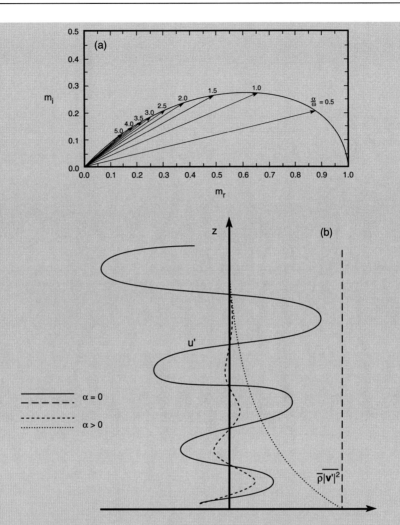

Figure 14.21 (a) Polar representation of complex vertical wavenumber for Rossby waves in the presence of Newtonian cooling with coefficient α. (b) Vertical structure and kinetic energy density of an individual wave component for $\alpha = 0$ (solid/dashed) and $\alpha > 0$ (short dashed/dotted).

where

$$m_0^2 = \left(\frac{N^2}{f_0^2}\right)\left[\frac{\beta_e}{\overline{u} - c_x} - |\mathbf{k}_h|^2\right] - \frac{1}{4H^2} \tag{14.87.2}$$

satisfies the dispersion relation in the absence of damping (14.77). Solving for the vertical wavenumber gives

$$m = m_r + im_i = \frac{e^{i\frac{\phi}{2}}}{\left[1 + \left(\frac{\alpha^2}{\omega^2}\right)\right]^{\frac{1}{2}}}m_0, \tag{14.88.1}$$

where (because $\omega < 0$)

$$\phi = tan^{-1}\left(\frac{\alpha}{|\omega|}\right). \tag{14.88.2}$$

Figure 14.21a plots m in the complex plane, as a function of cooling rate and intrinsic frequency. An individual wave component has vertical structure

$$e^{\left(\frac{1}{2H}+im\right)z} = e^{\left(\frac{1}{2H}-m_i\right)z} \cdot e^{im_r z}.$$

Because $m_0 > 0$ for downward phase propagation, (14.88) implies $m_i > 0$. Hence, thermal damping introduces an imaginary component to m that makes wave energy decay upward. It offsets the vertical amplification of wave activity that is introduced by stratification of mass (Fig. 14.21b). According to (14.88), m_i follows from the ratio of time scales for wave oscillation and damping. Under weakly damped conditions, it increases with increasing $\left(\frac{\alpha}{\omega}\right) m_0$. Wave amplitude then grows with height, but slower than under adiabatic conditions. Kinetic energy density (e.g., $\overline{\rho|v_h'|^2}$) therefore decreases upward, as progressively more wave activity is absorbed. For $\left(\frac{\alpha}{\omega}\right) m_0$ sufficiently large, even wave amplitude decreases with height. Components with small intrinsic frequency or short vertical wavelength (large m_0) therefore decay vertically above their forcing.

In practice, thermal dissipation is introduced whenever the circulation is driven out of radiative equilibrium. Wave motions do this naturally by displacing air from one radiative environment to another. In the stratosphere, the polar-night vortex is displaced out of radiative equilibrium by planetary waves. They drive air across latitude circles (cf. Fig. 1.10). Longwave cooling to space then acts on individual air parcels in proportion to their departure from local radiative equilibrium. It destroys anomalous temperature and, with it, the accompanying motion. Air that is displaced into polar darkness, where the radiative-equilibrium temperature is very cold, finds itself anomalously warm. It therefore cools through LW emission to space (Fig. 8.27). In the summer stratosphere, the vortex remains relatively undisturbed – because planetary wave propagation from below is blocked (Sec. 14.5). The summertime circulation therefore remains close to radiative equilibrium.

Newtonian cooling is proportional to temperature and hence to the vertical derivative of ψ'. Short vertical wavelengths are therefore damped faster than long vertical wavelengths. For vertical wavelengths comparable to H, the cool-to-space approximation breaks down. As shown in Fig. 8.29, LW exchange between neighboring layers then makes α itself strongly scale dependent. The cooling rate is much faster for short vertical wavelengths than for long vertical wavelengths.

Absorption can change along a ray because α varies spatially. According to Fig. 8.29, the time scale for thermal damping decreases from several weeks in the troposphere to only a couple of days in the upper stratosphere. Of greater significance is vertical shear, which can sharply reduce ω. If the intrinsic frequency is Doppler-shifted to zero, $\left(\frac{\alpha}{\omega}\right) m_0 \to \infty$. This is precisely what occurs near a critical line. There, oscillations become stationary relative to the medium, allowing wave activity to be fully absorbed.

14.7 NONLINEAR CONSIDERATIONS

Under inviscid adiabatic conditions, the equations governing wave propagation become singular at a critical line, where the mean velocity matches the trace speed:

$\overline{u} = c_x$. For stationary waves, this condition is achieved at the zero wind line ($\overline{u} = 0$). Dissipation removes the singularity by absorbing wave activity as the local wavelength shrinks and c_g approaches zero. Strong gradients of the wave field near its critical line then magnify thermal damping. They also magnify diffusion. By smoothing out gradients, both act to destroy organized wave motion.

These conclusions follow from considerations of the first-order equations, in which only simple forms of dissipation are represented. More generally, however, singular behavior near a critical line invalidates those equations. In the limit of small dissipation, wave stress exerted on the mean flow (14.5) becomes concentrated at the critical line and unbounded (Dickinson, 1970). Even with plausible dissipation, the convergence of momentum flux there implies large accelerations of $\overline{u}$ locally. Consequently, $\overline{u}$ cannot remain steady. Such behavior represents nonlinear interaction between wave activity and the mean flow. The same circumstances magnify nonlinear interaction between components of the wave spectrum.

Recall that the governing equations may be linearized so long as second-order terms such as $v' \cdot \nabla v'$ are much smaller than those first order in wave amplitude. For an individual wave component, the material derivative is

$$
\begin{aligned}
\frac{d}{dt} &= \frac{D}{Dt} + v' \cdot \nabla \\
&= -i\omega + i v' \cdot k \\
&= -ik\left[(c_x - \overline{u}) - v' \cdot \frac{k}{k} \right].
\end{aligned}
\tag{14.89}
$$

Ignoring second order behavior is therefore tantamount to requiring the intrinsic trace speed $|c_x - \overline{u}|$ to be large compared with the eddy velocity $|v'|$ that is induced by the wave. Near a critical line, this requirement breaks down because $|c_x - \overline{u}| \to 0$. Instead of experiencing a small oscillation during the passage of wave activity, an individual parcel then undergoes finite displacement. The latter can lead to a permanent rearrangement of air.[7]

Attention was first drawn to this issue by Kelvin (1880). He noted the peculiar behavior in Fig. 14.22a at the critical level of a gravity wave, under incompressible conditions and neutral stratification. For $N^2 = 0$, (14.40) implies a total streamfunction relative to the medium which is symmetric about $z = z_c$ and which contains closed streamlines within a neighborhood of the critical level.[8] Known as *Kelvin's cat's eye pattern*, this distribution of ψ describes a series of vortices. They overturn air inside the critical region, which is eventually wound up (Fig. 14.22d). Gradients that support organized wave motion are then sheared down to small scales, where diffusion destroys them efficiently (Fig. 12.4). The evolution culminates in the absorption of wave activity that is incident on the critical level.

In the limit of weak dissipation, another scenario is possible. It operates even under stable stratification, as supports gravity wave propagation. Mixing, implied by the motion in Fig. 14.22, will homogenize the distribution of potential temperature, driving $\frac{\partial \theta}{\partial z}$ and N^2 to zero (Sec. 7.6). Under the same conditions, vorticity u_z is also

---

[7] Dissipation keeps displacements bounded by introducing an imaginary component to $c_x - \overline{u}$.
[8] Rossby waves behave similarly, but with β_e in place of $-\overline{u}_{zz}$.

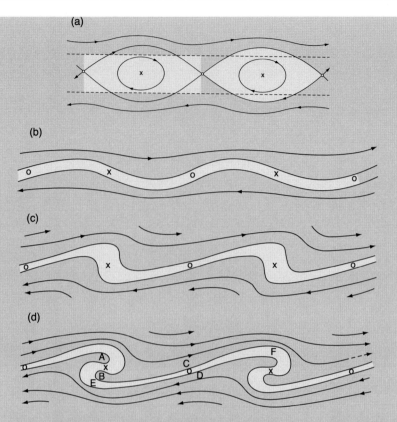

Figure 14.22 (a) Streamfunction at the critical level of a simple gravity wave under neutral stratification, which assumes the form of *Kelvin's cat's eye pattern*. (b) Evolution of a material contour initialized at the critical level. Adapted from Scorer (1978).

conserved. $\bar{u}_{zz}$ will thus likewise be driven to zero. Mixing thus culminates in the refractive index m^2 vanishing in some neighborhood of $z = z_c$ (14.40.2). Wave activity incident on the critical level is then not absorbed, but reflected.

Rossby waves produce such behavior through horizontal advection. Figure 14.23 shows planetary wave activity incident on a critical line at low latitude. Under weakly dissipative conditions (Fig. 14.23a), nonlinear advection near the critical line overturns material contours. The latter coincide with contours of potential vorticity Q, which is approximately conserved under these conditions. Air inside the critical region is rolled up in a series of vortices. They destroy organized wave motion on large scales by shearing it down to small scales, where diffusion is effective. Efficient horizontal mixing homogenizes Q near the critical line. It thus drives the refractive index there to zero (14.76). Wave activity is subsequently reflected from the critical line. This limiting behavior is implied by the weak phase tilt with latitude (dashed). The time required to establish this limiting behavior is relevant. If it is longer than the time scale for seasonal transience of $\bar{u}$ (e.g., in the stratosphere, where $\bar{u}$ reverses between winter and summer), reflection may never be achieved (Geisler and Dickinson, 1974).

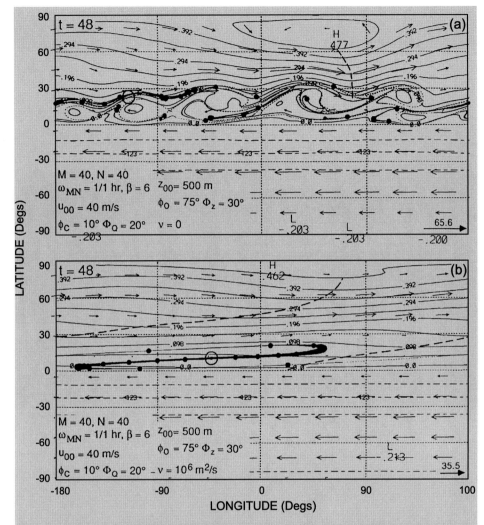

Figure 14.23 Nonlinear evolution of a planetary wave critical layer under barotropic conditions and with zonal wavenumber 1 incident on it, illustrated in terms of potential vorticity (contoured) and the position and thickness of a material contour (bold/broken curve and dots) that was initialized at the critical latitude. (a) In the presence of weak dissipation, potential vorticity is rolled into a series of vortices that mix air inside the critical layer, driving $\beta_e = \frac{\partial \bar{Q}}{\partial y}$ to zero. Wave activity is then reflected from the critical region, as manifested by the absence of phase tilt (dashed). (b) In the presence of strong dissipation, diffusion prevents the formation of sharp gradients that support vortex roll up, so $\frac{\partial \bar{Q}}{\partial y} > 0$ is maintained. The incident planetary wave then continues to propagate equatorward and be absorbed, as is manifested by its systematic meridional tilt. After Salby et al. (1990).

Under strongly dissipative conditions (Fig. 14.23b), the wave field evolves very differently. Diffusion then prevents the formation of sharp gradients that promote vortex roll-up. Wave activity therefore continues to propagate equatorward and be absorbed at the critical line. This limiting behavior is evident from the steep westward tilt of phase with decreasing latitude. It contrasts sharply with the limiting structure under weakly dissipative conditions (Fig. 14.23a), which exhibits little phase tilt.

Nonlinearity is also introduced by large wave amplitude. It likewise makes $|\boldsymbol{v}'| = O(|c_x - \overline{u}|)$. Figure 14.24 displays an amplified mountain wave over the Colorado Rockies, which was observed during an intense wind storm. Air moves along isentropic surfaces (solid), undergoing large vertical displacements. In the lee of the continental divide, it descends almost 5 km. The steep slope of isentropic surfaces reduces $\frac{\partial \theta}{\partial z}$ and static stability locally. In addition, the large amplitude exaggerates vertical shear. Both effects destabilize the organized motion, through the gradient Richardson number and mechanisms discussed in Sec. 13.3. Richardson numbers less than $\frac{1}{4}$ then produce intense turbulence. The latter damps the wave by cascading energy to small scales, where it is dissipated efficiently. Advected downstream, turbulence can be violent. It is visible only in the isolated rotor cloud at 600 hPa (Fig. 9.22). During the event shown, a Boeing 707 was on approach to Denver airport. It experienced impulsive accelerations in excess of 3 g's. Similar conditions were encountered during the 1950s in the Sierra Wave Project. The jet stream was then observed by instrumented sailplanes. The following excerpt from a chronicle of the project (Lincoln, 1972) describes an episode familiar to aviators:

> Editor's Note: On the day described, Larry Edgar was flying a Pratt-Read, one of the strongest sailplanes ever built, and he had the ultimate experience of flying turbulence in the rotor. This story has probably saved the lives of a number of aviators who have learned the indicated lesson and treat rotor clouds with respect.

> "Lloyd Licher and I examined the wreckage and compared notes with others. The nose pulled off just at the seats, by what seemed to be a tension failure of the steel tubing. Considering Larry's weight and the instrumentation, this should have required just in excess of 16 g's.... The wreckage showed the left wing to have broken at altitude. It broke downward near the root. The tail boom was broken cleanly from the fuselage pod, at altitude, and appears to have come off upward. The various control cables going from nose to tail were pulled apart completely – in a bunch. The force needed to do this should be far over 10,000 pounds."

> Joachim P. Kuettner

(The pilot survived.)

Large amplitude results naturally from vertical propagation. Upon reaching sufficient height, a gravity wave will have achieved a perturbation potential temperature θ' that is large enough to cancel the mean vertical gradient of potential temperature $\frac{\partial \overline{\theta}}{\partial z}$, which underpins vertical stability. At that level, the vertical gradient of total potential temperature is driven to zero

$$\frac{\partial \theta}{\partial z} = \frac{\partial \overline{\theta}}{\partial z} + \frac{\partial \theta'}{\partial z} = 0. \tag{14.90}$$

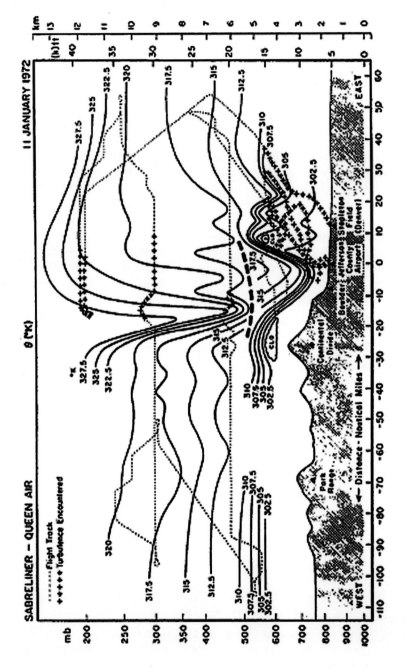

Figure 14.24 Cross section of isentropic surfaces in the mountain wave during a major wind storm over the Colorado Rockies. Severe turbulence (+) is found at elevations of the mountain cap cloud and below, as well as near the tropopause. It is visible, however, only in the isolated rotor cloud near 600 hPa. After Lilly (1978).

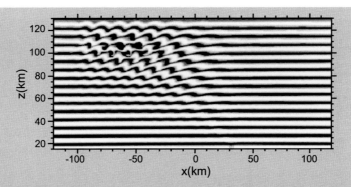

Figure 14.25 Potential temperature distribution in the presence of gravity waves that are excited in the troposphere by a moving bump analogous to a squall line. Positioned near $x = 30$ km, the disturbance is 10 km wide and migrates eastward. Gravity waves amplify with height sufficiently to break in the mesosphere, where they overturn θ surfaces. Note: Wave activity propagates eastward relative to the medium, so phase lines tilt eastward with height. Courtesy of R. Garcia.

Isentropic surfaces are then vertical. Such structure corresponds to neutral stability. The gravity wave then *breaks*, overturning the stratification. Such behavior is analogous to the breaking of a surface water wave that is incident on a shoreline. Air of low θ is folded over air of higher θ. Figure 14.25 displays such behavior for a gravity wave that has propagated to the mesosphere. By overturning isentropic surfaces locally, stable stratification is transformed into unstable stratification. In addition to being statically unstable, this structure violates the Ri criterion for dynamic stability. The ensuing stratification culminates in vigorous mixing. Wave momentum is then deposited at the breaking height, where wave activity is absorbed.

Planetary waves introduce similar behavior by rearranging air horizontally. Figure 14.26 shows the rearrangement of air by planetary waves under amplified conditions typical of the wintertime stratosphere. Anticyclonic motion has displaced the cyclonic polar-night vortex out of zonal symmetry. Mid-latitude air is then advected into the polar cap. To conserve potential vorticity, that air spins up anticyclonically. It establishes a reversed circulation about the pole, with circumpolar easterlies. Such behavior may be regarded as an expansion of the critical region in Fig. 14.23a, where $|v'| = O(|c_x - \overline{u}|)$. Both are characterized by closed streamlines that lead to a complex rearrangement of air.

Behavior such as that in Fig. 14.26 occurs sporadically in the stratosphere during northern winter (cf. Fig. 18.13). Planetary waves forced by large orography in the Northern Hemisphere often amplify sufficiently to form closed streamlines at middle and high latitudes. Termed *sudden stratospheric warmings*, such episodes are marked by an abrupt reversal of the zonal-mean flow. Through thermal wind balance, the latter is attended by abrupt warming at high latitude – as much as 50°K in just a couple of days. Air over the winter pole, which lies in perpetual darkness, then becomes warmer than air in the sunlit tropics, which experiences ozone heating. Figure 1.10a illustrates the 10-hPa circulation during a stratospheric warming. In concert with Fig. 14.26, it shows that this dramatic change in zonal-mean properties actually follows from a complex rearrangement of air.

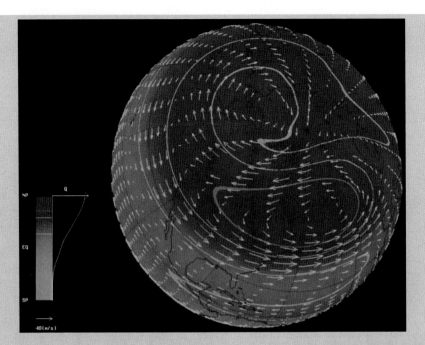

Figure 14.26 Distribution of potential vorticity Q and horizontal motion in a 2-dimensional calculation representative of the stratospheric circulation at 10 hPa under disturbed conditions. Color and velocity scales shown at left. High-Q polar air (blue), which marks the polar-night vortex, has been displaced well off the pole and distorted by an amplified planetary wave. Replacing it is low-Q air from equatorward (red), which has been advected into the polar cap. It spins up anticyclonically, forming a reversed circulation with easterly circumpolar flow at high latitude. These features are characteristic of a *stratospheric sudden warming* (cf. Fig. 18.13). See color plate section: Plate 11.

In the Southern Hemisphere, orographic features at the Earth's surface are comparatively small. Planetary waves are therefore weaker. For this reason, sudden stratospheric warmings and the disturbed conditions that accompany them are rare in the Southern Hemisphere. Less disturbed, the Antarctic polar-night vortex therefore remains closer to radiative equilibrium, much colder than its counterpart over the Arctic.

SUGGESTED REFERENCES

Atmosphere-Ocean Dynamics (1982) by Gill develops gravity waves at the graduate level. Advanced treatments are presented in *Atmospheric Waves* (1975) by Gossard and Hooke and *Waves in Fluids* (1978) by Lighthill.

An Introduction to Dynamic Meteorology (2004) by Holton develops Rossby waves. An advanced treatment, along with implications for transport, is presented in *Middle Atmosphere Dynamics* (1987) by Andrews et al.

PROBLEMS

1. Derive the zonal-mean momentum budget (14.5).
2. Lenticular cloud (Chap. 9) forms preferentially during winter. Explain its seasonality in relation to conditions that favor gravity waves.
3. Derive the group velocity (14.44) for simple gravity waves.
4. Demonstrate that, in mean westerlies, phase lines of a stationary gravity wave forced at the surface (e.g., the undulation of isentropic surfaces) must slope westward with height.
5. The Manti-La Sal mountains in southern Utah comprise an isolated range that has a characteristic width of 10 km and is oriented NS. If winds blow from the SW, the range's NS extent is much longer than 10 km, and static stability is characterized by $N^2 = 2.0 \ 10^{-4}$ s^{-2}, (a) at what polar angle will lenticular cloud be found from the center of the range? (b) What will be the orientation of individual cloud bands relative to the range?
6. A gravity wave of zonal wavenumber $k = 0.5$ km^{-1} propagates vertically through mean westerlies of uniform positive shear

$$\overline{u} = \overline{u}_0 + \Lambda z,$$

with $\overline{u}_0 = 10$ m s^{-1}, $\Lambda = 0.1$ m s^{-1}km^{-1}, and constant static stability $N^2 = 2.0 \ 10^{-4}$ s^{-2}. In the framework of the Boussinesq and WKB approximations, sketch the profiles of intrinsic frequency and vertical wavenumber if the gravity wave (a) is stationary, (b) propagates westward at 10 m s^{-1}, and (c) propagates eastward at 10 m s^{-1}.
7. A stationary gravity wave of the form

$$w'(x, z) = W(z)e^{ikx}$$

is excited by flow over elevated terrain. The horizontal and vertical scales are short enough for wave propagation to be controlled by local conditions. If the upstream flow $\overline{u}$ is independent of height but the stratification varies as

$$N^2(z) = N_0^2 \left(1 - \frac{z^2}{H^2}\right),$$

(a) characterize the propagation according to levels where wave activity is vertically propagating and external, (b) sketch the vertical wave structure, and (c) determine the range of $\overline{u}$ for which wave activity is trapped at the surface.
8. Barometric registrations following the eruption of Krakatoa revealed an oscillatory disturbance in surface pressure having an amplitude of 1 hPa. If the period of the oscillation was 1 hour, estimate the amplitude of (a) the pressure perturbation at 80 km, (b) the wind perturbation at 80 km.
9. In terms of the vorticity budget (12.36), discuss how volume heating excites planetary waves.
10. Stationary planetary wave activity is generated by westerly flow over elevated terrain. Show that such wave activity will be found downstream of its topographic forcing.
11. Show that the flux of zonal momentum transmitted vertically by simple gravity waves, $\rho_0 \overline{u'w'}$, is positive (negative) if their group velocity is upward and they propagate eastward (westward).

12. On approach to JFK airport in New York, winds are gusting nearly along the shoreline. Yet, white caps are observed to approach the shore virtually head on. Construct a simple model based on shallow water waves propagating in the $x - y$ plane, with depth decreasing linearly to zero at the shoreline, which coincides with $y = 0$. For a wavepacket characterized initially by $k = (k_0, l_0)$, (a) determine $k(y)$, (b) sketch phase lines corresponding to successive positions of the wavepacket for $l_0 > 0$, (c) plot rays that are initially oriented 60° and 30° from the shoreline.

13. Obtain the dispersion relation (14.34) for acoustic-gravity waves. (Hint: What condition must be satisfied for a homogeneous system of linear equations to have nontrivial solution?)

14. Show that wave activity for simple gravity waves propagates vertically one vertical wavelength for each horizontal wavelength that it propagates horizontally.

15. A gravity wave of wavenumber $k = (k, m)$ propagates into a region where potential temperature varies with height as

$$\frac{\theta}{\theta_0} = -\frac{(z - z_0)^2}{H^2},$$

with $H = $ const. (a) Describe propagation in the $x - z$ plane, sketching the corresponding ray. (b) Sketch the vertical structure of the wave. (c) Describe the behavior under fully nonlinear conditions.

16. *Inertio-gravity waves* are buoyancy waves with horizontal scales long enough to be influenced by rotation. Within the framework of the Boussinesq approximation, (a) provide a set of equations governing such motion on an f plane. (b) Show that these waves satisfy the dispersion relationship

$$\frac{m^2}{k^2} = \frac{N^2 - \omega^2}{\omega^2 - f^2},$$

which is a generalization of (14.36) and illustrates that inertio-gravity waves propagate vertically only for $|\omega| > |f|$. Atmospheric tides, which are generated by diurnal variations of heating, are inertio-gravity waves. Estimate the range of latitude for vertical propagation of (c) the diurnal tide: $\sigma = 1.0$ cpd, and (d) the semidiurnal tide $\sigma = 2.0$ cpd.

17. Show that the Lamb mode is the only nontrivial solution to the homogeneous boundary value problem (14.60).

18. Steady, barotropic, and nondivergent flow over an isolated mountain excites Rossby waves. The effect of the mountain may be treated as a vorticity source

$$Q(x, y) = \frac{Q_0}{2\pi L^2} e^{-\frac{x^2 + y^2}{2L^2}}.$$

(a) Provide a rationale for treating surface forcing in this manner. (b) Provide an equation governing the motion. (c) Express the streamfunction in terms of a Fourier integral to recover the wave field, subject to the condition of radiation away from the source region. (d) Plot the wave field.

19. Consider quasi-geostrophic motion in spherical geometry. (a) Provide an equation governing the propagation of planetary waves in latitude and height. For a stationary wave that is invariant with height ($\frac{\partial \psi'}{\partial z} = 0$) and propagates horizontally

through uniform westerlies of 20 m s^{-1}, estimate the polar turning latitude for (b) zonal wavenumber 1, (c) zonal wavenumber 4.

20. In light of Prob. 14.19, describe the wave propagation that could be established with a source at mid-latitudes if potential vorticity is homogenized at low latitudes by horizontal mixing (Sec. 14.7).

CHAPTER FIFTEEN

The general circulation

Thermal equilibrium requires that, for the Earth-Atmosphere system as a whole, net radiative heating must vanish. Although it applies globally, this requirement need not hold locally. In fact, net radiation (Fig. 1.34c during DJF) shows that low latitudes experience radiative heating: They receive more energy through absorption of SW than they lose through emission to space of LW. Conversely, middle and high latitudes experience radiative cooling, especially in the winter hemisphere: They lose more energy through emission to space of LW than they receive through absorption of SW. To preserve thermal equilibrium, these local imbalances in the radiative energy budget must be compensated by a poleward transfer of heat. Accomplished mechanically, the latter transfers energy from tropical regions, where it offsets the surplus of radiative energy, to extratropical regions, where it offsets the deficit of radiative energy. The poleward transfer of heat is accomplished by the general circulation of the Earth-atmosphere system, 60% of it by the circulation of the atmosphere (see Fig. 17.10).

The simplest mechanism to transfer heat poleward is a steady, zonally symmetric circulation between the equator and poles. Such motion is driven by atmospheric heating in the tropics and cooling in the extratropics. Atmospheric heating is concentrated at low latitude, where it derives from latent heat release inside centers of convection (Fig. 9.41b). Together with radiative cooling at higher latitude, it forces vertical motion across isentropic surfaces (Secs. 2.5, 3.6). The latter must be compensated by horizontal motion. The resultant overturning in the meridional plane transfers heat between regions of heating and cooling. Maintaining thermal equilibrium locally, it represents a *thermally direct circulation*. The Hadley circulation is of this form (Fig. 1.35). However, the observed Hadley circulation extends poleward to only about ±30°. It cannot achieve the global heat transfer required to maintain thermal equilibrium. The observed circulation is more complex, especially on individual days (Fig. 1.9).

How meridional heat transfer is actually accomplished in the atmosphere is influenced profoundly by the Earth's rotation. Net radiation is almost uniform in longitude. It therefore tends to establish thermal structure that is zonally symmetric, with temperature decreasing poleward. According to the hypsometric relation (6.12), isobaric height would then assume similar form, likewise decreasing poleward. Under these circumstances, contours of isobaric height would remain parallel to latitude circles. Geostrophic equilibrium would then require air motion to likewise be zonally symmetric: nearly zonal, with only a weak meridional component to transfer heat poleward. Such motion represents a skewed equator-to-pole circulation, with velocity deflected into the zonal direction. Air motion would then be nearly perpendicular to the thermally direct overturning hypothesized above. This peculiarity of the implied circulation is a manifestation of the geostrophic paradox (Sec 12.1). The zonally symmetric circulation would also be nearly perpendicular to the temperature gradient. It would therefore transfer little heat poleward.

With no offset to radiative heating and cooling, the meridional temperature gradient must steepen. It would continue to steepen until temperature in the tropics became sufficiently warm (increasing LW emission) and temperature in the extratropics became sufficiently cold (decreasing LW emission) to reduce net radiation in each region – just enough to be offset by the weak poleward heat transfer of the zonally symmetric circulation. This form of thermal equilibrium, however, would produce a meridional temperature gradient that is much steeper than observed (Fig. 10.1).

By thermal wind balance (12.11), the steep meridional temperature gradient must be accompanied by steep vertical shear and hence by a strong zonal jet overhead. The strength of the implied jet can be inferred from conservation of angular momentum. An air parcel drifting poleward in the zonally symmetric circulation conserves its angular momentum, $(u + \Omega a cos\phi)a cos\phi$ (11.29). If initially motionless at the equator, the parcel must then assume zonal velocity at latitude ϕ of

$$u = \Omega a \frac{sin^2\phi}{cos\phi}.$$

Upon reaching a latitude of $45°$, poleward-moving air must therefore attain wind speed in excess of 300 m s^{-1} – namely, of order Mach 1.

The observed circulation never even approaches such speed (Fig. 1.8). Long before it is attained, the circulation develops zonally asymmetric structure – large-scale disturbances, which completely alter the extratropical circulation. Those asymmetric motions transfer heat poleward far more efficiently than can a zonally symmetric circulation in the presence of rotation. Inherently unsteady, extratropical disturbances are fueled by reservoirs of atmospheric energy and exchanges amongst them, which shape the general circulation.

15.1 FORMS OF ATMOSPHERIC ENERGY

From the time it is absorbed until it is eventually rejected to space, energy undergoes a complex series of transformations between three basic reservoirs:

1. Thermal energy, which is represented in the temperature and humidity of air.
2. Gravitational potential energy, which is represented in the horizontal distribution of atmospheric mass.
3. Kinetic energy, which is represented in air motion.

Transformations from thermal and potential energy to kinetic energy are responsible for setting the atmosphere into motion. They maintain the circulation against frictional dissipation.

15.1.1 Moist static energy

The thermal energy of moist air has two contributions: Sensible heat content reflects the molecular energy of dry air, which is the dominant component of moist air. Latent heat content reflects the energy that can be imparted to the dry air component when the vapor component condenses. These contributions are collected in the specific enthalpy (Sec. 5.3)

$$h = c_p T + l_v r + h_0. \tag{15.1}$$

$c_p T$ measures the sensible heat content of a moist parcel, whereas $l_v r$ measures its latent heat content.

The First Law (2.15) implies that the specific enthalpy of an individual air parcel changes according to

$$\frac{dh}{dt} - \alpha \frac{dp}{dt} = \dot{q}_{net}, \tag{15.2.1}$$

where $\dot{q}_{net}$ is the net rate heat is absorbed by the parcel. Incorporating hydrostatic equilibrium allows (15.2.1) to be expressed

$$\frac{dh}{dt} + g \frac{dz}{dt} = \dot{q}_{net}$$

or with (15.1)

$$\frac{d}{dt}(c_p T + l_v r + \Phi) = \dot{q}_{net}$$
$$= \dot{q}_{rad} + \dot{q}_{mech}, \tag{15.2.2}$$

where $\dot{q}_{rad}$ and $\dot{q}_{mech}$ denote radiative and mechanical components of heat transfer. The quantity $(c_p T + l_v r + \Phi)$ defines the *moist static energy* of the air parcel. It includes thermal as well as potential energy. A parcel's moist static energy changes only through heat transfer with its environment. It is unaffected by vaporization and condensation, which involve only internal exchanges of sensible and latent heat contents.[1]

Mechanical heat transfer is dominated by turbulent mixing. It leads to exchanges of sensible heat (temperature) and latent heat (humidity)

$$\dot{q}_{mech} = \dot{q}_{sen} + \dot{q}_{lat}. \tag{15.3}$$

These components of turbulent heat transfer can be represented in terms of mean gradients of temperature and humidity and corresponding eddy diffusivities (13.13).

Equation (15.2.2) describes an air parcel's moist static energy in terms of advection and the distributions of sources and sinks. A parcel will acquire moist static energy at low latitude through absorption of radiative, sensible, and latent heats from the Earth's surface. Upon reaching middle and high latitudes, the parcel will lose moist static energy through radiative cooling to space. Completing a meridional circuit (symbolized

---

[1] This differs from the development in Sec. 10.7, which focuses on the dry air component. The latter must account for transfer of latent heat from the vapor component.

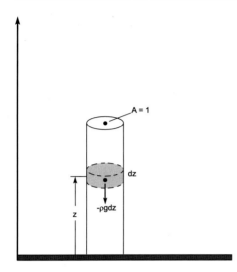

Figure 15.1 Incremental volume of unit cross-sectional area and height z inside an atmospheric column.

in Fig. 6.7) then results in a net poleward transfer of moist static energy. The latter compensates TOA radiative heating at low latitude and radiative cooling at middle and high latitudes. Integrating (15.2) over the entire atmosphere and averaging over time obtains the mean change of moist static energy. Under thermal equilibrium, it must vanish. Net heat transfer to the atmosphere must therefore likewise vanish

$$\overline{\dot{q}_{rad}} + \overline{\dot{q}_{sen}} + \overline{\dot{q}_{lat}} = 0, \qquad (15.4)$$

where overbar denotes global and time mean. Net radiative cooling of the atmosphere (Fig. 8.24) must then be balanced by transfers of sensible and latent heat from the Earth's surface (Fig. 1.32).

Inside the boundary layer, turbulent diffusion of heat and absorption of LW radiation from the Earth's surface increase an air parcel's sensible heat content. Absorption of water vapor from the surface of ocean increases its latent heat content. Both increase the moist static energy of the parcel. Outside the boundary layer and convection, turbulent diffusion is small enough to be ignored. Radiative cooling then becomes the primary diabatic process. It gradually depletes the parcel's store of moist static energy, which is subsequently replenished when it returns to regions of positive heat transfer.

15.1.2 Total potential energy

Thermal energy is reflected in the internal energy of dry air. Heat transfer from the Earth's surface increases an atmospheric column's temperature and hence its internal energy. The hypsometric relationship (6.12) then implies vertical expansion of the column. Because it elevates the column's center of mass, a process that increases the column's internal energy must also increase its potential energy.

Consider a dry atmosphere in hydrostatic equilibrium and in which air motion occurs on a time scale short enough to be regarded as adiabatic. An incremental volume of unit cross-sectional area (Fig. 15.1) has internal energy

$$d\mathcal{U} = \rho c_v T dz. \qquad (15.5.1)$$

Integrating upward from the surface gives the column internal energy

$$\mathcal{U} = c_v \int_0^\infty \rho T \, dz. \tag{15.5.2}$$

The potential energy of the incremental volume is

$$dP = \rho g z \, dz$$
$$= -z \, dp. \tag{15.6.1}$$

Then the column potential energy up to a height z is given by

$$P(z) = \int_{p(z)}^{p_s} z' \, dp', \tag{15.6.2}$$

where $z' = z'(p')$ and the integral is evaluated in the direction of increasing pressure. Integrating by parts transforms this into

$$P(z) = z' p' \big|_{p(z)}^{p_s} + \int_0^z p \, dz'$$
$$= -z p(z) + R \int_0^z \rho T \, dz',$$

which, for $z \to \infty$, reduces to

$$P = R \int_0^\infty \rho T \, dz. \tag{15.7}$$

A similar expression holds for the column internal energy (15.5.2). Thus,

$$\frac{P}{\mathcal{U}} = \frac{R}{c_v}$$
$$= \gamma - 1. \tag{15.8}$$

Equation (15.8) asserts that the internal and potential energies of an atmospheric column preserve a constant ratio. This interdependence is a consequence of hydrostatic equilibrium. It follows that energy can then be drawn from those reservoirs only in a fixed proportion. Adding the internal and potential energies of the column obtains

$$P + \mathcal{U} = \int_0^\infty \rho c_p T \, dz$$
$$= \frac{c_p}{g} \int_0^{p_s} T \, dp \tag{15.9}$$
$$= \mathcal{H},$$

which is just the column enthalpy. It defines the *total potential energy* (Margules, 1903).

Consider the atmosphere as a whole and, for the sake of illustration, in the absence of heat transfer at its boundaries. Because p vanishes at the top of the atmosphere and w vanishes at its bottom, no work is performed on this system as well. Then the First Law for the entire atmosphere, inclusive of kinetic energy $\mathcal{K}$, asserts that the change of total energy vanishes

$$\Delta(\mathcal{H} + \mathcal{K}) = 0. \tag{15.10.1}$$

Therefore

$$\Delta \mathcal{K} = -\Delta \mathcal{H}$$
$$= -\Delta(\mathcal{P} + \mathcal{U}),$$

(15.10.2)

where column properties are understood to be integrated globally. Under adiabatic conditions, the atmosphere's kinetic energy is drawn from its reservoir of total potential energy.

15.1.3 Available potential energy

Of the atmosphere's total potential energy, only a small fraction is actually available for conversion to kinetic energy. Consider an atmosphere that is initially motionless and barotropically stratified (Sec. 12.2). Isobaric surfaces then coincide with isentropic surfaces, as illustrated in Fig. 12.5a. Under adiabatic conditions, air parcels must move along θ surfaces. Consequently, no pressure gradient exists in the directions of possible motion. A barotropically stratified atmosphere thus possesses no means of generating motion internally. It follows that none of its potential energy is available for conversion to kinetic energy.

Suppose the above atmosphere is now heated at low latitude and cooled at middle and high latitudes, with net global heating of zero. By the First Law, the atmosphere's total energy is unchanged. If the atmosphere remains motionless, so is its total potential energy (15.10.1). Locally, however, total potential energy clearly does change. Increased temperature increases total potential energy at low latitude. Reduced temperature decreases total potential energy at middle and high latitudes. By the hypsometric relationship, the vertical spacing of isobaric surfaces is then expanded at low latitude but compressed at middle and high latitudes. Isobaric surfaces are thus tilted "clockwise" into the positions assumed in Fig. 12.5b. Under positive static stability, just the reverse occurs for isentropic surfaces. Heating increases potential temperature at low latitude. Because θ increases upward, it brings down greater values of θ from above. Conversely, cooling reduces θ at middle and high latitudes. It brings up smaller values of θ from below. Isentropic surfaces are thus tilted "counterclockwise." In this fashion, the nonuniform distribution of heating and cooling drives isobaric surfaces out of coincidence with isentropic surfaces. It thus drives thermal structure into baroclinic stratification.

Under such stratification, the atmosphere possesses a pressure gradient along isentropic surfaces (Fig. 15.2). An air parcel can then accelerate through the pressure-gradient force. Potential energy is thus available for conversion to kinetic energy. Even though no potential energy has been added to the atmosphere as a whole, that energy reservoir has been tapped by introducing a horizontal gradient of buoyancy. Analogous reasoning applies to uniformly heating the atmosphere. The latter clearly increases the atmosphere's store of total potential energy. However, it introduces none that is available for conversion to kinetic energy.

From hydrostatic equilibrium, the variation of pressure along a θ surface reflects a nonuniform distribution of mass above that surface. The situation is analogous to two immiscible fluids of different densities that have been juxtaposed horizontally (Fig. 15.3). Hydrostatic equilibrium implies a horizontal pressure-gradient force, directed from the heavier fluid to the lighter one. Motion will then develop internally to alleviate the mechanical imbalance. The nonrotating system in Fig. 15.3 accomplishes this by

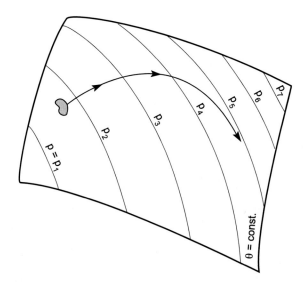

Figure 15.2 Variation of pressure along an isentropic surface under baroclinic stratification. The pressure variation vanishes under barotropic stratification.

rearranging mass so that heavier fluid undercuts and eventually comes to rest beneath lighter fluid. The system's final state is then hydrostatically stable. By lowering the center of gravity of the system, this process releases potential energy, which is converted into kinetic energy. The latter is eventually dissipated by viscosity, increasing the system's internal energy.

In the atmosphere, horizontal rearrangement of mass is inhibited by rotation. The Coriolis force deflects air motion parallel to isobars (into and out of the page in Fig. 15.3). Nevertheless, the pressure gradient along θ surfaces enables air motion to develop. Though more complex, it neutralizes the mechanical imbalance. The Coriolis force makes those motions highly rotational, a feature that favors horizontal mixing (Fig. 12.4). Such behavior is illustrated by interleaving swirls of tropical and polar air

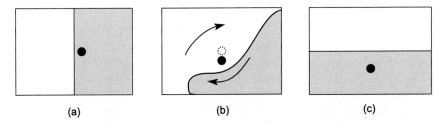

| (a) | (b) | (c) |

Figure 15.3 Schematic illustrating the rearrangement of mass in a system of two immiscible fluids of different densities that are initially juxtaposed horizontally. In the absence of rotation, heavier fluid (shaded) undercuts and comes to rest beneath lighter fluid. This process renders the system hydrostatically stable, while lowering its center of mass (solid circle). Rotation modifies this adjustment by deflecting upper motion out of the page (e.g., westerly) and lower motion into the page (e.g., easterly). Adapted from Wallace and Hobbs (1977).

in the cyclone off the coast of Africa in Fig. 1.18 (cf. Fig. 1.29). By mixing air horizontally, synoptic weather systems homogenize the horizontal distribution of mass over isentropic surfaces. This rearrangement of air drives isobaric surfaces back into coincidence with isentropic surfaces, restoring thermal structure to barotropic stratification. Such motion also results in a net poleward transfer of heat and moisture. Air drawn poleward from low latitude has greater moist static energy than replacement air that is drawn equatorward from middle and high latitudes. By making the horizontal mass distribution uniform, air motions lower the overall center of gravity of the atmosphere. Potential energy is therefore converted into kinetic energy. The latter, in turn, is dissipated by viscosity, converted into internal energy, and eventually rejected to space as heat.

Air motion responsible for this redistribution of mass is fueled by a conversion of potential energy to kinetic energy. It develops through *baroclinic instability*, wherein the source of energy is related to baroclinicity of the stratification. Temperature then varies along isobaric surfaces. By thermal wind balance, the energy source of baroclinic disturbances is also related to vertical shear.

Adiabatic adjustment

The potential energy available for conversion to kinetic energy is reflected in the departure from barotropic stratification. Consider an adiabatic redistribution of mass from a given baroclinic state. Because air must move along θ surfaces, horizontal mixing will eventually render the distributions of mass and pressure uniform over those surfaces. It therefore restores thermal structure to barotropic stratification. In that limiting state, the atmosphere possesses no more potential energy available for conversion, so $\mathcal{H} = \mathcal{H}_{min}$.

The *available potential energy* A (Lorenz, 1955) is defined as the difference between the total potential energy for the state under consideration and the minimum that would result through an adiabatic rearrangement of mass

$$A = \mathcal{H} - \mathcal{H}_{min}. \tag{15.11}$$

Available potential energy has the following properties:

1. $A + \mathcal{K}$ is preserved under adiabatic rearrangement. Hence $\Delta\mathcal{K} = -\Delta A$.
2. A is positive for baroclinic stratification and zero for barotropic stratification.
3. A is uniquely determined by the distribution of mass.

To derive an expression for A, consider an isentropic surface of area S. The average pressure on this surface

$$\bar{p} = \frac{1}{S} \int_S p \, dS \tag{15.12}$$

is conserved under adiabatic rearrangement (Prob. 15.8). The atmosphere as a whole has total potential energy

$$\mathcal{H} = \int_S dS \int_0^\infty \rho c_p T \, dz$$
$$= \frac{c_p}{g} \int_S dS \int_0^{p_s} T \, dp. \tag{15.13}$$

In terms of potential temperature, this becomes

$$\mathcal{H} = \frac{c_p}{g p_0^\kappa} \int_S dS \int_0^{p_s} \theta \, p^\kappa \, dp$$

$$= \frac{1}{p_0^\kappa (1+\kappa) \Gamma_d} \int_S dS \int_0^{p_s} \theta \, dp^{1+\kappa}, \tag{15.14}$$

where $\Gamma_d = \frac{g}{c_p}$ is the dry adiabatic lapse rate. Because the atmosphere is vertically stable, θ increases with height monotonically. It may therefore be interchanged with p for the vertical coordinate. Integrating (15.14) by parts obtains

$$\mathcal{H} = \frac{S}{p_0^\kappa (1+\kappa) \Gamma_d} \left[\overline{\theta \, p^{1+\kappa}} \Big|_0^{p_s} + \int_{\theta_s}^\infty \overline{p^{\kappa+1}} d\theta \right], \tag{15.15}$$

with θ_s denoting the potential temperature at the surface. The limit $p = 0$ in the first term in brackets vanishes. Defining $p = p_s$ for $\theta < \theta_s$ enables (15.15) to be expressed as

$$\mathcal{H} = \frac{S}{p_0^\kappa (1+\kappa) \Gamma_d} \int_0^\infty \overline{p^{\kappa+1}} d\theta. \tag{15.16}$$

When the stratification has been rendered barotropic, $\mathcal{H} = \mathcal{H}_{min}$ and p has become uniform over isentropic surfaces. Because p then equals $\overline{p}$,

$$\mathcal{H}_{min} = \frac{S}{p_0^\kappa (1+\kappa) \Gamma_d} \int_0^\infty \overline{p}^{\kappa+1} d\theta. \tag{15.17}$$

Hence, the state under consideration has available potential energy

$$\mathcal{A} = \frac{S}{p_0^\kappa (1+\kappa) \Gamma_d} \int_0^\infty \left[\overline{p^{\kappa+1}} - \overline{p}^{\kappa+1} \right] d\theta. \tag{15.18}$$

For small departures p' and T' from their areal averages, (15.18) is approximated by

$$\mathcal{A} = \frac{RS}{2g p_0^\kappa} \int_0^\infty \overline{p}^{1+\kappa} \left(\frac{\overline{p'^2}}{\overline{p}^2} \right) d\theta \tag{15.19.1}$$

and

$$\mathcal{A} = \frac{gS}{2} \int_0^{p_s} \frac{1}{N^2} \left(\frac{\overline{T'^2}}{\overline{T}^2} \right) dp, \tag{15.19.2}$$

demonstration of which is left as an exercise. The variance of pressure on an isentropic surface reflects the degree of baroclinicity. So does the variance of temperature on an isobaric surface. According to (15.19), each is a direct measure of available potential energy. Under typical conditions, as little as 0.1% of the atmosphere's total potential energy is available for conversion to kinetic energy (Prob. 15.4).

15.2 HEAT TRANSFER IN A ZONALLY SYMMETRIC CIRCULATION

Available potential energy is released by meridional air motion, which transfers heat poleward. Such motion is influenced importantly by the Coriolis force. To explore how rotation influences poleward heat transfer, we consider a simple model of the meridional circulation within the framework of the Boussinesq approximation (Sec. 12.5.3). A spherical atmosphere of finite vertical extent H is bounded below by a rigid surface and above by a free surface. The lower boundary is maintained at a zonally symmetric temperature $T_0(\phi)$, with T_0 decreasing poleward. The upper boundary is maintained at $T_0(\phi) + \Delta T$, with ΔT = const. If $\Delta T < 0$, this corresponds to heat being supplied at the atmosphere's lower boundary and rejected at its upper boundary.[2]

Following Charney (1973), we consider zonally symmetric motion $\overline{v} = (\overline{u}, \overline{v}, \overline{w})$ in the presence of turbulent diffusion. Scaling variables as in Sec. 11.4 casts the governing equations into nondimensional form. The prevailing balances can then be identified in terms of the dimensionless Rossby number and *Ekman number*:

$$Ro = \frac{U}{2\Omega a} \qquad E = \frac{K}{2\Omega H^2}, \qquad (15.20)$$

where $K = K_M = K_H$ is a constant eddy diffusivity (Sec. 13.1).

The limit of fast rotation and small friction makes $E \leq O(Ro) \ll 1$. To streamline the development, we follow the convention adopted in Sec. 12.5. Other than Ro and E, dimensionless factors are absorbed into the dependent variables. This enables the nondimensional equations to be expressed in a form nearly identical to their dimensional counterparts. Prevailing balances can then be identified in dimensional form because, at a specified order of Ro and E, those dimensionless parameters drop out. Proceeding as we did earlier casts the governing equations in log-pressure coordinates into

$$Ro\left\{\frac{\overline{v}}{a}\frac{\partial \overline{u}}{\partial \phi} + \overline{w}\frac{\partial \overline{u}}{\partial z}\right\} - f\overline{v} = EK\frac{\partial^2 \overline{u}}{\partial z^2} \qquad (15.21.1)$$

$$Ro\left\{\frac{\overline{v}}{a}\frac{\partial \overline{v}}{\partial \phi} + \overline{w}\frac{\partial \overline{v}}{\partial z}\right\} + f\overline{u} = -\frac{1}{a}\frac{\partial \overline{\Phi}}{\partial \phi} + EK\frac{\partial^2 \overline{v}}{\partial z^2} \qquad (15.21.2)$$

$$\frac{\partial \overline{\Phi}}{\partial z} = \frac{R\overline{T}}{H} \qquad (15.21.3)$$

$$\frac{1}{a\cos\phi}\frac{\partial}{\partial \phi}(\cos\phi\overline{v}) + \frac{\partial \overline{w}}{\partial z} = 0 \qquad (15.21.4)$$

$$Ro\left\{\frac{\overline{v}}{a}\frac{\partial \overline{T}}{\partial \phi} + \overline{w}\frac{\partial \overline{T}}{\partial z}\right\} = EK\frac{\partial^2 \overline{T}}{\partial z^2}, \qquad (15.21.5)$$

where factors like a, $f = sin\phi$, and K are understood to be dimensionless and $O(1)$. At the rigid lower boundary, no-slip and the imposed temperature require

$$\begin{aligned}\overline{u} = \overline{v} = \overline{w} = 0 \\ \overline{T} = T_0(\phi)\end{aligned} \quad z = 0. \qquad (15.21.6)$$

At the free surface, the stresses $\tau_{xz} = K\frac{\partial \overline{u}}{\partial z}$ and $\tau_{yz} = K\frac{\partial \overline{v}}{\partial z}$ must vanish. Further, for motion that is nearly geostrophic, vertical deflections of the free surface are small. The kinematic condition (Sec. 14.1) can then be replaced by the requirement of no

---

[2] It is through imposed temperature contrast that laboratory simulations are driven thermally. They will be seen in the next section to provide a mechanical analogue of this theoretical model.

vertical motion. With those approximations, the upper boundary conditions become

$$\frac{\partial \overline{u}}{\partial z} = \frac{\partial \overline{v}}{\partial z} = 0 \qquad \overline{w} = 0$$
$$\overline{T} = T_0(\phi) + \Delta T \qquad\qquad z = H. \qquad (15.21.7)$$

The dominant balances in (15.21) are geostrophic and hydrostatic equilibrium. Advective acceleration and friction introduce a small ageostrophic component into the momentum balance. As in Chap. 12, we consider an asymptotic series solution, one that is now valid in the limit $E \to 0$. Expanding dependent variables in power series of $E^{\frac{1}{2}}$, for example,

$$\overline{v} = \overline{v}^{(0)} + E^{\frac{1}{2}} \overline{v}^{(1)} \dots, \qquad (15.22)$$

enables the motion to be determined recursively to successively higher accuracy.[3]

To $O(E^0)$, the governing equations reduce to

$$f \overline{v}^{(0)} = 0 \qquad (15.23.1)$$

$$f \overline{u}^{(0)} = -\frac{1}{a} \frac{\partial \overline{\Phi}^{(0)}}{\partial \phi} \qquad (15.23.2)$$

$$\frac{\partial \overline{\Phi}^{(0)}}{\partial z} = \frac{R}{H} \overline{T}^{(0)} \qquad (15.23.3)$$

$$\frac{1}{a \cos\phi} \frac{\partial}{\partial \phi} \left(\cos\phi \overline{v}^{(0)} \right) + \frac{\partial \overline{w}^{(0)}}{\partial z} = 0, \qquad (15.23.4)$$

which are identical to their dimensional counterparts. Equations (15.23) imply that, to lowest order, the motion is zonal and in geostrophic balance. Cross-differentiating (15.23.2) and (15.23.3) obtains the thermal wind relation

$$\frac{\partial \overline{u}^{(0)}}{\partial z} = -\frac{R}{aHf} \frac{\partial \overline{T}^{(0)}}{\partial \phi}. \qquad (15.24)$$

With (15.23.1) and the boundary conditions, (15.23.4) gives

$$\overline{w}^{(0)} \equiv 0. \qquad (15.25)$$

Then the thermodynamic equation (15.21.5) implies

$$\frac{\partial^2 \overline{T}^{(0)}}{\partial z^2} = 0. \qquad (15.26.1)$$

According to (15.26.1) the vertical heat flux is nondivergent. The circulation is therefore in diffusive equilibrium. Heat absorbed at the atmosphere's lower boundary is transferred without loss to its upper boundary, where it is rejected. The solution of (15.26.1) satisfying the boundary conditions of imposed temperature is

$$\overline{T}^{(0)}(\phi, z) = T_0(\phi) + \Delta T \left(\frac{z}{H} \right). \qquad (15.26.2)$$

---

[3] Powers of $E^{\frac{1}{2}}$ are the appropriate form for the expansion because K enters the equations with second-order derivatives.

Then (15.24) implies that the E^0 motion intensifies with height linearly

$$\overline{u}^{(0)}(z) = -\frac{R}{aHf}\frac{\partial T_0}{\partial \phi}z. \tag{15.27}$$

The $O(E^{\frac{1}{2}})$ equations have the same form as (15.23). Proceeding along similar lines leads to

$$\overline{v}^{(1)} \equiv 0 \tag{15.28.1}$$

$$\overline{w}^{(1)} \equiv 0. \tag{15.28.2}$$

$$\overline{T}^{(1)} \equiv 0. \tag{15.28.3}$$

Thermal wind balance (15.24) then implies that the $O(E^{\frac{1}{2}})$ zonal flow is independent of height

$$\overline{u}^{(1)} \neq \overline{u}^{(1)}(z). \tag{15.29}$$

The solution accurate to $O(E^{\frac{1}{2}})$ has constant vertical shear. It therefore violates the upper boundary conditions of no stress on the free surface. This deficiency follows from the neglect of viscous terms, which reduces the order of the governing equations. Instead, the full solution must possess a boundary layer, wherein the motion adjusts sharply to meet boundary conditions. Viscous terms are then nonnegligible in a shallow neighborhood of the upper and lower boundaries. There, friction drives the motion out of geostrophic balance, which is inherent to the preceding equations.

To obtain an $O(E^{\frac{1}{2}})$ solution that is uniformly valid, we introduce a stretching transformation that accounts for sharp changes near the upper and lower surfaces. Boundary conditions can be satisfied by augmenting the first-order geostrophic motion with an ageostrophic correction

$$\overline{\boldsymbol{v}}^{(1)} = \overline{u}_g^{(1)}\boldsymbol{i} + \overline{\boldsymbol{v}}_a^{(1)}, \tag{15.30}$$

where $\overline{u}_g^{(1)}$ is the inviscid motion described by (15.29). The ageostrophic velocity $\overline{\boldsymbol{v}}_a^{(1)}$ imposes a secondary circulation onto the geostrophic flow. It is analogous to Ekman pumping (Sec. 13.4). The secondary circulation transports heat poleward by driving the motion across isotherms.

Inside the upper boundary layer, the stretching transformation

$$H - z = E^{\frac{1}{2}}\zeta \tag{15.31}$$

magnifies sharp changes. It makes viscous terms comparable to others in the momentum balance there, preserving the order of the governing equations. Then the $O(E^{\frac{1}{2}})$ momentum equations reduce to the Ekman balance

$$-f\overline{v}_a^{(1)} = K\frac{\partial^2 \overline{u}_a^{(1)}}{\partial \zeta^2} \tag{15.32.1}$$

$$f\overline{u}_a^{(1)} = K\frac{\partial^2 \overline{v}_a^{(1)}}{\partial \zeta^2}, \tag{15.32.2}$$

with the geostrophic contribution automatically satisfied by $\overline{u}_g^{(1)}$. Requiring the total solution accurate to $O(E^{\frac{1}{2}})$ to satisfy (15.21.7) provides boundary conditions at the free surface

$$\frac{\partial \overline{u}^{(0)}}{\partial z} - \frac{\partial \overline{u}_a^{(1)}}{\partial \zeta} = 0 \qquad \frac{\partial \overline{v}_a^{(1)}}{\partial \zeta} = 0 \qquad\qquad \zeta = 0. \tag{15.32.3}$$

For the solution to reduce to the inviscid behavior outside the boundary layer, the ageostrophic corrections must also satisfy

$$\overline{u}_a^{(1)} \sim 0 \qquad \overline{v}_a^{(1)} \sim 0 \qquad\qquad \zeta \to \infty. \tag{15.32.4}$$

The boundary value problem (15.32) has solution

$$\overline{u}_a^{(1)}(\zeta) = \frac{1}{2\gamma} \frac{\partial \overline{u}^{(0)}}{\partial z} e^{-\gamma \zeta} (\sin\gamma\zeta - \cos\gamma\zeta) \tag{15.33.1}$$

$$\overline{v}_a^{(1)}(\zeta) = \frac{1}{2\gamma} \frac{\partial \overline{u}^{(0)}}{\partial z} e^{-\gamma \zeta} (\sin\gamma\zeta + \cos\gamma\zeta), \tag{15.33.2}$$

where

$$\gamma = \sqrt{\frac{f}{2K}} \tag{15.33.3}$$

is the same parameter that defines the Ekman spiral in Sec. 13.2.1 (Figs. 15.4a,b).

Because $\overline{v}$ is concentrated inside the boundary layer, meridional transfer of mass is proportional to

$$\int_{\substack{H-\zeta \\ \zeta \to \infty}}^{H} E^{\frac{1}{2}} \overline{v}^{(1)} dz = E \int_0^\infty \overline{v}_a^{(1)} d\zeta$$

$$= -E \frac{KR}{aHf^2} \frac{\partial T_0}{\partial \phi}. \tag{15.34}$$

As $\frac{\partial T_0}{\partial \phi} < 0$, mass transfer along the upper surface is poleward and $O(E)$ (Fig. 15.4c). Integrating the continuity equation (15.21.4) gives the vertical motion just beneath the boundary layer

$$\overline{w}(H) = \int_{\substack{H-\zeta \\ \zeta \to \infty}}^{H} \frac{1}{a\cos\phi} \frac{\partial}{\partial \phi} (\cos\phi v) dz$$

$$= E \int_0^\infty \frac{1}{a\cos\phi} \frac{\partial}{\partial \phi} \left(\cos\phi \overline{v}_a^{(1)} \right) d\zeta.$$

It too is $O(E)$. Thus, vertical motion beneath the upper boundary layer is

$$\overline{w}^{(2)}(H) = -\frac{KR}{Ha^2 \cos\phi} \frac{\partial}{\partial \phi} \left(\frac{\cos\phi}{f^2} \frac{\partial T_0}{\partial \phi} \right). \tag{15.35}$$

At this order, $O(E)$, the zonal momentum equation in the interior reduces to

$$-f\overline{v}^{(2)} = K \frac{\partial^2 \overline{u}^{(0)}}{\partial z^2}$$

$$= 0. \tag{15.36}$$

Continuity then implies

$$\frac{\partial \overline{w}^{(2)}}{\partial z} = 0.$$

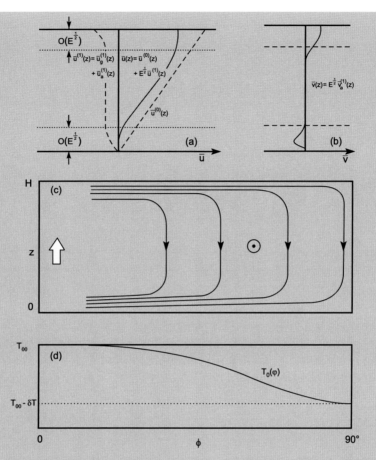

Figure 15.4 Zonally symmetric circulation in a Boussinesq spherical atmosphere of depth H that is driven thermally by imposing different temperatures along its rigid lower boundary and upper free surface. Motion is valid in the limit of fast rotation and small friction, which makes the *Ekman number E* small. The solution accurate to $O(E^{\frac{1}{2}})$: $\overline{\boldsymbol{v}} = \overline{\boldsymbol{v}}^{(0)} + E^{\frac{1}{2}}\overline{\boldsymbol{v}}^{(1)}$, is characterized by a strong zonal jet, with meridional motion confined to shallow boundary layers along the bottom and top. (a) Zonal motion. (b) Meridional motion. (c) Streamfield in a meridional cross section. (d) Meridional profile of imposed temperature. Adapted from Charney (1973) with permission of Kluwer Academic Publishers.

Hence, in the interior, poleward motion vanishes, whereas $\overline{w}^{(2)}$ is independent of height, equal to $\overline{w}^{(2)}(H)$. Air expelled from the upper boundary layer drives a gentle $O(E)$ subsidence in the interior. Subsiding air must be absorbed in another boundary layer at the bottom surface.

The lower boundary is a rigid surface. Consequently, motion there has the form of a conventional Ekman layer. Inside the lower boundary layer, the stretching transformation

$$z = E^{\frac{1}{2}}\xi \qquad (15.37)$$

magnifies sharp changes. It makes viscous terms comparable to others in the momentum balance there, preserving the order of the governing equations. Then, at $O(E^{\frac{1}{2}})$, the momentum equations reduce to the Ekman balance

$$-f\overline{v}_a^{(1)} = K\frac{\partial^2 \overline{u}_a^{(1)}}{\partial \xi^2} \tag{15.38.1}$$

$$f\overline{u}_a^{(1)} = K\frac{\partial^2 \overline{v}_a^{(1)}}{\partial \xi^2}, \tag{15.38.2}$$

with the geostrophic contribution again automatically satisfied by $\overline{u}_g^{(1)}$. Requiring the total solution accurate to $O(E^{\frac{1}{2}})$ to satisfy (15.21.6) and reduce to the inviscid behavior outside the boundary layer yields the boundary conditions

$$\overline{u}_g^{(1)} + \overline{u}_a^{(1)} = 0 \qquad \overline{v}_a^{(1)} = 0 \qquad \xi = 0 \tag{15.38.3}$$

$$\overline{u}_a^{(1)} \sim 0 \qquad \overline{v}_a^{(1)} \sim 0 \qquad \xi \to \infty. \tag{15.38.4}$$

The boundary value problem (15.38) has solution

$$\overline{u}_a^{(1)}(\xi) = -\overline{u}_g^{(1)} e^{-\gamma\xi} \cos\gamma\xi \tag{15.39.1}$$

$$\overline{v}_a^{(1)}(\xi) = \overline{u}_g^{(1)} e^{-\gamma\xi} \sin\gamma\xi, \tag{15.39.2}$$

which is analogous to the Ekman structure in Fig. 13.3 (Fig. 15.4a,b).

Because $\overline{v}$ vanishes in the interior, poleward mass transfer inside the upper boundary layer must be exactly compensated by equatorward mass transfer inside the lower boundary layer. This requires $\overline{v}_a^{(1)}$ to satisfy the condition

$$E\int_0^\infty \overline{v}_a^{(1)} d\xi = -E\int_0^\infty \overline{v}_a^{(1)} d\zeta$$

$$= E\frac{KR}{aHf^2}\frac{\partial T_0}{\partial \phi}. \tag{15.40.1}$$

Similarly, thermal wind (15.24) implies

$$\overline{u}_g^{(1)} = \frac{R}{aH\gamma f}\frac{\partial T_0}{\partial \phi} \tag{15.40.2}$$

for the $O(E^{\frac{1}{2}})$ geostrophic motion. Equations (15.40) complete the solution to this order.

Figure 15.4 illustrates the motion for the profile $T_0(\phi) = T_{00} - \delta T\sin^4\phi$ (Fig. 15.4d). This temperature structure is symbolic of the distribution of radiative heating (Fig. 1.34c). At low latitude, it is sufficiently flat for the solution to remain valid up to the equator. The motion characterizes a zonally symmetric Hadley cell, one that extends from the equator to the pole. Air heated near the equator rises, entering the Ekman layer at the upper boundary. It then spirals poleward in a strong westerly jet, which maximizes at $z = H$. At extratropical latitudes, air is expelled from the upper Ekman layer, sinking through the interior. It eventually enters the Ekman layer at the lower boundary, wherein it returns equatorward. In the interior, forces are in geostrophic balance. Horizontal motion there is zonal and parallel to isotherms. Meridional heat flux therefore vanishes. Poleward heat transfer is thus confined to the shallow Ekman layers at the upper and lower boundaries.

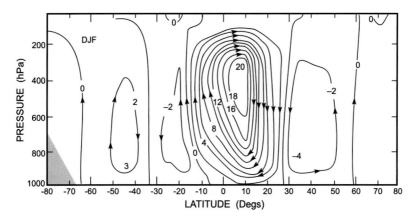

Figure 15.5 Zonal-mean meridional streamfunction observed during December–February. Meridional overturning equatorward of 30° comprises the Hadley circulation, which is strongest in the winter hemisphere. Poleward of downwelling is meridional overturning of opposite sense, Ferrell cells that prevail at mid-latitude. Adapted from Peixoto and Oort (1992).

The Ekman layers have thickness that is $O(E^{\frac{1}{2}})$. Meridional motion, $\bar{v}$, inside the Ekman layers is likewise $O(E^{\frac{1}{2}})$. Net poleward heat flux is thus $O(E)$. As Ω increases, $E \to 0$. So then does poleward heat transfer by the zonally symmetric circulation, which becomes increasingly inhibited or choked by rotation.

Plotted in Fig. 15.5, in zonal-mean motion, is the observed meridional circulation of the atmosphere. It qualitatively resembles the idealized Hadley cell – but only in the tropics. Rising motion appears near the equator in the summer hemisphere. It coincides with the ITCZ and organized latent heat release (Fig. 9.41). That rising motion is compensated in the subtropics of each hemisphere by sinking motion, chiefly in the winter hemisphere. Strong meridional motion appears near the tropopause and the Earth's surface, but only equatorward of 30°. Like subsidence, that ageostrophic motion is found chiefly in the winter hemisphere. It is there that the meridional temperature gradient and vertical wind shear are steepest (15.40); cf. Figs. 6.3, 6.4.

The limited extent of the Hadley cell can be understood from conservation of angular momentum, as discussed at the beginning of this chapter. Air moving poleward in the upper troposphere is deflected by the Coriolis force. To conserve angular momentum, it accelerates into strong westerlies (see Fig. 15.11a). The westerly acceleration, however, is observed only out to the poleward flank of the Hadley cell. The latter coincides with the subtropical jet, where westerlies maximize (cf. Fig. 1.8). At higher latitude, the thermally direct Hadley cell is replaced by a thermally indirect *Ferrell cell* (cf. Fig. 1.35). There, air sinks at low latitude and rises at high latitude.[4] In contrast to the Hadley cell, which is driven thermally by heating at low latitude, the Ferrell cell is driven mechanically by zonally asymmetric disturbances. They develop in strong

[4] This thermally indirect motion is actually an artifact of the Eulerian mean along latitude circles. In other representations, it disappears (see, e.g., Holton, 2004). For this reason, the Ferrell cell is not the principal mechanism of meridional heat transfer at extratropical latitudes.

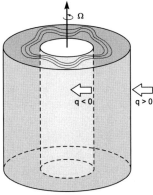

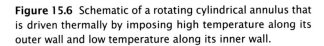

Figure 15.6 Schematic of a rotating cylindrical annulus that is driven thermally by imposing high temperature along its outer wall and low temperature along its inner wall.

westerlies that flank the Hadley cell. Associated with synoptic weather systems, those transient disturbances dominate the extratropical circulation, driving it out of zonal symmetry (Fig. 1.9).

15.3 HEAT TRANSFER IN A LABORATORY ANALOGUE

Insight into why the observed meridional circulation differs from the zonally symmetric model comes from laboratory simulations. A rotating cylindrical annulus (Fig. 15.6) is driven thermally by imposing high temperature along its outer wall, which represents the equator, and low temperature along its inner wall, which represents the pole. Figure 15.7 shows motion at the upper surface that is observed in the rotating frame, as a function of increasing rotation rate Ω. At slow rotation, the motion is zonally symmetric. It has the form of a thermally-direct Hadley cell, but with motion deflected into the zonal direction. Fluid along the upper surface gradually spirals inward, converging at the inner wall, where it sinks. As Ω increases, fluid moving inward experiences an intensified Coriolis force. Motion is therefore deflected increasingly into the zonal direction, parallel to isotherms. It forms a strong zonally symmetric jet in the interior, where forces are in geostrophic balance. Radial heat transfer then becomes confined to shallow Ekman layers that form along the upper and lower boundaries. With increasing Ω, that heat transfer decreases like E.

Beyond a critical rotation rate, zonally asymmetric disturbances appear. They develop from a steep temperature gradient that forms in the interior when radial heat transfer is choked. The zonally symmetric stratification is then strongly baroclinic. It therefore possesses available potential energy. By thermal wind balance (12.11), the radial temperature gradient is accompanied by strong vertical shear. Together, these features render the zonally-symmetric circulation *baroclinically unstable* (Chap. 16). Small disturbances to the flow then amplify by converting available potential energy of the zonal-mean state to eddy kinetic energy. Baroclinic disturbances accomplish this by transferring heat radially in *sloping convection*. Warm inward-moving fluid overrides heavier fluid and ascends. Simultaneously, cool outward-moving fluid undercuts lighter fluid and descends (cf. Fig. 15.3). The result is net heat transfer inward.

Behavior similar to that in Fig. 15.7 occurs at constant Ω if the imposed temperature contrast exceeds a critical value. It too renders the zonally symmetric circulation baroclinically unstable. Asymmetric heat transfer by baroclinic disturbances occupies much of the interior. Consequently, it is far more efficient than the $O(E)$ heat transfer

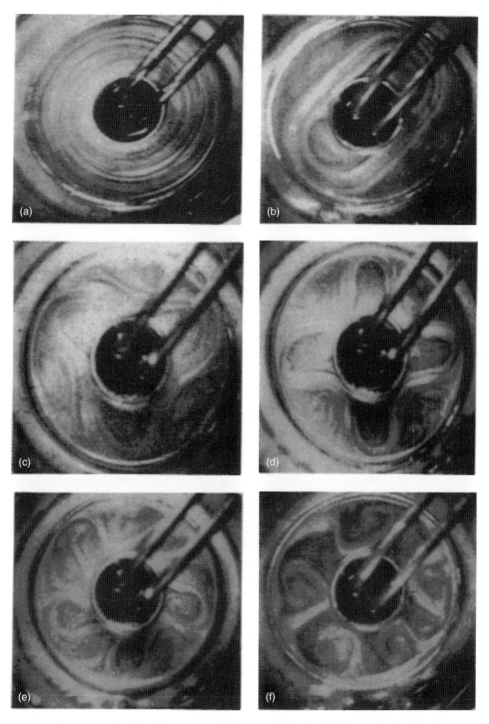

Figure 15.7 Surface flow pattern in the rotating cylindrical annulus for fixed temperature contrast and as a function of increasing rotation rate. Adapted from Hide (1966).

of the zonally symmetric circulation. Owing to rotation, the latter is confined inside shallow Ekman layers along the upper and lower surfaces. Sloping convection, although more complex, achieves the same result as the zonally symmetric circulation: it lowers the overall center of gravity, releasing available potential energy.

The dominant baroclinic disturbance has zonal wavelength that decreases with increasing rotation rate (Fig. 15.7b–e). At slow Ω, the circulation contains only wavenumbers 0 and 1; it is almost zonally symmetric. At intermediate Ω, wavenumbers 2 and 3 appear. Faster rotation leads to wavenumbers 5 and greater. In certain ranges of Ω, the annulus circulation resembles observed structure in the troposphere. The pentagonal structure in Fig. 15.7e is similar to patterns that are observed in the Southern Hemisphere (Fig. 2.10). There, the storm track is continuous, almost zonally symmetric (cf. Fig. 15.10b). Synoptic weather systems inside it often appear uniformly spaced in longitude, assuming baroclinic modal structure of the cylindrical annulus. At sufficiently fast rotation, the dominant wavelength in the annulus becomes small enough to make the disturbance itself unstable.[5] Undulations then wrap up, forming closed vortices. Attaining finite amplitude, those eddies transfer heat efficiently by mixing fluid across the annulus. This process acts to homogenize the horizontal distribution of temperature. It thus acts to restore barotropic stratification, eliminating strong vertical shear while exhausting available potential energy.

In the troposphere, the nonuniform distribution of radiative heating and cooling (Figs. 1.34c; 9.41b) continually drives thermal structure into baroclinic stratification. It therefore continually generates available potential energy, which fuels baroclinic disturbances. Baroclinicity is strongest at mid-latitudes, between the tropics, which experience radiative heating, and polar latitudes, which experience radiative cooling. Associated with steep meridional temperature gradient and steep vertical shear (Figs. 6.3, 6.4), strong baroclinicity coincides with the subtropical jet (Fig. 1.8). Synoptic disturbances along the poleward flank of the jet amplify through baroclinic instability. By transferring heat poleward in sloping convection, they limit the meridional temperature gradient, vertical shear, and hence the strength of the jet. In the stratosphere, the horizontal distribution of radiative heating and cooling is comparatively uniform. Baroclinicity therefore remains weak. Supported by strong static stability (15.19.2), this feature of the stratosphere limits available potential energy.

By transferring heat poleward, baroclinic disturbances and the Hadley circulation make the general circulation of the troposphere behave as a heat engine (Fig. 6.7). Air is heated at high temperature while near the equator and surface. After being displaced upward and poleward, it is cooled at low temperature. The Second Law then implies that, over a thermodynamic cycle (e.g., during a circuit between the tropics and extratropics), an individual air parcel performs net work. Kinetic energy produced in this fashion maintains the circulation against frictional dissipation. In the troposphere, about half of the dissipation of organized kinetic energy takes place inside the planetary boundary layer, where turbulent mixing is strong (Sec. 13.1.2). Kinetic energy is then cascaded efficiently to small scales, where it is absorbed by molecular

---

[5] Vertical shear of the disturbance reinforces that of the zonal-mean flow. This makes shear strong enough locally to render motion there baroclinically unstable. A parallel exists in the Northern Hemisphere. Planetary waves reinforce zonal-mean shear in the North Pacific and North Atlantic. Intensified shear there forms localized storm tracks, where conditions are favorable for the amplification of baroclinic weather systems (Sec. 15.4.2).

diffusion. The remainder takes place in the free atmosphere, through turbulence that is generated by dynamical instability and convection and, indirectly, through radiative damping of temperature anomalies (Secs. 14.7, 8.6).

15.4 QUASI-PERMANENT FEATURES

The observed circulation varies with horizontal position and time. In the troposphere, much of the variability is concentrated in synoptic weather systems, which dominate the extratropical circulation (Fig. 1.9). Averaging over a season removes those disturbances. The seasonal-mean circulation is smoother. Nonetheless, it too varies with horizontal position and time of year. The respective features characterize local climate and how it varies from one region to another and between seasons. Those variations are forced by the annual drift of solar insolation (Fig. 1.33). The latter modifies the TOA energy budget and hence local values of net radiation (Fig. 1.34c). Local conditions are also influenced by topography of the Earth's surface: elevated terrain and its interaction with the circulation. In addition, they are influenced by temperature of the Earth's surface, which dictates local heat transfer to the atmosphere.

15.4.1 Thermal properties of the Earth's surface

Surface temperature is shaped by optical and thermal properties, which determine the fate of energy that is supplied to or removed from the Earth's surface through radiation. The most important are the contrasting thermal properties of continent and ocean:

Ocean

- Water has a large specific heat capacity (Appendix B). A specified absorption of heat therefore results in a comparatively small increase of temperature (2.16). A specified removal of heat results in a comparatively small decrease of temperature.
- Because it is a fluid system, ocean is mobile. Heat that is supplied to or removed from the surface introduces anomalous energy which can therefore be redistributed downward efficiently through convection. Anomalous energy is thereby diluted over a deep volume.
- Radiative heating is offset by evaporative cooling, which transfers latent heat to the atmosphere.

These features combine to moderate the seasonal variation of ocean temperature. During summer, ocean becomes only modestly warmer than its annual-mean temperature. During winter, it becomes only modestly cooler.

Continent

- The specific heat capacity of land is comparatively small, less than 20% of that of water. The same absorption or removal of heat therefore results in a comparatively-large increase or decrease of temperature.
- Because it is solid, land is immobile. Heat that is supplied to or removed from the surface introduces anomalous energy which can therefore be redistributed downward

only through conduction. The latter is orders of magnitude slower than convection. Anomalous energy therefore remains concentrated in a shallow layer beneath the surface.

• Without evaporative cooling, radiative heating is offset only by sensible heat transfer to the atmosphere, which is comparatively small (Fig. 1.32).

These features combine to produce a large seasonal variation of continental temperature. During summer, continent becomes sharply warmer than its annual-mean temperature. During winter, it becomes sharply colder than its annual mean temperature – especially in the Northern Hemisphere, where landmasses extend into polar regions. How much warmer and colder it becomes is governed by the annual swing of solar insolation, which increases with latitude (Fig. 1.33).

These differences in the annual variation of temperature introduce large differences of temperature between continent and ocean during an individual season. During summer, continent becomes warmer than neighboring ocean. Overlying air then experiences anomalous heating. This makes it anomalously warm, while forcing upward motion across isentropic surfaces. By the hypsometric relation (6.12), the vertical spacing of isobaric surfaces in the lower troposphere is expanded. Isobaric surfaces near the ground are therefore displaced downward. This introduces low pressure over continent. Frictional convergence into low surface pressure reinforces upward motion (Sec. 12.3). During the same season, ocean is anomalously cool. It exerts the opposite influence. Overlying air then experiences anomalous cooling. This makes it anomalously cold, while forcing downward motion across isentropic surfaces. By the hypsometric relation, the vertical spacing of isobaric surfaces in the lower troposphere is pinched. Isobaric surfaces near the ocean surface are therefore displaced upward. This introduces high pressure over ocean. Frictional divergence from high surface pressure reinforces downward motion.

During winter, the horizontal pattern is reversed. Continent then becomes anomalously cold, favoring high surface pressure and downward motion. Conversely, ocean becomes anomalously warm, favoring low surface pressure and upward motion.

Plotted in Fig. 15.8 is the annual variation of surface temperature, displayed in the difference between the solstices: January–July. The annual variation is dominated by large changes at high latitudes of the Northern Hemisphere, where landmasses are prevalent. In the Siberian Arctic, wintertime temperature is as much as 60 K colder than summertime temperature. It is almost that much colder than the neighboring Pacific, where wintertime temperature is depressed comparatively little below its annual mean. Wintertime temperature is also sharply depressed in the Canadian Arctic. Like Siberia, it is therefore sharply colder than neighboring ocean. At lower latitudes, the annual variation of temperature becomes weak. This latitudinal dependence mirrors that in the annual swing of insolation (Fig. 1.33). Although weaker, distinct maxima appear in desert regions: the Sahara and Kalahari in Africa, the Atacama in South America, and central Australia. Situated in the subtropics, those regions experience strong warming during summer, when they remain relatively cloudless. In the Southern Hemisphere, the annual variation of temperature is conspicuously smaller than in the Northern Hemisphere. This distinction reflects the distribution of continent and ocean. Except for Antarctica, landmasses in the Southern Hemisphere are confined to lower latitude. Unlike Asia and North America, they are surrounded by ocean, which moderates temperature extremes.

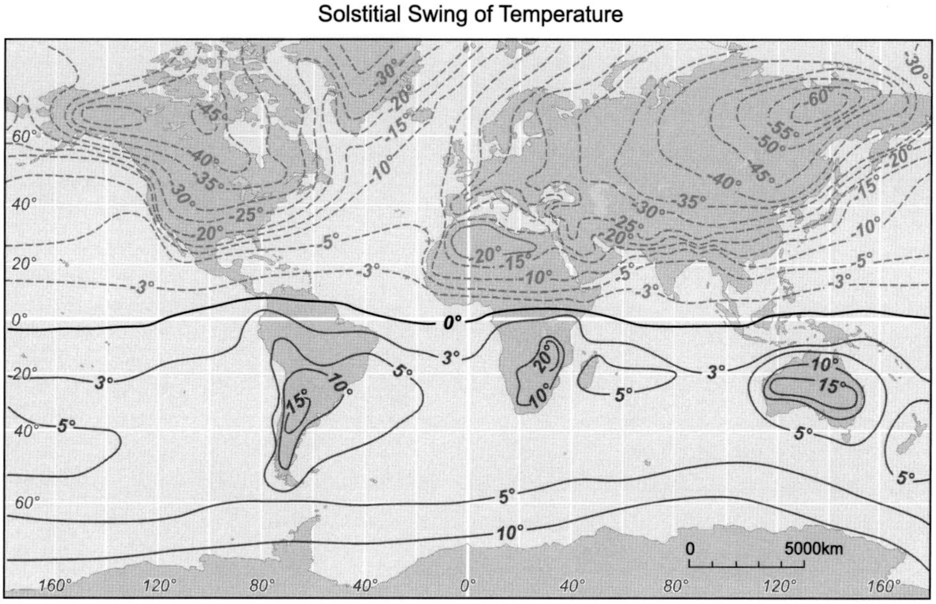

Solstitial Swing of Temperature

Figure 15.8 Annual swing of surface temperature: January–July. See color plate section: Plate 12.

15.4.2 Surface pressure and wind systems

Large anomalies of surface temperature introduce localized atmospheric heating and cooling, especially during the solstices. Augmenting them are anomalies of surface terrain, orographic features that displace the air stream vertically. Both force planetary waves. The latter are manifested at the surface by zonal asymmetries in the circulation.

Plotted in Fig. 15.9a is the distribution of SLP and surface motion during January. The Northern Hemisphere is punctuated by strong anomalies. They reflect anomalous surface temperature. Dominating Asia is the *Siberian High*. It coincides with anomalously cold Arctic surface, which cools overlying air. Over North America is the *Canadian High*, likewise coincident with cold Arctic surface. About both features is anticyclonic motion. It is accompanied by frictional divergence. The latter must be compensated aloft by subsidence. Contrasting with those continental regions is neighboring ocean. Comparatively warm, it is dominated by low-pressure anomalies: the *Icelandic Low* over the North Atlantic and the *Aleutian Low* over the North Pacific. Cyclonic motion about centers of low pressure is accompanied by frictional convergence. Over the subtropics are secondary centers of high pressure: the *Bermuda-Azores High* and the *Hawaiian High*. Those features are weaker than anomalies at higher latitude. They comprise the *subtropical high*, which forms beneath descending motion of the Hadley circulation. Divergent motion from those high-pressure centers, augmented by that from continental high pressure, fuels the north-easterly *trade winds* that prevail at low latitudes of the Northern Hemisphere. The trades converge into low pressure along the equator, which coincides with the ITCZ and nearly windless conditions (the *doldrums*). Positioned just south of the equator, the ITCZ lies beneath ascending motion of the Hadley circulation. Accompanying it is deep convection, with heavy

(a) January   **Surface Pressure and Motion**

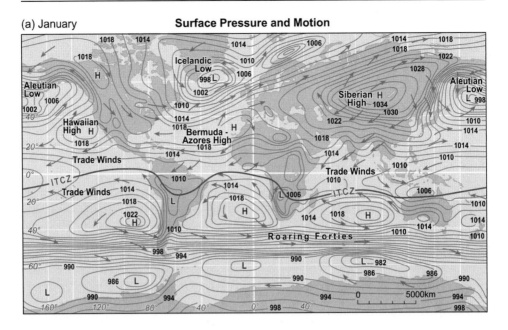

(b) July

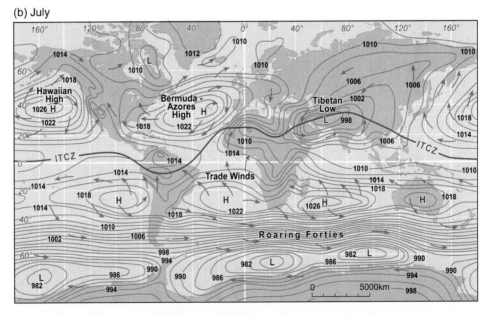

Figure 15.9 Climatological-mean SLP and surface motion for (a) January and (b) July. See color plate section: Plate 13.

precipitation (Figs. 1.30b; 9.41). Centers of low pressure form in the tropical Southern Hemisphere (in summer) over heated continent: equatorial Africa, South America, and northern Australia. In coastal regions, such as northern Australia, such structure defines the *monsoon low*; see Fig. 15.11c. These centers of convection contribute substantially to ascending motion of the Hadley circulation. Farther south are intensified centers of high pressure. Situated in the summer hemisphere, they form over comparatively cool ocean. Like counterpart features in the winter hemisphere, those centers comprise the subtropical high. Frictional divergence from them fuels the southeasterly trade winds in the Southern Hemisphere. Mirroring northeasterly trade winds opposite the equator, they converge into the ITCZ.

Plotted in Fig. 15.9b is the pattern for July. The distribution of highs and lows is now reversed. In the Northern Hemisphere (in summer), lows form over heated continent. Conversely, highs form over comparatively cool ocean. Asia is dominated by the *Tibetan Low*. Accompanied by frictional convergence, it is associated with the Asian monsoon. Counterparts are visible over heated regions of Africa and North America, albeit weaker. Over comparatively cool ocean are the *Bermuda-Azores High* and the *Hawaiian High*. They are sharply amplified, now prevailing across the North Atlantic and North Pacific. As during January, frictional divergence from those high-pressure centers fuels the north-easterly trade winds in the Northern Hemisphere, which converge into the ITCZ. During July, the ITCZ is displaced well north of the equator, as far as the Tibetan Low and the Asian monsoon. In the Southern Hemisphere (in winter), the subtropical high is almost zonally uniform. Because the Earth's surface in the Southern Hemisphere is chiefly maritime, it develops only weak temperature contrast (Fig. 15.8). This leads to centers of high pressure in the subtropics forming over ocean as well as continent. Frictional divergence from those centers fuels southeasterly trade winds of the Southern Hemisphere. After crossing the equator, they eventually converge into the ITCZ. Also exhibiting strong zonal symmetry is low pressure at polar latitude. Situated over the Southern Ocean, it circumscribes the much colder Antarctic.

Notice that, at mid-latitudes of the Southern Hemisphere, isobars remain zonal throughout the year. They are accompanied by circumpolar westerlies, the *Roaring Forties*. By contrast, at mid-latitudes of the Northern Hemisphere, the zonal circulation is disrupted by strong localized anomalies – especially during winter, when they extend to high latitude. Much stronger than counterpart features in the Southern Hemisphere, those anomalies represent planetary waves. They are amplified in the Northern Hemisphere by strong anomalies of surface heating and cooling and major orographic features.[6] These distinctions introduce hemispheric asymmetry into the general circulation.

Amplified planetary waves in the Northern Hemisphere yield a broken storm track. Plotted in Fig. 15.10, as a function of horizontal position, is the mean pressure of cyclonic depressions. It measures the average intensity of synoptic weather systems. During northern winter (Fig. 15.10a), synoptic weather systems in the Northern Hemisphere are concentrated over the North Atlantic and North Pacific. Those envelopes of cyclonic activity coincide with strong westerlies, steep meridional temperature gradient, and steep vertical shear (Figs. 1.9b, 6,3, 6.4). They represent the North Atlantic and North Pacific storm tracks. Closely related are the Icelandic Low and Aleutian Low, which reinforce westerlies to their south. Contrasting structure appears in the Southern Hemisphere (in summer). There, synoptic weather systems are distributed over longitude almost uniformly. That envelope of cyclonic activity represents the continuous storm track of the Southern Hemisphere, which circumscribes the Antarctic.

---

[6] During winter, these circulation anomalies tilt westward with height, signifying upward propagation of planetary wave activity (Sec. 14.5.2). In the stratosphere, the Aleutian Low is therefore replaced by the Aleutian High (Fig. 1.10b).

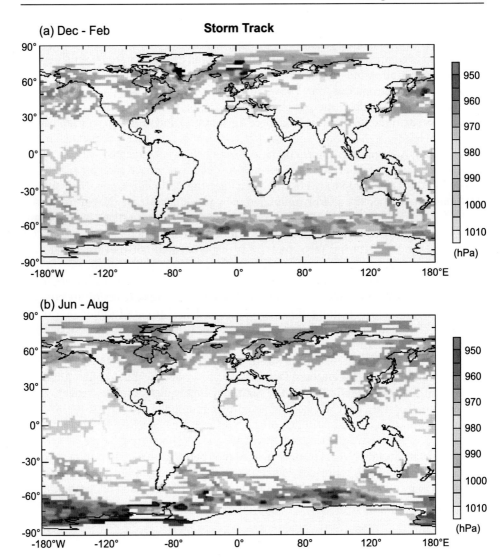

Figure 15.10 Mean intensity of cyclonic disturbances, measured by their depression of SLP, during (a) December–February 1984 and (b) June–August 1984. *Source*: http://data.giss.nasa.gov (10.07.10). See color plate section: Plate 14.

Much the same picture prevails during southern winter (Fig. 15.10b). Synoptic weather systems in the Southern Hemisphere thus develop at all longitudes. On individual days, they can assume a coherent global pattern, resembling baroclinic modal structure of the cylindrical annulus (Fig. 2.10).

15.4.3 Tropical circulations

At low latitude, the Coriolis force is weak. This feature of the momentum balance allows the tropical circulation to involve a greater contribution from the divergent component of motion and to be thermally direct (Sec. 12.1). In the tropics, the temperature distribution is relatively flat (Fig. 6.3). Implied is little available potential energy to drive large-scale motion. The primary source of energy for the tropical circulation is, instead, latent heat release inside organized convection. The latter, in turn, follows from evaporative cooling of the ocean.

Precipitation inside the ITCZ exceeds evaporation from the underlying surface (Fig. 9.41). Moisture must therefore be imported from surrounding areas. Much of it converges in the easterly and equatorward trade winds that prevail at low latitude. The trades comprise the lower branch of the zonal-mean Hadley cell in Fig. 15.5. They develop when equatorward-moving air is deflected westward by the Coriolis force. Air subsiding in the subtropics from the upper branch of the Hadley cell dries the troposphere. It also stabilizes the troposphere through mechanisms discussed in Sec. 7.4.4. By inhibiting convection, subsidence in the descending branch of the Hadley cell maintains desert conditions. It is no accident that all major deserts on Earth are found in the subtropics.

Although the zonal-mean overturning in Fig. 15.5 is intuitively appealing, the actual Hadley circulation is not zonally uniform. It varies with longitude according to surface temperature and humidity, which determine upward fluxes of sensible and latent heat. The local intensity of the Hadley circulation is reflected in upper-tropospheric divergence. Plotted in Fig. 15.11a is the streamfield at 200 hPa. Divergent motion is visible along much of the equator. However, it is concentrated in three major centers: equatorial Africa, equatorial South America, and, by far the largest, the equatorial Indian Ocean and western Pacific. The latter coincides with warm ocean surface, where SST exceeds 26°C (Fig. 5.1). Inside centers of convection, deep heating elevates isobaric surfaces in the upper troposphere (6.12). This favors anomalous high pressure above anomalous low pressure. Beneath each center of divergence at 200 hPa is convergent motion at 850 hPa (Fig. 15.11b). To conserve potential vorticity (12.36), that air motion spins up cyclonically (e.g., in the monsoon low over northern Australia). After being convected to the upper troposphere, it then diverges from the convective center, spinning up anticyclonically (Fig. 15.11a).[7] As air streams poleward at 200 hPa, it is deflected by the Coriolis force, forming strong westerlies that prevail at midlatitudes. Intensified jets appear to the east of Asia and North America, where isotachs exceed 45 m/s (shaded). Associated with steep vertical shear, they coincide with the North Atlantic and North Pacific storm tracks (cf. Fig. 15.10a).

In addition to the Hadley circulation, the tropics involve two other classes of circulation. They too are thermally direct, as well as zonally asymmetric. Monsoon circulations develop during summer over subtropical landmasses, such as India and northern Australia, which become warmer than neighboring ocean. Presented in Fig. 15.11c is instantaneous SLP and IR imagery under conditions characteristic of the Australian monsoon. Over the Northern Territory is the monsoon low. Much of that region is covered by cold high cloud. It marks divergent motion in the upper troposphere. Representing cirrus anvil from deep cumulonimbus, cold cloud spirals outward anticyclonically. It reflects diverging air that spins up anticyclonically to conserve potential vorticity. Convection in the monsoon circulation is fueled by moisture convergence at low levels. The latter is reinforced near the surface by frictional convergence into the

---

[7] Notice the reversal of curvature over South America where streamlines cross the equator.

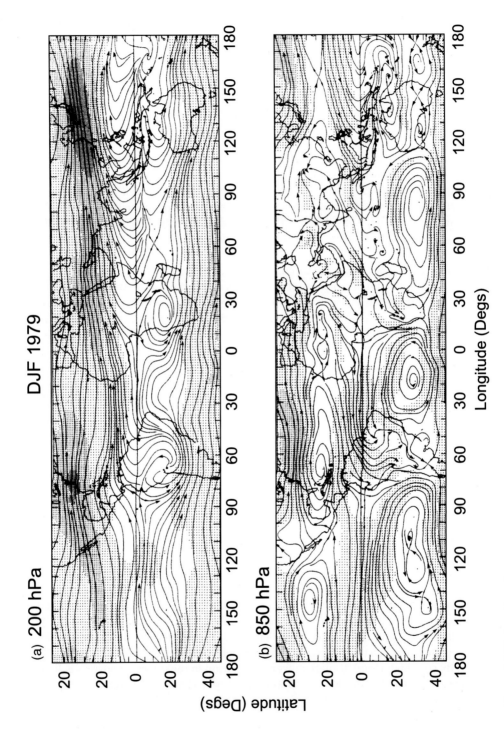

Figure 15.11 (*continued*)

496

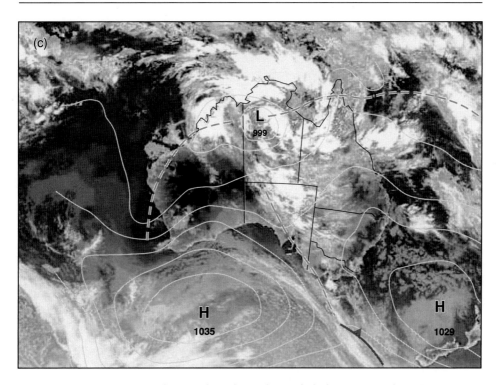

Figure 15.11 Streamlines and wind speed (stippled) during December 1978–February 1979 at (a) 200 hPa and (b) 850 hPa. *Trade winds* are manifested at 850 hPa by southwestward flow across the tropical western Pacific, Atlantic, and Indian oceans. Outflow at 200 hPa over tropical Africa, South America, and Indonesia assumes anticyclonic curvature as it diverges away from those convective centers. Note the reversal of curvature to conserve potential vorticity as streamlines cross the equator. Strong jets at 200 hPa east of North America and Asia mark the *North Atlantic* and *North Pacific storm tracks* (cf. Fig. 9.41), where the mean temperature gradient is steep and cyclone development is favored. Stipling of isotachs at $|v_h| = 15\text{-}30 \text{ m s}^{-1}$, $30\text{-}45 \text{ m s}^{-1}$, $> 45 \text{ m s}^{-1}$. Adapted from Lau (1984). (c) Instantaneous SLP and IR imagery under conditions typical of the Australian monsoon. Deep convection is organized by the monsoon low, as well as trough axes that emanate from it. Cirrus blow off at upper levels spirals outward anticyclonically. See color plate section: Plate 15.

monsoon low. The convergence of water vapor increases θ_e near the surface, rendering the lower troposphere potentially unstable (Fig. 7.12). Through mechanisms described in Sec 7.4.3, frictional convergence also leads to reduced vertical stability. These changes provide conditions favorable to deep convection, enabling air to reach the Level of Free Convection (LFC). From there, it accelerates upward through positive buoyancy, until reaching the Level of Neutral Buoyancy (LNB), which is typically between 250 and 150 hPa (Sec. 7.7).

The summer monsoon is an onshore circulation. It is analogous to a seabreeze circulation, which is likewise generated through a horizontal gradient of heating. Both may be understood from solenoidal production of horizontal vorticity, which occurs

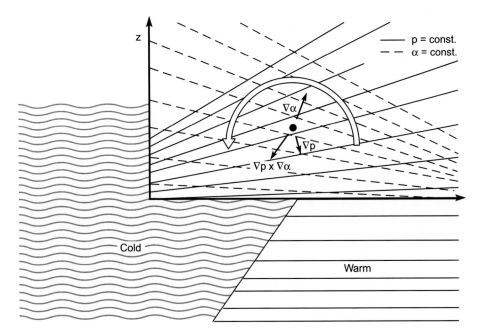

Figure 15.12 Schematic of a thermally-direct monsoon circulation that is established through solenoidal production of horizontal vorticity under baroclinic stratification. Heating over continent expands the vertical spacing of isobaric surfaces (solid), whereas it draws down isochoric surfaces (dashed) of larger α from aloft. The reverse occurs from cooling over ocean. The nonuniform distribution of heating drives isochoric surfaces out of coincidence with isobaric surfaces, producing baroclinic stratification.

under baroclinic stratification (Fig. 15.12). Nonuniform heating between continent and ocean introduces a horizontal variation of temperature. It drives isochoric surfaces ($\alpha = $ const) out of coincidence with isobaric surfaces. The budget of horizontal vorticity then includes a solenoidal production term, proportional to $\nabla p \times \nabla \alpha$. That source of horizontal vorticity is analogous to solenoidal production that appears in the budget of vertical vorticity (12.36). Derived from baroclinic stratification, it spins up horizontal vorticity as an onshore circulation, with upwelling over the region of heating and downwelling over the region of cooling.

The other class of circulation important in the tropics is the Walker cell, depicted in Fig. 1.36. The Walker cell is a zonally asymmetric overturning in the equatorial plane. It is characterized by upwelling over regions of heating and downwelling to the east and west. The Pacific Walker cell is forced by latent heating over the western Pacific, where ocean is anomalously warm. Low surface pressure in the equatorial western Pacific supports deep convection, in the same way it supports the monsoon circulation. Strong surface easterlies along the equator fuel convection with moisture that has been absorbed by air during its traversal of the western Pacific. There, SST in excess of 28°C provides an abundant source of water vapor (Fig. 5.1).

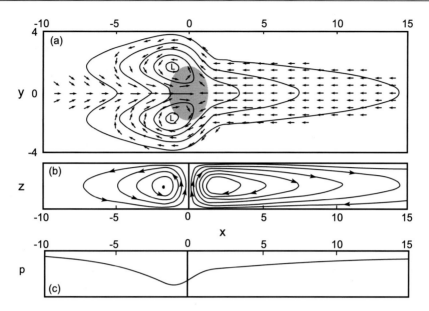

Figure 15.13 Zonally asymmetric circulation produced by a deep heating anomaly over the equator (shaded). (a) Planform view of surface motion. (b) Streamfield in a vertical cross section over the equator. (c) Surface pressure distribution. Adapted from Gill (1980).

Latent heating that drives zonal overturning in the Walker cell simultaneously drives meridional overturning. Associated with the Hadley circulation, it too is zonally asymmetric. Figure 15.13 plots the circulation produced by a localized and steady heating anomaly over the equator (shaded). At the surface, equatorial easterlies converge into the region of heating (Fig. 15.13a). Resembling the Pacific Walker cell, they are mirrored in the upper troposphere by equatorial westerlies (Fig. 15.13b). Surface low pressure forms just to the west of the heating anomaly (Fig. 15.13c). Ascending motion in the region of heating is flanked in the subtropics of each hemisphere by cyclonic gyres. Those features reinforce inflow along the equator to the west of the heating. They represent a Rossby wave mode that is trapped about the equator (see, e.g., Gill, 1982). Analogous structure is evident in the 200-hPa streamfield over the western Pacific, where deep convection is concentrated (Fig. 15.11a). Likewise for zonal flow to the east that comprises the Pacific Walker cell.

15.5 FLUCTUATIONS OF THE CIRCULATION

Beyond its annual variation, the general circulation also changes between years, as well as during individual seasons. Those changes comprise climate variability. They dominate the variation of regional properties (Fig. 1.38). Yet, because they are spatially incoherent, interannual changes need not contribute to the variation of global-mean properties. It is for this reason that the annual record of regional climate is generally incoherent with that of global-mean climate. Some components of climate

Anomalous SST

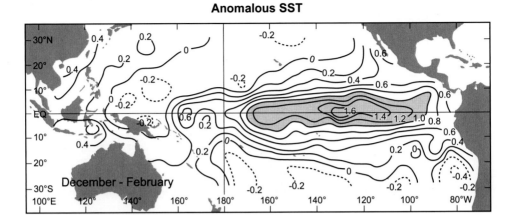

Figure 15.14 Anomalous SST (°K) during El Nino. After Rasmusson and Carpenter (1982).

variability, though mostly random, operate on a recurring basis and do possess spatial coherence.

15.5.1 Interannual changes

El Nino

The Pacific Walker cell fluctuates on interannual time scales. The respective changes characterize El Nino Southern Oscillation (*ENSO*). *El Nino* is a perturbation to the tropical circulation that occurs with a frequency of 3–5 years. It involves anomalously warm SST in the eastern and central Pacific (Fig. 15.14).[8] The equatorial Pacific is especially sensitive to temperature perturbations because, there, SST exceeds 26°C. Owing to exponential dependence in the Claussius-Clapeyron equation (4.39), small changes of SST produce large changes in evaporation and latent heat transfer to the atmosphere (Fig. 9.41). When the temperature anomaly in Fig. 15.14 is introduced, the warmest SST and convection, normally found in the western Pacific (Fig. 9.40a), shift eastward into the central Pacific.

El Nino involves a seesaw in surface pressure between the western and eastern Pacific: a longitudinal dipole. Known as the *Southern Oscillation*, this dipole shifts low pressure from the western Pacific and Indian ocean to the central and eastern Pacific. It is evidenced by the anti-correlation of pressure fluctuations in those regions (Fig. 15.15). During El Nino, SLP is anomalously low over the central and eastern Pacific, whereas it is anomalously high over the Indian ocean and western Pacific. The opposite phase of the Southern Oscillation is *La Nina* ("the girl"). SLP is then anomalously low over the Indian ocean and western Pacific, whereas it is anomalously high over the central and eastern Pacific.

ENSO upsets the pattern of tropical convection. Through anomalous latent heating, it perturbs the tropical circulation. In the Walker cell, upwelling is displaced from the

---

[8] The name El Nino ("the boy") refers to warm water off the coast of Peru that develops around Christmas.

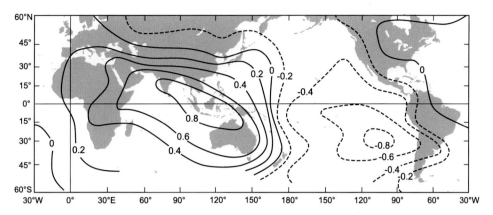

Figure 15.15 Correlation of monthly mean surface pressure in the Pacific with that at Jakarta. Oppositely phased variations in the eastern and western Pacific characterize the *Southern Oscillation*. Adapted from Berlage (1957).

equatorial western Pacific eastward into the central Pacific, along with downwelling to its east and west (Fig. 15.16); cf. Fig. 1.36. Subsidence is then positioned over the maritime continent and northern Australia. There, it inhibits monsoon activity during El Nino years. Those conditions favor drought and recurrent bush fire (McBride and Nichols, 1983); see Cover, discussion in Sec. 9.5.1, and color plate section.

ENSO is also felt at extratropical latitudes. Anomalous latent heating excites anomalous planetary wave activity that radiates poleward. Figure 14.20a plots anomalous

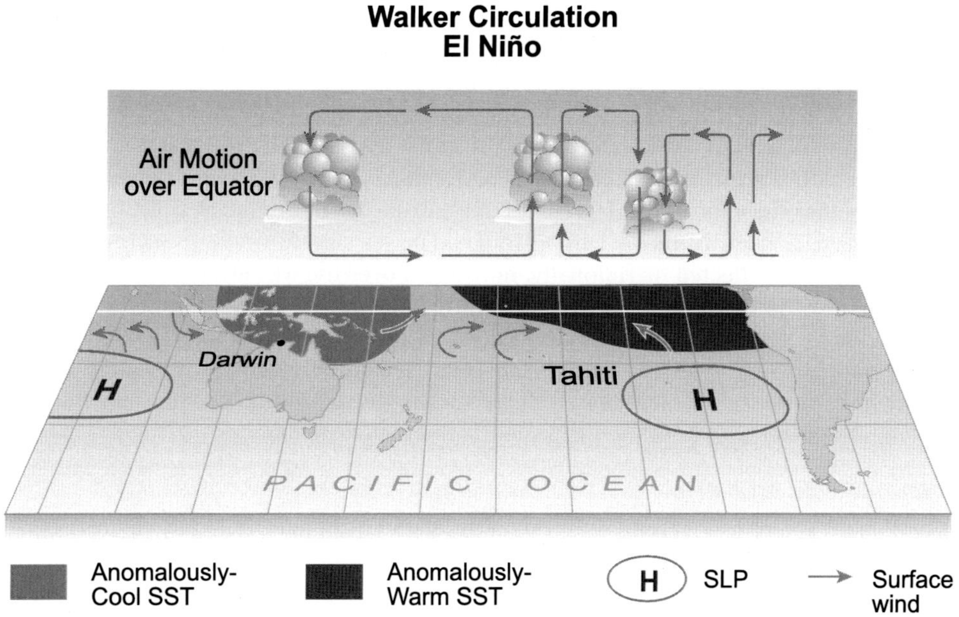

Figure 15.16 Disturbed Walker circulation during El Nino. Adapted from http://www.bom.gov.au (11.07.10). See color plate section: Plate 16.

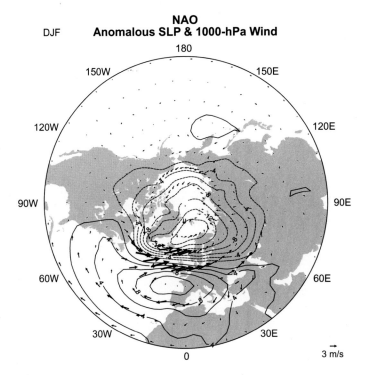

NAO
DJF   **Anomalous SLP & 1000-hPa Wind**

Figure 15.17 Anomalous SLP and 1000-hPa motion (scaled by 0.5) associated with the positive phase of the North Atlantic Oscillation (NAO). After Hurrel et al. (2003).

upper-tropospheric height for northern winters during El Nino, when convection is positioned in the central Pacific. A planetary wavetrain radiates poleward from anomalous convection. It is eventually refracted equatorward along a great circle route. Known as the *Pacific North American (PNA) pattern*, this perturbation to the height field modifies the extratropical circulation. It introduces a prominent ridge over western Canada, flanked to its east by a trough. The jet stream (wavy trajectory in Fig. 14.20a) is then deflected meridionally: northward over the western United States and Canada and southward over the eastern United States. The jet stream divides tropical and polar air (Fig. 6.3). Those deflections therefore introduce anomalously warm temperature over the western United States and Canada, but anomalously-cold temperature over the eastern United States. By perturbing the storm tracks, they also modify the development and movement of synoptic weather systems.

North Atlantic oscillation

Regional climate over Europe and Asia is moderated by maritime air, which is conveyed from the Atlantic by westerlies. This process is influenced importantly by the Icelandic Low and its equatorward complement, the Bermuda-Azores High (Fig. 15.9). The gradient between those features intensifies westerlies. Intensified with them is the advection of maritime air that moderates conditions over Europe and Asia.

The Icelandic Low and Bermuda-Azores High fluctuate between years, coherently but out of phase. These fluctuations form a meridional dipole in SLP. Known as the *North Atlantic Oscillation* (*NAO*), the dipole is plotted in Fig. 15.17. It is measured by the difference of SLP between the aforementioned sites. In its positive phase, the NAO reinforces the Icelandic Low and, simultaneously, the Bermuda-Azores High. Intensified westerlies between those features intensify the advection of maritime air, moderating conditions over northern Europe during winter and summer. In its negative phase, the NAO suppresses the Icelandic Low and the Bermuda-Azores High. This weakens westerlies across northern Europe, along with the advection of maritime air. In its place is colder air that advances southward from the North Sea and Arctic. The storm track is then displaced southward, toward southern Europe and the Mediterranean. So too are rain-bearing weather systems. Under these conditions, northern Europe and Asia experience winters that are anomalously cold and summers that are anomalously warm and dry.

The NAO is associated with much of the interannual variability over Europe and the northern Atlantic. It also influences the eastern seaboard of North America. There, anomalous southeasterlies advect maritime air. They also inhibit the outbreak and southward advance of cold air from the Canadian Arctic. Both changes moderate wintertime conditions.

Plotted in Fig. 15.18 is the anomalous wintertime temperature associated with positive phase of the NAO. Temperature is anomalously warm over northern Europe and Asia. The warm anomaly coincides with the region that experiences anomalous advection of warmer maritime air from the Atlantic (Fig. 15.17). Temperature is also anomalously warm over eastern North America. That warm anomaly bears a similar relationship to anomalous advection. Temperature is anomalously cold over north Africa and Labrador. Those regions experience anomalous advection of colder continental and Arctic air.

As for El Nino, zonal structure in the NAO involves changes of the planetary wave field. In its negative phase, the NAO weakens the Icelandic Low. Those conditions favor blocking highs over the North Atlantic, which impede the eastward migration of synoptic weather systems (Lejenas and Okland, 1983; Stein, 2008). They are associated with a modulation of planetary waves (Lejenas and Madden, 1992; Shabbar et al., 2001). The NAO also has implications for the tropical circulation. Divergent motion from the Bermuda-Azores High fuels southeasterly trade winds. The NAO should therefore interact with the Hadley circulation, which is coupled to the trades. Implied likewise is an influence on tropical cyclones. Some evidence suggests that tropical depressions track preferentially into the Gulf of Mexico during one phase of the NAO, but toward the eastern seaboard of the United States during the opposite phase. Although influences of the NAO on regional climate over Europe are well established, the mechanisms behind this fluctuation of the circulation remain poorly understood.

Arctic oscillation

Although defined from anomalies in the surface pattern, the NAO involves fluctuations with vertical structure. That structure links the NAO to a more general phenomenon, one that is closely related to planetary waves.

Fluctuations of SLP can be represented as a superposition of horizontally coherent structures that occur preferentially in the observational record. The structures

**NAO
Anomalous Temperature**

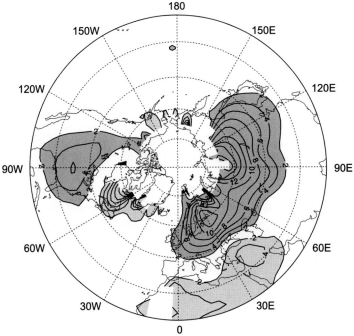

Figure 15.18 Anomalous wintertime surface temperature (scaled by 0.1 K) associated with the positive phase of the NAO. After Hurrel et al. (2003).

are called *Empirical Orthogonal Functions* (*EOF*s); Monahan et al. (2009) present an overview. Also known as *Principal Component Analysis*, the EOF description of fluctuations determines horizontal variations that operate coherently.[9] EOFs are ordered according to the fraction of overall variance that they represent. The first EOF represents the greatest variance, the second the next greatest, and so forth. Dynamical fluctuations are inherently red, with variance concentrated in the lowest wavenumbers (Sec. 14.1.4). For this reason, the leading EOFs, beyond representing the greatest variance, also represent the lowest wavenumbers and the largest spatial scales.

The *Arctic Oscillation* (*AO*) is defined as the leading EOF of SLP or its counterpart, 1000-hPa height. Plotted in Fig. 15.19b is anomalous structure associated with positive phase of the AO. It involves anomalous low pressure (height) over the Arctic, where the AO is almost zonally symmetric (wavenumber 0). Flanking it at subpolar latitudes are two centers of anomalous high pressure (wavenumbers 0 and 2): one over the North Atlantic, neighboring the Bermuda-Azores High, and another over the North Pacific, neighboring the Hawaiian High. Together with anomalous low pressure over the Arctic, which neighbors the Icelandic and Aleutian Lows, those features modulate

---

[9] By construction, EOFs are orthogonal: the product of two different EOFs vanishes when averaged over the Earth. Consequently, different EOFs are structurally independent. Nonetheless, different EOFs can represent the same phenomenon, which makes them physically interdependent.

Arctic Oscillation
Anomalous Height

(a) 50 hPa

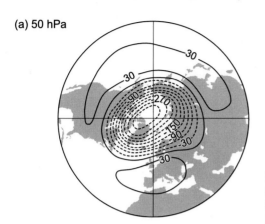

Figure 15.19 Anomalous height associated with the positive phase of the Arctic Oscillation (AO) at (a) 50 hPa and (b) 1000 hPa. Adapted from Baldwin and Dunkerton (1999).

(b) 1000 hPa

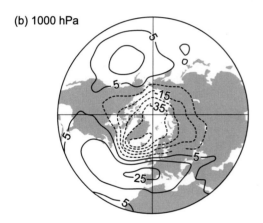

westerlies in the storm tracks over the north Atlantic and north Pacific. Accordingly, they exert influences analogous to those exerted on Europe by the NAO.

Vertical structure associated with the AO is illustrated by coherent fluctuations at higher levels. The pattern of anomalous height at 50 hPa is plotted in Fig. 15.19a. Zonally symmetric fluctuations over the Arctic are vertically coherent. They appear at all levels, nearly in phase across the troposphere and stratosphere. Those fluctuations amplify upward during winter. They have the horizontal structure of the polar-night vortex, which prevails in the wintertime stratosphere (cf. Fig. 1.10b). The polar-night vortex thus interacts with the tropospheric circulation. It is manifest in the AO as low as the Earth's surface (Fig. 15.19b). Owing to its deep vertical structure and zonal symmetry at polar latitudes, the AO is also referred to as the *Northern Annular Mode* (*NAM*).

In the zonal mean, the AO represents a meridional dipole, with anomalous height of one sign over the pole and of opposite sign at subpolar latitudes. Reflected in the high-latitude anomaly is the polar-night vortex. It develops through cold temperature that prevails over the Arctic during winter (Fig. 1.8). Plotted in Fig. 15.20 is anomalous

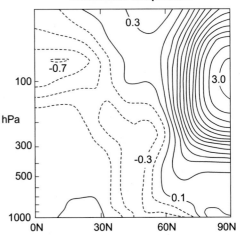

Arctic Oscillation Anomalous Temperature

Figure 15.20 Anomalous zonal-mean temperature associated with the *negative phase* of the AO (1 std deviation). Corresponds to Jan–Mar temperature over the Arctic that is anomalously warm. Adapted from Thompson and Wallace (2000).

zonal-mean temperature associated with negative phase of the AO (opposite to the phase in Fig. 15.19). Temperature is anomalously warm in the Arctic stratosphere. Through the hypsometric relation (6.12), that temperature anomaly corresponds to the height anomaly over the Arctic in Fig. 15.19 (but of reversed sign). The Arctic vortex is then anomalously warm and weak. In the opposite phase (positive AO, displayed in Fig. 15.19), the polar-night vortex is anomalously cold and strong.

The warm anomaly in Fig. 15.20 is a signature of anomalous downwelling and adiabatic warming. It penetrates well into the Arctic troposphere, as low as 500 hPa. Flanking it in the tropics is a cold anomaly. A signature of anomalous upwelling and adiabatic cooling, it maximizes just above the tropical tropopause. That temperature anomaly, in fact, has a form very similar to that of the cold tropical tropopause (Fig. 1.7). It represents an upward expansion of positive lapse rate from the troposphere, with commensurate cooling of the tropical tropopause (Sec. 18.8).

The changes over the tropics vary coherently but out of phase with the signature of anomalous downwelling over the Arctic. They imply the involvement of the Hadley circulation. Through upwelling and associated deep convection, the Hadley circulation controls the tropical tropopause (Secs. 7.7, 9.3.4). Varying in similar fashion is anomalous divergence near the tropical tropopause (Thuburn and Craig, 2000; Salby and Callaghan, 2005). Likewise for anomalous surface motion in the subtropics, which comprises the trade winds (Thompson and Wallace, 2000). Their involvement is consistent with the origin of the trades, outflow from the Bermuda-Azores High and the Hawaiian High, which are modulated by the AO (Fig. 15.9).

In the AO, fluctuations in the troposphere and stratosphere are coupled. On short time scales, amplifications of the AO appear first in the stratosphere and then descend. Plotted in Fig. 15.21 is a height-time section of the AO during an individual winter. Several amplifications appear in the upper stratosphere, descending over the course of a couple of weeks. The one in late winter, when the polar-night vortex is weak, descends into the troposphere. There, it modulates quasi-permanent features like the Icelandic Low and the Bermuda-Azores High. Such modulation perturbs the storm tracks over the north Atlantic and north Pacific, thereby influencing the development of synoptic weather systems.

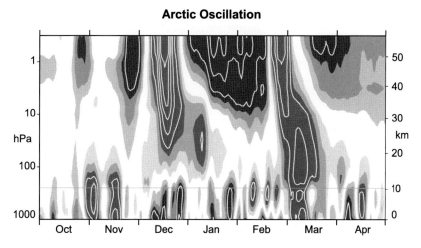

Figure 15.21 Time-height section of dimensionless AO index during 1998–1999. AO index represents the intensity of the leading EOF of geopotential height. After Baldwin and Dunkerton (2001). See color plate section: Plate 17.

The AO has varied systematically over consecutive years. Between the 1970s and 1990s, wintertime values intensified, reflecting an intensification of the polar-night vortex. The systematic change involves a trend in the Icelandic Low and Bermuda-Azores High. The accompanying trend in advection of warm maritime air (Fig. 15.17) accounts for about half of the observed warming over Eurasia (Hurrell, 1996; Thompson et al., 2000). Eventually, however, the systematic change collapsed, the AO then returning toward earlier values. The reversal echoes previous episodes in the observational record, when the AO varied systematically over even longer periods, before swinging in the opposite direction (IPCC, 2007).

The AO shares major structure with the NAO. Whether the two describe distinct phenomena is a matter of debate. Much of the uncertainty revolves about whether the anomalies over the north Atlantic and north Pacific operate coherently (Deser, 2000; Ambaum et al., 2001). The interdependence of those features hinges on the time scale of fluctuations, for example, manifest in daily versus interannual anomalies. Relevant as well is the discrimination to a single EOF. The AO is defined arbitrarily as the leading EOF of SLP. By ignoring neighboring EOFs, this definition can exaggerate spatial coherence.

It will be seen in Chap. 18 that the salient structure of the AO emerges as well from anomalous planetary wave activity that is transmitted upward to the stratosphere. The latter forces anomalous downwelling over the Arctic, which modulates the polar-night vortex. In the troposphere, anomalous planetary wave activity is associated with the fluctuation of quasi-permanent features, such as the Icelandic and Aleutian Lows. It is therefore accompanied by a perturbation of the storm tracks over the north Atlantic and north Pacific. Fluctuations of the planetary wave pattern are thus central to the AO. The mechanisms responsible for those fluctuations, like those behind the NAO, remain poorly understood.

Antarctic Oscillation
Anomalous SLP

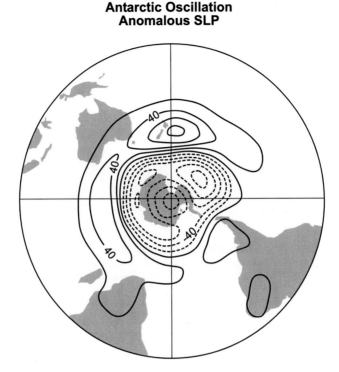

Figure 15.22 Anomalous SLP associated with of the Antarctic Oscillation (AAO) in its positive phase (arbitrary scaling). After Gong and Wang (1999).

Antarctic Oscillation

As it is a matter of definition, the AO has a counterpart over the Southern Hemisphere. The leading EOF of SLP (1000-hPa height) there defines the *Antarctic Oscillation (AAO)*. Plotted in Fig. 15.22 is anomalous structure associated with positive phase of the AAO. Like the AO, the AAO involves anomalous low pressure over the pole, flanked at subpolar latitudes by anomalous high pressure.[10] The AAO, however, is almost zonally symmetric at all latitudes. At subpolar latitudes, anomalous high pressure has crescent structure, zonally continuous except for a break between 0E and 90W. As for the AO, it shares zonal structure with the storm track (Fig. 15.10). Fluctuations of the AAO therefore perturb strong westerlies that comprise the jet stream. Associated with a displacement of the storm track, they have been related to fluctuations of rainfall over Australia (Hendon et al., 2007).

Zonally symmetric fluctuations over the Antarctic are vertically coherent. Like the AO, they operate in phase across the troposphere and stratosphere, amplifying upward during winter. That structure has the form of the Antarctic polar-night vortex. Owing to its deep vertical structure and zonal symmetry at polar latitudes, the AAO is also referred to as the *Southern Annular Mode (SAM)*.

---

[10] Owing to its high elevation, surface pressure over Antarctica is closer to 700 hPa.

The AAO has also varied systematically over consecutive years. It intensified between the 1970s and 1990s (Marshall, 2003). That trend in the AAO is concentrated in summer and autumn. Accompanying it is cooling over the Antarctic, with possible influence on Antarctic ice (IPCC, 2007). Mechanisms suggested for the trend include ozone depletion, which, through the ozone hole, favors an intensified polar-night vortex. However, the ozone hole has distinct seasonality that limits it to late winter and spring (Chap. 18). By summer, when the AAO trend appears, the ozone hole has dissolved. More significantly, the AAO trend that prevailed during closing decades of the twentieth century eventually reversed. Thereafter, the AAO returned toward earlier values. The reversal echoes previous episodes in the observational record, when a trend of the AAO persisted in one direction, before eventually swinging in the opposite direction (Jones and Widmann, 2004).

The global circulation also changes through interaction with other variations, some of which are inherent to the ocean, the stratosphere, and solar activity (Secs. 17.6, 18.8).

15.5.2 Intraseasonal variations

On time scales shorter than seasonal, variability of the circulation is dominated by synoptic weather systems, which prevail at mid-latitudes. In the tropics, another large-scale fluctuation operates. It leads to a global perturbation of convection and thermally direct circulations.

Madden-Julian Oscillation

Zonal wind over the equator fluctuates, conspicuously with periods of 1–2 months. Propagating eastward, this disturbance is named the *Madden-Julian Oscillation (MJO)*, after its discoverers (Madden and Julian, 1971). Figure 15.23a plots, as a function of period and wavenumber, the space–time spectrum of 850-hPa zonal wind over the equator. Fluctuations of zonal wind are concentrated at wavenumber 1 and eastward periods of 40–80 days. These fluctuations account for much of the intraseasonal variance in equatorial u_{850}. Plotted in Fig. 15.23b is the same information, but for OLR, which measures anomalous cloud cover and deep convection over the equator (Fig. 1.30). The space–time spectrum of convection is comparatively broad, with variance appearing in a red distribution over zonal wavenumber and eastward frequency (Sec. 14.1.4). Yet, like the spectrum of u_{850}, fluctuations of convection are concentrated at wavenumbers 1–2 and eastward periods of 40–80 days. The concentration of u_{850} and OLR variance at the same space and time scales indicates an interaction between the tropical circulation and the convective pattern.

Figure 15.24 illustrates anomalous motion and SLP at successive phases of the disturbance. Initially, the disturbance involves anomalous upwelling and low SLP over the Indian Ocean. Accompanying those perturbations is anomalous convection and latent heating. Thermally direct motion to the west and east has the form of a transient Walker circulation. Zonal structure is of planetary scale, principally wavenumber 1. These features amplify and migrate eastward into the Pacific. As they move into colder water of the eastern Pacific (Fig. 5.1), the convective anomaly collapses. Anomalous motion, however, continues eastward, eventually completing a circuit around the Earth.

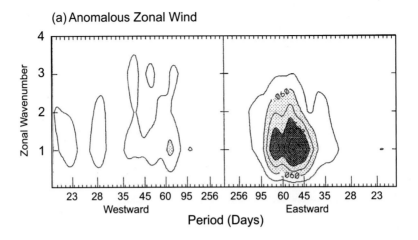

(a) Anomalous Zonal Wind

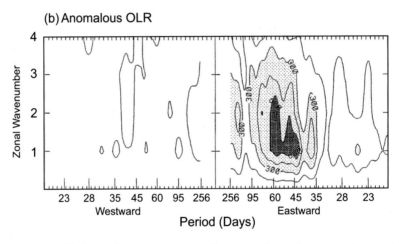

(b) Anomalous OLR

Figure 15.23 Space-time spectrum, as function of zonal wavenumber and period, of (a) anomalous equatorial zonal wind at 850 hPa and (b) anomalous OLR. After Salby and Hendon (1994).

Straddling anomalous zonal motion along the equator are gyres in the subtropics of each hemisphere (Madden, 1986; Hendon and Salby, 1994). They give the MJO horizontal structure which is similar to that produced by a localized heat source (Fig. 15.13), but which propagates eastward. In fact, moisture convergence by the MJO introduces anomalous convection and latent heating (Maloney and Hartmann, 1998). The latter operate sympathetically with the anomalous circulation: Anomalous motion leads to anomalous moisture convergence, which fuels anomalous latent heating, which in turn reinforces anomalous motion. Such feedback enables the MJO to sustain itself against frictional dissipation and amplify.

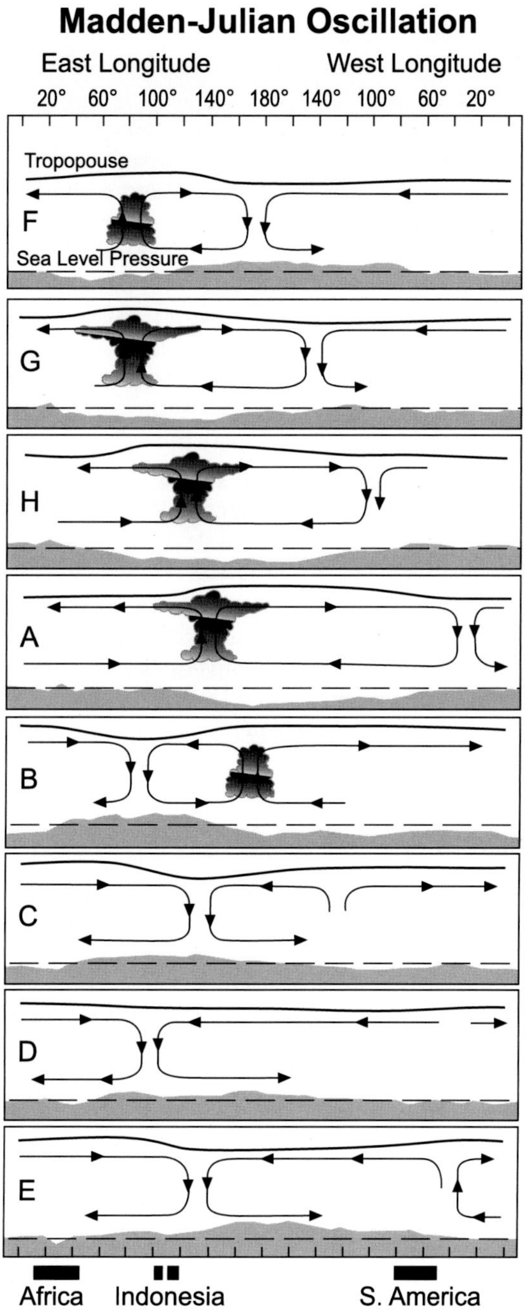

Madden-Julian Oscillation

Figure 15.24 Schematic of anomalous equatorial motion, SLP (shaded), and convection during the life cycle of the Madden-Julian Oscillation (MJO). After Madden and Julian (1972).

The positioning of moisture convergence is key to this process. It determines the phase of anomalous convection relative to anomalous temperature. If they are in phase, anomalous heating reinforces anomalously-warm temperature, amplifying the disturbance. If they are out of phase, anomalous heating suppresses anomalously-cold temperature, damping the disturbance. The distribution of divergent motion is therefore central (Sec 12.1.1). Moisture convergence is magnified near the equator. There, f is small enough for geostrophic balance to give way to frictional balance. Moisture then converges into anomalous low pressure along the equator. Flanked overhead by anomalous high pressure, the latter coincides with anomalously warm temperature (6.12). Frictional convergence into low surface pressure thus humidifies the lower troposphere at sites where temperature is anomalously warm. This feature enables the MJO to amplify through positive feedback with convection (ibid; Wang and Rui, 1990; Salby et al., 1994).

The MJO is also accompanied by coherent fluctuations of ocean temperature. Anomalous SST is of order 0.5 K in the equatorial Indian and Pacific oceans, sites where anomalous convection amplifies (Zhang, 1996; Shinoda and Hendon, 1998). Anomalous SST develops through (1) a modulation of evaporative cooling, which accompanies anomalous latent heat flux to the atmosphere, and (2) a modulation of SW heating, which is suppressed by cloud cover (Waliser, 1996; Maloney and Kiehl, 2002). These changes induce anomalously warm SST ahead of anomalous convection. It can reinforce moisture flux convergence and the organization of convection, providing additional feedback from the ocean (Wang and Xie, 1998). Like other elements of climate variability, these features form some of the challenges to large-scale simulation by GCMs.

SUGGESTED REFERENCES

An Introduction to Dynamic Meteorology (2004) by Holton treats the zonal-mean circulation and its relationship to baroclinic eddies.

Global Physical Climatology (1994) by Hartmann discusses eddy transports of heat and momentum, along with their roles in the general circulation.

Observations relevant to the general circulation are presented in *Physics of Climate* (1992) by Peixoto and Oort.

Tropical circulations are developed in *Atmosphere-Ocean Dynamics* (1982) by Gill.

An overview of GCMs is presented in *The Development of Atmospheric General Circulation Models* (2010), edited by Donner, Schubert, and Somerville.

PROBLEMS

1. Relate the change of moist static energy $h_{tot} = c_p T + l_v r + \Phi$ of an air parcel to the change of its equivalent potential temperature θ_e.

2. Consider an air parcel that moves equatorward at sea level, ascends adiabatically in the tropics, returns poleward at a constant altitude of 15 km, and then descends in the extratropics – as symbolized by the idealized thermodynamic circuit in Fig. 6.7. If heat transfer is restricted to upper and lower legs of the circuit and if, along the lower leg, the parcel's initial temperature and mixing ratio in the extratropics are $250°K$ and 1 g kg^{-1}, whereas it assumes a temperature and mixing

ratio in the tropics of 300°K and 20 g kg^{-1}, calculate the radiative heat transfer that occurs along the upper leg of the circuit.

3. Consider the following thermal structure, representative of zonal-mean stratification in the troposphere:

$$T(\phi, p) = \overline{T}(p) - T'(\phi, p),$$

where

$$\overline{T}(p) = \overline{T}_0 - \Gamma \bar{H} ln\left(\frac{p_0}{p}\right) + \frac{\Delta T}{4}\left(\frac{p}{p_0}\right),$$

$$T'(\phi, p) = \Delta T |sin\phi|^3 \left(\frac{p}{p_0}\right) - \frac{\Delta T}{4}\left(\frac{p}{p_0}\right),$$

and overbar denotes global average. For $\overline{T}_0 = 268°K$, $\Gamma = 4°K$ km^{-1}, $\bar{H} = 8$ km, $\Delta T = 70°K$, $\phi_0 = 45°$, and $p_0 = 1000$ hPa, (a) plot the thermal structure and zonal wind as functions of latitude and $\frac{z}{\bar{H}} = ln(\frac{p_0}{p})$, (b) calculate the available potential energy $\mathcal{A}$ of the atmosphere, and (c) plot the vertical concentration of available potential energy $\frac{\partial \mathcal{A}}{\partial z}$ as a function of $\frac{z}{\bar{H}}$.

4. For the atmosphere in Prob. 15.3, determine the fraction of total potential energy that is available for conversion to kinetic energy.

5. For the atmosphere in Prob. 15.3, determine a characteristic eddy velocity if thermal structure is driven adiabatically into barotropic stratification and if the kinetic energy produced is concentrated between $\pm 30°$ and $\pm 60°$ latitude.

6. (a) Use observed distributions of precipitation rate (Fig. 9.41) and total precipitable water vapor (Fig. 1.19) to calculate a characteristic time scale for the column abundance of water vapor in the tropics. (b) Discuss this time scale in relation to the efficiency of dehydration inside individual convective cells, their fractional coverage, and the efficiency with which water vapor is produced at the Earth's surface.

7. Atmospheric heating in Fig. 9.41b is concentrated about the equator. It is distinctly narrower than net radiative heating of the Earth-atmosphere system as a whole, represented in the TOA energy budget (Fig. 1.34c). Explain why and how.

8. Show that the pressure averaged over an isentropic surface $\overline{p}$ is preserved under an adiabatic rearrangement of mass.

9. Derive expressions (15.19) for available potential energy.

10. Zonal-mean flow in the troposphere is easterly at low latitude but westerly at middle and high latitudes (Fig. 1.8). Surface drag must therefore represent a sink of easterly momentum at low latitudes, which is equivalent to a source of westerly momentum. Conversely, it must represent a sink of westerly momentum at middle and high latitudes. On average, the atmosphere's angular momentum remains constant. Mechanical equilibrium therefore requires that the source of westerly momentum at low latitude and the sink of westerly momentum at middle and high latitudes be compensated by a poleward transfer of westerly momentum. This requirement is analogous to the poleward transfer of heat required by thermal equilibrium. (a) Demonstrate that transport of angular momentum by the zonal-mean Hadley circulation produces a momentum flux $\bar{u}\bar{v}$ of the correct sense to accomplish this transfer of westerly momentum. (b) Not all of the momentum transfer required to maintain mechanical equilibrium is accomplished by the

zonal-mean Hadley circulation. The remainder occurs through large-scale eddy transport. In terms of horizontal phase structure and group velocity, describe the meridional propagation of Rossby waves needed to accomplish the remaining momentum transfer between tropical and extratropical regions.

11. The length of day can vary through exchanges of momentum between the atmosphere and solid Earth. (a) Estimate the fluctuation of air velocity over the equator corresponding to observed fluctuations in the length of day of order 10^{-3} s, if the Earth has a mean density of order $5 \ 10^3$ kg m^{-3} and if the atmosphere compensates through uniform changes of angular velocity. (b) More generally, where would velocity fluctuations most effectively introduce changes in the length of day?

12. The height of convective towers is dictated by the moist static energy of surface air, in concert with the profile of environmental temperature, which determine CAPE (Sec. 7.4.1). Moist static energy, in turn, is determined by fluxes of sensible and latent heat from the surface (15.2). A surge of extratropical air enters the tropics, producing the temperature profile

$$
T(z) = \begin{cases} T_0 - \Gamma z & z < z_T \\ \\ T_T & z \geq z_T, \end{cases}
$$

where $T_0 = 290°$K, $\Gamma = 7.5°$K km^{-1}, $z_T = 12$ km, and $T_T = 200°$K. (a) If, through contact with an isolated landmass, surface air attains a temperature of 300°K and a mixing ratio of 23 g kg^{-1}, calculate the respective CAPE. (b) As in (a) but if the air is dry. (c) Estimate the height to which convection would develop under the conditions of (a) and (b).

SIXTEEN

Dynamic stability

Disturbances in the rotating cylindrical annulus are a laboratory analogue of synoptic weather systems, which prevail outside the tropics. Each develops through instability of the zonal-mean circulation. Instability was considered earlier in relation to the vertical stratification of mass (Chap. 7). If distributions of temperature and humidity violate the conditions for static stability, then small displacements lead to parcels accelerating away from their undisturbed positions. Unlike the response under stable stratification, this reaction leads to finite displacements of air. Fully developed convection then rearranges air, driving the stratification toward neutral stability.

Two classes of instability are possible. *Parcel instability* follows from reinforcement of air displacements by a negative restoring force, as occurs in the development of convection. *Wave instability* occurs in the presence of a positive restoring force, but one that amplifies parcel oscillations inside wave motions. Unstable waves amplify by extracting energy from the mean circulation, for example, from available potential energy that accompanies baroclinic stratification and vertical shear (Chap. 15). Strong zonal wind that follows from the nonuniform distribution of atmospheric heating, in concert with geostrophic balance, makes this class of instability the one most relevant to the large-scale circulation. Like parcel instability, it develops to neutralize instability in the mean flow. This objective is achieved through the rearrangement of air.

16.1 INERTIAL INSTABILITY

The simplest form of large-scale instability relates to the inertial oscillations described in Sec. 12.4. Consider disturbances to a geostrophically-balanced zonal flow $\bar{u}$ on an f plane. If the disturbances introduce no pressure perturbation, the total motion is governed by the horizontal momentum balance

$$\frac{du}{dt} - fv = 0 \tag{16.1.1}$$

$$\frac{dv}{dt} + f(u - \overline{u}) = 0, \tag{16.1.2}$$

where geostrophic equilibrium has been used to eliminate pressure in favor of $\overline{u}$.

Because a parcel's motion satisfies

$$v = \frac{dy}{dt}, \tag{16.2}$$

(16.1.1) implies

$$\frac{du}{dt} = f\frac{dy}{dt}. \tag{16.3}$$

Integrating from the parcel's initial position y_0 to its displaced position $y_0 + y'$ gives

$$u(y_0 + y') - \overline{u}(y_0) = fy'.$$

To first order in parcel displacement, this may be expressed

$$u(y_0) + \frac{\partial \overline{u}}{\partial y}y' - \overline{u}(y_0) = fy'$$

or

$$(u - \overline{u})\big|_{y_0} - \left(f - \frac{\partial \overline{u}}{\partial y}\right)y' = 0.$$

Then incorporating (16.1.2) yields

$$\frac{d^2 y'}{dt^2} + f\left(f - \frac{\partial \overline{u}}{\partial y}\right)y' = 0. \tag{16.4}$$

If the mean flow is without shear, (16.4) reduces to a description of the inertial oscillations considered previously. In the presence of shear, displacements either oscillate, decay exponentially, or amplify without bound. The system possesses unstable solutions if the absolute vorticity of the mean flow

$$f + \overline{\zeta} = f - \frac{\partial \overline{u}}{\partial y} \tag{16.5}$$

has sign opposite to the planetary vorticity f. Displacements then amplify exponentially. The zonal flow is said to be *inertially unstable*. These circumstances make the specific restoring force $f(f + \overline{\zeta})y'$ negative. The ensuing instability is therefore a form of parcel instability. Because $f + \overline{\zeta}$ is dominated by planetary vorticity, it usually has the same sign as f. The criterion for inertial instability is thus tantamount to the absolute vorticity reversing sign somewhere.

Inertial instability does not play a major role in the atmosphere. Extratropical motions tend to remain inertially stable, even locally in the presence of synoptic and planetary wave disturbances. However, the criterion for inertial instability is violated more easily near the equator, where f is small. There, even modest shear can lead to a reversal of absolute vorticity. Evidence of inertial instability exists in the tropical stratosphere. Horizontal shear that flanks strong zonal jets in the winter and summer stratosphere (Fig. 1.8) can violate the criterion for inertial stability.

16.2 SHEAR INSTABILITY

More relevant to the large-scale circulation is instability that derives directly from shear. Shear instability is a form of wave instability. It therefore requires a more involved analysis. Like the treatment of wave motions, the description of shear instability requires the solution of partial differential equations that govern perturbation or eddy properties. Closed form solutions can be found only for idealized profiles of zonal-mean flow $\overline{u}(y, z)$. However, an illuminating criterion for instability, due originally to Rayleigh, can be developed under fairly general circumstances.

16.2.1 Necessary conditions for instability

Consider quasi-geostrophic motion on a beta plane in an atmosphere that extends upward indefinitely and is bounded below and laterally at $y = \pm L$ by rigid walls. Disturbances to the zonal-mean flow $\overline{u}(y, z)$ are governed by first-order conservation of potential vorticity (14.75)

$$\frac{DQ'}{Dt} + v'\beta_e = 0,\tag{16.6.1}$$

where $\frac{D}{Dt} = \frac{\partial}{\partial t} + \overline{u}\frac{\partial}{\partial x}$,

$$Q' = \nabla^2\psi' + \frac{1}{\rho_0}\frac{\partial}{\partial z}\left(\frac{f_0^2}{N^2}\rho_0\frac{\partial\psi'}{\partial z}\right)\tag{16.6.2}$$

$$\beta_e = \beta - \frac{\partial^2\overline{u}}{\partial y^2} - \frac{1}{\rho_0}\frac{\partial}{\partial z}\left(\frac{f_0^2}{N^2}\rho_0\frac{\partial\overline{u}}{\partial z}\right)$$

$$= \frac{\partial\overline{Q}}{\partial y},\tag{16.6.3}$$

and z denotes log-pressure height. Requiring vertical motion to vanish at the Earth's surface (which is treated as an isobaric surface) gives, via the thermodynamic equation and thermal wind balance, the lower boundary condition

$$\frac{D}{Dt}\left(\frac{\partial\psi'}{\partial z}\right) - \frac{\partial\overline{u}}{\partial z}\frac{\partial\psi'}{\partial x} = 0 \qquad z = 0.\tag{16.6.4}$$

Physically meaningful solutions must also have bounded column energy. The latter provides the upper boundary condition

$$\text{finite energy condition}\quad z \to \infty.\tag{16.6.5}$$

At the lateral walls, v' must vanish, so $\psi' = \text{const}$. It suffices to prescribe

$$\psi' = 0 \qquad y = \pm L.\tag{16.6.6}$$

Equations (16.6) define a second-order boundary value problem for the eddy streamfunction $\psi'(\mathbf{x}, t)$ – one that is homogeneous. Involving no imposed forcing, (16.6) describes a system that is self-governing or autonomous. Nontrivial solutions (i.e., other than $\psi' \equiv 0$) exist only for certain *eigenfrequencies* that enable boundary conditions to be satisfied. Those eigenfrequencies are determined by solving the homogeneous boundary value problem for a given zonal flow $\bar{u}(y, z)$. They are, in general, complex.

Consider solutions of the form

$$\psi' = \Psi(y, z)e^{ik(x-ct)}, \tag{16.7}$$

where Ψ and $c = c_r + ic_i$ can assume complex values. Substituting (16.7) transforms (16.6) into

$$(\bar{u} - c)\left[\frac{\partial^2 \Psi}{\partial y^2} + \frac{1}{\rho_0}\frac{\partial}{\partial z}\left(\frac{f_0^2}{N^2}\rho_0\frac{\partial \Psi}{\partial z}\right) - k^2\Psi\right] + \beta_e\Psi = 0. \tag{16.8.1}$$

The lower boundary condition (16.6.4) becomes

$$(\bar{u} - c)\frac{\partial \Psi}{\partial z} - \frac{\partial \bar{u}}{\partial z}\Psi = 0 \qquad z = 0. \tag{16.8.2}$$

For c real, (16.8) is singular at a critical line where $\bar{u} = c$. The singularity disappears if $c_i \neq 0$. Behavior (16.7) then involves an exponential modulation in time. If the flow is stable, wave activity incident on the critical line is absorbed when dissipation is included (Sec. 14.3). The wave field then decays exponentially. If the flow is unstable, wave activity can be produced at the critical line. The wave field then amplifies exponentially.

Multiplying the conjugate of (16.8) by Ψ and (16.8) by the conjugate of Ψ and then subtracting yields

$$\left[\Psi\frac{\partial^2 \Psi^*}{\partial y^2} - \Psi^*\frac{\partial^2 \Psi}{\partial y^2}\right] + \left[\Psi\frac{1}{\rho_0}\frac{\partial}{\partial z}\left(\frac{f_0^2}{N^2}\rho_0\frac{\partial \Psi^*}{\partial z}\right) - \Psi^*\frac{1}{\rho_0}\frac{\partial}{\partial z}\left(\frac{f_0^2}{N^2}\rho_0\frac{\partial \Psi}{\partial z}\right)\right] - 2ic_i\frac{|\Psi|^2}{|\bar{u}-c|^2}\beta_e = 0 \tag{16.9.1}$$

and

$$\Psi\frac{\partial \Psi^*}{\partial z} - \Psi^*\frac{\partial \Psi}{\partial z} + 2ic_i\frac{|\Psi|^2}{|\bar{u}-c|^2}\frac{\partial \bar{u}}{\partial z} = 0 \qquad z = 0. \tag{16.9.2}$$

The chain rule enables terms in the first set of brackets in (16.9.1) to be expressed

$$\Psi\frac{\partial^2 \Psi^*}{\partial y^2} - \Psi^*\frac{\partial^2 \Psi}{\partial y^2} = \frac{\partial}{\partial y}\left[\Psi\frac{\partial \Psi^*}{\partial y} - \Psi^*\frac{\partial \Psi}{\partial y}\right].$$

Terms in the second set of brackets may be expressed similarly. Then (16.9.1) can be written

$$\frac{\partial}{\partial y}\left[\Psi\frac{\partial \Psi^*}{\partial y} - \Psi^*\frac{\partial \Psi}{\partial y}\right] + \frac{1}{\rho_0}\frac{\partial}{\partial z}\left[\frac{f_0^2}{N^2}\rho_0\left(\Psi\frac{\partial \Psi^*}{\partial z} - \Psi^*\frac{\partial \Psi}{\partial z}\right)\right] - 2ic_i\frac{|\Psi|^2}{|\bar{u}-c|^2}\beta_e = 0. \tag{16.10}$$

Integrating over the domain unravels the exterior derivatives in (16.10), leaving

$$\int_0^\infty \left[\Psi\frac{\partial \Psi^*}{\partial y} - \Psi^*\frac{\partial \Psi}{\partial y}\right]_{y=-L}^{y=L}\rho_0 dz + \int_{-L}^{L}\left[\frac{f_0^2}{N^2}\rho_0\left(\Psi\frac{\partial \Psi^*}{\partial z} - \Psi^*\frac{\partial \Psi}{\partial z}\right)\right]_{z=0}^{z=\infty} dy \tag{16.11}$$

$$- 2ic_i\int_{-L}^{L}\int_0^\infty\frac{|\Psi|^2}{|\bar{u}-c|^2}\beta_e dy\rho_0 dz = 0.$$

By (16.6.6), the first integral vanishes. The finite energy condition makes the upper limit inside the second integral also vanish. Then incorporating (16.9.2) for the lower limit yields the identity

$$c_i \left\{ \int_0^\infty \int_{-L}^L \beta_e \frac{\rho_0 |\Psi|^2}{|\bar{u} - c|^2} \, dy \, dz - \int_{-L}^L \left[\frac{f_0^2}{N^2} \frac{\rho_0 |\Psi|^2}{|\bar{u} - c|^2} \frac{\partial \bar{u}}{\partial z} \right]_{z=0} dy \right\} = 0. \tag{16.12}$$

Equation (16.12) must be satisfied for $\Psi(y, z)$ to be a solution of (16.6). Advanced by Charney and Stern (1962), it provides "necessary conditions" for instability of the zonal-mean flow $\bar{u}(y, z)$. If c is complex, (16.7) describes a disturbance whose amplitude varies in time exponentially

$$e^{ik(x-ct)} = e^{kc_i t} \cdot e^{ik(x-c_r t)},$$

with the growth/decay rate kc_i. Without loss of generality, k may be taken to be positive. The existence of unstable solutions then requires $c_i > 0$. Unstable solutions are thus possible only if the quantity in braces vanishes. If it does not, (16.12) requires that $c_i = 0$. Solutions to (16.6) are then stable.

16.2.2 Barotropic and baroclinic instability

Requiring (16.12) to be satisfied with $c_i > 0$ provides two alternative criteria for instability:

1. If $\frac{\partial \bar{u}}{\partial z}$ vanishes at the lower boundary, by thermal wind balance, so does the temperature gradient. Then $\beta_e = \frac{\partial \bar{Q}}{\partial y}$ must reverse sign somewhere in the interior. β_e is normally positive, dominated by the gradient of planetary vorticity β. A region of negative potential vorticity gradient, $\frac{\partial \bar{Q}}{\partial y} < 0$, is therefore identified as one where the mean flow is unstable.

2. If $\beta_e > 0$ throughout the interior, $\frac{\partial \bar{u}}{\partial z}$ must be positive somewhere on the lower boundary. By thermal wind balance, this implies the existence of an equatorward temperature gradient at the surface.

Other combinations are also possible, but these necessary conditions are the ones most relevant to the atmosphere. Neither represents a "sufficient condition" for instability. Satisfying criterion (1) or (2) does not ensure the existence of unstable solutions.

Criterion (1) defines the necessary condition for *free-field instability*, namely, instability for which boundaries do not play an essential role. From (16.6.3), the mean gradient of potential vorticity can reverse sign through strong meridional curvature of the mean flow or through strong (density-weighted) vertical curvature of the mean flow. It is customary to distinguish these contributions to $\frac{\partial \bar{Q}}{\partial y}$. If the necessary condition for instability is met through horizontal shear, amplifying disturbances are referred to as *barotropic instability*. If it is met through vertical shear (which is proportional to the horizontal temperature gradient and the departure from barotropic stratification), amplifying disturbances are referred to as *baroclinic instability*. Realistic conditions often lead to criterion (1) being satisfied by both contributions. Amplifying disturbances are then combined barotropic-baroclinic instability.

In the absence of rotation, criterion (1) reduces to Rayleigh's (1880) necessary condition for instability of 1-dimensional shear flow (Prob. 16.3). Criterion (1) is then equivalent to requiring the mean flow profile to possess an inflection point. Because β is everywhere positive, rotation is stabilizing. It provides a positive restoring force that inhibits instability and supports stable wave propagation.

Recall that, if $c_i = 0$, (16.8) is singular at a critical line, where $\bar{u} = c_r$. Exponential amplification removes the singularity by making $c_i > 0$. When boundaries do not play an essential role, amplifying solutions usually possess a critical line inside the unstable region, where $\frac{\partial \bar{Q}}{\partial y} < 0$ (e.g., Dickinson, 1973). Rather than serving as a localized sink of wave activity, as it does under stable conditions ($\frac{\partial \bar{Q}}{\partial y} > 0$), the critical line then serves as a localized source of wave activity (Sec. 14.5.4). Under these conditions, wave activity flux (14.81) diverges from the critical line, where it is produced by a conversion from the mean flow. Alternatively, wave activity incident on the critical line is "overreflected": More radiates away than is incident on the unstable region.

Criterion (2) describes instability that is produced through the direct involvement of the lower boundary. This criterion applies to baroclinic instability because it requires a temperature gradient at the surface and hence baroclinic stratification. Because air must move parallel to it, the boundary can then drive motion across mean isotherms. Through $v' \cdot \nabla \bar{T}$, such motion transfers heat meridionally (e.g., in sloping convection). If it transfers heat poleward, eddy motion weakens the zonal-mean gradient of temperature, which is directed equatorward. Eddy heat transfer then weakens baroclinicity, driving mean thermal structure toward barotropic stratification. It thus releases available potential energy, which is converted to eddy kinetic energy (Sec. 15.1). Such behavior is intrinsic to the development of extratropical cyclones that typify synoptic weather systems. Those systems amplify through baroclinic instability, which releases available potential energy. The latter is continually re-generated by the nonuniform distribution of atmospheric heating and cooling (Figs. 1.34c, 9.41b).

16.3 THE EADY MODEL

The simplest description of baroclinic instability is due to Eady (1949). Consider disturbances to a mean flow that is invariant in y, bounded above and below by rigid walls at $z = 0$ and H on an f plane, and within the Boussinesq approximation (Sec. 12.5). A uniform meridional temperature gradient is imposed. By thermal wind balance, it corresponds to constant vertical shear (Fig. 16.1)

$$\bar{u} = \Lambda z \qquad \Lambda = \text{const.} \tag{16.13}$$

Under these circumstances, $\frac{\partial \bar{Q}}{\partial y}$ vanishes in the interior. Instability can therefore follow solely from the temperature gradient along the boundaries.

Disturbances to this system are governed by the eddy potential vorticity equation in log-pressure coordinates

$$\frac{D}{Dt}\left[\nabla^2 \psi' + \frac{f_0^2}{N^2}\frac{\partial^2 \psi'}{\partial z^2}\right] = 0. \tag{16.14.1}$$

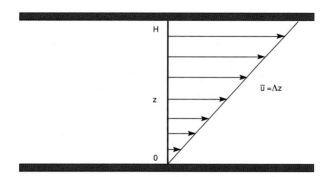

Figure 16.1 Geometry and mean zonal flow in the Eady problem of baroclinic instability.

With the thermodynamic equation, the boundary conditions become

$$\frac{D}{Dt}\left(\frac{\partial \psi'}{\partial z}\right) - \frac{\partial \bar{u}}{\partial z}\frac{\partial \psi'}{\partial x} = 0 \qquad z = 0, H. \qquad (16.14.2)$$

Considering solutions of the form

$$\psi' = \Psi(z)cos(ly)e^{ik(x-ct)} \qquad (16.15)$$

reduces (16.14) to the 1-dimensional boundary value problem

$$\frac{d^2 \Psi}{dz^2} - \alpha^2 \Psi = 0 \qquad (16.16.1)$$

$$(\bar{u} - c)\frac{d\Psi}{dz} - \Lambda \Psi = 0 \qquad z = 0, H, \qquad (16.16.2)$$

where

$$\alpha = \frac{N}{f_0}|k_h|, \qquad (16.16.3)$$

with $|k_h|^2 = k^2 + l^2$, is a stability-weighted horizontal wavenumber.[1] Solutions of (16.16.1) are of the form

$$\Psi = Acosh(\alpha z) + Bsinh(\alpha z). \qquad (16.17)$$

Substituting (16.17) into the boundary conditions (16.16.2) leads to a homogeneous system of two algebraic equations for the coefficients A and B. Nontrivial solutions exist only if the determinant of that system vanishes. Incorporating that condition yields the dispersion relation for Eady modes

$$\left(c - \frac{\Lambda H}{2}\right)^2 = \Lambda^2 H^2 \left[\frac{1}{4} - \frac{coth(\alpha H)}{\alpha H} + \frac{1}{(\alpha H)^2}\right]. \qquad (16.18)$$

If the right-hand side of (16.18) is positive, c is real. The system is then stable. If the right-hand side is negative, unstable solutions exist. Because $c - \frac{\Lambda H}{2}$ must then

---

[1] Considering structure of the form $cos(ly) = \frac{1}{2}\left(e^{ily} + e^{-ily}\right)$ implicitly presumes that disturbances are trapped meridionally (e.g., by rigid walls at $y = \pm\frac{\pi}{2l}$).

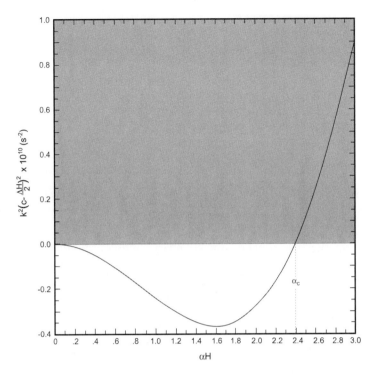

Figure 16.2 Frequency squared of Eady modes as a function of scaled horizontal wavenumber. Instability occurs for wavenumbers smaller than the cutoff α_c.

be imaginary,

$$c_r = \frac{\Lambda H}{2}. \tag{16.19}$$

Thus amplifying disturbances have phase speeds equal to the mean flow at the middle of the layer. They are advected eastward by $\bar{u}$ with its speed at the *steering level*: $z = \frac{H}{2}$. Although influenced by rotation, baroclinic disturbances are not Rossby waves in a strict sense. As the Eady model demonstrates, they can exist even on an f plane. Baroclinic disturbances can therefore exist in the absence of β, which provides the restoring force for Rossby waves.

The quantity $k^2 \left(c - \frac{\Lambda H}{2}\right)^2$ reflects the square of the complex frequency. It is plotted as a function of horizontal wavenumber, αH, in Fig. 16.2. For α greater than a critical value, $\alpha_c \cong 2.4$, $k^2 \left(c - \frac{\Lambda H}{2}\right)^2 > 0$ and solutions do not amplify. The system thus possesses a "short-wave cutoff" for instability. Amplifying solutions exist only for smaller α (larger horizontal scales). Wavenumbers $\alpha < \alpha_c$ are unstable: $k^2 \left(c - \frac{\Lambda H}{2}\right)^2 < 0$. A maximum growth rate kc_i is achieved at $\alpha H \cong 1.6$ for $l = 0$. This value of αH also maximizes the growth rate for waves of nonzero aspect ratio $\frac{l}{k}$, which have smaller kc_i. For square waves ($k = l$) and values typical of the troposphere, $\alpha H \cong 1.6$. It predicts a wavenumber of order 5, typical of extratropical cyclones (see Fig. 2.10).

The maximum growth rate is proportional to the vertical shear Λ. By thermal wind balance, (12.11), it is therefore proportional to the meridional temperature gradient.

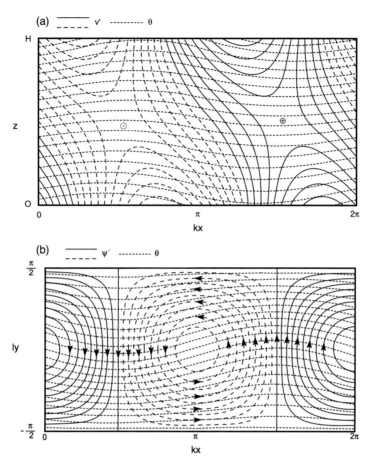

Figure 16.3 Structure of the fastest-growing square Eady mode ($k = l$). (a) Vertical section in the zonal plane at $y = 0$ of eddy meridional velocity v' (solid/dashed) and isentropic surfaces $\theta =$ const (dotted). Potential temperature increases upward. $\odot$ marks equatorward motion ($v' < 0$) and $\oplus$ marks poleward motion ($v' > 0$). (b) Horizontal section at $z = \frac{H}{2}$ of eddy streamlines (solid/dashed) and isentropes (dotted). Potential temperature increases equatorward.

Representative values give an e-folding time for amplification of a couple of days. The latter is broadly consistent with the observed development of extratropical cyclones.

Insight into how instability is achieved follows from the structure of the most unstable disturbance. The lower boundary condition (16.16.2) implies the following relationship between the coefficients in (16.17)

$$\frac{B}{A} = -\frac{\Lambda}{\alpha c}. \tag{16.20}$$

It yields the eddy thermal structure and motion in Fig. 16.3 for $k = l$. Plotted in Fig. 16.3a, on a vertical section passing zonally through the center of the disturbance ($y = 0$), is the eddy meridional velocity v' (solid/dashed). Superimposed is the distribution of isentropes: $\theta = \overline{\theta} + \theta'$ (dotted). The fastest-growing Eady mode is characterized by a westward tilt with height. It places streamfunction anomalies about 90° out of phase between the lower and upper boundaries, as is evident from $v' = ik\psi'$.

Eddy motion maximizes at $z = 0$ and H. It assumes the form of two edge waves that are sandwiched between the upper and lower boundaries. Owing to the temperature gradient along them, those boundaries drive air across mean isotherms. They also trap instability, enabling the disturbance to amplify. Each edge wave decays vertically with the Rossby height scale $H_R = \alpha^{-1}$. In the limit $|k_h| \to \infty$, $\alpha \gg \alpha_c$. H_R is then short enough for the two edge waves to be isolated from one another. Neither has vertical phase tilt. Hydrostatic balance (11.72.2) therefore implies a temperature anomaly that is in phase with the streamfunction anomaly (Prob. 16.4). The eddy velocity $v' = ik\psi'$ is therefore 90° out of phase with θ'. The two are said to be *in quadrature*. The eddy heat flux averaged over a wavelength $\overline{v'\theta'}$ then vanishes. Under these circumstances, eddy motion leaves the baroclinic stratification of the mean state unchanged. No available potential energy is released. It is for this reason that short wavelengths in the Eady model are stable.

For $\alpha < \alpha_c$, H_R is tall enough for the two edge waves to influence one another. Phase shifted, they produce the westward tilt apparent in Fig. 16.3a. The temperature anomaly is then no longer in phase with the streamfunction anomaly. Isentropes (dotted) actually tilt slightly eastward with height, making θ' positively correlated with v': Cold air ($\theta' < 0$) moves equatorward ($v' < 0$): out of the page. Warm air ($\theta' > 0$) moves poleward ($v' > 0$): into the page.[2] Both motions contribute positively to $\overline{v'\theta'}$. Consequently, they achieve poleward eddy heat flux. The latter releases available potential energy, which is converted into eddy kinetic energy. The mode's amplification is thus directly related to its westward tilt with height.

Plotted in Fig. 16.3b, for a horizontal section at $z = \frac{H}{2}$, is the eddy streamfunction (solid/dashed), along with the distribution of isentropes (dotted). At the center is a cyclonic anomaly, which moves eastward. Ahead of it, warm air (high θ) is displaced poleward. Behind it, cold air (low θ) is displaced equatorward. As the motion is adiabatic, air moves along isentropic surfaces. Plotted in Fig. 16.4, for a meridional section at $kx = \frac{3\pi}{2}$, is the eddy motion. Superimposed is the distribution of zonal-mean isentropes: $\overline{\theta}$ (dotted). Zonal-mean isentropes slope upward toward the pole (cf. Fig. 12.5). Eddy motion has similar slope. Consequently, poleward-moving air ahead of the cyclonic anomaly in Fig. 16.3b ascends. Equatorward-moving air behind the anomaly descends. Characteristic of sloping convection, this motion has the same form that it would were air to move along zonal-mean isentropic surfaces. However, eddy motion in Fig. 16.4 has only half the slope of those surfaces (Prob. 16.13). Eddy motion is therefore directed across zonal-mean isentropic surfaces. Its impact is manifest in the horizontal distribution of θ (Fig. 16.3b): Warm air is displaced poleward, whereas cold air is displaced equatorward. Both displacements represent a poleward transfer of heat. The latter releases available potential energy, enabling the disturbance

---

[2] At the steering level, $z = \frac{H}{2}$, v' and θ' are perfectly in phase.

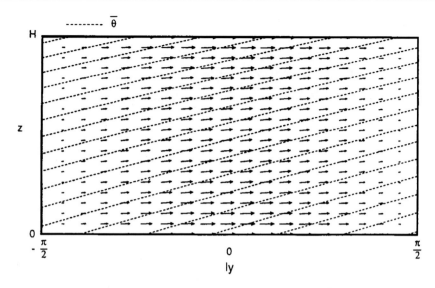

Figure 16.4 Vertical section in the meridional plane at $kx = \frac{3\pi}{2}$ of zonal-mean isentropic surfaces (dotted) and motion for the fastest-growing square Eady mode. Potential temperature increases upward and equatorward (cf. Fig. 12.5).

to amplify. Simultaneously, poleward heat transfer weakens the meridional gradient of zonal-mean temperature, which is directed equatorward. Through thermal wind balance, it must also weaken the strong zonal jet that accompanies that temperature gradient. These influences of the amplifying disturbance are equivalent to driving zonal-mean thermal structure toward barotropic stratification. By eliminating baroclinicity, they deplete available potential energy, which is converted into eddy kinetic energy.

Extratropical cyclones have qualitatively similar structure during their development. Figure 16.5 shows distributions of 700-hPa height and temperature for an amplifying cyclone situated off the coast of Africa on March 2, 1984. This disturbance is the precursor to the cyclone apparent in Figs. 1.18 and 1.29 two days later. During amplification, the system tilts westward. This transfers heat poleward and releases available potential energy – analogous to an unstable Eady mode with $\alpha < \alpha_c$. Eddy heat flux tends to maximize near 700 hPa, which typifies the steering level of observed cyclones. This is lower than the steering level predicted by the Eady model. However, an unbounded model treated by Charney (1947) reproduces the observed steering level while retaining the essential features captured by Eady's solution.

Isotherms in Fig. 16.5 (dashed) exhibit the characteristic signature of sloping convection: Analogous to behavior in the Eady model, a tongue of warm (high-θ) air is drawn poleward ahead of the closed low. Simultaneously, a tongue of cold (low-θ) air is drawn equatorward behind it (cf. Figs. 16.3b; 12.10). Those bodies of air have disparate histories, which are reflected in contemporaneous IR and water vapor imagery (Figs. 16.6a,b). A tongue of high cloud cover and humidity extends northwestward from the African coast. It marks the *warm sector* ahead of the cyclone (delineated by the warm and cold fronts in Fig. 16.5). A complementary tongue of cloud-free air and low humidity is simultaneously being drawn equatorward behind the cyclone. Sharp

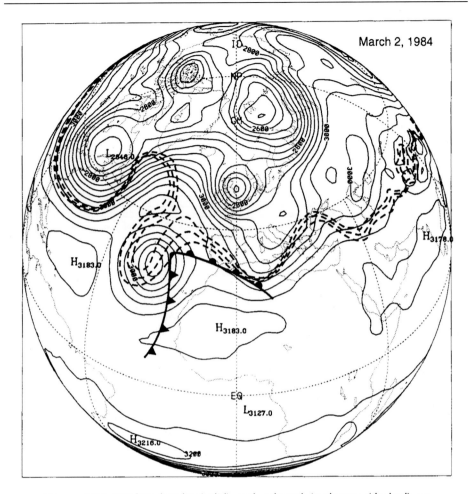

Figure 16.5 700-hPa height (solid) and selected isotherms (dashed) on March 2, 1984. A surface frontal analysis is superposed. Cold (warm) front marked by pointed (rounded) barbs. In the eastern Atlantic, isotherms have been deflected poleward ahead of the amplifying cyclone and equatorward behind it (cf. Figs 16.3b; 12.10).

gradients separating those bodies of air represent warm and cold fronts at the surface. Humid air comprising the warm sector can be traced, in water vapor imagery (Figs. 16.6d, 1.29), back to its source: the Amazon basin. There, tropical convection humidifies the troposphere (Sec. 9.3.4). Air advancing behind the cold front undercuts the warm sector in sloping convection. Humid air is then lifted, producing the extensive cloud shield in Fig. 16.6a.

16.4 NONLINEAR CONSIDERATIONS

While capturing the essential features of cyclone development, Eady's solution provides only a hint of their behavior at maturity. Conversion of available potential energy into eddy kinetic energy enables a baroclinic disturbance to intensify and eventually attain finite amplitude. Finite horizontal displacements then invalidate the

Infrared   Water Vapor

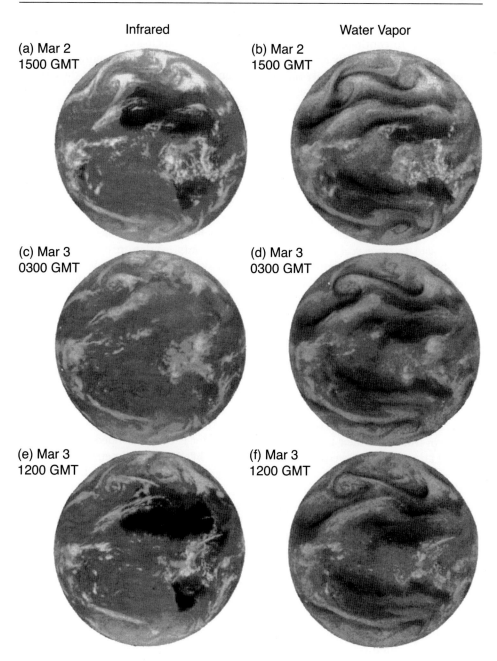

Figure 16.6 Infrared and water vapor imagery from Meteosat between March 2 and March 4, 1984, which reveals the evolution of an extratropical cyclone off the northwest coast of Africa. In the water vapor imagery, humid air is represented light and dry air dark. The warm sector ahead of the cyclone is marked by a tongue of high cloud and moisture that slopes northwestward on March 2. That air mass subsequently overturns near the juncture of cold and warm fronts that delineate it, forming an occlusion in which cold and warm air are entrained and mixed horizontally (cf. Fig. 9.21).

linear description on which Eady's model is based. The latter predicts exponential amplification to continue indefinitely. The situation is analogous to convective displacements under statically unstable conditions (Sec. 7.3). At finite amplitude, second-order effects enter. They modify the zonal-mean state, eventually limiting the amplification of a baroclinic disturbance.

As a baroclinic disturbance amplifies, horizontal displacements become increasingly exaggerated (Fig. 16.3b). Warm tropical air is eventually folded north of cold polar air. Such behavior is evident in the distribution of potential vorticity on March 2, 1984 (Fig. 12.10). Deformations experienced by those air masses steepen potential temperature gradients that separate warm and cold air (e.g., Fig. 12.4). The accompanying fronts are thus intensified. Continued advection leads to tongues of warm and cold air eventually encircling the low. The warm sector has then wound up at the junction of the warm and cold fronts, as is revealed by water vapor in Fig. 16.6 (see also cloud cover in Fig. 9.21). Upon reaching this stage, the cyclone in Fig. 16.5 is said to have *occluded*. Warm and cold fronts overlap in the center of the disturbance. There, the warm air mass has been swept over the cold air mass.[3] Characteristic of sloping convection, this structure marks the mature stage of the cyclone's life cycle. The surface low is then positioned beneath the upper-level low. This eliminates the disturbance's westward tilt, its poleward heat flux, and hence the release of available potential energy. The disturbance is then analogous to a stable Eady mode, with $\alpha > \alpha_c$. Cold and warm air drawn into the occlusion are subsequently wound together and mixed horizontally. By homogenizing temperature, this process drives thermal structure toward barotropic stratification. It thereby exhausts the supply of available potential energy.

Figure 16.6 shows sequences of IR and water vapor imagery while the disturbance in Fig. 16.5 matures. At 1500 UT on March 2 (Fig. 16.6a,b), the cold front is approaching the warm front near their junction. The warm sector is clearly defined, assuming the form of an inverted V. Twelve hours later, the 700 hPa low has deepened (not shown). The warm sector in IR and water vapor imagery (Fig. 16.6c,d) has then been sheared to the northwest. There, cold dry air is being entrained with warm humid air inside the occlusion that has formed. This process culminates in cold and warm air at the occlusion winding up into a spiral. It resembles behavior in the cylindrical annulus at fast rotation (Fig. 15.7f). By 1200 UT on March 3 (Figs. 16.6e,f), air inside the occlusion has wound up into a tight vortex, separating from the remaining warm sector to its south and east. Interleaving bands of cold and warm air are apparent in both cloud cover and humidity. They symbolize efficient horizontal mixing. One day later (Figs. 1.18, 1.28), that mass of air has become nearly homogeneous. The 700 hPa low has then weakened. Similarly, high cloud that developed earlier through sloping convection is dissipating. Only a broad spiral of equatorward-moving air remains. It is being drawn cyclonically around the now-diffuse anomaly of potential vorticity, wherein cold and warm air have been mixed.

Since θ (more generally, θ_e) is conserved for individual air parcels, horizontal mixing weakens the meridional gradient of temperature, rendering potential temperature nearly uniform. By thermal wind balance, it simultaneously weakens the strong zonal jet that accompanies baroclinic stratification (cf Figs. 6.3 and 6.4). The impact on the

---

[3] The occlusion actually forms as the surface low (not shown) separates from the junction of warm and cold fronts and deepens farther back into the cold air mass.

jet is illustrated in Fig. 16.7, which displays conditions that accompany an amplifying baroclinic system that has reached maturity. SLP is marked by a closed low over southern Australia (Fig. 16.7a). Extending equatorward from the low is a cold front, which divides warm/humid tropical air at the surface from cold/dry extratropical air.

The air masses are clearly distinguished in the contemporaneous water vapor image (Fig. 16.7b). Humid tropical air (light) has been drawn poleward, forming a warm sector with characteristic V shape. It wraps up inside the closed low, where the system has occluded. West of the warm sector is dry extratropical air (dark) that is being entrained with humid tropical air inside the occlusion. The deformation of warm and cold air masses exerts a similar influence on the horizontal temperature gradient. Through thermal wind balance, the distorted structure is conveyed into the horizontal distribution of vertical shear and, hence, into the horizontal structure of the jet stream. Superimposed in Fig. 16.7a are 300-hPa isotachs stronger than 20 m/s (shaded). The jet has been overturned, mirroring the interface between air masses.

Subsequent evolution homogenizes the meridional temperature gradient that existed initially, with a commensurate weakening of the jet. The amplifying system thus drives thermal structure toward barotropic stratification. Baroclinic instability that was present initially has then been neutralized. No more potential energy is available for conversion to eddy kinetic energy. The available potential energy has been exhausted. This process is a counterpart of vertical mixing by convection, which neutralizes static instability (Sec. 7.3). In each, unstable disturbances modify the mean state. Such interaction contrasts with wave propagation under stable conditions. Parcel displacements are then bounded. Outside regions of dissipation, where wave activity is absorbed (Sec. 14.5.4), they leave the mean state largely unaffected.

The preceding treatment applies formally to a zonally symmetric mean state. In practice, the results also lend insight into localized regions of instability. Amplified planetary waves in the Northern Hemisphere reinforce zonal-mean vertical shear, producing a broken storm track. The latter is marked by localized jets east of Asia and North America (cf. Figs. 15.11a). The North Atlantic and North Pacific storm tracks are strongly baroclinic. They are preferred sites of cyclone development (Fig. 15.10). Calculations in a zonally asymmetric basic state reveal that planetary waves concentrate unstable amplification in such zones (Simmons et al., 1983; Frederiksen, 2006). Through locally intensified shear, they make the fastest growth rate under zonally symmetric conditions even faster, while removing the shortwave cutoff for instability. The fastest-growing modes, which are shaped by the prescribed planetary wave pattern, then share features with teleconnection patterns (Sec. 14.5.3).

In the Southern Hemisphere, planetary waves are weaker. They leave the mean state close to zonal symmetry, producing a continuous storm track. This enables cyclones in the Southern Hemisphere to assume a distribution in longitude that is more uniform than is observed in the Northern Hemisphere. Occasionally, they assume the form of baroclinic modes in a rotating annulus, wherein the mean state is zonally symmetric (cf. Figs. 2.10, 15.7e).

The criteria for instability also have implications to planetary waves that have attained large amplitude. The situation is analogous to the breaking of gravity waves. The latter overturn isentropic surfaces (Fig. 14.25), folding high θ beneath

(a)

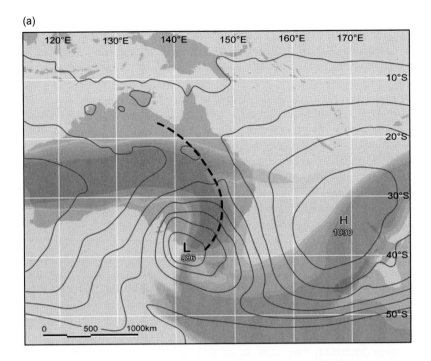

(b)

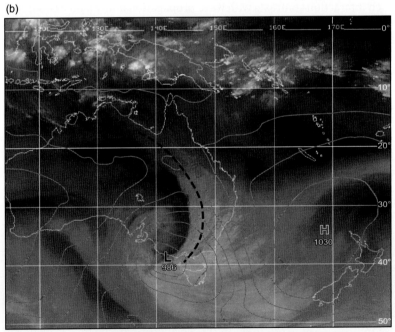

Figure 16.7 Conditions accompanying an amplifying baroclinic system that has reached maturity. (a) Distribution of SLP (contoured) and 300-hPa isotachs stronger than 20 m/s, which delineate the jet stream (shaded). (b) Contemporaneous water vapor image from MTSAT, representing humid air (light) and dry air (dark). See color plate section: Plate 18.

low θ. This makes $\frac{\partial \theta}{\partial z} < 0$ locally, rendering the region statically unstable. Convective mixing that ensues absorbs organized wave motion. Simultaneously, it drives $\frac{\partial \theta}{\partial z}$ to zero, neutralizing the instability. Planetary waves of sufficient amplitude exert an analogous influence on horizontal structure. They overturn the distribution of the conserved property Q (Fig. 14.26). High-potential vorticity is then folded south of low-potential vorticity. This makes $\frac{\partial Q}{\partial y} < 0$ locally, rendering the horizontal motion dynamically unstable (Sec. 16.2). At that point, the planetary wave field breaks. Small-scale eddies develop in the region of instability (Fig 14.23). They mix Q horizontally. This process absorbs organized wave motion. Simultaneously, it drives $\frac{\partial Q}{\partial y}$ to zero, neutralizing the instability.

SUGGESTED REFERENCES

An Introduction to Dynamic Meteorology (2004) by Holton provides a thorough treatment of baroclinic disturbances.

Atmosphere-Ocean Dynamics (1982) by Gill compares the Charney and Eady models of baroclinic instability.

A synoptic description of extratropical cyclones is presented in *Atmospheric Science: An Introductory Survey* (2006) by Wallace and Hobbs.

PROBLEMS

1. In the atmosphere of a newly discovered planet, the circulation is described by the barotropic zonal flow

$$\overline{u}(y) = -U \cos(2\phi),$$

   where $U = 0.55\Omega a$. (a) Characterize the inertial stability of this flow as a function of latitude. (b) Describe the zonal-mean circulation toward which inertial instability will drive the flow.

2. Consider westerly flow that corresponds to an angular velocity $A(z) = \frac{\overline{u}}{a\cos\phi}$ that varies only with height. (a) Within the framework of quasi-geostrophic motion and the Boussinesq approximation, what sign of curvature must the velocity profile $A(z)$ have for shear instability to develop? (b) At what latitudes is shear instability favored most if $A(z) = \epsilon(z) \cdot 2\Omega$, with $\epsilon \ll 1$?

3. Show that, in the absence of rotation, criterion (1) in Sec. 16.2.2 recovers Rayleigh's condition for instability: A barotropic flow must possess an inflection point in its interior. (b) Discuss the influence rotation has on shear instability.

4. Demonstrate that the limiting structure of Eady modes for $H \to \infty$ has v' in quadrature with θ'.

5. Derive an expression for the shortwave cutoff α_c for instability in the Eady problem.

6. Show that the maximum growth rate of Eady modes is achieved for $l = 0$. (b) Express the growth rate of the fastest-growing square Eady mode ($k = l$) in terms of that of the fastest growing Eady mode ($l = 0$).

7. Calculate the e-folding times of square Eady modes ($k = l$) for an f plane at $45°$, $N^2 = 10^{-4}$ s^{-2}, vertical shear of 3 m s^{-1} km^{-1}, a rigid lid at 10 km, and for zonal wavenumbers 1–8.

8. Use the Eady problem to explain (a) why mid-latitude cyclones develop and track along the jet stream, (b) how baroclinic instability would be altered if the jet were displaced equatorward.

9. Precipitation inside cyclones often assumes a banded structure. Discuss this feature in relation to the evolution in Fig. 16.6, the distribution of potential vorticity Q_g, and Ekman pumping (Sec. 13.4).

10. Unlike extratropical cyclones, tropical cyclones are driven by latent heat release. Suppose a typhoon drifts poleward from the ITCZ; see color plate section: Plate 0. (a) Describe its evolution as it migrates over colder SST. (b) What structure of mid-latitude westerlies will enable it to sustain itself?

11. Show that, for Eady modes, the eddy heat flux $\overline{v'\theta'}$ is positive and independent of height.

12. Derive a necessary condition for instability of a quasi-geostrophic zonal flow that is bounded vertically at $z = 0$ and H by rigid walls.

13. Show that, for the fastest-growing Eady mode, the maximum slope of motion in the meridional plane is only about half the slope of mean isentropic surfaces.

14. Within the framework of an initial value problem, describe the structure and evolution predicted by the Eady model if the flow in Prob. 15.7 is initialized with random structure having a broad wavenumber spectrum (a) under inviscid adiabatic conditions, (b) in the presence of linear dissipation[4] with a time scale equal to the shortest e-folding time of zonal wavenumber 2 under inviscid adiabatic conditions, (c) in the presence of linear dissipation with a time scale equal to the shortest e-folding time of zonal wavenumber 4 under inviscid adiabatic conditions.

15. A wave packet approaches its critical line with phase speed c_r equal to the real part of c in (16.8.1). Describe the wave packet's evolution in the presence of weak dissipation if, under inviscid adiabatic conditions, (a) c in (16.8.1) is real, (b) c in (16.8.1) is complex.

---

[4] Rayleigh friction and Newtonian cooling of the same time scale.

SEVENTEEN

Influence of the ocean

Regional climate is strongly influenced by thermal properties of the Earth's surface, in particular, by neighboring ocean that moderates extreme conditions (Chap. 15). An analogous influence is exerted on global-mean climate. Owing to its large heat capacity and capacity to hold substances in solution, the ocean serves as a reservoir of energy and carbon. It thereby provides thermal inertia to the climate system, figuring importantly in exchanges with the atmosphere of heat and carbon dioxide.

The Earth has only one ocean. It is compartmentalized in major basins, like the Atlantic and Pacific. Those bodies of water are interconnected by currents, circulation systems that exchange mass between ocean basins. Ocean circulations are driven by atmospheric wind stress, which transfers momentum to the ocean. They are also driven by transfers of heat and moisture, which, through density, influence the buoyancy of seawater.

17.1 COMPOSITION AND STRUCTURE

Seawater is a solution of pure water and a variety of salts, mostly sodium chloride, which accounts for more than 80% of its ionic composition. *Salinity S* is the relative concentration of dissolved material, measured in grams per kilogram or parts per thousand (‰). Each is approximately equal to Practical Salinity Units (PSU). Displayed in Fig. 17.1 is surface salinity. It varies between 30 and 40 g/kg. However, values are clustered about 35 g/kg. High salinity is found in the subtropical Atlantic and the Mediterranean Sea. There, as over the subtropical Pacific, salinity is increased by evaporation.

Figure 17.2 compares, in zonal-mean values, salinity (dashed) against net evaporation (solid). The latter equals the evaporation rate minus the precipitation rate. Maxima

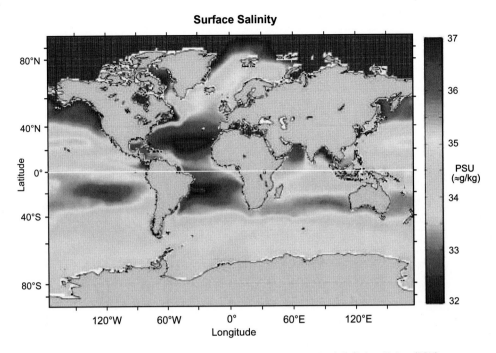

Figure 17.1 Annual-mean surface salinity, in Practical Salinity Units (PSU), approximately equal to g/kg. *Source:* Antonov et al. (2006). See color plate section: Plate 19.

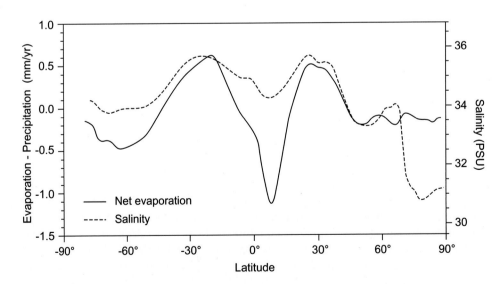

Figure 17.2 Zonal-mean surface salinity (solid) compared against net evaporation, equal to evaporation–precipitation (dashed). Adapted from Stewart (2008).

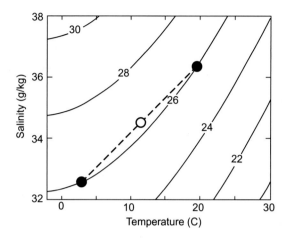

Figure 17.3 Density anomaly σ (kg/m^3), at sea level, as a function of salinity and temperature (solid). Superimposed are different mixing ratios of two bodies of water that have different values of salinity and temperature (solid circles) but identical $\sigma = 26$ g/kg (dashed). Mixing in equal proportion results in intermediate values (open circle).

of salinity appear in the subtropics. They coincide with positive maxima of net evaporation, which act to increase salinity. Those maxima reflect cloud-free conditions beneath the descending branch of the Hadley circulation, conditions that inhibit precipitation and maximize SW heating. Just the reverse appears near the equator, where salinity achieves a minimum. There, net evaporation attains a sharp negative minimum. It reflects net precipitation, which acts to reduce salinity. Concentrated inside the ITCZ, heavy precipitation lies beneath the ascending branch of the Hadley circulation (Fig. 9.41). The sharp minimum of S will be seen to follow also from upwelling of fresher water, which prevails in a neighborhood of the equator (Sec. 17.4.1).

The distribution of sea surface temperature (SST) is displayed in Fig. 5.1. It maximizes near the equator, exceeding 28°C in the Indian Ocean and western Pacific. Temperature decreases poleward, approaching 0°C in ice-laden regions at high latitude.

As seawater is a mixture, its state is determined by two thermodynamic properties plus salinity (Sec. 5.3). Density then has the form $\rho = \rho(T, p, S)$, analogous to $\theta = \theta(p, T, r)$ for moist air. Changes of density are small departures from the mean density of seawater (1027 kg/m^3). Discriminating to them is the *density anomaly*

$$\sigma(T, p, S) = \rho(T, p, S) - 1000 \, \text{kg/m}^3.$$

In the upper ocean, pressure remains small enough for seawater to be treated as incompressible. The density anomaly then depends only on temperature and salinity

$$\sigma = \sigma(T, S).$$

The forgoing represents the equation of state for seawater. Nonlinear, it is determined empirically. Figure 17.3 plots the dependence of density on T and S (at SLP). Density increases with decreasing temperature and increasing salinity. Isopleths of density (isochores or, in oceanographic vernacular, *isopycnals*) are concave in the same sense: toward decreasing temperature and increasing salinity. This nonlinear dependence

Ocean Stratification

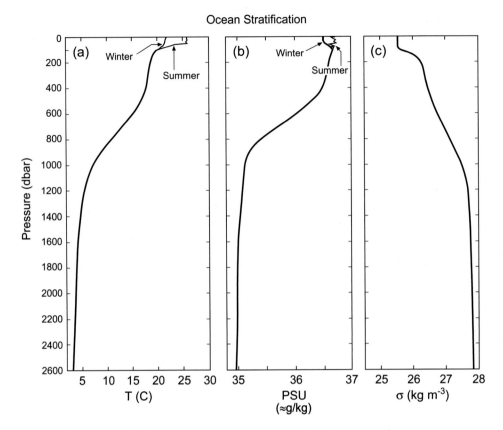

Figure 17.4 Profiles of (a) temperature and (b) salinity in Atlantic, near Bermuda, under summertime and wintertime conditions. (c) Corresponding mean profile of density anomaly. (1 dbar = 0.1 bar translates into approximately 1 m of depth.) *Sources:* Phillips and Joyce (2007), Java Ocean Atlas, http://www.epic.noaa.gov/epic (21.03.10).

has an important consequence for the buoyancy of seawater. Two water parcels of identical density but different salinity, upon being mixed, then achieve greater density. As indicated in Fig. 17.3, both parcels must initially lie on the same isopleth of σ. Because T and S are conserved, their ratio must be as well. The mixture must therefore lie on a line of constant $\frac{T}{S}$, intermediate to the two initial states. This positions the mixture at higher values of σ. Known as *cabbeling*, this process forms dense water in regions where surface currents converge. Such water, in turn, sinks to greater depth.

17.1.1 Stratification

As in the atmosphere, ocean properties are shaped by the vertical distribution of mass. Near the surface is a layer of 10–100 m in which vertical overturning prevails. Known as the *mixed layer*, it is the oceanic counterpart of the troposphere. Inside the mixed layer, temperature and salinity are invariant with depth (Fig. 17.4). So too then is density. Its uniform vertical distribution corresponds to neutral stability, which is achieved by efficient overturning.

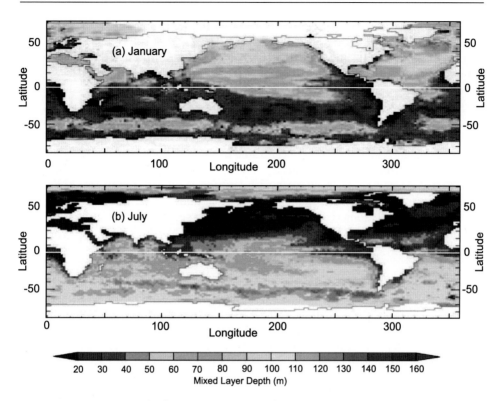

Figure 17.5 Depth of the mixed layer, as function of geographical position, during (a) January and (b) July. *Source:* http:/www./locean-ipsl.upmc.fr. See color plate section: Plate 20.

The mixed layer is maintained by two processes: Surface cooling destabilizes underlying ocean, driving thermal overturning (Fig. 7.11). Surface wind and accompanying wave motion drive mechanical overturning. Both influences are amplified during winter. As is evident from Fig. 17.4, the mixed layer is then deepest. During summer, surface cooling is replaced by surface heating. It stabilizes underlying ocean, while surface wind is weak. The mixed layer is then shallowest.

The depth of the mixed layer is plotted in Fig. 17.5. During January (Fig. 17.5a), the mixed layer is expanded in the north Atlantic and north Pacific. Driven by cooling and surface wind in the storm tracks, its depth there exceeds 100 m. At low latitudes, and in the summer hemisphere, it is shallower, with depth of 30–50 m. During July (Fig. 17.5b), the mixed layer is expanded adjacent to Antarctica, in the storm track of the Southern Hemisphere. Its depth then is again shallowest in the summer hemisphere, where values decrease to 20 m.

Beneath the mixed layer is the *thermocline*, in which temperature decreases and density increases downward (Fig. 17.4).[1] The downward increase of density makes the thermocline strongly stable. Together with continuity, which requires $\frac{d\rho}{dt} = 0$, this feature constrains water to move along isopycnal surfaces. Quasi-horizontal in the

---

[1] Salinity also influences the downward increase of density, which is known more generally as the *pycnocline*. However, in the open ocean, the variation of temperature dominates (cf. Fig. 17.4a,c). The pycnocline is then equivalent to the thermocline.

open ocean, isopycnal surfaces are the counterpart of isentropic surfaces in the atmosphere (for which $\frac{d\theta}{dt} = 0$). From the base of the mixed layer, where temperature is of order $20°C$, the thermocline extends downward to a depth of about 1000 m, where temperature is $2\text{--}4°C$.

Beneath the thermocline is the *deep ocean*. There, temperature and salinity vary with depth only gradually (Fig. 17.6). Characterized by temperatures of $\sim 4°C$, this layer represents 80% of the ocean's overall mass. Water in the deep ocean is largely sequestered from the surface. It therefore experiences limited interaction with the atmosphere. Such interaction is concentrated in those regions near the surface where *deep water* is produced and destroyed.

The submerged ocean layers vary with latitude. As illustrated in Fig. 17.6a, they are deepest in the tropics and shallowest at high latitude. Because SST decreases poleward (Fig. 5.1), the submerged layers eventually intersect the surface at high latitude, forming outcroppings. There, deep water and surface water interact.

17.1.2 Motion

Observations of the ocean circulation are mostly limited to a neighborhood of the surface. Plotted in Fig. 17.7 is the mean surface circulation. Large-scale motion is dominated by anticyclonic gyres that prevail in the subtropics of each hemisphere. Near the equator, they form strong easterly motion: the *North* and *South Equatorial Currents*. Compensating them at high northern latitudes is strong westerly motion: the *North Pacific* and *North Atlantic Currents*. These EW currents are bridged by coastal currents in the NS direction: In the Pacific, the *California Current* flows equatorward along North America, whereas the *Kuroshio Current* flows poleward along Asia. By transferring warm water poleward and cold water equatorward, they lead to poleward transport of heat. Similar motion prevails in other basins. In the Atlantic, the *Gulf Stream* flows poleward along North America. It eventually forms the *North Atlantic Current*, which moderates the temperature of air that is carried across Europe by westerlies. The same influence maintains most of the north Atlantic ice-free. The *Canary Current* flows equatorward along Europe and Africa. Mirroring the preceding features are coastal currents in the Southern Hemisphere: In the Pacific, the *Humboldt Current* flows equatorward along South America, whereas the *East Australian Current* flows poleward along Australia. In the Atlantic, the *Benguela Current* flows equatorward along Africa, whereas the *Brazil Current* flows poleward along South America.

The large-scale gyres in Fig. 17.7 have velocities of order 0.5 m/s. They are intensified along the western boundaries of ocean basins, where coastal currents transfer warm water poleward. There, motion is concentrated in jets, which are stronger and narrower than other legs comprising the gyres. The basin-scale circulations will be seen in Sec. 17.4 to be driven by wind stress that is exerted on the ocean surface.

Close to the equator, the easterly motion that prevails at low latitude reverses. The *Equatorial Counter Current* is westerly. Evident around the Earth, it will be seen to follow from certain features of equatorial dynamics. Crossing the equatorial Atlantic is northward motion. It transports water from the Southern Hemisphere to the Northern Hemisphere. That inter-hemispheric transport must be compensated by subsurface transport in the opposite direction.

At high latitudes are the Arctic Ocean and the Southern Ocean. The Arctic Ocean has limited interaction with surface water in other basins. Such interaction is concentrated

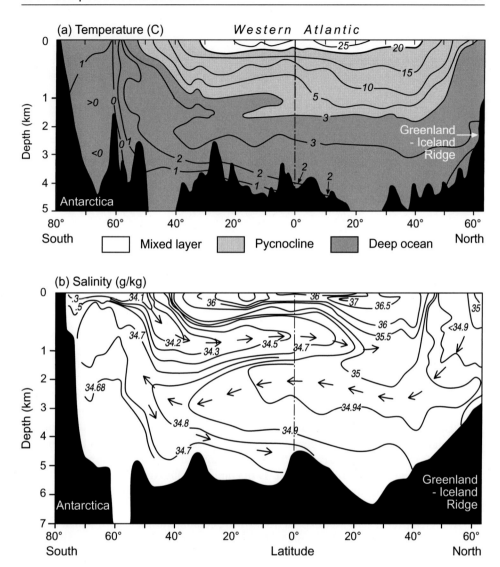

Figure 17.6 Latitude-depth cross section in western Atlantic of (a) temperature (°C) and (b) salinity (g/kg). *Sources:* Dietrich et al. (1980), Lynn and Reid (1968).

in the *Norwegian Current*, which pumps warm water from lower latitudes into the Arctic basin. Compensating that inflow is outflow of cold water along the coasts of Labrador and Greenland. At high southern latitudes is the *Antarctic Circumpolar Current*. Unbroken by continents, it is composed of a continuous westerly drift. The Antarctic Circumpolar Current coincides with SST of 4°C and colder. It therefore represents the motion of deep water, which is exposed at high latitude (Fig. 17.6a).

Observations of subsurface motion are hampered by limited coverage, as well as slow velocity that is characteristic of the deep ocean. Of order 1 mm/s, such motion is difficult to measure. At this speed, water that enters the deep ocean at one end

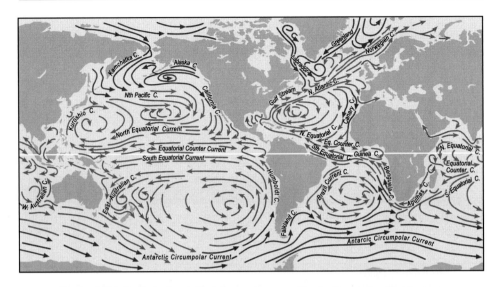

Figure 17.7 Surface circulation during January, illustrating water that is comparatively warm (red) and cold (blue). *Source:* US Navy Oceanographic Office. See color plate section: Plate 21.

of its circulation remains there for many centuries, before eventually emerging at the other end.

The deep ocean circulation is driven by buoyancy. Because density is determined by both temperature and salinity, the resulting motion is called the *thermohaline circulation* (also the *abyssal circulation* and *meridional overturning circulation*). The thermohaline circulation controls properties of the deep ocean and, therefore, most of the ocean's store of heat and carbon. Despite this pivotal role, observational limitations leave the thermohaline circulation poorly documented. Illustrated in Fig. 17.8, it must be inferred from the structure of conserved properties such as salinity, which serves as a tracer of motion (Fig. 17.6b).

At high latitude, seawater experiences strong cooling during winter, which renders it negatively buoyant. Further, some of it freezes, removing fresh water. Analogous to evaporation, this process enriches salinity, reinforcing negative buoyancy. Anomalously cold and salty, the resultant water has properties of deep water. It sinks along isopycnal surfaces, which become quasi-horizontal in the open ocean (Fig. 17.6a). The thermohaline circulation is thus initiated at sites where deep water is produced.

In the far north Atlantic, near Greenland, seawater experiences strong cooling by continental air from the Canadian Arctic (Fig. 15.9). Supported by freezing and enrichment of salinity, cooling produces *North Atlantic Deep Water* (indicated blue in Fig. 17.8). Anomalously dense, that water sinks through convection and then moves southward along isopycnal surfaces. It is marked in Fig. 17.6b by the tongue of salty water that extends southward near 3000 m. Upon reaching Antarctica, it overrides even denser *Antarctic Bottom Water*, which is produced through similar processes in the shallow Weddell Sea. Deep water then drifts eastward around Antarctica, eventually bifurcating: Part drifts northward into the Indian Ocean. The rest continues into the Pacific, where it drifts northward. Through mechanisms that remain obscure, both components of deep water become less dense, eventually finding their way to shallower

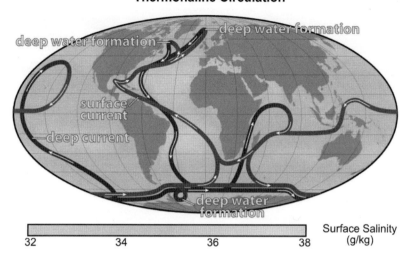

Figure 17.8 Schematic of the thermohaline circulation, along with surface salinity (shaded). Deep water (blue) forms in the far North Atlantic, near Greenland, and in the Weddell Sea, near Antarctica. After meandering through the global ocean, it eventually percolates upward, forming surface water (red). The time for the entire circuit to be completed is 1000–2000 years. *Source:* http://earthobservatory.nasa.gov. See color plate section: Plate 22.

depths (red).[2] There, the two streams from the deep ocean are thought to reunite in the Indian Ocean. After drifting into the surface circulation of the South Atlantic, the resultant water is carried north, first by the Benguela and Atlantic Equatorial Currents, then across the equator, and finally into the Gulf Stream, which returns it to the North Atlantic. The time for water to complete this circuit is longer than 1000 years.

Motion that comprises the thermohaline circulation is initiated by negative buoyancy. However, the circuit also involves vertical mixing. On long time scales, weak vertical mixing gradually diffuses deep water upward, enabling it to percolate through the stable thermocline and complete the circuit. Negative buoyancy is produced at high latitudes, especially in the far North Atlantic. There, cooling and freezing of seawater force downwelling. These diabatic processes must be attended by transfers to the atmosphere of sensible and latent heat. On long time scales, those transfers represent additional heating of air that is carried across Europe by westerlies. A weakening of the thermohaline circulation would therefore favor colder conditions over Europe. Such weakening has been hypothesized in relation to the Little Ice Age (Sec. 1.6.2). However, without a more concrete understanding of the deep ocean circulation, severely limited by available observations, that remains a speculation.

17.2 ROLE IN THE HEAT BUDGET

Because it covers 70% of the Earth, the ocean absorbs much of the overall SW energy that arrives at the Earth's surface. But even per m², ocean is particularly efficient as an

---

[2] On long time scales that characterize the thermohaline circulation, weak vertical mixing enables deep water to gradually percolate upward.

SW Heating

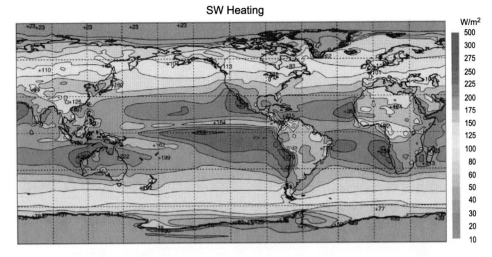

Figure 17.9 Annual-mean absorption of SW radiation at the Earth's surface. SW heating maximizes in the subtropics, especially in the eastern oceans, which are comparatively free of cloud and have large albedo. Derived from ERA-40 reanalysis. *Source:* Kallberg et al. (2005). See color plate section: Plate 23.

absorber. The darkest scenes in visible imagery are cloud-free ocean (Fig. 1.29). They lead to time-mean albedo of less than 20% (Fig. 1.34a).[3]

Plotted in Fig. 17.9 is the distribution of SW absorption at the Earth's surface. SW heating is concentrated in the subtropics. There, cloud cover is inhibited by down-welling of the Hadley circulation (Sec. 15.4.3). SW absorption is greatest over ocean, owing to its low albedo. It maximizes in the eastern oceans, where cloud-free conditions are maintained by cold SST and downwelling of the Walker circulation (Figs. 5.1, 1.36). In concert with small solar zenith angle and the low albedo of ocean, such conditions lead to column-integrated heating of the ocean in excess of 250 W/m^2.

The SW energy absorbed by the ocean is eventually transferred to the atmosphere, mostly through latent heat (Fig. 1.32). Plotted in Fig. 9.41b is the distribution of column-integrated heating of the atmosphere. Positive values mirror the pattern of precipitation (Fig. 9.41a). Latent heating thus provides a major source of energy for the atmosphere, energy transmitted from the ocean through evaporative cooling. Precipitation is concentrated inside the ITCZ, where atmospheric heating exceeds 400 W/m^2. Secondary maxima appear in the storm tracks over the north Atlantic and north Pacific. Elsewhere, values are negative. Atmospheric cooling is greatest in the subtropical eastern oceans. There, extensive coverage by low (comparatively warm) marine stratocumulus yields magnified albedo with strong LW emission, permitting efficient LW cooling to space.

During summer, the ocean acquires energy through enhanced absorption of SW, which then dominates emission of LW. The excess energy, which warms the ocean, is transferred downward. On seasonal time scales, it is redistributed across the mixed layer. During winter, the reverse operates. The ocean then loses energy through

---

[3] For overhead sun, instantaneous albedo is even smaller, approaching zero.

reduced absorption of SW, which leaves LW emission unbalanced. The energy lost is then drawn from the mixed layer, which, on seasonal time scales, transfers energy upward. The mixed layer thus serves as the principal repository of anomalous energy that is transferred to/from the ocean – at least on seasonal time scales. A similar process operates for continent. However, in continent, heat is transferred vertically only through conduction (Sec. 15.4). The affected depth is therefore shallow, limited to of order 1 m. These contrasting properties sharply differentiate the storage of heat in ocean and continent.

Columns of water and land have comparable density, each of order 10^3 kg/m^3. However, they differ in specific heat and, notably, in the depth that is affected by the annual variation of radiative heating and cooling. The ratio of ocean heat content to land heat content (per square meters^2) then involves the product of three ratios: (1) the ratio of specific heats, (2) the ratio of anomalous temperatures, and (3) the ratio of affected depths. The specific heat of water is four times that of most substances comprising land. The seasonal swing of ocean temperature, on the other hand, is less than one fourth that of land (Fig. 15.8). However, the depth over which it is distributed is two orders of magnitude greater. Additionally, coverage of the Earth's surface by ocean is twice that by land. Consequently, on seasonal time scales, the heat storage of ocean is two orders of magnitude greater than that of continent (Prob 17.4). It is concentrated in the mixed layer.

Most of the heat absorbed by the mixed layer during the annual cycle is transferred to the atmosphere. On time scales longer than seasonal, however, heat is exchanged with the deep ocean through convection and vertical diffusion. The latter is the same process that, in the gradual thermohaline circulation, eventually transfers deep water to the surface. Heat that is transferred into the deep ocean is sequestered for centuries, transported thereafter only by the slow drift of the thermohaline circulation. To the degree that it influences SST, such heat transfer imposes a long time scale on atmospheric climate. A stable average, one truly representative of climatological mean conditions, would then require averaging over centuries, if not millennia (Sec 1.6).

Absorption by the ocean of SW (Fig. 17.9) accounts for much of the positive net radiation that prevails at low latitudes in the TOA energy budget (Fig. 1.34c). What is not transferred to the atmosphere (Fig. 9.41b) must be transported poleward by the ocean to maintain thermal equilibrium.

Plotted in Fig. 17.10 is the northward transport of heat by the atmosphere and ocean, in petawatts (10^{15} W). Total poleward transport (solid) maximizes at latitudes near 30°, where it exceeds 5 PW. Most of that heat transport is accomplished by the atmosphere (dashed). Coincident with steep meridional temperature gradient and strong baroclinicity (Fig. 6.3), it is performed by baroclinic disturbances. Heat transport by the ocean is inferred as a residual between the total, which is required by the TOA distribution of net radiation (Fig. 1.34c), and heat transport by the atmosphere, which is better observed. Heat transport by the ocean (dotted) is generally smaller. However, it rivals that by the atmosphere equatorward of 20°, where the two are comparable. There, it reflects equatorial portions of the wind-driven gyres, which exchange warm tropical water with cooler extratropical water (Fig. 17.7). The thermohaline circulation, although poorly documented, may be equally important. Despite being comparatively weak, that handicap of the thermohaline circulation is offset by its vastly greater depth. In tandem, these features of the deep ocean can produce poleward heat transport as large as the surface circulation.

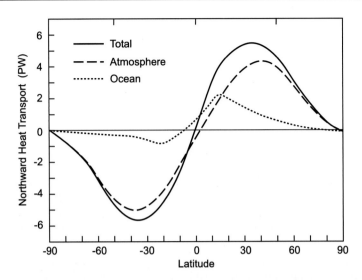

Figure 17.10 Northward transport of heat: total (solid), contribution from the atmosphere (dashed), and contribution from the ocean (dotted). Contribution from the ocean inferred as a residual between the total and atmospheric contributions. *Source:* Wunsch (2005).

17.3 ROLE IN THE CARBON CYCLE

Carbon dioxide dissolves readily in seawater. Its saturation mixing ratio increases with decreasing temperature. More CO_2 can therefore be absorbed into solution with cold water than with warm water. Alternatively, saturated water that is heated leads to outgassing of CO_2. Consequently, absorption of CO_2 is favored at polar latitudes, where it is carried into the deep ocean by the thermohaline circulation. Conversely, emission of CO_2 is favored at tropical latitudes, where deep water, rich in CO_2, percolates upward through weak vertical mixing (Fig. 17.8).

When it is absorbed, CO_2 combines with H_2O to produce carbonic acid

$$CO_2 + H_2O \rightleftharpoons H_2CO_3. \tag{17.1.1}$$

The formation of H_2CO_3 lowers the *pH* of seawater, which, owing to the presence of salt, is slightly alkaline: *pH* values of 7.5–8.5. Carbonic acid dissociates, producing bicarbonate ions

$$H_2CO_3 \rightleftharpoons H^+ + HCO_3^-. \tag{17.1.2}$$

They in turn dissociate, producing carbonate ions

$$H^+ + HCO_3^- \rightleftharpoons 2H^+ + CO_3^-. \tag{17.1.3}$$

The carbonate ion combines with other species to form organic matter that comprises marine organisms, for example, calcium carbonate.

Photosynthesis by phytoplankton has a similar effect by converting CO_2 into oxygen and carbohydrate. Phytoplankton account for as much as 50% of the emission of oxygen and absorption of CO_2. There is some evidence of a decline in phytoplankton during the the latter half of the twentieth century, by as much as 40% (Boyce et al.,

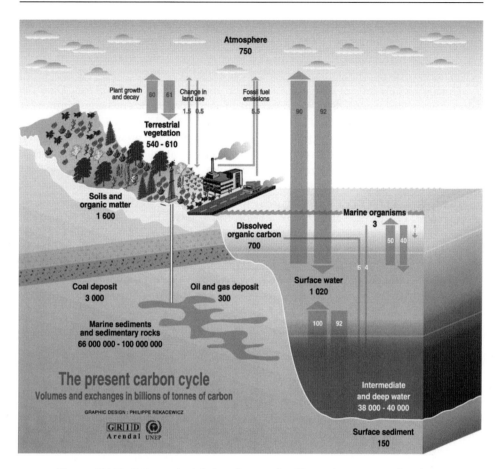

Figure 17.11 Estimated global carbon cycle, illustrating stores of carbon, in GtC, and transfers in GtC/yr, where 1 GtC = 10^9 tons of carbon. *Source:* http://maps.grida.no/go/graphic/the-carbon-cycle, design by Philippe Rekacewicz, UNEP/GRID-Arendal (11.07.10). See color plate section: Plate 24.

2010). Associated with increasing SST, such a decline would represent a significant decrease of CO_2 absorption. That would be equivalent to an anomalous source of CO_2 (cf. Sec. 8.7.1).

When marine organisms die, organic matter precipitates to the ocean floor, where it forms reduced carbon. A small fraction becomes sequestered in sediments, removing carbon from the ocean. However, most is reabsorbed into the ocean by living organisms.[4]

The storage of CO_2 is illustrated in Fig. 17.11. Except for deep sedimentary rock, which is sequestered, most of the carbon is stored in the ocean. It accounts for some 40,000 gigatons (10^{12} kg) of carbon (GtC), in the form of dissolved CO_2 and organic matter. Most resides in the deep ocean, where cold water supports the greatest observed

---

[4] The sequestration of organic matter in ocean sediments prevents its subsequent oxidation through decomposition. Otherwise, the latter process would systematically absorb oxygen from the climate system, depleting the atmosphere of O_2.

concentrations. There, dissolved CO_2 is controlled by the thermohaline circulation. Land and the adjoining biosphere account for only about 2000 GtC. The atmosphere contains less than 1000 GtC, concentrated in CO_2. Hence, the store of carbon in the ocean is two orders of magnitude greater than the store in the atmosphere.

Equally significant are transfers of carbon into and out of the ocean. Of order 100 GtC/yr, they exceed those into and out of land. Together, emission from ocean and land sources (~150 GtC/yr) is two orders of magnitude greater than CO_2 emission from combustion of fossil fuel. These natural sources are offset by natural sinks, of comparable strength. However, because they are so much stronger, even a minor imbalance between natural sources and sinks can overshadow the anthropogenic component of CO_2 emission (cf Secs 1.6.2, 8.7.1).

The values in Fig. 17.11 can be used to estimate the effective turnover time of atmospheric CO_2. At an absorption rate of 100 GtC/yr, the ocean will absorb the atmospheric store of CO_2 of 1000 GtC in about a decade. That absorption of CO_2, which is concentrated in cold SST at polar latitudes, is nearly offset by emission of CO_2 from warm SST at tropical latitudes. Warming of SST (by any mechanism) will increase the outgassing of CO_2 while reducing its absorption. Owing to the magnitude of transfers with the ocean, even a minor increase of SST can lead to increased emission of CO_2 that rivals other sources (Sec. 8.7.1). Further, if the increase of SST involves heat transfer with the deep ocean, the time for equilibrium to be reestablished would be centuries (Sec. 17.1.2).

17.4 THE WIND-DRIVEN CIRCULATION

The depth of the mixed layer maximizes in the storm tracks, where surface wind is strong (Fig. 17.5). This feature points to the importance of momentum transfer from the atmosphere, which is transmitted through wind stress on the ocean surface. By stirring the mixed layer, wind stress drives the surface circulation of the ocean.

17.4.1 The Ekman layer

A parallel of surface motion was treated earlier in the idealized zonally symmetric circulation (Sec. 15.2). There, the free surface is analogous to the ocean surface. Beneath the free surface is a shallow Ekman layer. Inside it, frictional-geostrophic balance enables the motion to adjust sharply to meet the boundary condition of zero stress. The ocean surface, on the other hand, experiences nonzero wind stress,

$$\boldsymbol{\tau}_0 = \tau_{0xz}\boldsymbol{i} + \tau_{0yz}\boldsymbol{j},$$

following from the drag exerted on it by surface wind. The wind stress is tantamount to a downward flux of horizontal momentum (Sec. 10.6). That flux must be continuous across the atmosphere-ocean interface.

Inside the Ekman layer, motion that is steady, in the presence of turbulent diffusion, but with the neglect of advective acceleration, is governed by the momentum balance

$$f\boldsymbol{k} \times \boldsymbol{v} = -\nabla\Phi + \frac{1}{\rho}\frac{\partial \boldsymbol{\tau}}{\partial z}, \tag{17.2.1}$$

where the geopotential Φ of isobaric surfaces follows from hydrostatic balance (14.7),

$$\boldsymbol{\tau} = \rho K \frac{\partial \boldsymbol{v}}{\partial z} \tag{17.2.2}$$

is the turbulent shear stress at level z, and K is the eddy diffusivity. Continuity requires

$$\nabla \cdot \boldsymbol{v} + \frac{\partial w}{\partial z} = 0. \tag{17.3}$$

Expressing motion in terms of a geostrophic component $\boldsymbol{v}_g$ plus an ageostrophic (Ekman) component $\boldsymbol{v}_E$, which varies sharply with depth, reduces the momentum balance inside the Ekman layer to

$$\begin{aligned} f\boldsymbol{k} \times \boldsymbol{v}_E &= \frac{1}{\rho}\frac{\partial \boldsymbol{\tau}}{\partial z} \\ &= K\frac{\partial^2 \boldsymbol{v}_E}{\partial z^2}, \end{aligned} \tag{17.4.1}$$

where z is understood to be a stretched vertical coordinate that renders frictional terms comparable to others in the momentum balance (Sec. 15.2). At the ocean surface, frictional stress must equal the imposed wind stress. At the bottom of the Ekman layer, it must vanish, requiring ageostrophic motion to disappear. Boundary conditions are thus

$$\begin{aligned} \rho K \frac{\partial \boldsymbol{v}_E}{\partial z} &= \boldsymbol{\tau}_0 & z &= 0 \\ \boldsymbol{v}_E &\sim 0 & z &\to -\infty \end{aligned} \tag{17.4.2}$$

If surface wind is northward,

$$\boldsymbol{\tau}_0 = \tau_0 \boldsymbol{j},$$

then (17.4) has solution

$$\begin{aligned} u_E &= v_0 e^{\gamma z} \cos\left(\gamma z + \frac{\pi}{4}\right) \\ v_E &= v_0 e^{\gamma z} \sin\left(\gamma z + \frac{\pi}{4}\right), \end{aligned} \tag{17.5.1}$$

where, analogous to (13.17.2),

$$\gamma = \sqrt{\frac{f}{2K}} \qquad v_0 = \frac{\tau_0}{\rho\sqrt{fK}}. \tag{17.5.2}$$

The ageostrophic motion has the form presented in Fig. 17.12; it is analogous to structure in Fig. 13.3. At the ocean surface, where

$$u_E = v_E = \frac{v_0}{\sqrt{2}},$$

Ekman current is deflected $45°$ to the right of wind in the Northern Hemisphere. This feature is in agreement with reports of mariners that sea ice drifts to the right of wind. With increasing depth, Ekman current veers increasingly to the right and weakens, until the total current $\boldsymbol{v} = \boldsymbol{v}_g + \boldsymbol{v}_E$ eventually becomes geostrophic. The depth of the Ekman layer is approximated by that depth where $\boldsymbol{v}_E$ first reverses:

$$d_E = \frac{\pi}{\gamma}. \tag{17.6}$$

It increases with eddy diffusivity, which, in turn, increases with wind stress. It decreases with latitude.

Motion at an individual level depends on the frictional formulation and, hence, on K. However, column-integrated transport within the Ekman layer does not. The

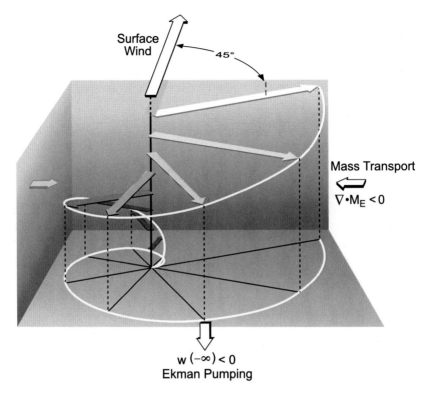

Figure 17.12 Ageostrophic current inside the Ekman layer. Mass transport M_E is orthogonal to surface wind, to the right in the Northern Hemisphere. Its convergence produces Ekman pumping from the frictional layer. Copyright American Meteorological Society, with permission.

ageostrophic *mass transport* is

$$M_E = \int_{-\infty}^{0} \rho \boldsymbol{v}_E dz.$$

Integrating the momentum balance vertically, together with boundary conditions (17.4.2), reduces (17.4.1) to

$$f \boldsymbol{k} \times \boldsymbol{M}_E = \boldsymbol{\tau}_0.$$

Operating across by $\boldsymbol{k} \times$ then obtains

$$\boldsymbol{M}_E = -\frac{1}{f} \boldsymbol{k} \times \boldsymbol{\tau}_0. \tag{17.7}$$

Ekman transport is determined entirely by the wind stress. More notably, it is directed just perpendicular to $\boldsymbol{\tau}_0$ – to the right in the Northern Hemisphere. This peculiarity of the surface current is another manifestation of the geostrophic paradox (Chap. 10).

Integrating the continuity equation in similar fashion reduces (17.3) to

$$\nabla \cdot \boldsymbol{M}_E + \rho \left[w(0) - w(-\infty) \right] = 0.$$

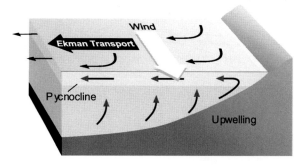

Figure 17.13 Coastal upwelling induced by along-shore wind through Ekman transport. Copyright American Meteorological Society, with permission.

At the ocean surface, steady vertical motion must vanish. With (D.7), vertical motion at the base of the frictional layer then follows as

$$w(-\infty) = \nabla \cdot \boldsymbol{M}_E$$
$$= \frac{1}{\rho} \boldsymbol{k} \cdot \nabla \times \left(\frac{\tau_0}{f} \right). \tag{17.8}$$

Representing Ekman pumping (Sec. 13.4), $w(-\infty)$ is proportional to the curl of the wind stress. The latter represents a torque that is exerted on the ocean surface. It creates relative vorticity, providing a parallel of Ekman pumping in the atmosphere (13.28). Wind stress that is anticyclonic produces Ekman pumping: $w(-\infty) < 0$. Mass is expelled from the Ekman layer at its base. Wind stress that is cyclonic produces Ekman suction: $w(-\infty) > 0$. Mass is absorbed by the Ekman layer at its base. As in the free atmosphere, vertical motion at the base of the Ekman layer imposes a secondary circulation on the geostrophic circulation of the deep ocean.

Together, Ekman transport and pumping have important implications for surface conditions. Along the eastern boundary of ocean basins (at the west coast of continents), surface wind has an equatorward component, induced by subtropical high pressure over ocean (Fig. 15.9). Both features are intensified during summer. Their influence on the ocean is exemplified by currents along the coast of California (Fig. 17.13). Northerly wind along the California coast drives Ekman transport in the ocean that is directed to the right and therefore offshore. The latter forces surface divergence along the coastal shelf. It must be compensated by upwelling of cold water from the deep ocean. Upwelling elevates the thermocline (Fig. 17.14), producing anomalously cold SST along the coast (Fig. 17.15).

Cold SST exerts a major influence on regional climate along the west coast of continent. By cooling the atmosphere from below, it stabilizes the troposphere, inhibiting deep convection (Fig. 7.11). Instead, cold SST favors surface fog, which forms when humidified air is cooled to saturation. Such conditions prevail over San Francisco during summer, when fog is swept inland by the onshore component of wind. They encourage semi-arid climate along the west coast of continent. There, regular rainfall is limited to winter, when coastal upwelling is weak and synoptic weather systems are active. Deep water, beyond being cold, is also nutrient rich.[5] For this reason, upwelling

---

[5] From the decomposition of dead organisms that have settled to the ocean floor.

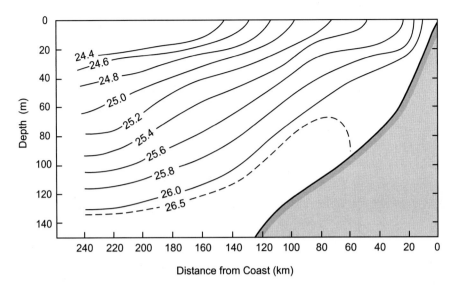

Figure 17.14 Cross section of density anomaly, orthogonal to Peruvian coast during May. *Source:* Strub et al. (1998).

simultaneously supports marine productivity. The west coasts of continents are sites of major fisheries. The catch off the coast of Peru alone accounts for a fifth of the global harvest.

Near the equator, subtropical highs in the atmosphere drive easterlies: the NE and SE trade winds (Fig. 15.9). As illustrated in Fig. 17.16a, Ekman transport is then northward in the Northern Hemisphere and southward in the Southern Hemisphere (17.7). This makes the equator a region of surface divergence. It too must be compensated by upwelling (Fig. 17.16b). Such vertical motion favors cold SST and low salinity along the equator, as is evident in Fig. 17.1. These features are amplified at the eastern boundaries of ocean basins, where equatorial upwelling is reinforced by coastal upwelling. Cold SST inhibits atmospheric convection. For this reason, the ITCZ tends to form to either side of the equator, but not directly over it (cf. Figs. 1.30b; 9.41).

When it deflects isopycnal surfaces, upwelling along the equator also introduces a deflection of isobaric surfaces. Vertical motion must satisfy

$$
\begin{aligned}
w &= \frac{dz}{dt} \\
&= \frac{1}{g}\frac{d\Phi}{dt}.
\end{aligned}
\tag{17.9.1}
$$

For steady motion, (17.9.1) reduces to

$$
\begin{aligned}
w &= \frac{1}{g}\boldsymbol{v}\cdot\nabla\Phi \\
&\cong \frac{1}{g}\boldsymbol{v}_g\cdot\nabla\Phi.
\end{aligned}
\tag{17.9.2}
$$

Upwelling associated with Ekman transport therefore implies anomalously high Φ in a neighborhood of the equator. Geostrophic balance (12.2) then forms current with low Φ to its left (right) in the Northern (Southern) Hemisphere. The result is westerly

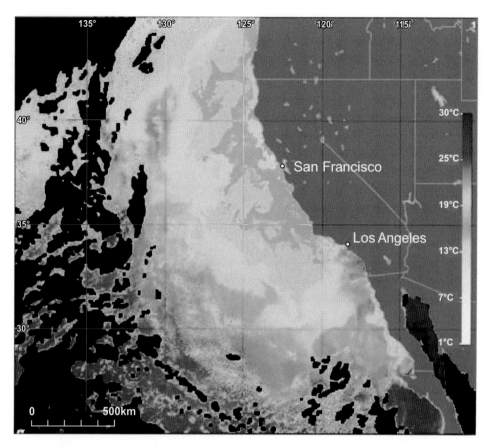

Figure 17.15 Surface temperature off the California coast. Temperature decreases shoreward, attaining values 10–15°C colder than in the open ocean. Black shading represents sites that are overcast, where data is unavailable. *Source:* http://oceanmotion.org. See color plate section: Plate 25.

current along the equator – just opposite to the easterly wind stress exerted by the trades. Reflecting the equatorial countercurrent (Fig. 17.7), it emerges naturally from the essential momentum balance governing the upper ocean.

17.4.2 Sverdrup balance

The upper ocean circulation is driven by the torque of wind stress, which forces vorticity. As in the treatment of geostrophic motion (12.47), applying $\boldsymbol{k} \cdot \nabla \times$ reduces (17.2.1) to the vorticity balance

$$\beta v + f \nabla \cdot \boldsymbol{v} = \frac{1}{\rho} \frac{\partial}{\partial z} \boldsymbol{k} \cdot \nabla \times \boldsymbol{\tau}. \tag{17.10}$$

With continuity (17.3), this becomes

$$\beta v - f \frac{\partial w}{\partial z} = \frac{1}{\rho} \frac{\partial}{\partial z} \boldsymbol{k} \cdot \nabla \times \boldsymbol{\tau}. \tag{17.11}$$

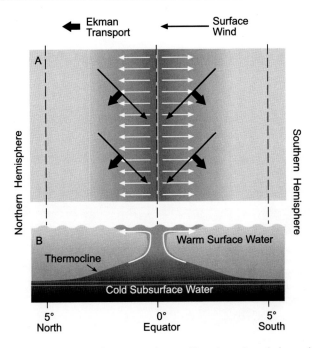

Figure 17.16 Schematic of equatorial upwelling introduced through Ekman transport. (a) Surface current (white arrows). Superimposed are surface trade winds (black arrows) and and Ekman transport (heavy black arrows). (b) Vertical section across equator, illustrating upwelling of cold subsurface water. Copyright American Meteorological Society, with permission.

Integrating vertically, introducing the column-integrated mass transport,

$$M = \int_{-\infty}^{0} \rho \boldsymbol{v} dz,$$

and requiring boundary conditions (17.4.2) and vertical motion to vanish at $z = 0$ and $z = -\infty$, then yields

$$\beta M_y = \boldsymbol{k} \cdot \nabla \times \boldsymbol{\tau}_0 \qquad\qquad (17.12.1)$$

$$\nabla \cdot \boldsymbol{M} = 0. \qquad\qquad (17.12.2)$$

Poleward transport of mass is determined entirely by the curl of the wind stress. Known as *Sverdrup balance*, (17.12.1) asserts that column motion can cross contours of planetary vorticity (e.g., basic-state potential vorticity) only through the creation of relative vorticity by the curl of the wind stress. Like (17.7), the column-integrated transport is independent of the frictional formulation.

Sverdrup balance describes how motion in the ocean interior is forced by surface wind through the Ekman layer. Ekman pumping leads to a downward displacement of isopycnals immediately beneath the surface layer. Vortex tubes in the ocean interior are therefore compressed, compensated by horizontal divergence. Conservation of potential vorticity then requires an anticyclonic adjustment of absolute vorticity (Fig. 12.9).

Such motion forces an equatorward drift (toward weaker planetary vorticity) and the anticyclonic gyres that prevail in the subtropics (Fig. 17.7).

In tandem with continuity (17.12.2), Sverdrup balance determines the horizontal distribution of ocean current from the distribution of wind stress. Equations (17.12), however, are only first order in M. Consequently, they enable only one boundary condition to be satisfied. Despite this limitation, Sverdrup balance provides great insight into the ocean circulation.

Requiring motion to be tangential to the coastline at the eastern boundary of ocean basins yields the distribution of mass transport in Fig. 17.17. Displayed is the streamfunction of equivalent volume transport, presented in units of Sverdrups (1 Sv = 10^6 m^3/s $\Leftrightarrow$ Mt/s). For its simplicity, Sverdrup balance captures the major features of the upper ocean circulation: Anticyclonic gyres prevail in the subtropics of each major basin. Represented likewise is an equatorial countercurrent, which was the original interest of Sverdrup (1947).

17.5 THE BUOYANCY-DRIVEN CIRCULATION

Descriptions of the thermohaline circulation, also known as the *abyssal circulation*, are mostly heuristic. Their qualitative nature reflects the dearth of quantitative documentation of motion in the deep ocean.

A simple model is due to Stommel and Arons (1960). It considers the deep ocean as a single homogeneous layer of depth H that extends from the ocean floor to the base of the thermocline. The layer is forced convectively at high latitude. There, injection of dense water from overhead (e.g., deep water formed through cooling at the ocean surface) serves as a concentrated mass source. Compensating it across the layer's upper surface is upwelling, which is weak but (except at sites of convection) horizontally uniform. Removing mass from the layer at the same rate, it serves as a distributed mass sink. Mass lost through upwelling eventually returns to the surface through vertical diffusion, which ventilates the thermocline with water from the deep ocean. Linking the mass source and mass sink is a western boundary current, which carries dense water equatorward to satisfy continuity (Fig. 17.18).

In the absence of turbulence, the vorticity balance (17.11) reduces to

$$\beta v = f \frac{\partial w}{\partial z}. \tag{17.13}$$

Advection of vorticity is then balanced by vertical stretching (Sec. 12.5). Vertical motion vanishes at the bottom of the layer, which is presumed flat. By construction, it is upward and horizontally uniform at the top of the layer. Integrating (17.13) vertically across the layer obtains

$$v = \frac{f}{\beta} \frac{w_0}{H}, \tag{17.14}$$

where w_0 is the vertical velocity at the top of the layer (i.e., at the base of the thermocline). Because $w_0 > 0$ throughout, meridional motion inside the layer must be everywhere poleward.

The implied circulation is illustrated in Fig. 17.18 for a sectoral basin symbolic of the North Atlantic. At the apex of the sector is a point source that represents the injection of North Atlantic deep water. Once introduced, dense water flows southward

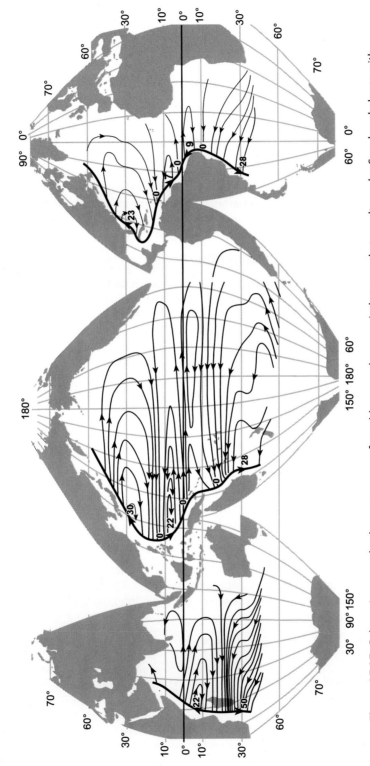

Figure 17.17 Column-integrated volume transport, forced by annual-mean wind stress, that results under Sverdrup balance with motion required to be tangential to the eastern boundaries of ocean basins. Transport streamfunction contoured at increments of 5 Sv (1 Sv = 10^6 m^3/s). Superimposed are the western-boundary currents required for continuity (heavy arrows). Adapted from Welander (1959).

Figure 17.18 Sectoral basin in Stommel and Arons (1961) model of the deep-ocean circulation. Superimposed is the point source of deep water, which forces the circulation, and the western-boundary current required by continuity.

along the western boundary of the basin. Compensating it in the interior is northward motion (17.14), which converges to form upwelling. Southward transport by the western-boundary current exceeds northward transport in the interior. Some of the northward stream must therefore peel off, rejoining southward motion at the western boundary in regions of recirculation.

The strength of the thermohaline circulation is determined by the horizontal gradient of surface buoyancy (e.g., by the formation of deep water in source regions). However, it is also determined by turbulent diffusion, which enables water from the deep ocean to percolate through the stable thermocline and, hence, return to the surface and complete the circuit. Tracer analyses such as that in Fig. 17.6b imply that deep water formation is concentrated in the Atlantic: in the North Atlantic near Greenland and in the South Atlantic at the Weddell Sea. Serving as mass sources for the deep ocean, those regions of dense water formation drive the gradual overturning that comprises the thermohaline circulation (Fig. 17.8). By contrast, the Pacific is virtually devoid of surface sources. It is supplied instead by deep water from the Atlantic, via the Southern Ocean.

Before returning to the surface, deep water must traverse the Indian and Pacific Oceans through a convoluted circuit. The overturning time depends on the path followed back to the surface. The time for water to reach the surface via the Indian Ocean is centuries. The time for it to return to the North Atlantic via the Pacific, which is devoid of major sources of deep water, is of order 2000 years.

17.6 INTERANNUAL CHANGES

As in the atmosphere, the circulation in the ocean varies. The wind-driven circulation evolves naturally with the annual cycle of surface pressure, wherein trade winds and the ITCZ migrate from one side of the equator to the other (Fig. 15.9). It also varies between years.

ENSO

El Nino Southern Oscillation involves a swing between extreme states of the circulation in the equatorial Pacific (Sec. 15.5.1). It is characterized by the *Southern Oscillation Index (SOI)*, which describes fluctuations of SLP in the eastern and western Pacific that are coherent but out of phase (Fig. 15.15). SOI is defined traditionally as SLP at Tahiti minus SLP at Darwin, Australia. Although alternative definitions exist, all measure the phase of the dipole oscillation across the equatorial Pacific. Negative SOI signifies SLP that is anomalously low in the eastern Pacific and anomalously high in the western Pacific. Easterly trade winds across the equatorial Pacific are then weak. Representing *El Nino* conditions, SOI < 0 corresponds to SST that is anomalously warm in the central and eastern Pacific and anomalously cold in the western Pacific. Positive SOI signifies SLP that is anomalously low in the western Pacific and anomalously high in the eastern Pacific. Easterly trade winds across the equatorial Pacific are then intensified. Representing *La Nina* conditions, SOI > 0 corresponds to SST that is anomalously cold in the central and eastern Pacific and anomalously warm in the western Pacific.

Figure 17.19 plots the record of normalized SOI, along with a measure of background variability that is unrelated to the southern oscillation. Contrary to its name, the southern oscillation is aperiodic. The greatest negative swing of SOI occurred during the 1982–1983 El Nino (Fig. 17.19a). SST in the eastern Pacific then warmed above its climatological mean by 7 K. That warm event, however, was not the strongest. During the 1997–1998 El Nino, SST in the eastern Pacific warmed by 8 K. Precipitation then was centered over the dateline (Fig. 17.20b). It was shifted well east of its climatological-mean position (Fig. 9.41a), leaving the western Pacific anomalously dry. Occurring simultaneously was conspicuous warming of global-mean temperature (Fig. 1.39). It reflects increased atmospheric heating, principally in the tropics. Through increased θ_e at the Earth's surface, increased heating is achieved through deeper convection and intensified cumulus heating of the troposphere (Sec. 7.7). Warm events (SOI < 0) also appear in SOI around 1987, 1993, 2005, and 2010. The protracted event around 2005 supported widespread forest fires in Indonesia. Dense haze produced by those fires led to health restrictions and public closures. Apparent in visible imagery (Cover; Plate 0), such haze is being drawn into one of two tropical cyclones that developed in the equatorial Pacific. Also punctuating the record of SOI are cold events (SOI > 0). La Nina conditions occurred around 1989, 1999, and 2007. During the 1998–1999 La Nina, precipitation was concentrated in the western Pacific (Fig. 17.20a). Its distribution then was not dissimilar to the climatological-mean pattern (Fig. 9.41a).

The warm and cold events must be considered in light of background variability, which is independent of ENSO. Such variability is reflected in the average of SLP at Tahiti and Darwin (Fig. 17.19b). Representing fluctuations across the Pacific that are coherent but in phase, the latter exceed 1-2 std deviations. Consequently, only strong events in SOI stand out clearly above random variability, which characterizes climate noise.

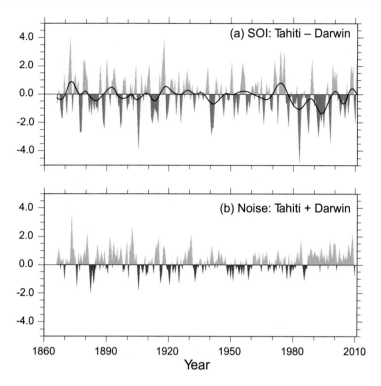

Figure 17.19 (a) Southern Oscillation Index (SOI), defined as SLP at Tahiti minus SLP at Darwin (shaded), along with the smoothed record (solid). Large deviations of SOI, which occur sporadically, represent fluctuations of SLP that are out of phase between those stations, characeristic of ENSO. (b) Fluctuations of SLP that are independent of SOI, defined by SLP at Tahiti plus SLP at Darwin. They represent fluctuations unrelated to ENSO. *Source:* http://www.cp.ncep. gov (14.07.10).

El Nino represents a perturbation to the annual cycle of SST. As depicted in Fig. 17.21, initial signs appear during the preceding northern summer, when anomalously warm SST appears off the coast of Peru. By autumn, the anomaly intensifies and expands westward, forming a tongue of warm SST that extends into the central Pacific. Reflecting the collapse of upwelling, the appearance of warm SST is attended by the disappearance of nutrient-rich water. The latter is followed by a rapid decline of fish populations off the coast of Peru. By winter, the warm anomaly has moved into the central Pacific. There, it forces anomalous convection and atmospheric heating (Fig. 17.20b). The influence of those anomalies is transmitted poleward through anomalous planetary waves (Fig. 14.20). During spring and the following summer, the warm SST anomaly decays, eventually replaced off the coast of Peru by a cold SST anomaly and the gradual return to climatological-mean conditions.

This perturbation to the annual cycle of SST depends intrinsically on the trade winds, whose EW structure comprises the Walker circulation. Under normal conditions, the mixed layer is warm and deep in the western Pacific, where the thermocline is depressed well beneath the surface (Fig. 17.22a). Conversely, the mixed layer is cold and shallow in the eastern Pacific. There, the thermocline is not far beneath the

(a) Dec – Feb 1999: La Nina Conditions

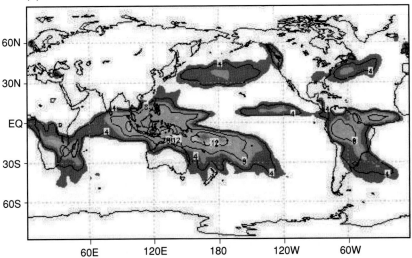

(b) Dec – Feb 1998: El Nino Conditions

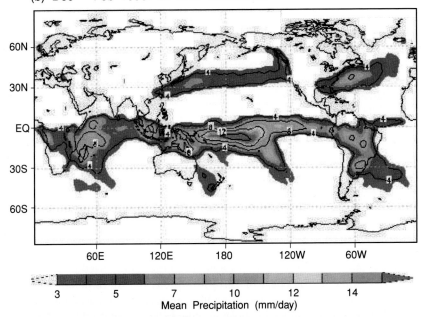

Figure 17.20 Global distribution of precipitation (a) during Dec–Feb 1999, under La Nina conditions, when rainfall was concentrated in the western Pacific, and (b) during Dec–Feb 1998, under El Nino conditions, when rainfall was displaced eastward into the central Pacific. Also evident during northern winter are the storm tracks over the North Atlantic and North Pacific, where precipitation is organized by amplifying synoptic weather systems. *Source:* http://www.esrl.noaa.gov (15.07.10). See color plate section: Plate 26.

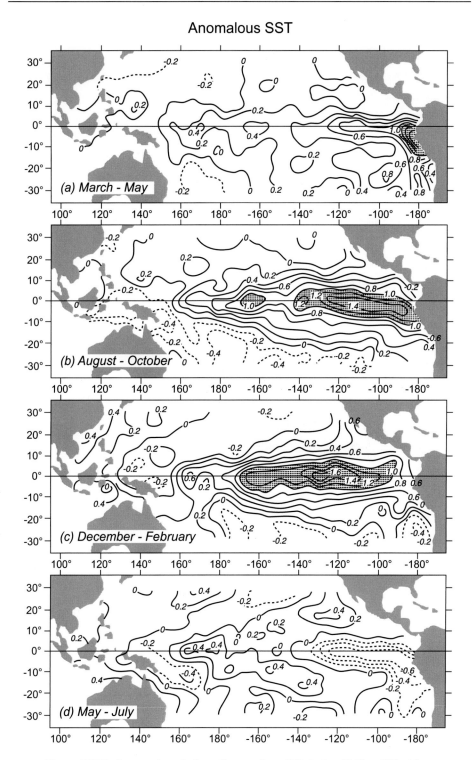

Figure 17.21 Seasonal evolution of anomalous SST during El Nino (°C). After
Stewart (2008), adapted from Rasmusson and Carpenter (1982).

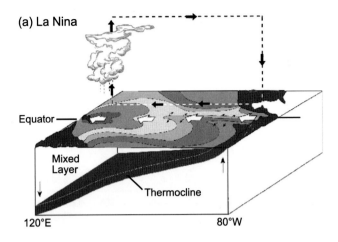

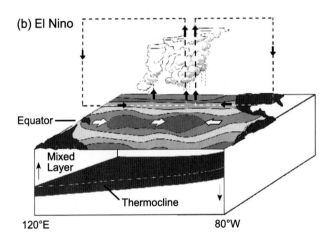

Figure 17.22 Ocean thermal structure, surface wind, SST (contoured, with red warmest), and Walker circulation under (a) La Nina conditions and (b) El Nino conditions. *Source:* http://www.pmel.noaa.gov. See color plate section: Plate 27.

surface, emerging in upwelling at the coast of South America. Characteristic of La Nina, these conditions are maintained by strong easterlies that prevail across the equatorial Pacific. The latter, through Ekman transport, induce equatorial upwelling that elevates the thermocline and produces cold SST across the eastern Pacific (Fig. 17.16). Warm SST and convection are then concentrated in the western Pacific. Warm water that has been dragged westward by the trades accumulates there, after being heated during its traversal of the equatorial Pacific. It forms a dome of high sea level in the western Pacific. About 0.5 m higher than sea level in the eastern Pacific, that elevation is maintained by easterly drag.

The onset of El Nino coincides with a weakening of the trades. This weakens upwelling that maintains cold SST in the eastern Pacific. It also weakens easterly stress that maintains the dome of warm water in the western Pacific.[6] Weakened easterlies and upwelling enable the mixed layer in the eastern Pacific to expand downward, lowering the thermocline there (Fig. 17.22b). Simultaneously, SST warms, flattening the EW gradient of temperature. This flattens the gradient of atmospheric heating, and hence the EW pressure gradient. The result is to further weaken easterly trade winds. The dome of warm water and the deep mixed layer in the western Pacific then surge eastward, behind an eastward-propagating wave front that traverses the equatorial Pacific in 2–3 months.[7] Upon arrival of the wave front, sea level, SST, and mixed layer depth all increase in the eastern Pacific. Accompanying those changes are low SLP and weak easterlies. Left behind the wave front in the western Pacific are westerlies. Along with easterlies in the eastern Pacific, they converge in the central Pacific, fueling upwelling and deep convection.

Together, these features represent a perturbed Walker circulation, with upwelling in the central Pacific and downwelling to the east and west (cf. Figs. 15.16; 1.36). As deep convection also comprises the upwelling branch of the Hadley circulation, the latter is likewise perturbed. The dynamical response to ENSO thus involves anomalous EW motion and anomalous NS motion, as results from a localized heating anomaly (Fig. 15.13). Especially sensitive to such changes are subtropical latitudes of the western Pacific (e.g., SE Asia and Australia). There, the regional Hadley circulation controls which sites experience upwelling and copious rainfall versus downwelling and cloud-free conditions (Fig. 15.11).

Figures 17.21 and 14.20a portray average conditions during warm events. In practice, however, ENSO varies considerably from one event to another. The 1982–1983 El Nino was accompanied by severe drought over eastern Australia. Yet the 1997–1998 El Nino, which was even stronger, was accompanied by normal rainfall. The El Nino of 2010 was accompanied by flooding. How such anomalous regional conditions develop clearly involves more than anomalous SLP and SST. The wide variance from these simple proxies underscores the role of dynamical changes that transmit those influences to individual regions. Those mechanisms remain poorly understood.

Pacific Decadal Oscillation

Another fluctuation of climate surrounds anomalous SST in the North Pacific. The *Pacific Decadal Oscillation (PDO)* is defined as the leading EOF of North Pacific SST (Sec. 15.5.1). Despite its name, it too is aperiodic. The PDO is displayed in Fig. 17.23 for opposite phases, along with accompanying anomalies of SLP and surface wind. In its positive phase, the PDO is characterized by SST that is anomalously cold in the northwestern Pacific and anomalously warm in the equatorial eastern Pacific (Fig. 17.23a). Accompanying anomalously cold SST in the North Pacific is anomalously low SLP and intensified westerlies in the North Pacific storm track. In its negative phase, the PDO is characterized by anomalies of opposite sign and weakened westerlies in the storm track (Fig. 17.23b). Like the AO, the change of circulation between opposite phases of the PDO influences the advection of maritime air, which moderates

---

[6] During sporadic episodes of the Madden-Julian oscillation (Sec. 15.5.2), the trades may even reverse. Equatorial westerlies then result in downwelling, compensated by the convergence of warm surface water. They also exert westerly drag on warm water to the west.

[7] Known as an *equatorial Kelvin wave*, it is a special form of deep water waves (Sec. 14.1).

Pacific Decadal Oscillation

(a) Positive Phase

(b) Negative Phase

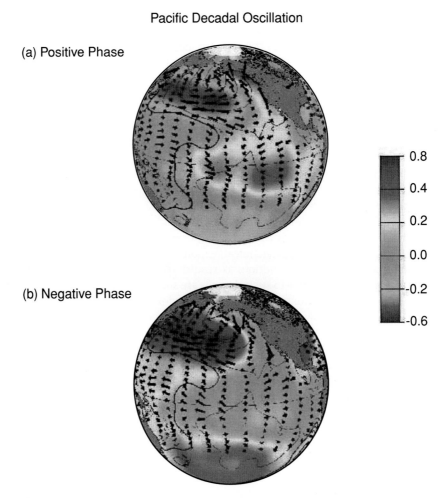

Figure 17.23 Anomalous SST (shaded), SLP (contoured), and surface wind associated with the Pacific Decadal Oscillation (PDO) during its (a) positive phase, when SST in the equatorial eastern Pacific is anomalously warm, and (b) negative phase, when SST in the equatorial eastern Pacific is anomalously cool. *Source:* http://jisao.washington.edu. See color plate section: Plate 28.

temperature extremes over neighboring continent (Sec. 15.5.1). Anomalous SST appears at high latitude, where isopycnals of the deep ocean intersect the surface. The PDO can therefore interact directly with the thermohaline circulation.

Cold temperature of the North Pacific, in concert with the exponential dependence on temperature of the Classius-Clapeyron relation (4.39), limits the PDO's interaction with the atmosphere through latent heat transfer. Perhaps relevant is the PDO's structure, which shares features with the extratropical response of El Nino (Fig. 14.20). Some fluctuations, in fact, operate coherently with those of SST in the tropics (Deser et al., 2004). There is also evidence of coherent changes in the Southern Hemisphere (Mantua and Hare, 2002).

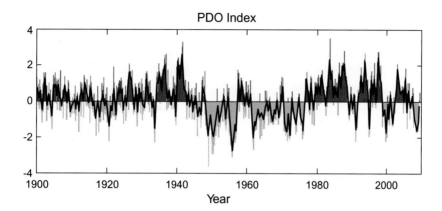

Figure 17.24 PDO index. The index fluctuates randomly, remaining of the same sign for several consecutive years, occasionally for longer than a decade. *Source:* http://jisao.washington.edu.

What distinguishes the PDO from ENSO is its time scale. Plotted in Fig. 17.24 is the PDO index, which reflects the sign and strength of the anomaly pattern. Although it fluctuates on time scales of years, the PDO remains of one sign for decades at a time. During the 1950s, 1960s, and 1970s, it was consistently negative. The PDO subsequently reversed phase. From the late 1970s until the close of the twentieth century, it was consistently positive. Since then, the PDO has again been mostly negative. A parallel exists in global-mean temperature (cf. Fig. 1.41). It varied consistently over similar periods, notably after the close of the 20th century. However, in view of the long correlation time of the PDO, the records are too short to establish a meaningful relationship. The mechanism behind this variation of SST, as well as how its influence is transmitted to other regions, remain poorly understood.

Other fluctuations

Other oceanic variations have been cited – more than a handful. A couple are worthy of mention. In the Atlantic is a counterpart of the PDO. The *Atlantic Multidecadal Oscillation (AMO)* is manifest in anomalous SST over much of the North Atlantic. Aperiodic, it varies on time scales of half a century or longer. The resolution of this feature in available records is therefore limited. Because it involves SST at high latitude, the AMO can also interact with the thermohaline circulation. There is some evidence that it influences Atlantic hurricanes, as well as drought in the United States.

The Indian Ocean exhibits an EW seesaw in SST, analogous to ENSO. Plotted in Fig. 17.25, the *Indian Ocean Dipole (IOD)* involves variations of SST in the eastern Indian Ocean, off the coast of Sumatra, and in the western Indian Ocean, off the coast of Africa, that are coherent but out of phase. The IOD emerges in the second EOF of SST, accounting for ~10% of the variance. Emerging in the leading EOF is ENSO, which accounts for ~30% of the variance.[8] Each perturbs the annual cycle of SST.

---

[8] ENSO (EOF 1) leads to anomalous SST that is in phase across the Indian Ocean. The IOD (EOF 2), which is out of phase across the Indian Ocean, thus has the structure of the first harmonic of ENSO.

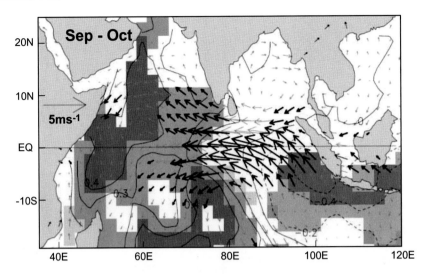

Figure 17.25 Anomalous SST (contoured), regions of >90% significance shaded, with red significant and warm and blue significant and cold, and surface wind associated with Indian Ocean dipole (IOD). After Saji et al. (1999). Reprinted by permission of Macmillan Publishers Ltd: *Nature* (Saji et al., 1999), Copyright (1999). See color plate section: Plate 29.

The IOD is aperiodic, fluctuating on a time scale of years. Anomalous surface wind that accompanies this variation of SST leads to an intensification and weakening of trade winds across the the equatorial Indian Ocean. The IOD has been associated with episodes of drought in Australia and excessive rainfall in Africa. However, like other fluctuations of SST, it represents only a fraction of the overall variance. This property limits the interpretation of individual episodes. The mechanism behind the IOD, as well as how its influence is conveyed to individual regions, suffer from considerations similar to those surrounding the AMO. They remain poorly understood.

SUGGESTED REFERENCES

The dynamics of ocean circulation are developed in *Atmosphere-Ocean Dynamics* (1982) by Gill and in *Introduction to Physical Oceanography* (2008) by Stewart.

Advanced treatments may be found in *Geophysical Fluid Dynamics* (1979) by Pedlosky and in *Atmospheric and Oceanic Fluid Dynamics* (2006) by Vallis.

Chemical considerations are discussed in *World's Oceans* (2003) by Sverdrup et al.

PROBLEMS

1. Suppose that warm salty water in the Gulf Stream, with temperature and salinity of 15°C and 36 g/kg, encounters water in the North Atlantic with temperature and salinity of 4°C and 33 g/kg. If the bodies of water mix in equal proportion, determine the (a) initial density anomaly of water in the Gulf stream, (b) initial density anomaly of water in the North Atlantic, (c) final density anomaly of the mixture; compare it to the average of the two initial density anomalies. What is the consequence of such mixing?

2. Surface salinity exceeds 36 g/kg in the subtropics of the south Pacific (Fig. 17.1). (a) If the mixed layer has a depth of 20 m, (a) what depth of water must be evaporated to increase its salinity from 35 g/kg, characteristic of the surroundings, to 36 g/km? (b) If achieved over a season, what is the attendant transfer of latent heat to the atmosphere in W/m^2? (c) If ocean temperature remains approximately constant, from where is that heat supplied? (d) Where is the latent heat that is transferred to the atmosphere eventually released? (e) Compare the area in (d) to that in (a); what is the implication?

3. Suppose that, in the North Atlantic, ambient temperature and salinity are 4°C and 33 g/kg, respectively. (a) Use the limiting behavior Fig. 17.4 to determine how much salinity would have to increase, with temperature fixed, to produce deep water. (b) If the mixed layer has a depth of 50 m, what depth of water must freeze to achieve that state? (c) If the freezing takes place over a season, how much heat transfer with the atmosphere must occur in W/m^2?

4. Demonstrate that, on seasonal time scales, heat storage of the ocean is two orders of magnitude greater than that of continent.

5. Suppose that heat transfer from the deep ocean leads to a gradual increase of SST in the tropics. Discuss the reestablishment of equilibrium between CO_2 in the ocean and atmosphere (cf. Fig. 1.43).

6. Show that the Ekman velocity satisfies the frictional momentum balance (17.4.1).

7. Consider a rectangular ocean basin of dimensions $L_x = a$ and $L_y = L_x/2$ that is narrow enough to be treated on an f plane at latitude 30°. Suppose that surface wind is geostrophic and controlled by a closed subtropical high, which is characterized by the 1000-hPa geopotential

$$\Phi_0(x, y) = 400 \cdot \cos\left(\frac{x}{X}\right) \cos\left(\frac{y}{Y}\right),$$

where $-\frac{L_x}{2} < x < \frac{L_x}{2}$, $-\frac{L_y}{2} < y < \frac{L_y}{2}$, and $X = L_x/\pi$, $Y = L_y/\pi$. If, in the spirit of Rayleigh friction, the surface stress can be described as

$$\tau_0 = \rho \alpha v_0$$

with $\alpha = 10^{-5}$ m/s, calculate and plot (a) the surface wind field, (b) the horizontal distribution of Ekman transport, (c) the horizontal distribution of Ekman pumping.

8. (a) For the conditions in Prob. 17.7, describe the horizontal distribution of SST implied by the motion. (b) As in (a), but if, opposite the equator, are a mirror-image ocean basin and surface wind.

9. With reference to the First Law (Sec. 10.7) and net radiation (Sec. 1.4.2), explain why northward transport of heat must assume the form in Fig. 17.10.

10. (a) For the conditions in Prob. 17.7, but on a β plane, calculate and plot the horizontal distribution of column-integrated mass transport under Sverdrup balance. (b) Compare the resulting transport against the Ekman transport in Prob. 17.7. (c) What additional feature is necessary to satisfy continuity?

CHAPTER EIGHTEEN

Interaction with the stratosphere

Above the tropopause, the horizontal gradient of heating is generally weak enough to leave thermal structure close to barotropic stratification. For this reason, baroclinic instability generated through conversion of available potential energy does not play an essential role. Further, strong westerlies in the winter stratosphere and easterlies in the summer stratosphere trap baroclinic disturbances in the troposphere. Their presence in the stratosphere is therefore limited to a neighborhood of the tropopause. These features leave the circulation in the stratosphere less disturbed than in the troposphere (Fig. 1.10). Exceptions are planetary waves. Unlike baroclinic disturbances, they propagate through strong westerlies of the polar-night jet (Sec. 14.5.2). Vertical amplification enables planetary waves to disturb the polar vortex. During major amplifications, they completely disrupt the circumpolar flow. Air then moves freely between tropical and polar regions, experiencing sharply different radiative environments. Global-scale disturbances such as the one in Fig. 1.10a play a key role in establishing mean distributions of radiatively active constituents.

With convection inhibited by strong static stability, the energy budget of the stratosphere is dominated by radiative transfer. It therefore remains closer to radiative equilibrium than the troposphere (Fig. 8.24). The radiative energy budget is controlled by SW heating due to ozone absorption, principally in the Hartley and Huggins bands (Sec. 8.3.1), and LW cooling due to CO_2 emission to space at 15 μm. Ozone, in concert with CO_2, thus shapes the thermal structure of the middle atmosphere. Through hydrostatic and geostrophic equilibrium, the latter controls the circulation. For this reason, ozone underpins much of the observed behavior of the middle atmosphere.

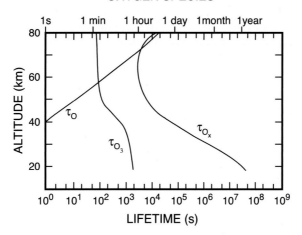

Figure 18.1 Photochemical lifetimes of oxygen species. *Source:* Brasseur and Solomon (1986).

18.1 OZONE PHOTOCHEMISTRY

The simplest treatment of ozone photochemistry is due to Chapman (1930). It considers a pure oxygen atmosphere. Composition is then governed by the following reactions

$$O_2 + h\nu \rightarrow O + O \qquad (J_2) \qquad\qquad (18.1.1)$$

$$O + O_2 + M \rightarrow O_3 + M \quad (k_2) \qquad\qquad (18.1.2)$$

$$O + O_3 \rightarrow 2O_2 \qquad (k_3) \qquad\qquad (18.1.3)$$

$$O_3 + h\nu \rightarrow O_2 + O \qquad (J_3), \qquad\qquad (18.1.4)$$

where rate coefficients, indicated parenthetically to the right, characterize the speeds of individual reactions.[1] Reactions (18.1.1) and (18.1.4) describe photodissociation or *photolysis* of O_2 by UV radiation in the Herzberg continuum near 242 nm and of O_3 in the Hartley and Huggins bands near 310 nm. Reaction (18.1.4) operates at all wavelengths shorter than 1 μm. Reactions (18.1.2) and (18.1.3) describe recombination of O_2 and O_3 with O. The molecule M represents a third body needed to conserve momentum and energy in the recombination of O and O_2 (e.g., air molecules, which are abundant). Atomic oxygen that is produced by photolysis of ozone in (18.1.4) recombines immediately with molecular oxygen in (18.1.2) to re-form ozone. Hence those reactions constitute a closed cycle, one that leaves O and O_3 unchanged yet absorbs UV radiation efficiently.

18.1.1 The chemical family

The rate coefficients determine the photochemical lifetimes of these species, which are presented in Fig. 18.1 as functions of altitude (Prob. 18.1). Lifetimes of the odd oxygen

---

[1] Dimensions of $\left(\frac{\text{molecules}}{\text{volume}}\right)^{-2}$ time^{-1} for k_2, $\left(\frac{\text{molecules}}{\text{volume}}\right)^{-1}$ time^{-1} for k_3, and time^{-1} for J_2 and J_3.

species O and O_3 are short by comparison with that of O_2, which is present in fixed proportion. Moreover, the lifetimes of O and O_3 differ by several orders of magnitude. This feature complicates their treatment in numerical calculations. For this reason, it is convenient to introduce the *odd oxygen family*

$$O_x = O + O_3. \tag{18.2}$$

Individual members of the family have lifetimes much shorter than that of O_x. They can therefore be treated as being in photochemical equilibrium with one another. Reactions (18.1) then describe comparatively slow changes in O_x that are attended by immediate adjustments of O and O_3 within the family to maintain photochemical equilibrium. Such changes are introduced whenever an air parcel is displaced from one radiative environment to another.

The relative abundance or *partitioning* of family members is determined by reactions (18.1.2) and (18.1.4), which operate much faster than the others. Those reactions preserve odd oxygen. Consequently, they represent a simple redistribution between O and O_3. The rate of destruction of O_3 in (18.1.4) is expressed by

$$\left.\frac{d[O_3]}{dt}\right|_{destruction} = -J_3[O_3], \tag{18.3.1}$$

where [ ] denotes the number density $\left(\frac{molecules}{volume}\right)$ and $\frac{d}{dt}$ represents the Lagrangian derivative. (The chemical budget should formally be expressed in terms of mixing ratio to account for expansion and compression of an air parcel (Sec. 1.2). For the moment, however, motion is ignored.) The rate of production of O_3 in (18.1.2) is given by

$$\left.\frac{d[O_3]}{dt}\right|_{production} = k_2[O_2][O][M]. \tag{18.3.2}$$

Adding yields the net rate of production of O_3 in (18.1.2) and (18.1.4)

$$\left.\frac{d[O_3]}{dt}\right|_{net} = k_2[O_2][O][M] - J_3[O_3]. \tag{18.3.3}$$

Photochemical equilibrium within the family requires the left-hand side to vanish. Therefore,

$$J_3[O_3] = k_2[O_2][O][M] \tag{18.4}$$

governs the partitioning of its member species. Below 60 km, O_3 is the dominant member of O_x. Consequently, ozone tracks the behavior of odd oxygen. The lifetime of O_x in the lower stratosphere (Fig. 18.1) is several weeks. It is an order of magnitude longer than the time scale of advection ($\sim$ 1 day). However, its lifetime decreases with height rapidly, shortening to less than 1 day above about 30 km.

18.1.2 Photochemical equilibrium

The remaining reactions (18.1.1) and (18.1.3) govern production and destruction of O_x. These reactions operate on time scales much longer than reactions for the individual member species. For photochemical equilibrium of the family as well, the net rate of production of $[O_x]$ must also vanish. The rate at which O_x molecules are produced in

(18.1.1) is expressed by

$$\frac{d[O_x]}{dt}\bigg|_{production} = 2J_2[O_2].$$
(18.5.1)

The rate at which they are destroyed in (18.1.3) is given by

$$\frac{d[O_x]}{dt}\bigg|_{destruction} = -2k_3[O][O_3].$$
(18.5.2)

Adding yields the net rate of production of $[O_x]$

$$\frac{d[O_x]}{dt}\bigg|_{net} = 2J_2[O_2] - 2k_3[O][O_3].$$
(18.5.3)

Photochemical equilibrium of O_x requires

$$J_2[O_2] = k_3[O][O_3].$$
(18.6)

Eliminating [O] from (18.4) and (18.6) then obtains the equilibrium ozone number density

$$[O_3] = [O_2]\left(\frac{k_2 J_2}{k_3 J_3}[M]\right)^{\frac{1}{2}},$$
(18.7)

which follows in terms of the number densities of molecular oxygen and air.

The profile of $[O_3]$ predicted by (18.7) is plotted in Fig. 18.2. For its simplicity, Chapman chemistry (solid) is remarkably successful. It reproduces the essential vertical structure of observed ozone, which is superposed for tropical (dashed) and extratropical latitudes (dotted). From TOA, ozone number density increases downward due to increasing $[O_2]$ and photolysis in (18.1.1). A maximum in $[O_3]$ is achieved near 30 km. At lower levels, UV flux and J_2 decrease sharply due to attenuation by photolysis.

Despite this general agreement, the behavior predicted by (18.7) deviates importantly from the observed distribution of ozone. Ozone number density is overpredicted at tropical latitudes, where much of ozone is concentrated (Fig. 1.20). For this reason, (18.7) gives total ozone that is too large, $\Sigma_{O_3} \cong 1000$ DU (Fig. 1.21). At extratropical latitudes, $[O_3]$ is underpredicted in the lower stratosphere and overpredicted in the upper stratosphere. Note the observed profile of $[O_3]$ is displaced downward from that in the tropics. Maximum concentration at extratropical latitudes is found below 20 km (cf. Fig. 1.20). At those latitudes, as much as 30% of total ozone lies in the troposphere. Tropospheric ozone increases during winter and spring (London, 1985; Logan, 1999). It also manifests an upward trend, long-term evolution that is not understood (WMO, 2006). When the horizontal distribution of total ozone is considered (Figs. 1.21, 1.22), the discrepancy with observed behavior is even more serious. Observed Σ_{O_3} at 60° is almost twice as great as in the tropics. By contrast, (18.7) predicts total ozone to minimize at high latitude – because there UV flux and hence J_2 are small. The latter vanish inside polar night, where observed values of Σ_{O_3} are, in fact, greatest.

These discrepancies are attributable to two key ingredients that are not accounted for by (18.7). First, ozone chemistry involves species other than oxygen. Catalytic cycles involving free radicals of hydrogen, nitrogen, and chlorine deplete odd oxygen. They make ozone dependent on a wide array of chemical species. The second factor not accounted for is motion. Below 30 km, where the ozone column is concentrated, the photochemical lifetime is long enough for O_3 to be transported by the circulation.

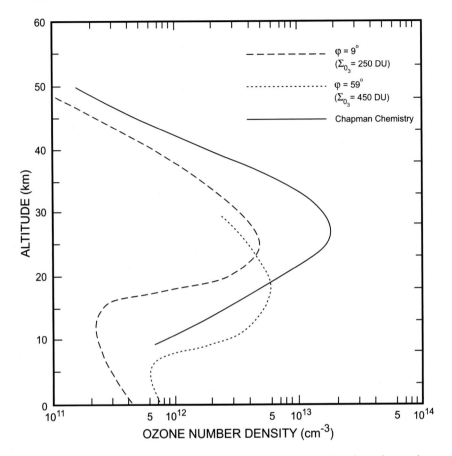

Figure 18.2 Vertical profile of ozone number density under photochemical equilibrium, as calculated from Chapman chemistry (solid), and observed at tropical (dashed) and extratropical (dotted) latitudes. Chapman chemistry was calculated with rate coefficients from Brasseur and Solomon (1986) and Nicolet (1980). It yields realistic vertical structure, but a column abundance of $\Sigma_{O_3} \cong$ 1000 DU. Observed data are from Hering and Borden (1965) and Krueger (1973).

It therefore passes between widely differing photochemical environments. Even at higher altitude, other species that participate in the complex photochemistry of ozone are influenced importantly by transport.

18.2 INVOLVEMENT OF OTHER SPECIES

Photodissociation by UV radiation produces a number of free radicals, derived from less reactive reservoir species of tropospheric origin. Photochemically active, the free radicals can then go on to destroy ozone in catalytic cycles that leave the free radical unchanged.

18.2.1 Nitrous oxide

The free radical nitric oxide NO represents an important link to human activities. Nitric oxide is a byproduct of inefficient combustion (e.g., in aircraft exhaust). In the stratosphere, the principal source of NO is dissociation of nitrous oxide N_2O through reaction with atomic oxygen

$$N_2O + O \rightarrow 2NO. \tag{18.8}$$

Nitrous oxide is produced in the troposphere by natural as well as anthropogenic sources (Chap. 1). Away from the surface, N_2O is long-lived, having a photochemical lifetime of order 100 years (Fig. 18.3). Further, nitrous oxide is not water soluble. It is therefore immune to normal scavenging mechanisms associated with precipitation (Sec. 9.5.2). These properties allow N_2O to become well-mixed in the troposphere. They also make it useful as a tracer of air motion in the stratosphere.

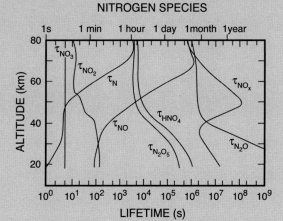

Figure 18.3 Photochemical lifetimes of nitrogen species. *Source:* Brasseur and Solomon (1986).

After entering the stratosphere, N_2O is photodissociated according to the reaction

$$N_2O + h\nu \rightarrow N_2 + O. \tag{18.9}$$

Reaction (18.9) is the primary destruction mechanism for nitrous oxide. It is responsible for the upward decrease of its mixing ratio r_{N_2O} near the tropopause (Fig. 1.23). The mixing ratio of N_2O is largest near the tropical tropopause (Fig. 18.4). In view of its long lifetime, this structure reveals how air enters the stratosphere. A plume of high values reflects upwelling in the tropical stratosphere, where N_2O-rich air is carried across the tropical tropopause from its tropospheric reservoir below.

The free radical NO forms via (18.8). Once produced, it can efficiently destroy ozone through the catalytic cycle

$$NO + O_3 \rightarrow NO_2 + O_2 \tag{18.10.1}$$

$$NO_2 + O \rightarrow NO + O_2. \tag{18.10.2}$$

Adding yields the net effect

$$O_3 + O \rightarrow 2O_2 \quad \text{(Net)}. \tag{18.10.3}$$

This closed cycle leaves $NO + NO_2$ unchanged. One molecule of nitric oxide can therefore destroy many molecules of ozone.

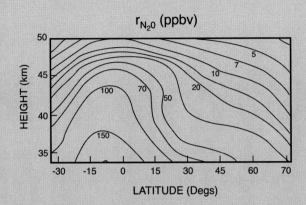

Figure 18.4 Zonal-mean mixing ratio of nitrous oxide N_2O on January 9, 1992, as observed by the ISAMS instrument on board the Upper Atmosphere Research Satellite (UARS). Adapted from Ruth et al. (1994).

Reactions with other nitrogen compounds make NO dependent on a number of species. Because reactions among those species are fast, it is convenient to introduce the *odd nitrogen family*

$$NO_x = N + NO + NO_2 + NO_3 + 2N_2O_5 + HNO_4, \tag{18.11}$$

the members of which may be treated in photochemical equilibrium with one another. Despite the short lifetimes of some its members, the lifetime of NO_x is much longer than the time scale for advection (Fig. 18.3). The abundance of NO then follows from $[NO_x]$ and from partitioning within the family. On a longer time scale, NO_x reacts with other families like O_x.

18.2.2 Chlorofluorocarbons

Free radicals of chlorine represent another important link to human activities. Atomic chlorine is produced naturally in the stratosphere through photodissociation of methyl chloride CH_3Cl, which is a product of ocean processes. Oxidation of CH_3Cl by reactive species likewise produces Cl. Atomic chlorine is also produced by photodissociation of CFCs, the sole source of which is industrial. Like N_2O, CFC-11 ($CFCl_3$) and CFC-12 (CF_2Cl_2) are long-lived – at least in the troposphere (Fig. 18.5). CFCs have photochemical lifetimes of several years. They are also water insoluble, which circumvents normal scavenging mechanisms. These properties allow CFCs to become homogenized in the troposphere (Fig. 1.23).

In the stratosphere, this picture breaks down. There, CFC-11 and CFC-12 decrease upward through photodissociation. This process releases free chlorine

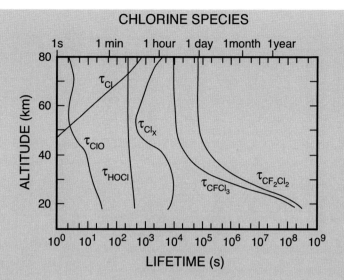

Figure 18.5 Photochemical lifetimes of chlorine species, including CFC-11 ($CFCl_3$) and CFC-12 (CF_2Cl_2). *Source:* Brasseur and Solomon (1986).

in the reactions

$$CFCl_3 + h\nu \rightarrow CFCl_2 + Cl \qquad (18.12.1)$$

$$CF_2Cl_2 + h\nu \rightarrow CF_2Cl + Cl \qquad (18.12.2)$$

that operate at wavelengths shorter than 225 nm. Once produced, free Cl reacts with ozone in the catalytic cycle involving chlorine monoxide ClO

$$Cl + O_3 \rightarrow ClO + O_2 \qquad (18.13.1)$$

$$ClO + O \rightarrow Cl + O_2. \qquad (18.13.2)$$

Adding yields the net effect

$$O_3 + O \rightarrow 2O_2 \quad \text{(Net)}. \qquad (18.13.3)$$

This closed cycle leaves $Cl + ClO$ unchanged. One atom of chlorine can therefore destroy many atoms of ozone.

Like free radicals of nitrogen, atomic chlorine can react with other species on widely varying time scales. It is therefore convenient to introduce the *odd chlorine family*

$$Cl_x = Cl + ClO + HOCl. \qquad (18.14)$$

The abundance of Cl then follows via the partitioning of member species. Unlike O_x in the lower stratosphere and NO_x, Cl_x is short-lived relative to air motion (Fig. 18.5). Therefore, it is not passively advected by the circulation. Nevertheless, Cl_x is much longer-lived than its members, streamlining the numerical treatment of chlorine species. Odd chlorine can interact with odd nitrogen through the reaction

$$ClO + NO \rightarrow Cl + NO_2. \qquad (18.15)$$

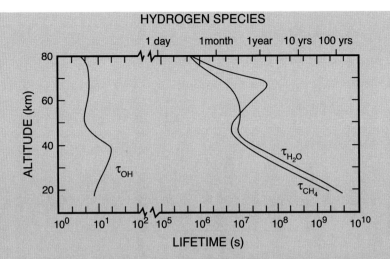

Figure 18.6 Photochemical lifetimes of hydrogen species. *Source:* Brasseur and Solomon (1986).

Reaction (18.15) regenerates free Cl, coupling the families NO_x and Cl_x. Other interactions among O_x, NO_x, and Cl_x bear importantly on ozone depletion in the polar stratosphere (Sec. 18.7).

18.2.3 Methane

Methane enters considerations of ozone because it interacts with NO_x and O_x. In addition to being radiatively active, CH_4 represents an important link between chemical constituents in the stratosphere and water vapor. For both, the troposphere serves as a reservoir. Like N_2O, methane has a photochemical lifetime of many years below 40 km (Fig. 18.6). It too is water insoluble. These properties allow CH_4 to become well-mixed in the troposphere (Fig. 1.23). They also make it useful as a tracer of air motion in the stratosphere. Figure 18.7 presents the zonal-mean

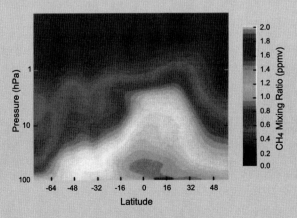

Figure 18.7 Zonal-mean mixing ratio of methane CH_4, as observed by the HALOE instrument on board UARS. Adapted from Bithell et al. (1994).

distribution of r_{CH_4} in the middle atmosphere. Mirroring structure in the N_2O distribution is a plume of high mixing ratio above the tropical tropopause. It reflects upwelling in the equatorial stratosphere.

Decreasing mixing ratio in the stratosphere follows from chemical destruction of CH_4. Methane is oxidized by the hydroxyl radical in the reaction

$$CH_4 + OH \rightarrow CH_3 + H_2O. \qquad (18.16.1)$$

It is also destroyed by atomic oxygen and chlorine in the reactions

$$CH_4 + O \rightarrow CH_3 + OH \qquad (18.16.2)$$

$$CH_4 + Cl \rightarrow CH_3 + HCl. \qquad (18.16.3)$$

The latter represents an important sink of reactive Cl_x. The free radical CH_3 produced in reactions (18.16) immediately combines with molecular oxygen to form CH_3O_2. Two reactions involving NO_x

$$CH_3O_2 + NO \rightarrow CH_3O + NO_2 \qquad (18.17.1)$$

$$NO_2 + h\nu \rightarrow NO + O \qquad (18.17.2)$$

then form a closed cycle. They have the net effect

$$CH_3O_2 + h\nu \rightarrow CH_3O + O \quad (Net). \qquad (18.17.3)$$

Methane oxidation thus leads to production of O_x. As the partitioning of O_x fixes the abundance of O_3, reactions (18.18) provide a potentially important source of ozone in the lower stratosphere.

The free radical OH initiates the above process. It is produced by dissociation of water vapor

$$H_2O + O \rightarrow 2OH. \qquad (18.18)$$

Above the tropopause, the lifetime of H_2O is many years (Fig. 18.6). Consequently, it too behaves as a tracer at these heights. Water vapor mixing ratio decreases sharply in the troposphere, attaining a minimum at the *hygropause* (Fig. 18.8). Found a couple of kms above the tropopause, the hygropause is characterized by mixing ratio of 1–2 ppmv. Notice that driest air is found at the altitude of highest cloud, cumulus overshoots that make data in Fig 18.8 unavailable (cf Fig 9.26). The steep decrease with height in the troposphere reflects convection, which dehydrates air by limiting r_{H_2O} to saturation values. Above the hygropause, r_{H_2O} increases due to oxidation of methane, as r_{CH_4} decreases. How much of the observed increase of water vapor actually follows from reactions (18.16) remains unclear.

18.3 MOTION

Photochemical considerations predict that ozone column abundance will maximize at low latitude, where UV radiation and photolysis maximize. The observed distribution, however, maximizes at high latitude (Fig. 1.21). Approaching 500 DU during late winter and spring, total ozone there increases from its fall minimum of about 270 DU by nearly 100%. This shortcoming of Chapman chemistry is unchanged by the addition

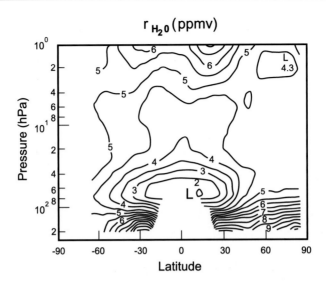

Figure 18.8 Zonal-mean mixing ratio of water vapor observed by Nimbus-7 LIMS. Masking beneath 60 hPa denotes sites in the tropics that are obscured by cloud, where data are unavailable. Adapted from Remsberg et al. (1984).

of other chemical species. The high-latitude maximum of Σ_{O_3} can be understood only through dynamical considerations.

18.3.1 The Brewer-Dobson circulation

In the winter stratosphere, ozone heating establishes an equatorward temperature gradient over a deep layer (Fig. 1.7). By thermal wind balance (12.11), this forces strong circumpolar westerlies. Stratospheric westerlies intensify upward along the *polar-night terminator*, where SW absorption vanishes. They form the polar-night jet (Fig. 1.8), which circumscribes the polar-night vortex (Fig. 1.10b). In the summer stratosphere there is a poleward gradient of heating, which follows from the distribution of daily insolation (Fig. 1.33). It produces a deep temperature gradient of the opposite sense, forcing strong circumpolar easterlies.

In each season, air motion is nearly zonal. Under radiative equilibrium, it is zonally symmetric, experiencing no net heating. By the First Law, individual air parcels then do not cross isentropic surfaces. This implies no net vertical motion and, by continuity, no meridional motion.

The situation is analogous to the zonally symmetric circulation in Sec. 15.2. Away from boundaries, where frictional drag is negligible, the motion is geostrophic and zonal. It accomplishes no transport. Meridional transport is confined to shallow Ekman layers along the upper and lower boundaries. There, frictional drag drives the motion out of geostrophic balance. It introduces an ageostrophic secondary circulation that transfers air across latitude circles. In the stratosphere, planetary waves play a similar role. When absorbed (Sec. 14.6), they exert a drag on the mean flow. It forces the motion out of geostrophic balance. An ageostrophic secondary circulation is then imposed onto the strong zonal motion. It forces air across latitude circles and therefore out of radiative equilibrium.

The nature of this secondary circulation is revealed by the structure of long-lived tracers (Figs. 18.4, 18.7). Distributions of N_2O and CH_4 imply upwelling in the tropics, where tropospheric air enters the stratosphere. Continuity then requires downwelling at middle and high latitudes. Connected by poleward motion, these regions of vertical motion comprise an equator-to-pole circulation (see Fig. 18.11). Although weak compared with zonal motion, this ageostrophic secondary circulation explains the large column abundance of ozone that is observed at high latitude. Ozone-rich air is drawn poleward from the chemical source region in the tropics, increasing r_{O_3} and total ozone at high latitude. When it descends, that air undergoes compression. This increases the absolute concentration ρ_{O_3} at high latitude (Fig. 1.20). Through adiabatic warming, it also increases temperature. Air descending over the pole warms by ~10 K for each kilometer that it sinks. During winter, it sinks several tens of kilometers. Adiabatic warming thus provides a major offset to LW cooling, which prevails inside polar darkness.

This gradual overturning in the meridional plane was first inferred by Dobson (1930) and Brewer (1949) to explain balloon observations of trace species. Known as the *Brewer-Dobson circulation*, it figures centrally in chemical considerations by shaping the distributions of many species. The Brewer-Dobson circulation also influences lower levels. Downwelling over the pole transfers stratospheric air into the troposphere (see Fig. 18.11). It must be compensated over the equator by upwelling that returns air to the stratosphere at the same rate. The Brewer-Dobson circulation thus couples the troposphere to the stratosphere.

18.3.2 Wave driving of mean meridional motion

Although characterized by a zonally symmetric overturning in the meridional plane, the Brewer-Dobson circulation actually develops through zonally asymmetric processes. A clue to its origin lies in the radiative-equilibrium state of the middle atmosphere (Fig. 18.9). During solstice, the sharp gradient of heating across the polar-night terminator produces radiative-equilibrium temperature T_{RE} colder than 150°K (Fig. 18.9a). This is much colder than observed temperature (Fig. 1.7). More uniform, observed temperature in the Northern Hemisphere remains above 200°K. With thermal wind balance (12.11), the deep layer of sharp temperature gradient implies radiative-equilibrium wind that intensifies upward – to more than 300 m s^{-1} (Fig. 18.9b). Analogous to temperature, this is much stronger than observed wind (Fig. 1.8).

The observed polar-night vortex is warmer than radiative equilibrium. It must therefore experience net radiative cooling (Sec. 8.5). Observed temperature and ozone imply a cooling rate inside polar darkness greater than 8°K day^{-1} (Fig. 8.27). By the First Law (2.36), that cooling must reduce the potential temperature of air inside the vortex. In concert with positive stability ($\frac{\partial \theta}{\partial z} > 0$), it implies downwelling across isentropic surfaces. Calculations in which this diabatic cooling is prescribed (Murgatroyd and Singleton, 1961) qualitatively reproduce the meridional circulation that was inferred from tracer behavior by Brewer and Dobson.

In the Southern Hemisphere, polar temperature is colder – as cold as 180°K. The Antarctic polar-night vortex, which is less disturbed by planetary waves, is thus colder and stronger than its counterpart over the Arctic. Closer to radiative equilibrium, it should therefore experience weaker radiative cooling. That, in turn, implies weaker downwelling at extratropical latitudes. Analogous considerations apply to the

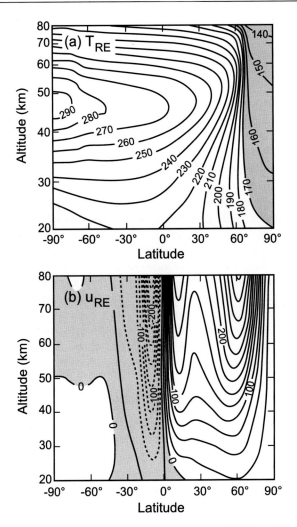

Figure 18.9 Radiative-equilibrium (a) temperature and (b) zonal wind in the middle atmosphere during solstice, as calculated in a radiative–convective–photochemical model. Thermal structure adapted from Fels (1985). Zonal wind calculated from thermal wind balance and from climatological motion at 20 km in Fig. 1.8.

summertime circulation (Fig. 8.27), in which strong easterlies block planetary wave propagation from the troposphere (Sec. 14.5).[2]

Understanding these inter-hemispheric differences requires an understanding of how the circulation is maintained out of radiative equilibrium. Observed motion during winter is rarely in a quiescent state of zonal symmetry. This is especially true in the Northern Hemisphere. There, the circumpolar flow is routinely disturbed by

---

[2] The absence of eddy motion during summer enabled the cloud of volcanic debris in Fig. 9.6 to remain confined meridionally. Once westerlies developed after equinox, planetary waves were able to propagate upward from the troposphere, disturbing the zonal flow. The volcanic cloud was then quickly dispersed across the hemisphere.

amplified planetary waves that propagate upward from the troposphere. Planetary waves in the Northern Hemisphere are stronger than those in the Southern Hemisphere, because of large orographic forcing and land-sea contrast (Sec. 15.4). Able to penetrate into strong westerlies of the winter stratosphere, they drive the circulation out of zonal symmetry and away from radiative equilibrium. As illustrated by Fig. 1.10a, air then flows meridionally between sharply different radiative environments: from the sunlit tropics, where it experiences net SW heating, into polar darkness, where it experiences only LW cooling. Such air motion operates on a time scale short compared with radiative adjustment. The heat transfer experienced by individual air parcels is therefore irreversible (Figs. 3.5, 3.6).

Figure 18.10a presents, from a numerical integration, the horizontal trajectory of an air parcel inside the polar-night vortex. Superposed are instantaneous distributions of motion and potential vorticity Q (Sec. 12.5). The parcel shown orbits about the disturbed vortex, which is sporadically displaced from zonal symmetry and distorted by planetary waves. In its thermodynamic state space (Fig. 18.10b), the parcel therefore cycles between sharply different radiative-equilibrium temperatures (solid). The short time scale of advection prevents the parcel from adjusting to local radiative equilibrium. Instead, it trails behind the ambient value of T_{RE} by a finite temperature difference. The parcel is therefore out of thermodynamic equilibrium. By the Second Law, the heat transfer it experiences is irreversible. This introduces a hysteresis into the parcel's thermodynamic state, one that accumulates with each circuit about the vortex. Net cooling during successive circuits then causes the parcel to drift across isentropic surfaces to lower θ (cf. Fig. 3.6). (Notice that the descent rate $\frac{d\theta}{dt}$ increases with the parcel's departure from local radiative equilibrium.) Superimposed is the parcel's evolution in the absence of mechanical forcing by planetary waves (dotted). The parcel then adjusts to radiative equilibrium, bringing vertical motion to a halt.

Downwelling is accompanied by the rejection of heat. The latter derives ultimately from work that is performed on the circulation by planetary waves when they experience dissipation. Analogous to paddle work, the latter makes the circulation of the stratosphere behave as a radiative refrigerator (Sec. 2.2.2). The work performed by planetary waves follows from drag that is exerted on the mean flow when wave momentum is absorbed. Measured by the convergence of wave activity flux (e.g., of EP flux in Sec. 14.5.4), planetary wave drag drives the motion out of geostrophic balance. Ageostrophic motion that is introduced then produces meridional transport. Involving a poleward drift of air (Fig. 18.10a), such transport is analogous to that produced by frictional drag in the zonally symmetric circulation of Sec. 15.2. Compensating it over the pole is downwelling. Accompanied by adiabatic warming, downwelling maintains the polar-night vortex warmer and weaker than it would be under radiative equilibrium – as much as 50 K warmer. The strength of downwelling follows in proportion to the absorption of wave activity. Amplified planetary waves of the Northern Hemisphere therefore drive the Arctic vortex farther from radiative equilibrium than the Antarctic vortex, for which planetary wave absorption is comparatively weak.[3]

---

[3] The Antarctic vortex is also reinforced by anomalous LW cooling, through exchange with the Antarctic plateau. Much colder than surrounding ocean. the Antarctic surface induces a strong meridional gradient of cooling in the stratosphere, one that is transmitted upward through the 9.6-μm band of ozone (Francis and Salby, 2001). Independent of planetary wave forcing, this radiative forcing of the Antarctic vortex accounts for about half of the observed difference with the Arctic vortex.

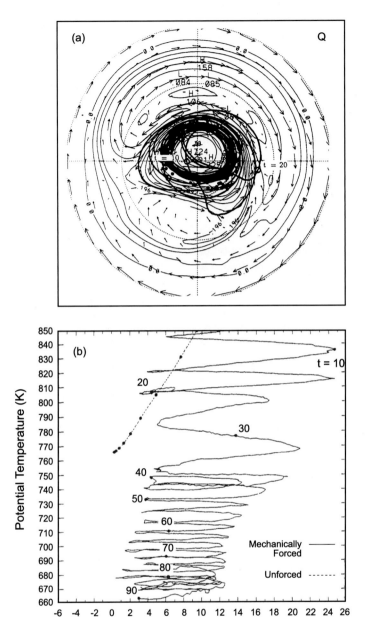

Figure 18.10 (a) Distributions of potential vorticity Q and horizontal motion in a 2-dimensional calculation representative of the polar-night vortex (cf. Fig. 14.26). Greenwich meridian is positioned at 0600 and International Dateline at 1200 on the clock dial. Bold line shows the trajectory of one of an ensemble of air parcels (solid circles) initialized inside the vortex. The folding of contours along the edge of the vortex and regions where the Q distribution has been homogenized by secondary eddies symbolize horizontal mixing. (b) Evolution of the parcel whose trajectory is shown in (a) in its thermodynamic state space (solid), represented in terms of the parcel's departure from local radiative equilibrium. Irreversible heat transfer introduces a hysteresis into the parcel's state with each circuit about the displaced vortex, which produces a steady drift of air to lower θ. In the same calculation, but without planetary waves to maintain the circulation out of radiative equilibrium (dotted), the parcel adjusts to radiative equilibrium, which halts subsequent cooling and descent inside the vortex. After Fusco and Salby (1994), copyright by the American Geophysical Union.

Planetary waves suffer absorption through thermal dissipation. Represented by cooling in Fig. 18.10, thermal dissipation damps perturbation temperature (Sec. 8.6). Planetary waves also suffer absorption through mechanical dissipation, notably, through quasi-horizontal mixing. By destroying large-scale gradients, it damps perturbation motion. As it operates on time scales short compared with radiative adjustment, mixing conserves θ, homogenizing all conserved properties along isentropic surfaces. Isentropic mixing is evidenced in Fig. 18.10a by the conserved property Q. Contours of potential vorticity have folded at the edge of the vortex (near 0 E; 0600 on the clock dial), where an anticyclonic eddy prevails and the distribution of Q has been homogenized. Those features neighbor the critical line of stationary planetary waves (cf. Fig. 14.23). There, large eddy displacements overturn the distribution of Q, folding contours of potential vorticity. By reversing the gradient of Q, this renders the local motion dynamically unstable (Sec. 16.2). Air is then wound up in secondary eddies. They cascade large-scale tracer structure to small scale, where it is absorbed by diffusion. Like thermal dissipation, mechanical dissipation of planetary waves acts to maintain the polar-night vortex warmer and weaker than it would be under radiative equilibrium.

18.3.3 Transformed Eulerian description

Although 3-dimensional, it is convenient to describe motion 2-dimensionally in terms of the zonal-mean circulation $\overline{\boldsymbol{v}} = (\overline{u}, \overline{v}, \overline{w})$. This representation is especially attractive in light of the numerous chemical species that must be described in the photochemistry of ozone. In this framework, zonal-mean distributions of motion and chemical species interact, with one another and with the wave field.

In log-pressure coordinates, motion is governed by

$$\frac{d\boldsymbol{v}}{dt} + w\frac{\partial \boldsymbol{v}}{\partial z} + f\boldsymbol{k} \times \boldsymbol{v} = -\nabla\Phi \qquad (18.19.1)$$

$$\nabla \cdot \boldsymbol{v} + \frac{1}{\rho_0}\frac{\partial}{\partial z}(\rho_0 w) = 0 \qquad (18.19.2)$$

$$\frac{dT}{dt} + \frac{N^2 H}{R}w = \frac{\dot{q}}{c_p}, \qquad (18.19.3)$$

where $\rho_0 = \rho_0(z)$, $\boldsymbol{v} = \boldsymbol{v}_h$,

$$\frac{d}{dt} = \frac{\partial}{\partial t} + \boldsymbol{v} \cdot \nabla, \qquad (18.19.4)$$

and T is related to Φ through hydrostatic equilibrium (11.72):

$$\frac{\partial \Phi}{\partial z} = \frac{RT}{H}. \qquad (18.19.5)$$

In the budget of zonal momentum (18.19.1), it is convenient to express the Lagrangian acceleration in flux form. Multiplying by ρ_0 and incorporating (D.5) obtains

$$\rho_0\frac{du}{dt} + \rho_0 w\frac{\partial u}{\partial z} = \rho_0\frac{\partial u}{\partial t} + \nabla \cdot (\rho_0 \boldsymbol{v}u) + \frac{\partial}{\partial z}(\rho_0 wu) - u\left\{\nabla \cdot (\rho_0 \boldsymbol{v}) + \frac{\partial}{\partial z}(\rho_0 w)\right\}.$$

With continuity (18.19.2), this reduces to

$$\rho_0 \frac{du}{dt} + \rho_0 w \frac{\partial u}{\partial z} = \rho_0 \frac{\partial u}{\partial t} + \nabla \cdot (\rho_0 \boldsymbol{v} u) + \frac{\partial}{\partial z} (\rho_0 w u). \tag{18.20}$$

Expanding in terms of zonal-mean and perturbation quantities, as in Sec. 14.5.4, applying the zonal average $\overline{(\ )}$, and then incorporating the zonal-mean continuity equation

$$\nabla \cdot \overline{\boldsymbol{v}} + \frac{1}{\rho_0} \frac{\partial}{\partial z} (\rho_0 \overline{w}) = 0$$

obtains

$$\rho_0 \overline{\frac{du}{dt}} + \rho_0 \overline{w \frac{\partial u}{\partial z}} = \frac{\overline{d}}{dt} (\rho_0 \overline{u}) + \frac{\partial}{\partial y} (\rho_0 \overline{u'v'}) + \frac{\partial}{\partial z} (\rho_0 \overline{u'w'}), \tag{18.21.1}$$

where

$$\frac{\overline{d}}{dt} = \frac{\partial}{\partial t} + \overline{v} \frac{\partial}{\partial y} + \overline{w} \frac{\partial}{\partial z}. \tag{18.21.2}$$

For quasi-geostrophic motion on a beta plane (Sec. 12.5.3),

$$\overline{w} = 0 \tag{18.22.1}$$

and

$$\overline{v} = \overline{\frac{\partial \psi}{\partial x}} \tag{18.22.2}$$
$$= 0,$$

as motion is cyclic in x. The zonal-mean Lagrangian acceleration then reduces to

$$\rho_0 \overline{\frac{du}{dt}} + \rho_0 \overline{w \frac{\partial u}{\partial z}} = \rho_0 \frac{\partial \overline{u}}{\partial t} + \frac{\partial}{\partial y} (\rho_0 \overline{u'v'}). \tag{18.23}$$

Applying the zonal average to (18.19) and incorporating (18.23) yield equations that govern the zonal-mean circulation:

$$\frac{\partial \overline{u}}{\partial t} - f\overline{v} = -\frac{\partial}{\partial y} (\overline{u'v'}) \tag{18.24.1}$$

$$f\overline{u} = -\frac{\partial \overline{\Phi}}{\partial y} \tag{18.24.2}$$

$$\frac{\partial \overline{v}}{\partial y} + \frac{1}{\rho_0} \frac{\partial}{\partial z} (\rho_0 \overline{w}) = 0 \tag{18.24.3}$$

$$\frac{d\overline{T}}{dt} + \frac{N^2 H}{R} \overline{w} = -\frac{\partial}{\partial y} (\overline{v'T'}) + \frac{\overline{\dot{q}}}{c_p}. \tag{18.24.4}$$

The zonal-mean circulation is forced by the convergence of eddy momentum flux, which forces zonal-mean wind, and the convergence of eddy heat flux, which forces zonal-mean temperature. For planetary waves, those forcings cannot modify zonal wind and temperature independently. Their impacts are coupled through geostrophic equilibrium (18.24.2) and hydrostatic equilibrium (18.19.5), which require the thermal wind balance

$$\frac{\partial \overline{u}}{\partial z} = \frac{R}{fH} \frac{\partial \overline{T}}{\partial y}. \tag{18.25}$$

A perturbation to the system (e.g., through convergence of wave momentum flux) must induce an ageostrophic circulation $(\overline{v}, \overline{w})$ that simultaneously modifies wind and temperature structure to restore thermal wind balance. The nature of the induced circulation is revealed by steady adiabatic conditions. The convergence of momentum flux in (18.24.1) must then be balanced by the Coriolis acceleration, which requires mean meridional motion $\overline{v}$. The convergence of eddy heat flux in (18.24.4) must be balanced by adiabatic cooling, which requires mean vertical motion $\overline{w}$. The two components of mean motion are coupled through continuity (18.24.3). These balances between eddy and zonal-mean terms are approximately satisfied even under unsteady conditions. Consequently, the induced mean circulation achieves a near cancellation with the eddy flux convergences of heat and momentum. The net circulation induced, the part that actually produces a tendency in (18.24), therefore follows as a small residual between those terms.

The *residual mean circulation* is defined as

$$\overline{v}^* = \overline{v} - \frac{R}{N^2 H} \frac{1}{\rho_0} \frac{\partial}{\partial z} \left(\rho_0 \overline{v'T'} \right) \tag{18.26.1}$$

$$\overline{w}^* = \overline{w} + \frac{R}{N^2 H} \frac{\partial}{\partial z} \left(\overline{v'T'} \right). \tag{18.26.2}$$

It represents that part of the induced meridional circulation that is not canceled by the convergence of eddy heat flux. Incorporating (18.26) reduces (18.24) to

$$\frac{\partial \overline{u}}{\partial t} - f\overline{v}^* = \frac{1}{\rho_0} \nabla \cdot \boldsymbol{F} \tag{18.27.1}$$

$$\frac{\partial \overline{v}^*}{\partial y} + \frac{1}{\rho_0} \frac{\partial}{\partial z} \left(\rho_0 \overline{w}^* \right) = 0 \tag{18.27.2}$$

$$\frac{d\overline{T}}{dt} + \frac{N^2 H}{R} \overline{w}^* = \frac{\overline{q}}{c_p}, \tag{18.27.3}$$

where

$$F_y = -\rho_0 \overline{u'v'} \tag{18.27.4}$$

$$F_z = \frac{fR}{N^2 H} \rho_0 \cdot \overline{v'T'}. \tag{18.27.5}$$

The vector $\boldsymbol{F}$ is recognized as the EP flux that was defined in Sec. 14.5.4. In this framework, eddy flux of heat enters only through the zonal momentum equation, where it translates into an upward flux of momentum.

The system (18.27) comprises the *Transformed Eulerian Mean* equations (Andrews and McIntyre, 1976). Unlike the system that governs the conventional Eulerian mean (18.24), it is forced exclusively in the momentum equation by the divergence of EP

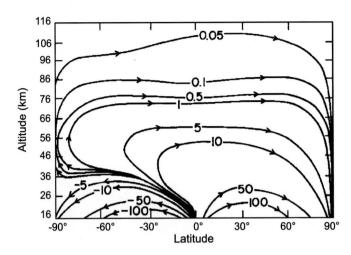

Figure 18.11 Streamlines of the residual mean circulation. Adapted from Garcia and Solomon (1983), copyright by the American Geophysical Union.

flux. As developed in Chap. 14, $\nabla \cdot \boldsymbol{F}$ describes the convergence of wave activity flux and (westward) momentum flux. It represents a westward drag that is exerted on the zonal-mean flow when planetary wave activity is absorbed.

These equations, supported by discussion in Sec. 14.5.4, show that residual mean motion is generated only if $\nabla \cdot \boldsymbol{F}$ is nonzero. This occurs if the waves are transient or if they experience dissipation, which results in the absorption of wave activity. As illustrated by (14.83) for potential vorticity, mean meridional transport of conserved properties then follows directly from $\nabla \cdot \boldsymbol{F}$.

Similar analysis applies to chemical species. If r is the mixing ratio of a long-lived species, then its zonal mean is governed by the continuity equation

$$\frac{\partial \bar{r}}{\partial t} + \bar{v}^* \frac{\partial \bar{r}}{\partial y} + \bar{w}^* \frac{\partial \bar{r}}{\partial z} = 0. \tag{18.28}$$

In concert with (18.27), planetary waves are seen to influence the distribution of $\bar{r}$ only if $\nabla \cdot \boldsymbol{F}$ is nonzero.[4]

Plotted in Fig. 18.11 is the residual mean circulation, calculated under conditions of northern winter. In the lower stratosphere are two equator-to-pole cells. They transport ozone-rich air from the photochemical source region in the tropics to extratropical regions. Meridional transport, accompanied by downwelling, explains the large total ozone observed at high latitude (Figs. 1.21; 18.2). Strong when planetary waves are amplified, that transport also explains the pronounced seasonality of total ozone. Maximum values of Σ_{O_3} appear during late winter and spring, just after the period of amplified planetary wave activity. Contemporaneous with downwelling, this is also when ozone increases in the troposphere (London, 1985).

In the mesosphere, the residual mean circulation is transformed into a pole-to-pole cell. At those heights, air passes from the radiatively-heated summer

---

[4] If the species undergoes chemical production and destruction, a chemical source term appears on the right-hand side of (18.28). It involves zonal-mean production and destruction, as well as interaction between eddy motion and chemistry (Andrews et al, 1987).

hemisphere to the radiatively-cooled winter hemisphere. Despite the distribution of net heating (Fig. 8.27), the summer mesosphere is actually colder than the winter hemisphere (Fig. 1.7). Like the winter stratosphere, the mesosphere is driven out of radiative equilibrium by mechanical disturbances that propagate up from below. Planetary waves play a role at these altitudes. However, gravity waves play an even greater role in disturbing the mesosphere. Their influence is reflected in the departure from radiative equilibrium. Unlike the stratosphere, which is disturbed chiefly in the winter hemisphere, both hemispheres of the mesosphere are disturbed from radiative equilibrium. This feature is consistent with the bi-directional propagation of gravity waves. Westward-propagating gravity waves can propagate through westerlies of the winter stratosphere without encountering a critical level (Sec. 14.3.4). Eastward-propagating gravity waves can do the same through easterlies of the summer stratosphere – unlike planetary waves. Upon reaching mesospheric heights, gravity waves excited in the troposphere have amplified sufficiently to break (Fig. 14.25). Momentum absorbed from the waves then exerts a drag on the zonal-mean flow. To maintain thermal wind balance, that drag induces a meridional circulation, which in turn drives thermal structure out of radiative equilibrium.

18.4 SUDDEN STRATOSPHERIC WARMINGS

During northern winter, the circulation often becomes highly disturbed. Accompanying an amplification of planetary waves, the disturbed motion is characterized by an abrupt deceleration of zonal-mean westerlies, even by their reversal into zonal-mean easterlies. Simultaneously, temperature over the polar cap increases sharply – by as much as 50 K. The winter pole, in darkness, then becomes warmer than the sunlit tropics. This dramatic sequence of events takes place in just a couple of days. It comprises a *sudden stratospheric warming*.

A stratospheric warming is underway on the day shown in Fig. 1.10a. Zonal-mean wind on the same day (Fig. 18.12b) has reversed across much of the winter stratosphere, from 30 hPa upward. Zonal-mean temperature (Fig. 18.12a) is anomalously warm from 70 hPa upward. Polar temperature is some 40 K warmer than climatological-mean temperature (Fig. 1.7). It is warmer even than temperature over the equator. The meridional temperature gradient is therefore reversed. These dramatic changes in the zonal-mean circulation result from an amplification of planetary waves. Enhanced absorption of wave activity ($\nabla \cdot \boldsymbol{F}$) intensifies the drag exerted on the mean flow, along with the residual mean circulation driven by it (18.27).

Although defined in term of zonal-mean properties, the stratospheric warming is a phenomenon that is inherently zonally asymmetric. Figure 1.10a reveals that zonal-mean deceleration on this day actually follows from a displacement of the cyclonic low out of the polar cap. It has been replaced by an anticyclonic high, which has invaded the polar cap from mid-latitudes. The zonally-asymmetric flow that results corresponds to an amplification of planetary wavenumber 1. It rapidly exchanges air between tropical and extratropical latitudes. Rather than operating on the long time scale of the residual mean circulation, this exchange operates on the time scale of only a day. Figure 18.13 shows contemporaneous distributions of motion and potential vorticity near 10 hPa. The cyclonic vortex has been displaced well off the pole and has suffered a complex distortion. High-latitude air (blue) is being drawn equatorward around the anticyclone that has invaded the polar cap. Notice that large eddy displacements have

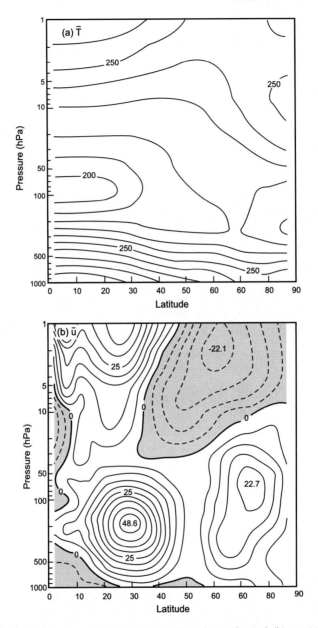

Figure 18.12 Zonal-mean temperature (a) and zonal wind (b) on March 4, 1984, during a stratospheric sudden warming (cf. Fig. 1.10a).

overturned the potential vorticity distribution near (135W,45N), reversing $\frac{\partial Q}{\partial y}$. Motion there is dynamically unstable (Sec. 16.2). The planetary wave field then breaks, forming secondary eddies that appear downstream (McIntyre and Palmer, 1983).

Secondary eddies in the region of instability lead to an irreversible rearrangement of air. Mid-latitude air (red) drawn poleward spins up anticyclonically to conserve potential vorticity. It thus forms easterly flow over the polar cap that prevails

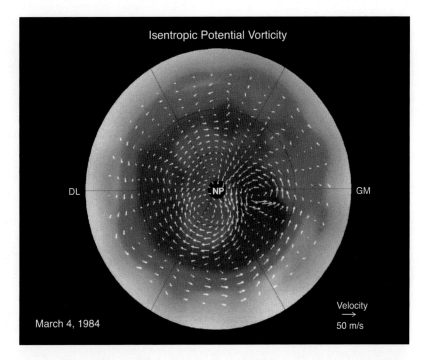

Figure 18.13 Distributions of potential vorticity Q and horizontal motion on the 850°K isentropic surface (near 10 hPa) for March 4, 1984. High-Q polar air (blue), which marks the polar-night vortex, has been displaced off the pole and distorted by a large amplitude planetary wave, with secondary eddies appearing along its tail. Low-Q air from equatorward (red) that is advected into the polar cap spins up anticyclonically, forming a reversed circulation with easterly circumpolar flow at high latitude (cf. 14.26). See color plate section: Plate 30.

in the zonal mean (Fig. 18.12). At the same time, polar air drawn equatorward degenerates into small anomalies of cyclonic vorticity; they are only suggested in the analyzed distribution of Q.[5] The foregoing behavior resembles an expanded critical layer, one that engulfs the winter stratosphere when the wave field amplifies and overturns the Q distribution (cf. Figs. 14.23; 14.26).

Motion during a stratospheric warming leads to a major rearrangement of air and chemical species. Ozone then flows freely from its chemical source region at low latitude to middle and high latitudes. Air advected poleward moves approximately along isentropic surfaces. In the lower stratosphere, where the ozone column is concentrated, isentropic surfaces slope downward toward the pole. Poleward-moving air then descends, undergoing compression. Through secondary eddies, it simultaneously experiences irreversible mixing. On a longer time scale, air displaced to high latitude also experiences radiative cooling, which produces downwelling across isentropic surfaces. These mechanisms are the same ones involved in residual mean motion. By

---

[5] The Q distribution must be derived from meteorological analyses, which, because they are based largely on temperature observations, have limited ability to resolve such features.

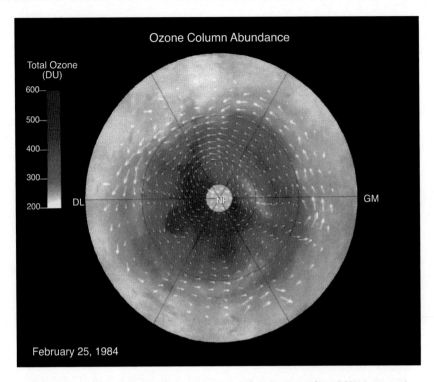

Figure 18.14 Total ozone Σ_{O_3} and horizontal motion on the 400°K isentropic surface in the Northern Hemisphere on February 25, 1984, during a stratospheric warming when zonal flow was disturbed down to 70 hPa. A tongue of enhanced Σ_{O_3} coincides with cross-polar flow between longitudes of 150 E and 30 W, where air descends along isentropic surfaces and undergoes compression. Air displaced equatorward in the Eastern Hemisphere ascends isentropically and undergoes expansion, introducing anomalously low Σ_{O_3}. See color plate section: Plate 31.

amplifying meridional displacements, a stratospheric warming therefore accelerates the residual mean circulation.

Stratospheric warmings are accompanied by a sharp increase of total ozone at high latitude. Plotted in Fig. 18.14 are distributions of Σ_{O_3} and 400 K isentropic motion on February 25, 1984, when zonal-mean motion was reversed down to 70 hPa. A tongue of enhanced total ozone ($\Sigma_{O_3} > 500$ DU) coincides with cross-polar flow. There, ozone-rich air from lower latitude descends along isentropic surfaces. It undergoes compression, increasing ρ_{O_3} and Σ_{O_3} (at the expense of ozone in the surroundings). In the Eastern Hemisphere, air is displaced equatorward. It experiences the reverse effect, leading to reduced Σ_{O_3}.

Stratospheric warmings occur primarily in the Northern Hemisphere, where planetary waves are strong. However, behavior like that described above closes the winter season of both hemispheres. Spring witnesses a weakening of the equatorward temperature gradient that supports strong zonal motion. Circumpolar westerlies then collapse in the *final warming*, replaced during summer by circumpolar easterlies.

18.5 THE QUASI-BIENNIAL OSCILLATION

The circulation in the tropics is also chiefly zonal. However, it is modified importantly by interannual variations. Figure 18.15 presents zonal wind averaged over equatorial stations, as a function of time and height. Prevalent at heights of 15–30 km is a cyclic variation of 24–36 months. Known as the *quasi-biennial oscillation* (QBO), this variation descends with time in an alternating series of easterlies and westerlies. They attain speeds of 20–30 m s^{-1}. At higher levels, the QBO gives way to the *semi-annual oscillation*, the second harmonic of the seasonal cycle. The QBO is symmetric about the equator, concentrated at latitudes of less than 15°.

Although nearly periodic, the QBO is not a harmonic of the seasonal cycle. It therefore cannot be explained merely as seasonality that is imparted to the equatorial stratosphere from other regions. Latent heating in the troposphere contains no preferred period on time scales of the QBO. Likewise, diabatic heating in the equatorial stratosphere is unable to explain the oscillation (Wallace, 1967). This leaves momentum transfer. During the westerly phase of the QBO, air over the equator moves faster than the Earth. Its specific angular momentum therefore exceeds that available from other latitudes. This feature rules out horizontal advection, which conserves angular momentum (11.29).

The QBO is thought to be driven mechanically by upward-propagating waves that transmit momentum from the troposphere. The particular waves involved remain under discussion. However, there is no debate over their origin. Unsteady latent heating inside tropical convection excites a spectrum of wave activity. Much of it propagates upward into the middle atmosphere. Upon being absorbed, that wave activity transfers momentum to the zonal-mean flow, which is thereby accelerated toward the phase speeds of individual waves.

As illustrated in Fig. 18.16, eastward- and westward-propagating waves have transmission and absorption characteristics that differ between easterly and westerly layers of the QBO. These characteristics are determined by the intrinsic frequencies of the waves, which in turn depend on zonal-mean wind (Sec. 14.6). The differences enable wave absorption to establish a westerly critical line. There, mean wind matches the phase speed of westerly waves, which are therefore absorbed. Flanking it overhead is an easterly critical line, where mean wind matches the phase speed of easterly waves that are likewise absorbed. Transfers of westerly and easterly momentum at those sites of wave absorption cause both critical lines to descend, as is observed in Fig. 18.15 (Lindzen and Holton, 1968; Holton and Lindzen, 1972).

The tropical QBO has an extratropical counterpart. During years when equatorial wind is easterly, the polar-night vortex is warmer, weaker, and more disturbed than during years when equatorial wind is westerly (Labitzke, 1982; Holton and Tan, 1980). Similar interannual variability is manifest by total ozone. A signature of the QBO appears in Σ_{O_3} at mid-latitudes, forced by the residual mean circulation of the QBO (Hasebe, 1983; Lait et al., 1989; Tung and Yang, 1994). The Antarctic ozone hole also varies with equatorial wind (Garcia and Solomon, 1987), although more so with planetary wave forcing of the residual mean circulation (Sec. 18.8).

The extratropical QBO results from a displacement of the critical line for planetary waves (e.g., the zero-wind line), to which the polar-night vortex is sensitive. The easterly phase of the QBO displaces the critical line into the winter hemisphere (green in Plates 32 and 34). In a neighborhood of the critical line, large eddy displacements

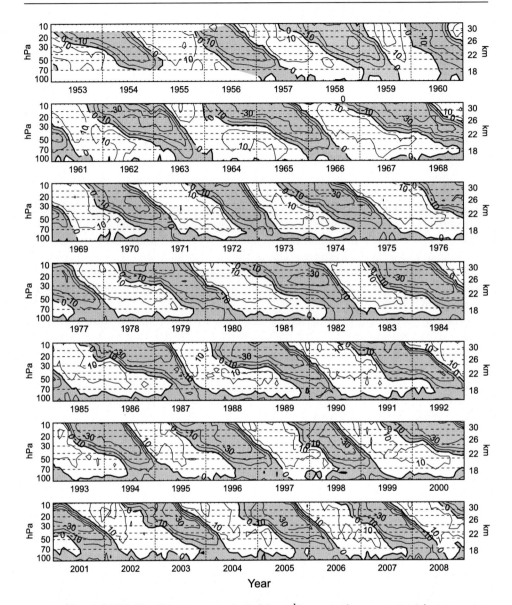

Figure 18.15 Monthly mean zonal wind (m s^{-1}) averaged over equatorial raw-insonde stations. Updated from Naujokat (1986). *Source:* http://www.geo.fu-berlin.de/en/met/ (10.07.10).

disturb the polar-night vortex, intensifying mean downwelling and adiabatic warming over the pole (Fig. 18.17a). Conversely, the westerly phase of the QBO removes the critical line into the summer hemisphere. This leaves the polar-night vortex comparatively undisturbed (Fig. 18.17b).

Because it occupies the lower stratosphere, the QBO influences the tropical tropopause, along with neighboring properties. Anomalous temperature can, by modifying stability, alter tropopause height and temperature. The latter, in turn, influence

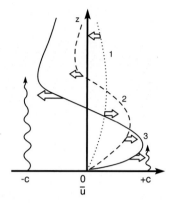

Figure 18.16 Schematic illustrating the differential transmission and absorption characteristics of vertically propagating eastward- and westward-traveling waves with phase speed c. Absorption of momentum from the eastward-traveling wave accelerates the zonal-mean flow toward the phase speed of that wave. Westerlies then form a critical level, where eastward-traveling wave activity is fully absorbed. The westward-traveling wave propagates through those westerlies without encountering a critical level, but it experiences absorption at higher levels in easterly shear. Absorption of its momentum accelerates the zonal-mean flow there toward the phase speed of the westward-traveling wave, producing a critical level where that wave is fully absorbed. Continued absorption of eastward and westward momentum at the critical levels of the two waves then causes the layers of easterlies and westerlies to descend. Adapted from Plumb (1982).

the height of overshooting cumulus (Collimore et al., 2003). Similarly, anomalous zonal wind can modify the steering level of tropical storms. The QBO's influence has been reported in relation to hurricanes (Shapiro, 1989).

18.6 DIRECT INTERACTION WITH THE TROPOSPHERE

The stratosphere is driven by the troposphere, indirectly by upward-propagating waves that transmit momentum. Vertical motion in the residual mean circulation also enables the stratosphere to interact directly with the troposphere. Chemical tracers like N_2O and CH_4 imply that tropospheric air enters the stratosphere across the tropical tropopause. By continuity, air must be returned to the troposphere at middle and high latitudes. The transfer of stratospheric air into the extratropical troposphere is explicit in the residual mean circulation (Fig. 18.11). Estimates of mass transfer suggest a residence time in the middle atmosphere of about 3 years (Holton, 1990). Exchanges of air between the troposphere and stratosphere are complex and poorly documented. However, tracer observations like those in Figs. 18.4 and 18.7 point to the involvement of tropical convection and mid-latitude weather systems.

Water vapor mixing ratio reaches a minimum at the hygropause, reflecting moisture in tropospheric air being limited to saturation values (Fig. 18.8). In the tropics, the hygropause is found as high as 18 km – a couple of kilometers above the mean elevation

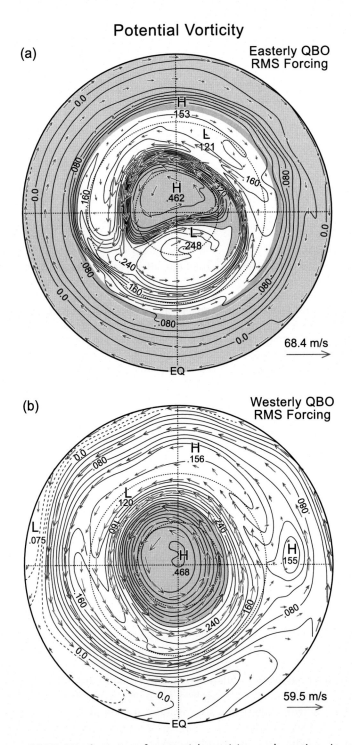

Figure 18.17 Distributions of potential vorticity and motion in a 2-dimensional calculation representative of the polar-night vortex (a) during the QBO easterlies (red), when the zero-wind line (green), which is the critical line for stationary planetary waves, invades the winter subtropics, and (b) during the QBO westerlies, when the critical line for stationary planetary waves recedes into the summer subtropics. Adapted from O'Sullivan and Salby (1990). See color plate section: Plate 32.

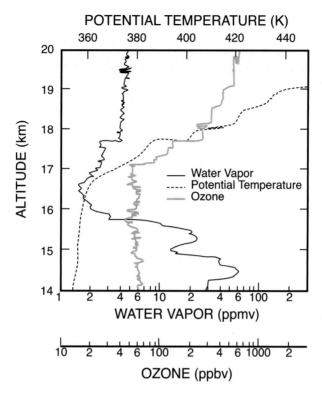

Figure 18.18 Vertical profiles of water vapor and ozone mixing ratio and potential temperature over Darwin Australia on January 13, 1987. *Source:* Kelly et al. (1993).

of the tropopause. Figure 18.18 plots the profile of water vapor mixing ratio over Darwin Australia, along with the mixing ratio of ozone and potential temperature. Well-mixed air below 18 km is marked by nearly uniform r_{O_3} and slowly increasing θ, the latter reflecting neutral stability under saturated conditions. Having been processed by deep convection, that air is characterized by r_{H_2O} that decreases exponentially to a broad minimum between 16 and 18 km. The minimum in water vapor mixing ratio extends about a kilometer above the local tropopause. Achieving values smaller than 2 ppmv, it is significantly drier than saturation values corresponding to the mean temperature of the tropical tropopause. These features indicate that tropospheric air enters the stratosphere in the tropics through overshooting cumulus, with temperatures of 190 K and colder. As discussed in Secs. 7.7 and 9.3, overshooting cumulus lead to mixing of tropospheric and stratospheric air. This process maintains the tropopause against radiative drive, which acts to extend the stratosphere downward. By mixing θ_e, it temporarily elevates the tropopause, extending neutral stability and positive lapse rate upward (Fig. 7.8).

Vertical mixing that occurs when cumulus updrafts invade stable air overhead can lead to upwelling across the tropical tropopause. A cumulus updraft that overshoots its Level of Neutral Buoyancy eventually collapses into cirrostratus anvil

(Fig. 7.7), cloud that is most extensive near the LNB (Fig. 9.25). There, tropospheric and stratospheric air are mixed.[6] Mixing of low-θ air from the overshooting updraft into high-θ air in its surroundings reduces the potential temperature of the environment. According to (2.36), this process serves as a mechanical heat sink for the environment at levels above the LNB. There, cumulus detrainment drives environmental air to temperatures colder than radiative equilibrium. By decreasing θ, it forces downwelling at sites of overshooting cumulus. Apparent in Fig. 9.26a, environmental downwelling extends some distance above the LNB (Gage et al., 1991; Salby et al., 2004a). After leaving those sites, environmental air that has been cooled through cumulus detrainment experiences radiative heating, which acts to restore it to radiative equilibrium. By increasing θ, radiative heating then produces upwelling across isentropic surfaces in the tropical stratosphere. Tropical upwelling must operate in unison with downwelling at extratropical latitudes. Forced by the absorption of planetary wave activity (Sec. 18.3), downwelling pumps stratospheric air into the extratropical troposphere. The latter must be compensated in the tropics by air being returned to the stratosphere at the same rate.

For this process to explain the observed distribution of water vapor, air introduced into the tropical stratosphere must be relatively free of water in condensed phase. Otherwise, that condensate would sublimate, increasing r_{H_2O} over the very lean values observed (Holton, 1984). Cumulonimbus anvil, in which tropospheric and stratospheric air have been mixed, is supplied with condensate by detrainment from cumulus updrafts. However, anvil is destabilized by radiative transfer: LW heating at its base and LW cooling at its top drive vertical overturning (Sec. 9.4). By encouraging particle growth, vertical motion enables condensate to precipitate out (Danielsen, 1982). In fact, airborne observations reveal a sink of total water at levels of cumulonimbus anvil (Kelly et al., 1993). This feature is consistent with the negative anomaly of mixing ratio found above the LNB, where anvil is most extensive (Fig. 9.26b). Appearing just above a layer of nearly saturated environmental conditions, it reflects efficient removal of ice particles that have grown large enough to precipitate to lower levels. This process acts to dehydrate air that has been mixed with stratospheric air and detrained into cumulonimbus anvil.

Stratospheric air enters the troposphere at high latitude. Vertical mixing is believed to be involved in this transfer as well, but on the larger scale of synoptic weather systems. The importance of those systems is underscored by the concentration of ozone column abundance between 10 and 20 km, where synoptic weather systems still influence motion (Figs. 1.20; 18.2). The daily distribution of total ozone (Fig. 1.22a) is, in fact, punctuated by anomalies of high Σ_{O_3}. Comparison with contemporaneous horizontal motion in Fig. 1.9a reveals that those anomalies are positively correlated with mid-latitude cyclones. The pressure on the 375 K isentropic surface (contoured in Fig. 1.22; Plate 1) indicates that, over mid-latitude cyclones, θ surfaces in the lower stratosphere are deflected downward to greater pressure. Ozone-rich air moving along isentropic surfaces must therefore experience compression. This magnifies Σ_{O_3} locally – by as much as 100% (Salby and Callaghan, 1993). Analogous behavior occurs on planetary dimensions during a stratospheric warming, when air is advected poleward along isentropic surfaces (Fig. 18.14; Plate 31).

---

[6] An updraft advances only with the entrainment of environmental air (Fig. 9.16), which, for an overshooting cumulus, is stratospheric.

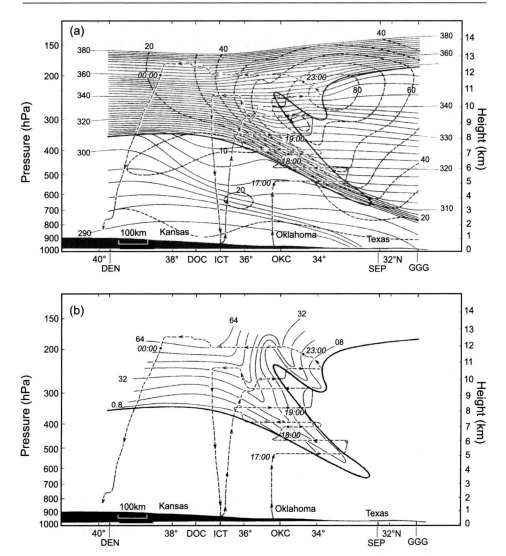

Figure 18.19 Vertical cross section through a tropopause fold. (a) Distributions of θ (solid) and wind speed (dashed). The jet is flanked by a baroclinic frontal zone in which isentropic surfaces descend into the troposphere from stratospheric levels. (b) Distribution of ozone mixing ratio. Tracer values advected isentropically decrease along the tropospheric intrusion, reflecting small-scale mixing and dilution of stratospheric ozone. Vertical arrows mark the flight path of airborne measurements from which the analyzed fields were inferred. After Shapiro (1982).

In meteorological analyses, isentropic surfaces undergo vertical deflections of a couple of kilometers. Airborne measurements, however, record stratospheric species as low as 700 hPa. Stratospheric air enters the troposphere inside *tropopause folds*, where isopleths of long-lived species are overturned (see Fig. 18.19). There, stratospheric air is drawn downward along the frontal zone that intensifies during the development of a baroclinic system. The impact of such transport is manifest in zonal-mean ozone mixing ratio (Fig. 1.20). $\bar{r}_{O_3}$ is deflected downward at high latitude. Such structure enhances ozone number density along the poleward flank of the subtropical jet – where baroclinic systems develop (Chap. 15).

Figure 18.19 plots a cross section of a tropopause fold. Isentropic surfaces (Fig. 18.19a) reflect stratospheric air in strong vertical stability (large $\frac{\partial \theta}{\partial z}$). They have been drawn downward, concentrated in a tilting frontal zone that divides cold and warm air. Strongly baroclinic, that region develops from deformation during the amplification of a baroclinic disturbance (Hoskins, 1982). Mirroring this structure is the distribution of ozone mixing ratio (Fig. 18.19b). Long lived at these altitudes (Fig. 18.1), r_{O_3} marks stratospheric air that has been drawn into the frontal zone. Anomalous mixing ratio is diluted by small-scale mixing, which reduces r_{O_3} along the intrusion. Once it enters the troposphere, O_3 is quickly transferred by convection to the Earth's surface, where it is destroyed through oxidation processes.

18.7 HETEROGENEOUS CHEMICAL REACTIONS

The chemical reactions considered in Secs. 18.1 and 18.2 involve species only in gas phase. During the 1970s and 1980s, these and other elements of gas-phase photochemistry were the focus of intensive investigation. Although an impact of anthropogenic species was suggested, it was not expected to emerge clearly before the twenty-first century. Even then, it was expected to reduce the column abundance of ozone by only about 5%. The timing and magnitude of anticipated changes were set by two considerations: (1) Reactive species had to increase to relatively high concentrations before catalytic destruction of ozone via (18.10) and (18.13) became sufficiently fast. (2) Those reactions were favored at higher altitudes, where only a small fraction of the ozone column resides.

For these reasons, the discovery of the Antarctic ozone hole in 1985 caught much of the scientific community by surprise. Moreover, the order-of-magnitude greater depletions observed were found at polar latitudes (Fig. 1.22b), where ozone was thought to be photochemically inert – because UV fluxes there are small. The key ingredient not considered previously was the presence of solid phase, which is normally excluded from the stratosphere by very low mixing ratios of water vapor.

Polar stratospheric cloud (Sec. 9.3) was already recognized to form over the Antarctic because of its very cold temperature. However, it was regarded largely as a curiosity. PSC appears far more frequently in the Antarctic stratosphere than in the warmer Arctic stratosphere (Fig. 18.20). During late Austral winter, when temperature is coldest, fractional coverage over the Antarctic exceeds 50%, chiefly by very tenuous cloud in the PSC I category. It is now widely accepted that PSC provides the surfaces on which certain reactions proceed much faster than they can in gas phase alone. Moreover, the presence of PSC shifts catalytic destruction of ozone from the upper stratosphere, where only a small fraction of the ozone column resides, to the lower stratosphere, where Σ_{O_3} is concentrated.

Figure 18.21 shows profiles of ozone concentration over Antarctica during Austral winter (solid) and shortly after equinox (dashed), when the sun rises above the horizon. A marked reduction of ozone has occurred between 10 and 20 km. Those are the levels where the ozone column is normally concentrated. Decreases in Σ_{O_3} of 50% are

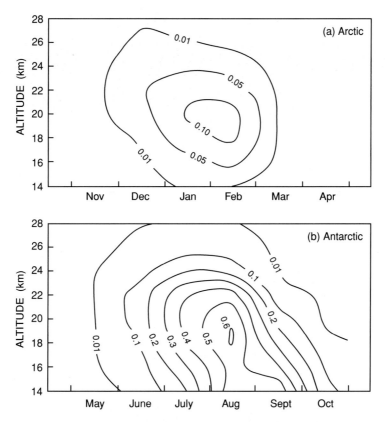

Figure 18.20 Relative frequency of polar stratospheric cloud sightings in satellite observations from SAM II, as a function of height and month, over (a) the Arctic and (b) the Antarctic. After WMO (1991).

observed at this time of year, with column abundances as low as 100 DU having been recorded. Superposed in Fig. 18.21 is the profile of ClO (stippled), which is produced by destruction of ozone (18.13). Consistent with the observed ozone depletion, r_{ClO} maximizes between 10 and 25 km. Those are the same levels where PSC is sighted (Fig. 18.20). A correspondence between reduced O_3 and increased ClO is also apparent across the edge of the polar-night vortex, which delineates the ozone hole (Fig. 18.22). Ozone decreases sharply where chlorine monoxide increases sharply. Both transitions are found where temperature becomes colder than 196 K, the threshold temperature for the formation of type I PSC.

The reactions now recognized to be primarily responsible for the ozone loss in Figs. 1.22b, 18.21, and 18.22 involve two stages: First, inactive chlorine species such as HCl and $ClONO_2$ are converted to reactive forms of Cl_x through heterogeneous reactions like

$$HCl(s) + ClONO_2(g) \rightarrow Cl_2(g) + HNO_3(s) \qquad (18.29.1)$$

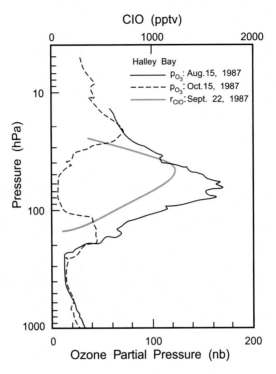

Figure 18.21 Profiles of ozone partial pressure over Antarctica during Austral winter (solid) and shortly after spring equinox (dashed). Profile of chlorine monoxide mixing ratio (stippled) after equinox is superposed. *Sources:* WMO (1988), Solomon (1990).

and

$$H_2O(s) + ClONO_2(g) \rightarrow HNO_3(s) + HOCl(g) \qquad (18.29.2)$$

that involve solid (*s*) as well as gas (*g*) phases. These reactions proceed rapidly on ice, but slowly in gas phase alone. Once produced, Cl_2 and HOCl are readily photolyzed by sunlight, releasing free chlorine

$$Cl_2 + h\upsilon \rightarrow 2Cl \qquad (18.29.3)$$

$$HOCl + h\upsilon \rightarrow OH + Cl, \qquad (18.29.4)$$

which is a reactive form of Cl_x. Then the sequence of reactions

$$Cl + O_3 \rightarrow ClO + O_2 \qquad (18.30.1)$$

$$2ClO + M \rightarrow Cl_2O_2 + M \qquad (18.30.2)$$

$$Cl_2O_2 + h\upsilon \rightarrow Cl + ClO_2 \qquad (18.30.3)$$

$$ClO_2 + M \rightarrow Cl + O_2 + M \qquad (18.30.4)$$

destroys ozone catalytically. It has the net effect

$$2O_3 + h\upsilon \rightarrow 3O_2 \ (Net). \qquad (18.30.5)$$

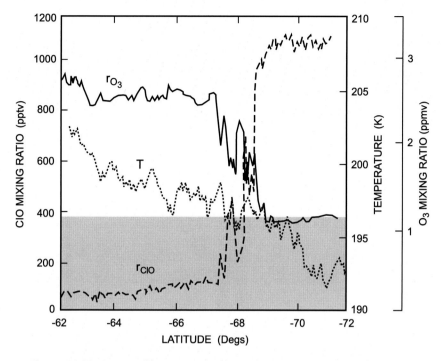

Figure 18.22 Ozone, chlorine monoxide, and temperature along a flight path into the Antarctic polar-night vortex. Temperatures colder than about 196°K (shaded) coincide with the formation of type I PSC. Source of O_3 and ClO profiles: Anderson et al. (1989).

A similar influence on ozone is exerted by bromine. Although less abundant, reactive bromine is some 50 times more effective at destroying ozone than reactive chlorine (WMO, 2006). With established reaction rates, observed mixing ratios of $r_{ClO} \cong 1$ ppbv are adequate to explain the observed ozone depletion rate of about 2% per day. The observed concentration of ClO is two orders of magnitude greater than that predicted by gas-phase photochemistry alone. However, it is consistent with calculations that include heterogeneous reactions like (18.30).[7]

The catalytic sequence (18.30) is initiated by free chlorine, which originates largely from photolysis of CFCs (18.12). Produced exclusively by industry, CFCs have led to steadily-increasing levels of atmospheric chlorine (Fig. 18.23). Above a natural background level of about 0.6 ppbv associated with ocean processes (Sec. 18.2.2), atmospheric chlorine has increased five fold since the 1950s – shortly after the introduction of CFCs in industrial applications. Tending in the opposite sense is the signature of ozone depletion over Antarctica. It accelerates after 1980, presumably when chlorine levels exceeded a threshold for reactions (18.30) to become an important sink of O_3 (Solomon, 1990).

---

[7] Those calculations rest on the photolysis rate of Cl_2O_2 (18.30.3), which has come into question (Pope et al., 2007; Schiermeier, 2007).

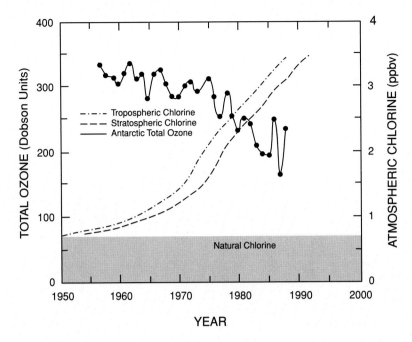

Figure 18.23 Time series of atmospheric total chlorine and total ozone over Halley Bay, Antarctica. Adapted from Solomon (1990).

While converting inert forms of chlorine into reactive Cl_x, heterogeneous reactions (18.29) have the opposite effect on reactive nitrogen. They convert NO_x into relatively inactive nitric acid. This bears importantly on ozone depletion because NO_x regulates the abundance of reactive chlorine. The principal means by which Cl_x is converted back to inactive forms is via reaction with nitrogen dioxide

$$ClO + NO_2 + M \rightarrow ClONO_2 + M. \qquad (18.31)$$

The abundance of NO_2 thus controls the duration over which reactive chlorine is available to destroy ozone in (18.30). Because PSC is composed of hydrated forms of nitric acid (Sec. 9.3), its formation removes NO_x from the gas phase. Should PSC particles become large enough to undergo sedimentation, NO_x is removed altogether. The stratosphere is then denitrified, leaving reactive chlorine available much longer to destroy ozone.

Figure 18.24 compares the seasonal cycle of Σ_{O_3} over Antarctica based on the historical record against that based on years after the appearance of the ozone hole (cf. Fig. 1.21). The two evolutions diverge near Austral spring. Solar radiation then triggers reactions (18.29.3) and (18.29.4) that release reactive chlorine, setting the stage for catalytic destruction of ozone in (18.30). Minimum column abundance is observed in October. By December, increasing values restore Σ_{O_3} toward historical levels, when ozone-rich air is imported from low latitudes during the final warming. But, even then, total ozone remains ~5% below historical levels, due to the dilution of

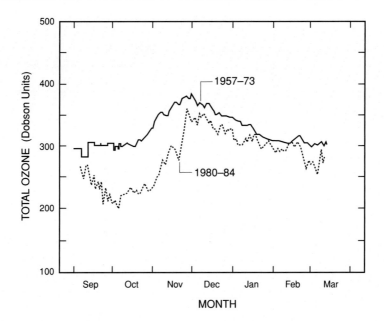

Figure 18.24 Seasonal cycle of total ozone over Halley Bay, Antarctica, based on the historical record since 1957 and on years since the appearance of the Antarctic ozone hole. After Solomon (1990).

subpolar air with ozone-depleted air from inside the polar-night vortex. As illustrated by Fig. 18.25, the breakdown of the vortex during Austral spring involves a complex rearrangement of air. Meridional displacements then allow ozone-depleted air (white), previously confined inside the vortex, to escape from the polar cap. Anomalies created in this fashion can survive for weeks before they are eventually destroyed by small-scale mixing and diffusion (e.g., Hess and Holton, 1985). Even after they dissolve into the wider air mass, the associated ozone deficit dilutes summertime ozone across much of the Southern Hemisphere (cf. Fig. 18.24).

Airborne observations have established that the same reactions which operate over the Antarctic operate over the Arctic. However, owing to the warmer temperature of the Arctic polar-night vortex, PSC is relatively infrequent. Moreover, free chlorine that is produced in isolated PSC via (18.29) is quickly acted on by dynamical effects. By elevating temperature, those effects reverse the process through other chemical reactions (e.g., Garcia, 1994). This limits the amount of reactive chlorine present when the sun rises over the Arctic, which in turn limits catalytic destruction of ozone via (18.30). Similar considerations apply to mid-latitudes, where ozone depletions of 5–10% have been reported (WMO, 1991). Temperature there is too warm to support cloud formation. Those depletions, however, may involve heterogeneous reactions that are supported by background aerosol in the Junge layer (Sec. 9.1.3).

The limiting factor in ozone depletion appears to be temperature, which controls the formation of PSC (Sec. 9.3). Indeed, the deepest reductions of Antarctic Σ_{O_3} are observed during the coldest winters. This dependence mirrors the spatial correspondence between temperature and perturbed photochemistry (Fig. 18.22). Consequently,

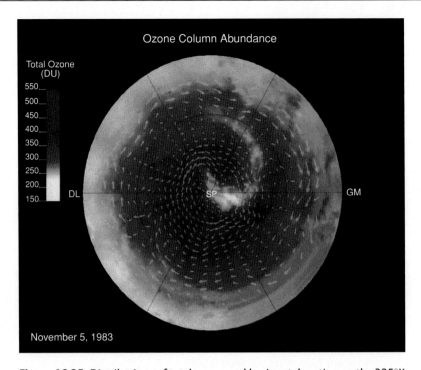

Figure 18.25 Distributions of total ozone and horizontal motion on the 325°K isentropic surface over the Southern Hemisphere on November 5, 1983, during the final warming. Ozone-depleted air (white), which had previously been confined inside the polar-night vortex, escapes in a tongue of anomalously-lean values that spirals anticyclonically into mid-latitudes to conserve potential vorticity. See color plate section: Plate 33.

dynamical disturbances that control temperature inside the vortex (Sec. 18.3) exert a major influence on ozone depletion at high latitude.

18.8 INTERANNUAL CHANGES

The stratosphere contains an internal mechanism for varying between years. In the QBO, the equatorial circulation swings between easterlies and westerlies. As noted in Sec. 18.5, those changes are also manifest in the polar-night vortex, which, through planetary waves, is sensitive to equatorial wind. Changes of $\nabla \cdot \boldsymbol{F}$ introduced by the QBO modify the residual mean circulation. It in turn modifies polar temperature and the strength of the polar-night vortex. The modulated influence of planetary waves (Fig. 18.17) makes stratospheric warmings more frequent during the easterly phase of the QBO than during its westerly phase (Naito and Hirota, 1997; Gray et al., 2004).

The QBO's influence on the vortex favors wintertime polar temperature that is anomalously warm during its easterly phase but anomalously cold during its westerly phase. Such dependence is manifest in polar temperature during many years – but not all of them. Plotted in Fig. 18.26 is anomalous wintertime temperature in the Arctic stratosphere, separated into years when wind in the equatorial stratosphere

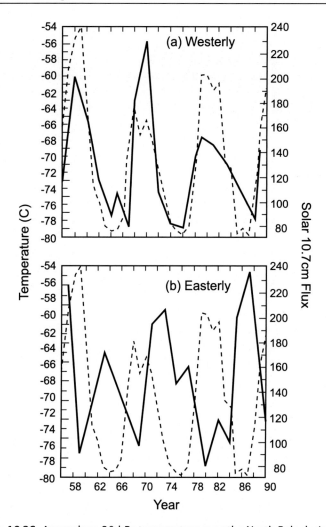

Figure 18.26 Anomalous 30-hPa temperature over the North Pole during Jan–Feb (solid) for years when the QBO in equatorial wind at 50 hPa is (a) westerly and (b) easterly. Superimposed is the 10.7-cm solar flux (dashed). After Labitzke and van Loon (1988). Reprinted from *J Atmos Terr Phys*, **50**, Labitzke K and H van Loon, Association between the 11-yr solar cycle, the QBO, and the atmosphere, 197–206, Copyright (1988), with permission from Elsevier.

is westerly and years when it is easterly (solid). During the westerly phase of the QBO (Fig. 18.26a), wintertime polar temperature evolves systematically, cycling on the time scale of a decade. Superimposed is 10.7-cm solar irradiance, a measure of net solar flux (dashed). It too evolves systematically – through the 11-year cycle of solar activity (Sec. 8.1.1). Polar temperature evolves coherently, in phase with solar flux. Much the same appears during the easterly phase of the QBO, (Fig. 18.26b). Polar temperature then also evolves systematically, coherently but out of phase with solar flux.

The change in overall flux during the solar cycle is small, of order 0.1% (Willson et al., 1986). This limits its direct impact at the Earths' surface, where most wavelengths in the SW spectrum are absorbed. In the stratosphere, however, the impact of the solar cycle can be substantially greater. The stratosphere is forced radiatively by ozone heating (Fig. 8.24). This feature makes it intrinsically sensitive to UV at wavelengths shorter than 200 nm. At those wavelengths, the 11-year variation of SW flux is of order 10%.

Variations of polar temperature in Fig. 18.26 imply the following relationship: For winters near solar min, polar temperature is anomalously warm during the easterly phase of the QBO, but anomalously cold during its westerly phase. This is the dependence implied by the QBO's influence on the extratropical circulation (Fig. 18.17). However, for winters near solar max, polar temperature obeys almost the reverse relationship: In those winters, polar temperature tends to be anomalously cold during the easterly phase of the QBO, but anomalously warm during its westerly phase.

The mechanism behind this decadal variation is poorly understood. One suggestion is a decadal modulation of planetary wave absorption, induced by differences of the polar-night jet between solar min and solar max (Kodera and Kuroda, 2002). Another involves the QBO directly. The QBO's period drifts between 24 and 36 months. That variation, however, is not altogether random. Presented in Fig. 18.27a, as a function year, is the anomalous frequency of the QBO, the deviation from its climatological mean. The QBO's frequency drifts, on the time scale of a decade. It increases near solar max, when its period approaches 24 months. It decreases near solar min, when its period approaches 36 months. The QBO induces its own residual mean circulation, with vertical motion of one sign over the equator flanked in the subtropics of each hemisphere by vertical motion of opposite sign (see, e.g., Andrews et al., 1987). Through adiabatic warming and cooling, that vertical motion introduces anomalous temperature. Plotted in Fig. 18.27b is the running power spectrum of temperature in the subtropical stratosphere. Variance associated with the QBO is concentrated about a period near 30 months. However, it drifts systematically on the time scale of a decade, cycling between shorter and longer period.[8]

The drift in frequency introduces a systematic drift of the QBO's phase with respect to winter months, when the extratropical circulation is sensitive to equatorial wind. Plotted in Fig. 18.28a, for years during the easterly phase of the QBO, is the correlation between February temperature and solar flux. The correlation is significant in the subtropical lower stratosphere and upper troposphere (shaded). Its implication is illustrated in Fig. 18.28b, which plots the respective time series at 100 hPa. February temperature near the tropopause evolves systematically, coherently and in phase with solar flux.

Thermal wind balance requires a corresponding drift in zonal wind – in the subtropical anomaly that is induced directly by the QBO. Neighboring the critical line of planetary waves, anomalous zonal wind there would introduce a modulation of $\nabla \cdot \boldsymbol{F}$. The latter would then force anomalous residual motion and polar temperature. Behavior analogous to that shaded in Fig. 18.28a, although weaker, is also evident at tropospheric levels (see Labitzke, 2005; Crooks and Gray, 2005; Salby and Callaghan, 2006).

---

[8] The signature of the QBO becomes obscure after 1980. That loss is contemporaneous with the introduction in meteorological analyses of satellite data, which have vertical resolution inadequate to resolve the QBO (Huesmann and Hitchman, 2003).

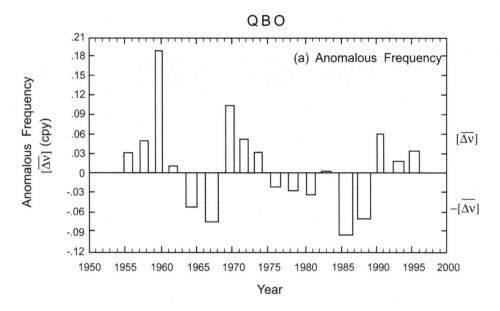

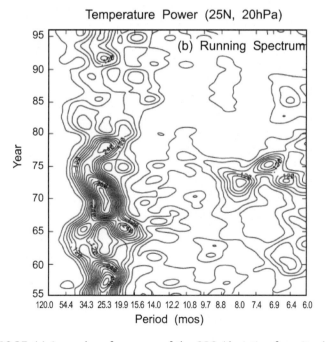

Figure 18.27 (a) Anomalous frequency of the QBO (deviation from its climatological mean: $0.41 \ yr^{-1} = (29.2 \ mos)^{-1}$), during westerly intervals of the QBO. Indicated at right is rms drift of frequency that varies coherently with the 11-year solar cycle. (b) Running power spectrum of temperature at 25N and 20 hPa, as function of year. After Salby and Callaghan (2000; 2006).

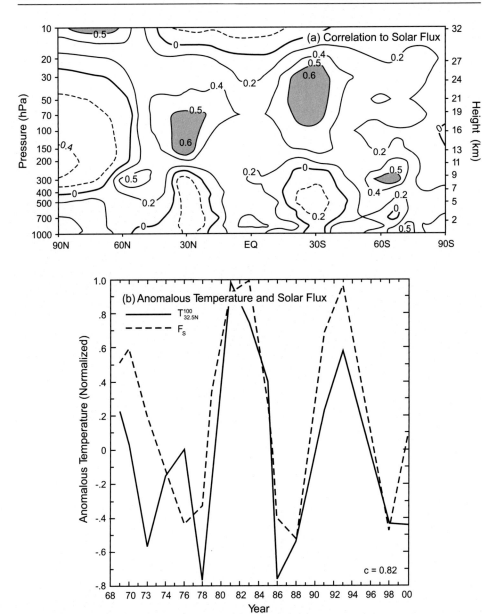

Figure 18.28 For years during the easterly phase of QBO, (a) correlation between February temperature and 10.7-cm solar flux (with regions of statistical significance shaded) and (b) as function of year, anomalous February temperature at 32N and 100 hPa (solid) and 10.7-cm solar flux (dashed). Correlation between the annual records is 0.82. After Labitzke (2005) and Salby and Callaghan (2006). Reprinted from *J Atmos Solar Terr Phys*, **67**, K Labitzke, On the solar cycle – QBO relationship: A summary, 45–54, Copyright (2005), with permission from Elsevier.

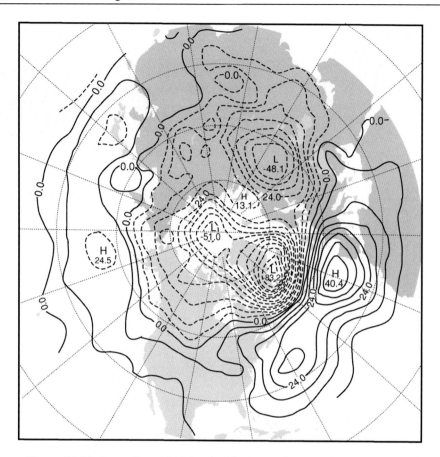

Figure 18.29 Anomalous 1000-hPa height (proxy for SLP) that varies coherently with the equatorial QBO. Compare with Figs. 15.17, 15.19b. After Holton and Tan (1980).

The QBO's influence on the extratropical circulation is conveyed poleward through a modulation of residual mean motion. As the residual circulation couples the stratosphere and troposphere, a change in $(\bar{v}^*, \bar{w}^*)$ must also exert an influence in the troposphere. Displayed in Fig. 18.29 is anomalous SLP that varies coherently with the equatorial QBO. It is marked by anomalously-low SLP over the Icelandic Low and anomalously-high SLP over the Bermuda-Azores High. These are the salient features of the NAO (Fig. 15.17). Closely related is the AO (Fig. 15.19). Indeed, power spectra associated with the AO evidence a peak at periods of the QBO (Coughlin and Tung, 2001).

The AO's relationship to the residual mean circulation is manifest in its structure. Annular at high latitude, AO structure extends upward like the polar-night vortex. The relationship is also manifest in signatures of vertical motion. Plotted in Fig. 18.30 is anomalous wintertime temperature that varies between years coherently with stratospheric temperature over the North Pole (open circle). Positive anomalous temperature over the winter pole is a signature of adiabatic warming and anomalous downwelling. From stratospheric levels, it extends downward coherently into the Arctic troposphere – as low as 500 hPa. Accompanying it at subpolar latitudes is negative

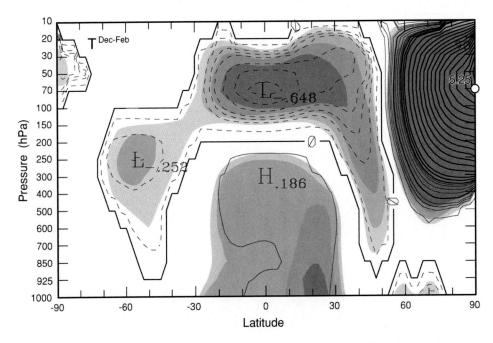

Figure 18.30 Anomalous wintertime tendency of temperature (Mar–Dec) that varies coherently with anomalous temperature at 70 hPa over the North Pole (open circle), which is a measure of anomalous downwelling. Values correspond to an rms intensification of downwelling in the Arctic stratosphere. Regions significant at the 90% and 95% levels are shaded. Compare with structure of the AO in Fig. 15.20. After Salby and Callaghan (2005).

anomalous temperature, a signature of adiabatic cooling and anomalous upwelling. Visible in the troposphere of both hemispheres, it joins a stronger signature of anomalous upwelling in the tropical stratosphere. The AO exhibits much the same structure (Fig. 15.20).

Inside the tropical troposphere, the signature of adiabatic cooling in Fig. 18.30 is interrupted by positive anomalous temperature. Although comparatively weak, it too is coherent with anomalous downwelling in the polar stratosphere. The warm anomaly appears equatorward of $30°$ and below 200 hPa – the LNB of cumulus updrafts (Sec. 7.7). At those levels, updrafts are positively buoyant. Cumulus detrainment there acts to heat the environment. Anomalous temperature becomes negative above 200 hPa. The cold anomaly maximizes just above the tropical tropopause. At those levels, updrafts are negatively buoyant. Cumulus detrainment there acts to cool the environment. Tropical anomalies at both levels operate coherently with anomalous downwelling over the winter pole. They imply an intensification of upwelling at upper levels of the Hadley circulation, one that accompanies an intensification of downwelling over the winter pole (Thuburn and Craig, 2000; Salby and Callaghan, 2005).

A similar relationship is implied for the tropical tropopause. The cold anomaly above the LNB acts to extend positive lapse rate upward. Its coherence with polar temperature implies a tropical tropopause that is anomalously cold and high when anomalous downwelling over the winter pole is anomalously strong. This relationship

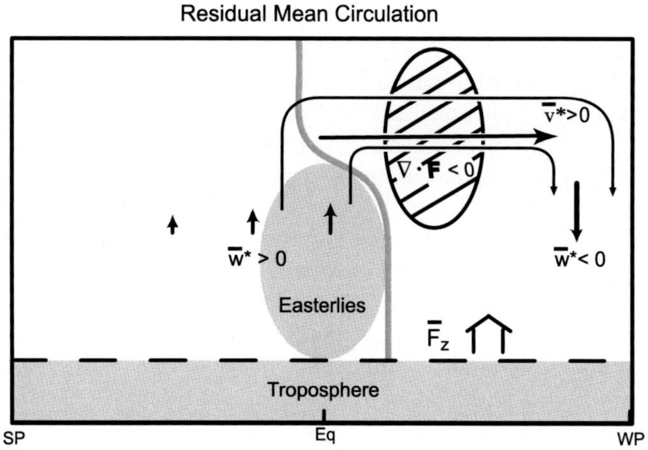

Residual Mean Circulation

Figure 18.31 Schematic representation of the residual mean circulation $(\overline{v}^*, \overline{w}^*)$. Driven by absorption of EP flux $(\nabla \cdot \boldsymbol{F})$ that is transmitted upward by planetary waves (18.27). The latter is measured by the upward EP flux at 100 hPa, integrated over latitude $\overline{F}_z$ (18.32). During QBO easterlies (red), the zero-wind line (green), which is the critical line for stationary planetary waves, invades the winter hemisphere (cf. Fig. 18.17a). See color plate section: Plate 34.

is consistent with the seasonality of the tropical tropopause. The latter is highest during northern winter, when planetary waves are amplified (Yulaeva et al., 1994).

Some insight into these changes follows from the mechanisms behind anomalous vertical motion, illustrated in Fig. 18.31. Those mechanisms are embodied in the transformed Eulerean mean equations. Under time-mean conditions, planetary wave drag is balanced by the Coriolis acceleration (18.27.1). At levels of wave absorption ($\nabla \cdot \boldsymbol{F} < 0$), wave drag induces a poleward drift

$$\overline{v}^* = -\frac{1}{\rho_0 f} \nabla \cdot \boldsymbol{F}$$

$$> 0.$$

$\qquad$ (18.32.1)

Averaging over the middle atmosphere, where EP flux is fully absorbed, and applying the divergence theorem recovers the poleward drift of mass in terms of the upward EP flux at 100 hPa integrated over latitude, $\overline{F}_z$. If $\nabla \cdot \boldsymbol{F}$ is sufficiently narrow in latitude to be treated on an f plane, then the poleward drift of mass reduces to

$$\langle \rho_0 \overline{v}^* \rangle = \frac{1}{f} \overline{F}_z,$$

$\qquad$ (18.32.2)

where $\langle \ \rangle$ denotes the integral over the region of wave absorption. The poleward drift of mass is proportional to the net transmission of momentum from the troposphere by planetary waves.

Integrating the continuity equation (18.27.2) vertically and incorporating (18.32.1) yields an expression for mean vertical motion:

$$\overline{w}^* = -\frac{1}{\rho_0} \frac{\partial}{\partial y} \left[\frac{1}{f} \int_z^\infty \nabla \cdot \boldsymbol{F} \, dz' \right].$$

$\qquad$ (18.32.3)

Vertical motion at height z is determined by the collective wave absorption overhead. The region of wave absorption is characterized by $\nabla \cdot \boldsymbol{F} < 0$. Along its equatorward flank, $\nabla \cdot \boldsymbol{F}$ decreases northward, so $\overline{w}^* > 0$. That region experiences upwelling and adiabatic cooling. Along the poleward flank of wave absorption, $\nabla \cdot \boldsymbol{F}$ increases northward, so $\overline{w}^* < 0$. That region experiences downwelling and adiabatic warming. Indicated in Fig. 18.31, these vertical motions imply the anomalous temperature structure evident in Fig. 18.30.

Anomalous temperature in Fig. 18.30 represents changes of thermal structure that are spatially coherent. Nearly identical structure characterizes anomalous temperature that varies coherently with $\overline{F}_z$ and the QBO (Salby and Callaghan, 2002). By modulating $\nabla \cdot \boldsymbol{F}$, the latter force an anomalous residual circulation (18.32.2). Through adiabatic warming and cooling, it induces anomalous temperature. Most of the temperature anomaly follows from $\overline{F}_z$, which modulates the magnitude of $\nabla \cdot \boldsymbol{F}$ (18.32.3).

Anomalous residual motion also influences the distribution of total ozone. By modulating poleward transport of ozone-rich air, anomalous $\overline{v}^*$ introduces anomalous total ozone at extratropical latitude. Plotted in Fig. 18.32, as a function of year, is the anomalous wintertime increase of NH total ozone (solid). It reflects the springtime maximum of total ozone; cf. Fig. 1.21. Superimposed is the component of ozone that varies coherently with anomalous $\overline{F}_z$ and the QBO, adjusted for sporadic enhancements of volcanic aerosol (dashed). Anomalous NH ozone closely tracks anomalous forcing of the residual circulation. Achieving a correlation to observed changes of 0.95, anomalous dynamical forcing accounts for nearly all of the interannual variance of NH ozone. Most of the ozone anomaly derives from changes of $\overline{F}_z$.

Similar considerations apply to polar ozone. However, those changes are complicated by the dependence on temperature of heterogeneous chemical destruction. The formation of PSC, which activates chlorine, is intrinsically sensitive to temperature (Sec. 9.3.1). Polar temperature, in turn, is determined by downwelling and, hence, by the residual mean circulation. Because the residual circulation also transports ozone to high latitude, the dynamics and chemistry of polar ozone are intertwined. This is especially true in the Northern Hemisphere, where amplified planetary waves sharply disturb the vortex (Fig. 1.10a).

In the Arctic stratosphere, wintertime temperature varies between years by as much as 10 K. The polar-night vortex thus varies from conditions representative of the Arctic, disturbed and warm, to those representative of the Antarctic, undisturbed and cold. The latter conditions support the formation of PSC and, hence, heterogeneous chemical destruction of ozone (18.29). Indeed, during cold winters, Arctic ozone is anomalously low (Muller et al., 2003; Tilmes et al., 2003; Rex et al., 2004). Reduced ozone then is consistent with accelerated chemical depletion. However, cold winters are also distinguished by diminished transport into the Arctic. With the latter, reduced ozone is likewise consistent. Unraveling contributions from anomalous transport and chemical destruction is complicated by their inter-dependence, through temperature.

A clue to the individual roles of transport and chemical destruction comes from the 3-dimensional structure of ozone. Compared in Fig. 18.33 is the distribution of r_{O_3} during March following warm winters against that following cold winters (solid). March r_{O_3} corresponds to the spring maximum of total ozone (Fig. 1.21). It is the culmination of increases during the winter season, when the vortex is disturbed and the residual circulation is active (Fig. 1.21). Superimposed in Fig. 18.33 are isentropic

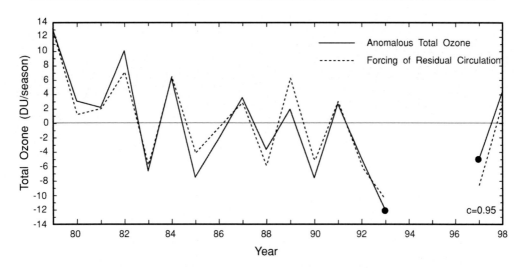

Figure 18.32 Anomalous wintertime tendency of NH total ozone, which dictates springtime ozone (solid). Superimposed is the component that varies coherently with anomalous forcing of the residual circulation, represented in $\overline{F}_z$ and the QBO, adjusted for sporadic enhancements of volcanic aerosol (dashed). Correlation between the wintertime records is 0.95. *Source:* Fusco and Salby (1999), Salby and Callaghan (2004b).

surfaces in the lower stratosphere (dashed), where total ozone is concentrated (cf. Fig. 18.2). Following cold winters (Fig. 18.33b), mixing ratio surfaces have been deflected downward across isentropic surfaces at extratropical latitudes of the winter hemisphere. That structure is a signature of downwelling by the residual mean circulation. It causes r_{O_3} to increase poleward along isentropic surfaces. Total ozone then also increases poleward – but only up to the edge of the vortex, where ozone accumulates. Poleward of 60N, r_{O_3} decreases along isentropic surfaces. Total ozone therefore decreases into the Arctic. During warm winters (Fig. 18.33a), r_{O_3} has similar structure – but not at high latitude. Poleward of 60N, mixing ratio surfaces have been laid flat. They have been driven into coincidence with isentropic surfaces. Total ozone then increases into the Arctic. It attains values some 60 DU greater than during cold winters, when ozone-rich air does not invade the Arctic.

The difference in total ozone follows from the structure of ozone mixing ratio surfaces. They are driven into coincidence with isentropic surfaces during warm winters but not during cold winters. Only one process achieves that structure. Isentropic mixing by planetary waves homogenizes long-lived species along θ surfaces. It offsets the deflection of mixing ratio surfaces across θ surfaces by the residual mean circulation (Holton, 1986). By exchanging air at low and high latitudes, isentropic mixing transfers ozone-rich air into the Arctic (Figs. 14.26; 18.13; 18.14). In concert with intensified downwelling, it increases Arctic ozone during warm winters over that during cold winters. In observed structure, these forms of transport account for some two thirds of the 60-DU anomaly that distinguishes Arctic ozone between warm and cold winters (Salby and Callaghan, 2007).

A similar conclusion follows from numerical simulations. Plotted in Fig. 18.34 is anomalous total ozone between simulations that were performed under conditions of

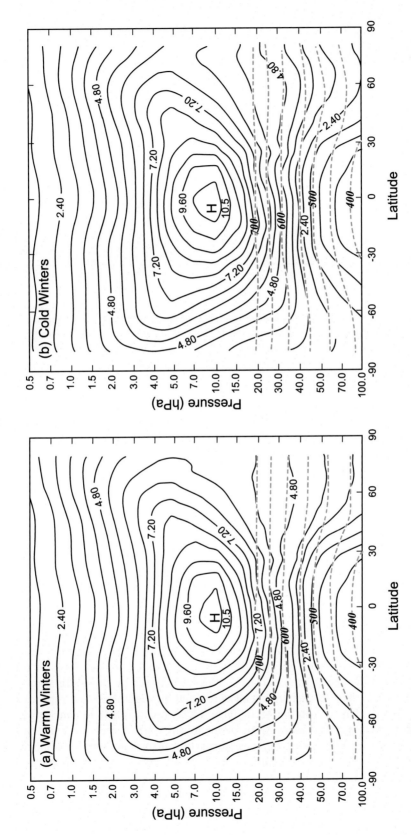

Figure 18.33 Zonal-mean ozone mixing ratio during spring following (a) warm winters and (b) cold winters (solid). Superimposed is the contemporaneous structure of isentropic surfaces (dashed). The difference corresponds to anomalous Arctic ozone between warm and cold winters of ∼60 DU. After Salby and Callaghan (2007).

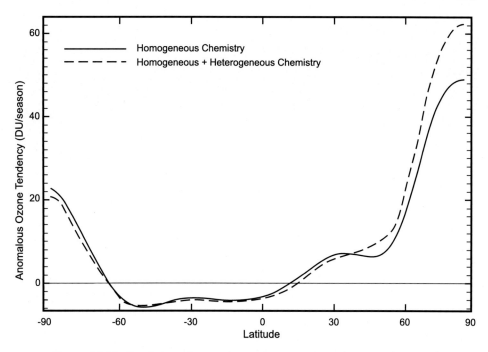

Figure 18.34 Simulation of anomalous springtime ozone between dynamical conditions representative of warm and cold winters. Simulated in the presence of homogeneous chemistry (solid) and inclusive of heterogeneous processes (dashed). Simulations performed in a 3-dimensional model based on the dynamical treatment of Callaghan et al. (1999) and the chemical treatment of Mozart-3 (Horowitz et al., 2003). Adapted from Salby (2011).

cold winters and warm winters. In one simulation, the polar-night vortex is strongly perturbed by amplified planetary waves, which are imposed from observed tropospheric structure during warm winters. In the other simulation, it is weakly perturbed by weak planetary waves, which are imposed from observed tropospheric structure during cold winters. Plotted in solid is anomalous total ozone (strongly perturbed – weakly perturbed) in the presence of only homogeneous (gas-phase) chemistry. Negative values appear in the tropics, where anomalous upwelling introduces ozone-lean air from the troposphere. From there, anomalous ozone increases poleward, exceeding 48 DU over the Arctic. Superimposed is the anomaly inclusive of full heterogeneous chemistry (dashed). Equatorward of 50N, it is nearly identical to the ozone anomaly that forms under homogeneous chemistry. At higher latitude, however, it diverges. Exceeding 60 DU over the Arctic, the computed anomaly is broadly consistent with the observed anomaly between warm and cold winters. Most of the contribution from heterogeneous chemistry enters through the simulation under weakly-perturbed conditions. Arctic temperature then decreases to 187 K. This magnifies PSC, accelerating chlorine activation and ozone depletion. The contribution from heterogeneous chemistry (ozone depletion) accounts for about 30% of the overall anomaly between warm and cold winters. The rest follows from changes of transport.

Over the Antarctic, similar considerations apply. However, weaker planetary waves of the Southern Hemisphere leave the Antarctic vortex cold enough to form

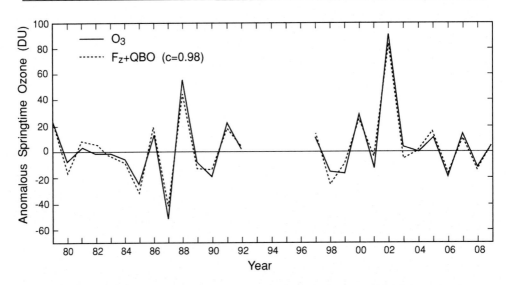

Figure 18.35 Interannual anomaly of springtime ozone (Sep–Nov) over the Antarctic (solid). Superimposed is the component that varies coherently with anomalous forcing of the residual circulation (dashed). Follows application of highpass filter that discriminates to periods shorter than a decade. Correlation between the records of springtime ozone is 0.98. After Salby et al., (2011).

PSC during all winters. How much forms depends on how cold temperature becomes. Plotted in Fig. 18.35 is the interannual anomaly of springtime ozone over the Antarctic (solid). Reflecting changes of the Antarctic ozone hole (Fig. 1.22b), it varies between neighboring years by 50–100 DU. This is almost as large as the mean depletion of ozone that defines the ozone hole. Superimposed is the component of ozone that varies coherently with anomalous $\overline{F}_z$ and the QBO (dashed). Changes of the ozone hole closely track anomalous forcing of the residual circulation. Achieving a correlation of 0.98, anomalous dynamical forcing accounts for nearly all of the interannual variance of the ozone hole. Most of the ozone anomaly derives from changes of $\overline{F}_z$, which modulate downwelling over the Antarctic (Huck et al., 2005; Salby et al., 2011). Through adiabatic warming, those changes shorten and prolong the period of coldest temperature which supports PSC. The latter, in turn, modulates the net activation of chlorine and hence springtime depletion of ozone (18.29).

The anomaly in Fig. 18.35 represents changes of springtime ozone between neighboring years. Introduced by anomalous forcing of the residual circulation, those changes describe the *climate variability* of Antarctic ozone (Sec. 1.6). Removing them leaves the component of anomalous ozone that is unrelated to dynamically-induced changes. Plotted in Fig. 18.36 is the record of springtime ozone that is independent of $\overline{F}_z$ and the QBO (solid). Representing the *secular variation* of Antarctic ozone, the part coherent over a decade and longer (Sec. 1.6), it declines over the 1980s and early 1990s, until that portion of the satellite record ends. By the late 1990s, when the satellite record resumes, the secular variation of Antarctic ozone manifests a gradual but systematic rebound. The upward trend is visible for almost as long as was the downward trend during earlier years. Reflecting ozone recovery, it follows

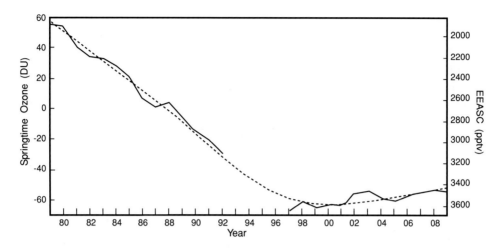

Figure 18.36 Component of springtime ozone over the Antarctic that is independent of dynamically-induced changes (solid). The upward trend after 1996 is significant at the 99.5% level. Superimposed is Equivalent Effective Stratospheric Chlorine for the Antarctic (with inverted scale), corresponding to a mean age of air of 5 years and a bromine scale factor of 60 (dashed). After Salby et al. (2011).

the elimination of CFC production by the Montreal Protocol, which has led to a gradual reduction of chlorine reservoir species (18.29). Superimposed in Fig. 18.36 (with inverted scale) is the record of Equivalent Effective Stratospheric Chlorine (EESC), inclusive of bromine (dashed). From its peak in the late 1990s, EESC has declined gradually, by ~15%. Tracking it is the secular variation of Antarctic ozone. In the full record of ozone, the systematic variation is obscured by dynamically-induced changes. Those changes, which represent climate variability, dominate the current evolution of Antarctic ozone. Owing to the gradual decline of chlorine, which is characteristic of secular changes of climate, this will remain the case for some time (WMO, 2006).

SUGGESTED REFERENCES

Interaction between planetary waves and the mean meridional circulation is developed in *An Introduction to Dynamic Meteorology* (2004) by Holton. An advanced treatment is presented in *Middle Atmosphere Dynamics* (1987) by Andrews et al.

Holton et al. (1995) presents an overview of stratosphere–troposphere exchange.

Atmospheric Physics (2000) by Andrews includes an overview of stratospheric chemistry and transport. A detailed treatment of gas phase photochemistry is presented in *Aeronomy of the Middle Atmosphere* (1986) by Brasseur and Solomon. Heterogeneous processes are reviewed in WMO (2006).

PROBLEMS

1. Photochemical lifetime is defined from reactions that destroy a species X as $\tau_X^{-1} = -\frac{1}{X}\frac{dX}{dt}$. Within the framework of Chapman chemistry, derive expressions for the photochemical lifetimes of O_3 and O_x. If below 50 km $[O_x] \cong [O_3]$, and if

the rate coefficients in cgs units are $k_2 = 6.0 \ 10^{-34} \left(\frac{300}{T}\right)[M]$, $k_3 = 8.0 \ 10^{-12} e^{-\frac{2060}{T}}$, where T is in Kelvin, and $[M]$ is the number density of air, $J_2 = 4.35 \ 10^{-15}$, $7.44 \ 10^{-11}$, $1.01 \ 10^{-9}$, and $J_3 = 4.47 \ 10^{-4}$, $7.80 \ 10^{-4}$, $6.96 \ 10^{-3}$ at 100 hPa, 10 hPa, and 1 hPa, respectively, evaluate τ_{O_3} and τ_{O_x} at (a) 100 hPa, (b) 10 hPa, (c) 1 hPa.

2. Ozone is useful as a tracer in the lower stratosphere, yet O_3 has a photochemical lifetime there of only hours (e.g., Fig. 18.1). Resolve this discrepancy.

3. Use observed temperature (Fig. 1.7), radiative-equilibrium temperature (Fig. 18.9), and the coefficient of Newtonian cooling (Fig. 8.29) to estimate the radiative cooling rate ($^\circ$K day^{-1}) inside the polar-night vortex at (a) 100 hPa, (b) 10 hPa, (c) 1 hPa.

4. According to the Brewer-Dobson circulation (Fig. 18.11), air enters the stratosphere from the tropical troposphere. Estimate the water vapor mixing ratio, in ppmv, for the tropical lower stratosphere if tropospheric air entering the stratosphere (a) evolves through a thermodynamic state corresponding to the mean temperature and height of the tropical tropopause (Fig. 1.7), (b) is actually introduced through isolated convective overshoots that have a mean height and temperature of 16.5 km and 185°K, (c) as in (b), but if, after entering the stratosphere, tropospheric air inside convective overshoots mixes with subtropical stratospheric air that has a mixing ratio of 5 ppmv.

5. Depleted ozone at high latitude introduces anomalous radiative transfer. Discuss how, by influencing the polar-night vortex, such changes can represent a positive feedback.

6. The residual mean circulation (Fig. 18.11) gives no indication of equatorward transport in the lower stratosphere. (a) How might subtropical air in Prob. 18.4c be supplied to the equator? (b) Why does such transport not appear in the residual mean circulation?

7. Methane oxidation is thought to be responsible for about 25% of stratospheric water vapor. Given the reactions

$$CH_4 + OH \rightarrow H_2O + CH_3 \quad (k_a)$$

$$H_2O + O \rightarrow 2OH \quad (k_b),$$

with the rate coefficients in cgs units: $k_a = 2.4 \ 10^{-12} e^{-\frac{1710}{T}}$ and $k_b = 2.2 \ 10^{-10}$, calculate the photochemical-equilibrium mixing ratio of water vapor produced by methane oxidation at 20 km, where $r_{CH_4} \cong 1.3$ ppmv, $r_{OH} \cong 1.5 \ 10^{-13}$, and $r_O \cong 4.9 \ 10^{-19}$. Compare this value with observed mixing ratios at this altitude.

8. Photochemical lifetime is treated in Prob. 18.1, for the family O_x. Consider now Chapman chemistry, but augmented by the catalytic cycle (18.13) involving Cl_x. Determine the photochemical lifetime of O_x at 100 hPa for (a) rate coefficients for (18.13.1) and (18.13.2) in cgs units of $k_a = 8.5 \ 10^{-12}$ and $k_b = 3.7 \ 10^{-11}$, respectively, and a ClO mixing ratio of 100 pptv, which are representative of gas-phase chemistry under unperturbed conditions, and (b) rate coefficients of $k_a = 8.5 \ 10^{-10}$ and $k_b = 3.7 \ 10^{-9}$, respectively, and a ClO mixing ratio of 1000 pptv, which are representative of heterogeneous chemistry involving (18.30) under chemically perturbed conditions. (c) Compare the photochemical lifetimes of O_x in (a) and (b) with those under pure oxygen chemistry calculated in Prob. 18.1.

Appendix A: Conversion to SI units

Physical Quantity	Unit	SI (MKS) Equivalent
Length	ft	0.305 m
	μm	10^{-6} m
	nm	10^{-9} m
Time	day	$8.64 \ 10^4$ s
Mass	lb	0.454 kg
Temperature	F	$273 + (F\text{-}32)/1.8$ K
Volume	liter	10^{-3} m^3
Velocity	mph	0.447 m s^{-1}
	Knots	0.515 m s^{-1}
	km hr^{-1}	0.278 m s^{-1}
	fps	0.305 m s^{-1}
Force	kg m s^{-2}	1 N
	lb	0.138 N
	dyne	10^{-5} N
Pressure	N m^{-2}	1 Pa
	bar	10^5 Pa
	mb	10^2 Pa
Energy	kg m^2 s^{-2}	1 J
	Nm	1 J
	erg	10^{-7} J
	cal	4.187 J
Power	kg m^{-2} s^{-3}	1 W
	J s^{-1}	1 W
	Langley day^{-1}	$4.84 \ 10^{-1}$ W m^{-2}
Specific Heat	cal gm^{-1}	$4.184 \ 10^3$ J kg^{-1}
Energy Flux	cal cm^{-2} min^{-1}	$6.97 \ 10^2$ W m^{-2}

Appendix B: Thermodynamic properties of air and water

Dry Air

Mean Molecular Weight	$M_d = 28.96$ g mol^{-1}
Specific Gas Constant	$R = 287.05$ J kg^{-1} K^{-1}
Density	$\rho = 1.293$ kg m^{-3} (at STP*)
Number Density (Loschmidt No.)	$n = 2.687\ 10^{25}$ m^{-3} (at STP)
Isobaric Specific Heat Capacity	$c_p = 1.005\ 10^3$ J kg^{-1} K^{-1} (at 273 K)
Isochoric Specific Heat Capacity	$c_v = 7.19\ 10^2$ J kg^{-1} K^{-1} (at 273 K)
Ratio of Specific Heats	$\gamma = \frac{c_p}{c_v} = 1.4$
	$\kappa = \frac{\gamma-1}{\gamma} = \frac{R}{c_p} = 0.286$
Coefficient of Viscosity	$\mu = 1.73\ 10^{-5}$ kg m^{-1} s^{-1} (at STP)
Kinematic Viscosity	$\nu = 1.34\ 10^{-5}$ m^2 s^{-1} (at STP)
Coefficient of Thermal Conductivity	$k = 2.40\ 10^{-2}$ W m^{-1} K^{-1} (at STP)
Sound Speed	$c_s = 331$ m s^{-1} (at 273 K)

Water

Mean Molecular Weight	$M_v = 18.015$ g mol^{-1}
	$\epsilon = \frac{M_v}{M_d} = 0.622$
Specific Gas Constant	$R = 461.51$ J kg^{-1} K^{-1}
Density (Liquid Water)	$\rho = 10^3$ kg m^{-3} (at STP)
Density (Ice)	$\rho = 9.17\ 10^2$ kg m^{-3} (at STP)
Isobaric Specific Heat Capacity (Vapor)	$c_p = 1.85\ 10^3$ J kg^{-1} K^{-1} (at 273 K)
Isochoric Specific Heat Capacity (Vapor)	$c_v = 1.39\ 10^3$ J kg^{-1} K^{-1} (at 273 K)
Ratio of Specific Heats (Vapor)	$\gamma = \frac{c_p}{c_v} = 1.33$
Specific Heat Capacity (Liquid Water)	$c = 4.218\ 10^3$ J kg^{-1} K^{-1} (at 273 K)
Specific Heat Capacity (Ice)	$c = 2.106\ 10^3$ J kg^{-1} K^{-1} (at 273 K)
Specific Latent Heat of Fusion	$l_f = 3.34\ 10^5$ J kg^{-1}
Specific Latent Heat of Vaporization	$l_v = 2.50\ 10^6$ J kg^{-1}
Specific Latent Heat of Sublimation	$l_s = l_f + l_v$
Specific Surface Tension	$\sigma = 0.076$ s^{-2} (at 273 K)

* Standard Temperature and Pressure (STP) = 1013 mb and 273 K.

Appendix C: Physical constants

Avogadro's Number	$N_A = 6.022 \ 10^{26} \ \text{mol}^{-1}$
Universal Gas Constant	$R^* = 8.314 \ \text{J mol}^{-1} \ \text{K}^{-1}$
Boltzmann Constant	$k = 1.381 \ 10^{-23} \ \text{J K}^{-1}$
Planck Constant	$h = 6.6261 \ 10^{-34} \ \text{J s}^{-1}$
Stefan-Boltzmann Constant	$\sigma = 5.67 \ 10^{-8} \ \text{W m}^{-2} \ \text{K}^{-4}$
Speed of Light	$c = 2.998 \ 10^{8} \ \text{m s}^{-1}$
Solar Constant	$F_s = 1.372 \ 10^{3} \ \text{W m}^{-2}$
Radius of the Earth	$a = 6.371 \ 10^{3} \ \text{km}$
Standard Gravity	$g_0 = 9.806 \ \text{m s}^{-2}$
Earth's Angular Velocity	$\Omega = 7.292 \ 10^{-5} \ \text{s}^{-1}$

Appendix D: Vector identities

$$A \times (B \times C) = (A \cdot C)B - (A \cdot B)C \tag{D.1}$$

$$A \cdot (B \times C) = (A \times B) \cdot C = B \cdot (C \times A) \tag{D.2}$$

$$(A \times B) \cdot (C \times D) = (A \cdot C)(B \cdot D) - (A \cdot D)(B \cdot C) \tag{D.3}$$

$$\nabla(fg) = f\nabla g + g\nabla f \tag{D.4}$$

$$\nabla \cdot (fA) = \nabla f \cdot A + f\nabla \cdot A \tag{D.5}$$

$$\nabla \times (fA) = \nabla f \times A + f\nabla \times A \tag{D.6}$$

$$\nabla \cdot (A \times B) = B \cdot (\nabla \times A) - A \cdot (\nabla \times B) \tag{D.7}$$

$$\nabla \cdot \nabla \times A = 0 \tag{D.8}$$

$$\nabla \times \nabla f = 0 \tag{D.9}$$

$$\nabla \cdot \nabla f = \nabla^2 f \tag{D.10}$$

$$\nabla \times \nabla \times f = \nabla(\nabla \cdot f) - \nabla^2 f \tag{D.11}$$

$$\nabla(A \cdot B) = A \cdot \nabla B + B \cdot \nabla A + A \times \nabla B + B \times \nabla \times A \tag{D.12}$$

$$\nabla(A \times B) = B \cdot \nabla A - A \cdot \nabla B + A(\nabla \cdot B) - B(\nabla \cdot A) \tag{D.13}$$

$$A \cdot \nabla A = \frac{1}{2}\nabla(A \cdot A) - A \times (\nabla \times A) \tag{D.14}$$

Appendix E: Curvilinear coordinates

Spherical Coordinates (λ, ϕ, r)

$$\nabla \psi = \frac{1}{r \cos \phi} \frac{\partial \psi}{\partial \lambda} \boldsymbol{e}_\lambda + \frac{1}{r} \frac{\partial \psi}{\partial \phi} \boldsymbol{e}_\varphi + \frac{\partial \psi}{\partial r} \boldsymbol{e}_r \tag{E.1}$$

$$\nabla \cdot \boldsymbol{A} = \frac{1}{r \cos \phi} \frac{\partial A_\lambda}{\partial \lambda} + \frac{1}{r \cos \phi} \frac{\partial}{\partial \phi}(\cos \phi A_\phi) + \frac{1}{r^2} \frac{\partial}{\partial r}(r^2 A_r) \tag{E.2}$$

$$\nabla \times \boldsymbol{A} = \frac{1}{(r^2 \cos \phi)} \left\{ r \cos \phi \left[\frac{\partial A_r}{\partial \phi} - \frac{\partial (r A_\phi)}{\partial r} \right] \boldsymbol{e}_\lambda \right.$$

$$\left. + r \left[\frac{\partial}{\partial r}(r \cos \phi A_\lambda) - \frac{\partial A_r}{\partial \lambda} \right] \boldsymbol{e}_\varphi + \left[\frac{\partial (r A_\phi)}{\partial \lambda} - \frac{\partial}{\partial \varphi}(r \cos \phi A_\lambda) \right] \boldsymbol{e}_r \right\} \tag{E.3}$$

$$\nabla^2 \psi = \frac{1}{r^2 \cos^2 \phi} \frac{\partial^2 \psi}{\partial \lambda^2} + \frac{1}{r^2 \cos \phi} \frac{\partial}{\partial \phi}\left(\cos \phi \frac{\partial \psi}{\partial \phi}\right) + \frac{1}{r^2} \frac{\partial}{\partial r}\left(r^2 \frac{\partial \psi}{\partial r}\right) \tag{E.4}$$

$$\nabla^2 \boldsymbol{A} = \left[\nabla^2 A_\lambda - \frac{A_\lambda}{r^2 \cos^2 \varphi} + \frac{2}{r^2 \cos \varphi} \frac{\partial A_r}{\partial \lambda} + \frac{2}{r^2} \frac{\sin \varphi}{\cos^2 \varphi} \frac{\partial A_\varphi}{\partial \lambda} \right] \boldsymbol{e}_\lambda$$

$$+ \left[\nabla^2 A_\varphi - \frac{A_\varphi}{r^2 \cos^2 \varphi} + \frac{2}{r^2} \frac{\partial A_r}{\partial \varphi} - \frac{2}{r^2} \frac{\sin \varphi}{\cos^2 \varphi} \frac{\partial A_\lambda}{\partial \lambda} \right] \boldsymbol{e}_\varphi$$

$$\left[\nabla^2 A_r - \frac{2}{r^2} A_r - \frac{2}{r^2 \cos \varphi} \frac{\partial}{\partial \varphi}(\sin \varphi A_\varphi) - \frac{2}{r^2 \cos \varphi} \frac{\partial A_\lambda}{\partial \lambda} \right] \boldsymbol{e}_r \tag{E.5}$$

Cylindrical Coordinates (r, ϕ, z)

$$\nabla \psi = \frac{\partial \psi}{\partial r} \boldsymbol{e}_r + \frac{1}{r} \frac{\partial \psi}{\partial \phi} \boldsymbol{e}_\varphi + \frac{\partial \psi}{\partial z} \boldsymbol{e}_z \tag{E.6}$$

$$\nabla \cdot \boldsymbol{A} = \frac{1}{r} \frac{\partial}{\partial r}(r A_r) + \frac{1}{r} \frac{\partial A_\phi}{\partial \phi} + \frac{\partial A_z}{\partial z} \tag{E.7}$$

$$\nabla \times \boldsymbol{A} = \left(\frac{1}{r} \frac{\partial A_z}{\partial \phi} - \frac{\partial A_\phi}{\partial z} \right) \boldsymbol{e}_r + \left(\frac{\partial A_r}{\partial z} - \frac{\partial A_z}{\partial r} \right) \boldsymbol{e}_\varphi + \left(\frac{1}{r} \frac{\partial (r A_\phi)}{\partial r} - \frac{1}{r} \frac{\partial A_r}{\partial \phi} \right) \boldsymbol{e}_z \quad (E.8)$$

$$\nabla^2 \psi = \frac{1}{r} \frac{\partial}{\partial r} \left(r \frac{\partial \psi}{\partial r} \right) + \frac{1}{r^2} \frac{\partial^2 \psi}{\partial \phi^2} + \frac{\partial^2 \psi}{\partial z^2} \qquad\qquad\qquad\qquad (E.9)$$

$$\nabla^2 \boldsymbol{A} = \left[\nabla^2 A_r - \frac{A_r}{r^2} - \frac{2}{r^2} \frac{\partial A_\phi}{\partial \phi} \right] \boldsymbol{e}_r + \left[\nabla^2 A_\phi - \frac{A_\phi}{r^2} + \frac{2}{r^2} \frac{\partial A_r}{\partial \phi} \right] \boldsymbol{e}_\varphi + \nabla^2 A_z \boldsymbol{e}_z \quad (E.10)$$

Appendix F: Pseudo-adiabatic chart

Adiabats (blue-solid), which are characterized by constant values of potential temperature θ; pseudo-adiabats (red-dashed), which are characterized by constant values of equivalent potential temperature θ_e; and isopleths of saturation mixing ratio with respect to water r_w (green), all as functions of temperature and pressure with the ordinate proportional to $\left(\frac{p_0}{p}\right)^{\kappa}$. See color plate section: Plate 35.

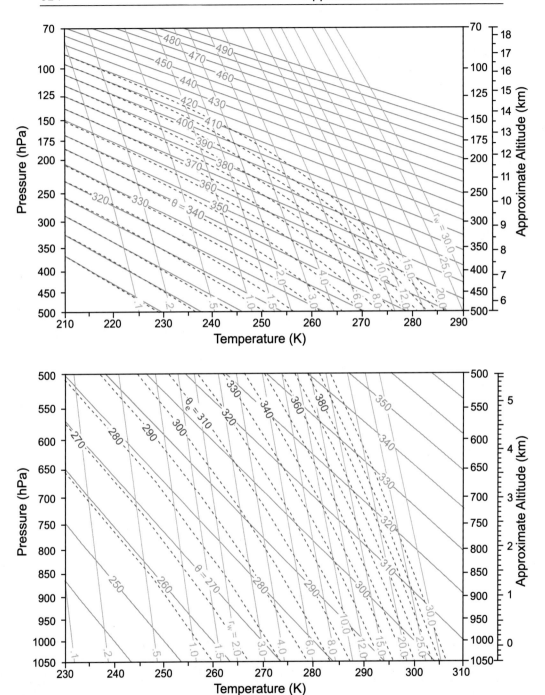

Appendix G: Acronyms

AAO	Antarctic Oscillation
AO	Arctic Oscillation
CAPE	Convective Available Potential Energy
CFC	Chlorofluorocarbon
DU	Dobson Units
EESC	Equivalent Effective Stratospheric Chlorine
ENSO	El Nino Southern Oscillation
EOF	Empirical Orthogonal Function
EP Flux	Eliassen-Palm Flux
ERBE	Earth Radiation Budget Experiment
GCM	Global Climate Model
GMT	Global Mean Temperature
HFC	Hydrofluorocarbon
IPCC	Intergovernmental Panel on Climate Change
IR	Infrared
ITCZ	Inter-Tropical Convergence Zone
LW	LongWave
LCL	Lifting Condensation Level
LFC	Level of Free Convection
LNB	Level of Neutral Buoyancy
MSU	Microwave Sounding Unit
MJO	Madden-Julian Oscillation
NAO	North Atlantic Oscillation
OLR	Outgoing Longwave Radiation
PDO	Pacific Decadal Oscillation
PL	Penetration Level
PNA	Pacific North America (pattern)
PSC	Polar Stratospheric Cloud

QBO	Quasi-Biennial Oscillation
RH	Relative Humidity
SLP	Sea Level Pressure
SST	Sea Surface Temperature
SOI	Southern Oscillation Index
SW	ShortWave
TOA	Top Of Atmosphere
UT	Universal Time (Local time at Greenwich England)
UV	Ultraviolet
WKB	Slowly Varying (short-wavelength) approximation, due to Wentzel, Kramers, and Brillouin
WMO	World Meteorological Organization

Answers to Selected Problems

CHAPTER 1

2. $\frac{V_i}{V_d} = \frac{r_i}{\epsilon_i}$.

3. $\frac{V_i}{V_d} = \frac{n_i}{n_d}$.

6. (a) $N_{H_2O} = 0.0231$, (b) $\overline{M} = 28.71$ g mol^{-1}, (c) $\overline{R} = 289.58$ J kg^{-1} K^{-1}, (d) $\rho_{H_2O} = 0.0173$ kg m^{-3}, (e) $r_{H_2O} = 0.0147$, (f) $\frac{V_i}{V_d} = 0.0236$.

8. $\Sigma_{H_2O} = \frac{1000}{\rho_w} \int_0^\infty r_{H_2O} \rho_d dz$ (mm).

9. (a) 47 mm, (b) 9.7 mm.

12. $\Delta T_e = 4.4$ K.

13. (a) 0.

15. $T(z) = \frac{gh^2}{2Rz}\left[1 + \left(\frac{z}{h}\right)^2\right]$.

16. 236 K.

CHAPTER 2

1. (a) 5 km, (b) 0.9 km.

2. (a) $\theta_{950} = 30$ C.

3. (a) $3.06 \ 10^3$ J, (b) $3.10 \ 10^3$ J, (c) $4.06 \ 10^4$ J.

6. 15.85.

CHAPTER 3

1. (a) 298 K, (b) -6.3 J K^{-1}, (c) 7.05 J K^{-1}, (d) 0.75 J K^{-1}.

2. (a) -20.7 J K^{-1}, (b) 21.5 J K^{-1}.

3. 0.90.

4. (a) 0.967.

5. (a) 277 K, (b) 285 K.

7. 199 J K^{-1} kg^{-1}.

9. (a) $1.3 \ 10^2$ J K^{-1} mol^{-1}, (b) $5.2 \ 10^4$ J mol^{-1}.

CHAPTER 4

2. (a) 95 C, (b) 86 C, (c) 71 C.

3. (a) 599 K.

4. (a) 188 K, (b) in the wintertime Antarctic stratosphere.

CHAPTER 5

1. (a) 33 C, (b) 10.2 g kg^{-1}, (c) 26%.

5. (a) 75%, (b) 25 C, (c) 325 K, (d) 770 hPa, (e) 400 K, (f) 22.5 g kg^{-1}, (g) 450 hPa.

6. 53%.

7. (a) 53%, (b) 100%.

8. (a) 20.1 C, (b) 32.6 C.

10. (e) 328 K.

11. 9 C.

12. (a) 920 hPa, (b) 610 hPa.

13. 9%.

14. (a) 5.4%, (b) 0.51 kg, (c) 7.22 10^6 J.

16. (a) 13.6 km.

17. 3.4 km.

22. (a) $T(z, t; x) = T_\infty(z - h(x - ct)) - \Gamma_d h(x - ct)$, (b) $r(z, t; x) = r_\infty(z - h(x - ct))$.

26. (d) 1.3 km.

CHAPTER 6

4. (a) $z = \frac{R}{g}\left\{(\overline{\overline{T}} + \overline{T}\cos\phi)\xi - T'\left(\frac{\xi_T}{\pi}\right)e^{-\frac{\lambda^2 + (\phi - \phi_0)^2}{l^2}}\sin\left(\frac{\pi\xi}{\xi_T}\right)\right\}$ $(\xi < \xi_T)$; $\frac{R}{g}\left\{(\overline{\overline{T}} + \overline{T}\cos\phi)\xi\right\}$ $(\xi \geq \xi_T)$.

5. 23 K.

6. (a) $z = \frac{R\overline{T}}{g}(\xi - \xi_s)\left\{1 - \frac{\xi_s\left[1 - e^{\alpha(\xi_s - \xi)}\right]}{(\xi_s - \xi)}\right\}$.

8. (a) 158 m.

10. $w = -R\oint T d\ln p = -RA$.

11. $w = c_p(T_2 - T_1)\left[1 - \left(\frac{p_{34}}{p_{12}}\right)\right] > 0$.

CHAPTER 7

2. (a) unstable: $z < z_c$, stable: $z > z_c$; $z_c = h_1 \ln\left(\frac{h_1}{h_2}\right)$, (b) $N^2 = g\frac{\frac{1}{h_1}e^{-\frac{z}{h_1}} - \frac{1}{h_2}}{e^{-\frac{z}{h_1}} - \frac{z}{h_2}}$.

4. 1.24 10^{-4} kg s^{-2}, (b) 5.41 10^{-4} 10^{-4} kg s^{-2}.

5. (a) $\frac{w_0}{N}$.

6. (a) $\theta = T_0\frac{a}{a+z}e^{\frac{\Gamma_d}{T_0}\left(z + \frac{z^2}{2a}\right)}$, (b) unstable: $z < z_c$, stable: $z > z_c$; $z_c = \sqrt{\frac{aT_0}{\Gamma_d}} - a$.

8. (a) $T_{trigger} = T_0 + \frac{3\Gamma_d}{2}$, (b) $\Delta t \cong 5$ hrs.

11. (a) 27 km.

12. 15 km.

15. (a) stable: $z < a$, unstable: $z > a$; (b) $\Delta z = 2a$ (exclusive of instability).

16. $\frac{d\theta_e}{dz} > 0$ (stable), $\frac{d\theta_e}{dz} = 0$ (neutral), $\frac{d\theta_e}{dz} < 0$ (unstable).

19. 57 m s^{-1}.

20. (a) $TI(z) = 4z - 20$, (b) 5 km, (c) $w_{max} = 58$ m s^{-1} at 5 km, $z_{max} = 9.5$ km.

21. (a) $K' > 0$ for $z < 9.8$ km, (b) $K' > 0$ for $z < 5.7$ km, (c) $K' > 0$ for $z < 5.3$ km.

23. $\theta'_e = 244$ K at 13 km.

24. (a) potentially unstable (stable) for $z < 1.6$ km ($z > 1.6$ km), (b) potentially unstable (stable) for $z < 4.8$ km ($z > 4.8$ km).

CHAPTER 8

1. (a) 254 K, (b) 290 K, (c) 270 K, (d) 240 K.

2. -32.4 K.

3. (a) 4.6 K.

4. (a) $u_\lambda(z) = \rho_0 r H e^{-\left(\frac{\lambda - \lambda_0}{\alpha}\right)^2 - \frac{z}{H}}$.

7. (a) 227 K.

8. (a) 95%, (b) 3.4.

9. (a) 393 K, (b) 393 K.

10. (a) $T_{RE} = \left(\frac{a_{SW}}{a_{LW}} \frac{F_s}{\sigma}\right)^{\frac{1}{4}}$, (b) 278 K.

11. $T(t) = \left[0.12\sigma t + \frac{1}{T_0^3}\right]^{-\frac{1}{3}}$.

12. $\frac{2 - a_{LW}^{atm} + a_{LW}^{atm} a_{LW}^{hood}}{2 - a_{LW}^{atm}} a_{SW}^{hood} F_s \cos\theta_s = a_{LW}^{hood} \sigma T_{hood}^4$; (a) 298 K, (b) 392 K.

13. (a) 324 K, (b) 212 K.

14. -97 K.

17. (a) 16 days, (b) 0.

18. (a) 4.6 K, (b) 17.6 K.

21. (a) $\left[a_{SW} + (1 - a_{SW}) a_{LW}\right] F_0 \sum_{n=0}^{\infty} \left(\frac{a_{LW}}{2}\right)^n$, (b) 252 K.

22. (a) 291 K, (b) 252 K, (c) 294 K.

23. (a) 5.0 hPa, (b) 6.4 hPa.

24. (a) $F_s = F_0 \frac{2}{2-a}$ ($N = 1$), $F_0 \frac{2+a}{2-a}$ ($N = 2$), ... $F_0 \frac{2+(N-1)a}{2-a} = F_0 \left[\frac{2}{2-a} + (N-1)\frac{a}{2-a}\right]$ (N).

CHAPTER 9

1. (a) $\frac{dS}{d(\log a)} = 4\pi a^2 \frac{dn}{d(\log a)}$, (b) $\frac{dM}{d(\log a)} = \frac{4}{3}\pi a^3 \rho \frac{dn}{d(\log a)}$.

3. $\Delta G = \dfrac{16\pi\sigma^3}{3\left[n_w KT \ln\left(\frac{e}{e_w}\right)\right]^2}$

4. (a) 0.60 μm, (b) 0.19 μm.

5. (a) $\sim$100 m.

6. (a) $w_t = \frac{2}{9} \frac{\rho_w g a^2}{\mu}$, (b) 87 μm, (c) 88 μm.

8. (a) 0.76 yrs, (b) 76 yrs.

11. (a) 0.013, (b) 0.051, (c) 0.301.

13. (a) 3.1 mm day^{-1}.

15. (a) 150 m, (b) 1500 m.

20. $\cos\Theta = \cos\theta \cos\theta' + \sin\theta \sin\theta'_2 \cos(\phi - \phi')$.

21. (b) $\beta_s(\pi) = \frac{64\pi^5}{\lambda^4} \frac{n}{\eta^2} a^6 \left(\frac{m^2 - 1}{m^2 + 2}\right)^2$.

23. (a) $a_e = 5.7 \ \mu$m, (b) $\Sigma_l = 785$, (c) $\tau_c = 208$.

24. $\tau_c = 410$.

25. (a) (9.40), but with $C = \dfrac{E_+\alpha_- e^{-\gamma\tau_c^*} - E_+\alpha_+ e^{-\frac{\bar{\mu}}{\mu_s}\tau_c^*} - (1-a)\alpha_+ E_-\left(e^{-\gamma\tau_c^*} - e^{-\frac{\bar{\mu}}{\mu_s}\tau_c^*}\right)}{\alpha_+^2 e^{\gamma\tau_c^*} - \alpha_-^2 e^{-\gamma\tau_c^*} + (1-a)\alpha_+\alpha_- \left(e^{-\gamma\tau_c^*} - e^{\gamma\tau_c^*}\right)}$

$D = \dfrac{E_+\alpha_- e^{-\frac{\bar{\mu}}{\mu_s}\tau_c^*} - E_-\alpha_+ e^{\gamma\tau_c^*} + (1-a)\alpha_- E_-\left(e^{\gamma\tau_c^*} - e^{-\frac{\bar{\mu}}{\mu_s}\tau_c^*}\right)}{\alpha_+^2 e^{\gamma\tau_c^*} - \alpha_-^2 e^{-\gamma\tau_c^*} + (1-a)\alpha_+\alpha_- \left(e^{-\gamma\tau_c^*} - e^{\gamma\tau_c^*}\right)}$.

27. $C_{SW} = -131$ Wm^{-2}, $C_{LW} = 125$ Wm^{-2}, $C = -6$ Wm^{-2}. (b) 3.1 mm day^{-1}.

28. Strong forward scattering implies that the reduction of direct transmission is offset by an enhancement of diffuse transmission.

30. (a) $\Sigma_l = 240$ g m^{-2}, (b) $\rho_l = 0.24$ g m^{-3}.

CHAPTER 10

2. (c) For a given time interval, the deformation experienced by a material volume decreases as its dimension decreases. Letting the volume's dimension go to zero averts deformation indefinitely.

3. (a) $\frac{d\rho}{dt} = \nabla \cdot \boldsymbol{v} = 0$, (b) $\frac{d\theta}{dt} = 0$.

4. $\boldsymbol{v}$ orthogonal to $\nabla \psi$.

6. (a) 6.9°, (b) 69.3°, (c) 693°, (d) 0.019, 0.19, 1.9.

CHAPTER 11

3. (b) $w = -\frac{1}{\rho g}\omega$, (c) -19 mm/s.

6. (a) $u_g = -\frac{1}{fa} \frac{\partial \Phi}{\partial \phi}$; $v_g = \frac{1}{fa\cos\phi} \frac{\partial \Phi}{\partial \lambda}$, (b) $u_g = -\frac{1}{fa} \frac{\partial \Psi}{\partial \phi}$; $v_g = \frac{1}{fa\cos\phi} \frac{\partial \Psi}{\partial \lambda}$.

7. $-\frac{\partial \boldsymbol{v}_g}{\partial t} \ln p = \frac{R}{f} \boldsymbol{k} \times \nabla_p T$.

12. $\frac{d}{dt}[(u + \Omega r\cos\phi)r\cos\phi] = -r\cos\phi \left[\frac{1}{\rho r\cos\phi} \frac{\partial p}{\partial \lambda} + D_\lambda \right]$.

15. $\frac{\partial \boldsymbol{v}_h}{\partial t} + f\boldsymbol{k} \times \boldsymbol{v}_h = -\nabla_\pi \Phi - \boldsymbol{D}_h$; $\frac{\partial \Phi}{\partial \pi} = -\theta$; $\frac{\partial}{\partial \pi}(\pi^{\frac{1}{\kappa}-1}\dot{\pi}) + \pi^{\frac{1}{\kappa}-1}\nabla_\pi \cdot \boldsymbol{v}_h = 0$; $\frac{d\theta}{dt} = \frac{\dot{q}}{\pi}$; $\frac{d}{dt} = \frac{\partial}{\partial t} + \boldsymbol{v}_h \cdot \nabla_\pi + \dot{\pi}\frac{\partial}{\partial \pi}$.

16. (a) $w = \zeta h\{1 - e^{-\frac{z^*}{h}}\}$, (b) Convergence inside boundary layer must be compensated by divergence aloft.

CHAPTER 12

1. 105 m s^{-1}.

2. (a) $\boldsymbol{v} = \frac{2}{L^2}e^{-\frac{x^2+y^2}{L^2}} \{-(\Psi y + Xx)\boldsymbol{i} + (\Psi x - Xy)\boldsymbol{j}\}$, (b) $\nabla \cdot \boldsymbol{v} = \frac{4}{L^2}Xe^{-\frac{x^2+y^2}{L^2}}\left[\frac{x^2+y^2}{L^2} - 1\right]$,

   (c) $\boldsymbol{k} \cdot (\nabla \times \boldsymbol{v}) = -\frac{4}{L^2}\Psi e^{-\frac{x^2+y^2}{L^2}}\left[\frac{x^2+y^2}{L^2} - 1\right]$.

3. (a) $\boldsymbol{v}_{500} = \frac{g}{f}\left\{\left[\frac{z_{500}}{a} - \frac{2\pi y}{L^2}z'_{500}\sin\left(\pi\frac{x^2+y^2}{L^2}\right)\right]\boldsymbol{i} + \left[\frac{2\pi x}{L^2}z'_{500}\sin\left(\pi\frac{x^2+y^2}{L^2}\right)\right]\boldsymbol{j}\right\}$,

   (c) $(\zeta + f) = \frac{g}{f}\frac{4\pi}{L^2}z'_{500}\left\{\sin\left(\pi\frac{x^2+y^2}{L^2}\right) + \pi\frac{(x^2+y^2)}{L^2}\cos\left(\pi\frac{x^2+y^2}{L^2}\right)\right\} + f_0$.

4. (a) 16 m s^{-1}, (b) 14.3 m s^{-1}.

5. (a) 768 m s^{-1}, (b) 45 m s^{-1}.

6. (a) 950-850: $\frac{\partial \overline{T}}{\partial x} = 0$, $\frac{\partial \overline{T}}{\partial y} = -.032$ K km^{-1}; 850-750: $\frac{\partial \overline{T}}{\partial x} = 0$, $\frac{\partial \overline{T}}{\partial y} = +.028$ K km^{-1}, (b) 950-850: 32 K day^{-1}; 850-750: -28 K day^{-1}, (c) 63 K km^{-1}.

7. $p = p_0 e^{\frac{\Omega^2 r^2}{2RT}}$.

8. (a) $\boldsymbol{v} = \pm\left(\frac{r}{L}\right)\sqrt{2\epsilon\overline{\Phi}}\boldsymbol{e}_\phi$.

9. 33°.

10. (a) $\zeta = -7.5 \cdot 10^{-5}$ s^{-1}, (b) $\zeta = -8.3 \cdot 10^{-5}$ s^{-1}.

11. (a) $v_\parallel = \frac{f^2 v_g}{K^2+f^2}$, $v_\perp = -\frac{fKv_g}{K^2+f^2}$ (b) $\delta = \tan^{-1}\left(\frac{K}{f}\right)$, (c) 24.2°, (d) 60.7°.

12. $\nabla \cdot \boldsymbol{v}_h = -\frac{K}{K^2+f^2}\nabla^2\Phi$.

13. (b) $\zeta = \frac{2y}{Y^2}e^{-\left(\frac{y}{Y}\right)^2}e^{i(kx-\sigma t)}$; $\nabla \cdot \boldsymbol{v}_h = ike^{-\left(\frac{y}{Y}\right)^2}e^{i(kx-\sigma t)}$;

   (c) $\nabla^2\chi = ike^{-\left(\frac{y}{Y}\right)^2}e^{i(kx-\sigma t)}$; $\psi = \frac{i}{k}\frac{\partial \chi}{\partial y}$.

14. (a) $\frac{\partial^2 y}{\partial x^2} + \frac{\beta}{u}y = 0$, (b) $y(x) = Ae^{i\sqrt{\frac{\beta}{u}}x} + Be^{-i\sqrt{\frac{\beta}{u}}x}$.

15. (b) Solenoidal production is absent in isobaric coordinates as well, but tilting remains.

CHAPTER 13

1. (a) .003, (b) 10^8, (c) $5 \ 10^{11}$.

2. (a) 20 m s^{-1} km^{-1}, (b) 50 m s^{-1} km^{-1}.

3. 87 km for H = 7.3 km.

4. 10^5 MW (!).

6. 8.9 mm s^{-1}

7. 0.08 N m^{-2}.

8. 4.5 days.

13. (a) $K\bar{u} - f\bar{v} = 0$; $K\bar{v} + f(\bar{u} - u_g) = 0$; $K = \frac{c_d |\bar{v}|}{h}$, (b) $\bar{v} = \frac{u_g}{1+\frac{K^2}{f^2}}(\mathbf{i} + \frac{K}{f}\mathbf{j})$,
(c) 0.46 days.

14. (a) $\frac{\partial r_t}{\partial x} = -\frac{1}{L}(r_t - r_a)a$, (b) $r_t(x) = (r_0 - r_a)e^{-\frac{x}{L}} + r_a$; $T(x) = (T_0 - T_a)e^{-\frac{x}{L}} + T_a$,
(c) $r_i(x) = \frac{\epsilon}{p}10^{10.55 - \frac{2667}{T_0 e^{-\frac{x}{L}} + T_a}}$.
(d) Dissolution occurs at $(r_0 - r_a)e^{-\frac{x_d}{L}} + r_a = \frac{\epsilon}{p}10^{10.55 - \frac{2667}{(T_0 - T_a)e^{-\frac{x}{L}} + T_a}}$, which can be solved graphically for x_d.

CHAPTER 14

8. (a) 0.0004 hPa, (b) 5.1 m s^{-1}.

12. (a) $\mathbf{k} = \left(k_0^2, \frac{\sigma^2}{gH(y)} - k_0^2\right)$.

18. (a) Through divergence, the mountain acts as a vorticity source,
(b) $\frac{D}{Dt}\nabla^2\psi' + \beta\frac{\partial\psi'}{\partial x} = \frac{Q_0}{2\pi L^2}e^{-\frac{x^2+y^2}{2L^2}}$,
(c) $\psi(x,y) = \frac{1}{2\pi}\int_{-\infty}^{\infty} dk \int_{-\infty}^{\infty} dl \Psi_{kl} e^{i(kx+ly)}$;

$\Psi_{kl} = \frac{i}{k}\left(\frac{Q_0}{2\pi}\right)\frac{e^{-\frac{(k^2+l^2)L^2}{2}}}{\bar{u}(k^2+l^2)-\beta}$; $xc_{gx} + yc_{gy} > 0$, with $\mathbf{c}_g = \frac{2\bar{u}^2 k}{\beta}\mathbf{k}$.

19. (a) $\frac{1}{a^2 \cos\phi}\frac{\partial}{\partial\phi}\left(\cos\phi\frac{\partial\psi'}{\partial\phi}\right) + \frac{1}{\bar{\rho}}\frac{\partial}{\partial z}\left(\frac{f^2}{N^2}\bar{\rho}\frac{\partial\psi'}{\partial z}\right) + \nu^2\psi' = 0$,
with
$\nu^2 = \frac{\beta_e}{\bar{u}-c} - k^2$
$\beta_e = \beta - \frac{1}{a^2\cos\phi}\frac{\partial}{\partial\phi}\left(\cos\phi\frac{\partial\bar{u}}{\partial\phi}\right) - \frac{1}{\bar{\rho}}\frac{\partial}{\partial z}\left(\frac{f^2}{N^2}\bar{\rho}\frac{\partial\bar{u}}{\partial z}\right)$,
(b) 74°, (c) 45°.

CHAPTER 15

1. $dh_{tot} = c_p T d\ln\theta_e$.

2. $-9.8 \ 10^4$ J kg^{-1}.

3. (b) $2.6 \ 10^{21}$ J.

4. 0.0013.

5. 40.5 m s^{-1}.

6. (a) 6 days.

10. (b) Phase lines tilt SW - NE in the Northern Hemisphere, so that wave activity propagates equatorward with $c_{gy} < 0$.

11. (a) 2.9 m s^{-1}.

12. (a) $9.2 \ 10^3$ J kg^{-1}, (b) $7.8 \ 10^2$ J kg^{-1}, (c) 13.3 km; 4.4 km.

CHAPTER 16

1. (a) unstable (stable) equatorward (poleward) of $25°$.

2. (a) positive, (b) $\phi = \cos^{-1}\left(\sqrt{\dfrac{\epsilon(z)}{1 - \frac{f^2}{N^2}a^2\frac{\partial^2\epsilon}{\partial z^2}}}\right)$.

5. $\dfrac{\alpha_c H}{2} = \coth\left(\dfrac{\alpha_c H}{2}\right)$.

7. 6.1 days, 3.2 days, 2.25 days, 1.86 days, 1.71 days, 1.77 days, 2.22 days, ∞.

CHAPTER 17

1. (a) $26.7\ \text{kg/m}^3$, (b) $26.25\ \text{kg/m}^3$, (c) $26.6\ \text{kg/m}^3$.

2. (a) -0.560 m, (b) $155\ \text{W/m}^3$.

3. (a) $S_f \cong 35.2$ g/kg, (b) -3.125 m, (c) $114.5\ \text{W/m}^3$.

7. (a) $v_0(x, y) = -\dfrac{200}{fY}\left\{\sin\left(\dfrac{x}{X} - \dfrac{y}{Y}\right) - \sin\left(\dfrac{x}{X} + \dfrac{y}{Y}\right)\right\}i - \dfrac{200}{fX}\left\{\sin\left(\dfrac{x}{X} - \dfrac{y}{Y}\right) + \sin\left(\dfrac{x}{X} + \dfrac{y}{Y}\right)\right\}j$

   (b) $M_E = -200 \cdot \dfrac{\rho\alpha}{f^2 X}\left\{\sin\left(\dfrac{x}{X} - \dfrac{y}{Y}\right) + \sin\left(\dfrac{x}{X} + \dfrac{y}{Y}\right)\right\}i$

   $+ 200 \cdot \dfrac{\rho\alpha}{f^2 Y}\left\{\sin\left(\dfrac{x}{X} - \dfrac{y}{Y}\right) - \sin\left(\dfrac{x}{X} + \dfrac{y}{Y}\right)\right\}j$ (c) $w(-\infty) = \dfrac{\alpha}{\rho f}\nabla^2\psi = -\dfrac{\alpha}{\rho f}\left(\dfrac{1}{X^2} + \dfrac{1}{Y^2}\right)\psi$

10. (a) $M_y = -200 \cdot \dfrac{\rho\alpha}{\beta f}\left(\dfrac{1}{X^2} + \dfrac{1}{Y^2}\right)\left\{\cos\left(\dfrac{x}{X} - \dfrac{y}{Y}\right) + \cos\left(\dfrac{x}{X} + \dfrac{y}{Y}\right)\right\}$

   $M_x = 200 \cdot \dfrac{\rho\alpha}{\beta f}\dfrac{X}{Y}\left(\dfrac{1}{X^2} + \dfrac{1}{Y^2}\right)\left\{-\cos\left(\dfrac{x}{X} - \dfrac{y}{Y}\right) + \cos\left(\dfrac{x}{X} + \dfrac{y}{Y}\right) + 2\sin\left(\dfrac{y}{Y}\right)\right\}$.

CHAPTER 18

1. $\tau_{O_3} = \dfrac{1}{J_3 + k_3[O]}$; $\tau_{O_x} = \dfrac{1}{2k_3[O]}$, (a) $\tau_{O_3} = 37$ mins; $\tau_{O_x} = 32$ yrs,

   (b) $\tau_{O_3} = 21$ mins; $\tau_{O_x} = 15$ days, (c) $\tau_{O_3} = 2.4$ mins; $\tau_{O_x} = 3.8$ hrs.

4. (a) 16.4 ppmv, (b) 1.6 ppmv, (c) 3.3 ppmv.

7. 1.26 ppmv.

8. (a) $\tau_{O_x} = \dfrac{1}{2k_3[O] + 2k_d[Cl]}$; $[O] = \dfrac{J_2[O_2]}{k_3[O_3] + k_b[ClO]}$, (a) 8 yrs, (b) 3.1 days.

References

Abramowitz and Stegun, 1972: *Handbook of Mathematical Functions.* Dover, New York, 1046 pp.

Ackerman, T, 1988: Aerosols in climate modeling. In *Aerosol and Climate*, P Hobbs and P McCormick eds., A. Deepak Publishing, Hampton, VA, pp. 335-348.

Adhikari and Kumon, 2001: MWP, JAPAN (wiki)

Ambaum, M, Hoskins, B, and D Stephenson, 2001: Arctic oscillation or North Atlantic oscillation? *J Climate*, **14**, 3495-3507.

Anderson, J, WH Bruce and MH Proffitt, 1989: Ozone destruction by chlorine radicals within the Antarctic vortex: The spatial and temporal evolution of $ClO - O_3$ anticorrelation based on *in situ* ER-2 data. *J Geophys Res*, **94**, 11, 465-11,479.

Andrews, D and M McIntyre, 1976: Planetary waves in horizontal and vertical shear: The generalized Eliassen-Palm relation and the mean zonal acceleration. *J Atmos Sci*, **33**, 2031-2048.

Andrews, D, 2000: *An Introduction to Atmospheric Physics.* Cambridge University Press, Cambridge, 228 pp.

Andrews, D, Holton, J, and C Leovy, 1987: *Middle Atmosphere Dynamics.* Academic Press, San Diego, 489 pp.

Antonov, J, Locarnini, R, Boyer, T, Mishonov, A, and H Garcia, 2006: *World Ocean Atlas 2005*, vol 2: Salinity, S Levitus, Ed. NOAA Atlas NESDIS 62, US Govt Printing Office, Washington, DC, 182 pp.

Araneda, A, Torrejon, F, Aguayo, M, Torres, L, Crueces, F, Cisternas, M, and R Urritia, 2007: Historical records of San Juan glacier advances: another clue to Little Ice Age timing in Southern Chile. *The Holocene*, **17**, 987-998.

Atmospheric Aerosols: Their Formation, Optical Properties, and Effects, 1982: A Deepak, Ed., Spectrum Press, Hampton, VA, 480 pp.

Aris, R, 1962: *Vectors, Tensors, and the Basic Equations of Fluid Mechanics.* Prentice Hall, Englewood Cliffs, NJ, 286 pp.

Arrhenius, S, 1896: On the influence of carbonic acid in the air upon the temperature of the ground. *Phil Mag*, **41**, 237-276.

Arrhenius, S, 1908: Das Werden der Welten. Academic Publishing House, Leipzig, 208 pp.

Bader, D, Covey, C, Gutowski, W, Held, I, Kunkel, K, Miller, R, Tokmakian, R, and M Zhang, 2008: *Climate Models: An Assessment of Strengths and Limitations.* Dept of Energy, Office of Biological and Environmental Research, Washington, DC, 124 pp.

Baldwin, M and T Dunkerton, 1999: Propagation of the Arctic Oscillation from the stratosphere to the troposphere. *J Geophys Res*, **104**, 30,937-30,946.

Baldwin, M and T Dunkerton, 2001: Stratospheric Harbingers of Anomalous Weather Regimes. *Science*, **294**, 581-584.

Battan, L, 1984: *Fundamentals of Meteorology*. Prentice Hall, Englewood Cliffs, NJ, 304 pp.

Banks, P and G Kocharts, 1973: *Aeronomy*. Academic Press, New York, 430 pp.

Barnola J, Raynaud, D, Korotkevitch, Y, and C Lorius, 1987: Vostok ice core: A 160,000 year record of atmospheric CO_2. *Nature*, **329**, 408-414.

Batchelor, G, 1977: *An Introduction to Fluid Dynamics*. Cambridge University Press, Cambridge, 615 pp.

Bauer, J, 2010: Molecular epidemiology of melanoma. *Melanoma Res*, **20**, doi: 10.1097/01.cmr.0000382758.93495.d6.

Berlage, H, 1957: Fluctuations in the general atmospheric circulation of more than one year: their nature and prognostic value. *Korte Ned Met Inst Meded Verh*, **69**, 1-152.

Bednarz, Z and J Ptak, 1990: Influence of temperature and precipitation on ring widths of oak in the Niepolomice forest near Cracow. *Tree Ring Bull*, **50**, 1-10.

Bithell, M, Gray, L, Harries, J, Russell, J, and A Tuck, 1994: Synoptic interpretation of measurements from HALOE. *J Atmos Sci*, **51**, 2942-2956.

Bleck, R, 1978a: Finite difference equations in generalized vertical coordinates. Part I: Total energy conservation. *Beiträge zur Physik der Atmosphäre*, **51**, 360-372.

Bleck, R, 1978b: Finite difference equations in generalized vertical coordinates. Part II: Potential vorticity conservation. *Beiträge zur Physik der Atmosphäre*, **52**, 95-105.

Bourke, W, 1974: A multi-level spectral model. I. Formulation and hemispheric integrations. *Mon Wea Rev*, **102**, 687-701.

Boyce, D, Lewis, M, and B Worm, 2010: Global phytoplankton decline over the past century. *Nature*, **466**, 591-596.

Branstator, G, 2002: Circumpolar teleconnections, the jet stream waveguide, and the north atlantic oscillation. *J Climate*, **15**, 1893-1910.

Brasseur, G and S Solomon, 1986: *Aeronomy of the Middle Atmosphere*. Reidel, Dordrecht, 2nd ed. 452 pp.

Brewer, A, 1949: Evidence for a world circulation provided by the measurements of helium and water vapor distributions in the stratosphere. *Quart J Roy Met Soc*, **75**, 351.

Briffa, K, 2000: Annual climate variability in the holocene: Interpreting the message of ancient trees. *Quaternary Sci Revs*, **19**, 87-105.

Brillouin, L, 1926: Remarques sur la mecanique ondulatoire. *J Phys Radium*, **7**, 353-368.

Broecker, W, 2000: Was a change in the thermohaline circulation responsible for the Little Ice Age? *Proc NAS*, 97, 1339-1342.

Callaghan, P, Fusco, A, Francis, G, and Salby, M, 1999: A Hough spectral model for 3-dimensional studies of the middle atmosphere. *J Atmos Sci*, **56** 1461-1480.

Carlson, T and S Benjamin, 1982: Radiative heating rates for a desert aerosol (Saharan dust). In *Atmospheric Aerosols: Their Formation, Optical Properties and Effects*, A. Deepak, Ed., Spectrum Press, Hampton, VA, pp. 435-457.

Cascinelu, N and R Marchesini, 2008: Increasing incidence of cutaneous melanoma, ultraviolet radiation, and the clnincian. *Photochem and Photobiol*, doi: 10.1111/1751-1097.1989tb05555.

Cess, R, 1976: Climate change: An appraisal of atmospheric feedback mechanism employing zonal climatology. *J Atmos Sci*, **33**, 1831-1843.

Chapman, S, 1930: On ozone and atomic oxygen in the upper atmosphere. *Phil Mag*, **10**, 369-383.

Charney, J, 1947: The dynamics of long waves in a baroclinic westerly current. *J Meteorol*, **4**, 135-163.

Charney, J, Fjortoft, R, and J von Neumann, 1950: Numerical integration of the barotropic vorticity equation. *Tellus*, **2**, 237-254.

Charney, J, 1973: Planetary fluid mechanics. In *Dynamical Meteorology*, P. Morel ed., Reidel, Dordrecht, 97-351.

Charney J and P Drazin, 1961: Propagation of planetary scale disturbances from the lower into the upper atmosphere. *J Geophys Res*, **66**, 83-109.

Charney J and M Stern, 1962: On the stability of internal baroclinic jets in a rotating atmosphere. *J Atmos Sci*, **19**, 159-172.

Christy, J Spencer, R, and D Braswell, 2000: MSU tropospheric temperatures: Dataset construction and radiosonde comparisons. *J Atmos Ocean Tech*, **17**, 1153-1170.

Cook, E, Palmer, J, and R D'Arrigo, 2002: Evidence for a 'Medieval Warm Period' in a 1100 year tree-ring reconstruction of past austral summer temperatures in New Zealand. *Geophys Res Lett*, **29**, doi: 10.1029/2001GL014580.

Collmore, C, Martin, D, and M Hitchman, 2003: On the relationship between the QBO and tropical deep convection. *J Climate*, **16**, 2552-2568.

Coughlin, K, and KK Tung, 2001: QBO signal found at the extratropical surface through northern annular modes. *Geophys Res Lett*, **28**, 4563-4566.

Coulson, K, 1975: *Solar and Terrestrial Radiation*. Academic Press, New York, 322 pp.

Cox, S, 1981: Radiation characteristics of clouds in the solar spectrum. In *Clouds: Their formation, Optical Properties, and Effects*, P Hobbs and A Deepak, Eds., Academic Press, New York, 241-280.

Crooks, S and L Gray, 2005: Characteristics of the 11-yr solar signal using a multiple regression analysis of the ERA-40 dataset. *J Clim*, **18**, 996-1015.

Danielsen, E, 1982: A dehydration mechanism for the stratosphere. *Geophys Res Lett*, **9**, 605-608.

Denbigh, K, 1971: *The Principles of Chemical Equilibrium*. Cambridge University Press, London, 494 pp.

Deser, C, 2000: On the teleconnectivity of the "Arctic Oscillation." *Geophys Res Lett*, **27**, 779-782.

DeWeaver E and S Nigam, 2004: On the forcing of ENSO teleconnections by anomalous heating and cooling. *J Clim*, **17**, 3225-3235.

Dickinson, R, 1968: Planetary Rossby waves propagating through weak westerly wind wave guides. *J Atmos Sci*, **25**, 984-1002.

Dickinson, R, 1970: Development of a Rossby wave critical level. *J Atmos Sci*, **27**, 627-633.

Dickinson, RE, 1973: Baroclinic instability of an unbounded zonal shear flow in a compressible atmosphere. *J Atmos Sci*, **30**, 1520-1527.

Dietrich, G, Kalle, K, Krauss, W, and G Siedler, 1980: *General Oceanography*, 2nd ed, Wiley-Interscience, New York.

Dikii, L, 1968: The terrestrial atmosphere as an oscillating system. *Izv Acad Sci USSR Atmos Oceanic Phys*, Engl. Transl., **1**, 469-489.

Dobson, G, 1930: Observations of the amount of ozone in the Earth's atmosphere and its relation to other geophysical conditions. *Proc Roy Soc London, Sec A*, **129**, 411-433.

Donner, L, Schubert, W, and R Somerville, 2010: *The Development of Atmospheric General Circulation Models*. Donner, L, Schubert, W, and R Somerville, Eds. Cambridge University Press, Cambridge, UK, 255 pp.

Douglas, B, 1997: Global sea rise: A redetermination. *Surveys in Geophysics*, **18**, 279-292.

Dowling, D, Radke, RR, and F Lawrence, 1990: A summary of the physical properties of cirrus clouds. *J Applied Meteorology*, **29** 970-978.

Drazin, P and W Reid, 1985: *Hydrodynamic Instability*. Cambridge University Press, Cambridge, 527 pp.

Dutton, J, 1986: *The Ceaseless Wind*. Dover, New York, 617 pp.

Eady, E, 1949: Long waves and cyclone waves. *Tellus*, **1**, 33-52.

Easterling, D, Horton, B, Jones, P, Peterson, T, Karl, T, Parker, D, Salinger, MJ, Raxuvayev, V, Plummer, N, Jamason, P, and C Folland, 1997: Maximum and minimum temperature trends for the globe. *Science*, **277**, 364-367.

Eckart, C, 1960: *Hydrodynamics of Oceans and Atmospheres*. Pergamon, New York, 290 pp.

Eisberg, R, 1967: *Fundamentals of Modern Physics*. John Wiley, New York, 729 pp.

Elsasser, W, 1938: Mean absorption and equivalent absorption coefficient of a band spectrum. *Phys Rev*, **54**, 126-129.

Ertel, 1942: Ein neuer hydrodynamischer Wirbelsatz. *Meteorol Z*, **59**, 271-281.

Etheridge, D, Steele, L, Langenfelds, R, Francey, R, Barnola, J, and V Morgan, 1996: Natural and anthropogenic changes in atmospheric CO_2 over the last 1000 years from air in Antarctic ice and firn. *J Geophys Res*, **101**, 4115-4128.

Fels, S, 1982: A parameterization of scale-dependent radiative damping rates in the middle atmosphere. *J Atmos Sci*, **39**, 1141-1152.

Fels, S, 1985: Radiative-dynamical interactions in the middle atmosphere. *Adv Geophys*, **28A**, 277-300.

Fischer, H, Wahlen, M, Smith, J, Mastrioianni, D, and B Deck, 1999: Ice core records of atmospheric CO_2 arouind the last three glacial terminations. *Science*, **283**, 1712-1714.

Fleming, E, Chandra, S, Schoeberl, M, and J Barnett, 1988: Montly-Mean Climatology of Temperature, Wind, Geopotential Height, and Pressure for 0-120 km. NASA TM-100697. Available from NASA Goddard Space Flight Center, Greenbelt, MD.

Friedli, H, Lotscher, H, Oeschger, H, Siegenthaler, U, and B Stauffer, 1986: Ice core record of the $^{13}C/^{12}C$ ratiio of atmospheric CO_2 in the past two centuries. *Nature*, **324**, 237-238.

Fujita, T, 1992: *The Mystery of Severe Storms*. WRL Research Paper No. 239, NTIS No. PB 92-182021, 298 pp.

Frederiksen, J, 2006: Instability of waves and zonal flows in two-layer models on a sphere. *Quart J Roy Meteoral Soc*, **104**, 841-872.

Francis, G and M Salby, 2001: Radiative influence of Antarctica on the polar night vortex. *J Atmos Sci*, **58**, 1300-1309.

Francois, R, Altabet, M, and R Goericke, 1993: Changes in the $\delta^{13}C$ of surface water particulate organic matter across the subtropical convergence inthe SW Indian Ocean. *Global Biological Cycles*, **7**, 627-644.

Fusco, A and M Salby, 1994: Relationship between horizontal eddy motions and mean meridional motions in the stratosphere. *J Geophys Res*, **99**, 20,633-20,695.

Gage, K, McAfee, J, Carter, D, Ecklund, W, Riddle, A, Reid, G, and B Balsley, 1991: Long-term vertical motion over the tropical Pacific: Wind profiling Doppler measurements. *Science*, **254**, 1771-1773.

Garcia, R, 1994: Causes of ozone depletion. *Physics World*, **7**, 49-55.

Garcia, R and M Salby, 1987: Transient response to localized episodic heating in the tropics. Part II: Far-field behavior. *J Atmos Sci*, **44**, 499-530.

Garcia, R and S Solomon, 1983: A numerical model of zonally-averaged dynamical and chemical structure of the middle atmosphere. *J Geophys Res*, **88**, 1379-1400.

Garcia, R and S Solomon, 1987: A possible interaction between interannual variability in Antarctic ozone and the quasi-biennial oscillation. *Geophys Res Lett*, **14**, 848-851.

Garrison, T, 2007: *Oceanography: Invitation to Marine Science*, 6th ed. Brooks Cole, Belmont, CA, 588 pp.

Geisler, J and R Dickinson, 1974: A numerical study of an interacting Rossby wave and barotropic zonal flow near a critical level. *J Atmos Sci*, **31**, 946-955.

Gill, A, 1980: Some simple solutions for heat-induced tropical circulation. *Quart J Roy Meteorol Soc*, **106**, 447-462.

Gill, A, 1982: *Atmosphere-Ocean Dynamics*. Academic Press, San Diego, 662 pp.

Goericke, R and B Fry, 1994: Variations of marine plankton $\delta^{13}C$ with latitude, temperature, and dissolved CO_2 in the world ocean. *Global Biol Cycles*, **8**, 85-90.

Gong, D and S Wang, 1999: Definition of the Antarctic Oscillaton. *Geophys Res Lett*, **26**, 459-462.

Goodridge, J, 1996: Comments on "Regional simulations of greenhouse warming including natural variability." *Bull Am Meteorol Soc*, **77**, 1588-1589.

Goody, R., 1952: A statistical model for water vapor absorption. *Quart Roy Meteorol Soc*, **78**, 165-169.

Goody, R and Y Yung, 1995: *Atmospheric Radiation: Theoretical Basis*. Oxford U Press, 544 pp.

Goody, R, West, R, Chen, L, and D Crisp, 1989: The correlated-k method for radiation calculations in nonhomogeneous atmospheres. *J Quant Spectroscopy and Radiative Transfer*, **42**, 539-550.

Gray, L, S Crooks, Pascoe, C, and S Sparrow, 2004: Solar and QBO influences on the timings of stratosphseric sudden warmings. *J Atmos Sci*, **61**, 2777-2796.

Grove, J and R Switsur, 1994: Glacial geological evidence for the medieval warm period. *Climate Change*, **26**, 143-169.

Gossard, E, and W Hooke, 1975: *Waves in the Atmosphere*. Elsevier, Amsterdam, 456 pp.

Green, G, 1838: On the motion of waves in variable canal of small depth and width. *Trans Cambridge Philos Soc*, in Mathematical Papers (1871), Macmillan, London, 223-230.

Greenspan, H, 1968: *Theory of Rotating Fluids*. Cambridge University Press, London, 327 pp.

Gruber, C and L Haimberger, 2008: On the homogeneity of radiosonde wind time series. *Meteorol Zeit*, **17**, 631-643.

Hall, B and G Denton, 2002: Holocene history of the Wilson Piedmont Glacier along the southern Scott Coast, Antarctica. *The Holocene*, **12**, 619-627.

Haltiner, G and R Williams, 1980: *Numerical Prediction and Dynamic Meteorology*. Wiley, New York, 477 pp.

Handbook of Geophysics and Space Environment, 1965: S. Valley, Ed., Air Force Cambridge Research Laboratory, Hanscom AFB, Mass., 692 pp.

Hansen, J and L Travis, 1974: Light scattering in planetary atmospheres. *Space Sci Rev*, **16**, 527-610.

Hartmann, D, 1993: The radiative effect of clouds on climate. In *Aerosol-Cloud-Climate Interactions*, P Hobbs, Ed., Academic Press, New York, 151-170.

Hartmann, D, 1994: *Global Physical Climatology*. Academic Press, San Diego, 408 pp.

Hartmann, D, Ramanathan, V, Berroir, A, and G Hunt, 1986: Earth radiation budget data and climate research. *Rev Geophys*, **24**, 439-468.

Hartmann, D and D Doelling, 1991: On the net radiative effectiveness of clouds. *J Geophys Res*, **96**, 869-891.

Hasebe, F, 1983: Interannual variations of global total ozone revealed from Nimbus 4 BUV and ground based observations. *J Geophys Res*, **88**, 6819-6834.

Hendon, H and M Salby, 1994: The life cycle of the Madden-Julian Oscillation. *J Atmos Sci*, **51**, 2225-2237.

Hendon, H and K Woodberry, 1993: The diurnal cycle of tropical convection. *J Geophys Res*, **98**, 16623-16637.

Hendon, H, Thompson, D, and M Wheeler, 2007: Australian rainfall and surface temperature variations associated with the southern annular mode. *J Clim*, **20**, 2452-2467.

Herzberg, G, 1945: *Molecular Spectra and Molecular Structure. II: Infrared and Raman Spectra of Polyatomic Molecules*. Nan Nostrand, New York, 616 pp.

Herzberg, L, 1965: In *Physics of the Earth's Upper Atmosphere*, C Hines, I Paghis, T Hartz, and J Fejer, Eds. Prentice Hall, Englewood Cliffs, 31-45.

Hering, W and T Borden, 1965: Ozone sonde observations over North America, vol. 3, AFCRL Report AFCRL-64-30. Air Force Cambridge Research Laboratories, Bedford, MA.

Hess, P and J Holton, 1985: The origin of temporal variance in long-lived trace constituents in the summer stratosphere. *J Atmos Sci*, **42**, 1455-1463.

Hide, R, 1966: On the dynamics of rotating fluids and related topics in geophysical fluid dynamics. *Bull Amer Meteorol Soc*, **47**, 873-885.

Holton, 1984: Troposphere-stratosphere exchange of trace constituents: The water vapor puzzle. In *Dynamics of the Middle Atmosphere*, J. Holton and T. Matsuno, Eds., Terrapub, Tokyo, 369-385.

Holton, J, 1986: Meridional distribution of stratospheric trace species. *J Atmos Sci*, **43**, 1238-1242.

Holton, J, 1990: On the global exchange of mass between the stratosphere and troposphere. *J Atmos Sci*, **47**, 392-395.

Holton, J, 2004: *An Introduction to Dynamic Meteorology*, 4th ed. Academic Press, San Diego, 535 pp.

Holton, J and R Lindzen, 1972: An updated theory for the quasi-biennial cycle of the tropical stratosphere. *J Atmos Sci*, **29**, 1076-1080.

Holton, J, Haynes, P, McIntyre, M, Douglass, A, Rood, R, and L Pfister, 1995: Stratosphere-troposphere exchange. *Rev Geophys*, **33**, 403-439.

Holton, J and HC Tan, 1980: The influence of the equatorial quasi-biennial oscillation on the global circulation at 50 mb. *J Atmos Sci*, **37**, 2200-2208.

Hoffman, D, 1988: Aerosols from past and present volcanic emissions. In *Aerosol and Climate*, P Hobbs and P McCormick, Eds., A Deepak Publishing, Hampton, VA, 195-214.

Hoffman, D and S Solomon, 1989: Ozone destruction through heterogeneous chemistry following the eruption of El Chichon. *J Geophys Res*, **94**, 5029-5041.

Horel, J and J Wallace, 1981: Planetary scale atmospheric phenomena associated with the inter-annual variability of sea-surface temperature in the equatorial Pacific. *Mon Wea Rev*, **109**, 813-829.

Horowitz, L, Walters, S, Mauzerall, D, Emmons, L, Rasch, P, Granier, C, Tie, X, Lamarque, J, Schultz, M, Tyndall, G, Orlando, J, and G Brasseur, 2003: A global simulation of tropospheric ozone and related tracers: Description and evaluation of Mozart version 2. *J Geophys Res*, **208**, 4784 doi:10.1029/2002JD002853.

Hoskins, B, 1982: A mathematical theory of frontogenesis. *Annu Rev Fluid Mech*, **14**, 131-151.

Hoskins, B, McIntyre, M, and A Robertson, 1985: On the use and significance of isentropic potential vorticity maps. *Quart J Roy Meteorol Soc*, **111**, 877-946.

Hoskins, B and D Karoly, 1981: The steady linear response of a spherical atmosphere to thermal and orographic forcing. *J Atmos Sci*, **38**, 1179-1196.

Houze, RA, 1982: Cloud clusters and large-scale vertical motions in the tropics. *J Meteorol Soc Jpn*, **60**, 396-410.

Houze, R, 1993: *Cloud Dynamics*. Academic Press, San Diego, 573 pp.

Huang, S and H Pollack, 1997: Late quaternary temperature changes seen in world-wide continental heat flow measurements. *Geophys Res Lett*, **24**, 1947-1950.

Huck, P, McDonald, A, Bodeker, G, and H Struthers, 2005: Interannaual variability in Antarctic ozone depletion controlled by planetary waves and polar temperature. *Geophys Res Lett*, **32**, L13819, doi:10.1029/2005GL022943.

Huesmann, A and M Hitchmann, 2003: Th 1978 shift in the NCEP reanalysis stratospheric quasi-biennial oscillation. *Geophys Res Lett*, **30**, 1048. doi:10.1029/2002GL016323.

Humphreys, W, 1964: *Physics of the Air*. Dover, New York, 661 pp.

Hurrell, J, Kushnir, Y, Ottersen, G, and M Visbeck, 1996: An overview of the North Atlantic Oscillation. In *The North Atlantic Oscillation: Climatic Signficance and Environmental Impact*. Geophysical Monograph 134, American Geophysical Union, 10.1029/134GM01.

Indemuhle, A, Monnin, E, Stauffer, B, and T Stocker, 2000: Atmospheric CO_2 concentration from 60 to 20 kyr BP fromthe Taylor Dome ice core, Antarctica. *Geophys Res Lett*, **27**, 735-738.

IPCC, 1990: *Climate Change. Intergovernmental Panel on Climate Change*. J Houghton, G Jenkins, and J Ephraums, Eds. Cambridge University Press, Cambridge, 365 pp.

IPCC, 2007: *Intergovernmental Panel on Climate Change. Climate Change 2007: The Physical Science Basis*. Cambridge University Press, Cambridge, 1008. pp.

Iribarne, J and W Godson, 1981: *Atmospheric Thermodynamics*. Reidel, Dordrecht, 259 pp.

Jackson, J, 1975: *Classical Electrodynamics*. Wiley, New York, 848 pp.

James, I, 1993: *Introduction to Circulating Atmospheres*. Cambridge University Press, New York, 422 pp.

Jensen et al, 2004: Ice supersaturations exceeding 100% at the cold tropical tropopause: Implications for cirrus formation and deydration. *Atmos Chem Phys*, **4**, 7433-7462.

Jin, F and B Hoskins, 1995: The direct response to tropical heating in a baroclinic atmosphere. *J Atmos Sci*, **52**, 307-319.

Johnson, R and D Kriete, 1982: Thermodynamic circulation characteristics of winter monsoon tropical mesoscale convection. *Mon Wea Rev*, **110**, 1898-1911.

Jones J and M Widman, 2004: Early peak in Antarctic oscillation index. *Nature*, **432**, 290-291.

Jouzel, J, Waelbroeck, B, Malaize, M, Bender, J, Petit, M, Stievenard, N, Barkov, J, Barnola, T, Kink, V, Kotlyakov, V, Lipenkov, V, Lorius, C, Raynaud,D, Ritz, Z, and T Sowers, 1995: Climatic interpretation of the recently extended Vostok ice records. *Climate Dynamics*, **12**, 513-521.

Kallberg, P, Berrisford, P, Hoskins, B, Simmons, A, Uppala, S, Lamy-Th'epaut, S, and R Hine, 2005: ERA-40 Atlas ERA-40 Project Report Series No 17. European Centre for Medium Range Weather Forcasting, Reading.

Keigwin, L, 1996: The Little Ice Age and Medieval Warm Period in the Sargasso Sea. *Science*, **274**, 1504-1508.

Kelly, K, Proffitt, M, Chan, K, Lowenstein, M, Podolske, J, Strahan, S, Wilson, J, and D Kley, 1993: Water vapor and cloud water measurements over Darwin during the STEP 1987 tropical mission. *J Geophys Res*, **98**, 8713-8723.

Kent, G and P McCormick, 1984: SAGE and SAM II measurements of global stratospheric aerosol optical depth and mass loading. *J Geophys Res*, **89**, 5303-5314.

Kiehl, J, and S Solomon, 1986: On the radiative balance of the stratosphere. *J Atmos Sci*, **43**, 1525-1534.

Kiehl, J, 1993: Atmospheric general circulation modeling. In *Climate System Modeling*. Cambridge University Press, Cambridge, 818 pp.

Kocharts, G, 1971: Penetration of solar radiation in the Schumann-Runge bands of molecular oxygen. In *Mesospheric Models and Related Experiments*, G Fiocco, Ed., Reidel, Dordrecht, 160-176.

Kodera, K and Y Kuroda, 2002: Dynamical response to the solar cycle: Winter stratopause and lower stratosphere. *J Geophys Res*, **107**, 4749.

Koschmieder, E and S Pallas, 1974: Heat transfer through a shallow horizontal convecting fluid layer. *Int J Heat Mass Transfer*, **17**, 991-1002.

Keeling, C, Bacastow, R, Carter, A, Piper, S, Whorf, T, Heimann, M, Mook, W, and H Rheloffzen, 1989: A three dimensional model of atmospheric $CO2$ based on observed winds: 1. Analysis of observational data. *Aspects of Climate Variability in the Western Pacific and Western Americas*, D Peterson, Ed., *Geophysical Monographs*, **55**, American Geophysical Union, Washington, DC, 165-236.

Keenan, J, 1970: *Thermodynamics*. MIT Press, Cambridge, 507 pp.

Kelvin, Lord, 1880: On a disturbing infinity in Lord Rayleigh's solution for waves in a plane vortex stratum. *Nature*, **23**, 45-46.

Knollenberg, RG, K Kelly, and JC Wilson, 1993: Measurements of measurements of number densities of ice crystals in the tops of tropical cumulonimbus. *J Geophys Res*, **98**, 8639-8664.

Kramer, H, 1926: Wellenmechanik und halbzhalige Quantisierung. *Z Phys*, **39**, 828-840.

Kramer et al, 2009: Ice supersaturations and cirrus cloud crystal numbers. *Atmos Chem Phys*, **9**, 3505-3522.

Kreutz, K, Mayewski, P, Meeker, L, Twickler, M, Whitlow, S, and I Pittalwala, 1997: Bipolar changes in atmospheric circulation during the Little Ice Age. *Science*, **277**, 1294-1296.

Krueger, A, 1973: The mean ozone distributions from several series of rocket soundings to 52 km at latitudes from 58 S to 64 N. *Pure Appl Geophys*, **106-108**, 1272-1280.

Lamb, H, 1965: Paleogeography, Paleoclimatology, Paleoecology. 1, 13-37.

Labitzke, K, 1982: On interannual variability of the middle stratosphere during northern winters. *J Meteorol Soc Japan*, **60**, 124-139.

Labitzke, K and H van Loon, 1988: Association between the 11-yr solar cycle, the QBO, and the atmosphere. *J Atmos Terr Phys*, **50**, 197-206.

Labitzke K, 2005: On the solar cycle – QBO relationship: A summary. *J Atmos Sol Terr Phys*, **67**, 45–54.

Lait, L, Schoeberl, M, and P Newman, 1989: Quasi-biennial modulation of Antarctic ozone depletion. *J Geophys Res*, **94**, 11559–11571.

Landau, X and Y Lifshitz, 1980: *Statistical Physics*, 3rd ed., Part 1. Buttenworth-Heinamann, Oxford, 544 pp.

Lamb, H, 1910: On atmospheric oscillations. *Proc Roy Soc London*, **84**, 551–572.

Lejenas, H and Okland, 1983: Characteristics of Northern Hemisphere blocking as determined from a ong time series of observational data. *Tellus*, **35A**, 350–362.

Lejenas, H and R Madden, 1992: Travelling planetary-scale waves and blocking. *Mon Wea Rev*, **120**, 2821–2830.

Lorius, et al, 1985: A 150,000 year climatic record from Antarctic ice. *Nature*, **316**, 591–596.

Lau, N, 1984: *Circulation Statistics Based on FGGE Level III-B Analyses*. NOAA Data Report ERL GFDL-5. NOAA Environmental Research Laboratories, Boulder CO.

Lee, J, Sears, F, and D Turcotte, 1973: *Statistical Thermodynamics*. Addison-Wesley, Reading, 371 pp.

Lighthill, J, 1978: *Waves in Fluids*. Cambridge University Press, Cambridge, 504 pp.

Lindzen, R and J Holton, 1968: A theory of the quasi-biennial oscillation. *J Atmos Sci*, **25**, 1095–1107.

Lindzen, R and Y Choi, 2009: On the determination of climate feedbacks from ERBE data. *Geophys Res Lett*, **36**, doi: 10.1029/2009GL039628.

Lilly, D, 1978: A severe downslope windstorm and aircraft turbulence event induced by a mountain wave. *J Atmos Sci*, **35**, 59–77.

Liou, K, 1980: *An Introduction to Atmospheric Radiation*. Academic Press, San Diego, 392 pp.

Liou, K, 1990: *Radiation and Cloud Processes in the Atmosphere*. Oxford University Press, New York, 487 pp.

Lincoln, C, 1972: *On Quiet Wings: A Soaring Anthology*. Northland Press, Flagstaff, AZ, 397 pp.

Liouville, J, 1837: Sur le développment des fonctions ou parties de fonctions en séries... *J Math Pure Apl*, **2**, 16–35.

List, R, 1958: *Smithsonian Meteorological Tables*, 6th ed. Random House, Smithsonian Institute Press, New York.

Loehle, C, 2007: A 2000-yr global temperature reconstruction based on non-tree ring proxies. *Energy & Env*, **18**, 1049–1058.

Loehle, C and J McCulloch, 2008: Correction to: A 2000-yr global temperature reconstruction based on non-tree ring proxies. *Energy & Env*, **19**, 93–100.

Logan, J, 1999: An analysis of ozonesonde data for the troposphere: Recommendations for testing 3D models and development of a gridded climatology for tropospheric ozone. *J Geophys Res*, **104**, 16115–16149.

London, J, 1980: Radiative energy sources and sinks in the stratosphere and mesosphere. *Proceedings of the NATA Advanced Study Institute on Atmospheric Ozone: Its Variation and Human Influences*. Rep. FAA-EE-80-20, A Aiken, Ed., 703–721.

London, J. (1985), The observed distribution of atmospheric ozone and its variations. In *Ozone in the Free Atmosphere*, R Whitten and S Prasad, Eds., van Nostrant-Reinhold, Princeton, 11–80.

Lorenz, E, 1955: Available potential energy and the maintenance of the general circulation. *Tellus*, **7**, 157–167.

Lynn, R and J Reid, 1968: Characteristics and circulation of the deep and abyssal waters. *Deep-Sea Res*, **15**, 577–598.

Madden, R and P Julian, 1971: Description of a 40-50 day oscillation in the zonal wind in the tropical Pacific. *J Atmos Sci*, **28**, 702–708.

Madden, R and P Julian, 1972: Description of global-scale circulation cells in the tropics with a 40-50 day period. *J Atmos Sci*, **29**, 1109–1123.

Madden, R, 1986: Seasonal variations of the 40-50 day oscillation in the tropics. *J Atmos Sci*, **43**, 3138-3158.

Madden, R and G Meehl, 1993: Bias in the global mean temperature estimated from sampling a greenhouse warming pattern with the current surface observing network. *J Clim*, **6**, 2486-2489.

Maloney, E and D Hartmann, 1998: Frictional moisture convergence in a composite life cycle of the Madden-Julian Oscillation. *J Clim*, **11**, 2387-2403.

Maloney, E and J Kiehl, 2002: MJO-related SST variations over the tropical eastern Pacific during Northern Hemisphere summer. *J Clim*, **15**, 675-689.

Manabe S and R Strickler, 1964: Thermal equilibrium of the atmosphere with convective adjustment. *J Atmos Sci*, **21**, 361-385.

Manabe, S and R Wetherald, 1967: Thermal equilibrium of the atmosphere with a given distribution of relative humidity. *J Atmos Sci*, **24**, 241.

Mantua, N and S Hare, 2002: The Pacific decadal oscillation. *J Oceanography*, **58**, 35-44.

Margules, M, 1903: Über die energie der sturme. *Jahrb Zentralanst Meteorol Wien*, **40**, 1-26.

Marland, G, Boden, T, and R Andres, 2008: Global, regional, and national fossil fuel CO_2 emissions. In *Trends: A Compendium of Data on Global Change*. Carbon Dioxide Information Analysis Center, Oak Ridge National Laboratory, US Dept of Energy, Oak Ridge, TN.

Marshall, J, 2003: Trends in the southern annular mode from observations and reanlyses. *J Clim*, **16**, 4134-4143.

Mastepanov et al, 2008: Large tundra methane burst during onset of freezing. *Nature*, **456**, 628-631.

Matveev, L, 1967: *Physics of the Atmosphere*. Israel Program for Scientific Translations. Jerusalem, 699 pp.

McBride, J and N Nichols, 1983: Seasonal relationships between Australian rainfall and the Southern Oscillation. *Mon Wea Rev*, **111**, 1998-2004.

McClatchey, R and J Selby, 1972: Atmospheric transmittance, 7-30 μm: Attenuation of CO_2 laser radiation. *Env Res*, Paper No. 419, AFCRL-72-0611.

McGann, M, 2008: High-resolution, forimaniferal, isotopic and trace element records from Holocene estuarine deposits of San Francisco Bay. *J Coastal Res*, **24**, 1092-1109.

McIntyre, M and T Palmer, 1983: Breaking planetary waves in the stratosphere. *Nature*, **305**, 593-600.

Mears, C, Schabel, M, and F Wentz, 2003: A reanalysis of the MSU channel 2 tropospheric temperature record. *J Clim*, **16**, 3650-3664.

Mie, 1908: Beigrade zur optik trüber medienspeziell kolloidaler metallösungen. *Ann Physik*, **25**, 377-445.

Miles, T and W Grose, 1986: Transient medium-scale wave activity in the summer stratosphere. *Bull Amer Meteor Soc*, **67**, 674-686.

Minnis, P and E Harrison, 1984: Diurnal variability of regional cloud and clear-sky radiative parameters derived from GOES data. III. November 1978 radiative parameters. *J Clim and Appl Meteor*, **23**, 1032-1051.

Mlawer, E, Taubman, S, Brown, P, Iacono, M, and S Clough, 1997: Radiative transfer for inhomogenous atmospheres: RRTM, a validated correlated-k model for the longwave. *J Geophys Res*, **102**, 16663-16682.

Moberg, A, Sonechkin, D, Holmgren, K, Datsenko, N, and W Karlen, 2005: Highly variable Northern Hemisphere temperatures reconstructed from low- and high-resolution proxy data. *Nature*, **433**, 613-617.

Monahan, A, Fyfe, J, Ambaum, M, Stephenson, D, and G North, 2009: Empirical orthogonal functions: The medium is the message. *J Clim*, **22**, 6501-6514.

Monin, A and A Yaglom, 1973: *Statistical Fluid Mechanics*. MIT Press, Cambridge, 769 pp.

Monnin, E, Indermühle, A, Dällenbach, A, Flückiger, J, Stauffer, B, Stocker, T, Raynaud, D, and J-M Barnola, 2001: Atmospheric CO_2 concentrations over the last glacial termination. *Science*, **291**, 112-114.

Morse, P and H Feschbach, 1953: *Methods of Theoretical Physics*, vols. I and II. McGraw Hill, New York, 1978 pp.

Muller et al, 2003: Chlorine activation and chemical ozone loss deduced from HALOE and balloon measurements in the Arctic during the winter of 1999-2000. *J Geophys Res*, **108**, 8302.

Murgatroyd, R and F Singleton, 1961: Possible meridional circulations in the stratosphere and mesosphere. *Quart J Roy Meteorol Soc*, **87**, 125-135.

Naito, Y and I Hirota, 1997: Interannual variability of the northern winter stratosphere circulation related to the QBOand solar cycle. *J Meteor Soc Japan*, **75**, 925-937.

Naujokat, B, 1986: An update of the observed quasi-biennial oscillation of he stratospheric winds over the tropics. *J Atmos Sci*, **43**, 1873-1877.

Nicolet, M, 1980: The chemical equations of stratospheric and mesospheric ozone. *Proceedings of the NATO Advanced Study Institute on Atmospheric Ozone (Portugal)*. US Dept. of Transportation, FAA, Washington, DC, Report No. FAA-EE-80-20.

Nigam, S, 2003: *Teleconnections*. In *Encyclopedia of Atmospheric Sciences*. Academic Press, San Diego, pp. 2243-2269.

Norris, J and M Wild, 2009: Trends in aerosol radiative effects over China and Japan inferred from observed cloud cover, solar dimming, and solar brightening. *J Geophys Res*, **114** doi: 10.1029/2008JD011378.

North, G, 1975: Theory of energy balance models. *J Atmos Sci*, **32**, 2033-2043.

Oort, A and J Peixoto, 1983: Global angular momentum and energy balance requirements from observations. *Adv Geophys*, **25**, 355-490.

Osterlind, A, Tucker, M, Stone, B, and Jensen, O, 2006: The Danish case control study of cutaneous malignant melanoma. II. Importance of UV-light exposure. *Int J Cancer*, **42**, 319-324.

O'Sullivan, D and M Salby, 1990: Coupling of the quasi-biennial oscillation and the extratropical circulation in the stratosphere through planetary wave transport. *J Atmos Sci*, **47**, 650-673.

Paeschke, W, 1937: Experimentelle Untersuchungen zum Rauhigkeits-und Stabilitaets-problem in der freien Atmosphaere. *Beitr Phys Atmos*, **24**, 163-189.

Paltridge, G and C Platt, 1981: Aircraft measurements of solar and infrared radiation and the microphysics of cirrus clouds. *Quart J Roy Meteorol Soc*, **107**, 367-380.

Parrenin et al, 2007: The EDC3 chrolonology for the EPICA Dome C ice core. *Clim Past Disc*, **3**, 575-606.

Patterson, E, 1982: Size distributions, concentrations, and composition of continental and marine aerosols. In *Atmospheric Aerosols: Their Formation, Optical Properties, and Effects*, A. Deepak, Ed., Spectrum Press, Hampton, VA, 1-23.

Pedlosky, J, 1979: *Geophysical Fluid Dynamics*. Springer-Verlag, New York, 624 pp.

Peixoto, J, and A Oort, 1992: *Physics of Climate*. American Institute of Physics, New York, 520 pp.

Peterson, T and R Vose, 1997: Overview of the global historical climatology network temperature data base. *Bull Am Meteorol Soc*, **78**, 2837-2849.

Petit et al, 1999: Climate and atmospheric history of the past 420,000 yrs from the Vostok ice core, Antarctica. *Nature*, **399**, 429-436.

Phillips, H and T Joyce, 2007: Bermuda's tale of two time series: Hydrostation S and BATS. *J Phys. Oceanography*, **37**, 554-571.

Platt, C, Dilley, A, Scott, J, Barton, I, and GL Stephens, 1984a: Remote sounding of high clouds. V: Infrared properties and structures of tropical thunderstorm anvils. *J Clim Appl Meteor*, **23**, 1296-1308.

Platt, C, Scott, J, and A Dilley, 1984b: Remote sounding of high clouds. IV: Optical properties of midlatitude and tropical cirrus. *J Atmos Sci*, **44**, 729-747.

Philander, S, 1983: El Nino Southern Oscillation phenomena. *Nature*, **302**, 295-301.

Phillips, N, 1966: The equations of motion for a shallow rotating atmosphere and the "Traditional Approximation." *J Atmos Sci*, **23**, 626-628.

Plumb, R, 1982: The circulation of the middle atmosphere. *Aust Meteorol Mag*, **30**, 107-121.

Pope, F, Hansen, J, Bayes, K, Friedl, R, and S Sander, 2007: UV absorption spectrum of chlorine peroxide, ClOOCl. *J Phys Chem A*, **111**, 4322-4332.

Pruppacher, H, 1981: The microstructure of atmospheric clouds and precipitation. In *Clouds: Their Formation, Optical Properties, and Effects.* P Hobbs and A Deepak, Eds., Academic Press, San Diego, pp. 93-186.

Pruppacher, H and J Klett, 1978: *Microphysics of Clouds and Precipitation.* Reidel, Dordrecht, 714 pp.

Queney, P, 1948: The problem of air flow over mountains: A summary of theoretical studies. *Bull Am Meteorol Soc,* **29,** 16-26.

Rayleigh, Lord, 1871: On the light from the sky, its polarization and colour *Phil Mag,* **41,** 107-120.

Rayleigh, Lord, 1880: On the stability or instability of certain fluid motions. *Proc London Math Soc,* **9,** 57-70.

Ramanathan, V, 1987: Atmospheric general circulation and its low frequency variance: Radiative influences. *J Meteorol Soc Jpn,* **65,** 151-175.

Rasmusson, E and T Carpenter, 1982: Variation of tropical sea surface temperature and surface wind fields associated with the Southern Oscillation/El Nino. *Mon Wea Rev,* **110,** 354-384.

Rektorys, K, 1969: *Survey of Applicable Mathematics.* MIT Press, Cambridge, 1369 pp.

Remsberg, E, Russell, J, Gordley, L, Gille, J, and P Bailey, 1984: Implications of the stratospheric water vapor distribution as determined from the Nimbus-7 LIMS experiment. *J Atmos Sci,* **41,** 2934-2945.

Rex, M, Salawitch, R, von der Gathen, P, Harris, N, Chpperfield, M, and B Naujoka, 2004: Arctic ozone loss and climate change. *Geophys Res Lett,* **31,** doi: 10.1029/2003GL018844.

Reynolds, D, Vonder Haar, T, and S Cox, 1975: The effect of solar radiation absorption in the tropical troposphere. *J Appl Meteor,* **14,** 433-443.

Robinson, A, Balliunas, S, Soon, W, and Z Robinson, 1998: Environmental effects of increased atmospheric carbon dioxide. *Med Sent,* **3,** 171-178.

Rodgers, C and C Walshaw, 1966: The computation of infrared cooling rate in planetary atmospheres. *Quart J Roy Meteor Soc,* **92,** 67-92.

Roper, R, 1977: Turbulence in the lower thermosphere, Chapter 7 in *The Upper Atmosphere and Magnetosphere,* F Johnson, Ed., U.S. National Academy of Science, Washington, DC, pp. 117-129.

Rosenfeld, D, 2006: Aerosol-cloud interactions control of Earth radiation and latent heat release budgets. *Space Science Revs,* **125,** 149-157.

Rossby, C, et al. 1939: Relations between variations in the intensity of the zonal circulation of the atmosphere and the displacements of the semipermanent centers of action. *J Mar Res,* **2,** 38-55.

Rotty, R, 1987: Estimates of seasonal variation in fossil fuel CO_2 emissions. *Tellus,* **39B,** 184-202.

Ruddiman, W and M Raymo, 2003: A methane based time scale for Vostok ice. *Quaternary Sci Rev,* **22,** 141-155.

Ruth, S, Remedios, J, Lawrence, B, and F Taylor, 1994: Measurements of N_2O by the Improved Stratospheric and Mesospheric Sounder during early northern winter 1991/92. *J Atmos Sci,* **51,** 2818-2833.

Saji, N, Goswami, B, Vinayachandran, P, and T Yamagata, 1999: A dipole mode in the tropical Indian Ocean. *Nature,* **401,** 360-363.

Salby, M, D O'Sullivan, R Garcia, and P Callaghan, 1990: Air motions accompanying the development of a planetary wave critical layer. *J Atmos Sci,* **47,** 1179-1204.

Salby M and P Callaghan, 1993: Fluctuations in total ozone and their relationship to stratospheric air motions. *J Geophys Res,* **98,** 2716-2727.

Salby, M and H Hendon, 1994: Intraseasonal behavior of clouds, temperature, and motion in the tropics. *J Atmos Sci,* **51,** 2207-2224.

Salby, M and P Callaghan, 2000: Connection between the solar cycle and the QBO: The missing link. *J Climate,* **13,** 2652-2662.

Salby, M and P Callaghan, 2002: Interannual changes of the stratospheric circulation: Relationship to ozone and tropospheric structure. *J Climate,* **15,** 3673-3685.

Salby, M, Sassi, F, Callaghan, P, Read, W, and H Pumphrey, 2003: Fluctuations of cloud, humidity, and thermal structure near the tropical tropopause. *J Climate,* **15,** 3428-3446.

Salby, M and P Callaghan, 2004a: Control of the tropical tropopause and vertical transport across it. *J Climate*, **17**, 965-985.

Salby, M, and P Callaghan, 2004b: Systematic changes of Northern Hemisphere ozone and their relationship to random interannual changes. *J Climate*, **17**, 4512-4521.

Salby, M and P Callaghan, 2005: Interaction between the Brewer-Dobson circulation and the Hadley circulation. *J Climate*, **18**, 4303-4316.

Salby, M and P Callaghan, 2006: Relationship of the QBO to the stratospheric signature of the solar cycle. *J Geophys Res*, **111**, D06110, doi: 10.1029/2005JD006012.

Salby, M, 2007: Influence of the solar cycle on the general circulation of the stratosphere and upper tropoposphere. *Solar Variability and Planetary Climates*, 287-303, Springer Verlag.

Salby, M and P Callaghan, 2007: On the wintertime increase of Arctic ozone: Relationship to changes of the polar-night vortex. *J Geophys Res* **112**, D06116, doi: 10.1029/2006JD007948.

Salby, M, 2011: Interannual changes of stratospheric temperatures and ozone: Forcing by anomalous wave driving and the QBO. *J Atmos Sci*, **68**, 1513-1525.

Salby, M, Titova, E, and L Deschamps, 2011: Rebound of Antarctic ozone. *Geophys Res Lett* (38, L09702, doi:10.1029/2011GL047266.).

Schiermeier, Q, 2007: Chemists poke holes in ozone theory. *Nature*, **449**, 382-383.

Schlicting, H, 1968: *Boundary Layer Theory*. McGraw Hill, New York, 744 pp.

Scorer, R, 1978: *Environmental Aerodynamics*. Ellis Horwood Ltd., West Sussex, 488 pp.

Scorer, R, 1986: *Cloud Investigation by Satellite*. Ellis Horwood Ltd., West Sussex, 23 chaps.

Shabbar, Huang, J, and K Higachi, 2001: The relationship between the wintertime north Atlantic oscillation and blocking episodes in the north Atlantic. *Int J Climatol*, **21**, 355-369.

Shapiro, L, 1989: The relationship of the quasi-biennial oscillation to Atlantic tropical storm activity. *Mon Wea Rev*, **117**, 1545-1552.

Shapiro, M, 1982: Nowcasting the position and intensity of jet streams using a satellite-borne total ozone mapping spectrometer. In *Nowcasting*, K. A. Browning, ed. Academic Press, 256 pp.

Shapiro, M and J Hastings, 1973: Objective cross-section analysis by Hermite polynomial interpolation on isentropic surfaces. *J Appl Meteorol*, **12**, 753-762.

Shea, D, K Trenberth, and R Reynolds, 1990: *A Global Monthly Sea Surface Temperature Climatology*. Tech Note TN 345+STR, National Center for Atmospheric Research, Boulder, CO, 167 pp.

Shinoda, T and H Hendon, 1998: Mixed layer modeling of intraseasonal variability in the tropical western Pacific and Indian Oceans. *J Clim*, **11**, 2668-2685.

Siddall et al, 2003: Sea level fluctuations during the last glacial cycle. *Nature*, **423**, 853-858.

Simmons, A, Wallace, J, and G Branstator, 1983: Barotropic wave propagation and instability and atmospheric telleconnection pattern. *J Atmos Sci*, **40**, 1363-1392.

Slingo, A, 1989: A GCM parameterization for the SW radiative properties of water clouds. *J Atmos Sci*, **46**, 1419-1427.

Slinn, W, 1975: Atmospheric aerosol particles in surface-level air. *Atmos Env*, **9**, 763-764.

Smith, W and D Gottlieb, 1974: Solar flux and its variations. *Space Sci Rev*, **16**, 771-802.

Soden, B, Wetherald, G, Stenchikov, G, and A Robock, 2002: Global cooling after the eruption of Mount Pinatubok: a test of climate feedback by water vapor. *Science*, **296**, 727-730.

Solomon, S, 1990: Progress towards a quantitative understanding of Antarctic ozone depletion. *Nature*, **347**, 347-354.

Sorbjan, Z, 1989: *The Structure of the Atmospheric Boundary Layer*. Prentice Hall, Englewood Cliffs, NJ, 317 pp.

Sowers et al, 1993: 135,000 year Vostok – SPECMAP common temporal framework. *Paleoceanogr*, **8**, 737-766.

Spichtinger, P, Gierens, K, and W Read, 2006: The global distribution of ice-supersaturated regions as seen by the Microwave Limb Sounder. *Q J Roy Meteorol Soc*, **129**, 3391-3410.

Stanford et al, 2006: Timing of the meltwater pulse 1A and climate responses to meltwater injections. *Paleoceanogr*, **21**, 4103.

Stein, O, 2008: The variability of Atlantic-European blocking as derived from long SLP time series. *Tellus*, **52**, 225-236.

Stewart, R, 2008: *Introduction to Physical Oceanography*. http://oceanworld.tamu.edu/ocng_textbook.

Stommel, H and A Arons, 1960: On the abyssal circulation of the world ocean (I). Stationary planetary flow patterns on a sphere. *Deep Sea Res*, **6**, 140-154.

Strub, P, Mesias, J, Montecino, V, Rutlant, J, and S Salinas, 1998: Coastal ocean circulation off western South America coastal segment. *The Sea*, **11**, 273-308.

Stull, R, 1988: *An Introduction to Boundary Layer Meteorology*. Kluwer, Boston, 666 pp.

Sun, W, and C Chang, 1986: Diffusion model for a convective layer. *J Clim Appl Meteorol*, **25**, 1445-1453.

Sverdrup, H, 1947: Wind-driven currents in a baroclinic ocean; with application to the equatorial currents of the eastern Pacific. *Proc Nat Acad Sci*, **33**, 318-326.

Taylor, G, 1929: Waves and tides in the atmosphere. *Proc Roy Soc London Ser A*, **126**, 169-183.

Tennekes, H and J Lumley, 1972: *A First Course in Turbulence*. MIT Press, Cambridge, 300 pp.

Thompson, D and J Wallace, 2000: Annular modes in the extratropical circulation. Part I: Month to month variability. *J Clim*, **13**, 1000-1016.

Thompson, D, Wallace, J, and C Hegerl, 2000: Annular modes in the extratropical circulation. Part II: Trends. *J Clim*, **13**, 1018-1036.

Thorpe, S, 1969: Experiments on the instability of stratified flows: Immiscible fluids. *J Fluid Mech*, **39**, 25-48.

Thorpe, S, 1971: Experiments on the instability of stratified flows: Miscible fluids. *J Fluid Mech*, **46**, 299-319.

Thuburn, J and G Craig, 2000: Stratospheric influence on tropopause height: The radiative constraint. *J Atmos Sci*, **57**, 17-28.

Tilmes, S, Muller, R, Groos, J, McKenna, D, Russel, J, and Y Sasano, 2003: Calculation of chemical ozone loss in the Arctic winter 1996-1997 using ozone tracer correlations. Comparison of ILAS and HALOE results. *J Geophys Res*, **108**, 4045.

Tsushima, Y and S Manabe, 2001: Influence of cloud feedback on annual variation of global mean surface temperature. *J Geophys Res*, **106**, 22635-22646.

Tung, KK and H Yang, 1994: Global QBO in circulation and ozone. II: A simple mechanistic model. *J Atmos Sci*, **51**, 2708.

Turco, R, Drdla, K, A Tabazadeh, and P Hamill, 1993: Heterogeneous chemistry of polar stratospheric clouds and volcanic aerosols. *The Role of the Stratosphere in Global Change*, NATO ASI Series I, **8**, M Chanin, Ed., Springer Verlag, Heidelberg.

Twomey, S, 1977: *Atmospheric Aerosols*. Elsevier, Amsterdam, 302 pp.

Understanding Climate Change, 1975: U.S. National Academy of Sciences, Washington, DC, 239 pp.

U.S. Standard Atmosphere, 1976: National Oceanic and Atmospheric Administration, National Aeronautics and Space Administration, United States Air Force. U.S. Govt. Printing Office, NOAA-S/T 76-1562, Washington, DC, 228 pp.

Vallis, G, 2006: *Atmospheric and Oceanic Fluid Dynamics*. Cambridge University Press, Cambridge, 745 pp.

Van Dyke, M, 1982: *An Album of Fluid Motion*. Parabolic Press, Stanford, 174 pp.

Vinnichenko, N, 1970: The kinetic energy spectrum in the free atmosphere – 1 second to 5 years. *Tellus*, **22**, 158-166.

Waliser, D, 1996: Formation and limiting mechanisms for very high sea surface temperatures: Linking the dynamics and thermodynamics. *J Clim*, **9**, 161-188.

Wallace, J, 1967: A note on the role of radiation in the biennial oscillation. *J. Atmos. Sci.*, **24**, 598-599.

Wallace, J and P Hobbs, 1977: *Atmospheric Science: An Introductory Survey*. Academic Press, San Diego, 467 pp.

Wallace, J and P Hobbs, 2006: *Atmospheric Science: An Introductory Survey*, 2nd ed. Academic Press, San Diego, 483 pp.

Wang, B and H Rui, 1990: Dynamics of the coupled moist Kelvin-Rossby wave on an equatorial beta plane. *J Atmos Sci*, **47**, 397–413.

Wang, B and X Xie, 1998: Coupled modes of the warm pool climate system. I: The role of air-sea interaction in maintaining the MJO. *J Clim*, **11**, 2116–2135.

Warner, J, 1969: The microstructure of cumulus cloud. Part I. General features of the droplet spectrum. *J Atmos Sci*, **26**, 1049–1059.

Warren, S and S Schneider, 1979: Seasonal simulation as a test for uncertainties in the parameter-izations of a Budyko-Sellers zonal climate model. *J Atmos Sci*, **36**, 1377–1391.

Webster et al, 2004: Drowning of the -150m reef off Hawaii: A causality of global meltwater pulse 1A? *Geology and GSA Today*, 249–252.

Welander, P, 1955: Studies of the general development of motion in a two-dimensional ideal fluid. *Tellus*, **7**, 141–156.

Welander, P, 1959: On the vertically integrated mass transport in the oceans. *The Atmosphere and the Sea in Motion*. B Bolin, Ed. Rockefeller Institute Press, Albany, 75–101.

Wentzel, G, 1926: Eine Verallgemeinerung der Quantenbedingung fur die Zwecke der Wellen-mechanik. *Z Phys*, **38**, 518–529.

Whipple, 1930: The great Siberian meteor, and the waves, seismic and aerial, which it produced. *Quart J Roy Meteorol Soc*, **56**, 287–298.

Wilson, A, Hency, C, and C Reynolds, 1979: Short-term climate change and New Zealand tempera-tures during the last millenium. *Nature*, **279**, 315–317.

Williams, M, Aydin, M, Tatum, C, and E Saltzman, 2007: A 2000 year atmospheric history of methyl chloride from a south pole ice core: Evidence for climate-controlled variability. *Geophys Res Lett*, **34**, doi:10.1029/2006GL029142.

Williamson, D, Kiehl, J, Ramanathan, V, Dickinson, R, and J Hack, 1987: *Description of the NCAR Community Climate Model*. NCAR/TN-285/STR. Available from NCAR, Boulder, CO.

Willson, RC, Hudson, HS, Frohlich, C, and RW Brusa, 1986: Long-term downward trend in total solar irradiance, *Science*, **234**, 1114–1117.

WMO, 1969: *International Cloud Atlas*. World Meteorological Organization, Geneva, 62 pp.

WMO, 1986: *Atmospheric Ozone: Assessment of our Understanding of the Processes Controlling its Present Distribution and Change*. Report No. 16, World Meteorological Organization, Global Ozone Research and Monitoring Project, Washington, DC.

WMO, 1988: *Report of the International Ozone Trends Panel 1988*. Report No. 18, World Mete-orological Organization, Global Ozone Research and Monitoring Project, NASA, Washington, DC.

WMO, 1991: *Scientific Assessment of Ozone Depletion: 1991*. Global Ozone Research and Monitoring Project Report 25, World Meteorological Organization, NASA, Washington, DC.

WMO, 1995: *Observed Chages in Ozone and Source Gases. Scientific Assessment of Ozone Depletion: 1994*. Rep 37. Available from NASA, Washington DC.

WMO, 2006: *Scientific Assessment of Ozone Depletion: 2006*. Global Ozone Research and Monitoring Project, Report No. 50, Geneva.

Wu, D, Read, W, Dessler, A, Sherwood, S, and J Jiang, 2005: UARS/MLS cloud ice measurements: Implications for H_2O transport near the tropopause. *J Atmos Sci*, **62**, 518–530.

Wunsch, C, 2005: The total meridional heat flux and its oceanic and atmospheric partition. *J Clim*, **18**, 4374–4380.

Yanai, M, Esbensen, S, and J Chu, 1973: Determination of the bulk properties of tropical cloud clusters from large-scale heat and moisture budgets. *J Atmos Sci*, **30**, 11–27.

Young, K, 1993: *Microphysical Processes in Clouds*. Oxford University Press, New York, 427 pp.

Yulaeva, E, Holton, J, and J Wallace, 1994: On the cause of the annual cycle in tropical lower-stratospheric temperature. *J Atmos Sci*, **51**, 169–174.

Zhang, C, 1996: Intraseasonal perturbations in sea surface temperatures of the equatorial eastern Pacific and their association with the Madden-Julian oscillation. *J Clim*, **14**, 1309–1322.

Index

Printed in the United States
By Bookmasters